D1602608

PLEASE STAMP DATE DUE, BOTH BELOW AND ON CARD

DATE DUE	DATE DUE	DATE DUE	DATE DUE
SEP 28 1992			
MAR 22 1993			
MAR 29 1994			
APR 12 1994			
JUN 13 1996			
MAR 4 1998			

GL-15

High-Resolution Transmission Electron Microscopy

High-Resolution Transmission Electron Microscopy
and Associated Techniques

EDITORS
Peter R. Buseck
John M. Cowley
Leroy Eyring

New York   Oxford
OXFORD UNIVERSITY PRESS
1988

Oxford University Press

Oxford New York Toronto
Delhi Bombay Calcutta Madras Karachi
Petaling Jaya Singapore Hong Kong Tokyo
Nairobi Dar es Salaam Cape Town
Melbourne Auckland

and associated companies in
Berlin Ibadan

Copyright © 1988 by Oxford University Press, Inc.

Published by Oxford University Press, Inc.,
200 Madison Avenue, New York, New York 10016

Oxford is a registered trademark of Oxford University Press

All rights reserved. No part of this publication may be reproduced,
stored in a retrieval system, or transmitted, in any form or by any means,
electronic, mechanical, photocopying, recording, or otherwise,
without the prior permission of Oxford University Press.

Library of Congress Cataloging-in-Publication Data

High-resolution transmission electron microscopy
and associated techniques.
Bibliography: p.
Includes index.
1. Electron microscope, Transmission. I. Buseck,
Peter. II. Cowley, J. M. (John Maxwell), 1923–
III. Eyring, LeRoy.
QH212.T7H54 1988   502′.8′25   87-14701
ISBN 0-19-504275-1

2 4 6 8 9 7 5 3 1

Printed in the United States of America
on acid-free paper

PREFACE

Scientific advances often follow improvements in the resolving power of measuring instruments. Transmission electron microscopes have yielded steady increases in spatial resolution and, consequently, improvements in our knowledge of solids. In recent years, new high-resolution instruments have become available that can provide information on a scale close to the atomic level. The main experimental and theoretical principles for the use of these instruments are now established, and there is a broad range of important new applications that has become possible in many fields of solid-state science. It is thus an appropriate time to review high-resolution transmission electron microscopy (HRTEM) and associated techniques and to take stock of their impact on related disciplines.

The goal of this book is to introduce the fundamental concepts and basic methods used for electron microscopy at high resolution in space, energy, or time. Our aim is to provide a volume that can be used by both advanced undergraduate students and established scientists who are entering or contemplating the field of HRTEM and who wish to obtain a sound, comprehensive overview of the subject and its possible impact on their own research.

HRTEM holds the possibility for a greatly improved understanding of the relationship between the atomic structure of materials and their chemical and physical properties. The instruments and techniques are applicable to problems in a wide range of disciplines, and thus the practitioners have greatly differing backgrounds. In particular, the familiarity with mathematics varies widely. With this in mind, the chapters on the basic theory of imaging and diffraction are written at two levels. The first and third chapters are in a descriptive style, with only the essential mathematical formulas included. The second and fourth chapters introduce the mathematical statements required for a more quantitative understanding, and thus they provide a basis for comprehending the more theoretical portions of the book and the literature.

All real crystals contain defects and local disorder, both chemical and structural. In addition to discussing the theoretical principles underlying HRTEM, in this volume we consider its role in understanding crystals and their complex structures, nonstoichiometry, interfaces, precipitates, and defects. Such local perturbations can have major influences on the electronic, mechanical, and chemical properties of their host crystals, and yet they are extremely difficult to study at spatial resolutions commensurate with their dimensions. Transmission electron microscopy offers the unique possibility to observe defects in crystals at close to the atomic level because defects scatter electrons differently from their surrounding matrices. Therefore, there has been much emphasis on the study of individual defects in crystals, and discussion of such features plays a major role in this book.

The various chapters provide accounts of the techniques and the ways in which they are yielding information in solid-state science. The high-resolution imaging of the structures and defects of thin crystals at the 2 to 4 Å level has become an important tool for solid-state chemistry and physics, mineralogy, metallurgy, ceramics, and other subjects involving non-amorphous solids. Results can be reliable and informative. Improved quantitative analysis and resolution, time-resolved information, and three-dimensional analysis are among the next important developments to be expected. Improvements in resolution, new variants on the experimental techniques, and new ways of using beams of fast (beam) electrons are appearing at a rapid rate. Knowledge of these advances will be a necessary prerequisite for forefront work in the study of solids.

The interpretation of images having the highest possible resolution (1 to 2 Å) and imaging of thicker crystals (more than 100 Å) can, in principle, give additional valuable information and extend the applicability of the technique to close-packed structures. Theoretically, thick crystals can yield information of greater sensitivity than thin crystals. However, the results are far more difficult to interpret. There are severe problems since many experimental variables are involved; these include beam defocus, optical aberrations, instrument alignment, crystal thickness, orientation, ionicity, and surface structure. It is difficult to determine values of the relevant parameters with sufficient accuracy to allow quantitative interpretation. Thus, future developments include on-line image analysis and automated adjustment of electron microscope parameters. Also needed are more effective systematic methods for assessment of specimen parameters.

The wealth of problems of managing relevant parameters parallels the wealth of important data that may be obtained, particularly for materials having small unit cells such as semiconductors and metals. Also of interest are amorphous and poorly crystalline materials, where interpretation of data at the limits of resolution or beyond offers the potential for major advances. Important unsolved problems include determination of the relationship of two-dimensional images to the three-dimensional real structures, taking into account the non-linearity of the imaging process.

The behavior of substances at interfaces and surfaces is critical for many reactions, and only a limited understanding is available for such behavior. Thus, increasing attention is being paid to the high-resolution imaging of surfaces and their structures. The presence of surface contamination or particular surface morphologies can influence contrast. The control of specimen environment can enhance the validity of results on thin films and facilitate the *in situ* study of transformations. The development of reflection electron microscopy, surface profile imaging, and special techniques for improving the contrast of surface structure are providing important new tools for surface studies that will be increasingly significant as the production of ultra-high vacuum and controlled environments progresses.

The observation of electron-diffraction patterns in parallel with HRTEM imaging has been of great importance. Recent developments of convergent-beam electron diffraction (CBED) for symmetry determination, lattice-parameter measurement, and some structural analysis are of increasing relevance. Diffraction from areas having dimensions smaller than unit cells is possible with scanning transmission electron microscopy (STEM) and can complement imaging in special cases.

Chemical analysis of sub-micron volumes using energy-dispersive, X-ray emission spectroscopy (EDS) or electron energy-loss spectroscopy (EELS) has developed to the stage that the accuracies and limitations of the methods are reasonably well defined. EELS is more compatible with HRTEM because the spectrometer and detector providing the signal are well removed from the specimen and so do not intrude on the critical imaging region. Additional development of spectrometers and computerized data handling will certainly allow the broader use of EELS imaging as an effective technique for identifying the atoms that are viewed in HRTEM images.

The study of inelastic energy-loss processes, and their interactions with dynamical diffraction, is introducing a number of new ideas for the use of TEM equipment for the study of matter on a submicron scale. The technique of atom location by channeling enhanced microanalysis (ALCHEMI) is useful for locating minority atoms and measuring their abundances. Studies of near-edge fine structure and cathodoluminescence can provide information on excited states and band structure. Combinations with surface-analysis techniques such as Auger electron spectroscopy and electron-stimulated desorption open up new possibilities.

The emerging application of techniques to obtain diffraction and image information at TV rates by means of videotape and framestore manipulation in real time extends the power of HRTEM. Such rapid resolution facilitates the detection of product nuclei and their evolution as reactions proceed, enabling explicit atomic-level mechanisms to be understood. Such information is valuable for understanding changes in materials as a function of time and changing external conditions. These techniques will also extend the usefulness of HRTEM to more beam-sensitive materials.

The literature relevant to HRTEM has come from several disciplines and thus a variety of nomenclature and sign conventions have been used, with the result that much confusion exists. In this volume we attempt to maintain consistency by establishing and adhering to an internally consistent set of sign conventions and nomenclature. This set is compatible with crystallographic usage and has been embodied in the International Tables for X-ray Crystallography. Unfortunately, the symbols differ from the conventions used in much of the physics literature and followed by some electron microscopists. In order to avoid error, it is important to establish and follow one consistent set of conventions. It would be convenient if all microscopists used the same set of conventions, and we hope that, by making the explicit suggestion, we can encourage the use of our chosen set.

We would like to acknowledge the following people who served as anonymous reviewers for various chapters: Drs. J. Barry, P. Bennett, C. Colliex, J. Gjønnes, T. Grove, D. Kohlstedt, S. Krause, C. Lyman, L. Marks, S. Nutt, M. Otten, D. Peacor, P. Rez, P. Self, D. Smith, J. Spence, and J. Vander Sande. Their help contributed significantly to this book. We are indebted to the many people who kindly shared their high-resolution images with the authors; the names are indicated with the respective illustrations. We also acknowledge, with thanks, the considerable technical or clerical help provided by J. Barry, C. Carson, D. Soo Hoo, J. Nickels, R. Kirkeide, J. Wheatley, L.-M. Kiessling, and the many people affiliated with the Arizona State University (A.S.U.) Facility for High Resolution Electron Microscopy, supported as a National Center by the National Science Foundation. This volume is an outgrowth of the A.S.U. Centennial Celebration, and it has received support from the A.S.U. Centennial Fund. We hope that our joint efforts will produce an understanding of the rich insights gained into the structures and properties of materials through high-resolution transmission electron microscopy.

Tempe, Arizona　　　　　　　　　　　　　　　　　　　　　　P. R. B.
June 1988　　　　　　　　　　　　　　　　　　　　　　　　　J. M. C.
　　　　　　　　　　　　　　　　　　　　　　　　　　　　　　　L. E.

CONTENTS

RECOMMENDED SYMBOLS, SIGN CONVENTIONS, AND ACRONYMS xvii

CONTRIBUTORS xx

1. IMAGING 3
   1.1 Introduction 3
     1.1.1 Electron-scattering and -imaging geometry 5
     1.1.2 Electron-microscopy specimens 7
   1.2 The imaging process 10
     1.2.1 Image formation 10
     1.2.2 Aberrations 13
   1.3 Phase contrast 15
     1.3.1 Thin specimens as phase objects 15
     1.3.2 The weak-phase-object approximation 18
     1.3.3 Imaging of weak phase objects 19
     1.3.4 The effects of partial coherence 24
   1.4 Images of periodic objects 28
   1.5 Dark-field images 30
   1.6 Scanning transmission electron microscopy (STEM) 33
   1.7 Resolution 34

2. IMAGING THEORY 38
   2.1 Waves and Scattering 38
     2.1.1 Scattering approximations 38
     2.1.2 Transmission of electron waves through matter 43
   2.2 Imaging 44
     2.2.1 Abbe theory 44
     2.2.2 Imaging of weak phase objects 46
     2.2.3 Imaging of phase objects 47
     2.2.4 Imaging with partial coherence 48
   2.3 Imaging of periodic objects 50
   2.4 Dark-field imaging 52
   2.5 Scanning transmission electron microscopy 54
   2.6 Conclusion 56

3. ELASTIC SCATTERING OF ELECTRONS BY CRYSTALS 58

3.1 General dynamical scattering 58
3.2 Kinematical scattering 60
    3.2.1 Kinematical diffraction from crystals: geometry 62
    3.2.2 Convergent-beam diffraction 67
    3.2.3 Kinematical diffraction from crystals: intensities 70
    3.2.4 Intensities for amorphous or microcrystalline specimens 72
3.3 Limitations of the simple approximations 74
    3.3.1 Kinematical-approximation limitations 74
    3.3.2 Phase-object-approximation limitation 77
3.4 Dynamical diffraction 78
    3.4.1 The Bloch-wave formulation 79
    3.4.2 The two-beam approximation 81
    3.4.3 The multislice formulation 86
3.5 Dynamical-diffraction symmetries 88
    3.5.1 Detection of symmetry elements 90
3.6 The imaging of crystals 93
    3.6.1 Imaging in the two-beam approximation 93
    3.6.2 Axial imaging of simple crystals 96
3.7 Diffraction and imaging of crystal defects and disorder 99
    3.7.1 The column approximation 100
    3.7.2 Local atom displacements: thermal vibrations 102
    3.7.3 Atomic disorder in crystals 103
    3.7.4 Stacking faults and twins: extended defects 105

4. ELASTIC-SCATTERING THEORY 109

4.1 Dynamical scattering 109
4.2 The kinematical approximation 111
    4.2.1 Diffraction by crystals 112
    4.2.2 Kinematical-diffraction intensities 113
4.3 Formulations for dynamical diffraction 115
    4.3.1 Bethe theory 115
    4.3.2 Progression of a wave through a crystal 119
    4.3.3 Basis for the multislice method 121
4.4 Images of crystals 124

5. INELASTIC ELECTRON SCATTERING: PART I 129

5.1 Introduction 129
5.2 Kinematics, single-event inelastic scattering, and the dielectric-response function 132
5.3 Plasmons, phonons, and single-electron excitations 142
5.4 Dynamical inelastic scattering 147

CONTENTS xi

6. **INELASTIC ELECTRON SCATTERING: PART II** 160

   6.1 Localization in inelastic scattering 160
   6.2 Inelastic electron imaging 166
   6.3 Absorption effects and parameters in HRTEM 173
   6.4 Multiple energy-loss effects and their removal 177
   6.5 Radiation damage in HRTEM 182

7. **TECHNIQUES CLOSELY RELATED TO HIGH-RESOLUTION ELECTRON MICROSCOPY** 190

   7.1 Introduction 190
   7.2 Extended electron-loss fine structure (EXELFS) 193
   7.3 Electron-loss, near-edge structure (ELNES) 198
   7.4 Orientation effects in EELS 211
   7.5 ALCHEMI 219
   7.6 Cathodoluminescence in STEM 224
   7.7 Microdiffraction 228
   7.8 Specimen preparation 235
   7.9 Real-time image acquisition and videorecording in HRTEM 237

8. **CALCULATION OF DIFFRACTION PATTERNS AND IMAGES FOR FAST ELECTRONS** 244

   8.1 Introduction 244
   8.2 Calculation of diffracted amplitudes and phases using multislice 246
       8.2.1 The transmission function 248
       8.2.2 The propagation function 251
       8.2.3 Multislice iteration 253
       8.2.4 Consistency tests 257
   8.3 Special systems 259
       8.3.1 Higher-order Laue zones 259
       8.3.2 Periodic continuation 265
       8.3.3 CBED and STEM 269
   8.4 HRTEM imaging 275
       8.4.1 Linear imaging 282
       8.4.2 Nonlinear imaging 286
       8.4.3 Limitations of the envelope functions 292
       8.4.4 Display techniques 295
       8.4.5 HRTEM-image processing 299
   Appendix A: The fast Fourier transform 303

9. **MINERALOGY** 308

   9.1 Introduction 308
   9.2 Reaction mechanisms 310
       9.2.1 Introduction 310
       9.2.2 Biopyriboles 311

	9.2.3 Graphite crystallization	318
	9.2.4 Cordierite transformation	320
	9.2.5 Biotite-chlorite reaction	322
9.3	Stacking disorder and polytypism	325
	9.3.1 Introduction	325
	9.3.2 Micas	327
	9.3.3 Chlorites	328
	9.3.4 Pyroxenes	330
	9.3.5 Pyrosmalite	331
	9.3.6 Other polytypic minerals	331
9.4	Intergrowth disorder and nonstoichiometry	332
	9.4.1 Introduction	332
	9.4.2 Sheet silicates	333
	9.4.3 Pyroxenoids	335
	9.4.4 Bastnaesite-synchysite	337
	9.4.5 Humites and leucophoenicite	337
	9.4.6 Oxysulfides	338
	9.4.7 Oxyborates and chemical twinning	340
9.5	Modulated structures and nonstoichiometry	340
	9.5.1 Introduction	340
	9.5.2 Antigorite and pyrrhotite	342
	9.5.3 Feldspars	346
	9.5.4 Other minerals	347
9.6	Characterization of minerals and structure determination	349
	9.6.1 Introduction	349
	9.6.2 Manganese oxides: fine-grained minerals	350
	9.6.3 Carlosturanite: a new type of chain silicate	354
	9.6.4 Other minerals (sursassite, takéuchiite, etc.)	355
9.7	Mineral definition and nomenclature	356
	9.7.1 Introduction	356
	9.7.2 Structural disorder and intergrowth structures	357
	9.7.3 Ordered structures	359
	9.7.4 Phases	361
9.8	Experimental techniques	362
	9.8.1 Introduction	362
	9.8.2 Special imaging to improve resolution (pyrrhotite)	362
	9.8.3 Radiation damage (biopyriboles, serpentines, and zeolites)	363
	9.8.4 "Controlled" heating by the electron beam (Cu–Fe sulfides)	364
	9.8.5 ALCHEMI and chemical disorder in minerals	365
9.9	Imaging artifacts and the role of calculations	367

10. SOLID-STATE CHEMISTRY — 378

10.1	Introduction	378
	10.1.1 Solid-state chemistry	378
	10.1.2 Historical aside	380
10.2	Application of HRTEM to solid-state chemistry	382
	10.2.1 The role of HRTEM in solid-state synthesis	382
	10.2.2 High-resolution microscopical analysis	383
	10.2.3 Nonstoichiometry and solid-state reactions	384

	10.2.4	Electron-beam-induced chemical change	387
	10.2.5	Image processing and structure determination	390
10.3	Structure and structural defects in the binary-tungsten oxides		391
	10.3.1	The {102} CS phase	394
	10.3.2	The {103} CS phase	396
	10.3.3	The $W_{24}O_{68}$ phase	396
	10.3.4	The $W_{18}O_{49}$ phase	397
	10.3.5	Defect structures	398
	10.3.6	Mechanisms of solid-state reactions	400
10.4	Doped WO_3 to give $(W, Ta, Nb)O_{3-\delta}$		402
	10.4.1	The ternary systems $(W, Nb)O_{3-\delta}$ and $(W, Ta)O_{3-\delta}$	402
	10.4.2	The quaternary $(Nb, Ta, W)O_{3-\delta}$ systems	404
	10.4.3	Pentagonal-tunnel structures and structural defects in the W-Nb-O system	407
10.5	The tungsten bronzes and related compounds		412
	10.5.1	The tungsten and vanadium bronzes	412
	10.5.2	Intergrowth tungsten bronzes (ITB)	414
	10.5.3	The ITB bronzoids	416
	10.5.4	Superstructures in ITB	417
	10.5.5	ITB structures in other systems	418
10.6	The $TiO_{2-\delta}$ phases		420
10.7	Compounds containing two-dimensional crystallographic shear		427
10.8	The structural characterization of zeolites		433
10.9	The anion-deficient, fluorite-related, rare-earth oxides		436
	10.9.1	The system	436
	10.9.2	Structural models for $Ce_{11}O_{20}$ and $Pr_{24}O_{44}$	440
10.10	The study of chemical reactions in thin films using HRTEM		446
	10.10.1	Introduction	446
	10.10.2	Chemical reactions studied by means of high-resolution diffraction methods	447
	10.10.3	Decomposition reactions	449
	10.10.4	The chemistry of thin-film reactions	450
10.11	*In-situ* HRTEM studies of chemical reactions in thin films		464
	10.11.1	The demands of technique development	464
	10.11.2	The recording of the formation of Au_3In *in situ*	466
10.12	Current and near-term extensions of HRTEM techniques applied to solid-state chemistry		468
	10.12.1	Structure elucidation in the bulk and in surfaces and interfaces	468
	10.12.2	Nanometer analytical electron microscopy	469

11. MATERIALS SCIENCE: METALS, CERAMICS, AND SEMICONDUCTORS   477

11.1	Introduction		477
	11.1.1	Materials	477
	11.1.2	Types of defects and processes	478

11.2	Imaging requirements		478
	11.2.1 Limitations and applications of lattice-fringe imaging		478
	11.2.2 Structure images and Fourier images		480
	11.2.3 Defects and crystal alignment		482
	11.2.4 Beam alignment		483
	11.2.5 Information finer than the Scherzer limit		484
11.3	Interfaces and grain boundaries		486
	11.3.1 Semiconductors		486
	11.3.2 Ceramics		490
	11.3.3 Metals		493
11.4	Planar defects		494
	11.4.1 Platelets in diamond		494
	11.4.2 Guinier-Preston zones		496
	11.4.3 Twin boundaries		496
	11.4.4 Extended crystallographic-shear defects		497
11.5	Line defects		498
	11.5.1 Dislocations		498
	11.5.2 Rodlike defects		499
11.6	Point defects		500
	11.6.1 Semiconductors		500
	11.6.2 Ceramics and metals		501
11.7	Small particles and surface-profile imaging		502
11.8	Metallic-alloy systems		504
11.9	Radiation damage		506
	11.9.1 Types of processes		506
	11.9.2 Metals		508
	11.9.3 Semiconductors		508
	11.9.4 Ceramics		509
	11.9.5 Effects of ion implantation and annealing		510
11.10	Complementary techniques		510
	11.10.1 Bright-field phase contrast		510
	11.10.2 Thickness contrast		511

12. PRACTICAL HIGH-RESOLUTION ELECTRON MICROSCOPY 519

12.1	Introduction		519
12.2	Instrumentation		520
12.3	Adjustment of the CTEM for high-resolution imaging		525
	12.3.1 Illumination		527
	12.3.2 Adjustment of the objective lens		532
	12.3.3 Aligning the rest of the microscope column		540
	12.3.4 Recording the high-resolution image		541
	12.3.5 Diffractogram analysis		543
	12.3.6 Automatic alignment		550
	12.3.7 Tilted-illumination imaging		552
	12.3.8 Nonlinear effects		553

12.4	Analytical electron microscopy	554
	12.4.1 Instrumental resolution	555
	12.4.2 Beam spreading	558
	12.4.3 Delocalization	558
	12.4.4 Statistical noise	561
	12.4.5 Optimum adjustment of the AEM	562
12.5	Conclusion	563

13. SURFACES — 568

13.1	Surface-sensitive methods	568
	13.1.1 Scanning electron microscopy (conventional SEM)	569
	13.1.2 Photoemission electron microscopy (PEEM)	571
	13.1.3 Scanning tunneling microscopy (STM)	572
	13.1.4 Field-ion microscopy (FIM)	573
	13.1.5 Low-energy, electron-reflection microscopy (LEERM)	573
	13.1.6 A short history of CTEM for surface studies	573
	13.1.7 STEM and SREM	575
13.2	Experimental techniques for surface CTEM	575
	13.2.1 Vacuum system	575
	13.2.2 In situ specimen treatment	576
	13.2.3 Analytical techniques	577
13.3	TEM studies of surfaces	577
	13.3.1 Image-formation process	577
	13.3.2 Surface-structure analysis by TEM-TED	581
	13.3.3 Observations of surface-dynamic processes	583
	13.3.4 HRTEM of surfaces	586
13.4	REM studies of surfaces	587
	13.4.1 Image-formation processes	587
	13.4.2 Observations of surface-dynamic processes	594
	13.4.3 Applications of REM to other fields of sciences	598
13.5	STEM, SREM, and microanalysis of the surface	599
	13.5.1 SREM and STEM	599
	13.5.2 Surface analysis	600
13.6	Concluding remarks	600

14. HIGHLY DISORDERED MATERIALS — 607

14.1	Significance and outline nature of the problem	607
14.2	Classical diffraction and structural description	608
	14.2.1 Radial-distribution function	608
	14.2.2 Interpretation of RDF data and structure modeling	610
14.3	Medium-scale structure	612
14.4	High-resolution, bright-field imaging	615
	14.4.1 Image theory and properties	615
	14.4.2 Image testing and assessment	618
	14.4.3 Projection effects and object reconstruction	619
	14.4.4 Results of high-resolution, bright-field imaging	620

14.5	High-resolution, dark-field imaging	622
	14.5.1 Dark-field image theory and properties	622
	14.5.2 Dark-field imaging of disordered alloys	624
	14.5.3 High-angle, incoherent, dark-field imaging	625
14.6	Other imaging modes	627
14.7	Conclusions	630

INDEX 633

RECOMMENDED SYMBOLS, SIGN CONVENTIONS, AND ACRONYMS

Symbols: notations*

Vectors, distances, coordinates
  Real space: $\mathbf{r}, r, xyz$ or $\mathbf{R}, R, XYZ$
  Reciprocal space: $\mathbf{u}, U, uvw$

Functions
  Real space: lower case: Roman or Greek
  Reciprocal space: capitals: Roman or Greek

Units   SI units except for Å; for uniformity we use Å throughout

Charge density $\rho(\mathbf{r})$

Defocus Δf (set positive for overfocus)

Electron wavelength $\lambda = h \left/ \left[2meE\left(1 + \dfrac{eE}{2mc}\right) \right]^{1/2} \right.$

Excitation error ζ_h or $\zeta(\mathbf{u})$

Extinction distance ξ_h

Interaction constant $\sigma = 2\pi me\lambda \left[1 + \dfrac{eE}{mc} \right] / h^2$

Potential
  Real space: $\phi(\mathbf{r})$
  Reciprocal space: $\Phi(\mathbf{u})$

Reciprocal lattice vectors $\mathbf{h}, \mathbf{g}$

Spread function $t(\mathbf{r}) = c(\mathbf{r}) + is(\mathbf{r})$

Transfer function $T(\mathbf{u}) = A(\mathbf{u}) \exp\{i\chi(\mathbf{u})\}$

Wave function
  Real space: $\psi(\mathbf{r})$
  Reciprocal space: $\Psi(\mathbf{u})$

Wave vector $\mathbf{k}$, magnitude $k = 2\pi/\lambda$; $\mathbf{k}_i - \mathbf{k}_0 = \mathbf{q}$

* Full definitions are found in Chapters 1 through 4.

Convolution	$f(x) \ast g(x) \equiv \int_{-\infty}^{\infty} f(X) g(x-X)\, dX$

Sign Conventions (Use the crystallographic convention, International Tables Vol. II p. 67.*)

Free space wave	$\exp\{i(\omega t - \mathbf{k} \cdot \mathbf{r})\}$
POA transmission function	$q(xy) = \exp\{-i\sigma\phi(x,y)\}$ where $\phi(xy) = \int \phi(xyz)\, dz$, and ϕ is a positive potential and therefore attracts electrons
Phenomenological absorption	$\sigma\phi(\mathbf{r}) \to \sigma\phi(\mathbf{r}) - i\mu(\mathbf{r})$
Propagation function	$\exp\{-2\pi i \zeta(\mathbf{u}) \cdot \Delta z\}$ where the excitation error $\zeta(\mathbf{u})$ is negative for reflections outside the Ewald sphere
Wave aberration function	$\exp\{i\chi(U)\}$ where $\chi(U) = \pi\lambda\Delta f U^2 + \pi C_s \lambda^3 U^4/2$ and $U^2 = u^2 + v^2$ Δf is positive for overfocus
Fourier transforms	
Real to reciprocal space:	$\Psi(\mathbf{u}) \equiv \mathscr{F}\psi(\mathbf{r})$ $= \int \psi(\mathbf{r}) \exp\{2\pi i \mathbf{u} \cdot \mathbf{r}\}\, d\mathbf{r}$
Reciprocal to real space:	$\psi(\mathbf{r}) \equiv \mathscr{F}^{-1}\psi(\mathbf{u})$ $= \int \psi(\mathbf{u}) \exp\{-2\pi i \mathbf{u} \cdot \mathbf{r}\}\, d\mathbf{u}$
Structure factors	$\Phi(\mathbf{u})$ or $F(\mathbf{u}) = \sum_j f_j \exp\{2\pi i \mathbf{u} \cdot \mathbf{r}_j\}$ f_j is scattering factor for atom j

Acronyms

Acronym	Definition
AEM	Analytical electron microscopy
AES	Auger-electron spectroscopy
AFF	Aberration-free focus
AFM	Atomic-force microscopy
ALCHEMI	Atom location by channeling enhanced microanalysis
APS	Appearance-potential spectroscopy
ARPES	Angle-resolved photoelectron spectroscopy
BF	Bright field
BIS	Bremsstrahlung isochromat spectroscopy
CBED	Convergent-beam electron diffraction
CL	Cathodoluminescence
CTEM	Conventional transmission electron microscopy

* The Alternative Sign Convention is that of Saxton et al., Ultramicroscopy *12*, 75 (1983).

CTF	Contrast transfer function
DF	Dark field
DOS	Density of states
DQE	Detector quantum efficiency
EBIC	Electron-beam-induced current
EDS	Energy-dispersive X-ray spectroscopy
EELS	Electron energy-loss spectroscopy
ELNES	Electron energy-loss near-edge structure
EXAFS	Extended X-ray-absorption fine-structure
EXELFS	Extended energy-loss fine-structure
FFT	Fast Fourier transform
FIM	Field-ion microscope
FOLZ	First-order Laue-zone
FWHM	Full width at half maximum
HOLZ	Higher-order Laue zone
HRTEM	High-resolution transmission electron microscopy
IPES	Inverse photoelectron spectroscopy
LEED	Low-energy electron diffraction
LEERM	Low-energy electron reflection microscopy
MD	Microdiffraction
NEXAFS	Near-edge X-ray-absorption fine-structure
PCDA	Projected-charge-density approximation
PCTF	Phase-contrast transfer function
PDOS	Partial density of states
PEEM	Photoemission electron microscopy
PL	Photoluminescence
PMT	Photomultiplier tube
POA	Phase-object approximation
RDF	Radial distribution function
REM	Reflection electron microscopy
RHEED	Reflection high-energy electron diffraction
SAED	Selected-area electron diffraction
SEM	Scanning electron microscope
SEXAFS	Surface extended X-ray-absorption fine-structure
SREM	Scanning reflection electron microscopy
STEM	Scanning transmission electron microscopy
STM	Scanning tunneling microscope
TED	Transmission electron diffraction
TEM	Transmission electron microscopy
UHV	Ultrahigh vacuum
UPS	Ultraviolet photoelectron spectroscopy
WPO	Weak phase object
WPOA	Weak-phase-object approximation
XANES	X-ray-absorption near-edge structure
XPS	X-ray photoelectron spectroscopy
ZOLZ	Zero-order Laue zone

CONTRIBUTORS

John C. Barry
Center for Solid State Science
Arizona State University
Tempe, AZ 85287

Peter R. Buseck
Departments of Geology and
 Chemistry
Arizona State University
Tempe, AZ 85287

John M. Cowley
Department of Physics
Arizona State University
Tempe, AZ 85287

LeRoy Eyring
Department of Chemistry
Arizona State University
Tempe, AZ 85287

A. Howie
Department of Physics
Cavendish Laboratory
University of Cambridge
Madingley Road
Cambridge CB3 0HE, England

Ondrej L. Krivanek
Gatan, Inc.
6678 Owens Dr.
Pleasanton, CA 94566

Michael A. O'Keefe
NCEM
Lawrence Berkeley Laboratories
University of California
Berkeley, CA 94720

Peter G. Self
CSIRO, Div. of Soils
Private Bag #2
Glen Osmond, S.A. 5064
Australia

David J. Smith
Center for Solid State Science
Arizona State University
Tempe, AZ 85287

John C. H. Spence
Department of Physics
Arizona State University
Tempe, AZ 85287

David R. Veblen
Dept. of Earth and Planetary
 Sciences
Johns Hopkins University
Charles & 34th Sts.
Baltimore, MD 21218

Katsumichi Yagi
Dept. of Physics
Tokyo Institute of Technology
Oh-Okayama, Meguro
Tokyo, Japan

High-Resolution Transmission Electron Microscopy

1
IMAGING
JOHN M. COWLEY

1.1 Introduction

Conventionally, the imaging with electrons in an electron microscope is discussed in the same terms and even with the same diagrams used for the imaging with light in optical microscopes. We follow this convention. In this way, we can build on familiar ideas and relate our treatments directly to common experience. For most purposes, this approach is satisfactory, although for neither electrons nor light is the description accurate or universally valid.

In electron microscopes, electrons are focused by use of the highly concentrated magnetic fields formed between soft-iron pole pieces and generated by currents flowing through annular coils. The trajectories of the electrons are complicated. Electrons diverging from a point are brought to a focus, but the electron paths spiral around the lens axis and are not confined to planes containing the lens axis, as are rays of light. The noticeable result of this spiraling is that the image is rotated with respect to the object. Subtle effects arise because this rotation is not constant but varies with the distance from the axis. However, in electron optics, we are usually concerned only with electrons making very small angles with the lens axis—that is, paraxial rays. Hence, these off-axis aberrations are negligible, and usually, a uniform rotation of the whole image is not important. Thus, we can describe the action of the magnetic lens in terms of the simple analog of Figure 1.1.

Light, of course, is electromagnetic radiation and, thus, should be described in terms of the electric- and magnetic-field vectors, rather than in terms of simple scalar amplitudes. The vector nature of light is manifest in such phenomena as polarization and birefringence. However, for most simple optical experiments, these phenomena can be ignored, and it is sufficient to consider scalar amplitudes as the basis for calculating observed intensities. Similarly, for electrons, some phenomena depend on the existence of the electron spin, but for our purposes, these special cases can be ignored. Consequently, we can discuss electron-beam intensities also in terms of simple scalar quantities, the wave amplitudes.

For both light and electrons, the concept of an ideal thin lens is so convenient that we tend to use it even though we are aware that it is not an accurate description of the actual lens configuration. Optical lenses are often quite thick compared with their focal lengths and usually are made from several components of different shape and composition. In electron

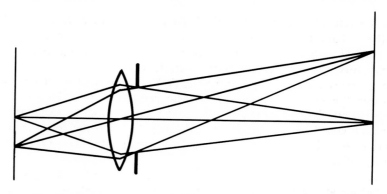

FIGURE 1.1 The focusing action of an ideal thin lens, with ray paths drawn from the object, through the lens and aperture, to the image plane.

lenses, likewise, the imaging field may extend over considerable distances. An electron-microscope specimen may be immersed within the extended magnetic field of the objective lens. Fortunately, for most purposes, it is possible to approximate the performance of a thick lens by considering it to be replaced by a small number of ideal thin lenses, with appropriate scaling factors applied to the distances and lens apertures concerned. This technique works especially well in electron microscopy, where we are concerned almost entirely with paraxial rays. In the diagrams and discussions in these introductory chapters, we therefore take advantage of this convenience. In doing so, we do not exclude any factors of importance for practical electron microscopy.

One further feature of the diagrams requires comment. In the conventional way, lines represent light rays or electron paths showing the passage of the radiation through the system. This is done even though most of our discussion is presented in terms of electron or light waves. In the even-numbered chapters (Chapters 2, 4), the theory is presented mostly in terms of the phases and amplitudes in wave fields. Then, individual and distinct rays or electron paths have no significance.

Of course, one can send a light ray of small diameter through an optical system, or one can trace the path of a fine electron beam. But there are limitations on the fineness of the beam that can be described in terms of the uncertainty principle. The lateral extent Δx of the beam and its lateral momentum Δp can be related by the expression

$$\Delta x \cdot \Delta p = h, \tag{1.1}$$

where h is Planck's constant.

Thus, as the diameter of the beam is made smaller, the uncertainty in its lateral momentum—and hence, the angular spread of the direction of the beam—becomes larger. In the limit, if Δx becomes equal to the wavelength λ, of the radiation, then the uncertainty of the lateral momentum becomes equal to h/λ, which is the total momentum of the electron or photon. Consequently, the direction of travel is undefined.

The geometric optics of ray paths is thus a convenience. It has validity only when the smallest dimensions considered are very much greater than the wavelength divided by the angular spread (in radians) of the paths about the optical axis. For any considerations of high-resolution imaging, the ray paths must be taken only as guides to the approximate direction of the flow of energy of the radiation.

To some extent, we conform to the conventions of light optics of the years before about 1950. At that time, geometric optics was the basis for discussion of optical systems, and the interference phenomena of wave optics were introduced as a necessary complication when the resolution of the imaging system was discussed. However, keep in mind the parallel theoretical treatment in terms of wave functions, which provides a more satisfactory, unified description of all aspects of imaging and diffraction. In this treatment, we follow the newer wave optics of light (see Goodman, 1968; Born and Wolf, 1980) or of other radiations (see Cowley, 1981). This approach is especially important for the consideration of crystalline specimens because interest here centers on the diffraction pattern formed in the back focal plane of the objective lens as a result of coherent interference of the radiation reacting with the specimen.

1.1.1 Electron-scattering and -imaging geometry

In the basic imaging arrangement of transmission electron microscopy (TEM), an electron beam from an electron gun illuminates the specimen, usually through an illuminating system of lenses. The radiation interacts with the specimen and is scattered. The scattered radiation is brought to a focus by the objective lens. Then, further magnifying (or projector) lenses produce an image of convenient size (see Figure 1.2).

For scanning transmission electron microscopy (STEM), the same diagram can be used, but the direction of the electron beam is reversed. The source replaces the TEM detection system. The projector lenses (sometimes called condenser lenses; see Figure 1.2) demagnify the source, which is then demagnified further by the objective lens to form a small probe at the specimen. This probe is scanned over the specimen by deflection coils, and the detector collects the electrons transmitted, or scattered, by the specimen. As shown later, these imaging modes are equivalent in principle, although they are different in their practical form and applications. For the moment, we concentrate on the more familiar TEM system.

Two important factors determine the essential geometry of the imaging system. The wavelength of the electrons determines the angular range of the scattering from the specimen. The aberrations of the electron lenses limit

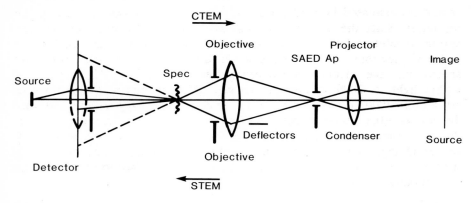

FIGURE 1.2 The main elements of CTEM and STEM systems. For CTEM, the electrons pass from left to right. For STEM, the direction is reversed. SAED Ap is the selected area electron diffraction aperture. Spec is the specimen.

the angular range of scattered radiation that can usefully contribute to the image. These factors are treated next in order.

The wavelength λ of an electron wave is given by

$$\lambda = \frac{h}{p} = h(2meE)^{-1/2}, \tag{1.2}$$

where h is Planck's constant, p is the momentum, E is the accelerating voltage of the electrons, and e and m are the electron charge and mass. The mass and effective voltage should include relativistic effects; but if these effects are ignored, the convenient approximate expression is $\lambda = (150/E)^{1/2}$, for E in volts and λ in angstroms (1 Å = 10^{-10} m).

For 100-keV electrons, $\lambda = 0.037$ Å = 3.7 pm. The scattering angles depend on the dimensions of the scatterers. Matter is composed of atoms having diameters d of the order of 1 to 2 Å. The average scattering angles are of the order of λ/d, or about 2×10^{-2} rad, or 1°. The scattering angles will be less by a factor of about 3 for 1 MeV electrons and greater by a factor of 3 for 10-keV electrons. In general, electron scattering from matter is strongly peaked in the forward direction, and except for very thick specimens, the scattered intensity is negligible for scattering angles beyond 10^{-1} rad.

In comparison with light-optics lenses, magnetic-electron lenses are very poor. They have inherent aberrations that are much worse than those of a simple, convex glass lens having spherical surfaces. Unfortunately, there is no way to correct aberrations by using multicomponent systems or other means, as has been done for light-optics systems. The effects of the aberrations increase very rapidly with the angular aperture α of the lens (with α^3 plus higher-order terms). So in practice, the angular aperture cannot be increased beyond about 10^{-2} rad without seriously limiting the resolution. Since the resolution of a microscope is on the order of λ/α, the current resolution limit of electron microscopes is of the order of 3 Å for 100-keV electrons. By comparison, in the optical case with corrected aberrations, the theoretical resolution limit approaches $\lambda/2$, which, for visible light, is 2000 to 3000 Å. Hence, although electrons can provide resolution 1000 times better than visible light—approaching atomic dimensions—the limitation resulting from lens aberrations is severe. But at the same time, this limitation implies that we need deal only with small scattering angles and can make small-angle, paraxial approximations, which simplify the description of the imaging and diffraction phenomena considerably.

1.1.2 Electron-microscopy specimens

Relative to other radiations (X-rays, neutrons) of comparable wavelength or energy, electrons interact very strongly with matter. For elastic scattering, the mean free path in solids varies from a few hundred angstroms for lighter elements to a few tens of angstroms for heavier elements for 100-keV electrons. In inelastic processes, having mean free paths of hundreds to thousands of angstroms, the incident beam loses energy to create excitation in the specimen. Decay of the excitation can lead to the emission of secondary radiation. Detection of this secondary radiation may give information concerning the specimen that is useful in some contexts, as outlined in later chapters.

For high-resolution imaging, the elastic scattering is of prime importance. To avoid serious complications, one must use a specimen thickness of the same order as the mean free path for elastic scattering. Special preparation techniques are needed to produce specimens of thickness in the required 100-Å range.

The scattering amplitudes and angular distributions for single, isolated atoms can be calculated from the so-called atomic-scattering factors (actually, Fourier transforms of the potential distributions of the atoms). These scattering factors are tabulated, for example, in the *International Tables for Crystallography*. From these data, one can derive the intensity distributions for images of single, isolated atoms. For example, Chiu and Glaeser (1975) calculated that a single mercury atom should give 15 percent contrast in a bright-field image obtained with an electron microscope currently available.

Such calculations are relevant only for special cases in which single-atom scattering can be approached, as for a heavy atom supported on a very thin, light-atom film. Scattering by real specimens is usually complicated, because the scattered-wave amplitudes are not given by the sums of the amplitudes of individual atoms.

The scattering process is readily understood by the analogy of the interaction of light with a transparent object. An incident wave traveling through a medium of refractive index different from unity suffers a phase change relative to the wave in vacuum. For electrons, there is a change of refractive index where the wave enters a region of different electrostatic potential, which changes the velocity and hence the wavelength of the electron waves. If the phase change is uniform, as for a plane sheet of glass for visible light, the wave-propagation direction does not change. But if there are lateral variations of the phase change (bumps in the glass plate), the radiation is scattered from its original direction of propagation.

For electrons, the scattering angles are small (about 10^{-2} rad), so in traversing a thin specimen (10^2 Å thickness), the electron wave does not suffer a sideways displacement of more than about 1 Å. Therefore, as a first approximation, we can assume that an electron wave traversing a thin specimen suffers a phase change that depends on the distribution of potential along a straight-line path through the object (see Figure 1.3). Thus, if the potential distribution in the object is represented by a function $\phi(xyz)$, a plane wave transmitted through the object in the z-direction suffers a phase change that is a function of the x,y-coordinates, proportional to the projection of the potential in the z-direction:

$$\phi(xy) = \int \phi(xyz)\, dz. \tag{1.3}$$

The phase change of the electron wave, relative to a wave transmitted through a vacuum ($\phi = 0$), is given by the product of $\phi(xy)$ and an interaction constant σ (equal to $\pi/\lambda E$), which defines the strength of the interaction of electron waves with matter. Mathematically, the effect on an incident wave of the phase change is given by multiplying the incident-wave amplitude by a transmission function

$$q(xy) = \exp\left[-i\sigma\phi(xy)\right]. \tag{1.4}$$

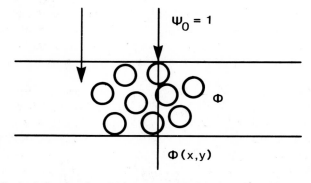

FIGURE 1.3 The basis for the phase-object approximation for a thin object.

This equation is called the phase-object approximation (POA). It is an approximation in that it ignores the sideways scattering of the waves, which can have significant effects, in practice, for thicknesses greater than a few tens of angstroms. It also ignores any inelastic-scattering processes—which, by preventing some electrons from contributing to the image, can have the effect of producing an absorption. If the absorption can be represented by a function $\mu(xyz)$ having a projection $\mu(xy)$, the transmission function can be written as

$$q(xy) = \exp[-i\sigma\phi(xy) - \mu(xy)]. \tag{1.5}$$

However, the contribution of absorption to the image is usually small. Thus, the error made in neglecting it is usually less than the other errors of the approximation.

In spite of its limitations, the POA is useful because it emphasizes the nonlinear nature of electron scattering. Unless the maximum value of $\sigma\phi(xy)$ is very much less than unity, changes in the projected potential do not give proportionate changes in the transmission function. For a single heavy atom, the phase differences from the center to the outside of the atom in projection may exceed π. If the electron wave passes successively through several atoms in a row, the phase changes add, and the nonlinear nature of the function (1.4) ensures that the scattering amplitudes are not added linearly.

For thicker specimens, it is not only the POA that fails. The whole concept of a transmission function becomes invalid. The effect of a specimen on an incident wave cannot be represented by a multiplicative function. Formally, the effect can be given by a matrix multiplication of an incident wave represented as an n-dimensional vector. This topic is discussed in Chapters 3 and 4.

For suitably thin specimens, the wave function at the exit surface of the specimen can be described as a two-dimensional function $\psi_1(xy)$, which can be represented as the product of the incident-wave amplitude $\psi_0(xy)$ and the transmission function of the specimen $q(xy)$. The exit wave has variations of phase and amplitude related to the specimen structure. Because of these variations, scattered waves diverge into the aperture of the objective lens. The aims of the imaging process are then to produce a magnified representation of $\psi_1(xy)$ and to record an intensity distribution from which some information about the structure of the specimen can be derived.

The derivation of the information about the specimen is often not a straightforward process. For example, if absorption effects are neglected, the only effect of a thin specimen is to change the phase of the incident wave. For a pure phase object of this type, the in-focus image with an ideally perfect microscope would show no contrast at all. The equivalent case is well known to biologists who use light microscopes to examine thin sections of biological tissue. To get appreciable contrast, one can defocus the microscope. But for light microscopes, this technique smears out the image and reduces the resolution. Better methods are to stain the specimens to produce absorption contrast or to use special phase-contrast equipment.

As a basis for the interpretation of electron micrographs, we must consider the imaging process in detail. When the best resolution of current microscopes is used, an intuitive interpretation of image contrast in terms of densities of scattering matter can often be completely misleading.

1.2 The imaging process

1.2.1 Image formation

Figure 1.4 suggests the basis for the production of an image following the wave-optics theory attributed to Abbe. A plane wave illuminates the specimen. The transmitted-wave function $\psi_1(xy)$ can be assumed to be composed of a forward-scattered, axial, transmitted wave plus other scattered waves that proceed in directions slightly inclined to it.

The ideal thin lens brings the parallel transmitted wave to a focus on the axis in the back focal plane. Waves leaving the specimen at the same angle ϕ_x with the axis, in the x, z-plane, are brought together at a point on the back focal plane at a distance $X \simeq f\phi_x$ from the axis, where f is the focal length of the lens. Thus, on the back focal plane, the waves from all parts of the specimen propagated in a given direction are added. So in this plane, the Fraunhofer diffraction pattern is formed, with maxima where waves from all parts of the specimen add constructively. The variables used for the distribution of amplitude or intensity in the diffraction pattern are the angular variables $u = (2/\lambda) \sin(\phi_x/2)$ and $v = (2/\lambda) \sin(\phi_y/2)$ or, in the

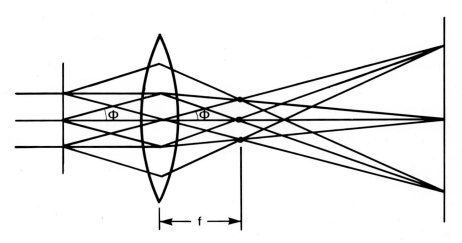

FIGURE 1.4 Ray diagram suggesting the formation of a Fraunhofer diffraction pattern on the back focal plane of a lens and the subsequent formation of an image.

small-angle approximation, $u = x/f\lambda$ and $v = y/f\lambda$. The amplitude of the wave in the back focal plane is written as $\Psi(uv)$.

Continuing the rays in Figure 1.4 beyond the back focal plane, we see that rays from a common point in the object converge on a common point in the image plane. For an ideally perfect lens, the image amplitude $\psi(xy)$ would be an inverted, magnified reproduction of ψ_1, namely, $\psi_1(-x/M, -y/M)$. Thus, the relationship of the wave function $\Psi(uv)$ in the back focal plane to that in the image is the inverse of the relationship of $\Psi(uv)$ with $\psi_1(xy)$. We can describe this relationship by saying that $\Psi(uv)$ is created by interference at infinite distance of the waves generated at the object. The back focal plane represents a plane at infinity because it is the plane on which parallel lines (in the object space) intersect. In propagation from the point at infinity, back to the plane conjugate to the object plane—that is, the image plane—the interference process is inverted and the object wave function is re-created.

Mathematically, the formation of the Fraunhofer diffraction pattern is described by a Fourier-transform operation, denoted by $\mathcal{F}$:

$$\Psi(uv) = \mathcal{F}\psi_1(xy).$$

The formation of the image is then described by two successive Fourier-transform operations, and apart from scaling factors, the equation is

$$\psi(xy) = \mathcal{F}[\mathcal{F}\psi_1(xy)] = \psi_1(-x, -y), \tag{1.6}$$

which therefore describes the two stages of the image-forming process. In summary, we may say that interference of the waves generated at the object gives the Fraunhofer diffraction pattern in the back focal plane. Interference of the waves from the back focal plane re-creates the object wave function in the image plane.

If all the waves leaving the object are brought together again with exactly the same relative amplitudes and phases, the image gives an exact reproduction of the transmission function of the object. This event can only happen with an ideally perfect lens having no limitation owing to a finite aperture. In practice, of course, lenses do have finite apertures, cutting off the waves traveling in directions at higher angles to the axis. They also have aberrations that distort the relative phases of the waves. The image is then not a perfect reconstruction of the object. The resolution of image detail is affected, and fine details of the intensity distribution may be distorted. A point source in the object plane does not give a point in the image plane. It gives a patch of wave amplitude of finite size in the image plane, described as the spread function that characterizes the lens. The image amplitude is thus obtained by using the spread function to smear out the amplitude distribution at the object with a consequent loss of fine detail—that is, a loss of resolution.

The spread function is related (by Fourier transform) to the function describing the imperfections in the action of the lens system: the limitation of amplitudes by the aperture and the distortion of relative phases by the aberrations. This latter function is known as the transfer function of the lens. It operates in the back focal plane. Thus, we say that the transfer function is the function characteristic of the lens that multiplies the wave function in the back focal plane. It modifies the wave amplitudes and phases of the Fraunhofer diffraction pattern of the object and therefore prevents the proper interference of these waves, which is needed to form an undistorted image.

For a lens having no aberrations, the transfer function includes only the aperture function. If the aperture is the usual circular hole in a screen, centered on the lens axis and of radius given by $(u^2 + v^2)^{1/2} = a/2$, we have the following result for the spread function:

$$t(xy) = \left(\frac{\pi a^2}{2}\right) \frac{J_1(\pi ar)}{\pi ar}, \qquad (1.7)$$

where $r = (x^2 + y^2)^{1/2}$ and J_1 is the first-order Bessel function. Thus, if the object wave function were a sharp point function, the image intensity would be $t^2(xy)$, which is known as the airy disk—a circular peak of intensity of radius $r = 1.22/a$ surrounded by rings of rapidly decreasing intensity.

In light optics, we assume that two point sources can be resolved by a microscope if the center of the airy disk for one falls no closer to the other than the first zero of its airy-disk function. This assumption is the Rayleigh criterion for resolution. The aperture radius is given (for $v = 0$) by $u = a/2$, with $u = x/f\lambda$ in the small-angle approximation, for which a/f is the aperture angle α. Then, the least resolvable distance between two point sources of light is

$$\Delta = \frac{1.22\lambda}{\alpha}. \qquad (1.8)$$

For electron microscopy, we cannot, in general, use this relationship. It is derived for the case of two point sources of radiation that are incoherent—i.e., that emit quanta of radiation independently so that no interference of the light from the two sources can take place. Then, one can add the intensities of the radiation from the two sources and derive the result in (1.8). For electron microscopy, we must follow the scheme of Figure 1.4, assuming that the wave function leaving the object is generated by transmission of an incident plane wave so that waves from all parts of the object are coherent and can interfere. Then, the image amplitude is given by adding the amplitudes of the waves from all parts of the object with their appropriate phases. The intensity distribution is obtained by multiplying the complex-wave amplitude by its complex conjugate or by taking the square of the modulus of the wave function. For two image points close together that have overlapping spread functions, the resultant intensity depends strongly on the relative phases of the waves at the two points (see Figure 1.5).

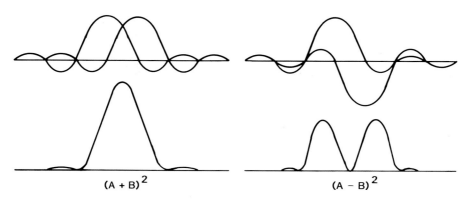

FIGURE 1.5 The addition of amplitudes in images of two point sources and the resulting intensity distributions for the cases when the point sources are in phase and out of phase.

The more general aspects of coherence and the special cases in electron microscopy for which the simple light-optics theory is a reasonable approximation are treated later in this chapter. First, we consider the effects of aberrations on the images.

1.2.2 Aberrations

In considering the distortion of the wave fronts by aberrations and the consequent loss of resolution, we are fortunate in electron microscopy that it is necessary to deal only with paraxial rays and small-angle scattering. It is usually possible to ignore the off-axis and higher-order aberrations that are important for light optics and can confine our attention to the simplest axial aberrations: defocus, spherical aberration, astigmatism, and chromatic aberration. The discussion of chromatic-aberration effects is deferred until the later presentation of effects of incoherence on imaging. Astigmatism refers to differences in focal lengths of a lens in various planes containing the optic axis and results from a lack of cylindrical symmetry of the magnetic field of the lens. This fault is common in electron microscopes but can be corrected, with care, by stigmators that supply correcting fields. Defocus and spherical aberration are treated together as providing the most significant, cylindrically symmetrical perturbations of the phase change produced by a lens.

In Figure 1.6, the geometric-optics and wave-optics representations of an imaging system with spherical aberration are compared. In geometric optics, paraxial rays are brought to a focus at the paraxial or Gaussian focus. Changes of the strength of the lens move this focus backward or forward by an amount Δf and so spread the image by amount $\Delta f \cdot \alpha$, where α is the angle the rays make with the axis. In the presence of third-order spherical aberration, the focal length decreases by an amount $C_s \alpha^2$ as α increases,

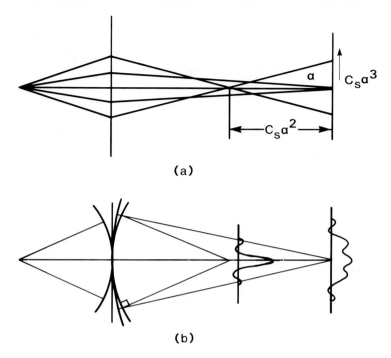

FIGURE 1.6 The effects of spherical aberration on the imaging of a point source. (*a*) In geometric optics. (*b*) In wave optics.

where C_s is the spherical-aberration constant, so that the crossover at the paraxial-focus position is spread by an amount $C_s\alpha^3$. For a point source, the minimum beam diameter in the image space is not at the paraxial-focus position but at some distance closer to the lens. Hence, the best resolution in an image is obtained by defocusing (weakening) the lens to bring this minimum beam size into the image plane.

In the wave-optics description, the action of an ideal lens is to transform the spherical wave emitted by a point in the object into a spherical wave that converges on the corresponding image point. The effect of the transfer function of the lens is to limit the angular range of, and to perturb the spherical shape of the convergent spherical wave. We consider only cylindrically symmetric perturbations, which can be expressed as a function of the angular variable α, a power series containing only even-ordered terms.

The relationship to the geometric-optics picture comes from the idea that the rays are perpendicular to the wave front. Spherical aberration, for example, perturbs the wave front by a fourth-order term, so the slope of the wave front—and hence, the deviation in the direction of the ray—is given by a third-order term in α. Instead of considering a change to a spherical wave front, we consider the equivalent picture of perturbations of phase on the back focal plane described by an even-ordered power series of the radial coordinate $U = (u^2 + v^2)^{1/2}$. The first two terms of this power series are included in the following expression for the phase factor:

$$\chi(U) = \pi \, \Delta f \, \lambda U^2 + \tfrac{1}{2}\pi C_s \lambda^3 U^4. \tag{1.9}$$

The second-order term represents the effect of a change of focus by an amount Δf. In the small-angle approximation, a spherical wave is represented by a parabolic variation of phase on a plane. Hence, the addition of a second-order perturbation changes the radius of curvature of the wave front and thus changes the focus position. The second term, containing the spherical-aberration coefficient C_s, changes the curvature of the wave point as a function of U, so the effective focal length decreases for the wave front further from the axis.

Interference of the waves from the back focal plane gives rise to the complex-amplitude distribution in the image plane. For a point source, this distribution is described by the spread function, which, for $C_s = 0$, has the form $J_1(x)/x$, as suggested in Figure 1.5. With spherical aberration, the spread function is modified into a more complicated and broader distribution. The spread function takes on the most desirable form of a narrow, single peak with small, subsidiary maxima and minima for some optimum defocus close to that suggested by the geometric-optics diagram.

1.3 Phase contrast

1.3.1 Thin specimens as phase objects

We stated earlier that very thin specimens can be approximated as phase objects for fast electrons. The thickness limitation depends on the experimental conditions: For 100 keV electrons, for example, the thickness must be less than about 50 Å if imaging with about 3-Å resolution is being considered. The transmission function represented by equation (1.4) describes a change of phase but no change of amplitude of the electron wave passing through the object. For an ideally perfect lens, the image wave function likewise has a phase modulation but no amplitude modulation, so the image-intensity distribution is a constant—that is, the image contrast is zero. To derive some information from the image concerning the object structure, we must use an imperfect lens, which means losing resolution, or resort to the use of analogs of one of the phase-contrast-imaging schemes used for light optics.

Any perturbation of the waves in the back focal plane affecting the contributions of these waves to the image will upset the relationship of phases and amplitudes that produces the zero image contrast. Hence, the finite lens aperture or any defocus or lens aberrations will give some image contrast. For example, a finite aperture will exclude those waves corresponding to higher-angle scattering from the object. As a first approximation, we may say that the loss of the intensity of the higher-angle diffracted beams will reduce the image intensity. Since more rapid variations of structure in the object produce scattering to higher angles, the image intensity will be less where the object transmission function changes more rapidly. Thus, the image will show dark lines around the edges of small particles. Also, the image intensity will be less for heavier-atom material—that is, there will be some amplitude contrast, with intensity decrease related to the amount of scattering material. The term *amplitude contrast* has been used here because of the concept that the aperture has removed some of the scattered radiation and therefore reduced the amplitude of the image signal.

Image contrast for thin objects can be enhanced by use of a small objective aperture. Biologists and others dealing with weakly scattering specimens often take advantage of this technique. However, use of a small aperture inevitably limits the resolution, as suggested by equation (1.8).

With the large apertures needed for high-resolution imaging, much stronger contrast can be produced by the phase changes due to defocus and aberrations. This type of contrast is known as phase contrast. For phase objects, the in-focus position shows minimum image contrast, and a major part of the contrast reverses with the change from overfocus to underfocus. For example, the main Fresnel fringe around the edge of a phase object is bright for underfocus (a weakening of the lens to increase its focal length) and is dark for overfocus.

In general, the image contrast produced by defocus and aberrations is not simply related to the amount of scattering matter. There is a relatively simple linear relationship with the projected-potential distribution of the object only for the special case, to be considered later, of weakly scattering, thin objects. For such objects there are only small differences in phase for the electron wave passing through neighboring parts of the object.

When moderately heavy atoms are present—and especially when crystals are viewed parallel to lattice planes or principal axes—the phase differences may be large, and this simple linear relationship no longer applies. However, a different and useful relationship does apply for small amount of defocus if the aberration effects are small. This relationship is the projected-charge-density approximation (PCDA). Cowley and Moodie (1960) showed that the second-order defocus term of equation (1.9) leads to an image contrast proportional to the second differential of the projected-potential distribution with respect to the coordinates, provided that the defocus Δf is sufficiently small. From Poisson's equation, the second differential of the projected potential is related to the projection $\rho(xy)$ of the charge-density distribution in the object. The charge-density distribution is made up of contributions from both the positive nuclei and the electron clouds of the atoms present. For an isolated atom, $\rho(xy)$ contains a central positive peak and a broader negative peak. When smeared out by thermal motion of the atom and the instrument's spread function, the result is an intensity peak, positive or negative, of strength proportional to the defocus, at the position of any isolated atom. Thus, the high-resolution image gives a direct representation of the projection of the structure because there is an intensity maximum or minimum wherever there is a peak in the projected charge density.

The use of this approximation for thin crystals has been discussed and evaluated, for example, by Anstis et al. (1973). It provides a useful guide in some cases but can be applied only for moderate resolutions. When there is an appreciable contribution to the image from the higher-angle scattering (higher U-values), the fourth-order term of equation (1.9) becomes important, and the simplicity of the relationship is lost.

For light optics, relying on defocus for phase-contrast imaging is usually not feasible. Because the objective aperture angles are large, the depth of focus is relatively small; and contrast is gained at the expense of resolution. Instead, phase contrast is often achieved by using the Zernike technique of putting a phase plate in the back focal plane to modify the relative phases of the waves.

The optics of the electron microscope appear to be ideally suited to application of the Zernike method. Because the illumination of the specimen can be assumed to be parallel and coherent radiation, the distribution of amplitude in the back focal plane of the objective lens is a sharp, central peak corresponding to the directly transmitted radiation and surrounded by the broad distribution of scattered radiation. It should be easy to insert a phase plate to change the phases of the central peak relative to that of the scattered radiation in such a way as to enhance the contrast.

Experiments along these lines have shown considerable success (Unwin, 1972), but there are practical difficulties. For example, one encounters problems in trying to insert a suitable thin object to act as a phase plate and maintain its phase change under the intense electron irradiation and possible contamination in the microscope. So the technique is not widely used.

The Schlieren technique used for special purposes in light optics can also be adapted to electron optics. In this method, an opaque half plane (or the edge of the objective aperture) is placed so that it cuts off half of the diffraction patterns in the back focal plane. The effect on the image contrast is to add an amplitude-contrast term described, very roughly, by the differential of $\phi(xy)$ in the direction perpendicular to the edge of the phase distribution. Thus, a peak in the projected-potential distribution gives a close black-white pair of peaks in the image. In this way, the visibility can be enhanced for small, weakly scattering objects. This technique has been applied effectively in electron microscopy, for example, by Cullis and Maher (1975). It is relatively easy to use and can be applied in many cases with appropriate specimens.

1.3.2 The weak-phase-object approximation

Even though the phase-object approximation is simple in principle, it can lead to complicated results in practice. The discussion of phase contrast imaging for thin specimens becomes simpler if it is possible to make the further assumption that the phase changes of the electron wave are small. Then, equation (1.4) can be approximated as follows:

$$q(xy) = \exp[-i\sigma\phi(xy)] = 1 - i\sigma\phi(xy). \qquad (1.10)$$

The constant 1 then represents the directly transmitted wave, unaffected by the object, which gives rise to the sharp, central peak of the central beam in the diffraction pattern in the back focal plane of the objective. The $i\sigma\phi(xy)$ is the scattering function, which gives rise to the distribution of scattered amplitude in the back focal plane. The i indicates a phase change of $\pi/2$. For weak elastic scattering, the scattered wave is always 90° out of phase with the incident beam. The interaction constant is given by

$$\sigma = \frac{\pi}{\lambda E} = \frac{2\pi me\lambda}{h^2}, \qquad (1.11)$$

and so decreases slowly with the incident-electron energy. The most important factor is thus the projected-potential distribution, which depends on the atomic number of the atoms present and the number of atoms superimposed in the beam direction.

The condition for validity of the weak-phase-object approximation (WPOA) is $\sigma\phi(xy) \ll 1$. This condition is not satisfied for single, very heavy atoms. At the center of the projection of a uranium atom, for example, $\sigma\phi$ may be of the order of 1. For a light atom such as carbon, nitrogen, or oxygen the maximum is of the order of 0.1. However, in a specimen 50 to 100 Å thick, especially for a crystal, 10 or more atoms may be aligned in the beam direction. In this situation, values of $\sigma\phi$ for the projection through the line of atoms may greatly exceed unity. Hence, the range of validity of the WPOA is strictly limited.

One factor is helpful in some cases. In the determination of the value of the projected potential $\phi(xy)$, an arbitrary constant potential may be added or subtracted. A constant added to the phase change can represent nothing more than a change of the reference position relative to which phases are measured. Hence, the maximum excursions of $\phi(xy)$ may be minimized by subtraction of the average value $\bar{\phi}$. In practice, this average is taken over the area of the object where waves contribute to the intensity at any one point of the image. The effect of subtracting this constant term is to greatly extend the range of validity of the WPOA, particularly for disordered materials or for crystals that are not aligned with principal axes parallel to the electron beam. This effect is illustrated in Figure 1.7.

For a crystal aligned with the beam parallel to some crystallographic axis, the projected potential consists of well-separated sharp peaks. The average

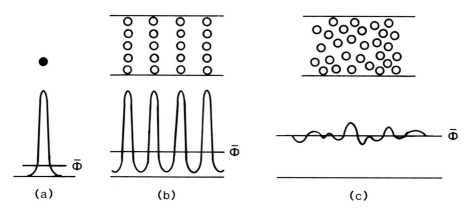

FIGURE 1.7 The form of the projected potential and the resulting deviations from the average projected potential $\bar{\phi}$. (a) A single, heavy atom. (b) A crystal composed of light atoms viewed in an axial direction. (c) A thin film of amorphous material.

potential $\bar{\phi}$ is relatively small, and the deviations from the average, $\phi(xy) - \bar{\phi}$, are large. For an amorphous material, the average potential may be the same; but because the atoms overlap randomly, the deviations from the average are much less. For a rough estimate, we may say that if the number of atoms within a region of diameter equal to the resolution limit is N, the fluctuation from this number is approximately $N^{1/2}$. Hence, the values of $\phi(xy) - \bar{\phi}$ are a factor of $N^{1/2}$ smaller than $\phi(xy)$.

The WPOA (and the POA, in general) is particularly useful for the small-angle-scattering conditions involved in most electron microscopy for which the resolution is much greater than the radius of the atoms. When high-angle-scattering is considered, as for diffraction patterns from amorphous materials or in the limit of very high resolution, the projection approximation fails. Then, for light atoms, the kinematic approximation is more appropriate. In this technique, the scattering amplitudes for all the individual atoms are added, with the phases factors corresponding to their positions in three-dimensional space.

1.3.3 Imaging of weak phase objects

Provided that the WPOA holds, the image wave function is derived by applying the spread function characteristic of the lens in order to smear out the distribution given by equation (1.10). The spread function is in general complex, with real and imaginary parts, $t(xy) = c(xy) + is(xy)$, because it is derived from a lens transfer function that includes phase changes. Thus, the $i\sigma\phi(xy)$ term of (1.10) must be brought into phase with the constant (1) term so that it can modulate the intensity of the image and give an image contrast proportional to $\phi(xy)$. To do so, we must combine this term with the imaginary part of the transfer function, $s(xy)$.

Thus, we write

$$\psi(xy) = [1 - i\sigma\phi(xy)] * t(x),$$

where the $*$ sign represents the smearing-out process corresponding to the loss of resolution (a convolution integral), or

$$\psi(xy) = 1 + \sigma\phi(xy) * s(xy) - i\sigma\phi(xy) * c(xy).$$

Then, the image intensity is found by multiplying ψ by its complex conjugate. If we ignore terms of second order in $\sigma\phi$, we obtain

$$I(xy) = |\psi(xy)|^2 = 1 + 2\sigma\phi(xy) * s(xy). \qquad (1.12)$$

The smearing function $s(xy)$ comes from a Fourier transform of the imaginary part of the lens transfer function. The effect of the lens aperture may be represented by multiplying the wave function in the back focal plane by the aperture function $A(U)$, which is unity for $U \leq a$ and zero elsewhere. The phase changes given in equation (1.9) resulting from the lens defocus and aberration may be represented by multiplying this wave function by $\exp[i\chi(U)]$. Since only the imaginary part of this term is involved in producing the WPOA intensity distribution (1.12), we can assume that the transfer function of the lens is

$$T(U) = A(U)\sin\chi(U) = A(U)\sin(\pi \Delta f \lambda U^2 + \tfrac{1}{2}\pi C_s \lambda^3 U^4). \qquad (1.13)$$

Thus, the effect of the lens aperture and aberrations can be assessed by finding the effect of multiplying the diffraction amplitude in the back focal plane by expression (1.13).

Expression (1.12) represents a simple smearing out of the function $\sigma\phi(xy)$, a projection of the object's potential distribution, which is a highly favorable result of the WPOA. However, its simplicity has led to some confusion of terms and concepts in the literature. For the incoherent-imaging process common in light optics, the image intensity may be represented in a way analogous to (1.12). A function representing the object-intensity distribution is smeared out by a smearing function to give the image intensity. In that case, the smearing function is the Fourier transform of a function known as the contrast transfer function (CTF). Function (1.13) used for coherent imaging of weak phase objects is sometimes referred to as a CTF, but this usage is incorrect. That is, for incoherent imaging, the smearing function is $c^2(xy) + s^2(xy)$, and the CTF is quite different from (1.13).

The effect of the lens transfer function (1.13) on the WPOA images can be calculated by deriving the smearing function $s(xy)$ and applying it as in (1.12). The best images obtainable with a lens of given aberration coefficient C_s are found by choosing the defocus Δf such that $s(xy)$ is as close as possible to a single, sharp peak with no subsidiary maxima. The forms of

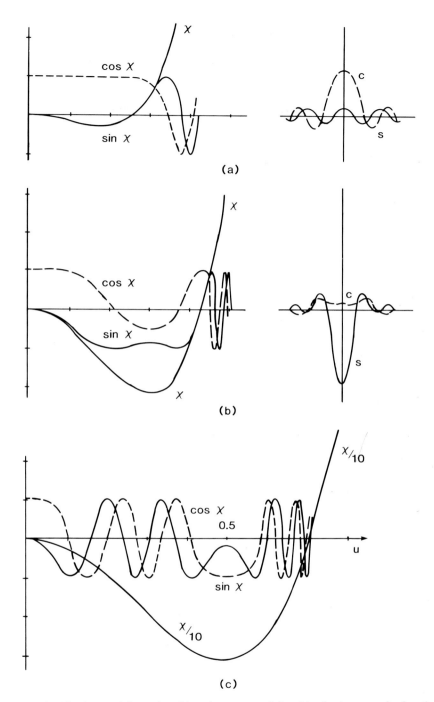

FIGURE 1.8 The forms of the real and imaginary parts of the objective-lens transfer function, $\cos\chi(u)$ and $\sin\chi(u)$, and the corresponding real-space spread functions $c(r)$ and $s(r)$ (right-hand side) for $C_s = 1$ mm and $E_0 = 200$ kV. (a) For a defocus value of 170 Å. (b) For a defocus value of 580 Å (Scherzer defocus). (c) For a defocus value of 1575 Å.

$s(xy)$ for various values of defocus in one particular case are shown in Figures 1.8a–c. There is a particular negative value of Δf, representing an optimum underfocus value, for which $s(xy)$ has the most desirable form, as illustrated in Figure 1.8b. For this defocus, $s(xy)$ has a sharp negative peak, so isolated atoms, or other features giving sharp maxima of $\phi(xy)$, appear in the image as small, dark spots.

The best way of assessing the effect of the lens parameters on the imaging process, however, is to consider the form of $T(U)$ of equation (1.13), rather than $s(xy)$. The forms of $\chi(U)$ and $\sin \chi(U)$, with $U = (u^2 + v^2)^{1/2}$, for various values of defocus Δf, are illustrated in Figure 1.8 for parameters appropriate for some contemporary microscopes ($C_s = 1$ mm; $\lambda = 2.5 \times 10^{-2}$ Å). The wave amplitude of the diffraction pattern in the back focal plane is $\sigma\Phi(uv)$, given by the Fourier transform of $\sigma\phi(xy)$. Multiplication by $\sin \chi(U)$ determines the extent to which these components of the scattered radiation in the diffraction pattern can contribute to the image. Thus, that $\sin \chi(u)$ is close to zero for small values of U implies that the contribution to the image of the inner parts of the diffraction pattern is suppressed. These amplitudes for small U correspond to slowly varying parts of the potential distribution. Under the conditions represented by any of the cases in Figure 1.8, for example, a modulation of the projected potential of the object that has a periodicity of 100 Å ($U = 0.01$) would have such low contrast that it would be invisible in the image.

For the outer parts of the diffraction pattern (large U-values), the $\sin \chi$ function oscillates rapidly in each case. Thus, for the fine detail of the object (the high-spatial-frequency components), the representation in the image is highly confused and, in general, uninterpretable. Usually, the objective aperture is used to eliminate this part of the diffraction pattern, since it may add fine detail to the image but does not add anything to the information that can be derived concerning the specimen.

For the ideal imaging conditions, the magnitude of $\sin \chi$ increases rapidly to near unity for small U-values and then stays close to unity for U-values as large as possible. Then, the greatest possible proportion of the diffraction-pattern amplitudes contribute to the image with the same maximum magnitude and the same phase, and the image is as accurate a representation as possible of $\phi(xy)$. The closest one can get to this ideal situation is for the optimum defocus position, first described by Scherzer (1949) and illustrated in Figure 1.8b. Here, the defocus value Δf is chosen so that the maximum negative value of $\chi(U)$ is about $-2\pi/3$. Then, $\sin \chi$ is close to -1 for $\chi(U)$ between $-\pi/3$ and $-2\pi/3$ and so is close to -1 for a wide range of U-values.

The value for this Scherzer optimum defocus is, for $d\chi/dU = 0$ when $\chi = -2\pi/3$,

$$\Delta f = -(\tfrac{4}{3} C_s \lambda)^{1/2}. \tag{1.14}$$

Under these conditions, the $\sin \chi$ curve does not cross the axis until

$U = 1.51 C_s^{-1/4} \lambda^{-3/4}$. Since for higher values of U, the signs of the contributions to the image are reversed and then oscillate rapidly, one usually inserts an objective aperture to limit the diffraction pattern for this U-value. Alternatevely, one may consider this U-value as determining the limit of the fineness of the meaningful contributions to the image—that is, the resolution limit. On this assumption, the resolution limit—or in common parlance, the "resolution"—is given by the inverse of this U-value as

$$\Delta x = 0.66 C_s^{1/4} \lambda^{3/4}. \tag{1.15}$$

The exact value of the constant in this expression depends on the assumptions made. It is lower if a greater deviation of $\sin \chi$ from unity is allowed so that the optimum defocus is greater. It is higher if the maximum value for U is taken to be that for which $\sin \chi$ is -0.5 or -0.87 instead of 0. In practice, the zero crossing is usually taken as the relevant point because it is easier to determine experimentally.

For this optimum defocus, as indicated previously, the spread function applied to form the image is a sharp negative peak, as illustrated in Figure 1.8b. Around this peak are positive and negative ripples; thus, the image of a single atom, for example, is a small, dark spot with bright and dark fringes around it. The first bright fringe can, in fact, be quite important. For the boundary of an extended object, for example, these fringes sum to the bright Fresnel fringe along the edge.

For zero U-values, $\sin \chi = 0$, which implies that the integral of $s(xy)$ over all x and y must be zero. The sum of all the bright parts must equal the sum over the negative, dark parts.

The existence of these subsidiary fringes around atom images may lead to artifacts. Krakow (1976) pointed out, for example, that in the images of some small molecules, the effect may be to create the appearance of an additional, nonexistent atom.

Thus, for a weak phase object, the bright-field image can give a reasonably good representation of the projected-potential distribution of the object in many cases. But the well-defined limitations on this representation must be borne in mind.

For special purposes or for special classes of objects, the considerations may be different. For example, as discussed later, a thin, perfect crystal gives a diffraction pattern consisting almost entirely of a few sharp peaks of amplitude—the Bragg reflections. For the imaging of such crystals, only the values of $\sin \chi$ for these few special U-values are important. The image is unaffected by any number of oscillations of $\sin \chi$ in between. For example, if the crystal has its first strong diffracted beams at $U = 0.5 \text{ Å}^{-1}$, corresponding to a lattice plane spacing of $d = 2.0$ Å, the defocus of Figure 1.8c is preferable to the optimum defocus of Figure 1.8b.

In practice, other factors besides the defocus and aberrations we have considered so far influence the formation of the images. We have been assuming the ideal case of a monochromatic plane wave incident on the specimen—that is, the case of perfectly coherent radiation. Now, we can discuss the more realistic situation of partially coherent radiation.

1.3.4 The effects of partial coherence

Under the assumption of perfectly coherent radiation, the imaging can be described in terms of a single wave function $\psi(\mathbf{r}, t)$. For high-resolution electron microscopy of thin specimens, this assumption is a good approximation for most purposes, because the spread of the electron energies in the incident beam—and, hence, of electron wavelengths—can be made very small. Also, the sources of electrons used have very small diameters and thus can be focused to give good approximations to incident plane waves. However, there are appreciable modifications of image detail at the current limits of resolution because the ideally coherent conditions cannot be achieved in practice, and the effective transfer function for weak phase objects is modified as indicated in Figure 1.9.

The discussion of imperfect or partial coherence can be complicated. However, for our present purposes, all the effects can be understood in terms of two assumptions:

1. For a source giving a range of wavelengths of radiation, the observed intensity is the sum of the intensities for all different wavelengths considered separately.
2. For a source of radiation of finite diameter, the observed intensity is the sum of the intensities produced by all points of the source considered separately. Even if the radiation were focused to give a wave as close to a plane wave as possible, there would be some convergence because of the finite source size.

In an electron microscope, a change in the energy—and, hence, in the wavelength—of the incident electrons will modify the strength of the interaction of the waves with matter (equation 1.4), the scattering angles, and the phase factor of the lens transfer function (equation 1.9). However, with the usual spread of wavelength of about 10^{-6}, these effects are negligible. The important factor is the change of the objective-lens focal length, which can cause a relatively large change in the defocus (equation 1.9).

The focal length of the lens may also be affected, of course, by fluctuations in the current I in the lens coils. The approximate result, often quoted, is that because of fluctuations ΔE in the energy of the electrons from the electron gun and of fluctuations ΔI of the lens current, the image intensity distribution is smeared out by a spread function of width

$$\Delta_c = C_c \alpha \left(\frac{\Delta E}{E} + 2 \frac{\Delta I}{I} \right), \tag{1.16}$$

where C_c is the chromatic-aberration constant and α is the objective-lens

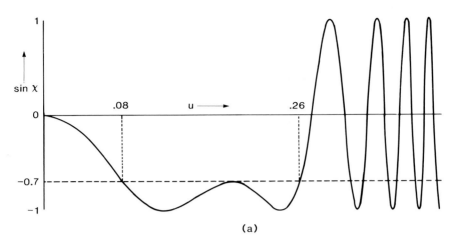

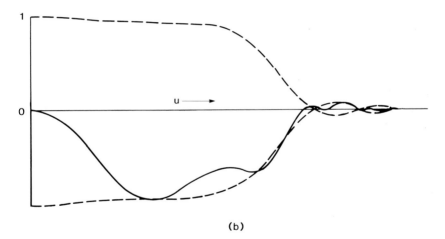

FIGURE 1.9 The effect of the envelope functions on the transfer function $\sin \chi(u)$ for a weak phase object resulting from beam-convergence and chromatic-aberration effects. (a) The function $\sin \chi$ for defocus -1005 Å for 100 keV and $C_s = 2.2$ mm. (b) With the effect of an envelope function. (After O'Keefe and Buseck, 1979.)

aperture angle. The corresponding variation in the defocus value Δf is approximately $\Delta_c \alpha^{-1}$. For microscopes having resolutions in the 2- to 3-Å range, this spread of defocus values may be 100 to 300 Å, which is significant in comparison with the Scherzer optimum defocus values of -500 to -1000 Å.

For weak phase objects, the effective transfer function (equation 1.13) is replaced by the sum of a range of transfer functions for different defocus Δf

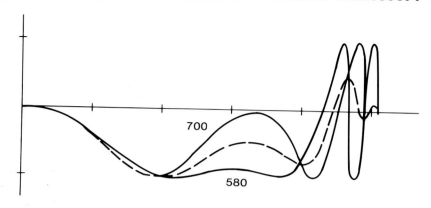

FIGURE 1.10 The effect of a focal-length variation due to high-voltage or lens-current fluctuations on the WPOA transfer function. The sin $\chi(u)$ curves are drawn for defocus values of -580 and -700 Å, and the average value is indicated by the dashed curve.

values. The effect is slight for the inner parts of the sin χ curves of Figure 1.8, which vary slowly with Δf. But for the outer, strongly oscillating parts of the curves, the effect is to average over a range of positive and negative values and greatly reduce the value of the effective transfer function (see Figure 1.10). The overall effect is usually represented by multiplying the transfer function by a so-called envelope function that is unity for small U-values but falls off rapidly for higher U-values (see Figure 1.9).

For a point source of monochromatic radiation, the wave incident on the specimen may be a plane wave for an ideally perfect condenser-lens system. Or it may be a coherent spherical wave converging on a point near the specimen if the condenser lens is focused to provide maximum intensity at the specimen. Thus, the wave irradiating the specimen is coherent, with a convergence angle varying from zero to a maximum value determined by the condenser-aperture size. To a good approximation, this situation is present in modern microscopes when a field-emission gun is used to produce a very small (a diameter of about 40 Å) and very bright electron source. In the more usual case, when a hot-tungsten or lanthanum hexaboride source is used, the effective source size is much larger. In most cases, one can assume that the illumination of the specimen is the same as it would be if the condenser aperture were an incoherent source illuminating the specimen. Then, the image intensity is equal to the sum of the image intensities given by taking each point within the condenser aperture as an independent, incoherent source.

For a weak phase object, the main effect of tilting the incident-beam direction away from the axis of the microscope is to displace the diffraction pattern sideways so that its zero point no longer coincides with the zero point of the function sin $\chi(u)$ (equation 1.13). The effect is much the same

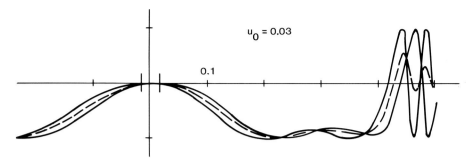

FIGURE 1.11 The effect of incident-beam convergence on the WPOA transfer function. Two curves, $\sin \chi(u)$, are drawn with a lateral displacement of $u_0 = 0.03 \text{ Å}^{-1}$, and the average value is indicated by the dashed curve.

as the effect of moving the $\sin \chi(u)$ function sideways. Thus, the effect of a finite, incoherent source is represented by applying a range of $\sin \chi(u)$ functions displaced sideways along the u-axis by various amounts, as suggested in Figure 1.11. The result is small where $\sin \chi(u)$ is a slowly varying function; but where $\sin \chi(u)$ oscillates rapidly, the oscillations are averaged out. Again, the effect can be simulated by multiplying the transfer function by an envelope function that reduces its value for large values of u by an amount that depends on the defocus and aberration coefficients (see Figure 1.9).

As in the case of the chromatic-aberration effect discussed earlier, these arguments apply only for the weak-phase-object approximation. For a phase object with large phase changes, we cannot use a simple envelope function to represent the effects of either the spread of focal lengths or the finite convergence angle. A more detailed discussion of this case for crystalline specimens is given in Chapter 8. For specimens that are not very thin, the effect of incident-beam convergence is complicated even further because electron waves coming from different directions see different projections of the atom positions. This effect is important when the product of the specimen thickness and the convergence angle become comparable to the microscope resolution. In this situation, the only way to calculate the image intensity is to add the image intensities calculated separately for all the directions of incidence that occur.

The discussion of beam convergence may be used to illustrate the importance of another basic assumption of the simple imaging theory, namely, that the incident beam is aligned exactly along the axis of the objective lens. Thus, the assumption has been that the incident-beam direction defines the central, zero spot of the diffraction pattern and that this direction coincides with the origin of the transfer function [$\sin \chi(u)$ for the WPOA]. If the microscope is misaligned, this assumption will no longer be valid. Then, the imaging properties of the lens may be seriously affected

(Smith et al., 1983). Methods for detecting and avoiding this defect are described in Chapter 12.

1.4 Images of periodic objects

Much of this book is concerned with the imaging of thin crystals, which approximate periodic objects. The interactions of electrons with crystals is the subject of following chapters. But at this stage, we can usefully review the special considerations relating to the imaging process when the wave function in the object plane is periodic because it represents the wave emerging from a crystal.

In terms of the Abbe imaging theory, the essential point about periodic objects is that their diffraction patterns consist of sets of sharp spots, the diffracted beams. The beams are regularly spaced in terms of the angular variables u, v, which represent distances in the back focal plane of the lens. The waves from this regular array of spots in the back focal plane then recombine to form periodic amplitude and intensity distributions in the image plane.

The effect of the lens defocus and aberrations is then to change the relative phases of the finite set of regularly spaced diffraction spots contained within the objective aperture. If the periodicity of the object is large—as for a crystal having a large unit cell, of dimensions 10 to 50 Å, say—the diffraction spots are closely spaced in the back focal plane (Figure 1.12a). Then, the imaging conditions are not very different from those for a diffraction pattern that is is a continuous function—that is, for a nonperiodic object. The equivalent statement relating to the image is that if the resolution limit of the microscope is much smaller than the unit-cell dimensions, each point within the unit cell is imaged independently. Consequently, the image intensity at any point is not influenced appreciably by the fact that the object is periodic.

In contrast, if the periodicity of the object is comparable to the resolution limit, the wave functions from adjoining unit cells overlap and interfere, giving an image intensity that is strongly affected by the periodicity. Correspondingly, in this case, the diffraction spots are well separated on the scale of the transfer function (Figure 1.12b). So only a few points of the transfer function are relevant, and the modifications of the phases of the diffracted beams may depend strongly on the lattice spacing and the defocus. Thus, the imaging of periodic objects can represent special cases.

Considering, for simplicity, the case of one-dimensional, weak phase objects, as in Figure 1.12b, we see that the behavior of the transfer function $\sin \chi$ between the diffraction spots clearly has no influence on the image. There may be any number of maxima, minima, and zeros in between them. There are several consequences of the fact that only particular values of $\sin \chi$ contribute to the image.

One result is that even though a diffraction spot may fall far outside the

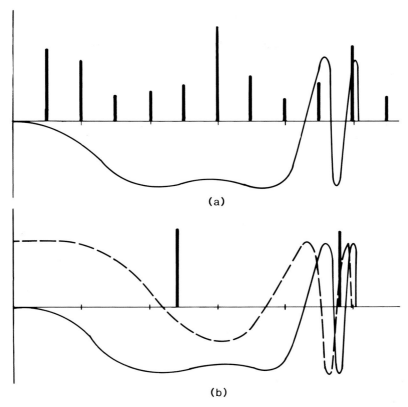

FIGURE 1.12 Relationship of a set of diffracted beams to the WPOA transfer function. (*a*) For a crystal with large unit-cell dimensions. (*b*) For a crystal with small unit-cell dimensions.

inner, slowly varying part of the transfer function, it will have an important contribution to the image if it happens to fall at a maximum or minimum of the oscillating transfer function. The provision is that the amplitude of the transfer function must not be reduced too much by the envelope functions owing to incoherence. Hence, provided that sufficiently coherent radiation is used, periodic components in the image can be produced with spacings much less than the theoretical point-to-point resolution, given conventionally by the position of the first zero of the transfer function. If a sharp diffraction spot is given a large weighting factor by the local value of $\sin \chi$, it produces a fringe pattern in the image by interference with the central beam. This result explains why electron microscopes having a nominal resolution limit of greater than 3 Å can be used to observe crystal-lattice fringe spacings as small as 0.6 Å (Matsuda et al., 1978). It also explains why two-dimensional images of thin crystals obtained with similar microscopes have seemed to show detail on the subangstrom scale (Hashimoto et al., 1977; Izui et al., 1977).

When a number of diffraction spots are involved, as in two-dimensional cases, the relative phases of the reflections vary strongly with defocus. Consequently, although the image shows periodicities with very fine detail, the intensity distributions vary greatly and do not necessarily have any obvious relationship to projections of the structure. When, as in most cases, the crystal specimens are not extremely thin and the relative phases of the diffracted beams are also strongly affected by dynamical-diffraction effects, the image-intensity distributions are perturbed in other ways. Then, the whole concept of simple envelope functions is invalid, as discussed in later chapters.

The functional dependence of the phase factors on defocus and spherical aberration leads to special imaging conditions for one-dimensional periodic objects and for two-dimensional periodicities of high symmetry. The same relative phase relationship of diffracted beams can occur for sets of defocus values separated by fixed defocus intervals or for regularly spaced values of the spherical-aberration coefficient (Hashimoto et al., 1977; Kuwabara, 1978).

This observation can be considered as an extension of the phenomenon of Fourier images, revived, after long neglect, by Cowley and Moodie (1957, 1960). These authors observed and explained that when a periodic object is irradiated by a plane, coherent incident wave, the transmitted-wave function is exactly re-created at defocus intervals of $R = 2na^2/\lambda$, where n is an integer, a is the periodicity of the object, and λ is the wavelength of the radiation. For $R = (2n + 1)a^2/\lambda$, the wave function is exactly re-created but with a lateral shift of $a/2$. For other R-values, the wave function is periodic but with relative phase shifts of the diffracted waves, which implies that the periodicity is the same but the intensity distribution is different.

With reasonably parallel illumination, the set of Fourier images with regularly spaced values of the defocus can be observed in any electron microscopy of thin crystal specimens. These images, though, can be confusing to the extent that the true in-focus position is difficult to select in the absence of nonperiodic objects. For nonperiodic objects such as crystal edges or defects, the normal Fresnel-diffraction effects appear, giving clear indications of underfocus and overfocus conditions.

Examples of the appearance of Fourier images have been shown both in the observations and in the computed, simulated images of crystals, for example, by Iijima and O'Keefe (1979) and Spence et al. (1977).

1.5 Dark-field images

For many purposes, a convenient technique is to use images formed from the electrons scattered by an object and excluding the directly transmitted beam, the central spot of the diffraction pattern. The simplistic interpretation of the image is that it shows the distribution of scattering matter in the specimen without the intense background due to the directly transmitted beam. For low-resolution imaging, this interpretation can often be useful. In studies of crystals, for example, the objective aperture may admit only one strong diffraction spot and its surrounding region of the diffraction pattern. The image then shows that portion of the object for which the crystal lattice is correctly oriented to give that particular Bragg reflection. Such images are useful in displaying the presence of defects or distortions that perturb the crystal lattice and change the diffracted intensity.

For high-resolution imaging of thin crystalline or noncrystalline materials, the origin of the dark-field-image contrast must be examined with more care. For a noncrystalline object, the diffraction amplitude falls off smoothly with scattering angle (Figure 1.13a). For a crystalline object, there are sharp diffraction spots, owing to the crystal lattice, surrounded by weaker scattering containing information on the size, shape, and defects of the crystalline regions. If the objective aperture is displaced to exclude the central beam, the various parts of the diffraction pattern will contribute to the image, with relative phases given by the value of the phase factor $\chi(u)$, as indicated. Clearly, for some particular value of the defocus, the value of $\chi(u)$ may be reasonably constant over the region of the objective aperture. But for other defocus values, it will not be constant, and severe distortion of the image intensities results.

The situation can be improved if, as in Figure 1.13b, the region of the diffraction pattern of interest is translated to the axis of the lens, the origin of the U-coordinate. The procedure is to tilt the incident-beam direction so that the central beam of the diffraction pattern is off-axis. Then, the phase factor $\chi(\mathbf{U})$ is reasonably constant over some range of $\mathbf{U}$ for any defocus. An optimum defocus, close to $\Delta = 0$, can be found that will maximize the range of $\mathbf{U}$ for which χ is uniform and close to zero—that is, for which the largest possible region within the objective aperture is imaged with the correct phase relationships of the diffracted beams. This defocus corresponds to the so-called high-resolution, dark-field condition.

Even so, for this configuration, the imaging does not necessarily give a good representation of the object. Because the atom-scattering factors—and in general, the amplitudes of scattering—fall off rapidly with scattering angle, some asymmetry is introduced into the imaging process. Also, only a small part of the diffraction pattern is included, and it may not contain some of the essential information concerning the object structure. Hence, a direct representation of the details of the object cannot be expected.

In principle, a more satisfactory method of dark-field imaging, which retains the correct symmetry of the imaging process, is the central-stop method. In this method, the incident beam is axial but is removed by a small stop placed on the axis in the back focal plane (Figure 1.13c). On the assumption that a negligible proportion of the diffracted electrons are removed by this stop, one can show that the image is then represented by

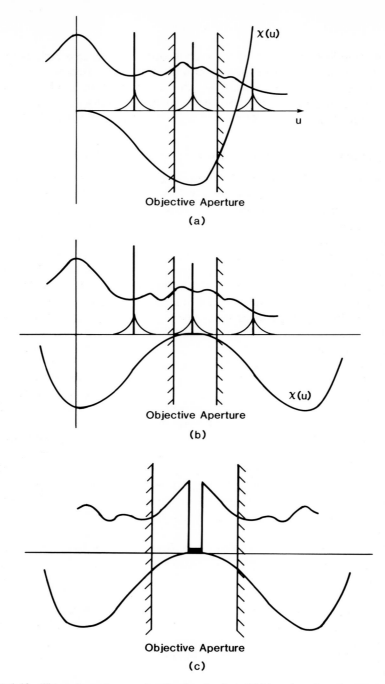

FIGURE 1.13 The reciprocal-space situation for the dark-field imaging of an amorphous object (amplitude of scattering indicated by the continuous curve) or of a finite, defective crystal (delta functions plus surrounding scattering) for the indicated objective-aperture position and phase function $\chi(u)$. (a) Dark-field imaging with a displaced aperture. (b) Dark-field imaging with a tilted incident beam. (c) Dark-field imaging with an axial, central-beam stop.

1.6 Scanning transmission electron microscopy (STEM)

In the early days of electron microscopy, experiments were made with the alternative method of imaging in which a very fine electron probe was scanned over the specimen and the transmitted electrons were collected to form an image signal. The image was displayed by modulating the beam intensity in a cathode-ray tube. The success of this STEM technique depended on the use of a source of sufficient brightness to give an electron current of sufficient magnitude in a very fine beam. So that a reasonable signal is produced, a large number of electrons must be scattered from each picture element of the image within a very short period of time. This result became feasible for high-resolution imaging only with the introduction of the field-emission gun (Crewe and Wall, 1970).

In a field-emission gun, electrons are drawn from a fine metal tip by an intense electrostatic field. The effective source diameter is 30 to 40 Å; so with a relatively small amount of demagnification and loss of intensity, an electron probe of diameter 5 Å or less can be formed on the specimen. The source brightness is up to 10^4 times greater than it is for the usual hot filament, and the energy spread of the electrons is less.

The image intensity in STEM can be related to that for conventional TEM (CTEM) imaging by application of the principle of reciprocity, first formulated for light waves by Helmholtz but more strictly valid for electrons than for light. The relationship is based on the observation, as suggested in Figure 1.2, that the STEM instrument is similar to that for TEM but with the electrons going in the opposite direction.

The reciprocity principle refers to the idealized situation of emission and detection at points. For any system involving scalar waves, scalar fields, and elastic-scattering processes, it may be stated as follows: The amplitude of radiation at a point B due to a source at a point A is identical to the amplitude at point A produced by the same source placed at point B.

For electron waves, the principle can be extended to include vector fields such as the magnetic fields of electron lenses or deflector coils if the reversal of direction of the electron beam is accompanied by reversal of the magnetic-field direction. The same relationship in terms of wave intensities rather than amplitudes can be applied to include inelastic-scattering processes, provided that the energy changes involved are sufficiently small (Pogany and Turner, 1968).

Application of this principle suggests that for point sources and detector elements, the image-intensity distributions should be identical for STEM and CTEM, provided that the geometries of the imaging systems are identical (Cowley, 1969). Extensions to finite sources and detectors can be made on the assumption that the sources and detectors are ideally incoherent—that is, that the image intensity is given by adding the intensities for all points of the source and detector separately. This equivalence of STEM and CTEM imaging has been adequately tested and provides a useful guide in many cases. In practice, however, the techniques and their applications differ widely.

The important limitations and advantages of STEM relative to CTEM derive from practical instrumental considerations. These issues are related to the serial nature of collection of the STEM-image signal as compared with the parallel-image recording in CTEM.

In STEM, one must obtain image signals separately and serially from a very large number (that is, 10^6) of image points in a reasonably short period of time (10 s). Even with a field-emission gun, obtaining a good signal-to-noise ratio for electron-beam diameters of 5 Å or less is difficult, so the quality of the bright-field images tends to suffer. For dark-field imaging, however, the scattered electrons can be detected with much greater efficiency in STEM, than in CTEM since an annular detector can be used to collect all electrons scattered outside the directly transmitted beam. In this configuration, the symmetry of the imaging geometry is maintained. The image-signal intensity can be given—to a good approximation and valid for all but the highest resolutions—by a simple incoherent-imaging approximation for weak phase objects. The image intensity is given, in fact, by the square of the object transmission function, with a smearing function equal to the intensity distribution of the incident electrom beam (Cowley, 1976).

The STEM instrument configuration provides great flexibility in that the many different signals generated when an incident electron beam strikes a specimen may be detected with convenient experimental geometry. The instruments provide either image signals or data for the microanalysis of small regions. Image signals may be collected from the whole central spot, from any part of the central spot of the diffraction pattern, or from any part of the array of diffraction spots. Either elastically or inelastically scattered electrons may be used. Images may also be generated by detection of the emitted X-rays, visible light, secondary electrons, Auger electrons, or conduction electrons in solids. Discussion and examples of the use of some of these modes are given in later chapters.

1.7 Resolution

So far, we have been using the terms *resolution, resolution limit,* and so on, in the imprecise and ill-defined manner conventional among electron microscopists. These terms refer vaguely to the ideas of classical light optics, involving incoherent imaging of amplitude objects. We have seen, however, that in general, the conditions of incoherent imaging do not apply in electron microscopy. We should therefore define more clearly what is implied in the use of these terms for electron microscopy and add more precise terminology when required.

We must distinguish two measures of performance of an electron microscope: the fineness of the detail that can be seen in an image and the amount of information that can be provided regarding the detailed structure of the object. For the classical incoherent imaging in light optics, this distinction is not necessary: Each point in the object plane produces an intensity distribution in the image, and the intensity distributions are added. To determine whether or not two small objects are resolved, we must ask whether the corresponding maxima or minima of intensity in the image can be distinguished as being separate. This judgment is formalized by use of the Rayleigh criterion (see equation 1.8) or some suitable variant.

To the extent that imaging in the electron microscope can be approximated by an incoherent-imaging assumption, the same resolution criteria may be applied, leading to what is usually referred to as the *point-to-point resolution*. This term, however, is also applied on occasion to the more precise definition of resolution based on the assumption of the weak phase object and the Scherzer optimum defocus (equation 1.15). This definition now represents the basis for a standardized and reproducible evaluation of the performance of a microscope and is being increasingly used by microscope manufacturers to describe the performance of their instruments.

To determine the resolution figures, one obtains the Fourier transform of the intensity distribution of the image of a thin, amorphous material. For a photographic recording of the image, an optical diffractometer is used to obtain the Fraunhofer diffraction pattern, which represents this Fourier transform. Alternatively, the transform may be calculated by a computer if the image is digitized. The diffraction pattern of the image allows the function $\sin^2 \chi$ to be deduced. The first zero of this function—and thus, of $\sin \chi$—is found for the Scherzer optimum defocus. The inverse of the U-value for this first zero is taken as the resolution or the resolution limit. With this procedure, the following precautions must be taken:

1. The object must be very thin and scatter weakly so that the weak-phase-object approximation applies.
2. The alignment and stigmation of the microscope must be carefully adjusted.
3. The recording (photographic or digital) must be linear over the intensity range of the image.
4. The defocus must be adjusted to the Scherzer optimum defocus, usually taken as the defocus for which the maximum phase change (before the first zero) is no more than $2\pi/3$.
5. The coherence of the incident radiation must be sufficient to ensure that the diffraction-pattern intensities are not reduced too much by the envelope functions. The first zero of the transfer function must be seen clearly.

It has been pointed out earlier that, especially for thin crystals, contributions from strong diffracted beams occurring far beyond the position of the first zero of the transfer function may produce very fine detail in the image. To some extent, the fineness of the detail observed may be taken to characterize the performance of the microscope, but the situation is complicated because the visibility of fine detail depends on the nature of the specimen.

For weak phase objects, one can define envelope functions by modifying the phase-contrast transfer function, and the form and the extent of the envelope functions can be deduced from optical diffraction patterns obtained from images of thin, amorphous materials. The maximum diameter in the optical diffraction pattern for which the intensity oscillations of $\sin^2 \chi$ are visible may be taken to indicate the limit of fineness of image detail. In principle, at least, the detail on this scale can be interpreted by comparison with calculations based on models of the structure of the specimen, provided that the phase-contrast transfer function is known with sufficient accuracy. This limit has been referred to as the information limit. It depends on the chromatic-aberration terms, involving the high-voltage and lens stabilities, and on the incident-beam convergence and effective source size. Other factors that may be relevant include mechanical vibrations of the microscope and any electrical or magnetic stray fields or other interference.

However, most crystalline samples are not weak phase objects. So the imaging cannot be described in terms of a simple transfer function, and the concept of an envelope function is irrelevant. Lattice-fringe spacings given by thicker crystals can be much smaller than the information limit, as judged on the basis of weak-phase-object theory. This result is evident from the common appearance of spots in optical diffraction patterns, which are due to lattice fringes and appear far beyond the region of appreciable intensity of the transfer function.

The merit of observing lattice-fringe spacings is that it gives a good measure of the mechanical and electrical stability of the instrument and the lack of specimen drift in directions that are not parallel to the fringes. However, the lattice-fringe visibility may not depend strongly on incident-beam convergence or focus variations, particularly for a defocus such that the strong reflections occur where the phase factor χ is almost constant. Hence, the usefulness of fringes as a measure of microscope performance is limited.

In this book, we will use the terms *resolution* and *resolution limit* to refer to the position of the first zero of the transfer function for weak phase objects for Scherzer optimum defocus. Otherwise, we will refer specifically to the *information limit* or the *minimum lattice-fringe spacing*, as appropriate.

REFERENCES

Anstis, G. R., Lynch, D. F., Moodie, A. F., and O'Keefe, M. A. (1973). *n*-Beam lattice images. III. Upper limits of ionicity in $W_4Nb_{26}O_{77}$. *Acta Crystallogr., Sect. A*, **29**, 138.
Born, M., and Wolf, E. (1980). *Principles of optics*. Pergamon Press, London.
Chiu, W., and Glaeser, R. M. (1975). Single-atom image contrast: Conventional dark-field and bright-field electron microscopy. *J. Microsc.*, **103**, 33.
Cowley, J. M. (1969). Image contrast in transmission scanning electron microscopy. *Appl. Phys. Lett.*, **15**, 58.
——— (1976). Scanning transmission electron microscopy of thin specimens. *Ultramicroscopy*, **2**, 3.
——— (1981). *Diffraction physics*. 2nd ed. North-Holland, Amsterdam.
———, and Moodie, A. F. (1957). Fourier images. I. The point source. *Proc. Phys. Soc., London, Sect. B*, **70**, 486.
———, and Moodie, A. F. (1960). Fourier images. IV. Phase gratings. *Proc. Phys. Soc., London, Sect. B*, **76**, 378.
Crewe, A. V., and Wall, J. (1970). A scanning microscope with 5-Å resolution. *J. Mol. Biol.*, **48**, 375.
Cullis, A. G., and Maher, D. M. (1975). Topographical contrast in the transmission electron microscope. *Ultramicroscopy*, **1**, 97.
Goodman, J. M. (1968). *Introduction to Fourier optics*. McGraw-Hill, New York.
Hashimoto, H., Kumao, A., and Endoh, H. (1978). Single atoms in molecules and crystals observed by transmission electron microscopy. In *Electron microscopy 1978*, Vol. III, ed. J. M. Sturgess, 244. Microscopical Society of Canada, Toronto.
———, Endoh, H., Tanji, T., Ono, A., and Watanabe, E. (1977). Direct observation of fine structure within images of atoms in crystals by transmission electron microscopy. *J. Phys. Soc. Jp.*, **42**, 1073.
Iijima, S., and O'Keefe, M. A. (1979). Determination of defocus values using "Fourier images" for high-resolution electron microscopy. *J. Microsc.*, **117**, 347.
International tables for crystallography, published for the International Union of Crystallography. D. Reidel Publishing Company, Boston.
Izui, K., Furuno, S., and Otsu, H. (1977). Observation of crystal structure images of silicon. *J. Electron Microsc.*, **26**, 129.
Krakow, W. (1976). Computer experiments for tilted-beam, dark-field imaging. *Ultramicroscopy*, **1**, 203.
Kuwabara, S. (1978). "Nearly aberration-free crystal images in high-voltage electron microscopy. *J. Electron Microsc.*, **27**, 161.
Matsuda, T., Tonomura, A., and Komada, T. (1978). Observation of lattice images with a field emission electron microscope. *Jpn. J. Appl. Phys.*, **17**, 2073.
O'Keefe, M. A., and Buseck, P. R. (1979). Computation of high-resolution TEM images of minerals. *Trans. Am. Cryst. Assoc.*, **15**, 27.
Pogany, A. P. and Turner, P. S. (1968). Reciprocity in electron diffraction and microscopy. *Acta Crystallogr., Sect. A*, **24**, 103.
Scherzer, O. (1949). The theoretical resolution limit of the electron microscope. *J. Appl. Phys.*, **20**, 20.
Smith, D. J., Saxton, W. O., O'Keefe, M. A., Wood, G. J., and Stobbs, W. M. (1983). The importance of beam alignment and crystal tilt in high-resolution electron microscopy. *Ultramicroscopy*, **11**, 263.
Spence, J. C. H., O'Keefe, M. A., and Kolar, H. (1977). Image interpretation in crystalline germanium. *Optik*, **49**, 307.
Unwin, P. N. T. (1972). Electron microscopy of biological specimens by means of an electrostatic phase plate. *Proc. R. Soc. London, Ser. A*, **329**, 327.

2
IMAGING THEORY
JOHN M. COWLEY

2.1 Waves and scattering

2.1.1 Scattering approximations

In Chapter 1, we made use of the familiar ideas of optics, which in many cases are described in terms of ray diagrams. But even so, we emphasized that, in electron microscopy—and particularly in high-resolution imaging—the relevance of geometric optics is severely limited. For the mathematical treatment in this chapter, the purely wave-optics approach is followed, not only because it is more convenient but also because it describes more directly and succinctly the essential ideas.

Fortunately, for most electron optics, it is possible to make use of wave mechanics in a particularly simple form. For elastic scattering only, the time-independent wave equation can be used. Also, because the scattering angles are very small for fast electrons, the small-angle-scattering approximations can be used; only forward scattering needs to be considered, and only paraxial optics is involved. Hence, although we start with a more general wave equation, the treatment rapidly becomes specialized and simplified because of these conditions. The treatment is not rigorous. It is kept as simple as possible, sacrificing some rigor in order to provide an introduction to the theory needed for an understanding of the chapters.

The time-independent wave equation for scalar waves can be written as

$$\nabla^2 \psi(\mathbf{r}) + k^2 \psi(\mathbf{r}) = 0, \qquad (2.1)$$

where k ($=2\pi/\lambda$) is the magnitude of the wave vector $\mathbf{k}$. This equation may be used for electromagnetic radiation when the vector nature of the electric and magnetic fields is not relevant to the experiment. It may also be used when polarization effects may be taken into account simply by multiplying intensities by a polarization factor, as in the kinematical scattering of X-rays.

For electrons, when spin is not important, the same wave equation may be used with

$$k^2 = \frac{2me}{\hbar^2}[E + \phi(\mathbf{r})] \qquad (2.2)$$

where eE is the kinetic energy of the incident electrons in vacuum and $\phi(\mathbf{r})$ is the distribution of electrostatic potential. The refractive index for electrons is then given by

$$n^2 = \frac{E + \phi(\mathbf{r})}{E}$$

or

$$n \simeq 1 + \frac{\phi(\mathbf{r})}{2E} \tag{2.3}$$

The latter approximation is valid for the scattering of fast electrons by solids since the average value for ϕ in a solid is of the order of 10 V, but the accelerating voltages we will be considering for electrons are 10^4 to 10^6 V.

The solutions of the wave equation are the wave functions, including those for a plane wave,

$$\psi(r) = \exp(-i\mathbf{k} \cdot \mathbf{r}) = \exp(-2\pi i \mathbf{s} \cdot \mathbf{r}), \tag{2.4}$$

and a spherical wave centered at the origin,

$$\psi(r) = r^{-1} \exp(-ikr) = r^{-1} \exp(-2\pi i s r). \tag{2.5}$$

The choice of positive or negative sign in the exponent is arbitrary. The negative sign is chosen as consistent with the "crystallographic," rather than the "physics," convention for Fourier transforms and for the expressions for scattering, imaging, and so on, as will be evident in later chapters. The use of a vector $\mathbf{s}$ ($= \mathbf{k}/2\pi$) is also consistent with crystallographic practice, but the $\mathbf{k}$-notation will be retained when convenient.

The principle of superposition describes properties of solutions of equations such as (2.1)—namely, that if $\psi_n(\mathbf{r})$ are solutions of the wave equation, then $\psi(\mathbf{r}) = \sum_n \psi_n(\mathbf{r})$ is a solution. Common practice is to describe wave functions in terms of sums of plane waves. For the periodic wave fields such as those occurring in crystal structures, the sum of plane waves becomes a Fourier series:

$$\psi(\mathbf{r}) = \sum_{\mathbf{h}} \Psi_{\mathbf{h}} \exp(-2\pi i \mathbf{h} \cdot \mathbf{r}). \tag{2.6}$$

In the general three-dimensional case, $\mathbf{h}$ stands for a set of three integers h, k, l, and the Fourier series can be written in detail as

$$\psi(xyz) = \sum_h \sum_k \sum_l \Psi(hkl) \exp\left[-2\pi i \left(\frac{hx}{a} + \frac{ky}{b} + \frac{lz}{c}\right)\right],$$

where a, b, c are the periodicities in the directions of the x,y,z-axes.

For the discussion of scattering problems, a convenient technique is to use the integral form of the wave equation, which, for electrons, is written as

$$\psi(\mathbf{r}) = \psi_0(\mathbf{r}) + \mu \int \frac{\exp[-i\mathbf{k}(\mathbf{r} - \mathbf{r}')]}{|\mathbf{r} - \mathbf{r}'|} \phi(\mathbf{r}')\psi(\mathbf{r}') \, d\mathbf{r}', \tag{2.7}$$

where μ is an interaction constant to be defined later and $\mathbf{r}'$ is the vector

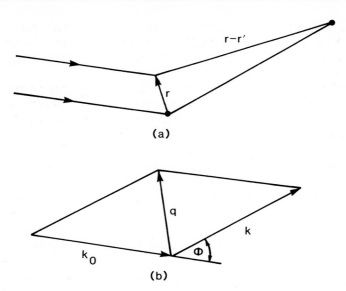

FIGURE 2.1 The scattering of radiation from an object. (*a*) In real space. (*b*) In reciprocal space.

from an origin in the scattering object (see Figure 2.1). The difficulty of solving this equation is that the solution $\psi(\mathbf{r})$ appears within the integral.

The equation states that the scattered wave that is added to the incident wave is found by adding spherical waves from each point of the scattering object with amplitudes depending on the wave amplitude and the scattering potential at the point. The usual method for obtaining a solution is by iteration to produce the Born series.

The first approximation is derived by assuming that the wave function $\psi(\mathbf{r})$ in the object may be approximated by the incident wave $\psi_0(\mathbf{r})$—that is, we assume that the modification of the incident wave by the scattering object is small.

The first Born approximation is then

$$\psi(\mathbf{r}) = \psi_0(\mathbf{r}) + \psi_1(\mathbf{r}),$$

where

$$\psi_1(\mathbf{r}) = \mu \int \frac{\exp[-i\mathbf{k}(\mathbf{r}-\mathbf{r}')]}{|\mathbf{r}-\mathbf{r}'|} \phi(\mathbf{r}')\psi_0(\mathbf{r}')\,d\mathbf{r}'. \tag{2.8}$$

The second approximation is then based on the assumption

$$\psi(\mathbf{r}) = \psi_0(\mathbf{r}) + \psi_1(\mathbf{r}) + \psi_2(\mathbf{r}),$$

where $\psi_2(\mathbf{r})$ is obtained by replacing $\psi_0(\mathbf{r}')$ by $\psi_1(\mathbf{r}')$ in (2.8). Subsequent approximations follow in the same way. The series can be interpreted as giving the sum of the unscattered wave, the singly scattered wave, the doubly scattered wave, and so on. This series provides the basis for one set of approaches to the dynamical scattering of electrons in crystals, as discussed in Chapter 4.

For weakly scattering objects, the first Born approximation is often sufficient. Then, on the assumptions that a plane wave $\psi_0(\mathbf{r}) = \exp(-i\mathbf{k}_0 \cdot \mathbf{r})$ is incident and the observation is made at a large distance R from the object, the expression (2.8) for the scattered wave is

$$\psi_1(\mathbf{R}) = \frac{\mu \exp(-i\mathbf{k} \cdot \mathbf{R})}{R} \int \phi(\mathbf{r}') \exp[+i(\mathbf{k} - \mathbf{k}_0) \cdot \mathbf{r}'] \, d\mathbf{r}'.$$

If we set $\mathbf{k} - \mathbf{k}_0 = \mathbf{q}$ and, to conform to crystallographic convention, $\mathbf{q} = 2\pi\mathbf{u}$, this equation becomes

$$\psi_1(\mathbf{u}) = \frac{\mu \exp(-i\mathbf{k} \cdot \mathbf{R})}{R} \int \phi(\mathbf{r}') \exp(+2\pi i \mathbf{u} \cdot \mathbf{r}') \, d\mathbf{r}'. \tag{2.9}$$

The scattering vector $\mathbf{u}$ has a magnitude of $2\lambda^{-1} \sin(\phi/2)$, where ϕ is the scattering angle. Hence, we obtain the important result that, for kinematical scattering (first Born approximation) in the far-field limit, the scattered-wave amplitude is given by a Fourier transform of the scattering potential.

For the transmission of a wave through an optical system, we consider how to derive the wave function on one plane (coordinates x, y) in the system, given the wave function on a prior plane (coordinates X, Y) (Figure 2.2). Then, the product $\phi(\mathbf{r}') \cdot \psi(\mathbf{r}')$ in equation (2.7) is, replaced by $\psi_1(XY) = [q(XY) - 1] \cdot \psi_0(XY)$, where $q(XY)$ is the object transmission function, so that $\psi_0(XY) \cdot q(XY)$ is the wave leaving the first surface. The distance $|\mathbf{r} - \mathbf{r}'|$ becomes

$$|\mathbf{r} - \mathbf{r}'| = [R^2 + (x - X)^2 + (y - Y)^2]^{1/2}. \tag{2.10}$$

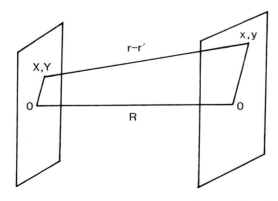

FIGURE 2.2 Coordinate systems for the consideration of Fresnel diffraction.

If, as is usual for electron scattering, we can make the small-angle approximation, $x - X, y - Y \ll R$, the usual expression for propagation of a wave in space by Fresnel diffraction follows:

$$\psi(x, y) = i\frac{\exp(-i\mathbf{k} \cdot \mathbf{R})}{R\lambda} \int \psi_1(XY) \exp\left\{-ik\frac{[(x-X)^2 + (y-Y)^2]}{2R}\right\} dX \cdot dY,$$

or

$$\psi(xy) = i\frac{\exp(-i\mathbf{k} \cdot \mathbf{R})}{R\lambda} \left\{\psi_1(xy) * \exp\left[\frac{-\pi i(x^2 + y^2)}{R\lambda}\right]\right\}, \quad (2.11)$$

where the $*$ sign represents a convolution integral defined, in one dimension, by

$$f(x) * g(x) \equiv \int_{-\infty}^{\infty} f(X)g(x - X) \, dX. \quad (2.12)$$

If, instead of using the small-angle approximation, we use the approximation that the object is very small compared with all other dimensions (that is, $X, Y \ll x, y, R$), then the approximation made in (2.10) is that

$$|\mathbf{r} - \mathbf{r}'| \simeq r = (R^2 + x^2 + y^2 - 2xX - 2yY)^{1/2}. \quad (2.13)$$

Substituting this expression in the integral gives the two-dimensional form of the Fraunhofer diffraction expression:

$$\psi(xy) = \frac{i \exp(-ikr)}{r\lambda} 0(\phi) \int \psi_1(X, Y) \exp[2\pi i(xX + yY)] \, dX \, dY,$$

where $0(\phi)$ is the obliquity factor. Setting $u = x/r\lambda \simeq 2\lambda^{-1} \sin(\phi_x/2)$ and $v = y/r\lambda \simeq 2\lambda^{-1} \sin(\phi_y/2)$ gives

$$\psi(xy) = C \int \psi_1(xy) \exp[2\pi i(uX + vY)] \, dX \, dY, \quad (2.14)$$

and the integral is the two-dimensional Fourier transform.

Thus, the passage of electrons through an electron optical system and the scattering of electrons by matter may be conveniently described in terms of Fourier transforms and convolution integrals. Fortunately, simple and convenient relationships exist between these operations. For example, the convolution theorem states (denoting the Fourier-transform operation by $\mathscr{F}$) that if

$$\mathscr{F}f(x) \equiv F(u)$$

and

$$\mathscr{F}g(x) \equiv G(u),$$

then

$$\mathscr{F}[f(x) * g(x)] = F(u) \cdot G(u). \quad (2.15)$$

The related multiplication theorem gives

$$\mathcal{F}[f(x) \cdot g(x)] = F(u) * G(u). \tag{2.16}$$

The convention for the Fourier transforms used is, in vector notation,

$$F(\mathbf{u}) = \mathcal{F}f(\mathbf{r}) \equiv \int_{-\infty}^{\infty} f(\mathbf{r}) \exp(2\pi i \mathbf{u} \cdot \mathbf{r}) \, d\mathbf{r}$$

and

$$f(\mathbf{r}) = \mathcal{F}^{-1} F(\mathbf{u}) \equiv \int_{-\infty}^{\infty} F(\mathbf{u}) \exp(-2\pi i \mathbf{u} \cdot \mathbf{r}) \, d\mathbf{u}. \tag{2.17}$$

Tables of Fourier transforms are given, for example, in Erdeyli (1954) or Gradshteyn and Ryzhik (1965). Detailed discussions on Fourier transforms in relation to diffraction phenomena are included in Cowley (1981), Goodman (1968), or other modern books on physical optics.

2.1.2 Transmission of electron waves through matter

For idealized optical systems, we assume that all changes of phase and amplitude of the waves occur on planes perpendicular to the optical axis. This assumption is appropriate for the electron microscopy of very thin specimens and serves as a starting point for the development of theories for real three-dimensional objects, as required.

An electron wave transmitted through a thin specimen undergoes a phase change relative to the vacuum wave of an amount given by the integral of the refractive index in the z-direction, the direction of propagation. Therefore, the wave emerging from the object is the incident wave multiplied by the transmission function,

$$q(xy) = \exp\left[-i2\pi\lambda^{-1}\int (n-1) \, dz\right] = \exp[-i\sigma\phi(xy)], \tag{2.18}$$

where $\phi(xy) = \int \phi(\mathbf{r}) \, dz$ and the interaction constant is $\sigma = \pi/\lambda E = 2\pi m e \lambda / h^2$. The approximation (2.18), known as the phase-object approximation (POA), neglects the lateral spread of the waves due to Fresnel diffraction. So it is a small-angle approximation, valid for very thin objects or for λ tending to zero, and hence, it may be also called the high-energy approximation. It may be derived from the Born series (2.8) by taking the limit for λ tending to zero for a two-dimensional object, giving

$$q(xy) = 1 - i\sigma\phi(xy) - \tfrac{1}{2}\sigma^2\phi^2(xy) + \cdots \tag{2.19}$$

Except for very thin objects, we cannot represent the transmission of waves through an object by multiplication by a transmission function. For finite thicknesses, the spreading of waves by Fresnel diffraction within the object, equation (2.11), prevents this representation. In general, however, we can express the action of an object on an incident wave in terms of an operation in Fourier-transform space.

If the incident and exit waves have the periodicity of a crystal specimen, they may be represented by Fourier series, and the coefficients of the Fourier series may be taken as the components of vectors $\boldsymbol{\Psi}_h^0$ and $\boldsymbol{\Psi}_h$. Then, the transformation of the incident-wave vector to the exit-wave vector by the crystal is represented by the action of a scattering matrix

$$\boldsymbol{\Psi}_h = \mathbb{S}\boldsymbol{\Psi}_h^0. \tag{2.20}$$

For nonperiodic objects, the dimensions of the vectors and matrix become infinite; but for many practical purposes, one may be able to use vectors of finite, although large, dimensions. This approach forms the basis for one class of solutions for the general problem of the dynamical scattering of electrons in crystals and may be extended to deal with other electron optical problems.

2.2 Imaging

2.2.1 Abbe theory

In the geometric-optics diagram (Figure 1.4), we see that rays leaving an object at the same angle ϕ are brought together at one point in the back focal plane of the lens. The back focal plane is, in effect, the plane at infinity. The wave amplitude on that plane is given, for an ideally perfect lens, by the Fourier-transform expression for Fraunhofer diffraction, equation (2.14):

$$\Psi(uv) = \mathscr{F}\psi_1(xy) \equiv \int \psi_1(xy) \exp\left[2\pi i(ux + vy)\right] dx\, dy, \tag{2.21}$$

where $\psi_1(xy)$ is the wave leaving the object. The variables u and v are given by the angles ϕ_x and ϕ_y at which the waves leave the object. From the geometry of Figure 1.4, we see that, in the small-angle approximation,

$$\psi(UV) = \mathscr{F}\Psi(XY).$$

The variables U and V are given in the same approximation as

$$U \equiv \frac{x}{(R-f)\lambda} \quad V \equiv \frac{y}{(R-f)\lambda},$$

so that

$$\psi(xy) = \psi_1\left(-\frac{x}{M}, \frac{-y}{M}\right), \tag{2.22}$$

Here, M is the magnification factor R/R_0, and the negative sign is introduced because, apart from the constants,

$$\psi(xy) = \mathscr{F}\Psi_1(uv) = \mathscr{F}[\mathscr{F}\psi_1(xy)] = \psi_1(-x, -y),$$

as can be deduced from the definitions in (2.17).

The magnification factor and image inversion are often ignored because the image information is usually referred back to the object. Hence, we will normally consider the process of imaging as being, ideally, the transition from the object exit-wave function $\psi_1(xy)$, to the Fourier transform $\Psi(uv)$, and back to the image function $\psi(xy)$.

In practice, the extent of the diffraction pattern contributing to the image is limited by the objective aperture in the back focal plane. The effect of the aperture is represented by multiplying the distribution $\Psi(uv)$ by the aperture function

$$A(uv) = \begin{cases} 1, & \text{if } (u^2 + v^2)^{1/2} < \tfrac{1}{2}a, \\ 0, & \text{otherwise.} \end{cases} \quad (2.23)$$

In the ideal case of a point object, the wave at the back focal plane would be a spherical wave converging on the image point. This spherical wave front is perturbed by defocus or aberrations of the lens. The perturbation is represented by multiplying $\Psi(uv)$ by the phase factor $\exp[i\chi(uv)]$. The phase change $\chi(uv)$ contains a second order term in u and v, proportional to the defocus, Δf. Astigmatism introduces different focal lengths in the u and v directions. Assuming astigmatism can be corrected, the important aberration term is of fourth order and derives from the third-order, spherical-aberration term (aberration coefficient C_s), so

$$\chi(uv) = \pi\lambda\,\Delta f(u^2 + v^2) + \tfrac{1}{2}\pi C_s \lambda^3 (u^2 + v^2)^2. \quad (2.24)$$

The function that multiplies $\Psi(uv)$ is thus $T(uv) \equiv A(uv)\exp[i\chi(uv)]$, known as the transfer function for the lens. The image amplitude is then

$$\psi(xy) = \mathscr{F}[\Psi_1(uv)\cdot T(uv)] = \psi_1(xy) \ast t(xy). \quad (2.25)$$

The intensity distribution of the image is

$$I(xy) = |\psi(xy)|^2 = |\psi_1(xy) \ast t(xy)|^2,$$

or, in terms of the two-dimensional vector $\mathbf{r}$,

$$I(\mathbf{r}) = |\psi_1(\mathbf{r}) \ast t(\mathbf{r})|^2. \quad (2.26)$$

The spread function $t(\mathbf{r})$, which produces the limitation of resolution and contrast of the image, is complex, with real and imaginary parts

$$c(r) = \mathscr{F}A\cos\chi(\mathbf{u}) \qquad s(r) = \mathscr{F}A\sin\chi(\mathbf{u}), \quad (2.27)$$

which are real, symmetrical functions of the two-dimensional vector $\mathbf{u}$ (components uv) because $\cos\chi$ and $\sin\chi$ are real and symmetrical.

Since, in general, both $\psi_1(\mathbf{r})$ and $t(\mathbf{r})$ are complex functions, it is not possible to deduce the effect of the lens aberrations on the image-intensity distribution except by calculation. By contrast, in the incoherent-imaging case, valid for the light-optics imaging of a luminous source, the intensity distribution is found by adding intensities for each point of the object. Since each object point gives a finite patch of light in the image with intensity distribution $|t(\mathbf{r})|^2$, the image intensity for incoherent imaging is then represented by

$$I(\mathbf{r}) = I_0(\mathbf{r}) \ast |t(\mathbf{r})|^2. \tag{2.28}$$

The spread function $|t(\mathbf{r})|^2$ for this incoherent-imaging case may be considered as being derived from a so-called contrast transfer function (CTF), given by

$$\mathscr{F}\,|t(\mathbf{r})|^2 = T(\mathbf{u}) \ast T^*(-\mathbf{u}). \tag{2.29}$$

The CTF can be measured experimentally and is used to characterize the performance of a lens for incoherent imaging. For no aberrations, $T(\mathbf{u})$ is the aperture function $A(\mathbf{u})$, and the spread function $t^2(\mathbf{r})$ is given by equation (1.7), leading, by application of the Rayleigh criterion, to the expression (1.8) for the resolution limit. We emphasize that this incoherent, light-optics case is not relevant for the usual coherent-imaging situation in electron microscopy.

2.2.2 Imaging of weak phase objects

For coherent imaging, equation (2.26) leads to an easily interpreted expression for the image intensity only in the limiting case of very thin, weakly scattering objects, for which the weak-phase-object approximation (WPOA) is valid. Then, for a plane wave of unit amplitude incident on the object, the transmitted wave is, from (2.19),

$$\psi(\mathbf{r}) = 1 - i\sigma\phi(\mathbf{r}),$$

and the image intensity is

$$I(\mathbf{r}) = |[1 - i\sigma\phi(\mathbf{r})] \ast [c(\mathbf{r}) + is(\mathbf{r})]|^2.$$

Since

$$c(r) \ast 1 = \int c(r)\,dr = 1$$

and

$$s(r) \ast 1 = \int s(r)\,dr = 0,$$

then, neglecting terms of second order in $\sigma\phi(\mathbf{r}) \ll 1$, we have

$$I(\mathbf{r}) = 1 + 2\sigma\phi(\mathbf{r}) \ast s(\mathbf{r}). \tag{2.30}$$

Thus, the image contrast gives a direct representation of the projected potential $\phi(\mathbf{r})$, smeared out by the spread function $s(\mathbf{r})$. By analogy with the incoherent-imaging case, we may say that the lens action is characterized by the multiplication of the scattered amplitude and a phase-contrast transfer function (PCTF), $A \sin \chi(\mathbf{u})$:

$$\mathscr{F}I(\mathbf{r}) = \delta(\mathbf{u}) + 2\sigma\Phi(\mathbf{u}) \cdot A(\mathbf{u}) \sin \chi(\mathbf{u}). \tag{2.31}$$

The form of the PCTF and the spread function, for various defocus values, is illustrated in Figure 1.8.

The Scherzer optimum defocus is that for which $|\sin \chi|$ is close to unity for the greatest possible range of $|\mathbf{u}|$ ($\equiv U$). If we allow a maximum value of $|\chi|$ of $2\pi/3$, so that $\sin \chi$ decreases to -1.0 and then increases only to -0.87, the optimum defocus is found by setting

$$\frac{d}{du} \sin \chi(\mathbf{u}) = 0 \quad \text{for} \quad \chi = \tfrac{2}{3}\pi,$$

which gives a defocus value of

$$\Delta f = -(\tfrac{4}{3}C_s\lambda)^{1/2}. \tag{2.32}$$

For this defocus, the $\sin \chi$ function crosses the axis for $\chi = 0$, or

$$U_m = 1.51(C_s\lambda^3)^{-1/4}.$$

Assuming that the oscillations in $\sin \chi$ for larger U-values imply that no useful information is obtained from the regions of the diffraction pattern beyond U_m, the value (2.32) gives the optimum size for the objective aperture. The least resolvable distance in the image is then taken to be U_m^{-1}, or the "resolution" is

$$\Delta x = 0.66(C_s\lambda^3)^{1/4}. \tag{2.33}$$

2.2.3 Imaging of phase objects

The range of validity of the WPOA is very small for electrons in the 10^5- to 10^6-V energy range. For light atoms, the most important deviation from the assumptions of the WPOA comes from Fresnel-diffraction effects, which become important for thicknesses of 50 to 100 Å, depending on the resolution being considered. For heavier atoms, the phases factor $\sigma\phi(\mathbf{r})$ may become comparable with unity for thicknesses of 20 to 50 Å. Hence, the POA, which neglectes Fresnel-diffraction effects but includes larger phase changes, has some limited range of validity. The main argument for considering imaging with the POA is that it provides an indication of the nature of the effects that may occur when the WPOA fails.

Since the central beam of the diffraction pattern may usefully be distinguished, we write

$$q(\mathbf{r}) = 1 + \{\exp[-i\sigma\phi(\mathbf{r})] - 1\},$$

so that the image intensity is given by (2.27) as

$$I(\mathbf{r}) = |\{1 + [\cos \sigma\phi(\mathbf{r}) - 1] - i \sin \sigma\phi(\mathbf{r})\} \ast [c(\mathbf{r}) + is(\mathbf{r})]|^2$$
$$= \{1 + [\cos \sigma\phi(\mathbf{r}) - 1] \ast c(\mathbf{r}) + \sin \sigma\phi(\mathbf{r}) \ast s(\mathbf{r})\}^2$$
$$+ \{[\cos \sigma\phi(\mathbf{r}) - 1] \ast s(\mathbf{r}) - \sin \sigma\phi(\mathbf{r}) \ast c(\mathbf{r})\}^2, \quad (2.34)$$

which involves the two transfer functions, $c(\mathbf{r})$ and $s(\mathbf{r})$. If, for simplicity, only terms of first and second order in $\sigma\phi$ are retained, we obtain

$$I(\mathbf{r}) = 1 + 2\sigma\phi(\mathbf{r}) \ast s(\mathbf{r}) - \sigma^2\phi^2(\mathbf{r}) \ast c(\mathbf{r}) + [\sigma\phi \ast s(\mathbf{r})]^2 + [\sigma\phi \ast c(\mathbf{r})]^2. \quad (2.35)$$

The relative importance of the various terms in this expression will vary with the defocus and objective-aperture size. Figure 1.8 gives the form of $\sin \chi(\mathbf{u})$, $\cos \chi(\mathbf{u})$, $s(\mathbf{r})$, and $c(\mathbf{r})$ for various values of defocus, including those near the Scherzer optimum defocus and for Δf small (actually, -170 Å).

For the Scherzer optimum defocus, $s(\mathbf{r})$ has a sharp negative peak, whereas $c(\mathbf{r})$ has no well-defined peak. For the imaging of fine detail, the $c(\mathbf{r})$ terms will contribute mostly low-resolution background fluctuation, which can be ignored. As a first approximation, $c(\mathbf{r}) = 0$, so

$$I(\mathbf{r}) = [1 + \sigma\phi \ast s(\mathbf{r})]^2, \quad (2.36)$$

and the image contrast gives a nonlinear representation of $\sigma\phi$.

Alternatively, for the low-resolution imaging given by the use of a small objective aperture, only the regions with small U-values will be significant. We note that, especially for $\Delta f \simeq 0$, $\cos \chi = 1$ and $\sin \chi = 0$, so $s(\mathbf{r})$ can be ignored. Thus,

$$I(\mathbf{r}) = 1 - \sigma^2\phi^2(\mathbf{r}) \ast c(\mathbf{r}) + [\sigma\phi \ast c(\mathbf{r})]^2. \quad (2.37)$$

Since, as illustrated in Figure 1.7, the value $\sigma\phi$ should be assumed to be $\sigma(\phi - \bar{\phi})$, which oscillates around zero, the last term of this expression will usually be small. Hence, the image intensity can again be considered as a nonlinear representation of $\sigma\phi$; but the squaring, in this case, implies also that the sign of $\sigma\phi$ is lost. Positive and negative deviations from the average projected potential can give similar contrast.

2.2.4 Imaging with partial coherence

The spread of focal lengths of the objective lens owing to the variations of the accelerating voltage for the electrons and the variations of objective-lens current can be represented by substituting $t(\Delta f, \mathbf{r})$ in equation (2.26) and integrating over all values of the defocus Δf, with a weighting function $W(\Delta f)$—that is,

$$I(\mathbf{r}) = \int W(\Delta f) \cdot |\psi_0(\mathbf{r}) \ast t(\mathbf{r}, \Delta f)|^2 \, d\Delta f. \quad (2.38)$$

We can reasonably assume that $W(\Delta f)$ is a normalized Gaussian,

$$W(\Delta f) = (\pi\varepsilon)^{-1/2} \exp\left[-\frac{(\Delta f - \Delta f_0)^2}{\varepsilon^2}\right], \qquad (2.39)$$

where Δf_0 is the average defocus value.

One can express the effect of the spread of defocus values in a simple form within the range of validity of the WPOA, for which (2.30) becomes

$$I(\mathbf{r}) = 1 + 2\sigma \int [\phi(\mathbf{r}) \ast s(\mathbf{r}, \Delta f)] W(\Delta f) \, d\Delta f.$$

Or (2.31) may be written as

$$\mathscr{F}I(\mathbf{r}) = \delta(\mathbf{u}) + \sigma\Phi(\mathbf{u}) \cdot A(\mathbf{u}) \int \sin \chi(\mathbf{u}, \Delta f) \cdot W(\Delta f) \, d\Delta f. \qquad (2.40)$$

The integral over Δf gives

$$\sin \chi(\mathbf{u}, \Delta f_0) \exp\left[\tfrac{1}{4}\pi^2 \lambda^2 (u^2 + v^2)^2 \varepsilon^2\right]. \qquad (2.41)$$

Hence, the effect is to multiply the PCTF by a so-called envelope function, as illustrated in Figure 1.9.

In general, expression (2.38) cannot be simplified so conveniently. In terms of the variables U and V, where $U^2 = (u^2 + v^2)$, equation (2.38) leads to

$$\mathscr{F}I(\mathbf{r}) = \iint \Psi_0(u - U) T(U, \Delta f_0) \Psi^*(-u - V) T^*(V, \Delta f_0)$$
$$\times \exp\left[-\pi^2 \lambda^2 (U^2 - V^2)^2 \varepsilon^2\right] dU \, dV, \qquad (2.42)$$

and the term in $U^2 V^2$ in the exponential prevents any simplification.

The effect of a variation of the angle of incidence of the electron beam on the specimen may be represented in a simple way if the exit wave is given by the multiplication of the entering wave by a transmission function. Then, for each direction of incidence, given by u', v', $\psi_0(xy)$ is multiplied by $\exp[2\pi i(u'x + v'y)]$. On the assumption that the illumination can be represented as coming from an incoherent source $C(u', v')$ in the condenser lens of the microscope, equation (2.26) becomes

$$I(r) = \iint C(u', v') |\{\psi_0(r) \exp[2\pi i(u'x + v'y)]\} \ast t(xy)|^2 \, du' \, dv'. \qquad (2.43)$$

Again, no simple expression can be found to represent the effect on the image in general. But for the WPOA, equation (2.31) becomes

$$\mathscr{F}I(\mathbf{r}) = \delta(\mathbf{u}) + \sigma\Phi(\mathbf{u})[C(\mathbf{u})A \cos \chi(\mathbf{u}) \ast A \sin \chi(\mathbf{u})$$
$$- C(\mathbf{u})A \sin \chi(\mathbf{u}) \ast A \cos \chi(\mathbf{u})].$$

If $C(\mathbf{u})$ is a sufficiently narrow distribution around $\mathbf{u} = 0$, then

$$\mathscr{F}I(\mathbf{R}) = \delta(\mathbf{u}) + \sigma\Phi(\mathbf{u})[C(\mathbf{u}) \ast A(\mathbf{u}) \sin \chi(\mathbf{u})]. \qquad (2.44)$$

The effect of the convolution with $C(\mathbf{u})$ can be approximated as the product of $A \sin \chi$ by the Fourier transform of $C(\mathbf{u})$ with respect to a complicated argument. So the effective envelope function for a circular source is usually written, including the focal-length variation, as

$$\frac{J_1(\eta)}{\eta} \quad \text{where} \quad \eta = \pi\lambda u_0[u\, \Delta f + \lambda u^3(C_s\lambda - i\pi\varepsilon^2)], \tag{2.45}$$

which clearly involves both the defocus value Δf and the angular width of the illumination u_0, as illustrated in Figure 1.9.

A second-order form for the modification of amplitudes and intensities, used for computation of images for the specific case of periodic objects when the WPOA is not valid, is given in Chapter 8.

2.3 Imaging of periodic objects

For transmission through periodic objects, an incident plane wave produces a periodic exit wave, which can be expressed as a Fourier series,

$$\psi_1(xy) = \sum_{h,k} \Psi_{h,k} \exp\left[2\pi i\left(\frac{hx}{a} + \frac{ky}{b}\right)\right], \tag{2.46}$$

If $\mathbf{h}$ is a two-dimensional, reciprocal-lattice vector with components $(h/a, k/b)$, then

$$\psi_1(\mathbf{r}) = \sum_{\mathbf{h}} \Psi_{\mathbf{h}} \exp(2\pi i \mathbf{h} \cdot \mathbf{r}). \tag{2.47}$$

The effects of the lens aberrations are incorporated by multiplying the coefficients $\Psi_{\mathbf{h}}$ by the appropriate values of the transfer functions, to give

$$\Psi'_{\mathbf{h}} = \Psi_{\mathbf{h}} \cdot A(\mathbf{h}) \exp\left\{i\pi\left[\lambda\, \Delta f\left(\frac{h^2}{a^2} + \frac{k^2}{b^2}\right) + \left(\frac{1}{2}\right)C_s\lambda^3\left(\frac{h^2}{a^2} + \frac{k^2}{b^2}\right)^2\right]\right\}. \tag{2.48}$$

The image will therefore be periodic with the same scaled periodicities as the object but with different phases for the Fourier coefficients. So the image amplitude and intensity within the periodically repeated unit will depend on the spherical-aberration coefficient and will vary with defocus.

The image-amplitude and image-intensity distributions may vary periodically with defocus. From (2.48), we see that the image will be the same for defocus values Δf_1 and Δf_2 if

$$\exp\left[i\pi(\Delta f_2 - \Delta f_1)\lambda\left(\frac{h^2}{a^2} + \frac{k^2}{b^2}\right)\right] = \exp(2\pi i m), \tag{2.49}$$

where m is an integer. This result will occur if $a^2/b^2 = n_1$ so that $(h^2/a^2 + k^2/b^2) = H/a^2$ for H an integer.

Then, condition (2.49) is satisfied if

$$\frac{(\Delta f_2 - \Delta f_1)\lambda}{a^2} = 2n,$$

or
$$\Delta f_2 - \Delta f_1 = \frac{2na^2}{\lambda}. \tag{2.50}$$

Thus, for square lattices—or for special cases of nonsquare lattices, including hexagonal—the whole sequence of image intensities will repeat as the focus is changed with the periodicity $2a^2/\lambda$. For these lattices, also, there is a repetition of the image intensity if the aberration coefficient can be varied for fixed defocus. The condition to be satisfied by the unit-cell dimensions is that it should be possible to write $(h^2/a^2 + k^2/b^2)^2 = H/a^4$. Then, the same image will be given for the two values C_{s1} and C_{s2} if

$$C_{s1} - C_{s2} = \frac{4ma^4}{\lambda^3} \tag{2.51}$$

for m any integer.

For square lattices, other special conditions exist. For example, if, instead of (2.50), the relation is

$$\Delta f_2 - \Delta f_1 = \frac{(2n+1)a^2}{\lambda},$$

then the exponential in (2.48) is multiplied by $\exp[\pi i(h^2 + k^2)] = \exp[\pi i(h+k)]$. And (2.6) can be written as

$$\psi^1(xy) = \sum_{h'k} \Psi^1_{hk} \exp\left\{2\pi i\left[\frac{b}{a}\left(x + \frac{a}{2}\right) + \frac{k}{a}\left(y + \frac{a}{2}\right)\right]\right\}, \tag{2.52}$$

so that the image at Δf_2 will be identical with that for defocus Δf_1 but shifted by half a unit cell in each direction. Hence, if this shift of the image is not taken into consideration, the periodicity for repetition of the images with defocus is given by

$$\Delta f_2 - \Delta f_1 = \frac{na^2}{\lambda}. \tag{2.53}$$

The repeated images of one in-focus image, or some other characteristic image, are generally known as Fourier images.

From the discussion in the previous section on the effects of partial coherence, the periodic repetition of image intensities with defocus or C_s will not occur with a finite incoherent source of electrons. For a finite beam convergence from an incoherent source, the envelope function for the WPOA will vary with the defocus, giving a progressive change of the image at successive Fourier-image positions.

For a periodic object illuminated by a point source rather than a plane wave, the Fourier coefficients of (2.46) are multiplied by the additional factor

$$\exp\left[\pi i R_0 \lambda \left(\frac{h^2}{a^2} + \frac{k^2}{b^2}\right)\right]. \tag{2.54}$$

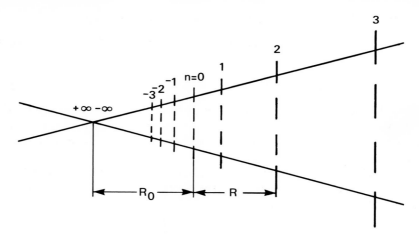

FIGURE 2.3 The Fourier-image positions for a periodic object placed at a distance R_0 from a point source.

It is derived by the Fourier transform of the expression $(\pi i/R_0)\exp[\pi i(x^2 + y^2)/(R_0\lambda)]$ for a spherical wave in the small-angle approximation. The condition that the Fourier coefficients of the image amplitude at a distance R beyond the object be the same as they are for $R = 0$ is given by a replacement of (2.53) by

$$\frac{1}{R} + \frac{1}{R_0} = \frac{\lambda}{na^2}. \qquad (2.55)$$

The image will then be the same as it is for $R = 0$, except for a magnification factor

$$M = \frac{R + R_0}{R_0}.$$

The spacing of the Fourier images is then as suggested in Figure 2.3, with $n = \infty$ or $-\infty$ for $R = -R_0$ (see Cowley and Moodie, 1957).

The special considerations that apply to the imaging of relatively thick crystals giving a small number of diffracted beams will be discussed in Chapters 3 and 4.

2.4 Dark-field imaging

Dark-field TEM images are formed when the central transmitted beam is excluded from contributing to the image, which is then formed only by scattered radiation. These images may thus be obtained either by displacing the objective aperture relative to an axial diffraction pattern in the back focal plane or by displacing the diffraction pattern relative to the axial aperture. In the first case, equation (2.26) may be written as

$$I(\mathbf{r}) = |\mathscr{F}\{\Psi_0(\mathbf{u})A(\mathbf{u} - \mathbf{u}_1)\exp[i\chi(\mathbf{u})]\}|^2. \tag{2.56}$$

In the second case,

$$I(\mathbf{r}) \simeq |\mathscr{F}\{\Psi_0(\mathbf{u} - \mathbf{u}_1) \cdot A(\mathbf{u})\exp[i\chi(\mathbf{u})]\}|^2. \tag{2.57}$$

In each case, the displacement $\mathbf{u}_1$ is such that $A(\mathbf{u}_1) = 0$. In neither case can one write a simple expression for the intensity even with the WPOA, although either expression can be evaluated readily by calculation for specific cases.

In the ideal, central-stop, dark-field technique, rarely approached in practice, the situation is much more favorable. With the WPOA, for example, the diffraction amplitude is

$$\Psi_0(\mathbf{u}) = \delta(\mathbf{u}) - i\sigma\Phi(\mathbf{u}),$$

and the introduction of the central stop removes $\delta(\mathbf{u})$ and also a portion $-i\sigma\Phi(0) \cdot \delta(\mathbf{u})$ of the scattered radiation. Hence, the image intensity becomes

$$I(\mathbf{r}) = |\mathscr{F}\{[\sigma\Phi(\mathbf{u}) - \sigma\Phi(0) \cdot \delta(\mathbf{u})]T(\mathbf{u})\}|^2 = |\sigma[\phi(\mathbf{r}) - \bar{\phi}] * t(r)|^2$$
$$= \{\sigma[\phi(\mathbf{r}) - \bar{\phi}] * c(\mathbf{r})\}^2 + \{\sigma[\phi(\mathbf{r}) - \bar{\phi}] * s(\mathbf{r})\}^2. \tag{2.58}$$

Here, we have set $\Phi(0) = \bar{\phi}$, the average value of the projected potential. For high-resolution imaging at about the Scherzer optimum defocus, $c(\mathbf{r})$ can be neglected as a first approximation. The image detail is then given by

$$I(\mathbf{r}) = \{\sigma[\phi(\mathbf{r}) - \bar{\phi}] * s(\mathbf{r})\}^2.$$

For low-resolution imaging with a small objective aperture, close to zero defocus, we have

$$I(\mathbf{r}) = \{\sigma[\phi(\mathbf{r}) - \bar{\phi}] * c(\mathbf{r})\}^2.$$

In each case, the contrast is of second order in the projected potential, so both positive and negative deviations from the average potential give positive intensity maxima. Arguments can be made that, for relatively small objective-aperture sizes and for large values of the displacement U_1 of the diffraction pattern, expression (2.57) for the so-called high-resolution, dark-field imaging reduces to (2.58), to a good approximation.

For a single, small-scattering object having dimensions much less than the resolution limit, the dark-field image will be approximated by $|t(\mathbf{r})|^2$, or $|s(\mathbf{r})|^2$ near the Scherzer optimum defocus. However, the bright-field-image contrast will show a peak represented by $s(\mathbf{r})$. Since $s^2(\mathbf{r})$ is a narrower peak than $s(\mathbf{r})$, we can say that the dark-field resolution is better than the bright-field resolution. In particular, if $s(\mathbf{r})$ can be approximated by a Gaussian, the $s^2(\mathbf{r})$ peak is narrower by a factor of $2^{1/2}$. So many approximations are involved in this argument, however, that it can, at best, be taken as only a very rough guide.

2.5 Scanning transmission electron microscopy

For a point source of electrons, the objective lens in a STEM instrument produces a spherical wave incident on the specimen that is perturbed by the transfer function of the lens. Hence, the wave incident on the specimen is

$$\psi_0(\mathbf{r}) = t(\mathbf{r}) = \mathscr{F}A(\mathbf{u}) \exp\left[i\chi(\mathbf{u})\right].$$

This incident beam is translated over the specimen by a deflection represented by a vector $\mathbf{R}$. Then, if the transmission through the specimen can be represented by a transmission function $q(\mathbf{r})$, the exit wave is

$$\psi_1(\mathbf{r}) = t(\mathbf{r} - \mathbf{R}) \cdot q(\mathbf{r}). \tag{2.59}$$

On the detector plane (see Figure 2.4), the intensity distribution is then

$$|\Psi(\mathbf{u})|^2 = |Q(\mathbf{u}) * T(\mathbf{u}) \exp(2\pi i \mathbf{u} \cdot \mathbf{R})|^2. \tag{2.60}$$

The image signal is obtained by integrating some portion of this intensity distribution selected by a detector aperture $D(\mathbf{u})$, giving an observed intensity as a function of the beam displacement $\mathbf{R}$,

$$I(\mathbf{R}) = \int D(\mathbf{u}) |Q(\mathbf{u}) * T(\mathbf{u}) \exp(2\pi i \mathbf{u} \cdot \mathbf{R})|^2 \, d\mathbf{u}. \tag{2.61}$$

For a very small axial detector, $D(\mathbf{u})$ can be represented by $\delta(\mathbf{u})$, and (2.61) then reduces to

$$I(\mathbf{R}) = \left| \int Q(\mathbf{u}) T(\mathbf{u}) \exp(2\pi i \mathbf{u} \cdot \mathbf{R}) \cdot d\mathbf{u} \right|^2 = |q(\mathbf{R}) * t(\mathbf{R})|^2, \tag{2.62}$$

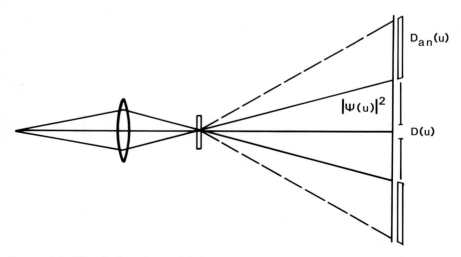

FIGURE 2.4 The placing of an axial detector $D(\mathbf{u})$ or an annular detector $D_{an}(\mathbf{u})$ in the detector plane of a STEM instrument.

which is identical with the equivalent expression for bright-field TEM imaging, expression (2.26). This case, relating to point sources and detectors, is an illustration of the application of the reciprocity principle, as stated in Chapter 1.

The use of a finite detector for STEM, represented by the more general expression of (2.61), is likewise seen to be identical with expression (2.43) for the TEM intensity distribution with a finite incoherent source. The variation of image contrast and resolution for STEM as a function of the diameter of the collector aperture has been explored by Cowley and Au (1978) and Butler and Cowley (1983). Other analogies between TEM and STEM can likewise be suggested. For example, the use of a finite incoherent source for STEM is equivalent to the convolution of the intensity distribution $I(\mathbf{r})$ given by (2.43) and a source function $S(\mathbf{r})$. Similarly, the equivalent of the high-resolution, dark-field, TEM configuration, equation (2.57), will be obtained in STEM by translation of the detector aperture.

For a weak phase object, the intensity distribution on the STEM detector plane is given, from (2.57), by

$$|\Psi(\mathbf{u})|^2 = |T(\mathbf{u})|^2 + \sigma^2 |\Phi(\mathbf{u}) \ast T(\mathbf{u}) \exp(2\pi i \mathbf{u} \cdot \mathbf{R})|^2. \qquad (2.63)$$

The first term $|T(\mathbf{u})|^2 \equiv A^2(\mathbf{u})$ is an image of the objective aperture, a round disk of intensity representing the directly transmitted electrons or, including the small contribution from the second term of (2.63), the central spot of the convergent-beam diffraction pattern obtained with the beam deflected to $\mathbf{R}$. The second term of (2.63) represents the electrons scattered through finite angles. Unless the objective aperture $A(\mathbf{u})$ is very large, most of this intensity will fall outside the central spot, and any or all of this intensity can be collected to form a dark-field image. Since the geometry is not equivalent to that for the usual dark-field-imaging modes of TEM, the dark-field image intensities will be different.

In the most common dark-field STEM configuration, an annular detector is used to collect all, or most of, the intensity scattered outside the central spot of the diffraction pattern. On the assumption that the intensity collected in this way is proportional to the total scattered radiation, the image intensity is given by

$$I(\mathbf{R}) = \sigma^2 \int |\Phi(\mathbf{u}) \ast T(\mathbf{u}) \exp(2\pi i \mathbf{u} \cdot \mathbf{R})|^2 \, d\mathbf{u},$$

which, by use of Parseval's theorem, is

$$I(\mathbf{R}) = \sigma^2 \int |\phi(\mathbf{r}) \cdot t(\mathbf{r} - \mathbf{R})|^2 \, d\mathbf{r},$$

or

$$I(\mathbf{R}) = \sigma^2 \phi^2(\mathbf{R}) \ast |t(\mathbf{R})|^2. \qquad (2.64)$$

Since $|t(\mathbf{r})|^2$ is the intensity distribution of the electron beam incident on the specimen, this result can be interpreted as the incoherent imaging of a specimen with scattering power $\sigma^2\phi^2(\mathbf{r})$, where, as before, $\phi(\mathbf{r})$ is interpreted as $\phi(\mathbf{r}) - \bar{\phi}$, the deviation from the average projected potential. This approximation has been shown to be reasonably accurate except near the resolution limits for the STEM instrument (Cowley, 1976).

The accessibility of the detector plane in a STEM instrument allows one to use a variety of detector configurations. Because STEM image signals are formed as electric signals in serial form, one can add, subtract, or otherwise manipulate and combine signals from various detectors. A review of the various ways in which specially shaped and multiple detectors have been used and expressions for the various image signals have been given, for example, by Cowley and Au (1978).

In various forms of analytical electron microscopy, the image is produced by the detection of secondary radiation produced by the inelastic scattering of the incident electron beam or by the corresponding energy losses of the incident electrons. For strongly localized inelastic interactions, such as the inner-shell excitations leading to the production of high-energy X-rays or Auger electrons, the resolution of the resulting images depends on the intensity distribution of the electron beam. For very thin specimens, this intensity is given by

$$I(\mathbf{r}) = |t(\mathbf{r})|^2. \tag{2.65}$$

For thicker specimens, the electron-beam intensity distribution is spread by multiple elastic and inelastic scattering. For nonlocalized, inelastic interactions, such as those associated with low-energy excitations of electrons in a solid, the situation is more complicated, as will be discussed in Chapters 5 and 6.

2.6 Conclusion

In this chapter, only the most general considerations of imaging have been included. The enormous variety of image-contrast effects, which make electron micrographs scientifically useful and aesthetically pleasing, comes from the wide range of diffraction conditions for particular specimens. For crystals, in particular, the diffraction effects determine the values for the Fourier coefficients (generally complex) used to describe the exact wave from the specimen, equation 2.46. The way in which these Fourier coefficients are determined is the subject of the next two chapters. The variety of image-contrast effects that result is well illustrated in the remainder of this volume.

The treatment of imaging given here relates only to elastic scattering of electrons, but it has relevance also for considerations of inelastic scattering. The inelastic-scattering processes introduce considerations of nonlocalized sources of electrons having changed wavelengths, incoherent with the primary, generating radiation. Once established, however, the inelastically scattered electron waves proceed through the specimen and form exit-wave fields, which are then imaged by the electron optical system in exactly the same way as the elastically scattered wave, although with a change of wavelength. The other effect of inelastic scattering is to modify the elastic-scattered wave in a way that can usually be described by the addition of an absorption function, out of phase with the phase-modifying potential function. This effect also is accommodated readily in the general treatment of imaging by a suitable change of the Fourier coefficients of the wave function at the exit surface of the specimen.

The applications of the imaging theory will therefore be considered and illustrated in subsequent chapters as the interactions of the electron waves with the specimen and the particular specimen structures are described in more detail.

REFERENCES

Butler, J. H., and Cowley, J. M. (1983). Phase contrast imaging using a scanning transmission electron microscope. *Ultramicroscopy* **12,** 39.

Cowley, J. M. (1976). Scanning transmission electron microscopy of thin specimens. *Ultramicroscopy* **2,** 3.

────── (1981). *Diffraction physics.* 2nd ed. North-Holland, Amsterdam.

──────, and Au, A. Y. (1978). Image signals and detector configurations in STEM. In *Scanning electron microscopy,* Vol. I, ed. Om Johari, 53. Scanning Electron Microscopy Inc. A.M.F., O'Hare, Ill.

──────, and Moodie, A. F. (1957). Fourier images. I. The point source. *Proc. Phys. Soc., London, Sec. B* **70,** 486.

Erdeyli, A. (1954). *Tables of integral transforms.* Vol. I. Bateman Mathematical Project. McGraw-Hill, New York.

Goodman, J. W. (1968). *Introduction to Fourier optics.* McGraw-Hill, New York.

Gradshteyn, I. S., and Ryzhik, I. M. (1965). *Tables of integrals, series and products.* Academic Press, New York.

3
ELASTIC SCATTERING OF ELECTRONS BY CRYSTALS
JOHN M. COWLEY

3.1 General dynamical scattering

Several simple approximations are convenient for describing the interaction of electrons with solids and the subsequent formation of an image. The phase-object approximation, mentioned in the previous chapters, is useful for discussing the imaging of very thin objects. The weak-scattering, kinematical approximation, developed for use in X-ray diffraction, is convenient for describing the geometry of electron diffraction patterns and images of thick crystals. Inevitably, our introduction to the subject of the scattering of electrons by crystals must be based on such approximations, because the dynamical theory of electron scattering is relatively complicated. The simple approximations are needed to establish the essential concepts of diffraction geometry and scattering amplitudes.

Before discussing these simpler approaches, however, we emphasize that they are approximations that fail badly in most cases of experimental, high-resolution imaging of crystals. Dynamical-scattering effects are so strong, except in special circumstances, that attempts to use the simple approximations result in gross error. Hence, we begin this chapter with a more realistic description of the scattering process. Because we are concerned solely with the transmission of reasonably high energy electrons through thin crystals, we can ignore the complications of backscattering, which are important for low-energy electron diffraction (LEED) and reflection high-energy electron diffraction (RHEED), and deal with forward scattering only.

The transmission of an electron beam through a crystal can be described in two ways: by referring to the electron-beam picture or to the picture of an electron wave being progressively modified as it travels through the structure. These are descriptions in reciprocal space and real space, respectively. In the first, we consider an incident beam in a direction s_0 entering a crystal, as in Figure 3.1a, and giving rise to a number of diffracted beams. Typically, for electrons, the angles of diffraction are 10^{-2} rad or less; the scattered amplitudes become appreciable for distances into the crystal from 5 to 50 Å, depending on the atomic numbers of the elements present; and the number of diffracted beams occurring simultaneously for such a thickness may be anything from about 10 to 1000. As it progresses through the crystal, each diffracted beam is diffracted again,

ELASTIC SCATTERING OF ELECTRONS BY CRYSTALS

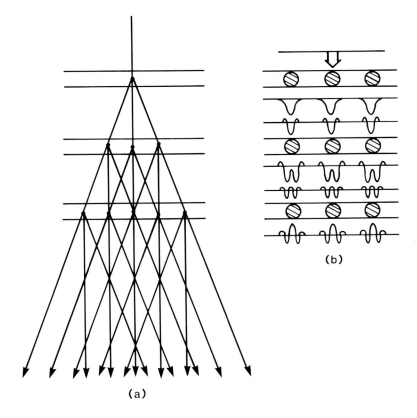

FIGURE 3.1 Representations of the many-beam, dynamical diffraction process. (a) In terms of the multiple scattering of electron beams. (b) In terms of the progressive modulation of an electron wave.

giving rise to a set of doubly diffracted beams. For small scattering angles, each set of doubly diffracted beams is in the same set of directions as the singly scattered beams. Beams progressing in the same directions are added in amplitude so that the total number of diffracted beams remains the same but the amplitudes and intensities are modified.

Then, each doubly scattered beam is scattered again, and so on. In Figure 3.1a, we have separated the single-, double-, and triple-scattering processes, but all these scattering processes occur simultaneously. The better description is that a limited set of diffracted beams passes through the crystal, but each beam amplitude is constantly being changed because every other beam is continually scattering in that direction, giving contributions that add with continually varying phase and magnitude.

The amplitudes of the electron beams emerging from the crystal then have very little relationship to the single-scattering amplitudes. The diverging beams travel through the imaging system, have their relative phases further modified by the lens aberrations and defocus, and then are recombined to form an image. The fixed angles of scattering of the diffracted beams define the periodicities of the image, but the intensities in the image depend on all the phase modifications made during the processes of multiple scattering in the crystal and the imaging by the lenses.

The second description of the interaction of electrons with a crystal is illustrated in Figure 3.1b. For simplicity, we think of the crystal as made up of well-separated parallel planes of atoms. A plane wave passing through the first plane of atoms is modulated in phase because, from the ideas of the phase-object approximation, the phase of the wave is changed in proportion to the projection of the potential distribution of the atoms. Hence, the plot of the phase of the wave as a function of lateral distance shows a periodic array of phase bumps. Next, the wave is propagated through space to the next plane of atoms. The propagation involves an interference effect, Fresnel diffraction, which complicates the phase distribution further, spreading out each of the phase bumps. Then, as the wave passes through the next layer of atoms, the phases are modified further, and so on. Gradually, the phase bump corresponding to any one row of atoms spreads out and overlaps that from neighboring rows of atoms, and interference effects occur, making the whole phase distribution complicated. The wave leaving the crystal then has modulations very different from those for a single layer of atoms, although the periodicity is the same. When the exit wave is imaged, the intensity distribution may be correspondingly complicated.

Again, we have oversimplified the situation because the phase changes produced by transversing the potential field and the phase changes due to Fresnel diffraction actually occur simultaneously. The result remains, however, that unless the phase changes for the layers of atoms are very small and the crystal is very thin, we cannot assume that the effect on the wave will be simply proportional to the projected potential of the crystal.

This complication from dynamical diffraction effects arises because the interaction of electrons with atoms is so strong. But it is because of the strong interaction that strong signals can be obtained from small numbers of atoms and electron microscopy can be a powerful means for investigating the structure of matter on an atomic scale. The complication of the image interpretation is an unavoidable adjunct of that capability.

3.2 Kinematical scattering

For the extreme case of a very weakly scattering material (e.g., small crystals of solid hydrogen), we can assume that the amplitudes of the diffracted beams are very weak compared with the amplitudes of the incident beam. Then, the amplitudes of doubly or multiply scattered beams are very small compared with the amplitudes of the single-scattered beams

ELASTIC SCATTERING OF ELECTRONS BY CRYSTALS 61

and so can be neglected. The total scattered amplitude in any given direction is then given by summing the amplitudes scattered by all atoms individually.

The amplitude scattered in a particular direction by a single atom is given by the atom-scattering factor $f(U)$. Here, U indicates the angle of scattering, appropriately scaled, and, has a magnitude of $U = (2 \sin \theta)/\lambda$, where 2θ is the scattering angle. Or U can be considered as having two components, u and v, for scattering from the z-axis toward the x-axis and the y-axis, respectively, so that $u = (2 \sin \theta_x)/\lambda$, $v = (2 \sin \theta_y)/\lambda$. The atom-scattering factors are calculated by the Fourier transform of the potential distributions of the atoms, and values are tabulated, for example, in the *International Tables for Crystallography*.

The contribution to the scattering from an atom located at a position $\mathbf{r}_j$ is given by the amplitude value $f(U)$ plus a phase factor depending on the atom position. The phase factor is given by $2\pi/\lambda$ multiplied by the path difference for scattering from that atom relative to the scattering from an atom at the origin (see Figure 3.2). If $\mathbf{s}_0$ represents a vector of length λ^{-1} in

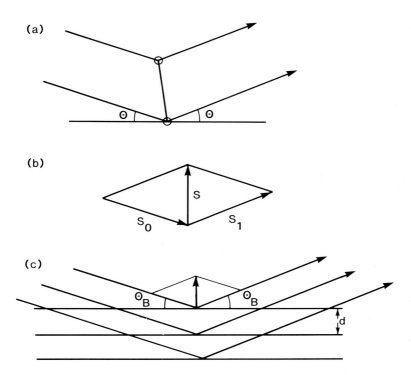

FIGURE 3.2 Diagrammatic ray-optics representations of kinematical scattering processes. (*a*) Kinematical scattering from two atoms. (*b*) The definition of the scattering vector **s**. (*c*) Bragg scattering from lattice planes.

the incident-beam direction and s_1 represents a vector of length λ^{-1} in the diffraction direction, then the path difference is given by the sum of the projections of r_j on these two directions (with s_0 reversed). Hence, the path difference is $(s_1 - s_0) \cdot r_j$. The vector $s_1 - s_0 = s$ is a vector indicating the change of direction—that is, the scattering vector. From Figure 3.2b, we can deduce that its magnitude is $(2 \sin \theta)/\lambda$, where θ is half the scattering angle, so that s is the same as the u defined earlier. Hence, the total scattered amplitude is given by

$$\Psi(\mathbf{u}) = \sum_j f_j \exp(2\pi i \mathbf{u} \cdot \mathbf{r}_j), \qquad (3.1)$$

and the phase factor $\mathbf{u} \cdot \mathbf{r}_j$ is given by the projection of the vector $\mathbf{r}_j$ on the scattering vector $\mathbf{u}$.

3.2.1 Kinematical diffraction from crystals: geometry

The summation in equation (3.1), taken over all atoms of the region of the specimen illuminated by the incident beam, is neither useful nor feasible in most cases, the possible exceptions being for the case of diffraction by very small particles containing no more than a few hundred atoms. The usual methods for evaluating diffraction intensities take advantage of some known or assumed relationship between the atom positions.

For crystals, for example, there is a periodic repetition of groups of atoms in three dimensions with periodicities described in terms of a real-space lattice of points defined by axes of length a, b, c with interaxial angles α, β, γ. One method for taking advantage of this regularity is to recognize that the lattice points lie on sets of parallel planes. These lattice planes are given the Miller indices h, k, l. For crystals having orthogonal axes, the distances between planes of the same set are given by d_{hkl}, where

$$\frac{1}{d_{hkl}^2} = \frac{h^2}{a^2} + \frac{k^2}{b^2} + \frac{l^2}{c^2}. \qquad (3.2)$$

More complicated expressions apply for nonorthogonal axes (see, for example, the *International Tables for Crystallography*).

Bragg introduced the idea of reflection from lattice planes. The waves reflected by successive planes are in phase, and the amplitudes add constructively if the path differences are integral numbers of wavelengths. The simple construction of Figure 3.2c gives the numerical relationship known as Bragg's law:

$$2d_{hkl} \sin \theta_{hkl} = \lambda, \qquad (3.3)$$

where θ_{hkl} is the Bragg angle for the sharp Bragg reflection from the h, k, l-planes.

Bragg's law implies more than the numerical relationship in (3.3) since the concept of reflection involves a geometric relationship. The diffraction vector $\mathbf{s} \equiv \mathbf{s}_1 - \mathbf{s}_0$ having magnitude $(2 \sin \theta_B)/\lambda$ must be perpendicular to the lattice planes and of length d_{hkl}^{-1}. The set of lattice planes, indices (h, k, l), define a set of points, indices h, k, l, when vectors of length d_{hkl}^{-1} are drawn from an origin in directions perpendicular to the planes. This set of points is the reciprocal lattice, well known and used long before the advent of X-ray or electron-diffraction techniques.

From equation (3.2), we see that, for a real-space lattice with orthogonal axes and unit-cell dimensions a, b, c, the reciprocal-lattice points having indices h, k, l are at distances $1/d_{hkl}$ from the origin of a reciprocal lattice having periodicities a^{-1}, b^{-1}, c^{-1}. Thus, the reciprocal lattice is defined by axial vectors $\mathbf{a}^*, \mathbf{b}^*, \mathbf{c}^*$, which, in this case, have lengths a^{-1}, b^{-1}, c^{-1}. The condition for a strong Bragg-diffracted beam being produced for an incident-beam vector $\mathbf{s}_0$ is that the diffraction vector, drawn from the origin of reciprocal space, should end on a reciprocal-lattice point, defined by a vector $\mathbf{h} = h\mathbf{a}^* + k\mathbf{b}^* + l\mathbf{c}^*$. Thus, Bragg's law may be written more completely as

$$\mathbf{s} \equiv \mathbf{s}_1 - \mathbf{s}_0 = \mathbf{h} \equiv h\mathbf{a}^* + k\mathbf{b}^* + l\mathbf{c}^*. \tag{3.4}$$

Equating the magnitudes of these vectors gives (3.3). A graphical representation of (3.4) is given by the Ewald-sphere construction (Figure 3.3). For a given incident-beam direction, the vector $\mathbf{s}_0$ of length λ^{-1} is drawn from a point L to the origin of the reciprocal lattice. Then, a sphere of radius λ^{-1} (the Ewald sphere) is drawn about the point L. If this sphere cuts the hkl-reciprocal-lattice point, then the vector from L to the reciprocal-lattice point will be the vector $\mathbf{s}_1$ for a diffracted beam, because the relationship $\mathbf{s} = \mathbf{h}$ (equation 3.4) will be satisfied.

The set of delta-function, reciprocal-lattice points corresponds to a perfectly periodic, real-space lattice. A periodic object is, by definition, infinite in extent. For real, finite crystals or for imperfect crystals, the diffracted beams will not be confined exactly to the directions defined by the reciprocal-lattice points. Correspondingly, the reciprocal-lattice points may

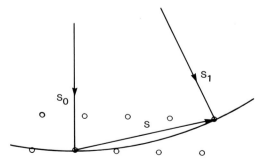

FIGURE 3.3 The condition for obtaining a diffracted beam from a crystal: The scattering vector $\mathbf{S}$ must end at a reciprocal-lattice point.

be considered to be spread out to form distributions of scattering power in reciprocal space. The Ewald sphere intersects the distribution of scattering power in reciprocal space to give scattered beams for a range of directions and not just for the sharp Bragg reflections corresponding to the reciprocal-lattice points. The scattered intensity for a scattering vector **s** corresponds to the value of the scattering power $I(\mathbf{u})$ for which the scattering vector **s** is equal to the reciprocal space vector **u** (coordinates u, v, w).

For a real crystal of finite size, the scattering power has sharp maxima around the reciprocal-lattice points. The widths of these maxima in reciprocal space are inversely proportional to the dimensions of the crystal. This result can be seen from the Bragg-diffraction diagram, Figure 3.2c. For an infinite set of parallel planes, the Bragg's law relation, equation (3.3), must be exactly satisfied if all planes are to scatter in phase. If there are only a few planes, the scattered beams add almost in phase, giving a high value for the total diffracted beam even if relation (3.3) is not exactly satisfied. A rough criterion is that the difference in phase between the first and last plane should not be more than π. Hence, the range of θ-values for which there is a strong constructive summation of scattered beams varies inversely as the number of planes.

The equivalent statement for reciprocal space is that if a crystal has a dimension D in some direction in real space, the region of scattering power around any reciprocal-lattice point is extended in the corresponding direction by an amount proportional to D^{-1}. Simple diffraction theory gives the scattering power of the form $(\sin^2 \pi D u)/(\pi u)^2$ for a crystal dimension D in the x-direction.

For example, the case of common interest in electron microscopy is a thin, single crystal lying perpendicular to the electron beam. For the ideal case of a perfect crystal of uniform thickness H, the reciprocal-lattice points are extended into regions of scattering power elongated in the w-direction (parallel to the real-space z-coordinate, which is parallel to the incident-beam direction) and having the form $(\sin^2 \pi H w)/(\pi w)^2$ (see Figure 3.4a). This peak has a height H^2 for $w = 0$ and a width, measured to the first zero, of H^{-1}. Integration over the peak gives an integrated intensity value proportional to the thickness H.

For electrons of wavelength about 0.037 Å corresponding to 100-keV energies, the radius of the Ewald sphere, λ^{-1}, is very large compared with the distance between reciprocal-lattice points, a^{-1} (for a real-space, unit-cell dimension a) or the distance H^{-1}. For the Ewald sphere drawn as in Figure 3.4a, the sphere cuts through several of the extended regions around the reciprocal-lattice points, giving rise to a number of diffracted beams, forming a cross-grating pattern of spots, as shown in Figure 3.4b. The intensities of the spots are proportional to $(\sin^2 \pi H \zeta_h)/(\pi \zeta_h)^2$, where ζ_h is the excitation error, or the distance in the w-direction between the reciprocal lattice point and the Ewald sphere.

If the incident beam is tilted with respect to the reciprocal-lattice plane,

ELASTIC SCATTERING OF ELECTRONS BY CRYSTALS 65

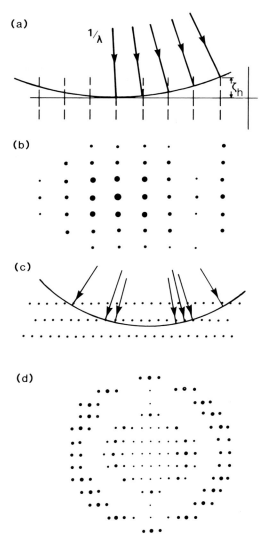

FIGURE 3.4 The intersection of the Ewald sphere with the extended regions of scattering power around the reciprocal-lattice points for a thin crystal and the resulting two-dimensional diffraction patterns. (a) and (b) Incident beam perpendicular to a reciprocal-lattice plane. (c) and (d) Incident beam tilted by a few degrees from this orientation.

the Ewald sphere intersects the reciprocal-lattice plane in a circle, the so-called Laue circle, giving a circle of strong reflections in the diffraction pattern (see Figures 3.4c and d). For higher angles of scattering, the Ewald sphere may intersect other planes of reciprocal-lattice points parallel to the reciprocal-lattice plane through the origin. The intersection with these

66 HIGH-RESOLUTION TRANSMISSION ELECTRON MICROSCOPY

planes gives further circles of strong reflection, concentric with the initial or zero-order Laue circle. In common parlance, the circle of strong spots through the zero spot, or incident-beam spot, is called the zero-order Laue zone (ZOLZ); the further circles of spots are associated with higher-order zones—that is, they are HOLZ reflections.

For example, in Figure 3.5, the diffraction pattern comes from a crystal of a superalloy with the incident beam almost parallel to the c-axis. The strong ZOLZ spots near the center come from the $hk0$ lattice points lying in

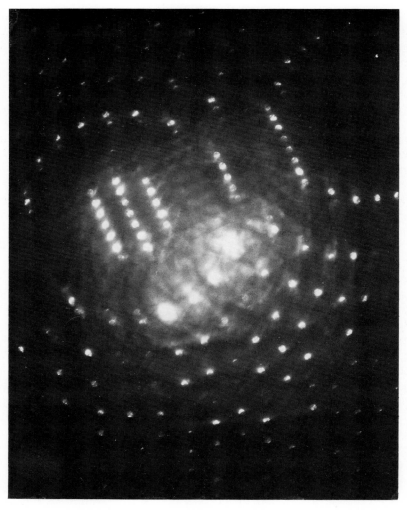

FIGURE 3.5 An electron diffraction pattern from a crystal of a nickel-base superalloy showing the spots lying on successive Laue zones. (Courtesy of Dr. Jing Zhu.)

the plane through the reciprocal-lattice origin. The other rings of spots, the HOLZ spots, come from the intersection of the Ewald sphere with the reciprocal lattice planes of $hk1, hk2, \ldots$, points.

The modulation of intensities by the $(\sin^2 \pi H\zeta)/(\pi\zeta)^2$ factor should, in principle, be evident in such patterns. It usually is not because thin crystals are often not perfectly flat, the incident beam is not a perfectly plane wave, and the crystal thickness is rarely a constant over large areas. All of these factors tend to smear out the intensity modulation.

3.2.2 Convergent-beam diffraction

For diffraction from very small crystals or from very small regions of larger crystals, the extension of the scattering power around the reciprocal-lattice points becomes appreciable in directions perpendicular to the incident beam. Then, the diffraction spots are no longer sharp but are spread out into patches having dimensions inversely proportional to the crystal dimensions or the diameter of the incident beam, whichever is smaller.

As a special case, we may consider convergent-beam microdiffraction. In this mode, a lens is used to focus an incident beam to a small spot on the specimen. If the beam is coherent, is coming from a point source, and is limited by a circular aperture in the lens, the amplitude distribution of the beam at the specimen level will be given by a Fourier transform of the aperture function as $J_1(\pi ar)/(\pi ar)$, as described in Chapter 1, for an angular aperture of width $u = a$. The intensity distribution at the specimen is given by the square, $J_1^2(\pi ar)/(\pi ar)^2$ and so is a peaked function of radius, to the first zero, $r = 1.22a$. The beam diameter at the specimen will be increased, of course, if the source of electrons is of finite size instead of being a point source.

In the diffraction pattern formed at a large distance beyond the specimen, the central spot is an image of the aperture, a bright disk of diameter $u = a$ (see Figure 3.6a), and each diffraction spot is likewise broadened into a disk. The intensity at each point of the zero-beam disk represents the transmitted-beam intensity for a particular direction of incidence, and the intensity at each point of the diffraction-spot disk depends on the intensity of the diffraction for the corresponding direction of incidence. This result may be described in terms of the Ewald-sphere construction in reciprocal space (Figure 3.6b). For each incident-beam direction within the cone, the Ewald sphere is drawn. The intersection of this Ewald sphere with the corresponding region of scattering power about the reciprocal-lattice point then gives the contribution to the intensity.

For example, if the sample is a thin, perfect crystal of uniform thickness H, the distribution of scattering power around the reciprocal-lattice point is proportional to $\sin^2(\pi H\zeta)/(\pi\zeta)^2$, and the intensity across the diffraction-spot disk varies in the same way, as suggested in Figure 3.6c. Since tilting of the incident beam in directions perpendicular to the plane containing both

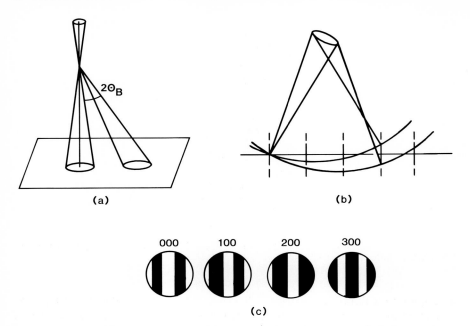

FIGURE 3.6 The formation of a convergent-beam electron diffraction (CBED) pattern. (*a*) Real-space diagram of the diffracting conditions. (*b*) Reciprocal-space diagram showing the intersection of a range of Ewald spheres with the extended reciprocal-lattice points for a thin crystal. (*c*) The CBED pattern corresponding to part (*b*).

the incident beam and the reciprocal-lattice vector does not change the excitation error, the modulation of the intensity in the diffraction-spot disks takes the form of bands at right angles to the diffraction vector. The zero beam shows dark bands corresponding to the loss of intensity to the diffracted beams.

This description of the formation of the convergent-beam electron diffraction (CBED) patterns is adequate for kinematical diffraction under a limited set of circumstances relevant to early experiments but is no longer valid for some modern instruments. Interference between electron waves scattered for different directions of the incident radiation can usually be neglected for hot-filament electron guns because the large, effective source size limits the coherence of the convergent beam. With some modern microscopes and particularly the dedicated STEM instruments, field-emission guns provide very small electron sources and so can give a high degree of coherence in the convergent beam. Then these interference effects can be prominent.

For perfect, thin, flat crystals for which the diffracted-beam disks do not overlap, the coherence of the incident convergent beam does not affect the intensities, because, for each direction of the incident beam, the diffracted

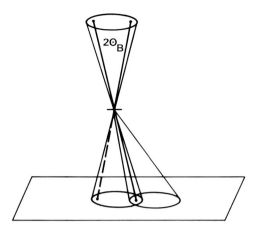

FIGURE 3.7 The formation of a CBED pattern with overlapping spots, showing the possibility of interference of beams in the region of overlap.

wave goes in a different direction. If the diffracted-beam disks overlap, the intensities in the regions of overlap are the result of interference between waves coming from different directions but diffracted by different sets of lattice planes (Spence and Cowley, 1978), as suggested in Figure 3.7. Thus, the intensities in the regions of overlap vary strongly with the relative phases of the two interfering waves. This variation is illustrated in the diffraction pattern shown in Figure 3.8. This effect may be the basis for the absolute determination of the relative phases of diffracted beams from crystals and hence the solution of the phase problem of structure analysis for kinematical intensities (Hoppe, 1969; Nathan, 1976) However, there are considerable complications involved. Even for the ideal-kinematical-scattering case, which is rarely achieved, the relative phases of the diffracted

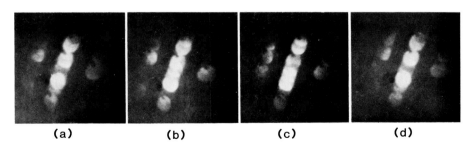

(a)   (b)   (c)   (d)

FIGURE 3.8 The CBED patterns from a thin crystal of the titanium-niobium oxide, $Ti_2Nb_{10}O_{29}$, obtained with an incident beam of diameter about 8 Å, which is smaller than the unit-cell periodicity (14 Å for the main line of spots). The diffraction spots overlap, and there are interference effects depending on the position of the beam in the unit cell. For (a) to (d), the incident beam is moved from one side of a unit cell to the other.

beams depend not only on the structure amplitude of the reflections (see section 3.2.3) but also on the defocus, lens aberrations, and alignment of the electron lenses. They also depend on the position of the axis of the electron probe relative to the origin of the unit cell of the crystal.

For small or imperfect crystals, the diffraction pattern for any one incident-wave direction is not confined to the set of sharp diffraction spots corresponding to the reciprocal-lattice points. Each incident-wave direction scatters into a continuous range of directions. Waves scattered from different incident-beam directions overlap and interfere. This result leads to a variety of interference phenomena that are visible in the diffraction patterns (Cowley and Spence, 1979; Cowley, 1980). For edges of crystals or planar faults, the effect is to produce a splitting of the diffraction spots into two or more components. Observation of this effect can be used in some cases to gain information concerning the nature of crystal defects (Zhu and Cowley, 1982).

3.2.3 Kinematical diffraction from crystals: intensities

Information relating to relative atom positions in a crystal is contained in the intensities of the diffraction spots. For kinematical diffraction, the relationship of intensities to the atomic positions is relatively simple. For the sharp Bragg reflections from a crystal, the scattered amplitudes from all real-space lattice points will add in phase. If there is an identical atom at each point of the lattice, the scattering from each atom is given by the atomic-scattering factor $f(\mathbf{u})$, so the total scattered amplitude in the direction defined by a reciprocal-lattice point is proportional to $Nf(\mathbf{u})$ if there are N atoms. Then, the intensity scattered in this direction, defined by $\mathbf{u} = \mathbf{h}$, is proportional to $N^2 f^2(\mathbf{u})$.

We have seen that, for a finite crystal, there is a region of scattering power around the reciprocal-lattice point. The peak value of the scattering power is proportional to the square of the crystal dimensions or, for a three-dimensional case, to the square of the volume, or to N^2. The width of the peak of scattering power is inversely proportional to the crystal dimensions. Thus, the integration over the peak of scattering power—and hence, the integrated intensity measured when the Ewald sphere is swept through the peak—is proportional to $Nf^2(\mathbf{u})$.

For most crystals, there is more than one atom associated with each real-space lattice point. The scattering amplitudes from all atoms within the unit cell of the crystal must be summed, with phase factors corresponding to the atom positions $\mathbf{r}_j$ with respect to the unit-cell origin. Thus, instead of an amplitude $f(\mathbf{u})$ from each lattice point, there is a contribution to the total scattered amplitude of $\Psi(\mathbf{u})$, proportional to

$$\Phi(\mathbf{u}) = \sum_j f_j(\mathbf{u}) \exp(2\pi i \mathbf{u} \cdot \mathbf{r}_j), \tag{3.5}$$

where the summation is over all atoms of the unit cell. The total integrated intensity for a Bragg reflection is then proportional to $N|\Phi(\mathbf{u})|^2$. The quantity $\Phi(\mathbf{u})$ is the structure amplitude. Usually, we note that the intensity is that of the **h**-reciprocal-lattice point and write

$$\Phi_\mathbf{h} = \sum_j f_j(\mathbf{h}) \exp(2\pi i \mathbf{h} \cdot \mathbf{r}_j),$$

or

$$\Phi(hkl) = \sum_j f_j(\mathbf{h}) \exp\left[2\pi i\left(h\frac{x_i}{a} + k\frac{y_i}{b} + l\frac{z_i}{c}\right)\right],$$

and

$$I(hkl) \propto |\Phi(hkl)|^2. \tag{3.6}$$

The actual intensities observed in practice depend on the experimental conditions and the method of measurement. Expressions appropriate for various experimental situations have been given, for example, by Vainshtein (1964).

We pointed out in earlier chapters that the electrons are scattered by the potential distribution of the atoms, $\phi(\mathbf{r})$. For a single atom, the atomic-scattering factor is given by

$$f(\mathbf{u}) = \mathscr{F}\phi(\mathbf{r}),$$

where the $\mathscr{F}$ symbol represents a Fourier-transform integral. Similarly, the structure amplitude $\Phi(\mathbf{u})$ is given by the Fourier transform of $\phi(\mathbf{r})$, the potential distribution in the whole of the unit cell. The Fourier-transform integral can be inverted. Thus, the potential distribution in the unit cell can be obtained from the scattered amplitude,

$$\phi(\mathbf{r}) = \mathscr{F}^{-1}\Phi(\mathbf{u}) = \int \Phi(\mathbf{u}) \exp(-2\pi i \mathbf{u} \cdot \mathbf{r}) \, d\mathbf{u}.$$

If $\Phi(\mathbf{u})$ is defined only around the reciprocal-lattice points, this integral can be replaced by the summation

$$\phi(\mathbf{r}) = \sum_\mathbf{h} \Phi_\mathbf{h} \exp(-2\pi i \mathbf{h} \cdot \mathbf{r})$$

$$= \sum_{hkl} \Phi(hkl) \exp\left[-2\pi i\left(\frac{hx}{a} + \frac{ky}{b} + \frac{lz}{c}\right)\right]. \tag{3.7}$$

Expression (3.7) is identical to the expression relating the electron-density distribution $\rho(\mathbf{r})$ to the structure amplitudes $F(hkl)$ for X-ray diffraction. This expression for the X-ray case is the basis for the structure analysis of crystals from X-ray-diffraction data. Similarly, (3.7) can be used as the basis for the structure analysis of crystals from electron-diffraction data. As in the case of X-rays, the outstanding problem for this process arises because it is the intensities, proportional to $|\Phi_\mathbf{h}|^2$, rather than the amplitudes $\Phi_\mathbf{h}$, that can be derived directly from kinematical-diffraction data. In the recording of the intensities, the phases of the structure amplitudes $\Phi_\mathbf{h}$, which are generally complex numbers, are lost. Solutions of the resulting phase problem are possible through the use of dynamical-scattering effects, as will be seen later (Chapter 4), but involve considerable complications within the context of kinematical scattering.

The methods and results for structure analysis from electron-diffraction data have been described, for example, by Vainshtein (1964) and Cowley (1967). To summarize the situation, the advantages of using electrons rather than X-rays for structure analysis include (1) the possibility of analyzing extremely small specimen regions; (2) improved sensitivity to light atoms such as H or H^+ in the presence of heavier atoms; and (3) the ease of observing the indications of disorder or defects in crystals given by diffuse scattering or streaking in the diffraction patterns.

The difficulties that have prevented electron-diffraction structure analysis from being more widely used include (1) the obvious failure of the assumption of kinematical scattering for most experimental, single-crystal patterns and the less obvious but important dynamical-diffraction effects for patterns from polycrystalline materials; (2) the difficulty of determining the experimental parameters (e.g., specimen thicknesses, crystal distortions) with sufficient accuracy in the case of very small specimen regions; and (3) the radiation-damage effects that severely restrict the applications of electron diffraction to organic, biological, and many inorganic materials.

The expressions for structure amplitudes, equation (3.5) or (3.6), suggest that the intensity $|\Phi_\mathbf{h}|^2$ is associated with a reciprocal-lattice point. From the results on the structure amplitudes and the considerations of the Ewald-sphere geometry, the intensities of the $hk0$ spots for a thin, perfect crystal and a beam incident in the direction of the c-axis are given by

$$I(hk0) = |\Phi(hk0)|^2 \sin^2 \frac{\pi H \zeta_{hk0}}{(\pi \zeta_{hk0})^2}, \qquad (3.8)$$

where $\Phi(hk0)$ is given by equation (3.6).

To get an intensity measurement proportional to $|\Phi|^2$, one must either determine the final geometric factor of (3.8) with high accuracy or obtain an integration over the excitation error ζ_h. The former method is rarely possible, so the latter procedure is normally used. The integration can be performed by rotating the crystal, as is usual in X-ray–diffraction experiments. Or it can be performed by using samples in which a well-defined range of orientations of the crystal exist with respect to the incident beam, as, for example, when a thin, single crystal is uniformly bent or when a polycrystalline sample is used with a random distribution of orientations of the crystallites (see Vainshtein, 1964).

3.2.4 Intensities for amorphous or microcrystalline specimens

When the dimensions of the individual diffracting crystals are very small or when amorphous materials are used, the intensities of diffraction from the individual small regions of the sample cannot be described in terms of peaks of scattering power centered on reciprocal-lattice points. Reciprocal-lattice points are derived from a real-space, periodic lattice. If such a real-space lattice cannot usefully be defined because there is no well developed periodicity of the atom positions, one can refer to the general expressions in (3.1). The Ewald-sphere construction can still be used. For any diffraction vector **u** corresponding to a point on the Ewald sphere, the diffracted intensity is given by

$$I(\mathbf{u}) = |\Psi(\mathbf{u})|^2 = \sum_j f_j \exp(2\pi i \mathbf{u} \cdot \mathbf{r}_j) \cdot \sum_i f_i^* \exp(-2\pi i \mathbf{u} \cdot \mathbf{r}_i)$$

$$= \sum_i \sum_j f_i f_j \exp[2\pi i \mathbf{u} \cdot (\mathbf{r}_j - \mathbf{r}_i)]. \tag{3.9}$$

Here, we have assumed that the atomic-scattering amplitudes are real numbers.

The important point to note is that the diffraction intensities do not depend on the individual atom positions $\mathbf{r}_j$. Thus, in the diffraction pattern, as distinct from the image, all information on absolute atom positions is lost. Only the vectors $\mathbf{r}_j - \mathbf{r}_i$, which are the vectors between atoms positions, are relevant.

For a normal diffraction specimen containing thousands or millions of atoms, the number of interatomic vectors is enormous. Equation (3.9) is useful only if the statistics of interatomic separations is known (or is found from observed intensities). For example, if one assumes that all the atoms are identical, (3.9) may be written as

$$I(\mathbf{u}) = f^2 \sum_i n_i \exp(2\pi i \mathbf{u} \cdot \mathbf{r}_i),$$

where n_i is the number of interatomic vectors of length $\mathbf{r}_i$. Or (3.9) may be written as

$$I(\mathbf{u}) = f^2 N \sum_i p(\mathbf{r}_i) \exp(2\pi i \mathbf{u} \cdot \mathbf{r}_i), \tag{3.10}$$

where $p(\mathbf{r}_i)$ is the probability that if any atom is taken as origin, there will be another atom at the position $\mathbf{r}_i$.

For gases, liquids, and amorphous materials, one can assume that, in a sufficiently large sample volume, interatomic vectors will occur equally often in all directions. Hence, equation (3.10) can be reduced to a one-dimensional equation for intensities as a function of U (where $U^2 = u^2 + v^2 + w^2$) having the form

$$I(U) = Nf^2(U) \sum_i p(r_i) \frac{\sin \pi U r_i}{\pi U r_i}, \tag{3.11}$$

where $p(r_i)$ is the probability that there will be an atom at a radial distance r_i from any atom chosen as origin.

The sum of weakly oscillating $(\sin \pi U r_i)/(\pi U r_i)$ terms in (3.11) is added to a uniformly falling background term $Nf^2(u)$, given for $r_i = 0$. The radially symmetric distribution of scattering power in reciprocal space, when intersected by the Ewald sphere, gives a weakly oscillating intensity distribution—a pattern of diffuse halos on a sharply falling background. However, this intensity distribution can sometimes be measured with high accuracy. Equation (3.11) is a one-dimensional, radial form for the Fourier transform and can be inverted. The probability function $p(r_i)$ can be obtained from the observed intensities

$$p(r) = 4\pi r \int_0^\infty \frac{UI(U)}{Nf^2(U)} \sin 2\pi Ur \cdot dU. \qquad (3.12)$$

This expression gives the radial-distribution function, the function that provides information on the interatomic distances and the relative frequency with which they occur.

This analysis of amorphous materials is a well-known application of the concept that kinematical-diffraction intensities depend on the statistics of the occurrence of interatomic vectors. The more general formulation of the concept and applications to the study of disordered or faulted crystals can be found, for example, in Cowley (1981).

3.3 Limitations of the simple approximations

The kinematical approximation and the phase-object approximation (POA) are relatively simple and provide useful guides for the preliminary assessment of diffraction and imaging phenomena. They accurately reproduce the geometry of the diffraction pattern and the periodicities of crystal images, except under special conditions. In many cases they provide a first, rough approximation for the intensities of diffraction spots and suggest the image intensities that may be produced in particularly favorable cases. There is a temptation to use these rough approximations for initial evaluation of diffraction patterns and images and then extend their application to regions where their validity is doubtful. Therefore at this stage it is useful to summarize the theoretical and experimental evidence on their domains of validity.

3.3.1 Kinematical-approximation limitations

The basic assumption of the kinematical approximation is that the ratios of the amplitudes of the first-order scattered radiation to the incident-beam amplitude are so small that second- and higher-order-scattering amplitudes may be neglected. The best way to test this assumption is to calculate the scattering amplitudes, using one of the approaches to the full, dynamical-scattering theory outlined later in this chapter. The results of such calculations suggest that the validity of the kinematical approximation depends strongly on the crystal orientation and on the values of scattering potential that is, on the atomic numbers and separations of the atoms concerned.

For a gold crystal with the incident beam parallel to a [100] axis, for example, the incident-beam intensity falls to a minimum of about 20 percent of its initial value, and the main diffracted beams rise to a maximum intensity for a thickness of about 20 Å for 100-keV electrons (Fisher, 1969). The kinematical limit then is surpassed for a thickness of about 10 Å. For a gold crystal in a different orientation, the limiting thickness may be 2 to 10 times as great.

For light-atom specimens such as organic crystals, the thickness for failure of the approximation is expected to be much greater. One may argue that if the unit-cell dimensions are large, as is the case for many organic materials,

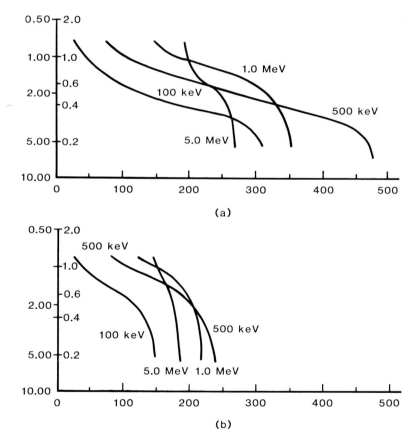

FIGURE 3.9 The approximate boundary lines for which the diffracted-beam magnitudes for kinematical scattering possess a reliability factor R; for crystals of two organic compounds as a function of crystal thickness and the electron-microscope resolution involved. The curves are for different accelerating voltages. (a) $R = 0.05$. (b) $R = 0.10$. (After Jap and Glaeser, 1980.)

the diffraction pattern will contain a very large number of very weak reflections, none of which will be intense for even large thicknesses of 100 to 1000 Å. In contrast, although double diffraction from one reflection may be very small, if there are N reflections, there will be N^2 double-diffraction contributions; and the total contribution of these effects to a diffraction pattern or image may be considerable.

Detailed calculations have been made by Jap and Glaeser (1980) for several organic materials of moderately large unit-cell dimensions (see Figure 3.9). They measured the validity of the kinematical approximation in terms of a reliability factor R similar to that used in X-ray–diffraction structure analysis. With an R factor of 10 percent as a reasonable limit, the

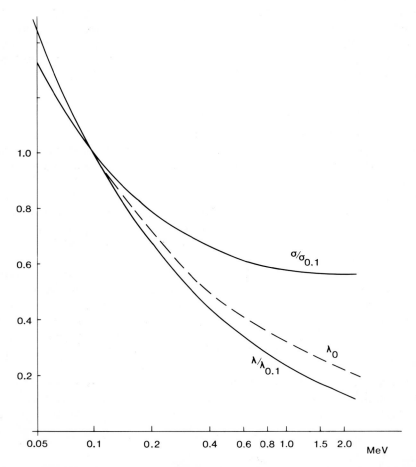

FIGURE 3.10 The variation with electron energy of the relativistic wavelength λ and the interaction constant σ, as compared with the values for 100 keV. The broken curve shows the nonrelativistic variation of λ.

kinematical approximation is a good guide for the calculation of diffraction intensities or amplitudes for thicknesses of about 100 Å for 100-keV electrons. The useful thickness increases with accelerating voltage to about 200 Å for 1 MeV.

The same increase, however, is not that expected from simple, nonrelativistic theory. The interaction constant σ, which describes the strength of the interaction of electrons with matter, may be written as

$$\sigma = \frac{\pi}{\lambda E} = \frac{2\pi m_0 e}{h^2} (\lambda^2 + \lambda_c^2)^{1/2}, \qquad (3.13)$$

where m_0 is the rest mass of the electron and λ_c is the Compton wavelength, $\lambda_c = h/m_0 c = 0.0242$ Å, corresponding to an accelerating voltage of 212 kV. This expression implies that, for very high accelerating voltages, the interaction constant tends to a constant value (see Figure 3.10), and the kinematical approximation does not improve as the electron energy increases. The value of σ comes within 7 percent of its limiting value for 1-MeV electrons.

3.3.2 Phase-object-approximation limitation

The limitations of the phase-object approximation (POA) have been discussed in general terms in Chapter 1. The approximation involved is the neglect of the spreading of the electron wave by Fresnel diffraction. From equation (2.11), the spread of the wave is on the scale of $(R\lambda)^{1/2}$ for a distance R in the beam direction. A rough estimate gives a spread of about 2 Å in a distance of 100 Å for 100-keV electrons. Since this spread is comparable with the distances between atoms, the POA will be poor for atomic-resolution imaging for such thicknesses and will be good only for much smaller thicknesses of 10 to 20 Å.

The detailed calculations of Jap and Glaeser (1980) for light-atom materials confirm this conclusion. For mediumweight and heavy-atom crystals, special considerations apply. There is growing evidence of a channeling effect that occurs when the incident beam is parallel to a zone axis (Marks, 1985; Tanaka and Cowley, 1986). The electron paths tend to be confined to atom rows, and the POA may be valid for thicknesses of up to 100 Å for 100-keV electrons.

Because of the dependence on the wavelength, the POA should improve with increased electron energy. In this case, the relativistic effects are such as to improve the approximation even further. The wavelength is given by

$$\lambda = h/(2m_0 e E^*)^{1/2},$$

where E^* is the effective, relativistic accelerating voltage given by

$$E^* = E + \left(\frac{e}{2m_0 c^2}\right) E^2. \qquad (3.14)$$

The decrease of wavelength due to the relativistic effect is thus about 5 percent for 100 keV, 41 percent for 1 MeV, and 230 percent for 10 MeV, with corresponding, but less spectacular, increases in the thickness for which the POA should apply.

The weak-phase-object approximation (WPOA) involves the assumptions of both the POA and the kinematical approximation and so has the limitations of both. It fails for 5- to 10-Å thicknesses for heavy atoms because the phase shift is not small. It fails for 10- to 20-Å thicknesses for light atoms because the spread of the waves due to Fresnel diffraction is appreciable. The region of usefulness for the WPOA is mostly in relation to low-resolution imaging of amorphous materials for which a spread of the waves by a few angstroms is not significant and the subtraction of the average potential $\bar{\phi}$, as in Figure 1.7, reduces the effective value of the relative changes of phase in the specimen.

3.4 Dynamical diffraction

The general description of the scattering of electrons by solids with which this chapter opened should now be expanded to include the actual methods by which the diffraction process has been formulated. These methods provide the basis for the calculation of diffraction amplitudes and image intensities. All the methods used are basically equivalent and must give the same results when correctly applied. The differences in approach, however, provide different insights into the diffraction process and allow particularly convenient descriptions of particular special cases.

In his original paper on the subject, Bethe (1928) set up the wave equation for electrons in the periodic potential field of a crystal and then applied the boundary conditions appropriate for a crystal surface with incident and diffracted waves in the vacuum. This general theory, aimed at the explanation of LEED experiments, was later developed more specifically for the transmission case (e.g., by MacGillavry, 1940), with the simplification of limiting the number of diffracted beams to one so that only the interactions of two beams (the incident beam and one diffracted beam) were considered. Later, this treatment was reexpanded to include many beams, usually with a matrix formulation, and has formed the useful basis for describing the medium-resolution electron microscopy of relatively thick crystals, as described in the books of Hirsch et al. (1965), Thomas and Goringe (1979), and others.

The alternative approach to the transmission-scattering problem is that implied in the introduction: namely, to consider the progressive modification of wave amplitudes or diffracted-beam amplitudes as the electron beam passes through the crystal. Various formulations are based on this concept, including the use of the Born series if the scattering is limited to the forward direction (Fujiwara, 1959). Essentially the same concept, but involving scattering by planar sections of the specimen rather than volume elements, is the basis of the multislice formulation of Cowley and Moodie (1957). The approach by Howie and Whelan (1961) was to write differential equations giving the progressive modification of diffracted-beam amplitudes by the scattering processes. In the semireciprocal formulation by Tournarie (1961), the uniqueness of the beam direction z is recognized. The two-dimensional wave equation is solved in the x,y-directions for which the periodicity of the crystal is assumed to be perfect, and the solution is then progressively modified by the propagation in the z-direction. Related developments, particularly appropriate for the consideration of the channeling of electrons through thick crystals, consider the scattering of electrons in progression along localized atomic strings through the crystal (see Buxton and Tremewan, 1980). There are also approaches based on the Bethe-type, Bloch-wave formulation that involve a progression through the crystal—notably, the scattering-matrix formulation, introduced by Sturkey (1957, 1961), which was the basis for the first attempts at many-beam, dynamical-intensity calculations. Of the many approaches, most have not found any considerable direct application to the subject matter of this book and so will not be discussed further here.

3.4.1 The Bloch-wave formulation

Following Bethe (1928), solutions are sought for the Schrödinger equation for electron waves in the periodic potential field of a crystal. A solution will consist of an incident beam, represented by a wave vector $\mathbf{k}_0$, directed to the reciprocal-lattice origin and many diffracted waves, represented by wave vectors $\mathbf{k_h}$, directed to the reciprocal-lattice points. All these wave vectors originate at the same point L (see Figure 3.11). The collection of incident and diffracted waves, constituting one solution of the wave equation, is known as a Bloch wave.

We note immediately that the $\mathbf{k_h}$-vectors do not have the same lengths. There is no restriction that all diffraction waves have the same wavelength, as in kinematical theory. Instead, with each wave vector $\mathbf{k_h}$, we associate an amplitude $\Psi_\mathbf{h}$; and this amplitude will be large if the $\mathbf{h}$-reciprocal-lattice point lies close to the Ewald sphere.

As the direction of the incident-wave vector $\mathbf{k}_0$ varies, its length also varies, depending on the modification of the wave by diffraction effects. Hence, instead of being a sphere of radius $2\pi/\lambda$, the locus of the point L is a surface, known as the dispersion surface, which is approximately a sphere when the diffraction effects are weak. For fast electrons, the deviations from the spherical shape are actually small in comparison with the radius; but for convenience, the derivations are usually drawn as if they were large.

In general, there are many solutions to the wave equation and hence many Bloch waves within a crystal. The number of Bloch waves is equal to the number of reciprocal-lattice points, which is infinite. But in practice, only a finite number of Bloch waves need be considered. And for some

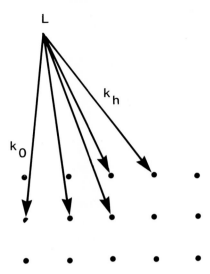

FIGURE 3.11 The set of wave vectors for the components of one Bloch wave in a crystal. The vector k_0 is the incident wave, drawn to the 000 reciprocal-lattice point. One vector k_h is drawn to each reciprocal-lattice point.

particular cases, one can deal with only two Bloch waves, corresponding to the two reciprocal-lattice points **0** and **h**, as discussed in the next section.

For each Bloch wave, there is a corresponding point $L^{(i)}$ and a corresponding surface that is the locus of $L^{(i)}$. These surfaces are referred to as the branches of the dispersion surface.

To determine what Bloch waves are generated in a crystal and what their relative amplitudes are for a given wave incident on the crystal from outside, one must apply the boundary conditions for the entrance face. The requirement is that the projection onto the crystal boundary of any incident-wave vector in the crystal must be equal to the projection on the boundary of the wave vector **K** of the incident wave in the vacuum. This condition implies that all points $L^{(i)}$ must lie on a line perpendicular to the surface, as in Figure 3.12. Since the positions of the points $L^{(i)}$ define all the wave vectors for all the Bloch waves, the problem is then to solve the wave equation to deduce the relative strengths of the various Bloch waves and the amplitudes of the diffracted waves within each Bloch wave. The formulation is usually written as a matrix equation, with these strengths and amplitudes as the eigenvalues and eigenvectors for the solution. This formulation is the basis for the computation of diffraction intensities and images using the Bloch-wave approach.

For a parallel-sided crystal slab, the exit surface adds no new boundary conditions. Each wave in the crystal becomes a wave in vacuum, with the wavelength corresponding to the vacuum electron energy. All waves

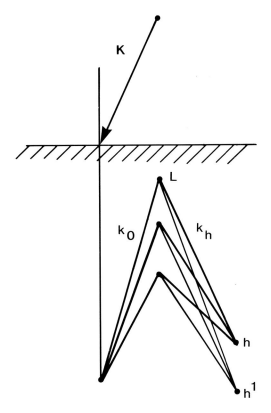

FIGURE 3.12 The relationship of wave vectors in vacuum and in a crystal for the O- and h-reciprocal-lattice points.

associated with a given reciprocal-lattice point emerge in the same direction, and their amplitudes are added to give the diffracted wave in vacuum. Since, inside the crystal, the **h**-waves have different $\mathbf{k_h}$-vectors and hence different wavelengths (different refractive indices), they emerge with relative phases depending on the crystal thickness. Hence, the diffracted-beam intensities will fluctuate, usually in a very complicated way, with thickness (see Figure 3.13).

3.4.2 The two-beam approximation

Except for special cases, in the dynamical theory of X-ray diffraction one may assume that only one diffracted wave is excited at any one time. Then, a two-beam theory is used to describe the interactions of the incident and diffracted waves. For electrons, in general the strength of the interaction with the crystal ensures that a large number of diffracted waves will be

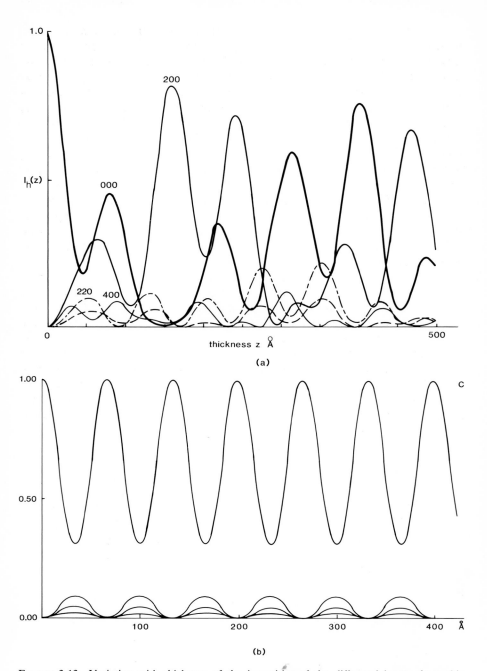

FIGURE 3.13 Variation with thickness of the intensities of the diffracted beams from thin crystals of disordered $CuAu_3$ alloy, calculated without absorption for 100-keV electrons. (*a*) With a tilt from [100] to satisfy the Bragg condition for the (020) reflection. (*b*) For an incident beam in a [100] direction. (From P. M. J. Fisher, 1969.)

excited and will interact coherently. There are some special circumstances, however, for which the two-beam approximation is good. For light-atom crystals having small unit cells and with the incident-beam direction far from an axial direction, one low-index reflection may be much stronger than all others. Calculations (e.g., Fisher, 1969) suggest that more than 99 percent of the energy may be in the two strong beams (**0** and **h**). Even for gold, only 2 or 3 percent of the energy may be in weak beams for 100-keV electrons.

This favorable circumstance has allowed the effective and valuable application of electron microscopes to the study of dislocations and other faults in metals, semiconductors, and ceramics with the relatively simple interpretations based on the two-beam theory (see Hirsch et al., 1965, and many later publications). Even though this approximation usually is not valid for most of the axial orientations used for high-resolution imaging or for crystals having larger unit cells, the basic phenomena of dynamical diffraction are most readily illustrated for the two-beam case and so will be reviewed briefly.

In the two-beam case for a thin crystal slab with the incident beam almost parallel to the lattice planes and perpendicular to the crystal surface, the dispersion-surface construction is represented in Figure 3.14. For the kinematical case with only those waves having $|\mathbf{k}| = 2\pi/\lambda$ in the crystal, the dispersion surface consists of two spheres centered on the **0**- and **h**-reciprocal-lattice points. With the interaction of the waves, the surfaces separate from the intersection point of these spheres with a gap of minimum width $2\sigma\Phi_\mathbf{h}$—that is, the modification of the wave-vector lengths depends on the structure amplitude for the reflection. There are two Bloch waves, each with components for the **0** and **h** reflections. The relative amplitudes of

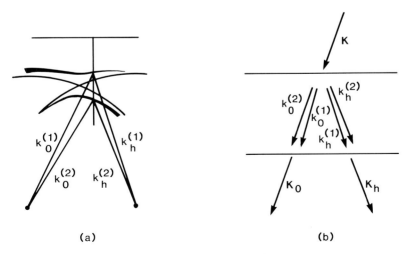

FIGURE 3.14 The wave vectors for the two Bloch waves for two-beam dynamical diffraction. (*a*) In reciprocal space. (*b*) Diagrammatically, in real space.

these components are suggested by the darkening of the lines of the dispersion surface. Well away from the Bragg angle, the incident-beam wave will be strong and the diffracted wave will be negligible. Whichever Bloch wave contains the incident beam, having the same $|\mathbf{k}_0|$ as for kinematical scattering, will predominate. Near the Bragg angle, both waves will be strong in both Bloch waves.

Because the two Bloch waves have $\mathbf{k}$-vectors of different length, they will, in effect, see the crystal as having different refractive indices, with $n-1$ proportional to $\Phi_0 + \Phi_\mathbf{h}$. When the two $\mathbf{k}_0$-waves or the two $\mathbf{k}_\mathbf{h}$-waves combine at the exit surface of the crystal to give the vacuum waves $\mathbf{K}_0$ or $\mathbf{K}_\mathbf{h}$ (see Figure 3.14b), they do so with a phase difference proportional to $\Phi_\mathbf{h}$ and the crystal thickness H. At the exact Bragg angle, the intensities of these beams then vary sinusoidally with thickness:

$$I_0 = \cos^2(2\pi\sigma\Phi_\mathbf{h}H) \qquad I_\mathbf{h} = \sin^2(2\pi\sigma\Phi_\mathbf{h}H). \tag{3.15}$$

This sinusoidal variation, shown in Figure 3.15 with absorption effects added, represents the "pendellosung," or pendulum solution of Ewald. The effects appear prominently as fringes of equal thickness in electron micrographs of wedge-shaped crystals for both bright-field (**0**-beam) and dark-field (**h**-beam) images. The periodicity of the oscillations is known as the extinction distance, $\xi_\mathbf{h} = (2\sigma\Phi_\mathbf{h})^{-1}$.

When the incident beam is not at the exact Bragg angle, one may deduce from Figure 3.14 that the difference in length of the $\mathbf{k}$-vectors increases so that the fluctuations of intensity with thickness have a smaller periodicity. However, the relative amplitudes of the components from the two Bloch waves are not equal, so the contrast of the modulation of the intensities is reduced.

The parameter w is used to measure the deviation from the Bragg angle, where $w = \zeta_\mathbf{h}/2\pi\sigma\Phi_\mathbf{h}$ and where $\zeta_\mathbf{h}$ is the excitation error of the kinematical

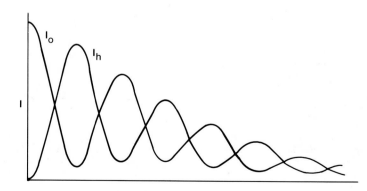

FIGURE 3.15 Variation of the intensities of the incident beam (I_0) and the diffracted beam ($I_\mathbf{h}$) for two-beam dynamical scattering with absorption.

theory. Then, the more general expression for the beam intensities is

$$I_0 = (1 + w^2)^{-1}\{w^2 + \cos^2[2\pi\sigma\Phi_\mathbf{h} H(1 + w^2)^{1/2}]\},$$
$$I_\mathbf{h} = (1 + w^2)^{-1} \sin^2[2\pi\sigma\Phi_\mathbf{h} H(1 + w^2)^{1/2}].$$
(3.16)

The variation with w or $\zeta_\mathbf{h}$ gives the "rocking curve," or the variation of intensity with deviation from the Bragg angle. This intensity variation is seen, for example, in the spots of a convergent-beam diffraction pattern. For w^2 much greater than unity, the expression for $I_\mathbf{h}$ clearly becomes proportional to $\sin^2(\pi H\zeta_\mathbf{h})/(\pi\zeta_\mathbf{h})^2$, which is the kinematical result. Thus, convergent-beam diffraction patterns give central lines and subsidiary fringes, as suggested in Figure 3.6c, except that, because (3.16) contains $(1 + w^2)^{1/2}$ rather than w, the intensities are modified for small w-values—that is, near the Bragg-reflection position. The amount of perturbation depends on the structure amplitude $\Phi_\mathbf{h}$. On this basis, MacGillavry (1940) attempted to derive structure-amplitude data from CBED patterns of mica, which represents the first practical use of dynamical-diffraction effects.

An alternative picture of the two-beam diffraction case, suggested in Figure 3.16, provides a real-space representation of the wave fields. Each Bloch wave established in a crystal is made up of two waves traveling in the **0**- and **h**-directions and separated by an angle $2\theta_B$. These waves interfere to give a standing wave having a lateral periodicity equal to the lattice plane spacing $d_\mathbf{h}$. There are two ways in which the wave field can have the same symmetry as the crystal lattice: The nodes may be either on the atom planes or halfway between them. These are the configurations for the Bloch waves.

The density of electrons is given by the square of the wave function. Hence, for Bloch wave number 1, the electrons most probably run along the planes of atoms and so experience a larger-than-average potential $(\Phi_0 + \Phi_\mathbf{h})$ and have a larger value for the refractive index. For Bloch wave number 2, the electrons are concentrated between the atom planes and so experience a

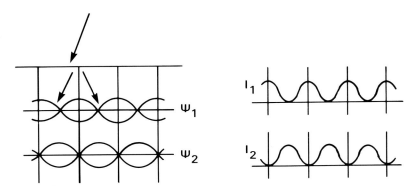

FIGURE 3.16 Real-space representation of the two Bloch waves in a crystal for two-beam dynamical diffraction.

smaller average potential ($\Phi_0 - \Phi_h$) and a smaller refractive index. Hence, the relative phases of the two Bloch waves change with increasing crystal thickness. When the Bloch waves combine at the exit face, the resultant wave amplitudes vary accordingly, giving rise to the pendellosung oscillations of the diffracted-beam intensities.

The effects of absorption may be included in this treatment. The origin of absorption effects for electrons in crystals will be discussed in later chapters. For present purposes, we may assume that the diffracted beams lose energy as a result of scattering by phonons and other excitations. The effect may be represented by use of an absorption function $\mu(r)$, so the POA is modified with an amplitude term:

$$q(xy) = \exp[-i\sigma\phi(xy)] \exp[-\mu(xy)]$$
$$= \exp\{-i\sigma[\phi(xy) - i\phi'(xy)]\}. \qquad (3.17)$$

As the second line of this equation demonstrates, we may represent the effect of absorption may by making the effective potential complex. The absorption function $\mu(r)$ can be described by a Fourier series with coefficients μ_0 and μ_h.

We assume that the absorption function has maxima at the atom positions. From Figure 3.16, Bloch wave number 1 is seen to concentrate the electrons on the planes of atoms and so has a higher-than-average absorption coefficient $\mu_0 + \mu_h$. Correspondingly, Bloch wave number 2 experiences a lower absorption coefficient $\mu_0 - \mu_h$. Then, even at the exact Bragg angle, the amplitudes from the two Bloch waves become unequal with increasing thickness, so, as suggested in Figure 3.15, the contrast of the pendellosung oscillations decreases with thickness. The average value for the intensity, however, decreases more slowly. When Bloch wave number 1 has been extinguished by its high absorption coefficient, Bloch wave number 2 remains. So for large thicknesses, the intensity decreases as $(\frac{1}{4}) \exp[-(\mu_0 - \mu_h)]$, as compared with the decrease $\exp(-\mu_0)$ for no diffraction.

This result is the origin of the Borrmann effect (Borrmann, 1950). For X-rays, μ_h may be almost as large as μ_0, since the absorption effects are highly localized. Thus, when a crystal is turned into a Bragg-diffraction orientation, the reduction of absorption and consequent increase in transmitted intensity is spectacular. For electron diffraction, the reduction of the absorption coefficient is not as dramatic, but the effect is significant for thick crystals and must often be taken into account for the interpretation of images of crystal defects.

3.4.3 The multislice formulation

The multislice approach was generated by Cowley and Moodie (1957) on the basis of the simple ideas of physical optics described in Chapters 1 and

2. Whereas the Bethe approach starts with the solution for the infinite periodic crystal, the multislice approach starts from the concept of the scattering by ideal, two-dimensional objects and so is often more appropriate for much of high-resolution electron microscopy, which invoves the scattering from very thin crystals.

The effect of a thin object on an incident electron wave may be approximated by considering only the change of phase experienced as the wave passes through the potential distribution. From the POA, the transmission function of the object is written as

$$q(xy) = \exp[-i\sigma\phi(xy)],$$

where

$$\phi(xy) \equiv \int \phi(\mathbf{r}) \, dz. \tag{3.18}$$

In this approximation, the spreading of the wave in the real, three-dimensional potential distribution by Fresnel diffraction (equation 2.11) is neglected. To give a correct representation of the scattering for three-dimensional object, one must introduce the simultaneous actions of the phase change due to the effect of the potential and the spreading of the wave by Fresnel diffraction. The two effects are first separated and then recombined. The phase changes are considered to be concentrated on a series of planes spaced a distance Δz apart (Figure 3.17). For each slice of the crystal for which the phase change is condensed on one plane, the phase change is given by equation (3.18), with the integration of $\phi(r)$ from z_n to $z_n + \Delta z = z_{n+1}$. The Fresnel-diffraction effects are considered to take place between these planes and are represented by equation (2.11), with the distance R replaced by Δz. Then, if the number of planes N tends to infinity and their separation Δz tends to zero—so that $N \Delta z = H$, where H is the crystal thickness—the result should represent the simultaneous operation of these two effects.

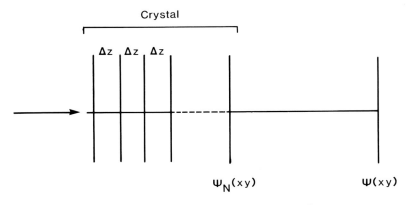

FIGURE 3.17 Scheme for multislice formulation of dynamical diffraction.

When the scattering object is a crystal, the potential distributions $\phi_n(xy)$ for the individual slices are periodic in two dimensions. A convenient technique is to deal with the set of Fourier coefficients $\Phi_n(h, k)$ of the Fourier series that describes $\phi_n(xy)$. The effect of Fresnel diffraction is introduced by multiplication of the amplitudes of the h,k-beams by relatively simple phase factors.

On this basis, Cowley and Moodie arrived at a general solution of the many-beam, dynamical-diffraction problem in terms of a doubly infinite series. Summing the series in various ways gave a single series from which, under the appropriate assumptions, they were able to derive the kinematical approximation, the two-beam approximation, the POA approximation, and other, less useful approximations.

The main practical value for this multislice formulation arises from a different type of approximation. As a basis for computer calculations of diffracted amplitudes, one assumes that the slice thickness may be taken as finite and that the number of slices may be taken as finite. Then, the diffraction amplitudes are obtained from a limited number of iterations of the process of modulating the phases by use of the POA and multiplying the amplitudes by a Fresnel-diffraction phase factor, as described in Chapter 8. The POA is considered to be a reasonable approximation for crystal thicknesses of 10 to 20 Å for 100-keV electrons. If the slice thicknesses are made much less than, say, 3 to 4 Å, the errors involved in taking the finite number of slices are usually small.

3.5 Dynamical-diffraction symmetries

The amplitudes and intensities of diffracted beams, in general, have complicated variation with crystal thickness (Figure 3.13a) and with incident-beam orientation. There are certain, axial, incident-beam directions for which the variation with thickness is particularly simple for some crystals (Figure 3.13b), but these cases are nontypical. The only generally valid statements that can be made concerning the observable intensities are those relating to the symmetries of the diffraction patterns. The symmetries are best seen in convergent-beam, electron-diffraction (CBED) patterns or by the use of related techniques giving equivalent information. The relationship of the CBED pattern to the crystal symmetry is more complicated than for the kinematical case but can provide correspondingly more information.

For kinematical scattering, with notable exceptions, the diffraction-pattern symmetries reflect the presence of the point-group symmetries of the crystal, the rotation axes, rotation-inversion axes, and mirror planes. The systematic absences of particular sets of reflections indicate the existence of the translational symmetries, the screw axes, and glide planes. The notable exceptions arise from the operation of Friedel's law, which implies that the presence or absence of a center of inversion (a "center of symmetry") cannot be deduced from kinematical-diffraction intensities. As a consequence, one cannot, for example, deduce the existence of a threefold rotation axis, as distinct from a sixfold axis. The determination of crystal symmetry from kinematical, X-ray–diffraction data is therefore severely limited. There are 230 space groups, but only 122 space groups, or groups of space groups, can be distinguished by such diffraction methods.

The fact that Friedel's law does not apply for dynamical diffraction, except in the two-beam approximation, was realized long ago (Miyake and Uyeda, 1950). But the possibility for systematical application of many-beam dynamical theory and CBED methods for symmetry determination is a recent development made possible by later advances in understanding of the theory and advances in electron optical instrumentation. Today, with few exceptions, all space groups may be distinguished by the systematic application of CBED techniques.

Symmetry information from the dynamical scattering of electrons is, however, fundamentally different in nature from that of kinematical scattering. The diffraction-pattern symmetries for transmission through a thin-crystal specimen reflect the symmetry of the wave function emerging from the exit face of the crystal. The symmetry of this wave function reflects the whole history of the transmission of the wave through each part of the crystal. The symmetry observations thus relate to the symmetry of the illuminated part of the crystal, rather than to the symmetry of the unit-cell contents.

For example, the symmetry of the diffraction pattern from a wedge-shaped crystal will usually be lower than the symmetry for a parallel-sided slab, because the wedge has, at best, a mirror-plane symmetry. Similarly, a parallel-sided slab crystal will show lower symmetry if the slab is not perpendicular to the electron beam. This effect was demonstrated by Goodman (1974).

The termination of the crystal at its surfaces may also affect the crystal symmetry. If the unit cell contains several distinct layers of atoms, termination of the crystal at one or another layer of atoms may produce an integral or nonintegral number of unit cells in the incident-beam direction; and, in general, the symmetry of the crystal will be different. If there is no center of symmetry in the unit cell, the variation of crystal symmetry will be different for fractional unit cells on the entrance and exit surfaces.

For thin crystals, the effect of crystal termination can be important. For thin, gold crystals viewed in the [111] direction, the addition or subtraction of one atomic layer from the $3n$ layers that constitute n unit cells removes the (almost complete) extinction of reflections that is associated with the rhombohedral stacking. The use of these nonextinguished spots to form dark-field images of the crystal then provides an effective way of imaging atom-high surface steps with good contrast (Cherns, 1974).

One or more stacking faults that are not parallel to the incident beam can also change the crystal symmetry. For example, Johnson (1972) observed a

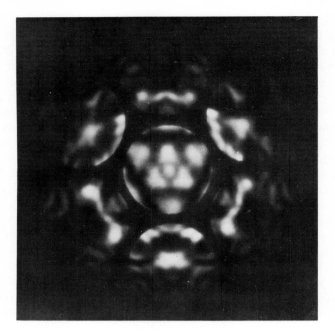

FIGURE 3.18 Convergent-beam, electron diffraction pattern from a faulted crystal of graphite showing a threefold symmetry. (From Johnson, 1972.)

strong threefold symmetry in CBED patterns from a thin, graphite crystal (Figure 3.18) having one or more stacking faults perpendicular to the beam. Perfect, thin, graphite crystals, however, give almost perfect hexagonal symmetry.

Information on the space-group symmetries of crystals can be found from CBED; but in view of the previous examples, the arguments and experiments must apply to parallel-sided, perfect crystal slabs with entrance and exit faces almost perpendicular to the incident beam. The slabs must be sufficiently thick, or suitably terminated, so that effects due to fractional unit cells are not significant.

3.5.1 Detection of symmetry elements

In relation to their influence on diffraction intensities, the symmetry elements of a crystal may be divided into two categories, types I and II. The type I symmetries are those whose operation leaves both the crystal and the z-axis (in the incident-beam direction) unchanged. These symmetries include the rotation axes (two-, three-, four-, or sixfold) parallel to the incident beam ("vertical") and mirror planes parallel to the beam. The corresponding symmetries are immediately evident in the diffraction patterns.

ELASTIC SCATTERING OF ELECTRONS BY CRYSTALS

The type II symmetries are those whose operation leaves the crystal unchanged except that the z-axis is inverted. These symmetries include the horizontal diad and screw axis, horizontal mirror and glide planes, and the center of inversion. The effects of these symmetries on diffraction intensities can be assessed by application of the reciprocity relationship.

An example illustrates the nature of the arguments involved. We consider a plane wave (from a distinct point source) incident so that the Ewald sphere comes close to the $hk0$-reciprocal-lattice point for an infinitely extended, plane lamellar crystal, with an excitation error ζ_{hk0}. The diffracted beam is detected at a point at infinity. If the source and detector points are interchanged (Figure 3.19a), the Ewald sphere will again pass through the $hk\zeta$-point. If the z-axis is inverted (i.e., if either the z-components of the beam wave vector or the reciprocal lattice is inverted), the Ewald sphere will intersect the $hk\bar{\zeta}$-point (Figure 3.19b). The reciprocity relationship then implies that the $hk\zeta$- and $hk\bar{\zeta}$-reflections will have the same amplitudes if the inversion of the z-axis leaves the crystal unchanged—that is, if the crystal has a mirror plane perpendicular to the z-axis.

The $hk\zeta$- and $hk\bar{\zeta}$-intensities will appear on either side of the Bragg position in the CBED disk for the $hk0$-spot. Hence, if the crystal has a mirror plane parallel to its surfaces, the $hk0$-spots will show intensities symmetrical about the central line that corresponds to the Bragg angle. Similar arguments may be made in relation to other type II symmetries.

Since the reciprocity relationship does not depend on the number or nature of the diffraction processes involved, the relationship between

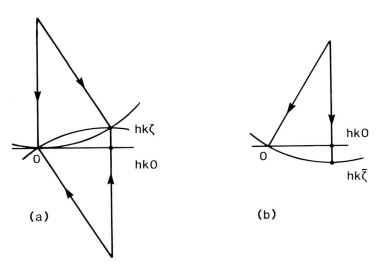

FIGURE 3.19 The use of a reciprocity argument to establish symmetry relationships for CBED patterns.

observed intensities is generally valid for any dynamical-diffraction conditions, involving any number of coherent, multiple interactions of diffracted waves.

On the basis of such arguments, one can formulate the sets of experiments needed to determine the presence of the various symmetry elements of the crystal and, under well-defined conditions, the space-group symmetry of the crystal structure. In principle, parallel-beam diffraction patterns of sharp spots could also be used for this purpose, but the intensities are then extremely sensitive to the exact orientation of the beam relative to the crystal. The method is practical only with convergent-beam diffraction in which the variation of diffracted-beam intensities is displayed for a range of incident-beam directions (see Figure 3.6). Then, much of the information on the type-II-symmetry elements is conveyed by the intensity distribution within the individual diffraction-spot disks.

In general, the type I symmetries are determined from whole-pattern symmetries, relating intensities of the distribution of all the convergent-beam spots. Type II symmetries are derived from observations of the internal symmetries within individual spots. Procedures have been outlined for the determination of point groups by Buxton et al. (1976). Procedures for resolving problems of space-group determination have been discussed by

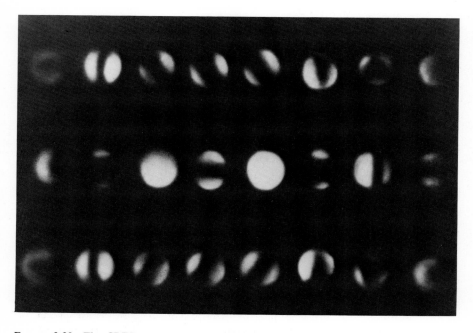

FIGURE 3.20 The CBED pattern from CdS close to the [1010] direction, showing the black G–M lines across the odd-order, 000l reflections (along the central, horizontal line) due to the glide-line symmetry of the planar group (pg). (After Goodman and Lehmpfuhl, 1968.)

Goodman and Secomb (1977) and Goodman and Johnson (1977). The analysis by Tanaka et al. (1983) includes tables giving exhaustive listings of dynamical-extinction data for the various space-group symmetries. Much of this information is derived from the original analysis by Gjønnes and Moodie (1965) of the appearance of dark extinction lines in CBED patterns corresponding to the systematic extinctions of reflections in kinematical scattering.

In kinematical scattering, such extinctions indicate the presence of screw axes or glide planes and provide important indications for the process of space-group determination. For dynamical scattering, the extinctions apply for ranges of orientations and so result in dark lines across the CBED disks, either parallel or perpendicular to the diffraction vector (Figure 3.20).

Strictly, these extinctions can occur only if the screw axis or glide planes conform to the conditions for type I symmetries. This limitation removes the possibility of extinctions for any but two-fold screw axes. In practice, however, the deviations from the conditions for type I symmetry may often be small and the extinction lines may appear, although not with complete extinction of the intensities, depending on experimental factors such as the crystal thickness.

3.6 The imaging of crystals

In Chapters 1 and 2, some account was given of the special considerations relevant for the imaging of periodic objects in general. There is little more to be said for the case of thin, perfect crystals for which the POA is valid. However, for thicker crystals and for crystals having defects, the combined effects of the dynamical diffraction within the crystal and the perturbations of the phases and amplitudes by the transfer function of the objective lens can lead to a variety of effects. These effects can be prominent in high-resolution images and confusing or misleading if not properly appreciated.

Although two-beam-diffraction conditions are rarely of interest for the high-resolution imaging of crystals, this relatively simple case may be considered as an initial guide to the kind of effects to be expected.

3.6.1 Imaging in the two-beam approximation

From the description of section 3.4.2 and Figure 3.16, we can derive a simple picture of the wave at the exit face of a crystal in the symmetrical, exact, two-beam situation. The two Bloch waves can be described as $\psi_1 = \cos \pi x/a$ and $\psi_2 = \sin \pi x/a$, respectively, since their maximum amplitudes are either at $x = 0$ (on the atomic planes) or at $x = \frac{1}{2}$ (between the atomic planes). The phase difference of the waves is $\pi/2$ at the entrance surface, where the intensity distribution is

$$I(x) = \left| \cos\left(\frac{\pi x}{a}\right) + i \sin\left(\frac{\pi x}{a}\right) \right|^2 = \cos^2\left(\frac{\pi x}{a}\right) + \sin^2\left(\frac{\pi x}{a}\right) = 1.$$

Then, the relative phase changes by $2\pi\sigma\Phi_\mathbf{h}H$ for a crystal thickness H. Thus, the intensity distribution at the exit face is

$$I(x) = \left| \cos\frac{\pi x}{a} + i \exp(2\pi i\sigma\Phi_\mathbf{h}H) \cdot \sin\frac{\pi x}{a} \right|^2$$

$$= 1 - \sin(2\pi\sigma\Phi_\mathbf{h}H) \cdot \sin\frac{2\pi x}{a}. \tag{3.19}$$

Thus, for a perfect imaging system, the image would show fringes of periodicity a, but the contrast would vary with the thickness and go to zero, changing sign for thicknesses $H = n/2\sigma\Phi_\mathbf{h}$ for integer n. The maxima and minima of intensity are at $x = a/4$ and $x = 3a/4$ and not on the atomic planes.

If the crystal is imaged with a nonideal lens and the incident beam is parallel with the lens axis, the transfer function adds a phase factor $\exp[i\chi(\mathbf{u})]$ for $\mathbf{u} = \mathbf{h}$ to the diffracted beam relative to the zero beam. This phase factor translates the fringe pattern sideways so that the positions of the fringes relative to the lattice planes depend on the defocus and the aberration coefficients. The fringes also are translated sideways by deviations from the Bragg condition. The relative phases and amplitudes of the zero and diffracted beams vary in a more complicated way with the excitation error and the thickness. For a wedge-shaped crystal, where the thickness varies, or for a distorted crystal, where the orientations of the lattice planes vary, the fringes become bent and are no longer parallel to the lattice planes, as pointed out by Hashimoto et al. (1961). Thus, in general, it is not possible to relate the fringes in the images to the lattice planes of the crystal. Usually, except in highly distorted crystal regions, the spacings of the fringes are nearly the same as the lattice plane spacing. But where there are rapid changes of thickness or of lattice strain, as in the vicinity of a dislocation or other defect, the bendings and even apparent terminations of fringes have no direct relationship to the actual atom displacements in the crystal (Cockayne and Gronsky, 1981).

For perfectly coherent, plane-wave illumination, the contrast of the fringes does not vary with defocus. However, since the fringe positions in the image vary with the phase factor of the transfer function, the range of defocus values due to the chromatic-aberration effect and the range of angles of incidence due to incident-beam convergence cause the fringes to be smeared out. Then, the contrast of the fringes is reduced.

These adverse effects of incoherence factors can be minimized if the incident beam is tilted at the Bragg angle to the objective-lens axis (Figure 3.21a). Then, in the back focal plane of the lens, the incident beam is at

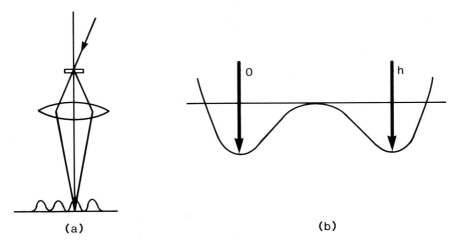

FIGURE 3.21 The real-space (a) and reciprocal-space (b) conditions for tilted-beam, high-resolution imaging of two-beam lattice fringes.

$u = -h/2a$, and the diffracted beam is at $u = +h/2a$; so the phase factor $\chi(u)$ is the same for both (Dowell, 1963). Then, variations of defocus will not change the relative phases of the two beams. Variations of angle of incidence will, in effect, translate the $\chi(u)$-curve relative to the two beam positions. The effect on the relative phases of the beams is minimized if the u-values of the two beams correspond to the positions of zero slope of the $\chi(u)$-curve—that is, at the minima, as in Figure 3.21b. Consequently, there is an optimum defocus for the imaging of two-beam lattice fringes given by

$$\frac{d\chi}{du} = 0 \quad \text{for} \quad u = \frac{h}{2a},$$

or

$$\Delta f = C_s \lambda^2 u^2 = \frac{C_s \lambda^2 h^2}{4a^2}. \tag{3.20}$$

Even under the ideal, tilted-beam conditions for two-beam lattice fringes, the correlation between perturbations of the lattice and perturbation of the lattice fringes in the image is not good. One well-documented example is for the images of an Au–Ni alloy for which spinodal decomposition resulted in a periodic segregation of the atoms and a consequent periodic fluctuation of the lattice plane spacing with a periodicity of 28 Å (Sinclair et al., 1976). The fluctuation in the fringe spacing in the image was approximately twice as great as the fluctuation in the lattice plane spacing which could occur, even if the variation was from pure gold to pure nickel.

3.6.2 Axial imaging of simple crystals

When the incident beam is exactly parallel to a principal axis of a crystal, the electron wave at the exit face reflects the symmetry of the crystal about that axis. In general, the same considerations of dynamical-diffraction symmetry apply as for the case of CBED, discussed previously. But for very thin crystals, the symmetry appearing in the image is the symmetry of the projection of the structure in the incident-beam direction.

The simpler, essentially one-dimensional case is that of the incident beam exactly parallel to a set of lattice planes but not close to a principal axis so that the diffraction pattern is a single line of diffraction spots (a systematic set). For a crystal of simple centrosymmetric structure and small unit cell, the strongest spots are the symmetrically equivalent **h**- and **h̄**-reflections. Those spots, together with the zero spot, provide a good approximation to the symmetrical three-beam case (Figure 3.22).

The wave function emerging from the crystal in this case has a center of symmetry on the lattice planes, so the positions of the planes of atoms are indicated in the image by either a maximum or a minimum of intensity—that is, a white line or a black line. There are two periodicities in the wave function. Interference of the **0**- with the **h**-beam, or the **0**- with the **h̄**-beam, gives the periodicity a/h. Interference of the **h**- with the **h̄**-beam gives the periodicity $a/2h$. The relative amplitudes and phases of the components having these two periodicities vary with crystal thickness, and either may dominate the intensity distribution. Thus, for an ideal imaging system, there will be some thickness for which only the half spacing $a/2h$ will be seen.

The components of the wave function having the two periodicities are affected differently by the lens transfer function. The **h**- and **h̄**-reflections have exactly the same value of the phase factor $\chi(u)$ for any defocus.

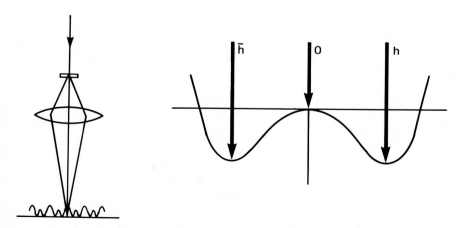

FIGURE 3.22 The real-space (a) and reciprocal-space (b) conditions for symmetrical, three-beam, lattice-fringe imaging.

Hence, the component of the wave function having the periodicity $a/2h$ is not changed by defocus. For the interference with periodicity a/h, however, the relative phases of the **0**- and **h**- or **h̄**-beams vary as $\chi(h/a)$, so this component varies with defocus and with the Fourier-image periodicity (equation 2.53). Thus, for a regularly spaced set of defocus values, the intensity of the images will be dominated by the strong periodicity a/h. But in between, when this periodicity is absent, a weaker $a/2h$-periodicity will remain.

The two components of the wave function are affected differently by the incoherence factors. A spread of defocus values due to chromatic-aberration effects tends to decrease the contrast of the a/h-periodicity but not that of the $a/2h$-periodicity. A variation of angle of incidence in an incident beam from a incoherent source of finite diameter affects both components. For each, the effect is minimized if the **h**- and **h̄**-reflections come at u-values where the $\chi(u)$-function has a minimum so that

$$\frac{d\chi}{du} = 0 \quad \text{for} \quad u = \frac{h}{a},$$

or for a defocus Δf given by

$$\Delta f = C_s \lambda^2 u^2 = \frac{C_s \lambda^2 h^2}{a^2}. \tag{3.21}$$

The optimum defocus for this case is thus four times as great as for the two-beam case, equation (3.20). Even for this optimum defocus, however, the difference in phase between the **h**- and **h̄**-reflections is less than between the **0**- and **h**- or **h̄**-reflections for a small lateral shift of the $\chi(u)$-curve. Hence, for both the main incoherence factors, the $a/2h$-periodicity will be less strongly affected than the a/h-periodicity.

Similar considerations apply to the case of a two-dimensional diffraction pattern given when the incident beam is parallel to a principal axis. Figure 3.23 suggests the case of an incident beam parallel to the [001] direction for a face-centered cubic crystal where the strongest spots are the 200, 2̄00, 020, and 02̄0 reflections, and the 220 and related spots are present although weaker. For the exact axial orientation, all the 200-type spots are equivalent, with equal amplitudes, excitation errors, and values of the phase factor $\chi(\mathbf{u})$; and all the 220-type spots are similarly equivalent.

The 200-type spots give lattice fringes in the image of spacing $a/2$ and $a/4$ in both the x- and y-directions and fringes of spacing $a/2\sqrt{2}$ in directions at 45° to these axes by interference of, for example, the 200 and 020 reflections. The 220-type spots give fringes of spacing $a/2\sqrt{2}$ by interference with the **0**-beam and $a/4$ and $a/4\sqrt{2}$ by interference between each other, in the directions at 45° to the x- and y-axes. Interaction of the 200-type and 220-type reflections gives component fringes having spacing of $a/2$ and $a/5\sqrt{2}$ in appropriate orientations. The relative amplitudes of all these

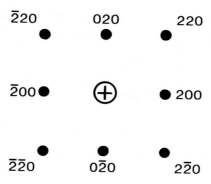

FIGURE 3.23 The inner reflections for a face-centered cubic structure with the incident beam in the [001] direction.

fringe components vary with crystal thickness, so the image-intensity distribution varies in a complicated way with thickness and with defocus but maintains the fourfold symmetry of the projection of the structure. Any small tilt will disrupt the symmetry and have drastic effects on the image.

As in the simpler three-beam, one-dimensional case, the various contributions to the image intensity are affected differently by the incoherence factors of chromatic spread of defocus and incident-beam convergence. Since all the 200-type beams are equivalent, the variations of the $\chi(\mathbf{u})$-factor have much less effect on their interactions with one another than on the fringes produced by interaction of these beams with the zero beam. Hence, the fringe components least affected by the incoherence factors are those of spacing $a/2\sqrt{2}$ and $a/4$ from the 200-type reflections and those of spacing $a/4$ and $a/4\sqrt{2}$ from the 220-type reflections. These components can predominate, especially for crystal thicknesses for which the **0**-reflection is relatively weak and the 200 and 220 are strong. The relative strengths of the 200 and 220 contributions depend on the crystal thickness and the defocus. By a suitable combination of parameters, one can produce many configurations for the images, including, for example, the appearance of the large, dark spots at the atom positions with small, bright spots in the middle of them, as seen by Hashimoto et al. (1977).

Similar considerations in more complicated form apply to the axial imaging of more complicated crystals having larger unit cells and many more diffracted beams of appreciable amplitude appearing in the diffraction pattern. For small thicknesses, for which the **0**-beam is of much larger amplitude than any diffracted beam, interactions of the type **0–h** predominate; the relative phases are determined by the $\chi(\mathbf{u})$-function. If a defocus is chosen close to the Scherzer optimum defocus so that $\chi(\mathbf{u})$ is almost the same for all beams, the resulting image is a reasonable representation of the projected potential. For larger thicknesses, as the amplitudes of the diffracted beams become large, the interactions of diffracted beams (which are favored by the incoherence factors) may become increasingly important. In some cases, these interactions can actually give the appearance of a better imaging of the crystal (Smith and O'Keefe, 1983); but in general, the appearance of the image will be highly variable with the experimental parameters and not directly related to the projected potential, except in symmetry.

3.7 Diffraction and imaging of crystal defects and disorder

Often, the main objective of high-resolution imaging is to investigate the defects or disorder in crystals rather than the structure of the perfect crystal. For the past thirty years, since the first systematic studies of dislocations and other defects in metal crystal films by Hirsch and co-workers (see Hirsch et al., 1965), electron microscopy has become the major tool for the study of extended defects. For much of this work, a moderate resolution ($\simeq 20$ Å) is sufficient, although in later refinements, such as the weak-beam-imaging technique (Cockayne et al., 1969), better resolution can be used. The diffraction theory needed to interpret most of the images deals with the modification of Bragg-reflection intensities by the strain fields that change the crystal-lattice orientation around the defect.

The approach to the study of defects by observation of diffraction patterns is essentially different. The basis for the derivation of information about defects is that the perfect crystal gives only sharp Bragg reflections, but deviations from the perfect periodicity give rise to other scattering. The defect scattering may take the form of a broadening or streaking of the sharp reflections, continuous sharp lines or broad lines, or general diffuse scattering in the background of the diffraction pattern. Electron diffraction is particularly well suited for the observation of defect scattering, because, relative to X-ray or neutron scattering, it is easy with electrons to use intense beams and high-scattering cross sections to see relatively weak, diffuse-scattering effects. Also, patterns can be obtained from very small regions in which the density of defects is high. In fact, very fine electron beams of diameter much less than 10 Å can be used to obtain diffraction patterns from individual defects (Cowley et al., 1985).

The information gained from images and from diffraction patterns is largely complementary, but there are important areas of overlap. The theory of diffraction by defects is an essential basis for the calculation of defect images, especially under high-resolution conditions.

The reciprocal relationship between dimensions in the image and the diffraction pattern defines the main areas of distinct application. For slowly varying strain fields, the effect on the image intensity can be great because the relative phases of the diffracted beams are modified by the variations of the excitation errors, but the effects in the diffraction patterns are not usually detectable. For localized defects such as point defects or the universal smearing of the structure by the thermal vibrations of atoms around their mean positions, the images show little effect except to a limited extent for the highest resolutions. However, the diffraction patterns show the broad distributions of diffuse scattering that can often be interpreted in terms of the statistics of the defect structures.

To provide a complete account of the theory or even of the general phenomena of diffraction effects due to crystal defects is beyond the scope of this book. For general treatments in the kinematic approximation, the reader is referred to the books of Guinier (1963), Warren (1969), or Cowley (1981). For electron diffraction, including dynamical-diffraction effects, see Cowley and Fields (1979) and Cowley (1981). Here, we give only a brief summary of the general principles and considerations for various forms of defect as they relate, in particular, to high-resolution imaging.

3.7.1 The column approximation

Because the scattering of high-energy electrons involves only small angles, one can deal with many problems of defect scattering by use of the so-called column approximation. It is the wave function at the exit face of a specimen that determines both the diffraction pattern and the image intensities. The wave amplitude at a point P on the exit face (see Figure 3.24) is assumed to depend only on the content of a column of material of width W extending through the specimen in the incident-beam direction.

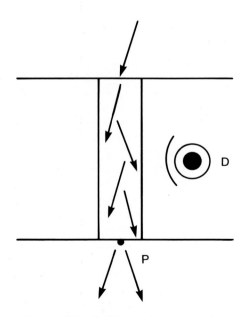

FIGURE 3.24 The channeling condition leading to the concept of a column approximation for calculating images of defects.

For a crystal with a defect D producing a strain field, for example, the amplitude at P is affected only by the displacements of the atoms within the column of width W. If the strain field varies sufficiently slowly, the variation of the strain field across the column can be neglected. Then, the amplitude at P can be calculated by assuming that the crystal is perfectly periodic in the x,y-directions perpendicular to the incident beam, but there are lateral displacements of the atoms (in the x- and y-directions) as a function of z. Calculations of the amplitude at P then involve only the sharp Bragg reflections. The effects of the lateral displacements due to the strain fields are to change the relative phases of the diffracted-beam amplitudes. The calculations can be made by use of the Howie–Whelan difference equations in which the beam amplitudes are progressively modified by the diffraction effects and by the strain fields. This procedure has been systematized by Head et al. (1973) to provide rapid computer simulations for images of extended defects. The column approximation can be used with confidence if the column width is smaller than the resolution limit of the microscope.

The width W of the column that can be assumed is often quite small. For a crystal in a symmetrical, two-beam orientation, for example, the incident beam travels at an angle θ_B to the z-axis. But after a distance of half the extinction length into the crystal, all the intensity is in the direction of the Bragg reflection at an angle of $-\theta_B$ to the z-axis. After a one-extinction distance, the intensity is again in the zero beam and directed at an angle θ_B to the z-axis (Figure 3.24). Thus, the energy flow is confined within a column of width given roughly by the product of the extinction distance and the Bragg angle. For a strong inner reflection from a simple metal, the Bragg angle may be 10^{-2} rad and the extinction distance about 400 Å, so the column width is $W \simeq 4$ Å, or the width of two or three lattice planes. Hence, for most medium-resolution imaging of crystal defects, the column approximation is sufficient, except for the regions of rapidly varying strains such as the core regions of dislocations.

For noncrystalline specimens or for crystals having large unit cells, the argument in terms of extinction distances is not appropriate. One basis for estimating the useful column width is the consideration of the spread of the waves by Fresnel diffraction. For a specimen thickness H, this spread is of the order of $(H\lambda)^{1/2}$. For a thickness of 400 Å, the width is, again, $W \simeq 4$ Å; for a thickness of 100 Å, the width is 2 Å. Thus, for thin specimens, the use of the column approximation becomes equivalent to the use of the phase-object approximation in which Fresnel diffraction is neglected and only a projection of the structure is considered.

For strongly scattering crystals viewed in an axial direction, an interesting approach is to consider that electrons may be channeled along the atom rows in the same way that high-energy ions are channeled through crystals, effectively yielding a column width equal to the interatomic spacings. The effects of crystal defects are then to block or distort the channels and, hence, modify the local intensities at the exit face.

3.7.2 Local atom displacements: thermal vibrations

Electrons pass through specimens so rapidly that vibrating atoms are seen as if stationary. The electron diffraction pattern and images are the sums of the intensities for the many instantaneous pictures of displaced atoms. Put another way, the energies of the thermal vibrations are so small compared with the incident-electron energies ($2 \cdot 10^{-2}$ as compared with 10^5 eV) that changes of the electron energies by the inelastic scattering from phonons are negligible.

If diffraction patterns are obtained from large numbers of atoms, as in selected-area electron diffraction (SAED) patterns, the diffraction effects of averaging over many small, static displacements of atoms, such as the displacements of atoms around point defects, will be much the same as the time average over the displacements of atoms in thermal vibration. The diffuse scattering will be much the same.

In images, the effects can be significantly different. The time averaging of thermal vibrations will be similar for all atoms, and all atom images will be smeared by an amount that may be significant for the best resolutions now available. For static displacements, the modifications of the transmitted-wave amplitudes will vary from place to place with the local concentration of defects, and the images will show a more or less random fluctuation of intensity.

Calculating the resulting intensity distributions in diffraction patterns by using the kinematical approximation or, in a crude way, by using the column approximation for dynamical scattering for thin crystals is straightforward. The sharp Bragg reflections are reduced in intensity by the Debye–Waller factor of the form $\exp(-4\pi^2 \overline{\varepsilon^2} U^2)$, where $\overline{\varepsilon^2}$ is the mean square displacement of the atoms. For the Bragg reflections, an absorption function should also be applied to take account of the loss of energy to the diffuse scattering. The continuous background of thermal diffuse scattering will be smooth and uniform for the case of independent vibrations of the atoms. Because the atom motions are correlated, the diffuse scattering is sharply peaked around the Bragg reflections but can also form diffuse bands and broad maxima between spots (see Figure 3.25). Because diffusely scattered electrons can undergo diffraction by the periodic-average structure, the diffuse scattering from thicker crystals shows the array of Kikuchi lines and bands.

For imaging, the effects of thermal vibrations are less obvious. The image is formed by recombination of all the diffracted beams transmitted by the objective aperture, including both the sharp Bragg reflections and the diffuse scattering. The Debye–Waller factor on the Bragg reflections will produce a broadening of the atom-image spots. The loss of some diffuse scattering beyond the objective aperture will produce the effect of an absorption function, but the diffuse scattering within the objective aperture will add some high-resolution detail to the image. Even in the relatively

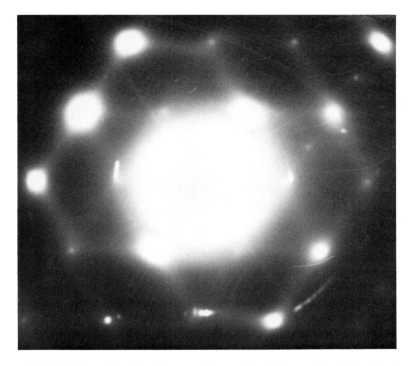

FIGURE 3.25 Electron diffraction from a thin crystal of zinc oxide, ZnO, showing the diffuse-scattering lines attributed to thermal vibrations.

simple POA, none of these effects will be simply related to the amplitudes of the atom displacements. A complete treatment of the problem, with fewer approximations, is not yet available.

3.7.3 Atomic disorder in crystals

For many crystals, particular sites within the unit cell may be occupied by any one of two or more types of atom or, in some cases, by vacancies. If the occupancy of the sites is the same over large distances (hundreds or thousands of angstroms), there is long-range order. In this case, there may be an alternation of a large region, or domains, in which the occupancy differs, and these regions are often clearly distinct in images as a result of the differences of scattering power for the different atoms. If the periodicities of the ordered structure are greater than the periodicities of the average structure, sharp superlattice reflections will appear in the diffraction pattern.

In many cases, the occupancy of a particular lattice site is almost random. Any one type of atom may tend to have the same type or a different type of atom in its neighboring lattice sites. This state of short-range order occurs for example, for many binary alloys at sufficiently high temperatures. Diffraction patterns from such materials contain diffuse maxima that occur between the sharp Bragg reflections of the average lattice if like atoms tend to avoid each other. There are diffuse spots around the Bragg reflections if like atoms tend to clump together.

There is an intermediate state of medium-range ordering for which the correlations between the occupancies of lattice sites extend over distances of typically 10 to 50 Å. The diffuse scattering produced in diffraction patterns is then often confined to relatively well defined diffuse lines, arcs, and maxima, as in Figure 3.26, which was obtained from the oxide TiO_x in which x is variable from 0.75 to 1.15 and there are about 16 percent of vacancies on the Ti or O sites. An adequate description of the state of medium-range order is, in most cases, not available and represents a challenge of the future for imaging and diffraction methods.

For kinematical scattering, diffraction from the various models for disordered or partially ordered systems can be calculated readily since this calculation involves a linear addition of scattering amplitudes, with appropriate phase factors, for each deviation from the average periodic lattice. For dynamical scattering, this simplicity vanishes. Variations of the wave function at the exit face of a crystal are not linearly related to the numbers of different kinds of atom on the lattice sites in the crystal, even if a column

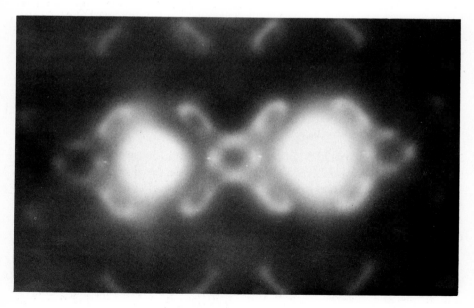

FIGURE 3.26 Electron diffraction pattern from a crystal of the titanium oxide, TiO_x (with $x = 1.19$) showing diffuse scattering owing to the local ordering of vacancies.

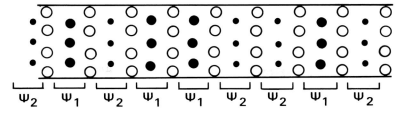

FIGURE 3.27 A thin crystal having disorder in one dimension, resulting in an irregular alternation of two types of local exit-wave function, ψ_1 and ψ_2.

approximation is made. For diffraction patterns, one often can make use of a dynamical factor, a smoothly varying function that is multiplied by the kinematical-diffuse-scattering intensity to take account of dynamical-scattering effects (Fisher, 1965; Cowley and Fields, 1979). However, this dynamical factor depends strongly on crystal orientation and thickness. For high-resolution images there are variations of intensity at the atom positions which depend on the numbers of the various atoms in the rows in the beam direction, but not linearly.

A more favorable situation exists when the disorder occurs only in one or two directions, perpendicular to the electron beam, and the crystal is well ordered in the beam direction, as suggested for a simple case in Figure 3.27. Use of the column approximation gives the conclusion that, in this case, the wave function at the exit face will be perturbed in one of two ways indicated by ψ_1 and ψ_2. The image will show two distinct intensity levels at the corresponding set of points. The diffraction pattern will show the same diffuse scattering as in the kinematical approximation except that the overall intensity distribution of the pattern will be different. Favorable cases of this sort are not uncommon for complex oxides such as the niobium-tungsten oxide studied by Iijima and Allpress (1974).

3.7.4 Stacking faults and twins: extended defects

In general, an extended defect such as a dislocation, a dislocation loop, or a small cluster of interstitials or vacancies involves a combination of additional atoms, missing atoms, and large atom displacements at a central core as well as an extended strain field. The strain field can be made clearly visible with moderate-resolution electron microscopy, and the calculations of the images involve only the phase changes of the Bragg reflections. It is the strongly perturbed core of the defect that is of interest for high-resolution electron microscopy. Here, the column approximation is not usually adequate. The diffuse scattering that conveys the information about the defect in the diffraction pattern is spread widely between the Bragg reflections. Then, complete dynamical-diffraction calculations using the method of periodic continuation (Chapter 8) are needed.

The only method now available for interpretation of images of defect core regions is that of postulating models, calculating image intensities for given crystal orientations, thickness, and instrumental parameters, and comparing the results with observed images. The number of parameters needed to describe a defect is, in general, so large that this process is impossibly arduous and confidence in the results cannot be assured. The cases treated to date are those for which the important simplifying assumption can be made that the defect is of high symmetry or is linear or planar and parallel to the incident beam, as in the studies of dislocation core structures by Bourret and Colliex (1982).

The simplest case to be considered is that of planar defects such as faults or twin planes parallel to the incident beam. Use of a column approximation allows the image on either side of the planar defect to be calculated; but if the exact nature of the lattice misfit or disjunction is to be investigated, a more careful analysis is required.

In the diffraction pattern, the effect of a planar discontinuity is to produce sharp continuous lines of scattering in the direction perpendicular to the planar fault. The information regarding the nature of the planar discontinuity is contained in the amplitude distributions along these lines, far from the Bragg reflections. Hence, no column-approximation treatment that considers only Bragg-reflection amplitudes can be adequate. Again, full dynamical-diffraction calculations are needed.

REFERENCES

Bethe, H. A. (1928). Theorie der Beugung von Elektronen an Kristallen. *Ann. Phys.* **87,** 55.
Borrmann, G. (1950). Die Absorption von Rontgenstrahlen im Fall den Interferenzen. *Z. Phys.* **127,** 297.
Bourret, A., and Colliex, C. (1982). Combined HREM and STEM microanalysis on decorated dislocation cores. *Ultramicroscopy* **9,** 183.
Buxton, B. F., and Tremewan, P. T. (1980). The atomic string approximation in cross-grating high-energy electron diffraction. I. Dispersion surface and Bloch waves. *Acta Crystallogr., Sect. A* **36,** 304.
———, Eades, J. A., Steeds, J. W. and Rackham, G. M. (1976). The symmetry of electron diffraction zone axis patterns. *Philos. Trans. R. Soc. London* **281,** 171.
Cherns, D. (1974). Direct resolution of surface atomic steps by transmission electron microscopy. *Philos. Mag.* **30,** 549.
Cockayne, D. J. H., and Gronsky, R. (1981). Lattice fringe imaging of modulated structures. *Philos. Mag.* **A44,** 159–175.
———, Ray, I. L. F., and Whelan, M. J. (1969). Investigations of dislocation strain fields using weak beams. *Philos. Mag.* **20,** 1265.
Cowley, J. M. (1967). Crystal structure determination by electron diffraction. *Prog. Mater. Sci.* **13,** 269.
——— (1980). Interference effects in a STEM instrument. *Micron* **11,** 229.
——— (1981). *Diffraction physics.* 2nd ed. North-Holland, Amsterdam.
———, and Fields, P. M. (1979). Dynamical theory for electron scattering from crystal defects and disorder. *Acta Crystallogr., Sect. A* **35,** 28.
———, and Moodie, A. F. (1957). The scattering of electrons by atoms and crystals. I. A new theoretical approach. *Acta Crystallogr.* **10,** 609.

———, and Spence, J. C. H. (1979). Innovative imaging and microdiffraction in STEM. *Ultramicroscopy* **3**, 433.

———, Osman, M., and Humble, P. (1985). Nanodiffraction from platelet defects in diamond. *Ultramicroscopy* **15**, 311.

Dowell, W. C. T. (1963). Das elektronenmikroskopische bild von netzebenenscharen und sein kontrast. *Optik* **20**, 535.

Fisher, P. M. J. (1965). The application of the Gjønnes theory of diffuse electron scattering to the problem of short-range order diffraction of electrons. In *Proceedings of the conference on electron diffraction and the nature of defects in crystals*. Australian Academy of Science, Canberra, paper I H–4.

——— (1969). Development and applications of an n-beam dynamic methodology in electron diffraction. Ph.D. diss. University of Melbourne.

Fujiwara, K. (1959). Application of higher-order Born approximation to multiple elastic scattering of electrons in crystals. *J. Phys. Soc. Jpn.* **14**, 1513.

Gjønnes, J. K., and Moodie, A. F. (1965). Extinction conditions in dynamical theory of electron diffraction. *Acta Crystallogr.* **19**, 65.

Goodman, P. (1974). Role of upper-layer interactions in electron diffraction symmetries. *Nature* **251**, 698.

———, and Johnson, A. W. S. (1977). Identification of enantiomorphically related space groups by electron diffraction—a second method. *Acta Crystallogr., Sect. A* **33**, 997.

———, and Lemhpfuhl, G. (1968). Observation of the breakdown of Friedel's law in electron diffraction and symmetry determination from zero-layer interactions. *Acta Crystallogr., Sect. A* **24**, 339.

———, and Secomb, T. W. (1977). Identification of enantiomorphously related space groups by electron diffraction. *Acta Crystallogr., Sect. A* **33**, 126.

Guinier, A. (1963). *X-ray diffraction in crystals, imperfect crystals and amorphous bodies.* Freeman, San Francisco.

Hashimoto, H., Mannami, M., and Naiki, T. (1961). Dynamical theory of electron diffraction for the electron microscopic image of crystal lattices. I. Images of single crystals. *Philos. Trans. R. Soc. London* **253**, 459.

———, Endoh, H., Tanji, T., Ono, A., and Watanabe, E. (1977). Direct observation of fine structure within images of atoms in crystals by transmission electron microscopy. *J. Phys. Soc. Jpn.* **42**, 1073.

Head, A. K., Humble, P., Clarebrough, L. M., Morton, A. J., and Forword, C. T. (1973). *Computed electron micrographs and defect identification.* North-Holland, Amsterdam.

Hirsch, P. B., Howie, A., Nicholson, R. B., Pashley, D. W., and Whelan, M. J. (1965). *Electron microscopy of thin crystals.* Butterworth, London.

Hoppe, W. (1969). Beugung im inhomogenen Primarstrahlwellenfeld. I. Prinzip einer Phasenmessung von Elektronenbeugungsinterferenzen. *Acta Crystallogr., Sect. A* **25**, 495.

Howie, A., and Whelan, M. J. (1961). Diffraction contrast of electron microscope images of crystal lattice defects. II. The development of a dynamical theory. *Proc. R. Soc. London, Ser. A* **263**, 217.

Iijima, S., and Allpress, J. G. (1974). Structural studies by high-resolution electron microscopy: Tetragonal tungsten bronze-type structures in the system Nb_2O_5–WO_3. *Acta Crystallogr., Sect. A* **30**, 22.

International tables for crystallography, published for the International Union of Crystallography. D. Reidel Publishing Company, Boston.

Jap, B. K., and Glaeser, R. M. (1980). The scattering of high-energy electrons. II. Quantitative validity domains of the single-scattering approximations for organic crystals. *Acta Crystallogr., Sect. A* **36**, 57.

Johnson, A. W. S. (1972). Trigonal symmetry in electron diffraction patterns from faulted graphite. *Acta Crystallogr., Sect. A* **28**, 89.

MacGillavry, C. H. (1940). Zur Prufung der dynamischen Theorie de Elektronenbeugung am Kristallgitter. *Physica* **7**, 329.

Marks, L. D. (1985). Direct observation of diffractive probe spreading. *Ultramicroscopy* **16**, 261.

Miyake, S., and Uyeda, R. (1950). An exception to Friedel's law in electron diffraction. *Acta Crystallogr.*, **3**, 314.

Nathan, R. (1976). Computer synthesis of high-resolution electron micrographs. In *Digital processing of biomedical images,* ed. K. Preston, Jr., and M. Onoe, 75. Plenum, New York.

Sinclair, R., Gronsky, R., and Thomas, G. (1976). Optical diffraction from lattice images of alloys. *Acta Metall.* **24**, 789.

Smith, D. J., and O'Keefe, M. A. (1983). Conditions for direct structure imaging in silicon carbide polytypes. *Acta Crystallogr., Sect. A* **39**, 139.

Spence, J. C. H., and Cowley, J. M. (1978). Lattice imaging in STEM. *Optik* **50**, 129.

Sturkey, L. (1957). The use of electron diffraction intensities in structure determination. *Acta Crystallogr.* **10**, 858.

——— (1961). The calculation of electron diffraction intensities. *Proc. Phys. Soc. London* **80**, 321.

Tanaka, N., and Cowley, J. M. (1986). Electron-microscope imaging of short-range order in disordered alloys. *Acta Crystallogr., Sect. A* **43**, 337.

———, Sekii, H., and Nagasawa, T. (1983). Space-group determination by dynamic extinction in convergent-beam electron diffraction. *Acta Crystallogr., Sect. A* **39**, 825.

Thomas, G., and Goringe, M. J. (1979). *Transmission electron microscopy of materials.* Wiley, New York.

Tournarie, M. (1961). Théorie dynamique rigoureuse de la propagation cohérente des électrons à travers une lame cristalline absorbante. *C. R. Acad. Sci.* **252**, 2862.

Vainshtein, B. K. (1964). *Structure analysis by electron diffraction* (translation). Pergamon Press, Oxford.

Warren, B. E. (1969). *X-ray diffraction.* Addison-Wesley, Reading, Mass.

Zhu, J., and Cowley, J. M. (1982). Microdiffraction from stacking faults and twin boundaries in FCC crystals. *J. Appl. Crystallogr.* **16**, 171.

4
ELASTIC-SCATTERING THEORY
JOHN M. COWLEY

4.1 Dynamical scattering

For the range of electron energies commonly employed for electron microscopy, we can use the time-independent Schrödinger equation to derive the elastic scattering of electrons by crystals. The required wave amplitudes are obtained by solution of the equation in the form

$$\nabla^2 \psi + \frac{2me}{\hbar^2}[E + \phi(\mathbf{r})]\psi = 0. \tag{4.1}$$

Although the energy range is such that relativistic effects are important, Fujiwara (1961), for example, demonstrated that no appreciable errors arise if equation (4.1) is used with the relativistically corrected values of the electron mass and wavelength:

$$m = m_0\left(1 - \frac{v^2}{c^2}\right), \tag{4.2}$$

$$\lambda = h\left[2m_0 eE\left(1 + \frac{eE}{2m_0 c^2}\right)\right]^{-1/2}. \tag{4.3}$$

Equation (4.1) may be used to obtain the wave function at the exit face of the specimen. The derivation of the observed intensities in the image or diffraction pattern follows from this wave function by application of the imaging theory of Chapter 2.

In the original formulation of the theory for diffraction by a crystal by Bethe (1928), the potential distribution $\phi(\mathbf{r})$ inserted in the wave equation was the infinite, periodic potential distribution of an ideal crystal. The possible solutions for the electron wave in the crystal for a given incident-wave vector are the Bloch waves, each representing a wave field having the periodicity of the crystal lattice. The relative amplitudes of the various Block waves are determined by the boundary conditions. At the entrance face of the crystal, the wave inside the crystal must be matched to the waves in the vacuum, which, in the case of transmission of high-energy electrons, include only the incident plane wave, with no appreciable reflected waves. A similar matching of the waves at the exit surface then gives the exit-wave function and introduces the dependence on the finite crystal thickness. So that complications are avoided, the crystal is usually assumed to be a thin, parallel-sided plate having entrance and exit surfaces almost perpendicular to the incident beam. Thus, the specimen periodicity in the x,y-directions within the plane of the plate is maintained, and the spatial limitations are applied only in the z-direction, the direction of the incident beams.

For high-energy electrons, we can make the simplifying assumptions that there is no backscattering and that all diffracted beams make small angles with the incident-beam direction. For the consideration of transmission electron microscopy and transmission, high-energy electron diffraction, the formulation of the dynamical-scattering problem may be made more transparent by taking these assumptions into account from the beginning. For an incident beam almost parallel to the z-axis, which is taken as the normal to the entrance face, the wave function is periodic in the x- and y-directions, and the modification of this wave function is traced as it progresses through the crystal in the z-direction. The progression of the wave can be expressed by differential equations or by multiplication of the column vector, whose elements are the Fourier coefficients of the wave function, by a scattering matrix. Such approaches offer a number of new insights into the consequences of the strong scattering of electrons by crystals.

A distinctly different approach to the dynamical-diffraction problem makes use of the integral form of the wave equation normally used for scattering problems, given in equation (2.7). From this form is generated the Born series,

$$\psi = \psi_0 + \psi_1 + \psi_2 + \cdots, \tag{4.4}$$

where ψ_0 is the incident wave, ψ_1 is the singly scattered wave, and the subsequent terms represent waves scattered two and more times.

For our purposes, the Born series in its original form has little relevance since, for strongly scattering objects, the convergence of the series is not good and each additional term of the Born series involves an additional integration over all volume elements of the specimen. An appropriate technique is to simplify the problem by taking advantage of the forward-scattering geometry of the diffraction of fast electrons. Then, thin slices of the specimen perpendicular to the incident beam may be considered as the scattering elements, and multiple scattering takes place only in progression through the crystal, with no backscattering. This procedure is the basis of the multislice approach, to be considered later. For the moment, however, we consider the use of the first Born approximation as a basis for the formulation of the kinematical approximation in terms of Fourier transforms. From equation (2.9), the single-scattering term is

$$\Psi_1(\mathbf{u}) = \mu \frac{\exp(-i\mathbf{k} \cdot \mathbf{R})}{R} \int \phi(\mathbf{r}) \exp(2\pi i \mathbf{u} \cdot \mathbf{r}) \, d\mathbf{r}, \tag{4.5}$$

or if we neglect the constant terms and those of modulus unity, the scattering amplitude is given by

$$\Psi_1(\mathbf{u}) \propto \Phi(\mathbf{u}) = \int_{-\infty}^{\infty} \phi(\mathbf{r}) \exp(2\pi i \mathbf{u} \cdot \mathbf{r})\, d\mathbf{r} = \mathscr{F}\phi(\mathbf{r}). \quad (4.6)$$

The Fourier-transform operation, denoted by $\mathscr{F}$, is thus the basic relationship between the real-space potential distribution $\phi(\mathbf{r})$ and the diffraction amplitude $\Phi(\mathbf{u})$ expressed as a function of the vector $\mathbf{u}$ in reciprocal space.

4.2 The kinematical approximation

The Fourier-transform relationship of (4.6) allows us to derive a distribution in reciprocal space, the space of the vector variable u, for any real-space distribution. The relationship of this reciprocal-space distribution to the observed intensities must then be considered as an associated aspect (to be treated in the next section of this chapter).

For single atoms, the atomic-scattering amplitude is given by

$$f_i(\mathbf{u}) = \int \phi_i(\mathbf{r}) \exp(2\pi i \mathbf{u} \cdot \mathbf{r}) \cdot d\mathbf{r}. \quad (4.7)$$

This scattering amplitude for electrons is related to that for X-rays and is derived from Poisson's equation

$$\nabla^2 \phi(\mathbf{r}) = -\rho(\mathbf{r})/\varepsilon_0 \quad (4.8)$$

where $\rho(\mathbf{r})$ is the charge-density distribution including both the nuclear charge $+Ze$ and the electron-density distribution $\rho_e(\mathbf{r})$; ε_0 is the permittivity in a vacuum. Since the atomic-scattering amplitude $f_X(u)$ for X-rays is given by the Fourier transform of $\rho_e(\mathbf{r})$, in electron units, the Fourier transform of both sides of (4.8) gives

$$f_e(\mathbf{u}) = \frac{e[Z - f_X(\mathbf{u})]}{4\varepsilon_0 \pi^2 u^2}. \quad (4.9)$$

This expression is the Mott formula. Values of $f_e(\mathbf{u})$ are tabulated in the *International Tables for Crystallography*, but they may also be derived from the tabulations of $f_X(\mathbf{u})$ by use of (4.9).

For assemblies of atoms, the potential distribution is expressed as the sum of the potential distributions of individual atoms:

$$\phi(\mathbf{r}) = \sum_i \phi_i(\mathbf{r}) * \delta(\mathbf{r} - \mathbf{r}_i), \quad (4.10)$$

where the convolution of $\phi_i(\mathbf{r})$ with the delta function shifts the origin of the atom distribution for the ith atom to the endpoint of the vector $\mathbf{r}_i$. A Fourier transform then gives the reciprocal-space-structure amplitude as

$$\Phi(\mathbf{u}) = \sum_i f_i(\mathbf{u}) \exp(2\pi i \mathbf{u} \cdot \mathbf{r}). \quad (4.11)$$

The $f_i(\mathbf{u})$ are usually taken to be the atomic-scattering amplitudes for isolated neutral atoms. This assumption is not necessarily a good one. For an isolated ionized atom, the value of $f_e(u)$ tends to $\pm\infty$ for $u \to 0$, since the X-ray–scattering factor f_X tends to $Z \pm n$, where n is the degree of ionization. For ions in crystals, no such infinities exist, although a local excess or defect of charge around the atoms or in bonds between atoms can have an effect on electron-scattering amplitudes, which is considerably greater than the effect for the X-ray case.

4.2.1 Diffraction by crystals

A crystal may be described as a periodic repetition of the group of atoms, $\phi_0(\mathbf{r}) = \sum_i \phi_i(\mathbf{r}) \ast \delta(\mathbf{r} - \mathbf{r}_i)$, with translations given by the real-space lattice axes $\mathbf{a}, \mathbf{b}, \mathbf{c}$,

$$\mathbf{R} = m\mathbf{a} + n\mathbf{b} + p\mathbf{c},$$

so that

$$\phi(\mathbf{r}) = \sum_i \phi_i(\mathbf{r}) \ast \delta(\mathbf{r} - \mathbf{r}_i) \ast \sum_j \delta(\mathbf{r} - \mathbf{R}_j). \tag{4.12}$$

The Fourier transform of the real-space lattice is the reciprocal lattice:

$$\mathcal{F} \sum_j \delta(\mathbf{r} - \mathbf{R}_j) = \sum_\mathbf{H} \delta(\mathbf{u} - \mathbf{H}), \tag{4.13}$$

where $\mathbf{H} = h\mathbf{a}^* + k\mathbf{b}^* + l\mathbf{c}^*$.

The vectors $\mathbf{H}$ define the reciprocal lattice that has axes

$$\mathbf{a}^* = \frac{(\mathbf{b} \times \mathbf{c})}{\mathbf{a} \cdot (\mathbf{b} \times \mathbf{c})} \qquad \mathbf{b}^* = \frac{(\mathbf{c} \times \mathbf{a})}{\mathbf{b} \cdot (\mathbf{c} \times \mathbf{a})} \qquad \mathbf{c}^* = \frac{(\mathbf{a} \times \mathbf{b})}{\mathbf{c} \cdot (\mathbf{a} \times \mathbf{b})}.$$

In each case, the denominator is the unit-cell volume Ω.

Hence, for an infinite, periodic-crystal structure, the reciprocal-space distribution is

$$\Phi(\mathbf{u}) = \sum_i f_i \exp(2\pi i \mathbf{u} \cdot \mathbf{r}_i) \cdot \sum_\mathbf{H} \delta(\mathbf{u} - \mathbf{H}), \tag{4.14}$$

where the summation over i is for all atoms within the unit cell of the crystal. This summation gives the structure amplitude $V_\mathbf{H}$. For the reciprocal-lattice points $\mathbf{u} = h\mathbf{a}^* + k\mathbf{b}^* + l\mathbf{c}^*$ and for atom coordinates given as fractions of the unit-cell dimensions, $\mathbf{r} = x_i\mathbf{a} + y_i\mathbf{b} + z_i\mathbf{c}$, this summation gives

$$V(hkl) = \sum_i f_i(\mathbf{u}) \exp[2\pi i(hx_i + ky_i + lz_i)], \tag{4.15}$$

since $\mathbf{a} \cdot \mathbf{a}^* = \mathbf{b} \cdot \mathbf{b}^* = \mathbf{c} \cdot \mathbf{c}^* = 1$ and $\mathbf{a} \cdot \mathbf{b}^* = \mathbf{a} \cdot \mathbf{c}^* = \mathbf{c} \cdot \mathbf{b}^* = 0$.

Thus, the reciprocal-space distribution consists of the set of delta functions forming the reciprocal lattice, each with a weighting $V_\mathbf{H}$. Real crystals are, of course, not infinite and not always exactly periodic. The finite dimensions of the crystal are usually taken into account by multiplying the periodic potential function by a shape function $s(\mathbf{r})$, which is unity inside the crystal and zero elsewhere. Then, the reciprocal-space distribution of (4.14) is modified by convolution with $S(\mathbf{u})$, the Fourier transform of $s(\mathbf{r})$. For example, if $s(\mathbf{r})$ represents a crystal having boundaries parallel to the unit-cell faces so that

$$s(\mathbf{r}) = \begin{cases} 1, & \text{for } |x| \leq \tfrac{1}{2}Ma, \ |y| \leq \tfrac{1}{2}Nb, \ |z| \leq \tfrac{1}{2}Pc, \\ 0, & \text{elsewhere,} \end{cases}$$

then $S(\mathbf{u})$ has the $(\sin x)/x$ form in each dimension: that is,

$$\Phi(\mathbf{u}) = \sum_{\mathbf{H}} V_{\mathbf{H}} \delta(\mathbf{u} - \mathbf{H}) * MNP \frac{\sin \pi Mau}{\pi Mau} \cdot \frac{\sin \pi Nbv}{\pi Nbv} \cdot \frac{\sin \pi Pcw}{\pi Pcw}. \quad (4.16)$$

Thus, around each reciprocal-lattice point, there is a distribution in $\Phi(\mathbf{u})$ having a central maximum and falling to zero for $u = 1/Ma$, $v = 1/Nb$, $w = 1/Pc$, with oscillations decreasing rapidly in amplitude in all directions.

For other shapes, the form of the shape transform $s(\mathbf{u})$ is written less readily, but the general principle remains. The extension of the function $\Phi(\mathbf{u})$ around the reciprocal-lattice points will be by an amount inversely proportional to the corresponding dimensions of the crystal.

4.2.2 Kinematical-diffraction intensities

The intensity of a diffracted beam in the kinematical approximation will be proportional to $|\Phi(\mathbf{u})|^2$, for $\Phi(\mathbf{u})$ defined by (4.6), provided $\mathbf{u} = \mathbf{s}_1 - \mathbf{s}_0$, where $\mathbf{s}_1$ and $\mathbf{s}_0$ are vectors of length $1/\lambda$ in the directions of the diffracted and incident beams, respectively. If, as in Figure 3.4, the incident-beam vector is drawn from a point P to the reciprocal-lattice origin, the locus of the endpoints of all vectors $\mathbf{s}_1$ drawn from P is the Ewald sphere. Then, diffracted intensity appears for all diffraction-beam directions for which $|\Phi(\mathbf{u})|^2$ is nonzero. For a crystal of large extent, the condition for an intensity maximum is that the Ewald sphere pass through a reciprocal-lattice point so that

$$\mathbf{s}_1 - \mathbf{s}_0 \equiv \mathbf{u} = \mathbf{H} = h\mathbf{a}^* + k\mathbf{b}^* + l\mathbf{c}^*. \quad (4.17)$$

The reciprocal-lattice geometry is such that the distance of the hkl-reciprocal-lattice point from the origin, which is the magnitude of $\mathbf{H}$, is equal to d_{hkl}^{-1}, where d_{hkl} is the distance between the lattice planes in real space having the Miller indices h, k, l. The magnitude of the vector $\mathbf{s}_1 - \mathbf{s}_0$ is seen from Figure 3.2b to be $2\lambda^{-1} \sin(\phi/2)$. So if $\phi/2 = \theta_{hkl}$, the Bragg angle for reflection from the h, k, l-planes, we have Bragg's law:

$$2d_{hkl} \sin \theta_{hkl} = \lambda. \quad (4.18)$$

The extension of the scattering amplitude $\Phi(\mathbf{u})$ around the reciprocal-lattice point by the shape transform, as in (4.16), becomes important for small crystal dimensions. Then, appreciable values of the intensity, given by $|\Phi(\mathbf{u})|^2$, can be produced simultaneously for many different reciprocal-lattice points, as suggested by Figure 3.4.

The usual approximation made for diffraction from crystals containing many unit cells is that the maxima of scattering amplitude around the reciprocal-lattice points are small, are isolated, and do not overlap. So the squares of their amplitudes may be considered as independent, and in the case of (4.16), for example,

$$|\Phi(\mathbf{u})|^2 \simeq \sum_{\mathbf{H}} |V_{\mathbf{H}}|^2 \, \delta(\mathbf{u} - \mathbf{H}) \ast |S(\mathbf{u})|^2,$$

where

$$|S(\mathbf{u})|^2 = (MNP\Omega)^2 \cdot \frac{\sin^2 \pi M a u}{(\pi M a u)^2} \cdot \frac{\sin^2 \pi N b v}{(\pi N b v)^2} \cdot \frac{\sin^2 \pi P c w}{(\pi P c w)^2}. \qquad (4.19)$$

This approximation will not be valid when the crystal dimensions correspond to only a few unit cells. For small crystals, one must take into consideration that the number of unit cells may not be an integral number for nonprimitive unit cells. Also, the use of a shape function multiplying a periodic function will not correctly represent the termination of the potential distribution at a crystal surface where the interatomic distances may be modified and where the potential function will vary smoothly from the value inside the crystal to the vacuum level.

To represent the intensity distributions for very small crystals, one may use the approximation of representing the potential distribution as the sum of potential distributions due to individual atoms, as in (4.10), giving

$$|\Phi(\mathbf{u})|^2 = \sum_i \sum_j f_i f_j^* \exp[2\pi i \mathbf{u} \cdot (\mathbf{r}_i - \mathbf{r}_j)]. \qquad (4.20)$$

However, a more strictly correct technique is to derive the scattering power from the potential distribution $\phi(\mathbf{r})$ by using (4.6).

For extended crystals that are imperfect because of the presence of strains, defects, or disorder, the distribution of scattering power in reciprocal space is not confined to the sharp maxima around the reciprocal-lattice points. The diffraction patterns may show various types of diffuse patches or diffuse or sharp streaks or spots, characteristic of the perturbation of the structure. Such effects are often clearly visible in electron diffraction patterns. For kinematical-scattering conditions, the form of the scattering function can be derived by consideration of the Patterson function, or autocorrelation function:

$$P(\mathbf{r}) = \mathcal{F}^{-1} |\Phi(\mathbf{u})|^2 = \phi(\mathbf{r}) \ast \phi(-\mathbf{r}) = \int \phi(\mathbf{R}) \phi(\mathbf{r} + \mathbf{R}) \, d\mathbf{R}. \qquad (4.21)$$

In terms of the contributions to $\phi(\mathbf{r})$ from individual atoms, $P(\mathbf{r})$ is given by the Fourier transform of (4.20) as

$$P(\mathbf{r}) = \sum_i \sum_j \phi_i(\mathbf{r}) * \phi_j(\mathbf{r}) * \delta[\mathbf{r} - (\mathbf{r}_i - \mathbf{r}_j)], \qquad (4.22)$$

and so consists of peaks of the form $\phi_i * \phi_j$ placed at the ends of the interatomic vectors $\mathbf{r}_i - \mathbf{r}_j$. The Patterson function is thus a weighted probability distribution giving the relative frequencies for the occurrence of the various interatomic vectors.

The methods by which this function—and consequently, the intensity distribution—may be derived for the various forms of crystal imperfection have been described by Cowley (1981). Such methods are important when the number of defects is so large that the diffraction pattern reflects the statistics of their occurrence and form, as is often the case in selected-area diffraction patterns. In relation to the imaging of individual defects using high-resolution electron microscopy or for the calculation of microdiffraction patterns from the region of individual defects, one must derive the scattering amplitude or intensity from a localized perturbation of a crystal structure. Within the limitations of the kinematical approximation, this derivation is most conveniently done by writing

$$\phi(\mathbf{r}) = \phi_0(\mathbf{r}) + \Delta\phi(\mathbf{r}), \qquad (4.23)$$

where $\phi_0(\mathbf{r})$ is the periodic, unperturbed potential distribution of a perfect crystal, contributing only the sharp Bragg reflections, and $\Delta\phi(\mathbf{r})$ is the perturbation due to the defect that is nonperiodic and hence gives a continuous scattering function $\Delta\Phi(\mathbf{u})$. Because the amplitude $\Delta\Phi(\mathbf{u})$ at the positions of sharp Bragg reflections is usually small compared with the scattering in the peaks due to the periodic structure, we may write

$$|\Phi(\mathbf{u})|^2 = \sum_{\mathbf{H}} |\Phi_{\mathbf{H}}|^2 \cdot \delta(\mathbf{u} - \mathbf{H}) * |S(\mathbf{u})|^2 + |\Delta\Phi(\mathbf{u})|^2, \qquad (4.24)$$

where $S(\mathbf{u})$ is the shape transform of the coherently illuminated portion of the crystal. The information concerning the defect is thus contained mostly in the diffuse scattering between the sharp diffraction spots.

4.3 Formulations for dynamical diffraction

4.3.1 Bethe theory

From the original ideas of Bethe (1928), the wave equation of (4.1) is written for the periodic potential field of a crystal, defined in order to absorb the constants, as

$$\frac{2me}{\hbar^2} \phi(\mathbf{r}) = v(\mathbf{r}) = \sum_{\mathbf{h}} v_{\mathbf{h}} \exp(2\pi i \mathbf{h} \cdot \mathbf{r}) = 2K_v \sigma\phi(\mathbf{r}), \qquad (4.25)$$

where $2meE/\hbar^2 = K_v$ and K_v is the magnitude of $\mathbf{K}_v$, the wave vector of the wave in vacuum incident on the crystal. A wave function satisfying the wave equation in a periodic potential field must have the periodicity of the crystal latice and so must have the form of a Bloch wave, written as

$$\psi(\mathbf{r}) = \sum_\mathbf{h} \Psi_\mathbf{h} \exp(2\pi i \mathbf{h} \cdot \mathbf{r}) \cdot \exp(i \mathbf{k}_0 \cdot \mathbf{r}) = \sum_\mathbf{h} \Psi_\mathbf{h} \exp(i \mathbf{k}_h \cdot \mathbf{r}), \quad (4.26)$$

where $\mathbf{k}_0$ is the incident-wave vector in the crystal and $\mathbf{k}_h = \mathbf{k}_0 + 2\pi \mathbf{h}$, where $\mathbf{h}$ is a reciprocal-lattice vector.

There are, in principle, an infinite number of such Bloch waves in the crystal. For the Bloch wave of index i, there is a set of plane waves having wave vectors $\mathbf{k}_h^i = \mathbf{k}_0^i + 2\pi \mathbf{h}$. The amplitudes of these waves are $\Psi_\mathbf{h}^i$. To determine the magnitudes of these vectors and these amplitudes, we derive a dispersion equation by inserting the periodic potential of (4.25) and the periodic wave function of (4.26) into equation (4.1), rewritten as

$$\nabla^2 \psi + [K_v^2 + v(\mathbf{r})]\psi = 0 \quad (4.27)$$

and equating each Fourier coefficient individually to zero, giving

$$(K_v^2 - k_\mathbf{h}^2)\Psi_\mathbf{h} + \sum_\mathbf{g} v_{\mathbf{h}-\mathbf{g}}\Psi_\mathbf{g} = 0.$$

From the summation, the term $v_0 \Psi_\mathbf{h}$ is extracted and combined with the first term, and $\kappa = (K_0^2 + v_0)^{1/2}$ is seen to be the magnitude of the wave vector inside a crystal for which $v_\mathbf{h} = 0$ for $\mathbf{h} \neq 0$. Then, the dispersion equations for the wave within the crystal are

$$(\kappa^2 - k_\mathbf{h}^2)\Psi_\mathbf{h} + \sum_\mathbf{g} v_{\mathbf{h}-\mathbf{g}}\Psi_\mathbf{g} = 0. \quad (4.28)$$

There is an ambiguity in sign of k_n because of the squared term $k_\mathbf{h}^2$ that appears in (4.28). Thus, the treatment to this stage includes waves going in all directions, backscattered as well as forward-scattered, as is essential if the treatment is to be applied to LEED or RHEED. For the transmission of high-energy electrons through thin crystals, there is no appreciable back-scattering. All diffracted beams of appreciable intensity occur within a narrow angular range of a few degrees around the incident-beam direction. Then, one may make the forward-scattering and small-angle approximations.

We consider an incident beam almost perpendicular to the entrance face of a thin, parallel-sided crystal plate. The boundary conditions that must be satisfied at the entrance surface are the usual ones of conservation of the component of the wave vector parallel to the surface. Then, the $\mathbf{k}_0^i$-vectors for all Bloch waves must have the same projection on the entrance surface as $\mathbf{K}_v$. If these vectors $\mathbf{k}_0^i$ are drawn to the origin of the reciprocal lattice, as in Figure 4.1, their starting points must lie on the same normal to the entrance surface as the starting point of κ.

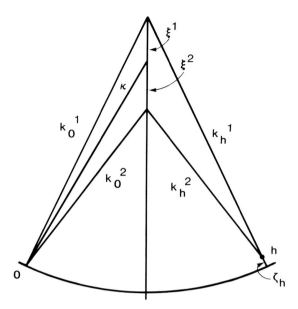

FIGURE 4.1 The excitation error $\zeta_\mathbf{h}$ and the anpassung values $\xi^{(1)}$ and $\xi^{(2)}$ for the two Bloch waves in two-beam dynamical diffraction.

Each Bloch wave is therefore characterized by a so-called tie point L^i on this surface normal. The distance of L^i along this line from the origin of the κ-vector is the anpassung ξ^i. The magnitudes of the significant values will be very much less than $|\kappa|$. If an Ewald sphere is drawn about L^0, the origin point of κ, the distance of the sphere from the $\mathbf{h}$-reciprocal-lattice point is the excitation error $\zeta_\mathbf{h}$; and since all ξ^i-values are small, this excitation error will be the distance along the direction of the $\mathbf{k}_\mathbf{h}^i$-vector for all i.

Then, in the forward-scattering, small-angle approximation,

$$\mathbf{k}_\mathbf{h}^i = \kappa - \xi^i - \zeta_\mathbf{h},$$

which may be inserted in (4.28), giving

$$\kappa^2 - k_\mathbf{h}^2 = \kappa^2 - (\kappa - \xi^i - \zeta_\mathbf{h})^2 = 2\kappa\xi^i + 2\kappa\zeta_\mathbf{h} - \zeta_\mathbf{h}^2. \tag{4.29}$$

This expression may be written as $x^i + p_\mathbf{h}$, where $x^i = 2\kappa\xi^i$ is a magnitude characteristic of the Bloch wave and $p_\mathbf{h} = 2\kappa\zeta_\mathbf{h} - \zeta_\mathbf{h}^2$ is a quantity associated with the reciprocal-lattice point, dependent on the geometry.

Then, (4.28) may be written as a matrix equation

$$\mathbb{M}\mathbf{\Psi} = x\mathbf{\Psi}, \tag{4.30}$$

where the matrix $\mathbb{M}$ is written, with $v_{\mathbf{h-g}} = v_{\mathbf{hg}}$, as

$$\mathbb{M} = \begin{pmatrix} p_0 & \cdots & v_{\mathbf{0h}} & \cdots & v_{\mathbf{0g}} \\ \vdots & & \vdots & & \vdots \\ v_{\mathbf{h0}} & \cdots & p_{\mathbf{h}} & \cdots & v_{\mathbf{hg}} \\ \vdots & & \vdots & & \vdots \\ v_{\mathbf{g0}} & \cdots & v_{\mathbf{gh}} & \cdots & p_{\mathbf{g}} \end{pmatrix} \quad (4.31)$$

The vector $\mathbf{\Psi}$ has components $\Psi_{\mathbf{h}}$. Then, the x^i and $\mathbf{\Psi}^i$ are the eigenvalues and eigenvectors and can be found, in principle, by use of the standard numerical methods available for the solution of matrix equations. In practice, the number N of coefficients $\Psi_{\mathbf{h}}^i$ having appreciable magnitude is finite, and the matrix $\mathbb{M}$ is an $N \times N$ square matrix that is Hermitian if the $v_{\mathbf{hg}}$ are the Fourier coefficients of a real potential function $v(\mathbf{r})$. Often, the effects of inelastic scattering on the elastic-diffraction process are included by assuming the existence of a phenomenological absorption function, which, when added $\pi/2$ out of phase with $v(\mathbf{r})$, makes the effective potential complex; and $\mathbb{M}$ is no longer Hermitian.

The boundary conditions at the entrance face determine the extent to which the various Bloch waves are excited; that is, they determine the coefficients α^i in the expression for the total wave function

$$\psi(\mathbf{r}) = \sum_i \alpha^i \psi^i(\mathbf{r}).$$

For the condition that the Bloch waves be orthogonal,

$$\alpha^i = \Psi_0^{i*}. \quad (4.32)$$

At the exit face of the parallel-sided crystal, the crystal waves are combined to give the set of vacuum waves of amplitude $\Psi_{\mathbf{h}}$ with wave vectors $\mathbf{K}_{\mathbf{h}}$ in the directions as given by the kinematical theory. Each vacuum amplitude $\Psi_{\mathbf{h}}$ is given by summing the $\mathbf{h}$-contributions from all Bloch waves in the crystal. For a crystal thickness Z, each Bloch-wave component will have changed in phase by an amount determined by the magnitude of the wave vector $\mathbf{k}_{\mathbf{h}}^i$—that is, by the effective value of the refractive index, which will be different for each component. If we include these phase terms and the coefficients α given by (4.32), the amplitude of the vacuum wave is then

$$\Psi_{\mathbf{h}}^v = \sum_i \Psi_0^{i*} \Psi_{\mathbf{h}}^i \exp\left(\frac{iZx^i}{2\kappa}\right), \quad (4.33)$$

and numerical evaluation of these amplitudes can be made by solution of the matrix equation in (4.30). More detailed derivations of (4.33) are given by Fujimoto (1959) and Cowley (1981).

For crystals having small unit cells, such as simple metals, and for incident-beam directions well removed from the principal axes, the diffraction pattern may contain only two beams of appreciable intensity, the

0- and **h**-beams, where **h** represents one of the reciprocal-lattice points close to the origin. Then, a good approximation is to assume that each Bloch wave has only two components, with amplitudes and wave vectors Ψ_0^i, Ψ_h^i, $\mathbf{k}_0^i$, $\mathbf{k}_h^i$ and only two Bloch waves, $i = 1, 2$, need be considered. The dispersion equation of (4.30) is then

$$\begin{pmatrix} 0 & v_{0h} \\ v_{h0} & p_h \end{pmatrix} \begin{pmatrix} \Psi_0 \\ \Psi_h \end{pmatrix} = x \begin{pmatrix} \Psi_0 \\ \Psi_h \end{pmatrix}. \quad (4.34)$$

The solution of this equation leads to the intensity expressions of (3.16), and the form of these expressions provides a basis for the description of many phenomena in the electron microscopy and diffraction from crystals, as outlined in Chapter 3.

4.3.2 Progression of a wave through a crystal

For the purposes of high-resolution, transmission electron microscopy and the associated electron-diffraction modes, the dynamical theory is greatly simplified by the use of the forward-scattering and small-angle approximations. If these approximations are applied from the beginning, instead of being inserted at a later stage of the general theory, they can provide descriptions of the diffraction process that are conceptually more satisfying and can lead more directly and logically to procedures for the numerical evaluation of intensities.

In the direction of the incident beam, taken to be close to the z-axis, the wave function will vary much more slowly than in the x, y-directions. The component k_z of the wave vector in the z-direction is changed very little by the scattering. Hence, the wave function can be replaced by

$$\psi \exp(i\mathbf{k}_z \cdot \mathbf{z}).$$

Substituting this expression into the wave equation of (4.1) and neglecting $\delta^2 \psi / \delta z^2$ then gives

$$\frac{\delta \psi}{\delta z} = i \left[\frac{1}{2\mathbf{k}_z} (\nabla_{x,y}^2 + k_x^2 + k_y^2) + \sigma \phi \right] \psi, \quad (4.35)$$

where

$$\nabla_{x,y}^2 = \frac{\delta^2}{\delta x^2} + \frac{\delta^2}{\delta y^2}.$$

Equation (4.35) has the form of a two-dimensional, time-dependent Schrödinger equation except that the z-coordinate replaces the time. Various techniques are available whereby the exit wave from a crystal may be deduced by integrating this expression through the crystal. Portier and Gratias (1981), for example, have expressed the equation in terms of a causal evolution operator that may be used as the basis for an iteration procedure. By a Fourier transform of (4.35), one reproduces the semi-reciprocal formulation of Tournarie (1961).

Alternatively, the action of the crystal of the incident wave may be represented by the action of a scattering matrix on the column vector $\boldsymbol{\Psi}$ whose elements are the Fourier coefficients $\Psi_\mathbf{h}$ in the expansion of the wave function:

$$\boldsymbol{\Psi} = \mathbb{S}\boldsymbol{\Psi}_0, \qquad (4.36)$$

where $\boldsymbol{\Psi}_0$, representing the incident plane wave, has the first element unity and the rest zero.

The form of the scattering matrix $\mathbb{S}$ was first derived from the Bethe theory by Sturkey (1962), who showed that, for a crystal thickness Z,

$$\mathbb{S} = \exp\left(\frac{iZ\mathbb{M}}{2\kappa}\right), \qquad (4.37)$$

where $\mathbb{M}$ is the matrix of (4.31). Then, for successive slices of the crystal, bounded by planes perpendicular to the z-direction, the accummulated effect of the diffraction by the slices is found by multiplying the corresponding matrices $\mathbb{S}_1, \mathbb{S}_2, \mathbb{S}_2, \ldots$. Or if the crystal is made up of n identical slices, the final exit-wave vector is given by

$$\boldsymbol{\Psi} = \mathbb{S}^n \boldsymbol{\Psi}_0$$

and

$$\mathbb{S}^n = \exp\left(\frac{inZ\mathbb{M}}{2\kappa}\right). \qquad (4.38)$$

From a different starting point, Howie and Whelan (1961) derived a difference-equation formulation similar in principle to that used by Darwin (1914) in his first treatment of X-ray diffraction from a crystal for the two-beam case. The change of each scattered-beam amplitude with an increase in z is described in terms of the contributions from the scattering from all other beams in the crystal, with phase factors determined by their excitation errors. In the many-beam form, with $\boldsymbol{\Psi}$ a column vector having elements equal to the amplitudes $\Psi_\mathbf{h}$, the basic equation is written as

$$\frac{d}{dz}\boldsymbol{\Psi} = 2\pi i(\mathbb{A} + \mathbb{B})\boldsymbol{\Psi}, \qquad (4.39)$$

where $\mathbb{A}$ is a square matrix with elements

$$A_{hh} = \zeta_\mathbf{h} + \frac{i\sigma\Phi_0'}{4\pi},$$

$$A_{hg} = \frac{\sigma(\Phi_{\mathbf{h}-\mathbf{g}} + i\Phi_{\mathbf{h}-\mathbf{g}}')}{4\pi}, \qquad (4.40)$$

and $\mathbb{B}$ is a diagonal matrix having elements

$$\beta_{\mathbf{h}} = \frac{d}{dz}[\mathbf{h} \cdot \mathbf{R}(z)].$$

The inclusion of the matrix $\mathbb{B}$ allows for the treatment of crystals in which there are distortions, with displacements $\mathbf{R}(z)$ of the planes of atoms away from the positions for the perfect crystal. With the application of the column approximation discussed in Chapter 3, equation (4.39), in a finite-difference form, can then be used widely as a basis for the calculation of electron-microscope images of crystals having distortions and extended defects such as dislocations and stacking faults. The terms $\Phi'_{\mathbf{h}}$ in (4.40) are the imaginary parts of the structure amplitudes, introduced to take account of the absorption effects due to inelastic-scattering processes.

The basis for the formulation in (4.39) is perhaps best illustrated by writing the form for the two-beam case for a perfect crystal without absorption:

$$\frac{d}{dz}\Psi_0 = i\sigma\Phi_{\mathbf{h}}\Psi_{\mathbf{h}},$$

$$\frac{d}{dz}\Psi_{\mathbf{h}} = i\sigma\Phi_{\mathbf{h}}\Psi_0 + 2\pi i\zeta_{\mathbf{h}}\Psi_{\mathbf{h}}. \qquad (4.41)$$

This form may be interpreted as follows: The amplitude of the **h**-beam is modified by a scattering from the **0**-beam of strength $\sigma\Phi_{\mathbf{h}}$ and by a phase change depending on the excitation error $\zeta_{\mathbf{h}}$ arising because the wave vector $\mathbf{k_h}$ differs in length from $\mathbf{k_0}$ by this amount.

Most of the calculations of electron-micrograph contrast for crystal defects have been made by use of the two-beam approximation, with equation (4.41) modified to include the absorption effects and the distortion of the crystal defined by the function $\mathbb{R}(z)$. The systematic computing methods of Head et al. (1973) have provided a basis for the rapid calculation of simulated images and a reliable means for defect identification, which has been valuable in many areas of materials science.

Most calculations of high-resolution images of crystal structures and crystal defects in recent years have, however, been based on the so-called physical-optics approach to a multislice formulation (Cowley and Moodie, 1957). This approach is also based on the idea of following the progressive modification of a wave through a crystal, but it is sufficiently different in both its origins and the means for its practical application to warrant a somewhat more extended treatment in the following section.

4.3.3 Basis for the multislice method

As it passes through a crystal, an incident electron wave may be considered to be modified in two ways. The phase of the wave is affected by traversing a medium of varying refractive index n, with $n-1$ proportional to the potential $\phi(\mathbf{r})$. Also, the wave is spread by Fresnel-diffraction effects. In the physical-optics approach, these effects are separated and are considered to occur successively, once within each thin slice of the crystal taken perpendicular to the incident-beam direction. In the limiting case that the slice thickness tends to zero, the effect of the simultaneous application of these two processes is correctly simulated.

For each slice of thickness Δz, the phase change due to the potential $\phi(\mathbf{r})$ is considered to take place on one plane and is found by multiplying the incident-wave function by

$$q_n(xy) = \exp[-i\sigma\phi_n(xy)],$$

where

$$\phi_n(xy) = \int_{Z_n}^{Z_n+\Delta z} \phi(\mathbf{r})\, dz. \tag{4.42}$$

The effect of Fresnel diffraction is assumed to be represented by convolution with the propagation function $p_n(xy)$, which acts over the distance Δz between slices. In the small-angle approximation, from equation (2.11), ignoring unimportant constants, the propagation function is

$$p(xy) = \exp[-\pi i(x^2+y^2)/\Delta z \cdot \lambda]. \tag{4.43}$$

Then, the wave function after the nth slice is given in terms of the wave function after the $(n-1)$ slice as

$$\psi_n(xy) = [\psi_{n-1}(xy) * p_{n-1}(xy)] \cdot q_n(xy). \tag{4.44}$$

The wave function at the exit face of the specimen can then be obtained by performing the series of alternating convolutions and multiplications of (4.44) for the successive slices of the specimen, starting from the incident-wave function

$$\psi_0(xy) = \exp[2\pi i(u_0 x + v_0 y)]. \tag{4.45}$$

For a crystal, the transmission functions $q_n(xy)$ are periodic:

$$q_n(xy) = \sum_{hk} Q_n(h,k) \exp\left[-2\pi i\left(\frac{hx}{a} + \frac{ky}{b}\right)\right].$$

In the limiting case that the slice thickness tends to zero, one can make the first-order approximation to $q_n(xy)$, letting

$$q_n(xy) = 1 - i\sigma\phi(xyz_n)\,\Delta z$$
$$= 1 - i\,\Delta z \cdot \sigma \sum_h \sum_k \sum_l \Phi(hkl) \exp\left[2\pi i\left(\frac{hx}{a} + \frac{ky}{b}\right)\right] \exp\left(2\pi i\frac{l}{c} z_n\right). \tag{4.46}$$

For a periodic object such as a crystal a more convenient method is to deal with the set of discrete structure amplitudes, rather than the continuous function of (4.44). The Fourier transform of (4.44) is

$$\Psi_n(uv) = [\Psi_{n-1}(uv) \cdot P_n(uv)] * Q_n(uv), \quad (4.47)$$

where

$$\Psi_n(uv) = \sum_h \sum_k \Psi_n(h, k)\delta\left(u - \frac{h}{a}, v - \frac{k}{b}\right).$$

The Fourier transform of the propagation function $p_n(xy)$ is

$$P_n(uv) = \exp\left[+\pi i \, \Delta z \cdot \lambda(u^2 + v^2)\right]. \quad (4.48)$$

And from (4.46),

$$Q_n(uv) = \delta(uv) - i \, \Delta z \cdot \sigma \sum_h \sum_k \left[\sum_l \Phi(hkl) \exp\left(2\pi i \frac{l}{c} z_n\right)\right]\delta\left(u - \frac{h}{a}, v - \frac{k}{b}\right). \quad (4.49)$$

Cowley and Moodie (1957) showed that the iteration of (4.46) through the N slices of a crystal can be expressed in a form analogous to that of a Born series. When the series is expressed as a power series in terms of the quantities $\Delta z \cdot \sigma \Phi(hk)$, the first-order terms represent scattering from only one slice of the crystal—that is, a single-scattering approximation leading to the kinematical-scattering approximation. The second-order terms represent scattering from any two slices; and so on.

The result may be expressed formally as a summation over n, the number of scatterings (Cowley and Moodie, 1962), as

$$\Psi(\mathbf{h}) = \sum_{n=1}^{\infty} \mathbb{E}_n(\mathbf{h}) \cdot Z_n(\mathbf{h}). \quad (4.50)$$

The operator $\mathbb{E}_n(\mathbf{h})$ represents sums over all scattering amplitudes involved:

$$\mathbb{E}_n(\mathbf{h}) = \sum_l \sum_{\mathbf{h}_1} \sum_{\mathbf{h}_2} \cdots \sum_{\mathbf{h}_{n-1}} \Phi(\mathbf{h}_1) \cdot \Phi(\mathbf{h}_2) \cdots \Phi\left(\mathbf{h} - \sum_{r=1}^{n-1} \mathbf{h}_r\right), \quad (4.51)$$

where h stands for the triple indices h, k, l.

The functions $Z_n(\mathbf{h})$ involve only the geometric aspects of the scattering, as embodied in the sets of excitation errors for the successive scatterings:

$$Z_n(\mathbf{h}) = \sum_{r=0}^{\infty} \frac{(-2\pi i H)^{n+r}}{(n+r)!} h_r(\zeta, \zeta_1, \zeta_2, \ldots, \zeta_{n-1}), \quad (4.52)$$

where H is the crystal thickness and the ζ_n are the excitation errors for the cumulative scatterings—that is,

$$\zeta_1 = -\frac{\lambda}{2}\left(\frac{h_1^2}{a^2} + \frac{k_1^2}{b^2}\right) - \frac{l_1}{c} + \frac{\lambda}{s}\left(\frac{u_0 h_1}{a} + \frac{v_0 k_1}{b}\right),$$

$$\zeta_n = -\frac{\lambda}{2}\left(\frac{h^{(n)^2}}{a^2} + \frac{k^{(n)^2}}{b^2}\right) - \frac{l^{(n)}}{c} + \frac{\lambda}{2}\left(\frac{u_0 h^{(n)}}{a} + \frac{v_0 k^{(n)}}{b}\right),$$

where

$$h^{(n)} = \sum_{r=1}^{n} h_r \qquad k^{(n)} = \sum_{r=1}^{n} k_r \qquad l^{(n)} = \sum_{r=1}^{n} l_r.$$

The excitation-error terms thus have a cumulative nature in that they rely on the whole scattering sequence rather than only on the final scattering direction h. This cumulative effect introduces the essential complication of the dynamical-diffraction result.

The functions $h_r(\ldots)$ are the complete, homogeneous, symmetric polynomials of degree r. For example,

$$h_1(\zeta, \zeta_1, \zeta_2) = \zeta + \zeta_1 + \zeta_2,$$
$$h_2(\zeta, \zeta_1, \zeta_2) = \zeta^2 + \zeta_1^2 + \zeta_2^2 + \zeta\zeta_1 + \zeta_1\zeta_2 + \zeta_2\zeta. \qquad (4.53)$$

The general series in (4.50) has little direct utility; but for certain limiting cases, it gives useful approximations. For example, in the limit of zero thickness or small excitation errors (i.e., λ tending to zero), it gives the phase-object approximation of (4.42). For single scattering ($\sigma\Phi_\mathbf{h} H$ tending to zero), it gives the kinematical approximation. Limiting the number of $\mathbf{h}$-values to two gives the two-beam result; and so on.

The general solution in (4.50) is not useful as a basis for practical calculations of diffracted amplitudes. To make such calculations, one approximates the scattering by considering a finite number of thin slices. The diffracted amplitudes are calculated by the iteration of (4.47), once for each slice. The number of iterations thus depends on the crystal thickness.

When slices of finite thickness are considered, the first-order approximation to the scattering per slice, (4.46), is not adequate; and the full phase-object form of (4.42) must be used, with

$$Q_n(u) = \mathcal{F}[\cos \sigma\phi(xy) - i \sin \sigma\phi(xy)].$$

In practice, slice thicknesses of less than 3 Å are needed in order to achieve accuracy for 100-keV electrons. If the lattice periodicity in the z-direction is smaller than 3 Å, all slices may be assumed to be the same. For larger periodicities in the beam direction, a succession of slices within each periodicity is needed to provide adequate representation of the scattering. An account of the computing methods currently used to deduce diffraction-pattern or electron-microscope-image intensities is given in chapter 8.

4.4 Images of crystals

At the exit face of a crystal, the Fourier coefficients Ψ_{hk} of the wave function produced for a plane wave incident on the crystal may be calculated by use of an appropriate formulation of the many-beam dynamical diffraction. The intensities of the diffracted beams are given by

$$I_{nk} = |\Psi_{hk}|^2.$$

The intensity distribution in the image is given by

$$I(xy) = \left| \sum_{hk} \Psi_{hk} \exp\left[i\chi(h,k)\right] \exp\left[2\pi i\left(\frac{hx}{a} + \frac{ky}{b}\right)\right] \right|^2,$$

where $\chi(h,k)$ is the phase factor of the lens transfer function for the h,k-reflection. Or with **H** equal to the two-dimensional, reciprocal-lattice vector, components h/a, k/b, the intensity is

$$I(\mathbf{r}) = \sum_{\mathbf{G}} \sum_{\mathbf{H}} \Psi_{\mathbf{H}} \Psi_{\mathbf{H}-\mathbf{G}}^* \exp\left[i[\chi(\mathbf{H}) - \chi(\mathbf{H}-\mathbf{G})]\right\} \cdot \exp(2\pi i \mathbf{G} \cdot \mathbf{r}). \quad (4.54)$$

The effects of the incoherent-imaging factors may then be included. The use of a beam of finite convergence (assumed to come from an ideally incoherent source) and the variations of defocus due to fluctuations of the high-voltage or objective-lens currents may be included as a first approximation by writing

$$\int \int W(\Delta f) \cdot C(\boldsymbol{\varepsilon}) \cdot \exp\{i[\chi'(\mathbf{H}) - \chi'(\mathbf{H}-\mathbf{G})]\} \, d\,\Delta f \, d\boldsymbol{\varepsilon} \quad (4.55)$$

in place of $\exp\{i[\chi(\mathbf{H}) - \chi(\mathbf{H}-\mathbf{G})]\}$ in (4.54), with

$$\chi'(\mathbf{H}) = \pi \, \Delta f \, \lambda \, |\mathbf{H} + \boldsymbol{\varepsilon}|^2 + \tfrac{1}{2}\pi C_s \lambda^3 \, |\mathbf{H} + \boldsymbol{\varepsilon}|^4. \quad (4.56)$$

Here, the vector $\boldsymbol{\varepsilon}$ represents the value of **u** for an incident-beam direction, inclined to the lens axis.

From (4.55), it is possible to deduce expressions for the first-order incoherence corrections, the so-called envelope functions, which modify the transfer function in the case that $\Psi_0 \gg \Psi_\mathbf{H}$ for all **H** (see Chapter 2). Equation (4.55) also includes the more general expressions to be applied to the more usual case that some $\Psi_\mathbf{H}$ have appreciable values (see Chapter 8). Expression (4.55), however, does not take into account that, except for very thin crystals, the amplitudes $\Psi_\mathbf{H}$ are themselves functions of the angle of incidence and so of $\boldsymbol{\varepsilon}$.

To illustrate in a simple way some of the factors associated with the imaging of crystals, we treat the two-beam case in several variations. For a crystal thickness H, the amplitudes of the exit wave, derived either from (4.34) with suitable boundary conditions or from the scattering-matrix formulation (4.37) in the two-beam form, are

$$\Psi_0 = \cos\left(\frac{\rho H}{2K}\right) - i\left(\frac{p_\mathbf{h}}{2\rho}\right) \sin\left(\frac{\rho H}{2K}\right),$$

$$\Psi_\mathbf{h} = i\left(\frac{v_\mathbf{h}}{\rho}\right) \sin\left(\frac{\rho H}{2K}\right) \quad (4.57)$$

where $\rho^2 = k^2\zeta_\mathbf{h}^2 + v_\mathbf{h}^2$, $p_\mathbf{h} = 2\kappa\zeta_\mathbf{h}$, and $\zeta_\mathbf{h}$ is the excitation error for the **h**-reflection.

For an incident beam parallel to the axis of the objective lens of a microscope, the values of the transfer function for the two beams are

$$\chi(0) = 0,$$
$$\chi(u_\mathbf{h}) = \pi \, \Delta f \lambda u_\mathbf{h}^2 + \tfrac{1}{2}\pi C_s \lambda^3 u_\mathbf{h}^4. \tag{4.58}$$

The image amplitude is then given by

$$\psi(x) = \Psi_0 + \Psi_\mathbf{h} \exp\left[i\chi(u_\mathbf{h})\right] \exp(2\pi i \mathbf{h} \cdot \mathbf{x}).$$

If we insert the wave amplitudes in (4.57) for a real, positive $v_\mathbf{h}$-value, the image intensity is

$$I(x) = 1 - \left(\frac{v_\mathbf{h}}{\rho}\right) \sin\left(\rho \frac{H}{\kappa}\right) \sin\left[2\pi \mathbf{h} \cdot \mathbf{x} + \chi(u_\mathbf{h})\right]$$
$$- \left(\frac{p_\mathbf{h} v_\mathbf{h}}{\rho^2}\right) \sin^2\left(\frac{\rho H}{2\kappa}\right) \cos\left[2\pi \mathbf{h} \cdot \mathbf{x} + \chi(u_\mathbf{h})\right]. \tag{4.59}$$

For the special case $\zeta_\mathbf{h} = 0$, this equation reduces to

$$I(x) = 1 - \sin\left(\frac{v_\mathbf{h} H}{\kappa}\right) \sin\left[2\pi \mathbf{h} \cdot \mathbf{x} + \chi(u_\mathbf{h})\right]. \tag{4.60}$$

Thus, fringes appear in the image with periodicity $|\mathbf{h}|^{-1}$. For $\chi = 0$—that is, for an ideal perfect lens in focus—the maxima and minima of intensity appear at $\tfrac{1}{4}$ or $\tfrac{3}{4}$ the distance between lattice planes; and the contrast varies sinusoidally with thickness, reversing in sign for thickness increments $H = \kappa/2\pi v_\mathbf{h}$. The phase factor χ gives a lateral shift of the fringe positions that varies linearly with the defocus Δf. For an angle of incidence different from the Bragg angle, $\zeta_\mathbf{h} \neq 0$, the last term of (4.51) introduces a lateral shift of the fringes depending on both H and $\zeta_\mathbf{h}$. Hence, for a wedge-shaped crystal for which H varies or for a distorted crystal for which $\zeta_\mathbf{h}$ varies locally, the fringes in the image will no longer be parallel with the lattice planes; and the correlation of image-fringe positions with lattice-plane positions is even more remote (see Hashimoto et al., 1961).

For a perfect, parallel-sided crystal, the dependence of the fringe position on defocus can be removed by arranging that the incident beam is inclined at an angle to the axis of the objective lens represented by $u = -u_\mathbf{h}/2$. Thus, the phase factors for the transfer function for the **0**- and **h**-beams are $\chi(-u_\mathbf{h}/2)$ and $\chi(u_\mathbf{h}/2)$, which are equal under the usual assumption that χ is a symmetrical function. Then, the image amplitude is

$$\psi(x) = \exp\left[i\chi\left(\frac{u_\mathbf{h}}{2}\right)\right] \exp(-\pi i \mathbf{h} \cdot \mathbf{x})[\Psi_0 + \Psi_\mathbf{h} \exp(2\pi i \mathbf{h} \cdot \mathbf{x})], \tag{4.61}$$

and the image intensity is

$$I(x) = 1 - \left(\frac{v_\mathbf{h}}{\rho}\right) \sin\left(\frac{\rho H}{\kappa}\right) \sin 2\pi \mathbf{h} \cdot \mathbf{x} - \left(\frac{p_\mathbf{h} v_\mathbf{h}}{\rho^2}\right) \sin^2\left(\frac{\rho H}{2\kappa}\right) \cos(2\pi \mathbf{h} \cdot \mathbf{x}). \tag{4.62}$$

Thus, the fringe contrast is completely independent of focus and is not affected by aberrations of the lens, provided that only even-ordered aberration factors are involved.

This result depends on the assumption that the wave incident on the crystal is a plane wave. For the more realistic case that the illumination is assumed to come from a finite, incoherent source, the intensities are added for each incident-beam direction:

$$I(x) = \int s(\varepsilon) I_\varepsilon(x) \, d\varepsilon, \tag{4.63}$$

when $s(\varepsilon)$ is the source function and

$$\begin{aligned}I_\varepsilon(x) &= |\Psi_0 \exp[i\chi(\tfrac{1}{2}u_\mathbf{h} - \varepsilon)] + \Psi_h \exp[i\chi(\tfrac{1}{2}u_\mathbf{h} + \varepsilon)] \exp(2\pi i u_\mathbf{h} x)|^2 \\ &= |\Psi_0|^2 + |\Psi_\mathbf{h}|^2 + 2\,\mathrm{Re}\,[\Psi_0^* \Psi_\mathbf{h} \exp\{i[\chi(\tfrac{1}{2}u_\mathbf{h} + \varepsilon) \\ &\quad - \chi(\tfrac{1}{2}u_\mathbf{h} - \varepsilon) + 2\pi i u_\mathbf{h} x]\}. \end{aligned} \tag{4.64}$$

If we can assume that the dependence of Ψ_0 and Ψ_h on ε is negligible, the integral over ε is

$$\int s(\varepsilon) \exp[2\pi i \lambda (\Delta f\, u_h + \tfrac{1}{4} C_s \lambda^2 u_\mathbf{h}^3)\varepsilon] \, d\varepsilon = S(\Delta f\, \lambda u_h + \tfrac{1}{4} C_s \lambda^3 u_\mathbf{h}^3). \tag{4.65}$$

Here, $S(U)$ is the Fourier transform of the source function $s(\varepsilon)$. For a Gaussian source, $S(U)$ is Gaussian with a maximum value for

$$U \equiv \Delta f\, \lambda u_\mathbf{h} + \tfrac{1}{4} C_s \lambda^3 u_\mathbf{h}^3 = 0.$$

Thus, the contrast of the lattice fringes in the image will be a maximun for the defocus value

$$\Delta f = -\tfrac{1}{4} C_s \lambda^2 u_\mathbf{h}^2, \tag{4.66}$$

and will decrease more rapidly with deviation from this defocus value as the effective, incoherent-source size increases.

A more accurate representation of the variation of lattice-fringe contrast with defocus is given by taking into account that both Ψ_0 and Ψ_h vary in amplitude and phase with $\zeta_\mathbf{h}$ and so with ε, as is evident from (4.57). This variation is minimized for $\zeta_\mathbf{h} = 0$ and for small thicknesses, but the effect can be evaluated analytically for small ε for this two-beam-diffraction situation.

The treatment of many-beam cases, which are of greater interest in relation to the high-resolution imaging of crystals, is considerably more complicated. However, similar dependences on defocus and relative values of the phase factor $\chi(u_\mathbf{h})$ influence the contributions to the image of all interacting pairs of diffracted beams.

REFERENCES

Bethe, H. A. (1928). Theorie der Beugung von Elektronen an Kristallen. *Ann. Phys.* **87**, 55.
Cowley, J. M. (1981). *Diffraction physics*. 2nd ed. North-Holland, Amsterdam.
——, and Moodie, A. F. (1957). The scattering of electrons by atoms and crystals, I. A new theoretical approach. *Acta Crystallogr.* **10**, 609.
——, and Moodie, A. F. (1962). The scattering of electrons by thin crystals. *J. Phys. Soc. Jpn.* Suppl. B2, 210-1–201-8.
Darwin, D. G. (1914). Rontgen-ray reflection. I and II. *Philos. Mag.* **27**, 315 and 675.
Fujimoto, F. (1959). Dynamical theory of electron diffraction in Laue-case. I. General theory. *J. Phys. Soc. Jpn.* **14**, 1558.
Fujiwara, K. (1961). Relativistic dynamical theory of electron diffraction. *J. Phys. Soc. Jpn.* **16**, 2226.
Hashimoto, H., Mannami, M., and Naiki, T. (1961). Dynamical theory of electron diffraction for the electron microscopic image of crystal lattices. I. Images of single crystals. *Philos. Trans. R. Soc. London* **253**, 459.
Head, A. K., Humble, P., Clarebrough, L. M., Morton, A. J., and Forwood, C. T. (1973). *Computed electron micrographs and defect identification*. North-Holland, Amsterdam.
Howie, A., and Whelan, M. J. (1961). Diffraction contrast of electron microscope images of crystal lattice defects. Part II. The development of a dynamical theory. *Proc. R. Soc. London, Ser. A* **263**, 217.
International tables for crystallography, published for the International Union of Crystallography. D. Reidel Publishing Company, Boston.
Portier, R., and Gratias, D. (1981). Une description unifiée de la theorie dynamique de la diffraction élastique aux petit angles des électrons. In *Microscopie électronique en science des Matériaux* 209. Editions du CNRS, Paris.
Sturkey, L. (1962). The calculation of electron diffraction intensities. *Proc. Phys. Soc.* **80**, 321.
Tournarie, M. (1961). Théorie dynamique rigoureuse de la propagation cohérente des électrons à travers une lame cristalline absorbante. *C. R. Acad. Sci.* **252**, 2862.

5
INELASTIC ELECTRON SCATTERING: PART I
JOHN C. H. SPENCE

5.1 Introduction

Although it is a powerful technique for the study of the defect strucure of crystals, high-resolution transmission electron microscopy (HRTEM) is subject to several limitations. Two of the most important are that (1) the method provides little information on the atomic number of the elements present and (2) HTREM images reveal only a projection of the crystal or specimen structure. In this and the following two chapters, we review the theory and applications of a variety of techniques based on inelastic electron scattering, which offer the possibility of overcoming these limitations of HRTEM.

Both of the limitations mentioned previously can, however, be overcome to some extent by using purely elastic electron scattering—for example, lattice images formed from reflections that are sensitive to differences in structure factors have been used to emphasize compositional variations in semiconductor multilayer structures (Hetherington et al. 1985; Ourmazd et al., 1986). Higher-order Laue-zone patterns also give three-dimensional information on crystal structure in convergent-beam microdiffraction patterns. But it is the promise of techniques that depend on inelastic scattering—such as EXAFS (extended X-ray–absorption fine-structure), XANES (X-ray–absorption, near-edge structure) and, more particularly, microdiffraction in STEM (scanning transmission electron microscopy) in combination with electron energy-loss spectroscopy (EELS)—that offers the best hope of overcoming these fundamental limitations of HRTEM. Since both EXAFS and XANES have their electron-beam equivalents (EXELFS—Extended Energy-Loss Fine-Structure—and ELNES—Electron Energy-Loss Near-Edge Structure—respectively), we devote the next three chapters to these and similar inelastic-scattering techniques and to the relevant theoretical background. All of these electron-beam techniques are instrumentally compatible with the HRTEM mode. In addition to providing atomic-structure information, EELS can provide fundamental insights into the electronic structure of solids not available from other techniques. Since the atomic and electronic structures of crystals are not independent, it is not surprising that the disentanglement of these effects on EELS data has proven most difficult. The distinction is therefore somewhat artificial.

There are, however, other reasons why an understanding of inelastic scattering is useful to a research worker in HRTEM. First, the depletion of the elastic wave field that is used to form the HRTEM image by inelastic electron scattering can cause dramatic changes in the appearance of lattice images at specimen thicknesses greater than a few hundred angstroms, since this scattering may affect the phases (and amplitudes) of the diffracted beams. Thus, inelastic scattering must be quantified if there is to be any hope of obtaining agreement between computed and experimental micrographs from thicker crystals, even for known structures. For materials of unknown structure, the inelastic scattering parameters greatly increase the number of adjustable parameters in an HRTEM structure refinement. The use of thicker samples is important if one wishes to minimize the effects on defects of surface relaxation. Second, progress in minimizing radiation damage in HRTEM depends on our understanding of the inelastic process responsible for it.

Inelastic scattering is not a negligible effect. In fact, the total single-electron inelastic scattering exceeds the total elastic scattering for elements lighter than copper. Since HRTEM-imaging conditions include virtually all the inelastic scattering but only a portion of the elastic scattering, HRTEM images must be expected to include a large inelastic background.

The literature on the inelastic scattering of electrons is vast, and a comment is needed to place these chapters in perspective. The very large amount of work on the inelastic scattering of low-energy (<1 kV) electrons in reflection from the vibrational modes of crystal surfaces will not be discussed here. Nor at the other end of the scale, will we be concerned with the classical and relativistic quantum theories of stopping power for high-energy particles described in texts on electrodynamics. Neither is the electron-energy-loss spectroscopy of gases discussed. We confine ourselves to the most probable inelastic processes for kilovolt electrons traversing thin specimens that have a bearing on the problem of extracting atomic- and electronic-structure information from small volumes. Several historically distinct areas are important. These areas include the theory of plasmon excitation based on the work of Bohm and Pines (see Raether, 1980, for a review), the theory of free, isolated-atom ionization by kilovolt electrons (originated by Bethe, 1930) and the theory of inelastic phonon scattering in crystals (developed by Takagi, 1958). Work on dynamical inelastic scattering from the elementary excitations of crystals is also relevant, as started by Kainuma (1955), using wave functions for both the beam and crystal electrons that have the full symmetry of the crystal. The tight-binding approximation provides a link between the free atom and crystal excitation theories. It is this last description of inelastic scattering that must be used to expose the crystal-structure information in electron-energy-loss spectra. The theory of EXAFS, and its electron equivalent EXELFS, and that of the near-edge fine structure are also relevant.

The material of the next three chapters has been covered in several review articles. For plasmons, the article by Raether (1980) is comprehen-

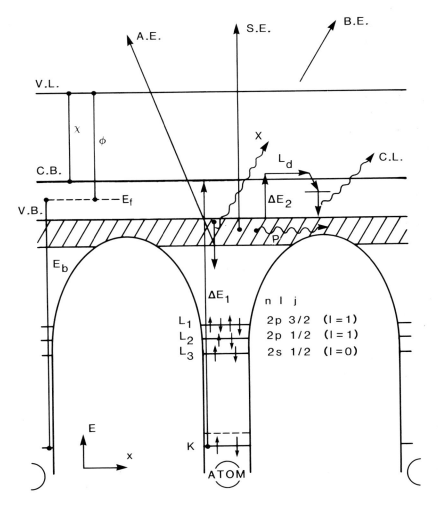

FIGURE 5.1 Schematic one-electron energy E plotted against distance x for the important excitations in a semiconductor in analytical electron microscopy. Here, χ is the electron affinity; ϕ the work function; ε_f the Fermi level; E_b a binding energy, C.B., V.B., and V.L. the conduction-band minimum, valence-band maximum, and vacuum-level, respectively; ΔE_1 a K-shell excitation; ΔE_2 a single-electron excitation; L_d a diffusion length; C.L. a cathodoluminescence photon; A.E. an Auger electron; X an X-ray; B.E. an elastically backscattered electron; S.E. a secondary electron; P a plasmon. In practice, in EELS, one measures differences in total energy between many-electron states. The one-electron atomic orbitals shown therefore provide only an approximate labeling scheme (see the discussion of Koopman's theorem in section 7.3).

sive. The effect of inelastic processes on elastic scattering was first described by Yoshioka (1957), and modern extension of his theory can be found in the articles by Serneels and Haentjens (1980) and Humphreys (1979). The theory of free-atom ionization, in particular, and a clear general review of EELS has recently been given in the text by Egerton (1986). Work on phonon scattering is summarized in the articles by Rez et al. (1977). Recent developments in the study of EXAFS and EXELFS spectra are described in the text by Teo and Joy (1981); work on ELNES (the electron equivalent of XANES or Near Edge X-ray Absorption Fine Structure (NEXAFS)) has been reviewed by Colliex et al. (1985). These last two topics are treated more fully in Chapter 7. A review of the spatial localization expected in the various energy-loss signals has been given by Howie (1979). General reviews emphasizing solid-state effects on EELS spectra can be found in the work of Silcox (1978) and Cowley (1981). Several comparisons of the EXELFS and EXAFS techniques have been published; these comparisons may be traced through the article by Stern (1982). Crystallographic-orientation effects on EELS spectra are briefly reviewed in Chapter 7 (see also Taftø and Krivanek (1982b).

Figure 5.1 indicates the main inelastic processes of interest. The elastic scattering includes large-angle Rutherford scattering from the nucleus. The decay of plasmons (with energies of a few tens of electron volts) results in the emission of UV light. The cathodoluminescence technique is based on the detection of the visible light, which results when an electron in a higher-energy state (usually at an impurity) fills a hole in a lower state that has been created by the fast electron. A similar process for holes created in deep-core states gives rise to X-ray emission, which is used for chemical identification in the X-ray–microanalysis technique (energy-dispersive X-ray spectroscopy, or EDS). Auger electrons may also be created in the de-excitation process; this process becomes the predominant decay mechanism for light elements. For each of these decay products, there is a corresponding electron-energy-loss event in the transmitted-electron beam that may be recorded by using a suitable energy-loss spectrometer.

An important unresolved question is whether the detection of soft X-rays by the windowless EDS technique will provide an more useful method of light-element microanalysis than does the direct recording of electron-energy-loss peaks from the corresponding K-shell excitations. In polyatomic specimens, the presence of the many unwanted low-energy peaks from heavy elements present frequently makes the identification of light elements difficult in EDS. In contrast, the large background in energy-loss spectra places a serious limitation on that technique for microanalysis.

5.2 Kinematics, single-event inelastic scattering, and the dielectric-response function

In this section, we apply the principles of conservation of energy and momentum to inelastic electron scattering, discuss the expression for the strength of this scattering, and relate it to the dielectric constant for the scatterer.

Consider a kilovolt electron scattered inelastically while traversing a thin crystal. Any of the elementary excitations of the crystal may be responsible for the scattering. If the Bloch wave-vector of the incident fast electron is $\mathbf{k}^i$ and that of the scattered electron is $\mathbf{k}^{j'}$, and if the wave vectors of the initial and final states of the crystal electron are χ_i and χ_f, respectively, then the most general requirements for conservation of energy and crystal momentum are

$$E_i(\mathbf{k})^i + \varepsilon_i(\chi_i) = E_f(\mathbf{k})^{j'} + \varepsilon_f(\chi_f) \tag{5.1}$$

$$\mathbf{k}^i + \chi_i = \mathbf{k}^{j'} + \chi_f - g, \tag{5.2}$$

where $E(\mathbf{k})$ and $\varepsilon(\chi)$ are the dispersion relations for the beam and crystal electrons, respectively (subscripts i and f denote initial and final states) and q is any reciprocal-lattice vector. Here χ_i and χ_f are restricted to the first two-dimensional Brillioun zone parallel to the specimen surface; however, $\mathbf{k}^i$ and $\mathbf{k}^{j'}$ are not restricted in this way, nor is q (see the following discussion). The superscripts i and j' refer to different branches of the dynamical-dispersion surfaces. The sign of g has been chosen to be consistent with the conventional Ewald-sphere construction, so that hg represents the momentum gained by the fast electron (not the crystal) as a result of the Umklapp process. This Umklapp involves a whole-body translation of the crystal. In general, these kinematics depend on the choice of basis functions used in the inelastic matrix element and will emerge naturally from its evaluation, and so they should not be imposed a priori. Thus, for example, χ_i is not a good quantum number for transitions from deeply bound, core-level excitations. It is also customary to define the total momentum transfer to the crystal:

$$\mathbf{q} = \mathbf{k}^i - \mathbf{k}^{j'} = \chi_f - \chi_i - \mathbf{g}. \tag{5.2a}$$

The crystal momentum $(-\chi_i - \mathbf{g})$ of the initial state corresponds to the term $\mathbf{p}/\mathbf{h}$ in equation (5.25) describing the recoil momentum of the ion in treatments of single-atom ionization (Meekison and Whelan, 1984). In the general case, $E(\mathbf{k})$ represents the dispersion surface of dynamical electron diffraction, for which the fast-electron wave function is written as a sum of Bloch functions. This case, with relativistic corrections, is described in section 5.4. Here, for simplicity, we specialize to the nonrelativistic case and represent the fast electron by plane waves of wave vectors $\mathbf{K}_i$ and $\mathbf{K}_f$, thereby neglecting multiple elastic scattering of both the incident and inelastically scattered fast electron. Then, $|K| = (2m_0 E/h^2)^{1/2}$; and if a crystal excitation (plasmon, phonon, valence, or core-level electron transition, etc.) of wave vector q and total energy ΔE is created, equations (5.1) and (5.2) become

$$\frac{h^2(K_i^2 - K_f^2)}{2m_0} = \Delta E(q), \tag{5.2b}$$

$$h\mathbf{K}_i = h(\mathbf{K}_f + q), \tag{5.2c}$$

for normal processes. Here, $\Delta E(q) = \varepsilon_f(\chi_f) - \varepsilon_i(\chi_i)$. For the flat (dispersionless), core-level, atomic-like energy bands in crystals, $\varepsilon_i(\chi_i)$ is independent of χ_i. In our later discussion of EXAFS, we will see that the kinetic energy of the final state can be expressed in terms of the binding energy E_b and the asymptotic value χ of χ_f (see equation 7.4).

By forming dK/dE and using $q(\perp) \simeq |K_i|\,\theta$ and $q(||) = |K_i|\,\theta_E$, we find, for small θ and ΔE,

$$q_f^2 = K_i^2(\theta^2 + \theta_E^2), \tag{5.3}$$

where

$$\theta_E = \frac{\Delta E}{2E_i}, \tag{5.3a}$$

as shown in Figure 5.2. Equation (5.3) is relativistically correct if θ_E is defined as $\Delta E/(\gamma m_0 v^2)$. For a given energy loss ΔE, there is a minimum value of q (we assume normal processes for the remainder of this chapter) given by

$$q_{min} = K_i \theta_E \qquad (\theta = 0), \tag{5.3b}$$

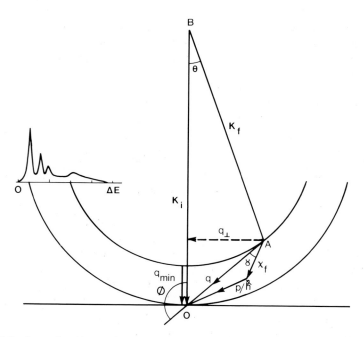

FIGURE 5.2 Scattering kinematics for a plane-wave beam electron. The angle ϕ is used in Figure 5.7. Here, p is the ion-core recoil momentum for single-atom ionization, or the crystal momentum of the initial crystal-electron state for ionization in a crystal, and χ_f is the asymptotic, crystal-electron wave vector after ejection. The energy-loss spectrum is indicated schematically at the left.

which should be compared with the corresponding values for the optical-absorption case (see section 7.2). For given E_i, with given (small) source position (i.e., a prepared initial state) and specified energy loss ΔE, q must connect the elastic and inelastic Ewald spheres shown in Figure 5.2. A feature corresponding to a particular excitation will then be seen in the EELS spectrum for those values of ΔE and q that satisfy the dispersion relation for this excitation. Alternatively, fixing E_i, $\mathbf{K}_i$, and the detector position restricts q to terminating on the line AB with corresponding energy losses given by (5.3) and $|q| < q_{max}$. For all energy losses, q must terminate within the elastic Ewald sphere; otherwise, energy gain would result.

Equation (5.3) allows the wave vector of the crystal excitation to be determined from the scattering angle and energy-loss measurement ΔE. For inner-shell excitations, the momentum transfer $(\pm hq)$ gives very approximately the direction in which the ejected photoelectron will travel (but see section 5.4). Since the final state of the fast electron depends on the diffraction of this photoelectron, both the near edge and extended fine structure of inner-shell edges have been shown to depend on the scattering geometry. In this way, by choice of scattering geometry, the EELS spectra can be used, in principle, to determine the near-neighbor crystallographic environment around chemically identified species. Applications of this method are described in Chapter 7. Note that $\delta\phi \simeq \delta\theta/\theta_E$, so the illumination semiangle and, hence, the available electron intensity are limited. Similar considerations apply to the detector solid angle, which, for a given energy resolution, will be limited by the luminosity (etendu) or throughput of the spectrometer. Some typical experimental inner shell excitation spectra are shown in Figure 5.3b,c.

For a dispersive excitation, the energy loss is a function of q, $[\Delta E = \Delta E(q)]$, and it is this dispersion relation that characterizes the excitation. The dispersion relation indicates the way in which the specimen can absorb energy and is characteristic of the electronic structure and hence the chemical behavior of the sample. It is used to distinguish the various elementary excitations in solids and can be determined from measurements of electron-loss spectra at various scattering angles.

In real specimens, each fast electron may excite several (possibly different) inelastic processes and, in addition, be subject to multiple, elastic Bragg scattering. Kikuchi lines are examples of the elastic Bragg scattering of inelastically scattered electrons. Because of the strength of the elastic interaction, it is probably impossible (except perhaps through the use of dynamically forbidden reflections) to devise experimental conditions that will completely exclude elastic scattering from an inelastic channel or detection window. This fundamental point is often overlooked; experimental evidence for these effects is given in Chapter 7. For large specimen thicknesses, these multiple inelastic and multiple elastic processes become heavily intermixed. For certain incident-beam directions, however, the penetration of an electron beam into a crystal is found to be greatly

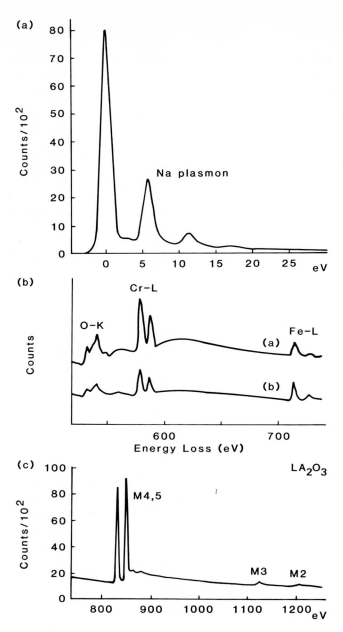

FIGURE 5.3 Electron-energy-loss spectra. (*a*) The transmission bulk-plasmon spectrum of sodium (the plasmon energy is 5.9 eV). The importance of multiple scattering is clear. The specimen thickness is less than 1000 Å, and the specimen a single crystal (Shindo and Spence, 1985). (*b*) The L_2- and L_3-edges of iron in a mixed-valency chromite spinel. The L_3-peak actually consists of two peaks, the lower one due to transitions from $2p$-states to empty d-states on the Fe^{2+} tetrahedral ions, the other to similar transitions on the Fe^{3+} octahedral ions. Spectra are shown for two channeling conditions of the incident-electron beam (see section 7.4 for a discussion.) (From Taftø and Krivanek (1982(a).) (*c*) The spectrum of lanthanum (from La_2O_3), showing the dipole forbidden M_3- and M_2-peaks (Ahn and Krivanek, 1983).

increased owing to reduced absorption. This result is known as the chaneling effect, and its relationship to the theory of dynamical inelastic scattering is described in the text by Ohtsuki (1983). The prediction of Kikuchi-pattern intensities (and backscattered channeling patterns) remains one of the most difficult and challenging problems in electron diffraction (Humphreys, 1979). These topics are discussed further in the next chapter.

The general expression for the scattered-electron intensity is given by equation (5.27). However, for sufficiently thin samples in orientations that avoid the strong excitation of Bragg beams, equation (5.27) gives the simple expressions of the Bethe theory for the inelastic-scattering cross sections, assuming single inelastic and no elastic scattering. For a general crystal excitation, the cross section, assuming a Columbic interaction between fast and crystal electrons, is

$$\frac{d^2\sigma}{d\Omega\, d\,\Delta E} = \left(\frac{e^4}{E_i\, \Delta E}\right) \frac{df(\theta, \Delta E)}{d\,\Delta E} \left(\frac{1}{\theta^2 + \theta_E^2}\right), \qquad (5.4)$$

where

$$\frac{df(\theta, \Delta E)}{d\,\Delta E} = \left(\frac{2m\,\Delta E}{h^2 q^2}\right) \left|\langle f| \sum \exp(i\mathbf{q}\cdot\mathbf{r}_i) |i\rangle\right|^2 \qquad (5.5)$$

is the oscillator strength for the interaction between ground state $|i\rangle$ and final state $\langle f|$ of the crystal (or free-atom) electrons, normalized per unit energy loss. A clear derivation of this result can be found in the text by Landau and Lifshitz (1978). The crucial approximations used in deriving equation (5.5) are the representation of the initial and final state of the fast electron by plane waves, the use of the first Born approximation ($\Delta E \ll E_0$), and the assumption that the fast and crystal electrons are distinguishable, so that their wave functions can be written as a simple product. The use of plane waves in place of the true, dynamical, fast-electron wave function eliminates the possibility of diffraction-contrast or lattice-image effects in inelastic images in this approximation. The factors independent of q incorporate a plane-wave density of states for the beam electron. Here, the r_i are the crystal electron coordinates.

We note that there is no contribution to the interaction from the nucleus in equation (5.4) (owing to the orthogonality of the atomic eigenfunctions); so unlike elastically scattered electrons (which are sensitive to the *total* crystal charge density), inelastically scattered electrons (like X-rays) are scattered only from the crystal electrons. Thus, for light elements, the total inelastic scattering is (1) stronger than the elastic scattering, (2) delocalized (see section 5.4), and (3) insensitive to the nuclear contribution to the crystal potential. The terms before the matrix element in equation (5.4) describe the Rutherford scattering of an electron by a single free electron (proportional to q^{-4}), but additional q-dependence is contained in the electron contribution from the crystal-electron wave functions in the matrix element (also known as the inelastic form factor).

If the oscillator strength is assumed to be independent of q (reasonable, perhaps, for small q), equation (5.4) shows that the angular distribution of inelastic scattering is Lorentzian, with half width (full width at half maximum height, or FWHM) $\theta_E = E_i/2\,\Delta E$. However, the mean angle of scattering may be as large as $10\theta_E$ (Egerton, 1986). For energy losses much larger than the ionization threshold, the energy-loss intensity becomes concentrated around the Bethe ridge, which occurs at a (nonrelativistic) scattering angle of $\theta_R = (2\theta_E)^{1/2}$. To progress further, one needs a detailed knowledge of crystal, excited-state wave functions, and estimates of these functions may be obtained by a variety of computational methods. Recent work, including a comparison of calculations for free atoms based on Hartree-Slater wave functions and those based on the modified hydrogenic model, is summarized by Rez (1982) and Egerton (1986), where a fuller discussion can be found. Figure 5.4 shows the results of these calculations for single atoms as used in problems of microanalysis by EELS (Egerton, 1982). This topic is further discussed in Chapter 7. The observation of these inner-shell losses in the reflection mode using bulk samples are described in Wang and Cowley (1988).

An alternative and more general approach to inelastic electron scattering is based on the dielectric formulation, and this approach has proved useful both in comparing the EXAFS and EXELFS techniques described in Chapter 7 and in relating EELS to optical-absorption and reflectance studies. The differential cross section per electron can be written (in cgs units; see Daniels et al., 1970) as

$$\frac{d^2\sigma}{d\Omega\, d\,\Delta E} = \left(\frac{1}{2\pi^2 n a_0 E_i}\right)\left(\frac{1}{\theta^2 + \theta_E^2}\right)\left[\operatorname{Im}\frac{-1}{\varepsilon(\omega, q)}\right], \tag{5.6}$$

where n is the number of electrons per unit volume, a_0 is the Bohr radius, and $\varepsilon(\omega, q) = \varepsilon_1 + i\varepsilon_2$ is the complex frequency- and wave-factor-dependent dielectric constant. In addition, we have

$$\operatorname{Im}\frac{-1}{\varepsilon(\omega, q)} = \frac{\varepsilon_2}{\varepsilon_1^2 + \varepsilon_2^2} = \left(\frac{4\pi^2 e^2}{q^2}\right) S(q, \omega),$$

where $S(q, \omega)$ is known as the dynamic-form factor ($\Delta E = hw$). Since the real and imaginary parts of $\varepsilon(\omega, q)$ are related through the Kramers-Kronig relations, transmission electron-energy-loss spectra can be used to determine the optical properties of solids (Daniels, 1969; Johnson, 1974; Raether, 1980; Isaacson, 1981).

The optical absorption coefficient for photons of energy ΔE is

$$\mu(\Delta E) = \frac{2\,\Delta E}{hc}\frac{1}{2}[|\varepsilon(\omega)| - \varepsilon_1(\omega)]^{1/2}, \tag{5.7}$$

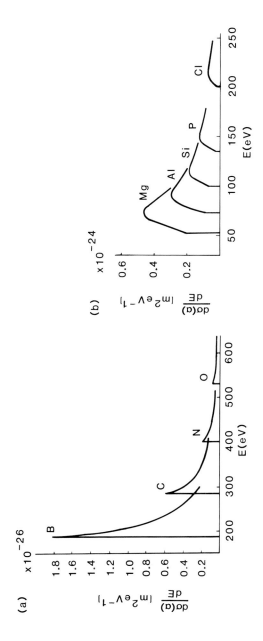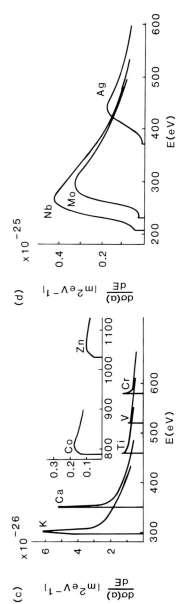

FIGURE 5.4 Computed atomic line shapes in EELS (from Leapman et al. (1980). (a) The K edges (3 mrad collection aperture, 80 keV). (b), (c) The $L_{2,3}$ edges (10 mrad). (d) The $M_{4,5}$ edges (10 mrad). 'White' lines not included in (c).

which gives an approximate photoabsorption cross section for inner-shell excitations

$$\sigma(\Delta E) \simeq \Delta E \frac{\varepsilon_2(\Delta E)}{hcN}, \tag{5.8}$$

where N is the number of atoms or molecules per unit volume and we have taken $\varepsilon_1 \approx 1$, $\varepsilon_2 \ll 1$ (Isaacson and Utlaut, 1978). The extraction of $\varepsilon_1(\omega, q)$ and $\varepsilon_2(\omega, q)$ from EELS data is, however, a lengthy procedure and involves the following considerations. The unwanted effects of multiple, fast-electron elastic scattering on the angular dependence of scattering must be removed if $\varepsilon(\omega, q)$ is to be determined. This removal has been undertaken in very few cases (Batson et al., 1976). For a recent review of the theory of angular broadening due to multiple, incoherent scattering (i.e., elastic scattering without Bragg diffraction), see Rez (1983). For the determination of $\varepsilon(\omega)$ alone, the unwanted effects of multiple energy-loss events can readily be removed by logarithmic deconvolution using the exact inversion of the multiple inelastic-scattering problem outlined in the next chapter (see section 6.4). This method involves no adjustable parameters (t/λ_{av} need not be known) and applies to any number of component-loss processes; however, it requires that data be collected over a large energy range and integrated over all angles for which appreciable inelastic scattering occurs. This angular integration may violate the $q = 0$ condition required for comparison with optical data.

Modified procedures requiring the collection of data over a smaller energy range have also been proposed that can be tested against the exact solution provided by logarithmic deconvolution. Finally, the Kramers-Kronig relations can be applied by using a simple and efficient algorithm that avoids numerical integration near poles (Johnson, 1975). The way in which noise and instrumental broadening propagate through the data analysis must be carefully considered. Figure 5.5 shows the result of a typical analysis (Johnson, 1974).

In a comparison of optical-absorption studies with EELS, several points must be borne in mind. Optical absorption is a single interaction in which a photon is completely destroyed, unlike inelastic light scattering, which is a second-order effect in which one photon is destroyed and another (of lower energy) created. The fast electron, however, may lose any portion of its energy. EELS covers the 0 to 3 kV range with an energy resolution of, at best, 0.5 eV for conventional instruments, although millivolt resolution has been obtained over a smaller energy range in purpose-built instruments (Geiger, 1981). The best recently reported resolution for X-ray work at low energies is about 0.75 eV at 900 eV. Only for small q can the EELS results be directly compared with optical studies, since only then can the dipole approximation be made in equation (5.5). For core losses, this approximation is valid if $q_{max}a_0 < 0.2$, where a_0 is the radius of the core state; and these conditions can be met by choice of detector geometry (see section

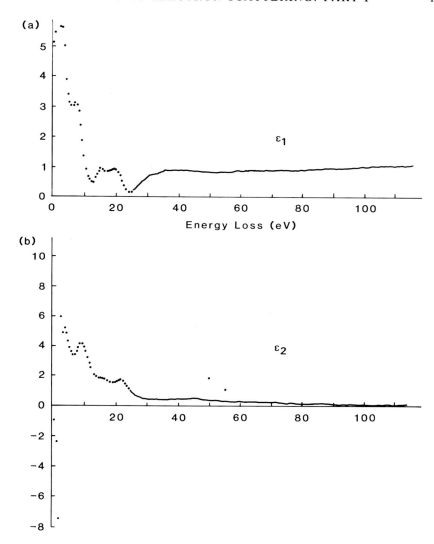

FIGURE 5.5 (a) Real (ε_1) and (b) imaginary (ε_2) parts of the frequency-dependent dielectric function for $BaTiO_3$ obtained from transmission electron-energy-loss data (Johnson, 1976) using the single-scattering analysis of section 6.4 and the Kramers-Kronig relations.

7.2). For plasmons, θ_E is sufficiently small to ensure that the optical response is a good approximation. Optically induced transitions are normally described as vertical on an energy against wave-vector diagram (i.e., $q = 0$). EELS provides the opportunity of obtaining more information by exploring the momentum dependence of crystal excitations (Chen and Silcox, 1977). By comparison, inelastic neutron diffraction probes small-energy phonon excitations over a wide range of q. The dielectric function given by Ritchie and Howie (1977) is of some interest, since it incorporates both the collective and single-particle behavior of the crystal electrons.

5.3 Plasmons, phonons, and single-electron excitations

In its simplest form, a plasmon is a collective excitation of the free-electron gas in a metal. Since typical plasmon energies are in the range $10\,\text{eV} < hw_\text{p} < 20\,\text{eV}$, plasmons are not excited optically; however, surface plasmons of lower energy may be. An extensive review of work on both is provided by Pines (1963) and Raether (1980). The simplest theory (exactly analogous to that used to account for the reflection of radio waves from the ionosphere) gives the plasmon energy as

$$hw_\text{p} = \left(\frac{ne^2}{\varepsilon_0 m}\right)^{1/2}, \tag{5.9}$$

in SI units, where n is the electron density, e its charge, and m its mass. The plasmon energy corresponds to the frequency ω_p for which $\varepsilon(w) = 0$. Modern work has concentrated on the observation of plasmons in insulators, on surface plasmons and their role in surface-enhanced Raman scattering, on the mechanisms of plasmon decay and dispersion, and on the use of surface-plasmon-decay emission for large-screen, optical-display devices. A comprehensive review of the effects of the lattice periodicity on plasmons and on the theory of plasmon line width in the dielectric formulation has been given by Sturm (1982). The occurrence of plasmons in semiconductors can be qualitatively understood by assuming that, since $hw_\text{p} \gg E_\text{g}$ (the band-gap energy), plasmons do not "see" this band gap. Changes in plasmon energy have also been used as the basis for a microanalytical technique through the variation of electron density with position (Williams and Eddington, 1976).

Ferrell (1956) has shown that the differential cross section for single plasmon excitation is

$$\frac{d\sigma_\text{p}}{d\Omega} = \left(\frac{G^{-1}(\theta)}{2\pi na_0}\right)\left(\frac{\theta_E}{\theta^2 + \Theta^2}\right), \tag{5.10}$$

where $G(\theta)$ is a function that is close to unity up to a critical cutoff angle $\theta_\text{c} = q_\text{c}/q$, where q_c is the cutoff wave vector for plasmon excitation (q_c is about 10^{-2} rad at 100 kV). From equation 5.10, a mean free path $\lambda_\text{p} = (n\sigma_\text{p})^{-1}$ can be obtained by integration, giving

$$\lambda_\text{p} = \frac{a_0}{\theta_E \ln(\theta_\text{c}/\theta_E)}. \tag{5.11}$$

For 100-kV electrons traversing a thin metal foil, λ_p is about 1000 Å (Spence and Spargo, 1970). We note that, for K-shell excitation in carbon,

$\lambda_k = 3 \times 10^4$ Å, so plasmons are generally the strongest feature in an EELS spectrum. Figure 5.3a shows the low-loss plasmon region of an EELS spectrum from sodium metal; Figure 5.3c shows a typical inner-shell-excitation spectrum.

Equations (5.10) and (5.11) were derived by neglecting elastic Bragg scattering, the effect of which is discussed in detail by Howie (1963) (see section 5.4). The simplest model for plasmon dispersion gives

$$E_p = \hbar\omega_p \left[1 + \left(\frac{3E_f}{5\omega_p^2 m}\right) q^2 \right], \tag{5.12}$$

where E_f is the Fermi energy. This slight dispersion was first confirmed experimentally by Watanabe. Beyond the cutoff wave vector q_c, single-electron excitations became kinematically permissible decay modes for plasmons; other modes such as exciton creation and ultraviolet-light emission also exist. However, there is evidence of collective-mode oscillation for wave vectors greater than q_c (Stiebling and Raether, 1978). Our understanding of plasmon decay is far from complete (Kliewer and Raether, 1974). The probability of exciting n plasmons in a thickness t of material is given by the Poisson distribution as

$$P_n = \frac{1}{n!} \left(\frac{t}{\lambda_p}\right)^n \exp\left(\frac{-t}{\lambda}\right), \tag{5.13}$$

This relationship has been confirmed experimentally. Since $\sum P_n = 1$, multiple-loss peak intensities increase with increasing specimen thickness. Ashley and Ritchie (1970a) have calculated the cross section for the second-order, double-plasmon process (an energy loss $2\hbar\omega_p$ occurring in a single interaction), and experimental evidence for this effect has been given by Spence and Spargo (1971). The thickness dependence of plasmon losses is demonstrated experimentally in Figure 5.6.

Surface plasmons have recently become increasingly important to electron microscopists in view of the renewed interest in the reflection mode for the study for surfaces (Raether, 1980). A complete theory of dynamical, coupled, elastic and inelastic electron scattering including surface plasmons does not yet exist; however, some important features of inelastic scattering in the Bragg (reflection) case have been noted (Howie, 1982). Energy losses due to transition radiation from the diffracted fast electron have also been observed in STEM in the reflection mode (Cowley, 1982). For the recently developed technique of surface-profile imaging at atomic resolution (Marks and Smith, 1983), the contribution of surface plasmons to the absorption potential remains to be determined. The dramatic effects of these and other new loss modes on EELS spectra from small particles have recently been studied in STEM (Marks, 1981; Batson, 1982; Wang and Cowley, 1987).

The thermal motion of atoms in a solid is customarily described by a set of time-dependent displacement waves known as phonons. The effects of

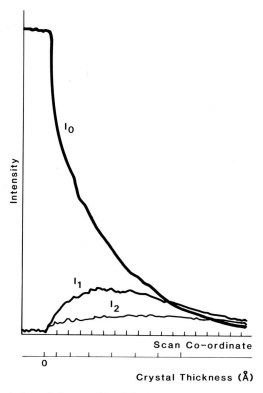

FIGURE 5.6 The variation of the zero-loss (I_0), first-plasmon-loss (I_1), and second-plasmon-loss (I_2) transmitted-electron intensities as a function of crystal thickness in aluminum. These experimental curves follow the Poisson law given in the text. The crystal thickness was calibrated from bright-field images using the known extinction distances (Spence and Spargo, 1971).

phonons on neutron diffraction and X-ray diffraction have been studied for many years by using a single-scattering theory, which, to a limited extent, can be used to understand the effects of phonons and temperature on electron diffraction patterns. Dynamical electron scattering from phonons is described in section 5.4. Here, we give a brief qualitative picture of the effects of phonons on electron diffraction in the single-scattering approximation.

Two extremes of approximation have been used to describe thermal-lattice vibration—the Einstein model in which each atom vibrates independently and the plane-wave-displacement model. We consider the Einstein model first. If the quantum nature of phonons is ignored, then in a simplified classical model the kinematic elastic scattering (as would be recorded in an energy-filtering electron microscope) from an assembly of vibrating atoms is given by the modulus squared of the Fourier transform of the time average of the crystal potential (Cowley, 1981). This time average may be expressed by convoluting the individual-atom potentials with a Gaussian function, which in one dimension may be written as

$$g(x) = (\pi b^2)^{-1/2} \exp\left(\frac{-x^2}{b^2}\right), \tag{5.14}$$

where b is the root-mean-square deviation of an atom from its lattice site. Since this time-averaged potential is periodic, the resulting elastic diffraction pattern shows only the sharp peaks of the undisplaced lattice attenuated by the square of the Fourier transform of equation (5.14), which is

$$D(q) = \exp(-2\pi^2 b^2 q^2) \tag{5.15}$$

This is known as the Debye-Waller factor. The total diffracted intensity, including inelastic scattering from the vibrational modes of the solid, is then given by this attenuated, elastic Bragg scattering plus a diffuse, inelastic background that occurs between the sharp Bragg reflections given by

$$|F(q)|^2 [1 - \exp(-2\pi^2 b^2 q^2)], \tag{5.16}$$

where $F(q)$ is the atomic-scattering factor and $q = \theta/\lambda$. Thus, the phonon diffuse scattering is zero at the origin and rises to a maximum (producing a bright ring of intensity in diffraction patterns recorded at high temperature) before falling off for large θ as $F(q)$ tends to zero.

At the other extreme of approximation, we can consider the electron scattering from a one-dimensional lattice subject to small displacements

$$\Delta = A \cos 2\pi \left(\frac{x}{L} - vt\right),$$

corresponding to a longitudinal wave of period L and frequency v. The phonon-dispersion relation gives the relationship between the phonon wave vector $k = 2\pi/L$ and its angular frequency $\omega = 2\pi v$. For such an idealized case, the method of generalized (time-dependent) Patterson functions gives the kinematic, elastic electron scattering as equal to that from the undisplaced lattice (i.e., a set of Bragg peaks) but attenuated by a factor $\exp(-4\pi^2 \langle \Delta^2 \rangle q^2)$, which is just the contribution to the Debye-Waller factor from this lattice wave. The inelastic scattering, corresponding to electrons that have gained or lost energy in the annihilation or creation of a phonon, can be shown to consist of satellite peaks displaced from the main Bragg peaks by $\Delta q = \pm L^{-1}$. The intensity of these peaks is proportional to $q^2 F^2(q)$ and to the energy hv of the lattice wave and so rises to a maximum at moderate scattering angles. Since there is one such pair of satellites for each lattice wave, and since the dispersion relation is linear for the long-wavelength phonons that contribute close to the Bragg peaks, one observes sharply peaked, inelastic, phonon diffuse scattering around the Bragg reflections. The energy resolution available in conventional EELS does not allow the determination of phonon dispersion curves from measurements of diffuse scattering, as is the case in inelastic neutron diffraction. A typical phonon energy is 20 meV at room temperature.

Interest in single-electron excitations from core levels in electron microscopy has greatly increased recently in view of the possibilities they offer for light-element microanalysis and EXELFS analysis. A discussion of both these techniques and a brief review of the theory of the fine structure that appears at core-loss peaks is deferred until Chapter 7. However, low-energy, single-electron excitations are also important since they make a substantial contribution to the background in EELS spectra. This background is generally believed to arise from multiple-phonon and low-energy, single-electron excitations. These excitations are divided into two groups—interband transitions between different bands of the crystal-electronic-band structure and, in metals where unfilled bands exist, intraband transitions occurring within the same band. These transitions should be clearly distinguished from the inter- and intraband transitions of the fast-electron, Bloch-wave states. Since the energy resolution of most conventional EELS is an appreciable fraction of the band gap for most materials, the direct observation of interband transitions at the low-energy end of EELS spectra is not usually practical; however, some interesting orientation effects have been observed in diamond by M. Brown and others (see section 7.4). These transitions could be treated by using equations (5.4) and (5.5) in a single-scattering approximation if suitable wave functions were available.

Multiple-scattering treatments of low-energy, single-electron excitations have been given (Cundy and Howie, 1969; Humphreys and Whelan, 1969), and these studies conclude that, certainly for losses of less than 100 eV and scattering angles of less than 10^{-3} rad, diffraction-contrast effects in images will be preserved. However, there has been little experimental study of these excitations since conventional spectrometers do not have the required angular or energy resolution, and a featureless spectrum is expected in many cases, similar to that due to crystal electrons promoted above the vacuum level. Exceptions include the study of interband transitions in aluminum by Chen and Silcox (1977) and work on the $4d$-shell excitations by Franck and Schnatterly (1982). The dominance of multiple-plasmon losses in this spectral region make the use of a method for removing these effects essential.

The preceding discussion has ignored relativistic effects in inelastic electron scattering. For plasmons, a detailed treatment of relativistic effects can be found in Ashley and Ritchie (1970b). For inner-shell excitations, the authoritative and comprehensive review of Bethe theory, including relativistic effects, is that of Inokuti (1971).

5.4 Dynamical inelastic scattering

A question of fundamental importance that must be resolved is whether the incident fast electron is more likely to suffer elastic Bragg scattering in a crystal or to be inelastically scattered. The mean free path for inelastic scattering gives an inverse measure of this probability. It might therefore be argued that, since these inelastic path lengths are usually much longer than typical two-beam-extinction distances (for a crystal set at the Bragg condition), multiple elastic scattering is of greater importance than inelastic scattering, however thin the specimen.

If we recall Bragg's idea of crystal planes as half-silvered mirrors, then the extinction distance is the distance the incident-electron wave field propagates through the crystal before being entirely diffracted away into a new Bragg direction. Thus, the use of plane waves to represent the fast electron (which eliminates the possibility of multiple elastic scattering) together with a single-inelastic-scattering theory (such as that given in section 5.2) may be entirely misleading in strongly diffracting orientations. The situation is complicated, however, by the fact that the extinction distance is not a very useful concept in a general orientation, since then the individual beams do not vary periodically in intensity with increasing thickness. The concept of an elastic mean free path is also not relevant to the case of coherent, multiple Bragg scattering, which occurs when a collimated, incident-electron beam is used.

A complete description of inelastic electron scattering in crystals must allow both for multiple energy-loss events and for multiple elastic Bragg scattering of those electrons that have lost energy to crystal excitations. This general treatment can then be used to estimate the validity of the approximations made in sections 5.2, 7.2, and 7.3, in which multiple, fast-electron elastic scattering is neglected. Only such a theory can predict the detailed intensity distribution of Kikuchi lines and energy-filtered images and diffraction patterns.

A general multiple-scattering theory is also required to predict orientation-dependent effects in EELS—in which the intensity of inner-shell peaks in EELS spectra is found to depend on the Bragg condition of the incident illumination—if a collimated beam is used. We have seen that the total elastic- and inelastic-scattering cross sections may be comparable. The neglect of the elastic scattering of inelastically scattered electrons must therefore be considered a severe approximation appropriate only to the thinnest specimens and lightest elements. This ratio of total elastic to inelastic scattering (excluding plasmon and phonon scattering) has been estimated very roughly as $n = 26/Z$, with Z the atomic number. However, in thin crystals, the elastic wave field may be concentrated by diffraction processes on particular sites, so total cross sections are hardly relevant.

We commence with a review of the kinematics of coupled, multiple elastic and inelastic scattering. If relativistic effects are also included, equations

(5.1) and (5.2) become (for $\chi_i = 0$)

$$q = \mathbf{k}^i - \mathbf{k}^{j'} \tag{5.17a}$$

$$E_i - E_f = \Delta E(q), \tag{5.17b}$$

$$hcK_i = E^i \beta, \tag{5.17c}$$

$$K_i^2 = \frac{E_i^2 - m_0^2 c^4}{h^2 c^2}, \tag{5.17d}$$

$$K_f^2 = \frac{E_f^2 - m_0^2 c^4}{h^2 c^2}, \tag{5.17e}$$

$$K_f^2 - \mathbf{k}^{(j')2} = -\lambda^{j'}, \tag{5.17f}$$

$$H_i^2 - \mathbf{k}^{(i)2} = -\lambda^i, \tag{5.17g}$$

where $\beta = v/c$, $\mathbf{k}^i$ and $\mathbf{k}^{j'}$ are the wave-vector labels of the ith and jth Bloch waves of the initial and final state of the fast electron, and $\Delta E(q)$ is the dispersion relation of the crystal excitation. Here, $\lambda^{j'}$ and λ^i describe the deviation of the dynamical electron dispersion surface from that of a plane wave for the final and initial states of the fast (beam) electron. They may be obtained from numerical solutions of the elastic, dynamical-electron-diffraction problem and are related to the extinction distances. The orientation and energy of the final state (and therefore its eigenvalues) are fixed in energy-loss spectroscopy by the experimental measurement of E_f and k_f. Each Bloch wave k^i of the initial state corresponds to the *same* total energy E_i of the fast electron (and similarly for the final state). The dispersion surfaces give the allowed wave vectors for fast electrons of energy E_i and E_f, and these wave vectors are coupled through the crystal-excitation dispersion relation. All quantities in equations (5.17a) to (5.17g) are laboratory frame quantities.

There have been broadly two main approaches to the problem of predicting the *intensities* of coupled, multiple, elastic and inelastic electron scattering in crystals. The first stems from the work of Kainuma (1955), Yoshioka (1957), and Howie (1963) and is based on coupled, Bloch-wave solutions to the Schrödinger equation; the second approach is described in a papers by Gjønnes (1964) and developed numerically by Cowley and Pogany (1968). We commence with a brief review of the Bloch-wave approach, which as been comprehensively reviewed by Kambe and Moliere (1970).

To obtain a rough physical picture of the effect of elastic scattering on inelastically scattered electron intensities, one may imagine that the effect of inelastic scattering is to broaden an initially collimated incident beam into a cone of semiangle $\theta_E = \Delta E / 2E_i$. Thus, diffraction effects such as pendellosung and lattice images resulting from the subsequent diffraction of this cone by the lattice will be washed out to a greater or lesser extent depending on the magnitude of the loss ΔE. The formation of lattice images by using plasmon-loss electrons (Craven and Colliex, 1977) can be understood from this simple beam-divergence model. Similarly, the loss of diffraction-contrast effects in energy-filtered images of increasing energy loss can be understood quantitatively as follows: The pendellosung fringes that form on the exit face of a wedge-shaped crystal in two-beam conditions owing to beating between strongly excited Bloch waves give rise to satellite peaks in the diffraction pattern whose spacing is inversely proportional to the extinction distance ξ_g. By recording these satellites as a function of crystal rotation, one may obtain a map of the dispersion surfaces (Ishida et al., 1975). The effect of inelastic scattering (for idealized plane-wave illumination) is to broaden these satellite peaks into disks subtending an approximate semiangle θ_E. If we imagine the pendellosung fringes in the image to arise from interference between these satellite reflections, then strong fringes can only be expected if these disks do not overlap. Alternatively, we might expect that diffraction contrast effects in inelastic images will be lost if θ_E exceeds the angular width of the two-beam rocking curve $(\xi_g g)^{-1}$. This condition leads to

$$\Delta E < \frac{2E_i}{g\xi_g} \qquad (5.18)$$

as the condition for the preservation of diffraction contrast. However, there is a difficulty here. The angular distribution of inelastic scattering has been seen to be approximately Lorentzian, having rather extended tails. Thus, criteria such as (5.18) must be considered very approximate, and the detail seen in inelastically filtered images will depend greatly on the background present and on the contrast-enhancement properties of the detector. The channeling effects on localized excitations described in section 7.4 depend on the use of scattering at angles considerably larger than θ_E.

In the Bloch-wave picture, the wave function $\phi_n(r)$ of the fast electron inelastically scattered from state m to n with energy $E_n = E_m - \Delta E_n$ satisfies the Schrödinger equation (Howie, 1963),

$$-\left(\frac{h^2}{2m\nabla^2} + H_{nn} - E_n\right)\phi_n = -\sum_{n \neq m} H_{nm}\phi_m \qquad (5.19)$$

where

$$H_{nm}(\mathbf{r}) = b_n^*(r_1, \ldots, r_n) H(r, r_1, \ldots, r_n) b_m(r_1, \ldots, r_n)\, dr_1, \ldots, dr_n. \qquad (5.20)$$

Here, b_m and b_n are crystal-electron wave functions before and after the energy loss ΔE; $H(r, r_1, \ldots, r_n)$ is the interaction energy between the fast electron at $\mathbf{r}$ and the crystal electrons $r_1, \ldots, r_n$. Howie (1963) has shown that the interaction potential $H_{nm}(\mathbf{r})$ may be expanded as

$$H_{nm}(\mathbf{r}) = \exp(-2\pi i \mathbf{q}_{nm} \cdot \mathbf{r}) \sum{}' H_g^{nm} \exp(2\pi i g \cdot r), \qquad (5.21)$$

where q_{nm} is the wave vector of the crystal excitation.

The solution to (5.19) is written as a Bloch wave,

$$\phi_n(r) = \alpha_n(z) \sum C_g(k_n) \exp[2\pi i(k_n + g) \cdot r], \quad (5.22)$$

and the aim becomes the determination of the depth dependence of the various Bloch-wave-excitation amplitudes $\alpha_n(z)$. The case of pure, elastic dynamical scattering is given by setting the right-hand side of equation (5.19) to zero. Then, $H_{00}(r)$ represents the ground-state crystal potential, and the solutions are those given in texts on electron diffraction (Hirsch et al., 1977). The strengths of the various Bloch waves describing electrons that have excited the various crystal excitations are coupled according to

$$\frac{d\alpha_n(z)}{dz} = \sum_{m \neq n} C_{nm}\alpha_m, \quad (5.23)$$

where

$$C_{nm} = -\left[\frac{2\pi i m}{h^2(k_n)_z}\right] \sum_{g,g'} C_g^*(k_n) H_{g-g'}^{nm} \cdot C_{g'}(k_m). \quad (5.24)$$

Equation (5.23) relates how, for example, the amplitude of a Bloch wave describing single-plasmon-loss fast electrons ($n = 1$) builds up with increasing thickness (as shown in Figure 5.5) at the expense of the elastic channel. The argument can be extended to cover wave functions consisting of linear sums of Bloch waves, each originating on a particular disperson surface. Thus, equation (5.23) describes the change in amplitude with depth of the Bloch wave associated with one of the dispersion surfaces of energy E_n owing to inelastic scattering, both interband, and intraband, from all other branches of different energy. We have ignored resonance-error effects that arise from the finite crystal thickness (Young and Rez, 1975); these are important in very thin specimens.

These equations have been solved for various inelastic processes by different authors. The case of plasmon losses has been treated by Howie (1963), who obtained agreement with the Poisson law for multiple losses (which may otherwise be derived by statistical rather than wave-mechanical arguments). Valence-band, single-electron excitations have been treated by Cundy et al. (1969) and Humphreys and Whelan (1969). Phonon excitation by kilovolt electrons in transmission has been treated by Takagi (1958) and Rez et al. (1977)—in all cases, incorporating multiple elastic scattering but treating multiple inelastic scattering to various degrees. A multiple-elastic/single-inelastic–scattering treatment [$n = 1$, $m = 0$ in equations (5.19) to (5.24)] is clearly appropriate to those excitations whose path length is much greater than the crystal thicknesses. Such a treatment has also been given for inner-shell, single-electron excitations in Maslen and Rossouw (1984) and Saldin and Rez (1986) which we now discuss.

Consider the ionization of an isolated atom by a fast electron through the promotion of a K-shell electron into the continuum. The asymptotic-crystal-electron wave vector is χ_f. Since nuclear recoil must be allowed for, the conservation-of-momentum equation of (5.2) becomes, for a free atom,

$$\mathbf{K}_i - \mathbf{K}_f = \chi_f + \frac{\mathbf{p}}{h} = q, \tag{5.25}$$

where **p** takes account both of the initial momentum $h\chi_i$ of the 1s electron and of nuclear recoil. Thus, **q** (whose magnitude and direction are known from the experimental geometry and energy loss) and χ_f may differ both in magnitude and direction. The relationship between the intensity distribution for the ejected atomic electron and the known value of q is thus of some interest, especially for directional EXELFS and other orientation-dependent EELS effects described in section 7.4.

In EELS experiments where the ejected atomic electron is not detected, an integration over all directions of χ_f is required to determine the energy-loss intensity. In certain coincidence experiments, however, one may detect both the fast and the atomic electrons. The results of theoretical calculations of Maslen and Rossouw (1983) for a hydrogenic atom are shown in Figure 5.7. Here, both $|\chi_f|$ and the energy loss are fixed (just beyond the carbon K-edge); the angular probability distribution for ejection of the atomic electron (which is rotationally symmetric about q) is shown for various values of $|q|$. The conclusion from this work is that, for small $|q|$ (within the limits of the dipole approximation), the amplitude variation resembles a distorted p-type wave with q as an axis of symmetry and a variation of 180° in the phase of the outgoing-beam electron as the angle γ between χ_f and q varies through a similar angle. For large q, both the amplitude and the phase become constant over γ. An important point concerns the large value of

$$\frac{d\phi}{d\theta} = \frac{K_f(K \cos\theta - K_f)}{q^2}$$

(see Figure 5.2), so extremely small detector-collection angles are needed for the fast electron to avoid significant averaging over the direction of q.

For inner-shell excitations in crystals, the situation is more complex. The complete expression for the inelastically scattered, fast-electron intensity involves products of the H_g^{nm}-terms of equation (5.21). They can be written in terms of free, single-atom, excited-state matrix elements in the tight-binding approximation (Whelan, 1965) thus:

$$H_g^{no} = \left(\frac{eN^{1/2}}{\varepsilon_0 V}\right) \exp[i(q-g)\tau] \frac{M_{no}(q-g)}{|q-g|}, \tag{5.26}$$

where

$$M_{no}(q) = \langle f| \exp(-2\pi i \mathbf{q} \cdot \mathbf{r}) |i\rangle.$$

These products render the result sensitive to the phase of $M(\mathbf{q})$, which

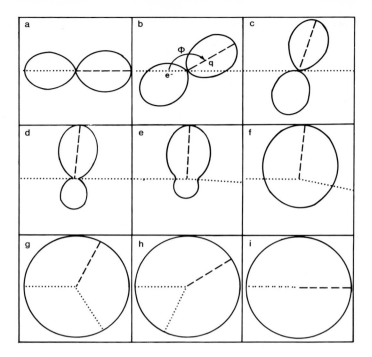

FIGURE 5.7 Polar diagrams showing the ejected-secondary-electron intensity per unit solid angle for an isolated carbon atom whose K-shell is ionized by a 100-keV-beam electron traveling in the direction of the dotted line. The kinetic energy of the secondary electron is 15 eV. The dashed line shows the direction of q, and the secondary electron has wave vector $|\chi_f| = 1.98$ Å^{-1}. The values of θ (in radians; see Figure 5.2) and $|q|$ (per angstrom) are as follows: (a) 0.0 and 0.0278. (b) 0.001 and 0.325. (c) 0.005 and 0.892. (d) 0.02 and 3.4. (e) 0.05 and 8.49. (f) 0.2 and 33.8 (g) 1 and 162.7. (h) 2 and 285.5. (i) π and 339.2. As q increases beyond the dipole approximation, the emission pattern changes from a p-wave to the form of an outgoing spherical wave. Nuclear recoil accounts for the difference between $|\chi_f|$ and $|q|$. (From Maslen and Rossouw, 1983.)

cannot therefore be obtained from the oscillator strength given in equation (5.5), as has been done in some simplified, dynamical-inelastic treatments. It has been shown (Meekison and Whelan, 1984) that the wave vector of the final state $\langle f|$ here is, in the tight-binding model, both the band-labeling wave vector and the wave vector of the electron ejected into the continuum.

A further important point concerns the question of Umklapp (the term g in equation (5.2a) associated with core-loss excitation. In crystals, the momentum-conversion law results from the evaluation of the matrix element in equation (5.26). This momentum conservation law will contain an Umklapp term if the corresponding crystal periodicity is included in the excited-state, crystal-electron wave function. In this sense, we might say that Umklapp is synonymous with EXELFS, since both correspond physically to diffraction of the excited-state crystal electron.

The difference $(\mathbf{q} - \chi_f)$ in equation (5.25) must be considered a crystal momentum of the initial state of the crystal electron [see also equation (5.2a)], and may perhaps be equivalent to single-atom nuclear recoil in an Einstein model of crystal vibrations at high temperatures. The criterion that this momentum difference should be attributed to single-atom recoil if the target atom would move off with a velocity v greater than that of sound, and to phonon emission if v is less than the velocity of sound, has also been suggested (Howie, 1984). The radiolysis mechanism described in section 6.4 differs from these processes in that it consists of two independent steps. The simultaneous emission of phonons and electronic excitation would, however, appear to require the use of a second-order theory involving products of matrix elements, by analogy with the theory of indirect, optical, band-gap transitions in semiconductors. The very short core-hole lifetimes probably make such effects very difficult to observe. A theoretical treatment of these matters can be found in Ohtsuki (1983) and Saldin and Rez (1987).

In an extension of the work of Kainuma (1955), by representing both the incident and inelastically scattered fast electrons by Bloch waves, Maslen and Rossouw (1984) obtain, for the (dimensionless) K-shell inelastically scattered intensity,

$$\frac{dI}{d\Omega} = \left(\frac{8NtK_f}{a^2 K_i}\right) \sum_{ghg'h'} [\text{site}][e, 2e] \sum_{iji'j'} [BW], \quad (5.27)$$

where N is the number of lattice sites per unit volume in a crystal of thickness t and

$$[\text{site}] = \sum_n \exp[2\pi i(\mathbf{Q}_1 - \mathbf{Q}_2) \cdot \tau_n] \quad (5.28)$$

for n atomic sites τ_n. Here,

$$\mathbf{Q}_1 = q + \mathbf{h} - g'$$

and

$$\mathbf{Q}_2 = q + g - \mathbf{h}'.$$

The Bloch-wave term is

$$[BW] = C_0^{i'} C_0^{j'*} C_0^j C_0^{i*} C_h^j C_g^{i*} C_{h'}^{j'} C_{g'}^{i'} X, \quad (5.29)$$

where C_g^i ($C_g^{i'}$) are eigenvector components of the initial (final) fast-electron state, and

$$X = \frac{\exp[2\pi i(k^j - k^i)t] - \exp[2\pi i(k^{i'} - k^{j'})t]}{it(k^j - k^i - k^{i'} + k^{j'})} \quad (5.30)$$

expresses pendellosung effects. The matrix-element term is

$$[e, 2e] = M_{n0}(\mathbf{Q}_1, \chi_f) M_{n0}^*(\mathbf{Q}_2, \chi_f) \rho(\chi_f) d^3 \chi_f, \quad (5.31)$$

where $\rho(\chi_f)$ is the density of final crystal-electron states. Since, in this model, M_{n0} depends on both q and χ_f, the $[e, 2e]$ term must be integrated over the possible directions of χ_f (the asymptotic, excited-state, crystal-electron wave vector) to obtain the inelastically scattered, fast-electron intensity.

We now discuss the main features of the general equation for dynamical inelastic scattering (equation 5.27) in order to understand how the various approximations to it (such as the Bethe theory used in ELNES) and the channeling effects on inner-shell, energy-loss peaks (both described in Chapter 7) are obtained.

First, we note that, in simple crystals in which the primitive unit cell contains only a single atom, [site] = 1. In order for the intensity of an inner-shell, electron-energy-loss filtered diffraction pattern (or the ELNES structure) to be sensitive to the crystallographic site τ_n of the target atom, one must have $Q_1 \neq Q_2$. This site sensitivity is lost in the random-phase approximation of Craven et al. (1978), in which the phases of $M_{n0}(Q_1)$ and $M_{n0}(Q_2)$ are assumed unrelated for simplicity. The inelastic-rocking-curve asymmetry described in Chapter 7 is also lost in this model. A second approximation that has been used (Spence, 1981) might be termed the oscillator-strength approximation. Here, the required integration over $[e, 2e]$ needed to include all angles of secondary-electron ejection is written as a product of separate integrals over $M_{n0}(Q_1)$ and $M_{n0}(Q_2)$. This procedure is physically equivalent to assuming a spherical waveform for the ejected secondary electron.

Second, we consider the case of extreme localization (a delta-function inelastic-scattering potential) for which $[e, 2e] = 1$ for all $Q_{1,2}$. This corresponds physically to the case where the outgoing *fast* electron is modeled as spherical wave. The inelastic intensity may then be integrated over thickness, to give

$$I(R) = A \sum |C_0^i|^2 |C_0^{j'}|^2 \sum_{g,h,g',h'} C_h^i C_g^{i*} C_h^{j'} C_g^{j'*} \exp[i(h - g + h' - g') \cdot R] \quad (5.32)$$

for a loss even localized at **R**, with A a constant. $I(R)$ is equal to the square of the current density $\psi(\mathbf{R})\psi(\mathbf{R})^*$ of the fast electron if q is parallel to $\mathbf{K}_j$. Equation 5.32 may then be used to interpret the orientation effects of inner-shell losses described in section 7.4. For off-axis points in an energy-filtered, inner-shell diffraction pattern, Kikuchi bands will be seen in which, for an axial incident beam, the edge of the excess band is associated with a final fast-electron state that moves current density away from heavy-atom positions. A deficient band will be observed from lighter atoms.

Finally, the Bethe theory [equations (5.4) and (7.6), on which the interpretation of ELNES is based] is obtained by taking $C_g^i = C_g^{i'} = 0$ for all i and $g \neq 0$ in equation (5.27). We must also make allowance for the fact that the final-state wave functions in the Bethe theory (equation 7.6) are normalized per unit energy range, whereas in equation (5.27) a delta-function normalization is used.

Equations (5.27) to (5.31) do not make the independent Bloch-wave approximation. Experimental evidence for the failure of this approximation at small thickness has been given by Cherns et al. (1973) and can be understood by using the phase-grating approximation.

Equation (5.25) is based on the tight-binding approximation. This approximation is well known to be a poor one for excited states well above the Fermi level. In addition, the assumption of "no overlap" between orbitals on neighboring atoms specifically excludes the possibility of EXELFS effects, as would the use of a plane-wave density of states $\rho(\chi_f)$. Nevertheless, this treatment contains all the essential features necessary for the quantitative prediction of dynamical inner-shell-excitation intensities. In particular, Kikuchi patterns due to localized inner-shell excitations, channeling and blocking effects, orientation effects in EELS core losses (see section 7.4), and the loss of diffraction-contrast effects with increasing energy loss may all be computed by using this treatment. The dependence of Kikuchi-band intensities on the crystal site of the inelastically scattering species is also revealed. The important parameters for such a calculation are the ground-state crystal potential and the excited-state wave functions of the crystal electrons. In this way, reasonable agreement has been obtained between experimental, orientation-dependent, X-ray and EELS spectra from spinel crystals and calculations (Maslen and Rossouw, 1984). However, these comparisons do not provide a very sensitive test of theory. Whereas the Maslen and Rossouw treatment uses atomic wave functions for the excited-state crystal electron, a more rigorous treatment would use the true wave functions for the elementary excitations of the crystal (Kainuma, 1955; Meekison and Whelan, 1984). Unfortunately, the accurate calculation of these wave-functions remains a difficult problem in theoretical solid-state physics. The recent work of Saldin and Rez (1987) should also be consulted in connection with this topic.

An alternative approach to dynamical-inelastic calculations is that of Doyle (1971), which gives a more physical picture of the scattering process. In this approach, instead of working with the component Bloch waves of the total wave function *inside* the crystal, one works with a wave function matched to the crystal-exit-face boundary condition, whose Fourier transform is therefore the observable two-dimensional distribution of diffuse inelastic scattering seen in an energy-filtered diffraction pattern. The multislice algorithm (see Chapter 7) is used to calculate the dynamical-elastic-scattering distribution and the effect of multiple elastic scattering on the inelastically scattered wave. An inelastically scattered wave function is introduced at a particular slice and crystallographic site, weighted by the incident, elastic, dynamical wave function for that slice. This inelastically scattered wave is then allowed to propagate dynamically through the remaining slices of the crystal, and the resulting wave function is stored. The entire process must then be repeated, allowing the inelastic event to occur in successive slices (and all nonequivalent crystallographic sites). The resulting stored amplitudes or intensities are then added. This method has also been used to calculate the elastic scattering from defects and short-range order scattering (Cowley, 1981). For delocalized excitations, complex amplitudes are added for every depth of the excitation; for localized excitations, intensities are added. The intermediate case may be treated by a running average (Doyle, 1971).

This method of calculation is very similar to the method that has been successfully used to compute simulated elastic microdiffraction patterns from coherent, subnanometer electron probes in STEM (see Chapter 7). Applications of this method to the problem of impurity-site occupancy determination are described in Chapter 7. As in the Bloch-wave method, the essential parameters on which the calculations depend are (1) an accurate knowledge of the crystal-ground-state potential $H_{00}(\mathbf{r})$, (2) a knowledge of the inelastic-scattering matrix element M_{n0}, and (3) for calculations of energy-filtered images of inner-shell losses, a knowledge of the phase of the inelastically scattered wave. This phase may be important if high spatial resolution is sought. The resulting calculations of energy-filtered, inner-shell-excitation diffraction patterns all show considerable sensitivity to the crystallographic site $\mathbf{R}_m$ of the excited species. Thus, by providing the chemical identification of these species, these patterns offer the prospect of overcoming one of the major limitations of the HRTEM method.

REFERENCES

Ahn, C. C., and Krivanek, O. L. (1983). *EELS atlas*. GATAN Corp., Warrendale, Pa., and Center for Solid State Science, Arizona State University, Tempe, Ariz.

Ashley, J. C., and Ritchie, R. H. (1970a). The double plasmon excitation. *Phys. Status Solidi* **38**, 425.

———, and Ritchie, R. H. (1970b). Mean free path of relativistic electrons for plasmon excitations. *Phys. Status Solidi* **40**, 623.

Batson, P. E. (1982). A new surface plasmon resonance in clusters of small aluminum spheres. *Ultramicroscopy* **9**, 277.

———, Chen, C. H., and Silcox, J. (1976). Plasmon dispersion at large wave vectors in Al. *Phys. Rev. Lett.* **37**, 937.

Bethe, H. (1930). Zur theorie des Durchgangs schneller Korpuskularstrahlen durch Materie. *Ann. Phys.* **5**, 324.

Chen, C. H., and Silcox, J. (1977). Direct nonvertical interband transitions at large wave vector in aluminum. *Phys. Rev. B* **16**, 4246.

Cherns, D., Howie, A., and Jacobs, M. (1973). Characteristic X-ray production in thin crystals. *Z. Naturforsch., Teil A* **28**, 565.

Colliex, C., Manoubi, T., Gasgnier, M., and Brown, L. M. (1985). Near-edge fine structures on EELS core loss edges. In *Scanning electron microscopy—1985*, p. 489 A.M.F. O'Hare, Chicago.

Cowley, J. M. (1981). *Diffraction physics*. 2nd ed. North-Holland, New York.

——— (1982). Energy losses of fast electrons at crystal surfaces. *Phys. Rev. B* **25**, 1401.

———, and Pogany, A. P. (1968). Diffuse scattering in electron diffraction patterns. I. General theory and computational methods. *Acta Crystallogr., Sect. A* **24**, 109.

Craven, A. J., and Colliex, C. (1977). The effect of energy-loss on lattice fringe images of dysprosium oxide. In *35th annual proceedings of the electron microscopy society of America*, ed. G. W. Bailey, 242. Claitor's Publishing, Baton Rouge.

———, Gibson, J. M., Howie, A., and Spalding, D. R. (1978). Study of single-electron excitations by electron microscopy. 1. Image contrast from localized excitations. *Philos. Mag., Ser. A* **38**, 519.

Cundy, S. L., Howie, A., and Valdre, U. (1969). Preservation of electron microscope image contrast after inelastic scattering. *Philos. Mag.* **20**, 147.

Daniels, J. (1969). Determination of optical constants of palladium and silver from 2 to 9 eV from energy-loss measurements. *Z. Phys.* **227**, 234.

———, Von Festenberg, H., Raether, H., and Zeppenfeld, D. (1970). Springer Tracts in Modern Physics, vol. 54. Springer-Verlag, New York.

Doyle, P. A. (1971). Dynamical calculation of electron-scattering by plasmons in aluminum. *Acta Crystallogr., Sect. A* **27**, 107.

Egerton, R. F. (1982). Electron energy-loss spectroscopy for chemical analysis. *Philos. Trans. R. Soc. London, Ser. A* **305**, 521.

——— (1986). *Electron energy loss spectroscopy*. Plenum, New York.

Ferrell, R. A. (1956). Angular dependence of the characteristic energy loss of electrons passing through metal foils. *Phys. Rev.* **101**, 554.

Franck, C., and Schnatterly, S. E. (1982). Excitation of 4d shell in Sb, Te and BaF_2 with inelastic electron scattering: collective or single particle? *Phys. Rev. A* **25**, 3049.

Geiger, J. (1981). Inelastic electron scattering with energy losses in the meV region. In *39th annual proceedings of the electron microscopy society of America*, ed. G. W. Bailey, 182. Claitor's Publishing, Baton Rouge.

Gjønnes, J. K. (1964). Influence of Bragg scattering on inelastic and other forms of diffuse scattering of electrons. *Acta Crystallogr.* **20**, 240.

Hetherington, C. J. D., Barry, J. C., Bi, J. M., Humprheys, C. J., Grange J., and Wood, C. (1985). High resolution electron microscopy of semiconductor quantum well structures. *Mat. Res. Soc. Symp. Proc.* **37**.

Hirsch, P. B., Howie, A., Nicholson, R. B., Pashley, D. W., and Whelan, M. J. (1977). *Electron microscopy of thin crystals*. Krieger, New York.

Howie, A. (1963). Inelastic scattering of electrons by crystals. I. The theory of small-angle inelastic scattering. *Proc. R. Soc. London, Ser. A* **271**, 268.

——— (1979). Image contrast and localized signal selection techniques. *J. Microsc.* **117**, 11.

——— (1982). Surface reactions and excitation. *Ultramicroscopy* **11**, 141.

——— (1984). Personal communication.

Humphreys, C. J. (1979). Scattering of fast electrons by crystals. *Rep. Prog. Phys.* **42**, 1825.

———, and Whelan, M. J. (1969). Inelastic scattering of fast electrons by crystals. I. Single electron excitations. *Philos. Mag.* **20**, 165.

Inokuti, M. (1971). Inelastic collisions of fast charged particles with atoms and molecules—the Bethe theory revisited. *Rev. Mod. Phys.* **43**, 297.

Isaacson, M. S. (1981). The potential of energy-loss spectroscopy for electronic characterization of structures on the nanometer scale. *Ultramicroscopy* **7**, 55.

———, and Utlaut, M. (1978). Comparison of electron and proton beams for determining the microchemical environment. *Optik* **3**, 213.

Ishida, K., Johnson, A. W. S., and Lehmpfuhl, Z. (1975). Bloch-wave analysis in electron-diffraction experiments with a CaF_2 single crystal wedge. *Z. Naturforsch. Teil A* **30**, 1715.

Johnson, D. W. (1974). Optical properties determined from electron energy-loss distributions. In *Electron microscopy—1974* (8th International congress), ed. J. V. Sanders and D. J. Goodchild, 388. Australian Academy of Science, Canberra.

——— (1975). A Fourier method for numerical Kramers-Kronig analysis. *J. Phys. A: Gen. Phys.* **8**, 490.

——— (1976). Ph.D. diss., University of Melbourne, Australia.

Kainuma, Y. (1955). The theory of Kikuchi patterns. *Acta Crystallogr.* **8**, 247.

Kambe, K., and Moliere, K. (1970). *Advances in structure research by diffraction methods.* Vol. 3, ed. R. Brill and R. Mason, 53. Pergamon Press, Oxford.

Kliewer, K. L., and Raether, H. (1974). Compton-effect cross sections for electron gas collective to single-particle transition. *J. Phys. C* **7,** 689.

Landau, L., and Lifshitz, L. (1978). *Nonrelativistic quantum mechanics* Pergamon Press, New York.

Leapman, R. D., Rez, P. and Mayers, D. F. (1980). K-, L-, and M-shell generalized oscillator strengths and ionization cross sections for fast-electron collision. *J. Chem. Phys.* **72,** 1232.

Marks, L. D. (1981). Imaging low-energy losses from MgO smoke particles. *Institute of physics conference series,* No. 61, 259. Institute of Physics, Bristol, England.

———, and Smith, D. J. (1983). Direct imaging of surfaces in small metal particles. *Nature* **303,** 316.

Maslen, V. W., and Rossouw, C. J. (1983). The inelastic scattering matrix element and its application to electron energy-loss spectroscopy. *Philos. Mag., Ser. A* **47,** 119.

———, and Rossouw, C. J. (1984). Implications of ($e, 2e$) scattering for inelastic electron diffraction in crystals. I. Theoretical. *Philos. Mag., Ser. A* **49,** 735.

Meekison, C. D., and Whelan, M. J. (1984). The matrix element for inelastic scattering by a crystal. *Philos. Mag.* **50,** L39.

Ohtsuki, Y. (1983). *Charged beam interactions with solids.* Taylor and Francis, London.

Ourmazd, A., Rentschler, J. R., and Taylor, D. W. (1986). Direct resolution and identification of the superlattices in compound semiconductors by high-resolution transmission electron microscopy. *Phys. Rev. Lett.* **57,** 3073.

Pines, D. (1963). *Elementary excitations in solids.* Benjamin, New York.

Reather, H. (1980). *Excitation of plasmons and interband transitions by electrons.* Springer Tracts in Modern Physics, vol. 88. Springer-Verlag, New York.

Rez, P. (1982). Cross sections for energy-loss spectrometry. *Ultramicroscopy* **9,** 283.

——— (1983). A transport equation theory of beam spreading. *Ultramicroscopy* **12,** 29.

———, Humphreys, C. J., and Whelan, M. J. (1977). The distribution of intensity in electron diffraction patterns due to phonon scattering. *Philos. Mag.* **35,** 81.

Ritchie, R. H., and Howie, A. (1977). Electron excitation and the optical potential in electron microscopy. *Philos. Mag.* **36,** 463.

Saldin, D., and Rez, P. (1987). The theory of the excitation of atomic inner shells in crystals by fast electrons. *Philos. Mag.* **55,** 481.

Serneels, R., and Haentjens, D. (1980). Extention of the Yoshioka theory of inelastic electron scattering in crystals. *Philos. Mag.* **42,** 1.

Shindo, D., and Spence, J. C. H. (1985). Unpublished work.

Silcox, J. (1978). Inelastic electron-matter interactions. In *Electron microscopy—1978* (9th international congress), ed. J. M. Sturgess, 259. Microscopical Society of Canada, Toronto.

Spence J. C. H. (1981). The crystallographic information in localized characteristic-loss electron images and diffraction patterns. *Ultramicroscopy* **7,** 59.

———, and Spargo, A. E. (1970). The plasmon mean free path in aluminum. *Phys. Lett. A* **33,** 116.

———, and Spargo, A. E. (1971). Observation of double-plasmon excitation in aluminum. *Phys. Rev. Lett.* **26,** 895.

Stern, E. A. (1982). Comparison between electrons and X-rays for structure determination. *Optik* **61,** 45.

Stiebling, J., and Raether, H. (1978). Dispersion of volume plasmon of silicon (16.7 eV) at large wave vectors. *Phys. Rev. Lett.* **40,** 1293.

Sturm, K. (1982). Electron energy loss in simple metals and semiconductors. *Adv. Phys.* **31,** 1.

Taftø, J., and Krivanek, O. L. (1982a). Site-specific valence determination by electron energy-loss spectroscopy. *Phys. Rev. Lett.* **48,** 560.

———, and Krivanek, O. L. (1982b). Characteristic energy-losses from channeled 100-keV electrons. *Nucl. Instrum. Methods* **194,** 153.

Takagi, S. (1958). On the temperature diffuse scattering of electrons. II. Application to practical problems. *J. Phys. Soc. Jpn.* **13,** 287.
Teo, B. K., and Joy, D. C. (1981). EXAFS spectroscopy. Plenum, New York.
Wang, Z. L., and Cowley, J. M. (1987). Generation of surface plasmon excitation of supported metal particles by an external electron beam. *Ultramicroscopy* **21,** 347.
——, and Cowley, J. M. (1988). Atomic inner shell excitations for EELS in the reflection mode. *J. Microsc.* (in press).
Whelan, M. J. (1965). Inelastic scattering of fast electrons by crystals. I. Interband excitations. *J. Appl. Phys.* **36,** 2099.
Williams, D. B., and Eddington, J. W. (1976). High-resolution microanalysis in materials science using electron energy loss measurements. *J. Microsc.* **108,** 113.
Yoshioka, H. (1957). Effect of inelastic waves on electron diffraction. *J. Phys. Soc. Jpn.* **12,** 618.
Young, A. P., and Rez, P. (1975). Resonance errors and partial coherence in inelastic scattering of fast electrons by crystal excitations. *J. Phys. C* **8,** L1.

6
INELASTIC ELECTRON SCATTERING: PART II
JOHN C. H. SPENCE

6.1 Localization in inelastic scattering

The previous chapter reviewed the main inelastic-scattering processes of importance for kilovolt electrons traversing thin crystals. A brief overview of the main theoretical approaches to the problem of multiple elastic scattering of inelastically scattered electrons was also given. We now discuss the important concept of inelastic localization, which places fundamental limits on the spatial resolution of inelastic images. We then review briefly the experimental work on energy-filtered, transmission-electron-microscope images. We end with short accounts of the calculation of absorption effects, the removal of multiple inelastic scattering from energy-loss spectra, and some notes on radiation damage as it affects HRTEM imaging.

We have seen that the total inelastic scattering of kilovolt electrons from atoms of the first three rows of the periodic table is greater than the elastic scattering, unaffected by the nuclear contribution to the atomic-charge density (unlike elastic scattering), and delocalized. In this and the following section, the meaning and implications of delocalization are discussed for experiments aimed at extracting structural and chemical information from inelastic excitations and their various products.

If the inelastic electron scattering is taken to be confined to a cone of semiangle $\theta_E = \Delta E / 2 E_0$ (see equation 5.4), then this radiation must originate from a specimen volume whose transverse dimension is $L = \lambda / \theta_E$, where L is known as the inelastic localization and can be taken as a measure of the size of the region in which the inelastic scattering is coherently generated. It is therefore also the size of the image element that would be formed if all this scattering were passed through a perfect lens and focused. This simple result for L is consistent with the results of more elaborate quantum-mechanical calculations (Craven et al., 1978) and with an argument based on the time and energy uncertainty principle (Howie, 1979).

This process can be understood from Figure 6.1, in which the classical impact parameter b is defined. A beam electron traveling at velocity v spends an amount of time $\Delta \tau = b/v$ in the neighborhood of the target atom. If we think of the pulse of electromagnetic energy associated with the beam electron's passage evaluated at this atom, then this pulse will contain Fourier components over a range $\Delta \omega = 2\pi / \Delta \tau$, for which the range of associated quantum-mechanical excitation energies (the beam-electron

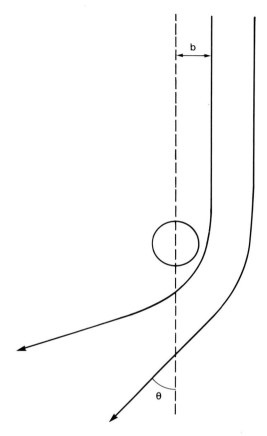

FIGURE 6.1 Trajectories for electrons with small and large impact-parameters b scattered by an atom through large and small asymptotic angles θ. Note that, unlike positrons, electrons are deflected toward the target atom.

energy loss) is $\Delta E = \hbar\omega = hv/b$. Using $v = \hbar k/m$ and $E_i = mv^2/2$, we then find that $L = \lambda/\theta_E = 2\pi b$, so the localization L is approximately equal to the classical impact parameter b. Alternatively, the inelastically scattered amplitude of equation (5.4) can be written as the Fourier transform of an inelastic-scattering potential in Born's approximation. The forces due to this potential affect the passing electron only when it is within a distance b of the atom and for times of the order of $\Delta \tau$. However, the theory of the photoelectric effect, on which the analysis of X-ray–absorption, near-edge structure (XANES) and energy-loss spectra (ELNES) are also based (see section 7.2), assumes instantaneous excitation of an atom (Feuerbacher et al., 1978). Here, the abrupt switching on of the core hole (within less than 10^{-17} s) is known as the sudden approximation. No evidence for the failure of this approximation has yet been produced in inelastic electron scattering.

TABLE 6.1
Time and Energy Relationships in Electron Microscopy ($\Delta\tau\,\Delta E = h$)

Result	Interpretation	Magnitude
$\Delta\tau_1 = \dfrac{b}{v}$ $\left(L = 2\pi b = \dfrac{\lambda}{\theta_E}\right)$	b = classical impact parameter L = inelastic localization v = beam electron velocity	$\Delta\tau_1 = 2.4 \times 10^{-18}$ s for $b = 4$ Å (100 kV) $\Delta E_1 = 1.7$ kV
$\Delta\tau_2 = \dfrac{\lambda(\text{plasmon})}{v'}$ (photoelectron)	ΔE_2 = energy width of an inner-shell peak due to photoelectron lifetime $\Delta\tau_2$ and velocity v'	$0.1 < \Delta E_2 < 10$ eV $(4 \times 10^{-14} < \Delta\tau_2 < 4 \times 10^{-16}$ s$)$
$\Delta\tau_3 = \dfrac{\lambda(\text{phonon})}{v}$ (beam electron)	ΔE_3 = line width of coherent bremsstrahlung or channeling radiation	$\Delta\tau_3 = 0.6 \times 10^{-15}$ s (100 kV) $\Delta E_3 = 6$ eV (nonrelativistic)
$\Delta\tau_4 = \dfrac{h}{\Delta E_4}$	ΔE_4 = electron-source energy spread; $\Delta\tau_4$ is coherence time; electron antibunching is observable if many electrons are generated per $\Delta\tau_4$ in the same mode; for electron-image recording times greater than $\Delta\tau_4$, the *intensities* of the images due to each component-source beam energy are added (incoherently)	$\Delta\tau_4 = 8.3 \times 10^{-15}$ s (for $\Delta E_4 = 0.5$ eV) $L_c = v\,\Delta\tau_4 = 1.3\,\mu$m (100 kV)

The hole switch-on time depends on the photoelectron's kinetic energy, reaching the adiabatic limit at the onset of ionization. Table 6.1 gives a few of the times and energies, related by the uncertainty principle, that are important in electron microscopy. A more rigorous definition of localization can be found in the work of Rossouw and Maslen referred to in Section 7.5.

Broadly, there are, for our purposes, three closely related classes of experiments in which localization effects became important: (1) the observation of diffraction-contrast effects in energy-filtered images; (2) the dependence of secondary-emission-product intensities (e.g., Auger electrons, X-rays, or cathodoluminescence) or those of the corresponding transmitted-electron, energy-loss peaks on the crystallographic orientation of an incident, collimated electron beam; and (3) energy-filtered images from inner-shell excitations formed from large-angle scattering intended to reveal high-resolution detail combined with chemical information.

We have known for many years that low-resolution, diffraction-contrast, energy-filtered images formed from plasmon loss or valence-band, single-electron excitations appear almost identical to the corresponding elastic images (i.e., they preserve contrast) (Howie, 1979). In the previous chapter, the simple approximate condition

$$\Delta E < \frac{2E_0}{g\xi_g} \tag{6.1}$$

was given for the preservation of diffraction contrast at energy loss ΔE. These experiments must be distinguished from lattice imaging or high-resolution, phase-contrast experiments performed by using energy-filtered, plasmon-loss electrons (Craven and Colliex, 1977). Here, the observation of high-resolution detail results from the prior or subsequent elastic scattering of the inelastically scattered electron. (The question of whether the scattering events can actually be sequentially ordered in time is subject to the uncertainty principle.) The case of energy-filtered, diffraction-contrast images formed from inner-shell excitations has been discussed in detail elsewhere (see Chapter 5). The characteristic asymmetry found in core-loss, energy-filtered rocking curves may also be used to identify the crystallographic site of an atomic species (Maslen and Rossouw, 1984). Experimental results, in rough accord with the these predictions, have been obtained by Taftø and Krivanek (1981), as discussed in the next chapter.

If we consider energy-filtered diffraction patterns of increasing energy loss (and therefore increasing $\theta_E = \Delta E/2E_0$ and decreasing localization L), the following qualitative trend might be expected for thin crystals. For small losses, the small scattering angle θ_E results in a point diffraction pattern of sharp Bragg reflections, which may be used to form diffraction-contrast images; and these images will be similar to those seen by using only elastically scattered electrons. With increasing energy loss and θ_E, these Bragg spots broaden to become overlapping disks, producing a pattern of diffuse inelastic scattering crossed by Kikuchi lines. The condition that these disks overlap is that

$$\theta_E > 2\theta_B \tag{6.2}$$

or that

$$\Delta E > \frac{2E_0 \lambda}{d} \tag{6.3}$$

for crystal planes of spacing d with Bragg angle θ_B. This condition, in turn, is equivalent to the requirement that the localization $L = \lambda/\theta_E$ become smaller than the interplanar spacing d. The Kikuchi lines result from the elastic scattering of electrons scattered inelastically into non-Bragg directions. Experimental energy-filtered patterns showing this general trend have been published (Egerton and Whelan, 1974); however, a quantitative analysis must also take account of absorption and thickness effects. In summary, however, we expect diffraction-contrast effects to be preserved for $\theta_E \ll 2\theta_B$; orientation effects and the possibility of higher-resolution imaging arise for $\theta_E \gg 2\theta_B$. A comparison of equations (6.1) and (6.3) shows that they are of similar form, with $W = 2\theta_B \xi_g$ (the column width or Takagi-triangle width) in (6.1) playing the role of d in (6.3).

We will see in Chapter 7 that the observation of orientation-dependent effects on either secondary-emission products or transmitted-electron, energy-loss peaks only become possible if the inelastic localization L is smaller than the lattice spacing responsible for elastic diffraction. This inelastic localization has another interpretation: It may be taken to be the approximate width of the absorption potential for the inelastic process of interest at the appropriate crystallographic site. This absorption potential $V'(\mathbf{r})$ is discussed more fully in section 6.3. However, Heydenreich (1962) showed that it gives rise to a rate of loss of electrons from the elastic channel that is proportional to $V'(\mathbf{r})|\psi(\mathbf{r})|^2$ at the point $\mathbf{r}$. For the simplest case of plasmon excitation, the function $V'(\mathbf{r})$ is roughly constant and has therefore a single zero-order Fourier coefficient that is inversely proportional to the plasmon path length. Thus, we speak of plasmon excitation as delocalized, whereas, at the other extreme of approximation, inner-shell excitations have been modeled as a delta function $V'(\mathbf{r})$ broadened only by thermal vibrations (Cherns et al., 1973).

The study of localized excitation in solids has developed historically along several different lines, resulting in a confusing amount of jargon in the literature, often specialized to particular research groups. Thus, for example, an analysis of the preservation of diffraction contrast in images and rocking curves at large energy losses (Cundy et al., 1969; Humphreys and Whelan, 1969) may treat exactly the same problem as a study of channeling and blocking effects on the orientation dependence of inner-shell excitations (Taftø and Krivanek, 1981). Both must consider the prior elastic scattering of the fast electron in the presence of other absorption processes (channeling), the inelastic event of interest with localization L, and the subsequent elastic scattering of the inelastically scattered electrons (blocking). For both, the independent Bloch-wave model must be expected to fail for thicknesses of less than a few hundred angstroms (Cherns et al., 1973). Useful comparisons between experiments can only be made by specifying the energy and direction of the incident and detected particles and the excitation spectrum, structure, and boundary conditions of the specimen, since these factors alone define the quantum-mechanical scattering problems.

In a series of recent experiments, Taftø and Krivanek (1981) have shown how the effective localization can be varied (and, for the first time, measured) by choice of scattering geometry. In this work, the height of the magnesium, aluminum, and oxygen edges in EELS spectra from spinel crystals were studied as a function of both the incident- (collimated) and scattered-beam directions. The researchers found that a large orientation effect (i.e., dependence on the incident-beam direction) was only obtained if the scattering angle θ shown in Figure 6.1 was also large. Classically, this effect can be understood by interpreting L as the impact parameter, which, as shown in the figure, is small for large-angle scattering. The orientation effect becomes strong for $L < d$ (see equation 6.3). For a wave-mechanical picture, we might imagine an annular detector of similar inner semiangle. This work is described in more detail in section 7.4.

We must emphasize that inelastic localization is a poorly defined concept; in view of the Lorentzian tails on the inelastic-scattering distribution,

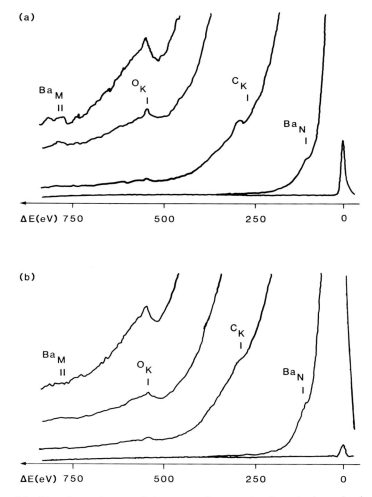

FIGURE 6.2 Experimental transmission-energy-loss spectra from barium aluminate. These data were obtained by using a subnanometer electron probe. (*a*) For probe located over a plane of barium atoms (beam direction in plane of atoms). (*b*) For probe located between the barium planes. The localized barium M-peaks change in intensity, but the more delocalized N-edges do not. Probe size was about 5 Å; interplanar spacing was 11.4 Å. A vacuum-generator HB5, field-emission STEM instrument was used at 100 kV. The spectrum ordinate has been rescaled for clarity.

experimental measurements of L are difficult and will depend greatly on any background present and on the characteristics of the detector. However, the experiments described above can be interpreted as approximate measurements of localization. A study of localization effects on characteristic X-ray emission and energy losses has also been made by Bourdillon and co-workers (see Bourdillon, 1984, and references therein), who find that the inelastic localization is smaller than the thermal-vibration amplitude for all the currently detectable, characteristic X-ray–emission lines. These authors also provide a useful comparison of localization effects on X-ray emission with those on the corresponding energy losses. More recent measurements of localization for both cases confirm equations (6.2) and (6.3) (thus eliminating a factor of two given earlier). Since K-shell X-rays may be produced by electrons promoted well above the top of the valence band, the use of a ΔE value from K-shell EELS spectra in equation (6.3) determines an upper limit on L (Self and Buseck, 1983). These authors find complete localization down to about 500 eV. At larger energy losses, the localization is found to be smaller than the mean thermal-vibration amplitude. More recent measurements of localization can be found in the work of Pennycook (1987).

A more direct impression of the effects of localization on transmitted-electron energy-loss spectra can be found in the work of Spence and Lynch (1982), who observed barium $M_{4,5}$- and N-edges from crystals of barium aluminate by using a field-emission STEM instrument with an electron-probe size of about 5 Å. This probe size is smaller than either the $M_{4,5}$-inner-shell-excitation localization L or the projected unit-cell dimension (11.4 Å). Thus, as shown in Figure 6.2, as the electron probe is moved from a plane of barium atoms (which contains the beam direction) to a position midway between the barium-atom planes (a distance of 5.7 Å), the height of the localized $M_{4,5}$-edges changes; but that of the lower-energy, more delocalized N-edge does not. Because these barium-atom planes coincide with mirror planes of symmetry in the structure, the researchers were able to position the probe accurately over the barium planes by observing the symmetry of the corresponding microdiffraction pattern. Movements of the probe by less than 2 Å may cause contrast reversals in portions of these microdiffraction patterns. It is the portion of the microdiffraction pattern that is used to form the STEM lattice image.

6.2 Inelastic electron imaging

The suggestion has frequently been made that one of the major limitations of high-resolution electron microscopy may be overcome by forming high-resolution images using only those transmitted electrons that have excited some characteristic inner-shell excitation at an identifiable atomic species. The resulting images could, in principle, then be used to locate known species in crystalline or disordered material.

There have been two main approaches to the attainment of this goal, the first using TEM instruments fitted with imaging electron-energy-loss filters of the type originally described by Castaing and Henry and the second using a magnetic-sector analyzer fitted to a STEM instrument. (For an interesting recent use of magnetic-sector analyzers for imaging, see Kondo et al. (1985)). Since high-resolution information is our main concern, we shall not discuss the work done on the energy filtering of diffraction-contrast images or other lower-resolution images. Large-angle scattering must be used to synthesize high-resolution images; therefore, these images cannot be formed from small energy losses (such as losses resulting from plasmons), for which θ_E is small. Furthermore, images formed from the elastic scattering of plasmon-loss electrons contain little useful chemical information other than the density of free electrons (see Williams and Edington, 1976).

Several imaging electron-energy-loss filters for TEM instruments have been built that are capable of chemical mapping, including those of Egerton (1976), Ottensmeyer et al. (1982), Zanchi et al. (1980), Krahl (1982), and Schuman and Somlyo (1982). Most of the applications of these devices have been to biological samples, where the improvement in contrast that results from the elimination of chromatic aberration is important. For a lens focused for the elastic image, the point resolution of the inelastic image due to electrons that have lost energy ΔE in the sample is degraded by an amount

$$\Delta X = \theta_0 C_c \left(\frac{\Delta E}{E_i} \right), \tag{6.4}$$

where θ_0 is the objective-aperture semiangle (assumed filled) and C_c is the chromatic-aberration constant of the objective lens. In high-resolution lattice images, this out-of-focus background from, for example, plasmon-loss electrons, usually appears as a slowly varying background and may, in principle, be removed by energy filtering. For certain focus settings and low-atomic-number specimens supporting a sharp plasmon excitation, though, the plasmon contribution to the unfiltered image may show some structure (Krivanek, 1985). However, the ultimate spatial resolution (which has yet to be reached) in these energy-filtering TEM instruments must be limited by the inelastic localization L.

STEM systems have also been developed for inner-shell-loss imaging, including those of Pearce-Percy and Cowley (1976), Jeanguillaume et al. (1978), Rez and Ahn (1982), Schuman and Somlyo (1982), and Leapman and Swyt (1983). Here, there is no analogous contrast reduction owing to chromatic aberrations. Important problems with all these systems, however, include the removal of thickness contrast, background subtraction and estimation in real time, the development of parallel or multiple detectors, and the minimization of radiation damage. Again, localization places a fundamental limit on the ultimate spatial resolution obtainable.

We consider now the problem of energy-filtered imaging using inner-shell excitations or microanalysis in *crystals* by STEM, a subject that has received little attention in the literature. The experimental arrangement is one in which a subnanometer, coherent electron probe traverses a thin crystal (thickness about 200 Å) and both the elastic microdiffraction pattern and energy-loss spectrum (containing inner-shell losses) are observable. In one recent design (Cowley and Spence, 1979), electrons pass through a small hole in the center of the microdiffraction screen into the spectrometer. The screen allows parallel recording of the majority of the microdiffraction pattern and subsequent storage in a computer through a suitable image intensifier and fast-frame store. This experimental arrangement appears to offer the best hope of obtaining both chemical and crystal structure information at atomic resolution in electron microscopy.

We first must review briefly the principles of STEM lattice imaging (Spence and Cowley, 1978). A ray diagram is shown in Figure 6.3. The *experimental* observation of interference effects near P_3 (Cowley, 1981) indicates that the objective aperture OA can be taken to be coherently filled—that is, that P_0 can be treated as a point source. Then, in the absence

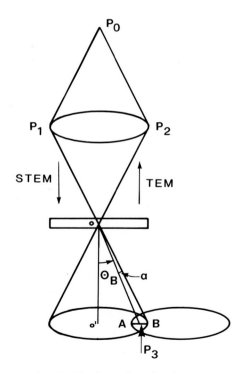

FIGURE 6.3 Ray diagram showing the formation of coherent overlapping orders in STEM. The illumination semiangle is slightly greater than the Bragg angle θ_B, allowing interference at P_3 between the optical paths $P_1 P_3$ and $P_2 P_3$.

of lens aberrations, a diffraction-limited probe of "size" a will result from the use of an objective-aperture semiangle $\theta_R = 0.61\lambda/a$ if the probe size is defined as half the width between the first minima of the probe-intensity profile. The Bragg angle for lattice planes of spacing d is $\theta_B = 0.5\lambda/d$, and thus, the condition that the probe be smaller than the lattice spacing is equivalent to the condition that $\theta_R > \theta_B$—that is, that the adjacent diffraction disks shown in Figure 6.3 overlap. A lattice image can only be obtained in STEM by placing a detector within the region of overlap of these disks and recording the variation of intensity with probe position. In this rather loose sense, lattice imaging becomes possible in STEM as the probe size approaches that of the lattice spacing. In reality, with $C_s \neq 0$, lattice imaging is normally undertaken with a probe substantially larger than the lattice spacing. The important requirement for the formation of lattice images in STEM is not that the electron probe be smaller than the lattice planes of interest but, rather, that the objective aperture be coherently filled over an angular range greater than θ_B. In terms of the reciprocity theorem, overlap of the orders allows interference at a single detector point along two optical paths (P_1P_3 and P_2P_3 in Figure 6.3) that differ by twice the Bragg angle, and this arrangement is exactly equivalent (if ray paths are reversed) to the tilted-illumination, two-beam case in TEM.

In general, for a perfect, parallel-sided crystal that allows only elastic scattering, Spence and Cowley (1978) have shown that (1) the intensity within all the microdiffraction disks is independent of the electron-probe position if these disks do not overlap (i.e., if the probe is larger than the projected unit cell); (2) the intensity within the overlap region is sensitive to the probe position within the unit cell, to the focus setting and spherical-aberration constant of the probe-forming lens, and to the crystal structure. In particular, the projected point-group symmetry of the microdiffraction pattern will be that of the projected crystal structure, as reckoned about the center of the probe, if the probe is smaller than the unit cell. Only as the probe is enlarged will the full-space-group-symmetry elements appear. This general result (true for dynamical scattering) provides a powerful method of locating absolutely the position of a STEM probe with respect to a crystal structure. Thus, Spence and Lynch (1982) were able to use the knowledge that the planes of barium atoms in the barium aluminate structure lie on mirror planes to locate a 5 Å-diameter STEM probe over (and between) these planes, where the energy-loss spectra shown in Figure 6.2 were recorded. This procedure is equivalent to determining whether atoms appear white or black in a high-resolution lattice image. (Using symmetry arguments, Olsen and Spence (1980) have shown that this contrast ambiguity can be resolved in favorable cases.) By displaying the barium $M_{4,5}$-edge intensity (shown in Figure 6.2) as a function of probe position (with suitable background subtraction), one may obtain a chemically specific lattice image. However, there are many difficulties, both fundamental and practical, with these experiments. The most important practical difficulty is obtaining a sufficiently stable probe to allow the required portion of the EELS spectrum to be recorded, including enough data points to allow a reliable estimate of the background. Inelastic localization provides a fundamental difficulty.

To understand filtered STEM lattice images in crystals more fully, we consider the form of an idealized energy-filtered microdiffraction pattern, bearing in mind that the portion of this pattern that falls on the detector will be used to form the image. Consider the case where the illumination takes the form of a plane wave. Then, each Bragg diffraction spot in the energy-filtered pattern will be broadened by approximately $\theta_E = \Delta E/2E_0$ for energy loss ΔE, and the entire pattern will be crossed by Kikuchi lines resulting from elastic scattering of the inelastically scattered electrons. For the barium edges shown, these disks will overlap, and thus the Kikuchi lines that result from localized scattering may be thought of as arising from interference between overlapping orders, as in the theory of elastic STEM lattice imaging. By analogy with that theory (Spence and Cowley, 1978), the intensity distribution of the Kikuchi lines is sensitive to the crystallographic site of the inelastic event (Gjønnes, 1966). (There is, however, an additional integration over crystal thickness in the Kikuchi line theory, not present in the STEM lattice imaging case.) For the case of plane wave illumination, however, there is no possibility of forming a STEM lattice image, and we must now consider the case in which the illumination takes the form of a focused STEM probe.

For the experimental conditions used to record the spectra shown in Figure 6.2, the lattice was resolved in the elastic channel (i.e., the probe was smaller than the lattice spacing, and the illumination semiangle was larger than the Bragg angle). The resulting complicated intensity distribution in the two-dimensional, energy-filtered microdiffraction pattern for an inner-shell excitation will thus show overlapping inelastic disks convoluted against the overlapping orders of the elastic channel. Some general comments about such patterns, formed from localized excitations in the axial orientation, from which STEM energy-filtered lattice images are formed, can be made:

1. The energy-filtered microdiffraction pattern intensity must be zero everywhere for a small probe at a large distance from the localized inelastic event (this point affects the normalization of wave functions in computer calculations).
2. The patterns will be highly sensitive to both the probe coordinate and that of the crystallographic site on the inelastic event.
3. Kikuchi lines should be seen.
4. For certain conditions of objective-aperture semiangle and inelastic-scattering angle θ_E, the intensity at the midpoint between Bragg points in the energy-filtered inelastic pattern will be independent of the electron-probe focus setting and of the spherical-aberration constant but

not of the crystallographic coordinate of the inelastic event. These considerations will suggest a choice of detector position which minimizes the sensitivity of filtered lattice images to instrumental parameters.

Dynamical calculations for inner-shell, energy-filtered microdiffraction patterns are shown in Figure 6.4, based on the algorithm of Doyle (see section 5.4) and using hydrogenic wave functions to describe the inelastic scattering. An incoherent addition of dynamical intensities over all equivalent lattice sites was used. The pattern is filtered for the beryllium K-loss electrons and shows the effects of moving a subnamometer probe that is smaller than the lattice spacing from the beryllium site to the gold atom site. Since there is no forward-scattered beam in inelastic scattering, the use of a small axial detector produces a dark-field, energy-filtered image. A STEM energy-filtered lattice image would be obtained by displaying a portion of the intensity shown in Figure 6.4 as a function of probe position. Only for very thin specimens ($t < 100$ Å) and at the optimum probe focus [$\Delta f = -0.75(C_s \lambda)^{1/2}$] is there a direct correlation between the intensity maxima in the filtered image and the site of the corresponding atoms in the lattice, as

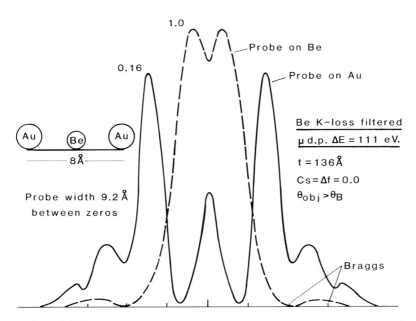

FIGURE 6.4 Multiple-scattering calculation for a beryllium K-shell, energy-filtered microdiffraction pattern in the artificial BeAu alloy structure shown. The electron-probe size is comparable with the lattice spacing, and both are smaller than the inelastic localization L. The dotted curve shows the case where the probe falls over the beryllium site; the continuous curve indicates the case where the probe lies over the gold atoms. The minimum in the continuous curve may be interpreted as a Kikuchi line, and these calculations suggest the optimum form of STEM detectors for chemical mapping by energy-filtered lattice imaging.

revealed in an elastic STEM lattice image of sufficiently high resolution. The choice of focus that gives the most compact probe in STEM has recently been analyzed in detail by Mory, Colliex, and co-workers.

Finally, we consider the effect of inelastic localization on energy-filtered electron images of inner-shell excitations in STEM or HRTEM and on their symmetry. Figure 6.5 shows the beryllium K-shell, energy-filtered, HRTEM image of a beryllium atom lying on a crystalline gold substrate in the axial orientation. The substrate thickness is 72 Å. This physically unlikely case has been chosen to illustrate the important point that, although the image has been energy-filtered for the beryllium atoms, the lattice planes of the gold-substrate crystal are also revealed. Similar effects have also been observed experimentally (Craven and Colliex, 1977) in plasmon-loss, energy-filtered lattice images. For localized inner-shell losses, however, a new feature arises. Dynamical calculations show that the inelastic localization L acts as a window, revealing the crystal structure in the image of all the unfiltered atomic species falling within a region of size L around the filtered species. This effect is a consequence of the elastic scattering of inelastically scattered electrons. The calculation of Figure 6.5 assumes single-event inelastic but multiple elastic scattering and uses an artificial superlattice larger than the inelastic localization.

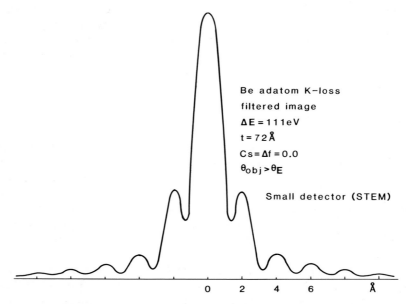

FIGURE 6.5 Dynamical calculation for a beryllium K-shell, energy-filtered lattice image of a single beryllium adatom on a crystalline gold substrate (thickness 72 Å). Lattice fringes from the gold substrate appear (owing to elastic scattering of the inelastically scattered electrons) within the inelastic localization range L despite filtering for the beryllium ionization edge. (From Spence and Lynch, 1982.)

The symmetry of inner-shell, energy-filtered diffraction patterns formed by using plane-wave illumination may be different from that of the crystal as a whole and may reveal just the local symmetry about the energy-filtered species (Spence, 1981). This result suggests the possibility of determining the site symmetries of selected species by a method that, unlike EXAFS, contains no adjustable parameters. This result is based on a simple argument that combines the column approximation with the reciprocity principle.

The properties of electron images formed by using only those fast electrons that have excited surface plasmons or interband excitations have recently been analyzed by Marks (1981) (for MgO), Cowley (1982) (for MgO), Kohl (1983), and Howie and Milne (1984) (for Cu and MgO). These workers find that the classical dielectric theory of energy loss at a surface gives a good account of the RHEED (reflection high-energy electron diffraction) energy-loss spectrum and that the surface-plasmon, energy-filtered electron image is most intense just outside the sample. They (Howie and Milne, 1984) also concluded that a surface-reconstructed layer of thickness a will affect only those surface modes for which $qa > 1$. Since the energy losses studied are all small, q is close to the minimum value allowed (see equation 5.3b), and for realistic values of a, they therefore suggest that high-energy electron beams in reflection produce energy-loss spectra that are rather insensitive to the details of surface structure. The corresponding energy-filtered images of surfaces would be similarly insensitive. At lower energy, however, considerable sensitivity to surface reconstruction has been observed experimentally (Ichinokawa et al., 1981). The new surface modes that occur in clusters of small metal particles can also be imaged in STEM (Batson, 1982).

6.3 Absorption effects and parameters in HRTEM

Microdiffraction patterns from regions of a crystal used to form high-resolution images show a considerable amount of scattering in non-Bragg directions, even from perfect crystals. In the absence of defects, this scattering arises from inelastic scattering of the fast electron by the various crystal elementary excitations, chiefly plasmons, phonons, inner-shell excitations, and valence-band, single-electron excitations. The question arises as to what contribution, if any, these scattered electrons make to a high-resolution image.

The usual assumption has been that electrons that lose energy ΔE in traversing the specimen will be out of focus by an amount $C_c(\Delta E/E_0)$ if the objective lens is correctly focused for the elastic or zero-loss electrons, owing to the chromatic aberration of the lens. They therefore contribute only a slowly varying blurred background to the HRTEM image, as indicated by equation (6.4). However, two independent effects influence this background—namely, the size of the objective aperture used and the strength of the inelastic scatting. Furthermore, in addition to contributing a low-resolution background to HRTEM images, inelastic scattering also depletes the elastic wavefield and so modifies the amplitudes and phases of the beams used for image formation. A full account of this process for high-resolution imaging has not been given, although several historically distinct bodies of literature have a bearing on the problem. These studies include the theory of the Debye-Waller factor, the justification for the use of an absorption or optical potential in electron diffraction, and work on the effects of absorption on diffraction-contrast images. As briefly outlined in section 5.3, the Debye-Waller factor expresses the attenuation of the elastic Bragg beams due to the excitation of phonons in a crystal according to the kinematic theory. The use of a complex optical potential to describe depletion of the elastic wave field by all inelastic processes in the dynamical theory was first justified by Yoshioka (1957). This use of a time-independent theory for inelastic processes is correct if all possible inelastic processes are included. The bulk of the work in the literature on imaging is concerned with estimating the effect of inelastic absorption on a particular elastic Bragg beam, which may then be related to a diffraction-contrast image formed from this beam alone (using a small objective aperture) through the column approximation. This approach is not directly applicable to HRTEM many-beam images, since these images are formed by using larger apertures at higher resolution.

In principle, the classical method of treating phonon-absorption effects in HRTEM is straightforward. The interaction time of a fast electron traversing a thin crystal is much shorter than that of any of the inelastic crystal excitations. Therefore, a time average is required of the intensity of the dynamical, many-beam, aperture-limited image for every configuration of the instantaneous crystal potential. This procedure requires a complete description of all atomic displacements resulting from the thermal motion of the atoms. The time averaging must be performed on the dynamical image rather than on diffracted beams or the crystal potential. A second aspect of such a calculation is the estimation of the distribution of diffuse inelastic scattering due to electronic excitations, from which the background due to the inelastic electrons that pass through the objective aperture can be estimated. The contribution of these electronic processes to the imaginary part of the optical potential must also be determined. The result gives the HRTEM image observed without the use of an energy-filtering microscope—that is, due to both elastic and inelastically scattered electrons.

While such a calculation is possible in principle, it would be impossibly laborious in practice; and a variety of approximations must be made for realistic calculations. In addition, this classical approach ignores the quantum nature of phonons, which absorb or give up a quantum of energy to the fast electron during their creation or annihilation.

Because of the presence of chromatic aberration, only electrons within a small range of energies around the accelerating potential will contribute to the in-focus HRTEM image. Thus, we may think of lattice imaging as a kind of energy filtering for elastically scattered electrons. Then, two effects must be considered: the modification to the amplitudes and phases of the elastic beams due to inelastic scattering and the contribution to the low-resolution background from inelastically scattered electrons that pass through the objective aperture.

For electronic excitations, the depletion of the elastic wave field may be represented by the addition of a small imaginary part V'_g to the Fourier coefficients of crystal potential. For example, for plasmon scattering, a single coefficient V'_0 is used, which is related to the path length λ_p for plasmon excitation by

$$2\sigma V'_0 = \lambda_p^{-1}. \tag{6.5}$$

For valence electrons, with the nomenclature of section 5.4 (and $m = 0$), the contribution to the imaginary part of the potential is (Whelan, 1965; Humphreys and Whelan, 1969).

$$V'_g = \frac{2m\Omega}{h^2 16\pi^2 k_z} \sum_{n \neq 0} H_g^{0n} H_0^{n0} \, d\sigma_n,$$

where Ω is the crystal volume and the integration is over the energy- and momentum-conserving states. In fact, virtual inelastic-scattering and exchange effects between the fast and crystal electrons also make a small contribution to V_g (the elastic potential); however, Rez (1977) has shown that these effects are negligible. We note here that these exchange effects between the fast electron and crystal electrons are fundamentally different from the exchange effects included in band-structure calculations, which take account of exchange among crystal electrons. Thus, the potential measured by electron diffraction, while sensitive to bonding effects, is not the same as that used for band-structure calculations. Values of these (amplitude) absorption coefficients

$$\mu_g = \sigma V'_g \tag{6.6}$$

have been given by several workers (Humphreys and Hirsch, 1968; Radi, 1970; Reimer and Wachter, 1980). However, the straightforward application of coefficients intended for use in diffraction-contrast-image calculations to HRTEM-image calculations is unlikely to be correct for several reasons:

1. Recent careful experimental measurements of V'_g for silicon (Voss et al., 1980) indicate that earlier theoretical estimates (Radi, 1970) are about three times too large. Here, the measured values of V'_g are found to be about a hundredth of V_g.
2. Inelastic processes that preserve diffraction contrast (such as plasmon excitation and small-energy-loss, single-electron excitation) may not be included in an estimate of V'_g or may be included in a way that makes V'_g

a function of the objective-aperture size used. This aperture will generally be much smaller than that used in HRTEM.
3. The case of phonon scattering raises special difficulties (see the discussion that follows).
4. The use of the independent Bloch-wave model in some of these calculations must be questioned for the thin specimens used in HRTEM work.

In summary, the few reliable values of V'_g useful for HRTEM calculations are those that have been measured experimentally by convergent-beam or similar diffraction methods (Goodman and Lehmpfuhl, 1967; Voss et al., 1980). Alternatively, the empirical relationship

$$V'_g = V_g(a\,|g| + bg^2) \tag{6.7}$$

may be used, where a and b are fitting parameters, and g a reciprocal lattice vector. This model includes the phonon contribution.

In correcting HRTEM-image calculations for the effects of inelastic scattering, the low-resolution-background contribution to the images from inelastically scattered electrons that pass through the objective aperture must also be estimated. This procedure does not appear to have been adopted in the past. In principle, a calculation of the type described in section 5.4 would be required, since the detailed angular redistribution of multiple inelastic scattering within the objective aperture must be determined. In practice, a simpler approach would be to include, as background, an appropriate fraction of the energy lost from the elastic wave field as a consequence of the use of an absorption potential.

For phonon scattering, the situation is less clear. Transmitted electrons that lose energy owing to the creation of phonons in a crystalline sample give up very small amounts of energy (less than 1 eV even for multiple losses) and so remain in focus. In addition, as discussed in section 5.3, the angular distribution of phonon-loss electrons shows a broad maximum for $0.5 < (q) < 1\,\text{Å}^{-1}$ and shows peaks at the Bragg positions. Much of this broad maximum falls within the resolution limit of modern HRTEM instruments. The question therefore arises as to what contribution this scattering makes to HRTEM images. There has been little study of this problem; however, Cowley (1983) has suggested that phonon-loss electrons will make a high-resolution contribution to the many-beam image.

To include both a Debye-Waller factor and a contribution to V'_g from phonon excitations in dynamical calculations for the elastic wave field is not inconsistent (Ohtsuki, 1967). In the quantum picture, the Debye-Waller factor describes the modification to the elastic wave field due to virtual inelastic scattering that results from the creation and annihilation of phonons in a single interaction (at the same time). The imaginary part of the potential V'_g, however, describes the redistribution of elastic scattering that results from real inelastic phonon scattering through non-Bragg angles.

In the classical picture, the Debye-Waller factor describes the time-averaged, periodic crystal potential responsible for the purely elastic Bragg scattering in Bragg directions. For a review of the electronic contribution to the absorption potential, see Ritchie and Howie (1977).

For completeness, the expression for a many-beam-structure image in the Bloch-wave picture including the absorption effect is (Pirouz, 1979)

$$I(\mathbf{r}) = I_B + \sum_j \sum_l B^{jl} \sum_{g \neq G} D^{jl}_{gG} \exp[2\pi i(\mathbf{g} - \mathbf{G}) \cdot \mathbf{r}], \qquad (6.8)$$

where

$$B^{jl} = C_0^j C_G^{*l} \exp(2\pi i t \, \Delta\gamma jl) \exp[-2\pi t(\mathbf{q}^j + \mathbf{q}^l)]$$

and

$$D^{jl}_{gG} = C_g^{(j)} C_G^{*l} \exp(-2\pi i[\chi(\mathbf{g} + \mathbf{q}_\perp) - \chi(\mathbf{G} + \mathbf{q}_\perp)])$$

and where t is the crystal thickness, I_B is a constant background term, and C_g and $\Delta\gamma^{jl}$ are the eigenvectors and eigenvalue differences of the crystal-structure matrix. In this derivation, the Bloch waves are assumed to be absorbed independently (a good approximation in thicker crystals), each with absorption coefficient q^j. Here, χ is the lens-aberration function, and $q_\perp$ is the scattering vector.

Detailed calculations for elastic HRTEM images of Au_4Mn have also shown that these images are drastically altered for thicknesses of greater than a few hundred angstroms by the inclusion of absorption effects (Van Dyck, personal communication).

6.4 Multiple energy-loss effects and their removal

The primary quantity of interest in the studies described in Chapters 7 and 5 on electron-loss, near-edge structure (ELNES), extended electron-loss fine structure (EXELFS), dielectric-function measurement, and microanalysis by energy-loss spectroscopy is the single-scattering, electron-energy-loss distribution function $F_1(E)$. The quantity recorded in experiments from all but the very thinnest samples is, however, the multiple-scattering, electron-loss distribution function $F_2(E)$ convoluted or broadened by the thermal-energy-spread function of the electron source $I(E)$, which we shall call the instrument response. The relative probability of exciting the various elementary excitations of a crystal (plasmons, phonons, single-electron excitations) is inversely proportional to the mean free path λ_i for that process. Thus, for example, plasmon satellites are commonly seen in ELNES spectra since λ_i for plasmons is smaller than that for inner-shell excitations. So it becomes very probable that a fast electron will excite both an inner-shell excitation and a plasmon in passing through a thin crystal in successive, independent events. This process leads to a peak in the ELNES spectrum at an energy equal to the sum of the individual process loss energies. This double-loss process must be distinguished from that described by the terms H_g^{12} of equation (6.5)—which refers to a single plasmon loss in a crystal in an excited state (e.g., already containing a plasmon)—and from double-plasmon excitation, which is a second-order process in which two plasmons are created at the same instant of time.

The problem of multiple, independent scattering events described by Poisson statistics occurs in many fields of physics, and a general solution to this problem was given by Landau (1944) for the closely related problem of energy losses by fast particles in cosmic-ray showers (see Whelan, 1976, for a discussion). Landau's solution may be applied to the EELS problem and inverted to give the required single-loss spectrum in terms of the observed multiple-loss distribution in a way that also correctly incorporates the instrument response. The exact result is (Johnson and Spence, 1974; Spence, 1979)

$$\bar{F}'_1(w) = \bar{I}'(w) \ln \left[\frac{C \, |\bar{F}_2(w)|}{\bar{I}(w)} \right], \tag{6.9}$$

where $\bar{F}_2(w)$ is the Fourier transform of the multiple-loss spectrum $F_2(E)$, $\bar{I}(w)$ is the transform of the instrument response $I(E)$, and $\bar{F}'_1(w)$ is the transform of the desired single-loss spectrum convoluted by a new instrument response $I'(E)$, whose transform is $\bar{I}'(w)$. In most cases, we may take $\bar{I}'(w) = \bar{I}(w)$ (see the following discussion). Here ln is the complex natural-logarithm function, and C is an unimportant constant that is approximately equal to the ratio of the peak height of the experimentally determined instrument response (which must be recorded in a separate experiment) to that of the unscattered peak at the origin in the multiple-loss spectrum $F_2(E)$. This result applies to any number of mixed energy-loss processes, each with mean free path λ_i. The total path length is

$$\lambda_T^{-1} = \sum_i \lambda_i^{-1}. \tag{6.10}$$

Equation (6.9) allows the exact recovery of the single-loss spectrum as it would have been recorded on the original instrument [if $I'(w) = I(w)$] without knowledge of the scattering parameter t/λ_T (or any of the components λ_i). Thus, for example, the application of equation (6.9) will, in principle, remove all the contributions to the background beneath an inner-shell edge due to multiple valence-electron excitations. In fact, the constant C is related to (t/λ_T); however, (6.9) may be rewritten as

$$\bar{F}'_1(w) = \bar{I}'(w) \ln C + \bar{I}'(w) \ln \left[\frac{|\bar{F}_2(w)|}{\bar{I}(w)} \right]. \tag{6.11}$$

This equation shows that, since $I'(w)$ is a broad function [$I'(E) = I(E)$ is a narrow-peaked function], changes in C will affect the recovered single-loss spectrum only in a narrow range near the origin. Thus, the retrieval is effectively independent of the value of t/λ_T assumed. An expression similar to equation (6.9) in terms of Fourier coefficients suitable for use in computers is given in Johnson and Spence (1974), which also contains a discussion of the effects of changes in the height, width, and position of the instrument response $I(E)$ between the time that it is recorded and the time that it acts on $F_2(E)$. By recording $I(E)$ in a separate experiment, one fully corrects for any misalignment of the spectrometer, since $I'(E)$ may be chosen to be a Gaussian peak whose width ΔE is equal to the desired energy resolution in $F_1'(E)$. However, from studies in many fields, we know that the penalty for resolution increase by deconvolution is noise amplification, an effect we now discuss briefly.

If the width of $I(E)$ is ΔE and that of the reconvolution function $I'(E)$ is $\Delta E'$, then this trade-off between resolution increase $\Delta E/\Delta E'$ and noise increase has the general form shown in Figure 6.6 (Johnson, 1975). For simplicity, we have ignored here the single-scattering retrieval and con-

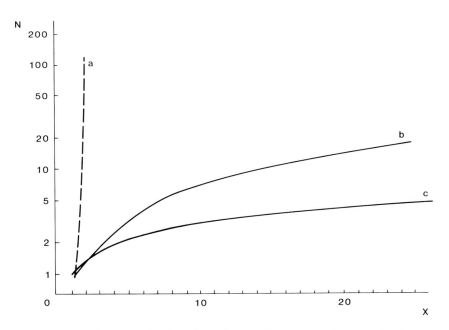

FIGURE 6.6 The noise penalty N of resolution increase by numerical deconvolution (see curve a), by use of a high-pass filter in conjunction with subsequent numerical deconvolution (see curve b), and by use of a narrower-source energy distribution, without deconvolution (see curve c). The ordinate is the ratio of the noise amplitude on the deconvoluted result to that on the unprocessed data (see text for details). The abscissa X is the ratio ($\Delta E/\Delta E'$) of the half width at half the height of the instrument response $I(E)$ to that of the reconvolution function $I'(E)$. For curves b and c, the loss of intensity due to the narrowing of the source energy distribution has been incorporated into the analysis in order to allow a meaningful comparison. (From Johnson, 1975.)

sidered the effect of simple deconvolution according to

$$\bar{R}(w) = \frac{\bar{I}'(w)\bar{F}(w)}{\bar{I}(w)}, \qquad (6.12)$$

where $\bar{I}'(w)$, $\bar{F}(w)$, and $\bar{I}(w)$ are the transforms of the reconvolution function $I'(E)$, the raw data $F(E)$, and the true instrument response $I(E)$, respectively. Curve a shows $N = \sigma(R(E))/\sigma(F(E))$, where σ represents the variance (assumed approximately independent of E) in the distribution of counts for $R(E)$ (the result of the deconvolution procedure) or $F(E)$. The quantity N is plotted against the ratio of the widths $\Delta E/\Delta E'$. Thus, the plot reveals how noise increases on the deconvoluted result $R(E)$ as the reconvolution function $I'(E)$ is made narrower. [$I'(E)$ is assumed noise-free, and Poisson statistics are otherwise assumed.] The important point is that the detailed shape of the curve depends greatly on the form of $I(E)$. In general, for a more favorable trade-off between noise amplification and resolution increase, one seeks an instrument response $I(E)$ containing sharp edges in order to enhance its high-order Fourier coefficients and so provide increased immunity to noise (which is also expressed by high-order Fourier coefficients). Note that a one-sided, high-pass filter gives almost as good a performance after numerical deconvolution as a narrower source-energy distribution, as shown in curve b. In these curves, $I'(E)$ has been chosen to give minimum ripple on the result (Johnson, 1975).

Additional points arise in the application of the single-scattering analysis (equation 6.9), as follows:

1. The method assumes that the EELS data have been collected over a sufficiently large range of scattering angles to include all appreciable inelastic scattering. At large thickness, this collection may not be possible, and the effects of angular truncation owing to the inevitable presence of apertures are not easily incorporated into the analysis. Note, also, that the failure of the dipole approximation must be considered (see section 7.3) for these angle-integrated spectra.
2. The signal-to-noise ratio for the peak corresponding to loss process λ_i in the single-loss spectrum is a maximum if this spectrum is retrieved from a multiple-loss spectrum (which may contain other processes) recorded at a specimen thickness $t = \lambda_i$. Thus, for plasmon losses, higher-order peaks in the multiple-loss spectrum fold back useful signals into the retrieved single-loss peak.

The effects of including a range of thickness in the data and of energy-gain processes are discussed elsewhere (Johnson and Spence, 1974; Spence 1979). The removal of a plasmon-satellite peak by using equation (6.9) on the K-edge, electron-energy-loss spectrum of beryllium is demonstrated in Figure 6.7 (from Spence, 1985). Recent calculations by the multiple-scattering XANES method (see section 7.3) indicate that the

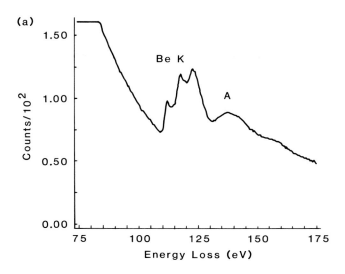

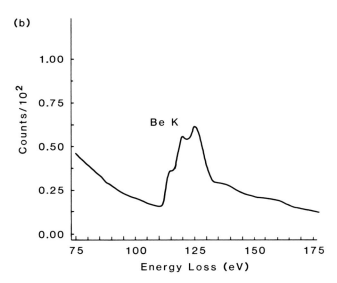

FIGURE 6.7 The effect of the multiple-scattering analysis (logarithmic deconvolution) given in equation (6.9) on an experimental beryllium K-edge spectrum. (*a*) The raw experimental data. (*b*) The single-scattering spectrum retrieved, which shows a marked reduction in the fourth peak, suggesting that this peak contained a contribution from successive K-shell and plasmon-excitation events. Subsequent ELNES calculations by the multiple-scattering method have confirmed this. The spectrum was obtained by using a lanthanum hexaboride electron source, a Philips EM 400 electron microscope, and a GATAN model 607 spectrometer from a region of specimen about 1000 Å in diameter and about 800 Å thick. (From Spence, 1985.)

remaining feature is not due to multiple scattering of the beam electron. Although the application of equation (6.9) to the low-energy end of the EELS spectrum (and so to dielectric-response-function studies) has proven straightforward, several difficulties arise in connection with its application to inner-shell processes occurring at larger energy losses. The most important is the need to collect data for both $F_2(E)$ and $I(E)$ over a very large energy range so that $F_2(E)$ falls to a small value and does not introduce artifacts into the periodic continuation implied by the use of Fourier series. For a solution to this problem, a number of approximate algorithms have been proposed (for reviews, see Egerton et al., 1984; Schattschneider and Solkner, 1984. The method of Misell and Jones (1969) uses a series of convolutions; however, this series converges only for small t/λ_T and does not correctly incorporate the instrument response. The alternative-matrix method of Schattschneider and Solkner (1984) is equivalent to this method; it has the advantage of ensured convergence for all t/λ_T. In addition, it allows the use of data collected over a smaller energy range. However, full computational trials on real data correctly including the effects of the instrument response have yet to be undertaken. Finally, the method of Leapman and Swyt (1981) consists in dividing the Fourier coefficients of the high-energy region of the spectrum (with background subtracted) by those of the low-energy region, with the elastic peak replaced by a numerical delta function of equal area. This *ad hoc* procedure has proven useful, but should be tested, in particular cases, against the exact result given by equation (6.9). This equation makes no approximation and treats the background, the effects of noise, and the instrument response correctly.

6.5 Radiation damage in HRTEM

In the study of electron-beam radiation damage in perfect crystals, we can conveniently distinguish between the effects of ionization that predominate at accelerating voltages below the atomic-displacement threshold and those that predominate above it. Above the threshold, direct collisions between the beam electron and an atom may displace the atom. A further useful classification is according to the type of bonding present in the material, such as metallic, covalent, ionic, or van der Waals. A sizable amount of literature exists on each of these topics. Two excellent recent reviews (Hobbs, 1979; Urban, 1980) contain extensive references to particular materials and processes; we restrict the following discussion to a qualitative review of principles and to recent work that has a particular bearing on high-resolution imaging. A third distinct area of study is that of radiation-induced defect reactions based on the chemical-reaction-rate theory. Radiation damage in organic materials is not considered here; for a review of this large area, see Zietler (1982). The review articles by Fryer et al. referenced therein are particularly recommended. In organic materials, ionization damage is the primary concern, and a reduction in specimen temperature is one of the more promising research directions (Glaeser and Taylor, 1978). Because of its chemical specificity, electron-energy-loss spectroscopy also appears to be a promising technique for this field (Egerton, 1980).

The displacement threshold for metals varies from about 150 kV for aluminum (at 300 K) to about 1 MeV for gold, and it has a weak temperature dependence. Below these accelerating voltages, direct atomic displacement from collisions between the fast electron and an atomic nucleus (knock-on) is unlikely; however, ionization of an atom may lead, through secondary processes, to the ultimate rearrangement of the atoms in the crystal. This process is known as radiolysis. The Bethe theory for single-electron excitation described in Chapter 5 gives an ionization cross section inversely proportional to the square of the fast-electron velocity, indicating that ionization damage should decrease at higher voltages. This cross section (for carbon) is of the order of 10^6 barns (b) at 100 keV ($1\,b = 10^{-24}\,cm^2$). Unfortunately, the elastic cross section falls off in a similar way, resulting in a reduction in image contrast at higher voltages that largely offsets the reduction of radiation damage for a specimen of given thickness. The essential requirements for the energy of excitation to appear as an atomic displacement are (1) that the excitation be localized (this requirement excludes plasmon excitation); (2) that there be sufficient energy in the excitation to move a nucleus through the crystal-potential saddle point; (3) that the lifetime of the excitation exceed typical atomic vibration times (about a picosecond).

Radiolysis is an unimportant process in metals because of screening and the delocalized nature of excitations; however, it may occur in semiconductors (particularly at point defects) and is very important in organic materials, in the alkali halides, and in many oxides, minerals, and other insulating crystals of ionic or covalent character. The energies needed to break bonds in these materials varies from about a tenth of an electronvolt for van der Waals bonding to tens of electronvolts for ionic crystals. For the simplest ionic materials, the criterion for displacement that the electron-hole recombination energy should exceed the lattice-binding energy is a useful guide; however, this simple rule fails for many covalent solids such as silicates. The most extensive studies of radiation damage by this mechanism have been done for the alkali halides. Note that, although the cross section for inner-shell ionization is small, the corresponding energy loss is large; so the rate of energy transfer to the specimen may be comparable to that from valence-band excitations.

The rate at which interstitials and vacancies are created owing to the knock-on process at accelerating voltages above the displacement threshold is given by the product of the displacement cross section and the local electron-current density. (The total dose may be more important for ionization damage.) Since this last factor is influenced by electron-channeling effects and diffraction conditions, increased damage has been seen inside bend contours (Fujimoto and Fujita, 1972). For most materials, these cross sections are in the range of 1 to 100 b at 100 kV. The displaced atom and remaining vacancy are known as a Frenkel pair. Displacement energies for many crystals are spread throughout the literature; typical values are 20 to 40 eV for metals, 80 eV for diamond, about 60 eV for magnesium and oxygen in magnesium oxide, and 11 to 22 eV for silicon. Thus, although in earlier HRTEM instruments operating below 120 kV, only first- and second-row elements were subject to the knock-on mechanism, a much larger range of materials is now vulnerable with modern HRTEM instruments operating between 200 kV and 1 MeV.

A wide variety of reaction products and mechanisms have been observed and studied, including the recombination of interstitials and vacancies, their interactions with surfaces, dislocations and grain boundaries, and the formation of vacancy clusters and dislocation loops. A great deal of work has been devoted to problems associated with swelling in nuclear-reactor materials. We emphasize, however, that the displacement threshold energy quoted for a perfect crystal is rarely relevant, since damage occurs initially at the weakly bonded atoms on surfaces or in line or point defects. Thus, for example, float-zone silicon remains undamaged (as evidenced by the HRTEM image) for extended periods in 400-kV microscopes, but CZ silicon at the same voltage damages heavily within a few minutes, presumably because of the presence of oxygen-related defects. Similarly, deformed crystals containing a high concentration of point defects are found to damage more readily than perfect crystals as one approaches the displacement threshold. In some high-voltage machines, the complication also arises of distinguishing electron-beam damage from damage due to ions generated in the gun (Werner and Paseman, 1982). Some machines are fitted with a suitable ion trap.

We now consider specifically the direct evidence for radiation damage seen in HRTEM structure images of inorganic materials. Figure 6.8 shows a structure image of $4Nb_2O_5 \cdot 9WO_3$ both before and after about 20-min irradiation by 1-MeV electrons at typical microscope beam currents (perhaps 1 A/cm^2) (Horiuchi, 1982). In Figure 6.8a, the dark spots are the cations; bright areas are pentagonal or square tunnels through the structure. The structure consists of corner-sharing octahedra, and the beam direction is parallel to the c-axis. The contrast changes at the arrowed tunnel positions are interpreted as knock-on events at the pentagonal-tunnel sites, and these changes are not seen at the square-tunnel sites. Some pentagonal tunnels contain a string of alternating metal-oxygen atoms, and it is assumed that oxygen release leaves a weakened metal-metal bond. Subsequent knock-on transfers one of these metals into a neighboring, empty pentagonal tunnel.

This study, and the many others like it that have appeared since, reveals the power of high-resolution imaging for the determination of the atomic mechanisms involved in radiation damage. In a similar study of Zr_3Al, Au_3Cd, and Au_3Mn alloys, Shindo et al. (1984) have used direct-structure

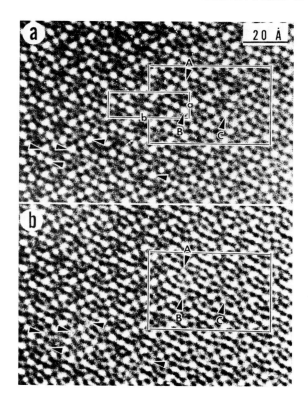

FIGURE 6.8 (a) An atomic-resolution micrograph of undamaged $4Nb_2O_5 \cdot 9WO_5$ with the unit cell indicated. (b) The altered image after 20-min irradiation of 1-MeV electrons. Arrows indicate pentagonal tunnels filled with metal atoms that reverse contrast as the atom is knocked-on to a neighboring pentagonal tunnel. (From Horiuchi, 1982.)

imaging to show that replacement disordering in these alloys owing to 1-MeV irradiation occurs in the atomic rows parallel to the beam. Dislocation loops are also observed with a diameter of less than 10 Å at higher intensities, and lighter atoms are observed to be displaced preferentially.

The study of radiation damage in quartz is important in both the geological and nuclear-waste-storage fields. Cherns et al. (1980) have found that dissociated dislocations in quartz become vitrified after irradiation; Pascucci et al. (1981) have observed a two-stage damage process. At low doses, ionization damage produces a heterogeneous nucleation of amorphous inclusions. Higher doses produce an additional homogeneous loss of correlation in the surrounding matrix. The role of water or OH radicals (which causes a marked softening of quartz) has yet to be fully elucidated. Other studies of radiation damage in inorganic materials by HRTEM include those of sodium β'' alumina by Hull et al. (1981) and the spectacular subnanometer hole drilling of Mochel et al. (1983). The well-established method of video-recording HREM images (Spence et al., 1982) should greatly assist in this work, since replay and freeze-frame reproduction of the first images of a new area of crystal frequently reveal details that cannot be seen in micrographs.

Much current high-resolution work is concentrated on the effects of radiation damage on surfaces. Thus, Sinclair et al. (1982) were able to observe radiation-induced, partial-dislocation motion in cadmium telluride by using structure images and other dynamic effects of atoms in surface positions. More recently, Bovin and co-workers (Bovin et al., 1985; Iijima and Ichihashi, 1985) have obtained some remarkable atomic resolution videorecordings of the radiation-induced motion of small metal particles (less than 100 Å in diameter) on silicon. Individual atoms are seen performing a random walk on the crystal surface. Radiation-induced effects are clearly seen in most of the recent HRTEM-profile images (Smith and Marks, 1985), and the usefulness of this effect for generating submicron metallic layers on dielectric metal-oxide crystals is one subject of current interest. The limitations that radiation damage impose on the microanalysis of a few hundred atoms by energy-loss spectroscopy are well illustrated in the work of Bourret et al. (1984) on oxygen segregation to dislocation cores in germanium. The considerable amount of work on damage in the rare-earth compounds can be traced through the recent study of vacancy ordering in scandium sulphide by Franzen et al. (1983).

REFERENCES

Batson, P. (1982). A new surface plasmon resonance in clusters of small aluminum spheres. *Ultramicroscopy* **9,** 277.
Bovin, J. O., Wallenberg, R., and Smith, D. (1985). Imaging of atomic clouds. *Nature* **317,** 47.
Bourdillon, A. J. (1984). The measurement of impact parameters by crystallographic orientation effects in electron scattering. *Philos. Mag., Ser. A* **50,** 839.
Bourret, A., Colliex, C., and Trebbia, D. (1984). Oxygen segregation on a dislocation core in germanium studied by electron energy loss spectroscopy. *J. Physique Lett.* **44,** L33.
Cherns, D., Howie, A., and Jacobs, M. H. (1973). Characteristic X-ray production in thin crystals. *Z. Naturforsch.* **28a** 565.
———, Hutchison, J. L., Jenkins, M. L., Hirsch, P. B., and White, S. (1980). Electron irradiation induced vitrification at dislocations in quartz. *Nature* **287,** 314.
Cowley, J. M. (1981). Coherent interference effects in STEM and CBED. *Ultramicroscopy* **7,** 19.
——— (1982). Surface energies and surface structure of small particles. *Surf. Sci.* **114,** 587.
——— (1983). Personal communication.
———, and Spence, J. C. H. (1979). Innovative imaging and microdiffraction methods in STEM. *Ultramicroscopy* **3,** 433.
Craven, A. J., and Colliex, C. (1977). High-resolution, energy-filtered images in STEM. *J. Microsc. Spectrosc. Electron.* **2,** 511.
———, Gibson, J. M., Howie, A., and Spalding, D. R. (1978). Study of single-electron excitations by electron microscopy. *Philos. Mag., Ser. A* **38,** 519.

Cundy, S. L., Howie, A., and Valdre, U. (1969). Preservation of electron microscope image contrast after inelastic scattering. *Philos. Mag.* **18**, 251.

Egerton, R. F. (1976). Inelastic scattering and energy filtering in the transmission electron microscope. *Philos. Mag.* **24**, 49.

——— (1980). Measurement of radiation damage by electron energy loss spectroscopy. *J. Microsc.* **118**, 389.

———, and Whelan, M. J. (1974). The contribution of inelastically scattered electrons to the diffraction pattern and images of crystalline specimen. In *Proceedings of the 8th international congress electron microscopy*, Vol. 1, 276, Australian Academy of Sciences, Canberra.

———, Williams, B. G., and Sparrow, I. G. (1984). Fourier deconvolution of energy loss spectra. *Proc. R. Soc. London, Ser. A* **398**, 395.

Feuerbacher, B., Fitton, B., and Willis, R. F. (1978). *Photoemission and the electronic properties of surfaces.* Wiley, New York, 111.

Franzen, H., Tuenge, R. T., and Eyring, L. (1983). Vacancy ordering in scandium sulphide. *J. Solid State Chem.* **49**, 206.

Fujimoto, F., and Fujita, H. (1972). Radiation damage induced by channeling. *Phys. Status Solidi* **11(a)**, K103.

Gjønnes, J. K. (1966). The influence of Bragg scattering on inelastic and other forms of diffuse scattering of electrons. *Acta Crystallogr.* **20**, 240.

Glaeser, R. M., and Taylor, K. A. (1978). Radiation damage relative to transmission electron microscopy of biological specimens at low temperature: A review. *J. Microsc.* **112**, 127.

Goodman, P., and Lehmpfuhl, G. (1967). Electron diffraction study of MgO (hoo) systematics. *Acta Crystallogr.* **22**, 14.

Heydenreich, R. D. (1962). Attenuation of fast electrons in crystals and anomalous transmission. *J. Appl. Phys.* **33**, 2321.

Hobbs, L. W. (1979). Radiation effects in analysis of inorganic specimens by TEM. In *Introduction to analytical electron microscopy*, ed. J. Hren, J. Goldstein, and D. C. Joy, 437. Plenum, New York.

Horiuchi, S. (1982). Detection of point defects accommodating nonstoichiometry in inorganic compounds. *Ultramicroscopy* **8**, 27; *Acta Crystallogr.* **15**, 323 (1982).

Howie, A. (1979). Image contrast and localized signal selection techniques. *J. Microsc.* **117**, 11.

———, and Milne, R. H. (1984). Interactions of electrons with surfaces. *J. Microsc.* **136**, 279.

Hull, R., Cherns, D., Humphreys, C. J., and Hutchison, J. L. (1981). Electron beam damage in sodium β'' alumina. *Institute of Physics Conference Series*, No. 61, Chap. 1, 23. Institute of Physics, Bristol, England.

Humphreys, C. J., and Hirsch, P. B. (1968). Absorption parameters in electron diffraction. *Philos. Mag.* **18**, 115.

———, and Whelan, M. J. (1969). Inelastic scattering of fast electrons by crystals. I. Single-electron excitations. *Philos. Mag.* **20**, 165.

Ichinokawa, T., Ishikawa, Y., Awaya, N., and Onoguchi, A. (1981). Analytical SEM in UHV. In *Scanning electron microscopy*, Part 1, ed. O. Johari, 271. IITRI, Chicago.

Iijima, S., and Ichihashi, T. (1985). Motion of surface atoms on small gold particles. *Jpn. J. Appl. Phys.* **24**, L125.

Jeanguillaume, C., Trebbia, P., and Colliex, C. (1978). About the use of electron energy loss spectroscopy for chemical mapping. *Ultramicroscopy* **3**, 237.

Johnson, D. (1975). Ph.D. diss., University of Melbourne, Australia.

———, and Spence, J. C. H. (1974). Determination of the single scattering distribution from plural scattering data. *J. Phys. D* **7**, 771.

Kohl, H. (1983). Image formation by inelastically scattered electrons. *Ultramicroscopy* **11**, 53.

Kondo, Y., Yoshioka, T., Oikawa, T., Kokubo, Y., and Kersker, M. (1985). Electron energy loss imaging. In *Proceedings of the 43rd electron microscopy society of America*, 1985, 410. San Francisco Press, San Francisco.

Krahl, D. (1982). Elemental analysis and chemical mapping. In *Electron microscopy—1982*

(10th international congress on electron microscopy), Vol. 1, 173. Deutsche Gesellschaft für Elektronmikroscopie, Hamburg.
Krivanek, O. L. (1985). Private communication.
Landau, L. (1944). On the energy loss of fast particles by ionization. *J. Phys. (Moscow)* **8**, 204.
Leapman, R., and Swyt, C. (1981). Electron energy loss spectroscopy. In *Analytical electron microscopy*—1981, ed. R. H. Geiss, 164. San Francisco Press, San Francisco.
———, and Swyt, C. (1983) Electron energy loss imaging in the STEM. In *Proceedings of the 41st electron microscopy society of America*, San Fransisco Press, San Francisco.
Marks, L. D. (1981). Imaging low-energy losses from MgO smoke particles. In *Institute of physics conference series*, No. 61, 259. Institute of Physics, Bristol, England.
Misell, D., and Jones, A. F. (1969). Determination of single-scattering line profile from observed spectrum. *J. Phys. A.* **2**, 540.
Mochel, M. E., Humphreys, C. J., Mochel, J. M., and Eades, J. A. (1983). Cutting of 20-Å holes and lines in β-aluminas. *Appl. Phys. Lett.* **42**, 392.
Ohtsuki, Y. H. (1967). Normal and abnormal absorption coefficients in electron diffraction. *Phys. Lett. A* **24**, 691.
Olsen, A., and Spence, J. C. H. (1980). Distinguishing shuffle and glide set dislocations by HREM. *Philos. Mag., Ser. A* **43**, 945.
Ottensmeyer, F. P., Arsenault, A. L., and Yu, A. (1982). Quantitative elemental mapping in electron images of biological specimens. In *Electron microscopy*—1982 (10th international congress on electron microscopy), Vol. 1, 597. Deutsche Gesellschaft für Electronmikroskopie, Hamburg.
Pascucci, M. R., Hobbs, L. W., and Hutchison, J. L. (1981). Lattice image of the metamict transformation in synthetic quartz. In *Proceedings of the 39th electron microscopy society of America*, 110. San Franscisco Press, San Francisco.
Pearce-Percy, H. T., and Cowley, J. M. (1976). On the use of energy filtering to increase the contrast of STEM images. *Optik* **44**, 273.
Pennycook, S. J. (1987). Impurity atom location using axial electron channeling. In *Scanning electron microscopy*—1987, ed. O. Johari. A.M.F. O'Hare, Chicago (forthcoming).
Pirouz, P. (1979). Effect of absorption on lattice images. *Optik* **54**, 69.
Radi, G. (1970). Complex lattice potentials in electron diffraction. *Acta Crystallogr., Sect. A* **26**, 41.
Reimer, L., and Wachter, M. (1980). Complex Fourier coefficients of the crystal potential. In *Electron microscopy*—1980 7th European congress on electron miscroscopy, Vol. 3, ed. P. Brederoo and G. Boom, 192. Electron Miscropie Foundation, Leiden.
Rez, P. (1977). Ph.D. diss., Oxford University, England.
———, and Ahn, C. (1982). Computer control for X-ray and energy loss line profiles and images. *Ultramicroscopy* **8**, 341.
Ritchie, R., and Howie, A. (1977). Electron excitation and the optical potential in electron microscopy. *Philos. Mag.* **36**, 463.
Rossouw, C. J., and Maslen, V. M. (1984). Implications of $(e, 2e)$ scattering for inelastic electron diffraction in crystals. *Philos. Mag., Ser. A* **49**, 743.
Schattschneider, P., and Solkner, G. (1984). A comparison of techniques for the removal of plural scattering. *J. Microsc.* **134**, 73.
Schuman, H., and Somlyo, A. P. (1982). Energy-filtered transmission electron microscopy of ferritin. *Proc. Natl. Acad. Sci USA* **79**, 106.
Self, P. G., and Buseck, P. R. (1983). Low-energy limit to channeling effects. *Philos. Mag. Lett., Ser. A* **48**, L21.
Shindo, D., Hiraja, K., Hirabayashi, M., and Aoyagi, E. (1984). *HREM of radiation defects in ordered alloys*. Science Reports of the Research Institutes, Tohoku University, A- Vol. 32, No. 1, Sendai, Japan.
Sinclair, R., Smith, D. J., Erasmus, S. T., and Ponce, F. A. (1982). Lattice resolution movie of defect modification in cadmium telluride. *Proceedings of the 10th international congress on electron microscopy*, 47. Deutsche Gesellschaft für Elektronmikroskopie, Hamburg.

Smith, D. J., and Marks, L. D. (1985). Direct imaging of atomic rearrangements on extended gold surfaces. In *Mat. res. soc. symp. proc.*, Vol. 41, 129. Materials Research Society, Boston.
Spence, J. C. H. (1979). Uniqueness in the inversion problem of incoherent multiple scattering. *Ultramicroscopy* **3,** 433.
——— (1981). The crystallographic information in filtered diffraction patterns. *Ultramicroscopy* **7,** 59.
——— (1985). The structural sensitivity of electron energy loss near edge structure. *Ultramicroscopy* **18,** 165.
———, and Cowley, J. M. (1978). Lattice imaging in STEM. *Optik* **50,** 129.
———, and Lynch, J. (1982). STEM microanalysis by ELS in crystals. *Ultramicroscopy* **9,** 267.
———, Disko, M., Higgs, A., Wheatley, J., and Hashimoto, H. (1982). A video system for HREM. In *Proceedings of the 10th international congress electron microscopy*, Deutsche Gessellschaft für Elektronmikroscopie, Hamburg. 519.
Taftø, J., and Krivanek, O. L. (1981). Characteristic energy losses from channeled 100-keV electrons. *Nucl. Instrum. Methods* **194,** 153.
Urban, A. (1980). Radiation damage in inorganic materials in the electron microscope. In *Electron microscopy*—1980, Vol. 4, 188. (Electron Microscopie Congress Foundation, Leiden.
Voss, T., Smith, P., and Lehmpfuhl, G. (1980) Influence of doping on the crystal potential in silicon. *Z. Naturforsch.* **35a,** 973.
Werner, P., and Paseman, M. (1982). Generation of radiation-induced defects in silicon. *Ultramicroscopy*, **7,** 267.
Whelan, M. J. (1965). Inelastic scattering of fast electrons by crystals. I. Interband excitations. *J. Appl. Phys.* **36,** 2099.
——— (1976). On the energy loss spectrum of fast electrons. *J. Phys. C* **9,** L195.
Williams, D. B., and Edington, J. W. (1976). High-resolution microanalysis in materials science using electron energy loss measurements. *J. Microsc.* **108,** 113.
Yoshioka, H. (1957). Effect of elastic waves on electron diffraction. *J. Phys. Soc. Jpn.* **12,** 618.
Zanchi, G., Sevely, J., and Jouffrey, B. (1980). Filtered electron image contrast in amorphous objects. *J. Phys. D.* **13,** 1589.
Zeitler, E. (1982). Cryomicroscopy and radiation damage. *Ultramicroscopy* **10,** special issues Nos. 1 and 2.

7
TECHNIQUES CLOSELY RELATED TO HIGH-RESOLUTION ELECTRON MICROSCOPY

JOHN C. H. SPENCE

7.1 Introduction

For each new detector that is fitted to an electron microscope, a new subdiscipline of electron microscopy is created. These subdisciplines are usually closely related to existing, well-established fields. Thus, for example, the theory of cathodoluminescence (CL) closely parallels that of photoluminescence (PL), energy-dispersive X-ray spectroscopy (EDS) has close similarities with X-ray fluorescence spectroscopy, and electron-energy-loss spectroscopy (EELS) has a good deal in common with X-ray–absorption studies, both with X-ray–absorption, near-edge-structure work (XANES) and with the extended X-ray–absorption fine structure (EXAFS). Both of these, in turn, have much in common with photoelectron spectroscopy, using either incident X-rays (XPS, or X-ray photoelectron spectroscopy) or ultraviolet light (UPS, or ultraviolet photoelectron spectroscopy). Thus, the student or research worker interested in any of the growing number of new spectroscopies found on modern electron microscopes would be well advised to consult the literature of the parent subject, always bearing in mind any differences that may be important. For example, the books by Cardona and Ley (1978), Teo and Joy (1981), and Egerton (1986) contain much relevant background material for this chapter.

Unlike high-resolution electron microscopy, which records the positions of atoms, spectroscopy is concerned with the measurement of energies. The bewildering variety of spectroscopic acronyms and techniques can be best classified according to the nature of the source particle, the detected particle, and the degree of collimation used for both. This taxonomy is shown in Table 7.1, which lists the main techniques based on electron or photon sources or detectors. Some techniques are more surface-sensitive than others. Angle-resolved variants of several are possible, such as ARPES (angle-resolved photoelectron spectroscopy), and inverse methods are sometimes fruitful [BIS (Bremsstrahlung isochromat spectroscopy) or IPES (inverse photoelectron spectroscopy)]. Auger-electron spectroscopy (AES), surface extended X-ray–absorption fine-structure SEXAFS) and appearance-potential spectroscopy (APS) complete the list of techniques given in Table 7.1.

TABLE 7.1
Some Closely Related Techniques in Surface Science and Electron Microscopy

Technique	Scattering geometry	Source (energy E; resolution ΔE)	Detector	Comments
		Photons	**Electrons**	
UPS	Bulk	$21\,eV = E$	$\Delta E \simeq 50\,meV$	Surface sensitive; beam size >1 mm
AES	Bulk	$2 < E < 5\,kV$	$\simeq 1\%$ of Auger energy	Surface sensitive; beam size >1 mm
XANES	Bulk	$2 < E < 5\,kV$	$1 < \Delta E < 5\,eV$	Beam size >1 mm; surface sensitive
ARPES	As for UPS			
XPS	Bulk, gases	$1.5\,keV = E$	$0.25\,eV$	
SEXAFS	Bulk	$2 < E < 5\,keV$	$5 < \Delta E < 10\,eV$	Surface sensitive; beam size >1 mm
		Electrons	**Photons**	
EDS	Transmission	$50\,kV < E < 1\,MeV$	$\Delta E \simeq 150\,eV$	Beam size >5 Å
WDS	Bulk	$10\,kV < E < 30\,keV$	$\Delta E \simeq 5\,eV$	Beam size $10\,\mu m$
CL	Transmission or bulk	$30\,kV < E < 300\,keV$	Grating	Diffusion length $\simeq 1\,\mu m$ or thickness divided by 2.4
IPES (KRIPES)	Bulk	$E < 100\,eV$	Grating	Beam size >1 mm
BIS	Bulk	$E \simeq 2\,keV$	Grating	Beam size >1 mm
ALCHEMI	As for EDS			
		Photons	**Photons**	
EXAFS	Transmission	$2 < E < 5\,keV$	$\Delta E \simeq 0.7\,eV$	Beam size >1 mm
PL	Bulk	Grating	Grating	Beam size >1 micron
		Electrons	**Electrons**	
AES	Bulk (or transmission)	$2 < E < 30\,keV$	$\simeq 1\%$ of Auger energy	Surface sensitive; beam size $\simeq 1$ mm or $\simeq 500$ Å or $\simeq 80$ Å
ELNES	Transmission	$30\,keV < E < 1\,MeV$	$0.3\,eV$	Beam size >5 Å
EXELFS	As for ELNES			
APS	Bulk $\Delta E \simeq 0.5\,eV$	$E < 5\,keV$	Surface sensitive	Beam size >0.1 mm

Of the four possibilities using photon or electron sources or detectors, the three groups most important historically are those that derive from the discovery of the photoelectric effect (such as XPS, UPS, ARPES, and some EXAFS and XANES work), those that derive from the original Franck–Hertz experiment (such as electron-loss, near-edge structure (ELNES) and extended electron-loss fine structure (EXELFS), and those based on optical-absorption measurements (such as conventional EXAFS and XANES). An important distinction must also be made between absorption and emission experiments (see Figure 7.1). The first probes all possible transitions that differ by a given energy. This energy may be that of the

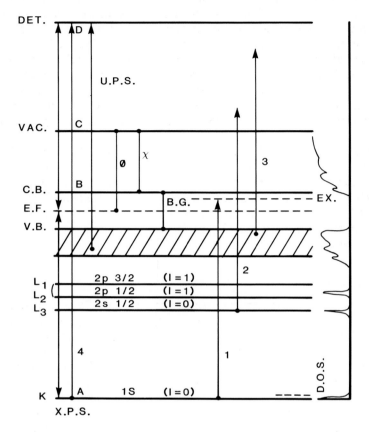

X.P.S., U.P.S.   E.L.N.E.S., E.X.E.L.F.S., E.X.A.F.S., X.A.N.E.S.

FIGURE 7.1 Simplified energy-level diagrams comparing XPS, UPS, XANES, EXAFS, ELNES, and EXELFS in a semiconductor. In the absorption experiments (XANES, EXAFS, EXELFS, ELNES), all transitions (e.g., 1, 2, 3) that *differ* by the energy ΔE to which either the source (for XANES or EXAFS) or detector (for ELNES or EXELFS) is tuned will contribute to the spectrum. Some of these transitions therefore contribute to an unwanted background. The emission techniques (UPS, XPS, ARPES) have the advantage that a single transition is selected (e.g., 4), although secondary-loss processes for the emerging photoelectron may not be negligible. For the XPS transition (4) shown, AB represents the binding energy, BC the work function ϕ, CD the kinetic energy of the ejected photoelectron, and AD the energy of the incident photon. The density of states (DOS) is suggested at the right. The energies of the vacuum level (VAC), (empty) conduction-band minimum (C.B.), (filled) valence-band maximum (V.B.), Fermi level (E.F.), exciton level (EX), optical band gap (B.G.), work function (ϕ), and electron affinity (χ) are all indicated. This ground-state, one-electron orbital diagram is a convenient approximation that may not correspond with the measured excited-state spectrum. The diagram also takes no account of the momentum-conservation constraint, which may be important for the electron-beam techniques.

incident X-ray beam in EXAFS or that of the energy loss in ELNES, for example. The second (emission experiments) may be of two types: those in which an electron in a particular state is ejected and detected (such as UPS and XPS) or those in which the secondary decay-products of an excited atom or crystal are detected (such as CL, EDS, or AES). All these techniques produce a signal that is proportional to certain matrix elements between electronic states of the crystal. Thus the underlying physics is similar in all cases, although the fine details of the spectra may be understood only in the light of particular mechanisms specific to each technique.

In the history of microanalysis by techniques such as EELS and EDS, the problem has usually been treated as one of free-atom ionization, ignoring solid-state effects altogether. Some of the new effects discussed in this chapter such as ELNES, atom location by channeling enhanced microanalysis (ALCHEMI), and EXELFS result directly from the imposition of the full target-crystal symmetry on either the crystal or beam-electron wave functions (or both).

7.2 Extended electron-loss fine structure (EXELFS)

EXELFS is the electron-beam counterpart of the EXAFS technique. For the present, we assume that the theory of EXAFS applies directly to EXELFS; differences between these techniques will be clarified in section 7.3 [compare also equations (5.5) and (7.1)]. We therefore commence with a very brief review of the principles of EXAFS analysis and that of the closely related X-ray photoelectron-spectroscopy (XPS) method.

In the simplest EXAFS experiments, the attentuation of a monochromatic photon beam passing through a thin sample is measured as a function of the photon energy. Since the surviving transmitted photons have the same energy as the incident photons, this technique must be distinguished from inelastic light scattering (a second-order effect). The most important absorption mechanism is the generation of photoelectrons; thus, the theory of EXAFS (and that of XPS and UPS) is the theory of the photoelectric effect in solids. The first-order cross section per atom for excitation of photoelectrons from core states $|i\rangle$ into continuum states $\langle f|$ is

$$\sigma_{pe}(E) = \left(\frac{4\pi^2 e^2 E}{\hbar c}\right) |\langle f| \exp(i k_p \cdot \mathbf{r}) \hat{\varepsilon} \cdot \mathbf{r} |i\rangle|^2 \, \delta(E - \varepsilon_f - \varepsilon_i) \quad (7.1a)$$

$$\approx \left(\frac{4\pi^2 e^2 E}{\hbar c}\right) |\langle f|\hat{\varepsilon} \cdot \mathbf{r} |i\rangle|^2 \, \delta(E - \varepsilon_f - \varepsilon_i). \quad (7.1b)$$

This cross section is related to the dielectric constant through equation (5.8). Here, $\langle f|$ and $|i\rangle$ are crystal-electron wave functions, with $\langle f|$ normalized per unit energy range; k_p is the wave vector of the incident photon; E is its energy; and ε is a unit vector giving the direction of the electric-field polarization. The approximation of equation (7.1b) is known as the dipole approximation. The intensity-absorption coefficient that is measured is

$$\mu(E) = n\sigma_{\text{pe}}(E), \tag{7.2}$$

where n is the number of atoms per unit volume.

In XPS and UPS (and some EXAFS work), the photoelectron itself is detected (this signal also contains EXAFS structure). In EXELFS, the transmitted-beam electron responsible for exciting the photoelectron (called a secondary electron in this case) is detected. The distribution of energy loss for the transmitted-electron beam in the region beyond an ionization edge then constitutes the EXELFS spectrum, which, under certain experimental conditions, may be given the same interpretation as an EXAFS spectrum. Although fast electrons may lose any amount of energy in an interaction, photons are either annihilated or survive without interaction. Thus, EXELFS is equivalent to an EXAFS experiment using incident, white X-ray radiation and a monochromator before the detector.

We now restrict the discussion to K-shell excitation so that $|i\rangle$ represents the 1s-atomic orbital; by the dipole-selection rule in equation (7.1b), the final state must therefore be of P-symmetry. The validity of this approximation is discussed in section 7.3. In equation (7.1), $\langle f|$ represents the one-electron wave function of the ejected photoelectron, which is an excited-state wave function of a crystal electron. The kinetic energy of this photoelectron is

$$E_{\text{kin}} = \hbar w - E_{\text{b}}, \tag{7.3}$$

where E_{b} is the binding energy of the 1s-state and hw is the photon energy. The ejection of the photoelectron from the core state leaves the atom in an excited state containing a core hole, and the energy E_{b} is subsequently released in other emission products such as X-rays or Auger electrons. In UPS and XPS, the photoelectron itself is detected, and a series of peaks are therefore seen in the energy spectrum of the ejected photoelectrons, corresponding to the various ionization energies E_{b} of the solid (see Figure 7.1). However, unlike EELS, which probes the density of unfilled final states (for K-shell excitations), XPS probes the density of occupied core-level states, since the final state of the ejected photoelectron is roughly a plane wave of large kinetic energy. (A plane-wave final state cannot always be assumed in UPS.) In EXELFS, each point ΔE in the energy-loss spectrum of the transmitted electrons beyond an absorption edge at E_{b} corresponds to a value of $(E_{\text{kin}} + E_{\text{b}})$. However, the intensity at ΔE will also contain a background due to the promotion of more weakly bound electrons high into the continuum and also from multiple energy-loss processes. Figure 7.1 provides a simplified energy-level scheme. The *asymptotic* photoelectron wave vector χ (called a secondary electron in EXELFS) is defined by

$$\frac{h^2\chi^2}{2m} = \Delta E - E_b, \quad (7.4)$$

where the binding energy E_b can be determined in various ways (see Figure 7.1).

Since the core state $|i\rangle$ is compact, equation (7.1) shows that the fine structure beyond an absorption edge ($hw > E_b$) will depend on the wave function of the ejected photoelectron evaluated at the site of the target atom. For energies well above the edge, a single-scattering approximation is commonly made for this photoelectron, and the EXAFS effect is attributed to the interference at the target atom between the reflected and outgoing components of the photoelectron wave function. The diffraction of this electron is a problem similar to that of low-energy electron diffraction (LEED); thus, EXAFS has been described as LEED with the source and detector at the same point (the target atom). From equation (7.1b), we see that only the overlap between the 1s and photoelectron wave functions is important, and this overlap depends both on the diffraction of the photoelectron wavefield and on its range (see section 7.3).

The success of EXAFS can be traced to the development of synchrotron sources and to the realization that equation (7.1) can be manipulated into a form such that a simple Fourier transform of the absorption function with respect to χ yields the radial distribution function about the target atom. The promise of both EXAFS and EXELFS is that it may provide both the number of and the distances to the nearest-neighbor atoms about a chemically identified species in crystalline, liquid, and amorphous materials. Applications to catalysis, biological materials, solution chemistry, alloys, and surfaces are too numerous to mention here; for a comprehensive review of EXAFS and EXELFS, see Lee et al. (1981) or Teo and Joy (1981). Outstanding problems include the transferability of scattering-atom phase shifts $\phi_j(k)$, the energy dependence of the absorption function $\lambda_j(k)$ for the photoelectron, the determination of bond angles in addition to distances, the importance of multiple scattering for small χ-values, the determination of the threshold energy E_b, and the form of the Debye–Waller factor M_j used (not exactly the same as that used in X-ray crystallography).

The use of EELS for EXAFS was first suggested by Ritsko et al. (1974) and Leapman and Cosslett (1976) and was applied to nanometer-sized areas by Batson and Craven (1979) soon after. In the single-scattering approximation, the EXAFS intensity is given by

$$I(\chi) = |\chi|^{-1} \sum_j \left(\frac{N_j}{r_j^2}\right) f_j(\chi) \sin\left[2\chi r_j + \phi_j(\chi)\right] \exp\left[-2M_j^2\chi^2 - \frac{2r_j}{\lambda(q)}\right], \quad (7.5)$$

where N_j is the number of neighboring atoms in the jth coordination shell of radius r_j with backscattering amplitude $f_j(\chi)$. The inversion of (7.5) by Fourier transform gives values of r_j with an accuracy of a small fraction of an angstrom in EXELFS.

Any comparison of the EXAFS and EXELFS techniques must include the considerations of reaction cross sections, source brightness, detector efficiency, and radiation damage. The cross sections are related through the dielectric-response function [equations (5.8) and (5.6)]. Recent comparisons (Stern, 1982) have shown that EXELFS using field-emission guns and parallel detection is highly competitive with EXAFS using synchrotron sources for edges below about 2 kV ($= E_b$), with the major uncertainty lying in the detailed comparison of radiation-damage mechanisms in the two cases. EXAFS work in this region is complicated by the need for windowless detectors, unless photoelectrons are detected. Perhaps more important than these considerations is the fact that EXELFS is instrumentally compatible with the HRTEM, microdiffraction, and EDS techniques. Thus, the EXELFS method can be applied to the same region (perhaps a few nanometers in diameter) from which an atomic-resolution, transmission-electron image has been obtained. This image provides all the

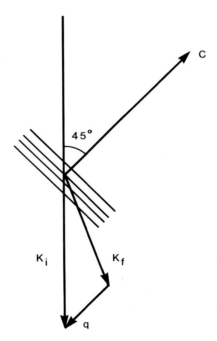

FIGURE 7.2 Scattering geometry used to observe directional effects on EXELFS spectra. The momentum transfer q is arranged to lie parallel to the crystal c-axis and normal to it in two successive experiments; K_i and K_f are wave vectors of the incident- and scattered-beam electron.

defect-structure information described in the remaining chapters of this book. The most important limitation of the EXELFS technique is that, within the limited energy range available, the overlap of EXELFS structure from different elements severely limits the resolution obtainable in polyatomic materials. The development of parallel-recording detectors has greatly improved the quality of EXELFS data.

The analysis of EXELFS data is based on the similarity between equations (5.5) and (7.1). Hence, small-angle-scattering conditions are required to allow the dipole approximation to be made in equation (5.5). For an axial detector, however, nondipole contributions in EXELFS are severely attenuated by the q^{-4}-term in equation (7.6) (following expansion of the exponential). Multiple electron-energy-loss effects, particularly plasmon peaks following the edge, can be removed by logarithmic deconvolution, as described in section 6.4. In addition, under the dipole approximation, q in EXELFS plays the role of the photon polarization in EXAFS. Thus, by choice of scattering geometry for certain favorable cases, one can

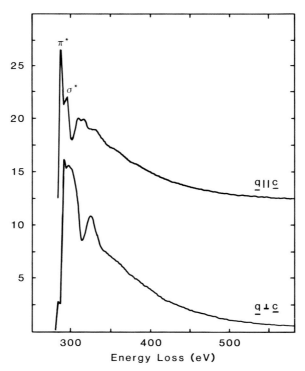

FIGURE 7.3 Carbon K-edges recorded by electron-energy-loss spectroscopy for the two orientations shown in Figure 7.2. The Fourier transforms of $I(\chi)$ [see equation (7.5) for these EXELFS spectra] show maxima near 3.6 Å for q parallel to c and at 1.4 Å for q perpendicular to c, in agreement with the known near-neighbor spacings in graphite.

determine interatomic distances in particular crystallographic directions by EXELFS, thereby overcoming one of the major limitations of the standard technique. As an example, we consider the recent application of this idea to the study of EXELFS in graphite (Disko et al., 1982). A diagram of the scattering kinematics used for these experiments is shown in Figure 7.2. Under the dipole approximation, the photoelectron wave function is heavily weighted in the directions of $\pm q$ (see Figure 5.7), which is constrained to lie within and normal to the graphite-sample planes in two successive EXELFS experiments. The results of the experiments are shown in Figure 7.3, giving good agreement with both the known in-plane and c-axis interatomic distances for graphite.

While the most important approximations involved in the application of equation (7.5) to EXELFS data are probably the neglect of multiple elastic scattering by both the photoelectron and the fast electron, Spence et al. (1981) have suggested that the channeling effects that result from diffraction of the fast electron (or X-ray) (see section 7.4) may be used to enhance EXELFS (or EXAFS) spectra. In crystals containing the same species in two nonequivalent crystallographic sites, conventional EXELFS spectra are difficult to interpret. However, the possibility arises of focusing the incident beam preferentially onto each of these sites in successive EXELFS experiments, using the channeling effect (described in more detail in section 7.4). The required general, theoretical treatment of dynamical EXELFS has not yet been developed and would involve considerable complexity, unless the simplifying experimental conditions described in section 7.4 were used.

7.3 Electron-loss, near-edge structure (ELNES)

ELNES is the electron-beam counterpart of X-ray–absorption, near-edge structure (XANES), or near-edge X-ray–absorption fine-structure (NEXAFS), as it is sometimes known. It concerns the study of the first few tens of electronvolts beyond an ionization threshold in optical or electron-beam absorption spectra. Reviews of ELNES research can be found in Colliex (1984) and in Colliex et al. (1984). These acronyms appear to have replaced the earlier terms of *Kossel structure* (for XANES) and *Kronig structure* (for EXAFS), honoring their first investigators.

The use of focused electron beams for XANES studies has several important advantages:

1. The ability to study the electronic structure of isolated, well-characterized defects or surface states when nanometer-sized probes are used (Pennycook, 1981; Baston, 1986).
2. The ability to distinguish similar species in nonequivalent sites by using the channeling effect on ELNES (see section 7.4).
3. The possibility of studying nonvertical (dipole-forbidden) transitions (see the discussion that follows).

4. The advantages of convenient access to the soft–X-ray region of the spectrum (0 to 3 kV).

As with EXELFS, the detected count rates and energy resolution possible in ELNES over the soft–X-ray region are comparable with those obtainable from synchrotron studies by XANES (Stern, 1982). Since near-atomic-resolution electron images can also be obtained from the same region as the ELNES, one may work with samples that have been "characterized" at the atomic level.

Both XANES and ELNES concern the first 40 eV or so beyond an X-ray–absorption edge. They may be distinguished from the EXAFS region beyond by the fact that much stronger oscillations are seen, by the fact that, in this region, the wavelength of the ejected photoelectron is larger than the interatomic distances, and by the fact that multiple rather than single scattering of the ejected electron occurs. It is this multiple scattering that offers the prospect of determining both interatomic distances and angles from the ELNES. But the gross features of ELNES edges are revealed by purely atomic calculations for isolated atoms; for example, K-edges show a characteristic sawtooth shape, while $M_{4,5}$-edges show a delayed hump, due to centrifugal-barrier effects, which increases with increasing atomic number. These generic atomic edge-shapes are shown in Figure 5.4.

It is perhaps misleading to classify ELNES as a "technique," since, for fundamental reasons, there is as yet no generally useful and predictive theory of XANES (or ELNES) that applies to all materials. For the ELNES obtained from thick specimens, the multiple scattering of the beam electron complicates the interpretation still further. By ELNES, we specifically refer therefore to electron-beam experiments whose interpretation follows similar lines to that used for XANES. The experimental conditions that allow such an interpretation are those that allow equation (5.21) (the general dynamical expression for electron-energy loss) to approximate equation (5.27) (the Bethe theory), and these conditions will be discussed further shortly. However, ELNES spectra are generally collected by using small, axial sources and detectors (i.e., at small values of $|q|$), so they represent only a very small fraction of the information present in a series of energy-filtered diffraction patterns that might be recorded over the same energy range. By providing angle-resolved information, such patterns must be expected to provide both a more sensitive test of theories of electronic structure and more direct information on the local coordination and symmetry of atoms in crystals. (For example, from section 5.4, we have seen that the fine structure of energy-filtered Kikuchi lines from inner-shell excitations contains angle-resolved XANES information compounded with the crystal rocking curve in a complicated way.) Nevertheless, we include ELNES as a technique here because a fair amount of evidence now indicates that the ELNES and XANES spectra are characteristic of the local arrangement of atoms about the excited atom (Knapp et al., 1982; Taftø and Zhu, 1982) and so may be used as fingerprints of the local atomic coordination. In addition, one may take advantage of the very large amount of theoretical work that has been done on XANES when interpreting ELNES spectra.

This theoretical work shows that the optical-absorption spectrum for an isolated atom is modified in a complicated way when a number of these atoms are assembled into a crystal. For example, in most cases, the K-edge ELNES for an isolated atom is predicted to be a simple sawtooth shape, as shown in Figure 5.4 (Rez and Leapman, 1981; Ahn and Rez, 1985). In a crystal, however, large modulations of the ELNES are seen, which depend both on the local crystalline environment (as demonstrated in Figure 7.4) and on more subtle many-electron interactions. These latter effects may be incorporated in both atomic calculations and band-structure calculations; however, they extend well beyond the scope of this book. One of them (the core exciton) dominates the ELNES of many insulators. Before very briefly summarizing recent research on XANES, we therefore introduce several important concepts: the core exciton, chemical shifts, relaxation and the definition of the lifetime of an excited state.

A core exciton is a bound, localized state formed between the excited core electron and the positively charged core hole that it leaves behind. Because less energy may be required to create such a state than is needed to promote a core electron into the lowest unfilled state (as predicted by ground-state, one-electron, band-structure calculation), core excitons frequently appear as sharp peaks just before the expected ionization energy of the solid, as shown in Figure 7.1. The expected ionization energy here refers to the level that would be occupied by an additional electron in the crystal in the absence of any core-hole interaction. While similar in principle to the familiar valence-band exciton in that this electrically neutral electron-hole pair cannot contribute to photoconduction, core excitons are found to lie much deeper into the band gap than valence excitons. In addition, core excitons are not mobile and have extremely short lifetimes (see Table 6.1). The presence of a core hole destroys the translational symmetry of the crystal. Since the ejection of the photoelectron leaves behind a positively charged atom, the eigenstates for the final state of the photoelectron may differ from those of the crystal ground state, thus giving rise to excitonic levels in the band gap. For emission spectra (such as XPS or EDS), many researchers use the potential for the ground state of the emitting atom, while for absorption (ELNES), the final state (containing a core hole) is relevant. Many-body effects can be incorporated by using the initial-state partial density of states (PDOS) for transitions into a nearly full band and the final-state PDOS for transitions into the initially empty band of an insulator (Stern and Rehr, 1983).

For insulators, XPS (or high-energy-resolution EDS) measures the difference between a given core level and the top of the valence band if the work function is known. Thus, the addition of the measured, optical band gap to this difference gives the expected onset energy of the corresponding

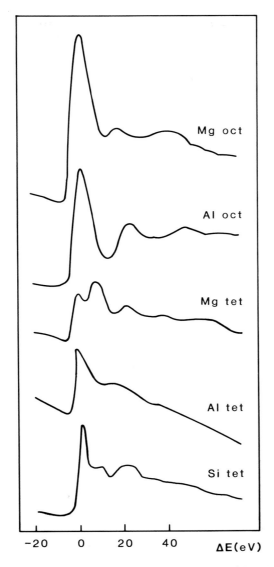

FIGURE 7.4 The *K*-edges in the ELNES from three atoms (Si, Al, Mg) in differing crystallographic environments. From the top: Mg in olivine (octahedral), Al in spinel (octahedral), Mg in spinel (tetrahedral), Al in orthoclase (tetrahedral), and Si in olivine (tetrahedral). A prominent broad peak is seen for the octahedral sites (white line); atoms on tetrahedral sites produce a more atomic sawtoothlike shape.

XANES edge. Any features at energies lower than this energy must be suspected as due to excitons (Pantelides, 1975). Note, however, that although the core electron is removed entirely from the solid in XPS, it remains in a localized excited state (possibly screening the core hole) in both ELNES and XANES. Core excitons were identified in graphite for the first time by ELNES (Mele and Ritsko, 1979).

Experimentally, the absolute calibration of the energy axis in EELS is vital, both for the identification of excitons and for the study of chemical shifts. These shifts constitute a separate effect, which lead to changes in the ionization energy of atoms as a result of differing chemical surroundings. This effect, due to changes in the atomic potential arising from altered bonding, can also be seen in XPS. Chemical shifts arise because the work needed to remove a core electron depends on the spatial distribution of valence charge and on that of the neighboring ion cores. Typical shifts are ±4 eV and increase with oxidation state. The situation is complicated, however, by an additional effect that is often difficult to disentangle from these effects: relaxation energy.

In EELS, we measure an energy loss $\Delta E = E_f^0 - E_i^0$, where E_f^0 and E_i^0 are the final and initial energies of the total (many-electron) crystal states. Methods such as the one-electron, Hartree–Fock, self-consistent scheme give a set of energy levels ε_i whose differences are only approximately equal to the allowed values of $E_f^0 - E_i^0$. Koopmans' theorem may be used to show that the ε_i could only be interpreted as ionization energies if no change in the other ε_i resulted from the removal of a particular electron. Thus, we customarily refer to any difference between ΔE and the coresponding ε_i loosely as a relaxation energy, and this energy may include many effects such as electron correlation and relativistic effects. Physically, the relaxation energy refers (in XPS) to the reduction in binding energy that results from the adiabatic relaxation of the uninvolved "spectator" electrons, which rush in to screen the abruptly created core hole, thereby reducing the energy needed to take the ejected core electron to a detector at infinity. In the chemical literature, a common method for going beyond the one-electron scheme is known as the configuration-interaction technique; by taking some account of the repulsion between the many electrons of an atom, this technique leads to more accurate predictions of spectroscopic-energy levels.

Another important concept in optical-absorption spectroscopy and ELNES is that of the excited-state lifetime. A core electron in a metal promoted by fast-electron ionization into a state more than about 15 eV above the Fermi level has sufficient energy to excite a plasmon or any of the other elementary excitations of the crystal. In doing so, the photoelectron wave field is depleted or absorbed, resulting in a limited photoelectron range or mean free path λ_{pe}. Figure 7.5 shows the approximate form of λ_{pe} as a function of the photoelectron's energy for two metals (Muller et al., 1982). Note that, at the ionization threshold, λ_{pe} becomes infinite; thus, the excited state for the photoelectron is here a Bloch wave of infinite extent.

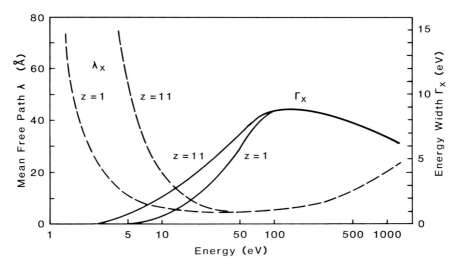

FIGURE 7.5 Mean free path λ_x for excited-crystal electrons as a function of their energy. Here, z is the number of valence electrons and Γ_x is the corresponding line width [evaluated from equation (7.6)]. Theoretical absorption spectra are commonly convoluted with a Lorentzian function whose width is equal to $\Gamma_x + \Gamma_c$, where Γ_c is the energy broadening due to the core-hole lifetime. These curves apply to metals and must be considered only very approximate indicators of the range of excited-crystal electrons. Energies are measured from the Fermi level and correspond roughly to E_{kin} in equation (7.3).

At onset, we would not expect the ELNES to be a useful probe of local structure (bound states are here ignored for simplicity). In the EXAFS or EXELFS region ($E_{kin} > 40$ eV), the shorter mean free path of the excited-crystal electron may well account for the success of the single-scattering approximation in EXAFS. As in LEED, a single-scattering theory may work surprisingly well under conditions of strong multiple scattering if there is sufficiently severe absorption or attenuation of the elastic wave field. Now, a crystal electron of energy E_{kin} traveling at velocity $V = (2E_{kin}/m)^{1/2}$ requires a time $\Delta \tau = \lambda_{pe}(m/2E_{kin})^{1/2}$ to travel a distance λ_{pe}. Thus, the excited-photoelectron state exists for a time $\Delta \tau$ and so, from the uncertainty principle, must have an associated energy width given by

$$\Delta E' = \left(\frac{2h}{\lambda_{pe}}\right)\left(\frac{2E_{kin}}{m}\right)^{1/2}.$$

These widths, which form one of the fundamental line-broadening mechanisms on ELNES spectra, are also shown on Figure 7.5 as derived from the λ_{pe}-curve. Note that the lifetime broadening increases with energy loss. Similar arguments apply to the core-hole lifetime, for which the associated energy width can be obtained from measurements of X-ray–emission line widths. A tabulation of core-hole natural line widths for all the spectra of interest in ELNES can be found in Keski–Rahkonen and Krause (1974). These values vary from a fraction of an electronvolt to tens of electronvolts and so are comparable with the energy resolution of modern electron spectrometers. These lifetime effects can be incorporated approximately by convoluting the unbroadened spectrum by a Lorentzian function of width $\Delta E'$. Since $\Delta E'$ increases with E_{kin}, the density-of-states (DOS) information is progressively lost with increasing energy loss beyond an edge. A complete treatment of inelastic effects in ELNES or XANES is beyond the scope of this chapter; a summary of the main effects can be found in Rehr and Stern (1983).

In summary, there are six main effects that must be considered when interpreting ELNES spectra:

1. Energy-broadening effects due to the finite resolution of the electron spectrometer.
2. Energy-broadening effects due to the core-hole and photoelectron lifetime.
3. The local density of empty states available for the excited-core electron (see the discussion that follows). Through this local density of states (PDOS), the ELNES spectrum is rendered sensitive to the local atomic structure.
4. Many-body effects, such as core excitons and those predicted in metals by Mahan et al. (for a review, see Citrin et al., 1979).
5. Phonon excitation, which may be important if the core-hole lifetime becomes comparable with that of a typical phonon period, resulting in a Gaussian line broadening.
6. Spin-dependent exchange effects, which are important in explaining the anomalous $L_{2,3}$-line-intensity ratios (Waddington et al., 1986).

Of these effects, 4 and 6 tend to be confined to the first few electronvolts beyond the onset of ionization.

A convenient classification divides the interpretation of XANES and ELNES spectra into three types: spectra from metals, spectra from highly ionic insulators (which necessarily have a large band gap), and spectra from covalent materials and semi-conductors of smaller band gap.

The best agreement between theoretical predictions and experimental XANES or ELNES spectra has been obtained for the transition metals. Here, a simple ground-state, one-electron band structure and PDOS calculation has frequently been found to give excellent agreement with experiment [see Muller et al. (1978) or Bianconi et al. (1983) and references therein, for XANES; Leapman et al. (1982), Figure 14, for ELNES]. The absence of excitonic effects in metals owing to screening may be understood qualitatively as follows: In insulators, the outgoing photoelectron experiences an additional potential $V(r) = -|e|^2/\varepsilon r$ due to the core hole, which may be taken to be stationary with a large effective mass. The hydrogenic levels of the excited electron in this potential then constitute the excitonic states. In metals, however, this core-hole charge is screened by the conduction electrons, reducing the potential to $V_m(r) = [-|e|^2/r]\exp(-r/\lambda)$, where λ is a small screening radius. This attenuated potential is then unable to bind a new localized state.

A comprehensive review of the influence of many-body effects on the soft–X-ray–absorption spectra of simple metals can be found in Citrin et al. (1979), who find slight effects for the $L_{2,3}$-edge onsets in sodium, magnesium, and aluminum. These effects are due to the adjustment of all the other crystal electrons to the excitation. Thus, equation (7.6) should also include products of overlap terms between the initial and final states of all the other crystal electrons. These and similar topics, such as the "shake-off" process, and the sudden and adiabatic approximations lie beyond the scope of this brief review. It has yet to be shown, however, that such complicated effects are truly important for metals, and the limits of the single-electron picture have yet to be fully explored. In view of the general success of one-electron, band-structure methods and the increasing power of small computers, ELNES studies of partially ordered alloys by STEM would appear to be a profitable field of research, since elastic microdiffraction patterns and images can also be obtained from the same subnanometer-sized region from which the ELNES has been collected (see section 7.7).

Sharp peaks at onset (known as "white lines") are also seen in the ELNES spectra from metals; these peaks are frequently accounted for by transitions into the high density of unfilled d-states. Since the $L_{2,3}$-edge initial states have p-symmetry, these edges probe final states of s- and d-symmetry. White lines on K- and L_1-edges (for which the final states must have p-symmetry) must therefore be due to a different mechanism such as excitons in insulators (Brown et al., 1977). These mechanisms can sometimes be distinguished by line-shape analysis. Some recent atomic calculations giving good agreement with experiment can be found in Waddington et al. (1986).

For ionic crystals such as the alkali halides and other crystals with large band gaps, the situation is entirely different; and the observed spectra may be almost entirely excitonic in nature, in some cases bearing no resemblance at all to the crystal-ground-state PDOS. Since any valence-electron charge that might otherwise be available for screening is confined predominantly to the halogen ion, the strongest excitonic effects are generally seen on the metal ion. These excitons are frequently *not* associated with critical points in the band structure (see below). By combining XPS (or high-energy-resolution EDS) information, however, one may be able to disentangle excitonic and band-structure effects. For a review of this work, see Pantelides (1975). The rich variety of features in the electron-loss, near-edge structure of magnesium oxide are believed also to be predominantly excitonic (Henrich and Dresselhaus, 1975); however, some limited success in matching ELNES spectra for magnesium oxide has recently been obtained through the multiple-scattering method by Lindner et al. (1985) and Neuman and Spence (1987).

For covalent insulators and semiconductors with smaller band gaps, rather less work has been done; and both DOS and excitonic states may be important. A valuable analysis of the X-ray–*emission* spectra from silicates and their relationship to bonding has been given by Dodd and Glen (1968), using a high-energy-resolution (wavelength-dispersive), electron-beam microprobe and the molecular-orbital picture for interpretation. The most detailed ELNES study yet undertaken of insulators is that of Leapman et al. (1982) on the transition metals and their oxides. They found that differences between the chemical shifts observed in XPS and ELNES could only be accounted for by screening and relaxation effects. Similarly, the oxygen ELNES edges in titanium dioxide and nickel oxide have been compared with theoretical calculations by Grunes et al. (1982). The molecular-orbital picture was found to be useful for TiO_2, but Hartree–Fock calculations gave poor agreement for NiO. The fine structure at the carbon K-edge in nucleic-acid bases has also been analyzed in detail by Isaacson (1979) in terms of the distribution of charge on atoms. For a theoretical analysis of the near-edge structure in boron nitride, see Robertson (1983b).

A few general comments about the interpretation of ELNES can be made. The width of a peak seen at onset may be due either to lifetime or band-structure effects. Although the latter contribute an asymmetrical line shape, the natural line width is Lorentzian in shape. If a peak is known to reflect the DOS, then its width gives a measure of the dispersion of that band, which, in a tight-binding picture, is proportional to certain overlap-matrix elements between near-neighbor-atomic orbitals. Thus, a broad peak in the ELNES suggests large overlap and a correspondingly delocalized final state (in real space), and vice versa. Also, the correlation between local coordination (octahedral or tetrahedral) seen in Figure 7.4 may result from the fact that the ELNES probes the angular-momentum content of the final (empty) states. Thus, for K-edges, these final states must have p-symmetry, and p-orbitals are consistent with the symmetry of an octahedron. Tetrahedral symmetry requires some mixing of s-states. Thus, we expect a higher density of states (and hence, the broad peak at onset) for the octahedral rather than the tetrahedral case, as observed experimentally. Since cubic crystals are isotropic [the dielectric susceptibility is simply related to the matrix element of equation (7.6)], the ELNES in these crystals is not expected to depend on the direction of q (for plane-wave, beam-electron states). Thus, a simple test for site symmetry would be to observe the ELNES, using a small axial detector, as a function of crystal orientation. For target species on cubic sites, there should be no dependence on crystal orientation, but some dependence should be observed for nonisotropic sites.

The case of graphite (whose ELNES has been studied in detail) is particularly instructive, since here two well-defined peaks have been identified as resulting from transitions to empty σ-orbitals that lie in the crystallographic planes normal to the c-axis, forming metallic bands, and to transitions (at slightly lower energy) into empty π-orbitals (see section 7.4).

Mele and Ritsko (1979) have shown that this π^*-transition is a localized excitonic state, reflecting the local excited density of states.

We now consider some of the features of ELNES interpretation that may distinguish it from the interpretation of XANES spectra. To interpret XANES intensities, we commonly assume neligible variation in the matrix element of equation (7.1) with energy, and we attribute the structure either to excitonic effects or to variations in the appropriate PDOS as specified by the dipole-section rules (for EXAFS, the opposite assumption is made). For ELNES, however, the detailed scattering kinematics must be considered; and in crystals, the fast-electron wave function should in general be modeled by a sum of Bloch waves, following the dynamical inelastic treatments outlined in section 5.4. In this way, channeling effects due to multiple scattering of the fast electron are included. The Bloch-wave kinematics of equations (5.17a) to (5.17g) would then also be used. For sufficiently thin samples in orientations that avoid the excitation of strong Bragg beams, however, we may represent the fast-electron states by plane waves, with the appropriate kinematics (equations 5.2a and 5.2b); so the cross section becomes, for a single atom,

$$\frac{d^2\sigma(q)}{d\,\Delta E\,d\Omega} = \left(\frac{4}{a_0^2 q^4}\right) |\langle f| \exp(2\pi i \mathbf{q}\cdot\mathbf{r}) |i\rangle|^2, \qquad (7.6)$$

where the final-state wave function is normalized per unit-energy range, and a_0 is the Bohr radius. Expressions similar to this one have recently been evaluated for applications in electron-energy loss for microanalysis (Ahn and Rez, 1985), using Hartree–Slater wave functions.

If a small axial detector is used, such that $|q| \ll r_c^{-1}$, where r_c is the radius of the initial state, the dipole approximation may be made in equation (7.6). This approximation can be shown to be equivalent to the requirement that $hq \ll (2mE_b)^{1/2}$, or that the scattering angles are less than $\theta_m = (E_b/E_i)^{1/2}$. For an axial detector, nondipole contributions in equation (7.6) are severely attenuated by the q^{-4}-term. Figure 5.3c, however, shows experimental evidence of the dipole-forbidden $M_{2,3}$-edges in La_2O_3, which may be removed by the use of a sufficiently small detector. If, in addition, the more severe approximation is made that the matrix element is independent of q over the angular range of the detector, then equation (7.6) becomes

$$\frac{d\sigma}{d\,\Delta E} = \left(\frac{4\pi e^4}{h^2 v^2}\right) \ln\left[1 + \left(\frac{\theta_m}{\theta_E}\right)^2\right] |\langle f| \boldsymbol{\varepsilon}_q \cdot \mathbf{r} |i\rangle|^2 \qquad (7.7)$$

for the total intensity falling within a detector of semiangle θ_m. Here, v is the fast-electron velocity, and $\boldsymbol{\varepsilon}_q$ is a unit vector along $\mathbf{q}$. This expression is similar to equation (7.1), suggesting similar shapes for ELNES and XANES spectra under the described experimental conditions. Since dipole-selection rules apply to equation (7.7), the final state must have p-symmetry for K-shell excitations.

Equations (7.6) and (7.7) refer to single atoms or possibly to an isolated but crystallographically aligned target atom whose orbitals lie in a known orientation with respect to q. For core excitations in crystals (again, ignoring multiple elastic scattering of the beam electron), the cross section may be written, for normal processes, as

$$\frac{d^2\sigma(q)}{d\,\Delta E\,dq} = \left(\frac{8\pi e^4}{h^2 q^3 v^2}\right) \sum_{\chi_f} |\langle \varepsilon_f, \chi_f | e^{2\pi i q \cdot r} | \varepsilon_i, \chi_i \rangle|^2 \delta_{\Delta\chi,q} \delta(\varepsilon_f - \varepsilon_i - \Delta E). \quad (7.8)$$

Here, box-normalized Bloch waves are used to represent the crystal electrons, and the dimensionless Kronecker-delta function $\delta_{\Delta\chi,q}$ indicates that the matrix element is zero unless crystal momentum is conserved—that is, unless equations (5.2) and (5.2c) are satisfied. Here, we see the essential feature that distinguishes ELNES from XANES: In ELNES, in addition to the energy-conservation requirement that occurs in XANES, there is also a momentum-conservation requirement, since the direction and energy of the final state of the fast electron are measured. Thus, in general, the ELNES intensity for a given $\mathbf{q}$ is obtained by summing (7.8) over all values of χ_f and ε_f in the three-dimensional, crystal-band structure that satisfy both the conditions

$$\Delta\chi = \chi_f - \chi_i = \mathbf{q}$$

and

$$\varepsilon_f - \varepsilon_i = \Delta E. \quad (7.8a)$$

This sum must be repeated, and the results added, for every point within a finite detector. Thus, we cannot conclude that, in general, the ELNES is proportional to the DOS used to interpret XANES spectra.

Simplifications occur, however, for the case where the primitive unit cell contains only one atom and where the initial state is not dispersive—that is, $\varepsilon_i(\chi_i)$ is a constant, independent of χ_i. Then, the number of pairs of initial and final states in the three-dimensional band structure that differ in energy by ΔE and by $\Delta\chi$ in wave vector is independent of $\mathbf{q} = \Delta\chi$, so the cross section becomes proportional to the simple optical DOS in energy. The fact that the momentum constraint in ELNES is not important for transitions from a nondispersive initial state may explain the observed close similarity between experimental XANES and ELNES spectra and the apparently weak dependence of ELNES spectra on the size of the (axial) detector used. In addition, for an axial detector, the term in q^{-4} in equation (7.6) severely attenuates nondipole contributions. For an off-axis point detector, the momentum constraint may be important, and this constraint allows nonvertical transitions to be studied (Chen and Silcox, 1977). In the simplest case of a crystal with a monatomic unit cell, a small axial detector, and a nondispersive initial state, the sum over energy-conserving delta functions becomes a DOS:

$$\rho(\Delta E) = \left(\frac{6}{2\pi}\right) \int \frac{ds}{|\nabla_\chi \varepsilon_f(\chi_f)|}, \quad (7.8b)$$

where ds is an element of surface in χ-space on the surface defined by $\varepsilon_f(\chi_f) - \varepsilon_i = \Delta E$. Thus, sharp peaks are expected in the ELNES from critical points in the band structure where $\nabla_\chi \varepsilon_f(\chi)$ is small. Here, we have assumed that there is negligible variation in the matrix element with the angle across the detector.

In multiatom crystals, under similar conditions, the ELNES is proportional to a local PDOS defined as

$$\rho(\Delta E, \mathbf{r}) = \sum_n |\psi_n(r)|^2 \, \delta(\Delta E - E_n), \tag{7.8c}$$

where $\psi_n(r)$ is an eigenstate evaluated at the site $\mathbf{r}$ of the core hole whose symmetry is that specified by the dipole-selection rule. Thus, there is a double projection on the unoccupied density of states, one in space at the site of the core hole and one in symmetry (Colliex, 1984). This DOS may be related to a multiple-scattering series (the first term of which represents EXAFS) through the Green's function formalism (Economou, 1983), giving

$$\rho(\Delta E, \mathbf{r}) = \operatorname{Im} G(\mathbf{r}, \mathbf{r}', \Delta E), \tag{7.8d}$$

where $G(\mathbf{r}, \mathbf{r}', \Delta E)$ is the appropriate Green's function. Physically, this function gives the response of the crystal to a spherical wave launched at the site of the core hole.

PDOS calculations have now been carried out by a number of workers for comparison with XANES experimental results [see Colliex (1984) for a review]. Generally, they entail very large calculations for each material. The two most successful current methods in use for XANES interpretation on the single-electron PDOS model are the augmented-plane-wave (APW) method [see Leapman et al. (1982) for references] and the Korringa–Kohn–Rostor (KKR) Green's-function method, as developed by Pendry and Durham from LEED work [see Durham (1985) for a review]. Both make the approximation of a "muffin tin" potential, which may fail for materials with strong directional bonding. All such one-electron DOS methods are formally equivalent, differing only in the way in which exchange and correlation effects are represented. However, the KKR technique uses the language of scattering theory and so provides a simpler connection with EXAFS in the single-scattering limit. It does not assume translational symmetry, and so it is well suited to disordered materials and to the computation of core-hole effects in the optical-ALCHEMI approximation (Hjalmarson et al., 1981). However, it may obscure the spectroscopic labeling of peaks, which is more readily apparent in atomic calculations. The convergence of this real-space, Green's-function method with increasing number of shells about the target atom will depend on the lifetime-broadening parameter, since it determines the photoelectron range. In APW calculations, the equivalent parameter is the number of sampling points in the brillioun zone.

An older approach has been to look for patterns in the XANES from elements in known coordination and to compare gas spectra with those from crystals containing similar coordination units in order to confirm that these effects are short-range. Molecular-orbital theory has also been used extensively, and group theory can be used to provide selection rules. Many crystal-field-splitting effects seen in optical work may, however, be unresolvable in ELNES. Striking similarities were pointed out, for example, by Van Nordstand [see Srivastara and Nigam (1972–1973) for a review], who obtained XANES spectra from 95 compounds of four elements in various chemical environments, which he was able to classify into four main types. In a similar way, Taftø and Zhu (1982) have noted similarities in the K-shell ELNES of magnesium, silicon, and aluminum when octahedral and tetrahedrally coordinated forms are compared, as shown in Figure 7.4. A single broad peak tends to be seen from octahedral sites, while tetrahedral coordination is indicated by several narrower peaks.

ELNES is well suited to these light elements for which the corresponding XANES would occur in the soft–X-ray regions. Differences in the ELNES from three forms of carbon have also been observed and compared with DOS calculations (Egerton and Whelan, 1974). ELNES contains atomic-number information because the initial state is a core state of atomic character and well-defined energy. But the extraction of site-symmetry information from the broader, final-conduction-band state of the excited-crystal electron has proven much more difficult. Indeed, from the many band-structure calculations published, no completely general rules relating local-site symmetry to PDOS have yet been extracted. Although symmetry arguments can be used to classify the likely ELNES peaks, they cannot predict their energies. The molecular-orbital approach has proved useful for the interpretation of ELNES, even for some crystals, and provides some qualitative rules, such as the energy reduction that results from orbital overlap and the formation of bonding and antibonding states.

Since general trends are clearly present in the experimental data, a theoretical approach that emphasizes these qualitative trends has been applied to ELNES data from Be_2C and Mg_2Si by Disko et al. (1986). This tight-binding approach is known to fail for excited states of large energy; however, for the ELNES, it may reveal general trends. In this work, the degree of ionicity was found to strongly affect the predicted edge shapes. Calculations for neutral atoms gave a poor fit to the data, but calculations by the multiple-scattering XANES method for the ionicity $Be_2^{1.5+}C^{3.0}$ gave a greatly improved fit. More sophisticated calculations have recently been completed for Cu_2O (Robertson, 1983a), using the optical-ALCHEMI approximation to indicate likely core-exciton levels; and fair agreement has been obtained between theoretical calculations and experimental ELNES for TiO_2 (rutile) and NiO, using tight-binding and Hartree–Fock methods, respectively (Grunes et al., 1982). Here, core excitons are identified in NiO but not in TiO_2.

Clearly, a considerable computational effort is required at present to interpret either the XANES or ELNES structure, and few general principles apply. The recent application of a $\chi_f \cdot \mathbf{r} = $ constant rule (Natoli, 1983) to experimental XANES and ELNES spectra (Stohr et al., 1984; Spence, 1985) for the determination of bond lengths does, however, appear promising in favorable cases. Nevertheless, the promise of ELNES is great, since, by comparison with XANES, it allows the analysis of submicron areas containing light elements in correlation with HRTEM images of the same region. Equally important are the insights it provides into the electronic structure of solids. Comparisons between XANES and ELNES can be made through the dielectric-response function, as for EXAFS (see Stern, 1982). By comparison with EXELFS, however, ELNES allows many atomic species to be analyzed with the limited-energy range available in EELS without the problem of edge overlap. In addition, channeling effects can be used to provide additional ELNES information, as described in the next section.

In summary, we may say that the problem of the multiple scattering of the beam electron in ELNES has proven tractable to computation; but further developments in the interpretation of ELNES depends on advances in methods for calculating the excited-state wave functions of crystal electrons, and these methods still present unsolved problems. The greatest success (with some exceptions) has been obtained for simple metals, and the difficulties seem to increase with the size of the band gap. The degree of charge transfer between atoms in insulators is clearly an important (often unknown) factor. The measurement of low-order-structure factors by convergent-beam electron diffraction would therefore appear to be a powerful complementary technique. In addition, no a priori method of band-structure calculation is at present able to predict correctly the crystal-band gap energies. Errors of at least comparable magnitude must therefore be expected in calculations of the onset energy for ELNES and XANES.

7.4 Orientation effects in EELS

Experimentally, inner-shell-excitation spectra, EXELFS spectra, and some band-gap spectra all vary appreciably with the crystallographic direction of either a collimated incident-electron beam or a small detector in the geometry of Figure 5.2. Here, we discuss briefly recent work on this effect and its interpretation in terms of the atomic and electronic structure of crystals.

Four distinct classes of experiments have been reported in this new field of research:

1. Experiments on the orientation dependence of the first few electronvolts of the EELS spectrum (the band gap) (Brown, 1985).
2. Experiments on the orientation dependence of EXELFS spectra (Disko, et al., 1982).

3. Experiments on the orientation dependence of ELNES in specimens sufficiently thick to produce channeling or blocking conditions and using sufficiently large values of **q** to ensure localization (Taftø and Lehmpfuhl, 1982).
4. Experiments at small q in thin specimens in which multiple elastic scattering is minimized (Leapman et al., 1983).

All of these experiments can be understood by taking the appropriate limit of the general equation for dynamical inelastic scattering, equation (5.32).

The work of Leapman et al. (1983) provides the most important example of this last type of experiment. Here, in the scattering geometry of Figure 7.2, carbon, boron, and nitrogen edges were studied from specimens of graphite and boron nitride with **q** arranged to lie successively parallel to and normal to the layers of the crystal. The dipole approximation was satisfied in all experiments; hence, the angular-momentum quantum number must change by one. Since the orbitals $\langle f|$ in equation (7.6) point in particular crystallographic directions, a variation of the transmitted-electron-loss intensity is expected as a function of detector position (and hence q) for each of the two crystal orientations. In graphite, the π^*-orbital extends along the c-axis, while σ^*-orbitals lie in the crystal planes. These two states produce two distinct peaks separated in energy in the ELNES, which can be identified from DOS calculations. Figure 7.6 shows the differing angular dependencies of the scattering from each state and a comparison with a simple no-overlap, tight-binding theory. We note the following:

1. By writing the initial and final states of the crystal electron in spherical harmonics, these authors obtain a simple expression for the angular variation of inelastically scattered beam electrons. For the scattering geometry of Figure 7.2, they obtain

$$\frac{d^2\sigma}{d\Omega\, d\,\Delta E} = \frac{A \sin^2\left[\tan^{-1}(\theta/\theta_E) - \gamma\right]}{\theta^2 + \theta_E^2}$$

for the σ^*-loss peak, and they obtain a similar expression with the sine replaced by a cosine for the π^*-peak. Here, A is a constant and γ is the angle between the incident beam and the c-axis of the graphite crystal. These functions are compared with the experimental results in Figure 7.6.
2. Unlike synchrotron studies, which probe states that are oriented predominantly perpendicular to the X-ray beam, the small-angle ELNES technique probes final states whose principal axis lies parallel to the beam (for a K-shell transition).
3. Very high angular resolution is needed for this work, and it must be a small fraction of θ_E (see Figure 7.4).
4. Since $|q| < |g|$, these losses represent delocalized excitations (large impact parameter), and this delocalization distinguishes this work from

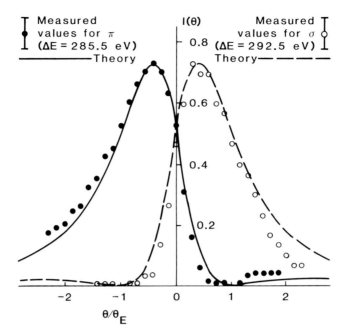

FIGURE 7.6 Measured angular variation of the intensity of electrons that have excited a K-shell-ionization event in traversing a thin crystal of graphite. The filled circles indicate the intensity of the low-energy π^*-peak; open circles represent data collected from the higher-energy σ^*-peak. The theory curves are discussed in the text. (From Leapman et al., 1983.)

that on the orientation dependence of localized losses under channeling conditions (to be described later).

5. Extremely thin ($t < 100$ Å) samples are required both to eliminate multiple elastic scattering and to eliminate multiple losses, such as the double scattering, which produces plasmon satellites in the ELNES structure (see section 5.5).
6. The core-peak shapes are not simply related to the crystal total DOS, since, for transitions from a 1s-state, only final states with p-character are selected. Secondly, the matrix element itself may be energy-dependent. Finally, the possibility of core excitons occurs (see section 7.3).

Several recent papers have described orientation effects in thicker samples owing to localized losses (Spence et al., 1981; Taftø and Krivanek, 1982; Taftø and Lehmpfuhl, 1982). In the simplest application of this effect, strong elastic scattering is used to concentrate the (thickness-averaged) beam electron's wave field onto particular crystallographic planes. The principle is thus similar to that of the ALCHEMI technique described in section 7.5. For example, in the [100] systematics orientation, the

ABAB... stacking sequence of crystal planes in the natural spinel structure contains all the aluminum and oxygen atoms on the *A*-planes (see Figure 7.10 shown later in the chapter) and all the magnesium atoms on the *B*-planes. If the crystal is traversed by a collimated, kilovolt electron beam in a direction approximately normal to [100], the standing wave established across these crystal planes by dynamical diffraction will result in different thickness-integrated electron intensities (the shaded areas I_A and I_B in Figure 7.10) on each of the *A*- and *B*-planes. Since the transition rate for ionization is proportional to this integrated intensity (see section 6.1) for ideally localized losses, and since I_A and I_B depend on the direction of the incident beam, we expect the magnitude of the magnesium and aluminum inner-shell-loss edges to vary with incident-beam direction. However, in the energy-loss case, the diffraction of the outgoing, inelastically scattered beam electron must also be considered (blocking). Thus, a quantitative treatment must be based on equation (5.27) and so requires an accurate knowledge of the crystal potential, absorption coefficients, and crystal-electron, excited-state wave functions.

Useful simplifications are possible in certain cases to assist interpretation, however. For example, in the spinel case (and many others), one may compare the orientation dependence of different atomic species on the *A*-plane (which show the same orientation dependence) with those on the *B*-plane. This comparison provides a simple method of distinguishing normal and inverse spinels, which is otherwise difficult. For normal spinel, the *A*-planes contain all the aluminum atoms (in octahedral sites) and oxygen atoms; the *B*-planes contain the magnesium atoms on tetrahedral sites. The inverse spinel is obtained by interchanging half of the aluminum atoms with the magnesium atoms. Thus, the spinels can be distinguished according to whether the magnesium *K*-loss peak follows the oxygen reference peak in its orientation dependence. The power of this method is that it is independent of the details of dynamical diffraction and can be applied to light elements and many oxides, since oxygen can be used as a reference atom for EELS but not easily for ALCHEMI. These light elements may be neighbors in the periodic table, making conventional X-ray–diffraction studies difficult. Finally, the method may be applied to very small areas and microcrystals. Figure 7.7 shows experimental results from natural spinel.

In these experiments, the detector is displaced in a direction normal to the systematics line (i.e., it is displaced in the *y*-direction in Figure 7.8). This arrangement allows the diffraction (channeling) conditions to be varied (in the plane whose normal is *y*) *independently* of the localization, since the most important finding of this work is that the localization can be increased through the use of a large inelastic-scattering angle and correspondingly small impact parameter (see section 6.1). The scattering angles and source and detector sizes can be judged from the (400) Kikuchi lines sketched in the inset on Figure 7.7.

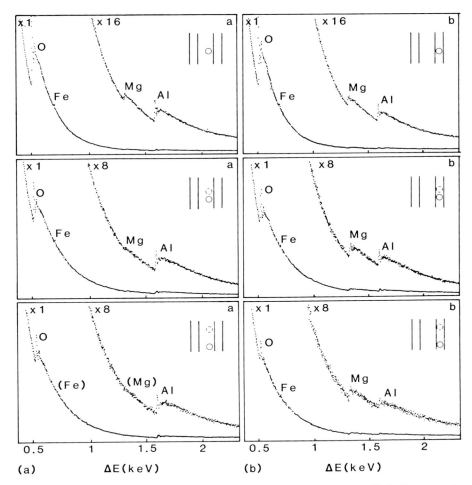

FIGURE 7.7 Energy-loss spectra from an $MgAl_2O_4$ spinel crystal in the $(h, 0, 0)$ systematics orientation. The $[h, 0, 0]$ direction runs normal to the Kikuchi lines [sketched in the inset, where the source (solid circle) and detector (dotted circle) are also drawn to scale]. Only for a large angular separation between source and detector (lower panels) is there a significant change in the ionization edges. (a) Spectra recorded with the (400) reflection just outside the Ewald sphere. (b) Spectra with (400) reflection just inside (see insets).

A second useful simplification results from the application of the reciprocity theorem, which has been shown experimentally to hold for all the energy losses important in EELS (Lehmpfuhl and Taftø, 1980). By using a sufficiently large axial detector subtending an angle much larger than the Bragg angle, diffraction effects in the final fast-electron state may be eliminated. Indeed, if all energy-loss electrons were collected, their dependence on the orientation of a collimated incident beam should be the same as

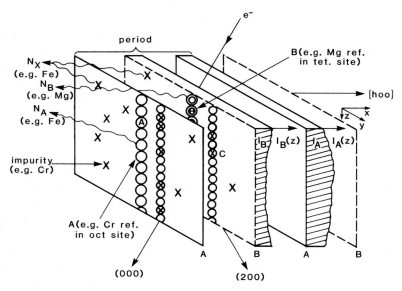

FIGURE 7.8 Principle of the ALCHEMI method. The $(h, 0, 0)$ systematics crystal planes, with stacking sequence $ABABAB \cdots$ are shown; $I_A(z)$ and $I_B(z)$ are the electron-beam intensities on any of the A- or B-planes, and these intensities are constant in the y-direction. Characteristic X-rays are emitted by species A, B and the substitutional impurity X.

that of the corresponding X-ray–emission products. By the reciprocity principle, this result also applies to the experimentally convenient case of a large source (convergent-beam illumination) and a small detector subtending an angle much smaller than the Bragg angle. In this case, a theoretical treatment similar to that used for ALCHEMI [and much simpler than equation (5.27)] can be applied if ideally localized losses are assumed. Equation (6.8) gives the dynamical intensity $I(\mathbf{r})$ inside the crystal, from which an expression for the depth-integrated intensity follows for an atom at $\mathbf{r}$ in a crystal of thickness t (Taftø, 1979):

$$\int_0^t I(\mathbf{r}, z)\, dz = t + \sum_j \frac{|C_0^j|^2}{2\mu}(\exp(-2\mu^j t) - 1) + \sum_{jj'hh'} C_0^j C_0^{j'*} C_h^j C_h^{j'*} \cdots$$

$$\frac{\exp[ir(h-h')]}{(-\mu^j - \mu^{j'}) + i(\gamma^j - \gamma^{j'})}(\exp[(-\mu^j - \mu^{j'})t + i(\gamma^j - \gamma^{j'})t] - 1) \quad (7.9)$$

where the instrumental parameters have been set to zero. Note that this result can also be obtained from the general equation for dynamical inelastic scattering (equation 5.27) by taking $[e, 2e] = 1$ for all $Q_{1,2}$ (extreme localization) and deleting all primed quantities in equation (5.32) with $C_0' = 1$ (corresponding to the neglect of diffraction for the initial beam-electron state). These expressions can readily be evaluated by computer, using the absorption coefficients discussed in section 6.3, and give a good approximation to the orientation dependence of localized losses under these experimental conditions. An even more severe approximation is to include only two beams and to make the independent Bloch-wave approximation (the neglect of interference terms between Bloch waves). If, in addition, absorption effects are also neglected, the resulting expression for the angular variation of the total, incident (thickness-integrated), fast-electron intensity at an atom site $\mathbf{r}$ becomes

$$I(W) = 1 - \frac{W \cos(2\pi g \cdot r)}{1 + W^2}. \tag{7.10}$$

Here, $W = S_g \xi_g$ denotes the deviation of the incident (or scattered, but not both) beam direction from the Bragg condition at $S_g = 0$. This function is shown in Figure 7.9 for atomic planes at $r = 0$ and $(2g)^{-1}$, such as the gallium and arsenic planes in gallium arsenide along [200] when using the (200) reflection. Equation (7.10) is in qualitative agreement with experimental results (Taftø and Lemhpfuhl, 1982). A more general two-beam result, including absorption, can be found in Spence and Graham (1986). By choosing higher-order reflections, one may further subdivide the crystal period along the systematics direction.

While the argument used to distinguish normal and inverse spinels is strictly independent of the choice of orientation and thickness, the best experimental conditions for producing the largest orientation effect must be found experimentally. For very thin crystals, we do not expect any orientation dependence, since, for example, the phase-grating approxima-

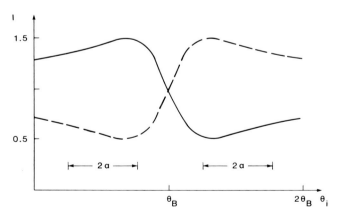

FIGURE 7.9 Theoretical intensity variation of either characteristic X-ray emission or energy-loss intensity from a localized process. A collimated incident beam and a detector semiangle much larger than the Bragg angle have been assumed (two-beam, independent, Bloch-wave theory). The curves refer to a systematics (or planar-channeling) orientation in a crystal with $ABABAB\ldots$ stacking sequence in the neighborhood of the first-order Bragg condition. The continuous curve shows the emission from the species on the A-planes (assumed lighter), and the dashed curve shows that from the species on the B-planes (heavier).

tion given in Chapter 2 indicates a constant intensity across the unit cell for all incident-beam directions. Experimental support for this idea is indicated in section 5.4. At very large thicknesses, the coherence and collimation of the dynamical wave field is destroyed by diffuse inelastic scattering, which degrades the channeling effect. In practice, a thickness of about 100 nm seems best (at 100 kV), with a collimation angle of about one-half of the Bragg angle, although larger angles have also given strong effects.

From equation (7.10), the two incident-beam directions giving the largest difference in inelastic core-loss intensity are seen to correspond, in the neighborhood of the Bragg condition, to

$$S_g = \pm \xi_g^{-1} \tag{7.11}$$

in the two-beam, independent, Bloch-wave approximation.

Two other recent studies have extended this work. In chromite spinel, three-quarters of the Fe atoms occupy tetrahedral sites and one-quarter occupy octahedral sites; these sites occur in planes that alternate along [400]. The distribution of Fe^{2+} to Fe^{3+} over the sites was not known; however, Taftø and Krivanek (1982) were able to use the orientation dependence of the Fe $L_{2,3}$-edge to determine it. The structure also contains O (all atoms on octahedral sites) and Cr atoms (all on octahedral sites), which may be used as reference signals. A 2-eV splitting of the Fe$L_{2,3}$-edge (see Figure 5.3b) was attributed to a chemical shift between states localized on the Fe^{2+} and Fe^{3+} atoms. Since the subpeak associated with the Fe^{3+} follows the Cr edge in its orientation dependence, these authors conclude that the octahedral site contains Fe^{3+} ions, while the tetrahedral site contains Fe^{2+} ions, a result unobtainable by other methods. In a similar way, channeling effects offer the possibility of separating the ELNES or EXELFS from the same species that may occur on nonequivalent sites in a crystal. Thus, Taftø (1984) has obtained separate ELNES data from Al atoms in sillimanite that occur with the same valence in both octahedral and tetrahedral sites. Figure 7.10 shows the separated ELNES for successive orientations that maximize the incident-electron intensity on the tetrahedral and octahedral sites. Since chemical shifts in the core levels are not expected (each Al occurs as Al^{2+}), these differences are attributed to differences in the local PDOS about each atom, owing to its differing crystalline environment. The experiments were performed in the [100] systematics orientation with a detector displayed normal to this direction; confirmation of the validity of the dipole approximation was therefore necessary.

In both of these last two studies, a further simplifying condition was used to make the interpretation of orientation-dependent ELNES possible. This simplification is that the effect of the elastic diffraction of the inelastically scattered fast electron can be incorporated by squaring equation (7.9) for the special case that the source and detector angles are equal (Lehmpfuhl and Taftø, 1980), so the initial and final fast-electron states are related by time reversal.

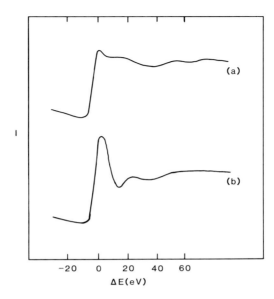

FIGURE 7.10 Separated electron-energy-loss, K-shell, near-edge structure from the aluminum atoms in sillimanite, which occur with the same valence in an octahedral [curve (b)] and a tetrahedral [curve (a)] site. Electron channeling has been used to isolate the contributions of the two sites. (From Taftø, 1984.)

Orientation effects have recently also been observed in EELS from optical band-gap transitions in diamond (Brown, 1985). At these low energies, the losses are delocalized; however, an orientation effect can be explained by considering how the number of states across the band gap that differ by a given q varies with crystal orientation.

7.5 ALCHEMI

ALCHEMI is a quantitative technique for identifying the crystallographic sites and distribution and types of substitutional impurities in many crystals. The technique requires a modern transmission electron microscopy (TEM) and EDS system. If characteristic X-rays from trace elements of interest in crystalline regions of a TEM sample can be detected by using the EDS system, then, in most cases, one can determine the crystallographic site of the impurity if it is substitutional. The method involves no adjustable parameters, the fractional occupancies of the substitutional impurities of interest being given in terms of measured X-ray counts alone. Since TEM is used, the method can be applied to areas as small as a few hundred angstroms in diameter; the detection sensitivity is limited by that of the EDS system to about 0.1 atomic percent. Elements that are neighbors in the periodic table can normally be readily distinguished. The method uses the collimated electron-beam-orientation dependence of characteristic X-ray emission and does not require either the specimen thickness or the precise beam orientation to be known. No dynamical-electron-diffraction calculations are required for the interpretation of this quantitative method. The classical problems of cation ordering in spinels, feldspars, and olivine minerals have all been studied by this method. The first application of the method was described by Taftø and Liliental (1982); tutorial reviews of the principles and applications of the technique can be found in Spence and Taftø (1983) and Krishnan (1988). These also contains a very brief historical summary of the related effects (such as the Borrman effect in X-ray diffraction) on which it is based.

The principle of the method is illustrated in Figure 7.8. The characteristic X-ray–emission intensities from both the impurity atom and the atoms of the host crystal are measured for two or three crystallographic orientations of the collimated, incident-electron beam. A specimen thickness of about 1000 Å is sought; however, the thickness is not critical. The illumination beam divergence should be a fraction of the Bragg angle for a first-order reflection. In many problems, one can limit the substitutional sites of the impurity atom to a few likely possibilities. The crystalline specimen is then oriented in the systematics or planar-channeling orientation, so crystal planes whose reciprocal-lattice vector is normal to the beam contain alternating candidate sites for the impurity and alternating species of the host crystal. For example, in the zinc blende structure, the $[00h]$ systematics orientation would contain all the Zn atoms on the A-planes of Figure 7.8, and the B-planes contain only S atoms. By measuring the Zn, S, and impurity X-ray emission for two incident-beam directions (both approximately normal to $[00h]$, one can determine the fractions of the impurity that lie on the A- and B-planes, respectively. By repeating this experiment for other sets of planes, one can determine the crystal site.

For each of these two incident-beam orientations, the fast electron sets up a standing wave in the crystal, whose intensity variation has the period of the lattice (e.g., from A to A, as shown in Figure 7.8). For ideally localized X-ray emission, the characteristic X-ray–emission intensity due to the ionization of the crystal atoms is proportional to the height of this standing wave at the atom concerned. The total X-ray emission from, say, the species on the A-planes of Figure 7.8 is then proportional to the thickness-integrated electron intensity I_A on the A-planes, shown shaded in the figure. This area changes with changes in the incident-beam direction. Because a systematics orientation has been chosen, the electron intensity is constant along the A- and B-planes. We let C_X be the concentration of impurities on the A-planes and $(1 - C_X)$ the concentration on the B-planes. The X-ray emission from the host atoms in known sites on the A- and B-planes will be used to provide an independent monitor of the electron intensities I_A and I_B, which also excite X-rays from the impurities.

Thus, the principle of ALCHEMI is to use the host atoms as reference atoms or detectors, to "measure" the thickness-integrated, dynamical-electron-intensity distribution. We let $N_A^{(1,2)}$, $N_B^{(1,2)}$, and $N_X^{(1,2)}$ be the six X-ray counts from elements A, B, and X for a channeling orientation (1) and a nonchanneling orientation (2) in which the electron intensities on the A- and B-planes are equal. Then, we have the six relationships

$$N_B^{(1,2)} = K_B I_B^{(1,2)}, \tag{7.12}$$

$$N_A^{(1,2)} = K_A I_A^{(1,2)}, \tag{7.13}$$

$$N_X^{(1,2)} = K_X C_X I_B^{(1,2)} + K_X (1 - C_X) I_A^{(1,2)}, \tag{7.14}$$

where the superscripts refer to X-ray counts obtained in two successive orientations. In addition, for the nonchanneling orientation, $I_A^{(2)} = I_B^{(2)}$. Here, K_A, K_B, and K_X are constants that take account of differences in fluorescent yield and other scaling factors. These equations can be solved for C_X in a way that eliminates I_A and I_B. If we define the ratio of counts for two orientations as

$$R\left(\frac{A}{X}\right) = \frac{N_A^{(1)}/N_X^{(1)}}{N_A^{(2)}/N_X^{(2)}}, \tag{7.15}$$

then

$$C_x = \frac{R(A/X) - 1}{R(A/X)(1 - \beta)},$$

where

$$\beta = \frac{N_B^{(1)} N_A^{(2)}}{N_A^{(1)} N_B^{(2)}}. \tag{7.16}$$

Equation (7.16) can thus be used to find the concentration of species X on the A-planes (see Figure 7.8) in terms of the measured X-ray counts $N_A^{(1,2)}$, $N_B^{(1,2)}$, and $N_X^{(1,2)}$ alone. In practice, orientation (1) is usually chosen as slightly greater than the Bragg angle, and orientation (2) is one that avoids the excitation of Bragg beams. Note that it is *not* assumed that the electron-beam intensity is the same for both spectra.

This technique has now been applied to determine the occupancies of chromium, aluminum, iron, and magnesium in samples of chromite spinel (Taftø, 1982) and to the occupancies of iron and trace elements in an Mg–Fe olivine (Taftø and Spence, 1982). Here, results in agreement with those obtained by the Mössbauer method were found. A second independent check on ALCHEMI is described in the work of Taftø and Buseck (1982), who compared results for the occupancy of aluminum on the T1-site in orthoclase feldspar with X-ray–diffraction results from the same sample. The occupancies derived by the two methods agreed to within less than 2 percent. The method has also been applied to uranium atoms in nuclear-waste materials (Taftø et al., 1983). More recently, a generalization of the method to a larger number of sites and orientations has been described (Krishnan and Thomas, 1984), and work on ceramics has also been reported (Chan, 1984). In semiconductor applications, a variant of this method that

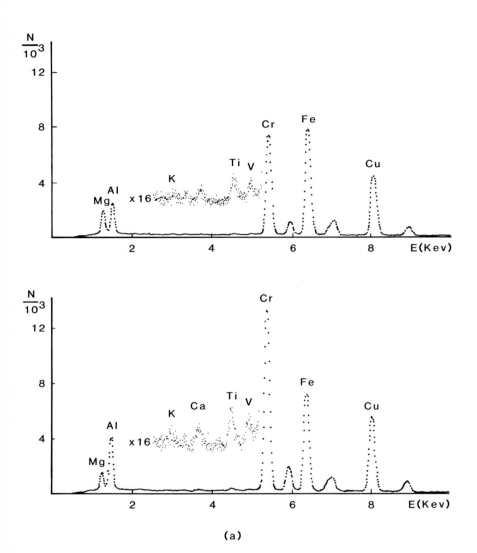

FIGURE 7.11 Electron channeling effects on X-ray emission in two different materials. (*a*) Characteristic X-ray–emission spectra recorded from a $(Cr, Fe, Al, Mg)_3O_4$ spinel in two different channeling orientations, near the (400) Bragg condition. Planes normal to $[h, 0, 0]$ contain, alternately, all the oxygen atoms and the octahedral cations, then the tetrahedral cations. The changes in peak-intensity ratios can be used to determine the crystallographic-site location of the trace elements Ti and V and the Mg, Cr, and Al species. (From Taftø, 1982.) (*b*) Characteristic X-ray emission from Si (circles) and As (crosses) as a function of beam direction for (220) planar-channeling conditions at 100 keV. The As concentration was 3×10^{19} atoms cm^3. The dotted curve shows the theory for interstitial As; the dashed curve is 80 percent substitutional (20 percent incoherent); and the continuous curve is 100 percent substitutional.

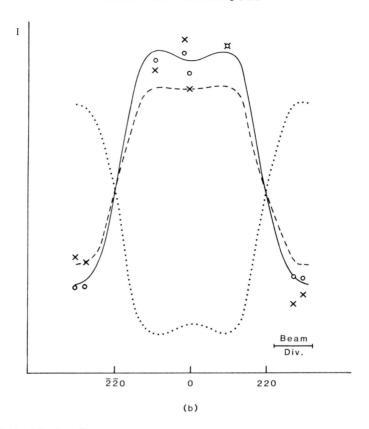

FIGURE 7.11 (*Continued*)

depends on the possibility of comparing absolute X-ray intensities for two different incident-beam orientations has been reported for the case of arsenic in silicon (Taftø and Spence, 1983). Results of this work are shown in Figure 7.11*b*.

The method is subject to all limitations of conventional EDS. Thus, it cannot easily be applied to elements of lower atomic number than sodium (unless a windowless detector is used), and the sensitivity is restricted. However, it has some important advantages in comparison with techniques such as EXAFS, Rutherford ion-backscattering (RBS), and X-ray diffraction; in particular, neighboring species in the atomic table may be distinguished, and the method may be applied to small areas in combination with TEM imaging. Thus, real crystals (such as meteorite fragments or fine-grained, polycrystalline, nuclear-waste materials) rather than crystals of specially grown synthetic analogs may be analyzed.

A crucial assumption of the method is that the X-ray–emission process be highly localized. This assumption has been tested by Self and Buseck

(1983), whose experimental results on localization in energy-loss spectra indicate that ALCHEMI using EDS can be expected to give reliable results down to X-ray energies of 500 eV or possibly less (see section 6.1). This result opens up the possibility of applying the ALCHEMI technique to elements such as oxygen, using thin-window EDS detectors. More extensive measurements of localization may be found in the work of Pennycook (1987).

Since crystal planes can be found that contain interstitial sites, the method may, in principle, be extended to include interstitial-occupancy-site determination. In addition, under the column approximation, any local change in diffraction conditions owing to crystal bending or thickness changes under the probe will not affect the ALCHEMI results, since these affect both the impurity emission and that from the reference host atoms in a similar way (Spence and Taftø, 1983). A final important assumption of the method is that the impurities be uniformly distributed in depth throughout the crystal. Thus, the method fails for a segregated sheet of impurities normal to the beam at a particular depth in the crystal.

In some cases, the "known" atomic species may segregate into distinct columns. Then, the much stronger axial-channeling effect may be used in a similar way (Otten and Buseck, 1987; Pennycook, 1987). The crystallographic constraints are, however, more severe in this case; because of the finer transverse oscillations of the wave function, localization effects become more important. For a detailed analysis of localization effects on axial ALCHEMI, see Rossouw and Maslen (1987).

As an example, Figure 7.11a shows the EDS spectra obtained from two orientations of a $(Cr, Fe, Al, Mg)_3O_4$ spinel crystal containing other trace elements (Taftø, 1982). The changes in peak height with incident-electron-beam orientation can be clearly seen. By repeating this experiment for many orientations along the same systematics row, the researchers found that the simple ratio of peak heights changes greatly between pairs of orientations, but all these ratios reduce to the same occupancy factor C_x given by equation (7.16) (see also Taftø and Spence, 1982). A full list of references to ALCHEMI work and measurements and analysis of the effect of temperature reduction on ALCHEMI can be found in Spence and Graham (1986). This work includes references to applications to problems of dopant-site location in semiconductors (Taftø and Spence, 1983), rare-earth additions to thin-film garnets (Krishnan et al., 1985), aluminum and silicon in feldspars (Taftø and Buseck, 1983) and calcium in barium titanate (Chan et al., 1984). A special issue of *Ultramicroscopy* is also planned for 1988 to report the proceedings of a conference on electron diffraction and channeling. It will contain several papers on ALCHEMI.

7.6 Cathodoluminescence in STEM

The study of the electronic properties of individual defects in crystals requires the development of a technique that is (1) sensitive to the very low concentrations of impurities that may be electrically important [usually present in concentrations too low to be detected by EDS, EELS, or high-resolution electron microscopy (HRTEM)]; (2) capable of sufficient spatial resolution to isolate individual defects; and (3) able to provide sufficiently high spectral-energy resolution to study the electronic states of interest. At present, the cathodoluminescence (CL) technique in scanning TEM (STEM) provides the most favorable combination of these properties. Here, the optical emission excited by the electron beam in passing through a thin sample is collected by a small mirror and passed to a conventional optical spectrometer for analysis, as shown in Figure 7.12. By forming a small electron probe and plotting the CL intensity within a small spectral range as a function of electron-probe position, one may also obtain a scanning-monochromatic CL image, giving a spatial map of the impurity or electronic state of interest. For example, Roberts (1981) has used this method to relate structural inhomogeneities to luminescence to ZnS phosphors.

The usefulness of the STEM CL technique for the study of semiconductor defects was first demonstrated by the pioneering work of Petroff and others

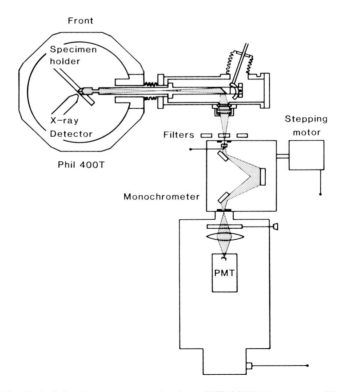

FIGURE 7.12 Cathodoluminescence apparatus for a STEM/TEM instrument. The optical path between the ellipsoidal light-collection mirror and detector is shown shaded.

(Petroff, et al., 1980). Here, a systematic study of dislocations in gallium arsenide was undertaken, using the increased accelerating voltage of a high-voltage STEM to allow the study of thicker specimens. A correlation between the electrical activity and type of dislocation was obtained in this work. More recently, systems have been fitted both to the HB5 field-emission STEM instrument and to conventional TEM–STEM machines.

The spatial resolution possible in STEM CL appears to depend greatly on the specimen thickness for thin samples. In general, it is expected to be given by

$$d_r = (d_p^2 + d_g^2 + d_D^2)^{1/2}, \qquad (7.17)$$

where d_p is the electron-probe diameter, d_g is the electron-hole, pair-generation volume, and d_D is the carrier-diffusion length (Pennycook, 1981). For thin samples, d_g (the beam spreading discussed in connection with EDS) and d_p are small compared with d_D, which is dominated by surface recombination. As a rough approximation, this diffusion length is expected to give (Brown, 1984).

$$d_r = \frac{t}{\sqrt{6}} \qquad (7.18)$$

for a thin specimen of thickness t. For such very thin specimens, the emission intensity is very small, requiring efficient light-collection optics. The spatial resolution of a monochromatic-scanning CL image is not influenced by the optical-resolution limit of the light-collection mirror or lens; however, this lens does affect the amount of stray light that is collected and, hence, the signal-to-noise ratio in the image. From such a small volume (it may be as small as a few thousand angstroms), a spectral resolution of perhaps 10 Å may be obtained from very low concentrations of impurities, and this spectral information may be collected together with the corresponding HRTEM image. In most applications, however, spectra are recorded from a region of about 1 μm in diameter in combination with diffraction-contrast images, and this ability to correlate CL spectra with high-resolution TEM images provides an important advantage of the CL–STEM technique over conventional scanning electron microscope (SEM) CL work.

In addition to the semiconductor work referred to earlier, there have been several studies of magnesium oxide and diamond by this method. Using a tapered silver tube as a light-collection element fitted to the VG HB5 STEM, Pennycook et al. (1980) obtained CL images of individual dislocations in type IIb diamond in correlation with their transmitted-electron images. In this material, almost all the optical luminescence arises from dislocations; however, not all dislocations are luminescent. Both 60° and screw dislocations were found to be luminescent. By subtracting a spectrum recorded from the defect of interest from one recorded nearby, the workers eliminated the effects of stray luminescence, filament light, and impurity luminescence. In development of this work, Yamamoto et al. (1984) have used a system based on that developed by Roberts (1981) to obtain weak-beam images and spectra from individual dislocations of known type in diamond. The apparatus (shown in Figure 7.12) has been fitted to a Philips EM 400 TEM–STEM instrument. This apparatus allows spectra to be collected over a range of temperatures down to about 20 K.

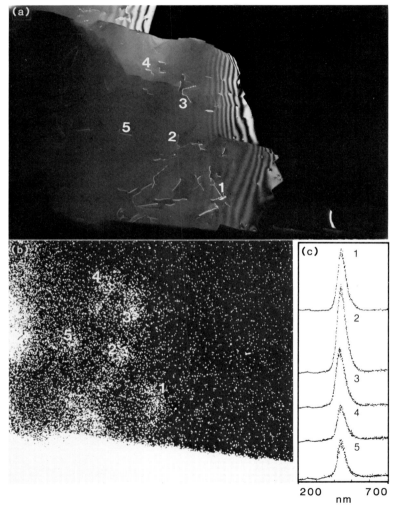

FIGURE 7.13 Correlated TEM and cathodoluminescence images and spectra from diamond. (a) A dark-field image of several dislocations of known type in a thin, type IIb diamond sample. (b) The scanning-monochromatic (425-nm) cathodoluminescence image. This identifies the dislocations responsible for emission. (c) The emission spectra from the individual dislocations numbered in the image in (a). Number 1 is a screw type, 2 is an edge, 3 is a screw, 4 is an edge, and 5 is a screw dislocation.

An example of results from this work are shown in Figure 7.13, where the TEM image of a single-line defect and its associated optical-emission spectrum are shown together with the scanning-monochromatic CL image used to identify the defect. The spectrum was recorded at 89 K. At low temperatures, CL in most materials is found to increase in intensity, and the resulting fine structure observed may be more readily correlated with theoretical models. In this work, in addition, the polarization of the CL measured from individual dislocations of known type and degree of dissociation. The dislocation emission (which occurs at 435 nm) is confirmed to be polarized along the dislocation-line (rather than Burgers-vector) direction and to form a broad band whose width is 0.412 eV. No clear correlation between CL activity and dislocation type or degree of dissociation was found; however, a one-dimensional donor-acceptor model for dislocation-core CL is suggested as consistent with all the experimental findings. A high kink density was found to reduce the CL intensity.

The limited aim of most of the STEM CL work undertaken so far has been to correlate a spectral signature with a particular defect that can be imaged at high resolution by TEM. More detailed quantitative interpretation of CL spectra has yet to be attempted. This interpretation must follow closely along the lines of the theory of photoluminescence. Only at liquid-helium temperatures and for very simple, well-characterized systems would there appear to be any possibility of obtaining quantitative agreement between the intensities of experimental CL spectra and theoretical predictions. The identification of defect levels and phonon lines according to their energy is, however, more straightforward. But since it gives most sensitive spectral signature obtainable in correlation with HRTEM images, the techniques shows considerable promise in many areas of microscopy from mineralogy to catalysis and semiconductor studies. For example, measurements of CL polarization may be related to the crystallographic site of impurities, and the extension of CL detectors into the infrared region has opened up the possibility of work on silicon and other small-band-gap semiconductors. As an example of STEM CL work the infrared, Graham et al. (1986) have recorded spectra in the 0.8 to 1.0 eV range from groups of straight dislocations in silicon, showing the recently discovered "D" lines.

7.7 Microdiffraction

A clear trend has recently emerged for the provision of convergent-beam electron diffraction (CBED) facilities on modern HRTEM instruments. To some extent, the information obtainable from microdiffraction patterns complements that provided by HRTEM, since if CBED patterns and HRTEM images can be obtained from the same region of specimen, these patterns can be used for many purposes, including the following:

1. To determine the crystal periodicity in the beam direction. This

periodicity is given by

$$C_0 = \frac{2}{\lambda U_0^2}, \qquad (7.19)$$

where λ is the electron wavelength and U_0 is the radius (in reciprocal angstroms) of the first-order Laue-zone (FOLZ) ring seen in the microdiffraction pattern. The unit-cell dimensions and angle normal to the beam can also be obtained from the zero-order, Laue-zone pattern.
2. To align the crystal-zone axis with the electron beam. Modern HRTEM instruments provide probe sizes as small as about 30 Å, so the alignment of the region of interest for HRTEM becomes possible if no mechanical alteration to the specimen height is required in changing from the microdiffraction to the HRTEM mode.
3. To determine specimen thickness. In thicker specimens of small unit-cell crystals, the two-beam CBED method of specimen of thickness determination may be applied (Blake et al., 1978).
4. To determine the space group of the crystal in favorable cases (Goodman, 1975; Eades et al., 1983; Steeds and Vincent, 1983). A thicker region of crystal will generally be required for the CBED pattern than for the HRTEM image, since CBED patterns from very thin crystals show little contrast. A possible solution to this problem lies in the use of the Tanaka wide-angle CBED method (Eades, 1984). This method allows convergent-beam patterns to be obtained over an angular range larger than the Bragg angle without overlap of orders. It may therefore be used to reveal contrast at higher angles in CBED patterns from very thin crystals if sufficiently large areas of uniform thickness are available. The ability of this technique to isolate orders without overlap is particularly valuable for the thin specimens of large-unit-cell crystals (with correspondingly small Bragg angles) commonly used for HRTEM.

In general, the HRTEM method is most useful for crystals with a short period in the beam direction and large cell dimensions normal to the beam, since these crystals give the maximum information in this projection. Such crystals are the least favorable for CBED work (since orders are likely to overlap), for which the opposite dimensions are preferable unless a wide-angle technique can be used. The HRTEM method is also preferable for the study of the defect structure of crystals and for the observation of superlattices, which may give rise to very weak reflections in a CBED pattern but to intense points in an HRTEM image. The great power of the CBED method lies in its ability to provide symmetry information on the specimen structure that is independent of the HRTEM-imaging parameters (e.g., astigmatism correction) and, in high-angle patterns, to provide information on the crystal structure in the electron-beam direction from a single-crystal setting. The limited-tilting facilities available on modern HRTEM machines makes this information otherwise difficult to obtain.

230 HIGH-RESOLUTION TRANSMISSION ELECTRON MICROSCOPY

Thus, for example, the component of a dislocation Burgers vector in the beam direction can be determined in favorable cases from CBED patterns (Carpenter and Spence, 1982). The tetragonal distortion associated with phase transitions in metals has been studied by Porter et al. (1983).

Methods for combining the CBED and HRTEM techniques in older instruments have been discussed in the literature (Goodman and Olsen, 1981), and a general review of the electron-microdiffraction technique has recently appeared (Spence and Carpenter, 1986). For recent examples showing the power of this combination of techniques, see Yamamoto and Ishizuka (1983), Fung and Yang (1984) (for the Philips EM400T), and Moodie and Whitfield (1984) (for the JEOL 200CX). Figure 7.14 shows, as an example, the lattice image formed from the forbidden ($\bar{4}22$)/3-type reflections due to an intrinsic stacking fault lying normal to the incident beam in silicon. A microdiffraction pattern from a similar fault is also shown, clearly revealing the forbidden reflections (Alexander et al., 1986).

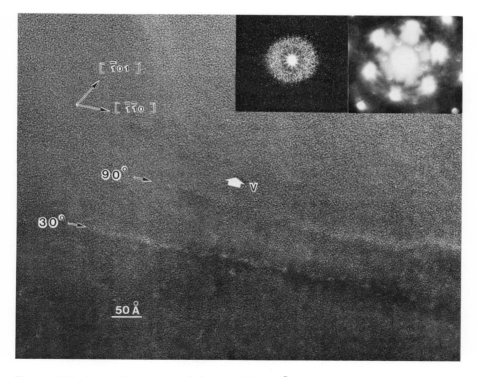

FIGURE 7.14 Lattice image formed from forbidden ($\bar{4}22$)/3 reflections and corresponding microdiffraction pattern (inset) showing these reflections due to the stacking fault, which lies normal to the beam. The arrow indicates the direction of motion of the partial dislocations that bound the fault. The ($\bar{4}22$)/3 fringes have the correct orientation and spacing with respect to bulk fringes and delineate the partial dislocation cores. The ($\bar{4}22$)/3 fringe spacing is 3.3 Å. (From Alexander et al., 1986.)

The probe size used was smaller than the separation of the partial dislocations.

We consider now the microdiffraction patterns formed by using field-emission STEM instruments, since the information these patterns contain is closely related to that of an HRTEM image if an electron probe of near-atomic dimensions is used. The size of the microdiffraction probe depends on the demagnification of the probe-forming lenses, on the electron source size, and on the focus setting, aperture size, and spherical-aberration constant of the probe-forming lens. For field-emission sources, these last three factors are the most important, with the probe-broadening effect of spherical aberration becoming most important for large aperture sizes. In fact, a useful approximation is to consider that the electron source in field-emission instruments is an idealized point emitter. Then the focused probe is diffraction-limited (i.e., it is the image of a point source as formed by an imperfect lens) and so, in the absence of spherical aberration, has a size given very approximately by

$$r_s = \frac{0.61\lambda}{\theta_c}, \qquad (7.20)$$

where θ_c is the semiangle subtended by the STEM objective aperture and shown in Figure 7.15. In the presence of spherical aberration, no simple expression for probe size exists; however, computer calculations have shown (Spence, 1978; Mory, 1985; Mory et al., 1987) that, except at the optimum focus setting, the probe-intensity distribution contains rather extended "tails" and oscillations. In practice, the size of the STEM probe can be roughly estimated from the resolution of a STEM-lattice image or from the formation of a probe image directly on certain machines. On modern machines, this probe size may be as small as 4 Å. Recent calculations by Colliex and coworkers (Mory, 1985) have shown that the optimum focus needed to form the most compact probe is approximately $\Delta f = -0.75 C_s^{1/2} \lambda^{1/2}$.

The assumption that the probe is diffraction-limited corresponds to the requirement that the objective aperture (C2 in Figure 7.15) be coherently filled. This is the case if the coherence width

$$X_a = \frac{\lambda}{2\pi\theta_s} \qquad (7.21)$$

in the plane of the aperture is larger than the aperture. Here, θ_s is the semiangle subtended by the geometrical electron-source image (formed on the specimen) at the aperture. Thus, the requirement for a coherently filled aperture depends on the electron-source size, which is proportional to θ_s. Figure 7.15 shows a ray diagram for coherent CBED. Since P in this diagram is conjugate to the set of points P', we do not expect that CBED patterns from perfect crystals should differ in the two extreme cases where

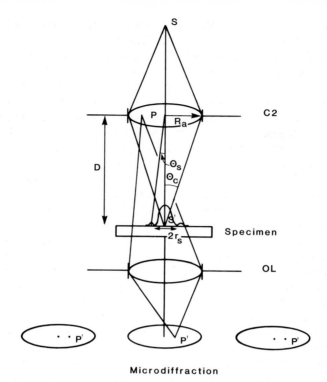

FIGURE 7.15 Ray diagram for coherent microdiffraction. Here, P is conjugate to the set of points P' (one in each diffraction disk); $C2$ is the probe-forming lens aperture, and OL is the objective lens. Also, θ_s determines the coherence X_a in the plane of $C2$. The probe is an aberrated image of the point source s.

$C2$ is either coherently or incoherently filled. However, these points are conjugate only at exact focus, so the through-focus behavior of coherent and incoherent CBED patterns differ greatly.

For a discussion of the close relationship between coherent microdiffraction and lattice imaging in a STEM, refer to section 6.2 and to Figure 6.3. In the coherent case, a method for the computer simulation of microdiffraction patterns has been described by Spence (1978) in which the probe wave function is used as the boundary condition within an artificial superlattice at the entrance surface of the specimen and a multiple Bragg-scattering algorithm used. In a series of papers, experimental microdiffraction patterns from various materials have been compared with the results of these calculations (e.g., see Zhu and Cowley, 1982). For the incoherent case, a method of microdiffraction-pattern simulation for defects has also been described; it is based on the use of the popular diffraction-contrast software under the column approximation and takes advantage of the reciprocity theorem (Carpenter and Spence, 1982).

The similarities between the problems involved in the interpretation of HRTEM images and coherent CBED patterns become clear when we consider the case of CBED patterns formed by using a probe that is smaller than the crystal unit cell. Then, as shown in section 6.2, the CBED orders will overlap. Interference effects will then be seen within the region in which they overlap, and these depend on the focus setting and spherical-aberration constant of the probe-forming lens (Cowley and Spence, 1979). Thus, the problem of determining atom positions by a comparison of computed and experimental coherent microdiffraction patterns involves a set of adjustable parameters (atom coordinates and instrumental parameters) similar to the set involved in HRTEM image matching, for both defects or, in the microdiffraction case, for a particular identified group of atoms within a unit cell. However, the determination of instrumental parameters from coherent microdiffraction patterns with overlapping orders (the "electron Ronchigram") is more straightforward and may be done in real time (Cowley, 1981). These Ronchigrams may also be interpreted as point-projection shadow images of the crystal lattice (at larger defocus) or as electron holograms. Figure 7.16 (from Cowley, 1981) shows such a pattern from $Ti_2Nb_{10}O_{29}$ in which, since the probe size is smaller than the unit-cell size, the symmetry of the pattern is not that of the crystal as a whole but, rather, possesses the local symmetry about the probe (Cowley and Spence, 1979). The patterns are seen to repeat as the probe is moved by a lattice-translation vector (28 Å in length). An analysis of the propagation of energy through a crystal in a diffraction-limited probe under conditions of multiple electron scattering can be found in Marks (1985).

Since the intensity within the region of overlapping orders is sensitive to probe position, mechanical movement of the specimen or probe movement due to electronic instabilities is as important for coherent microdiffraction as

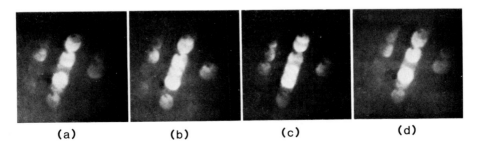

(a) (b) (c) (d)

FIGURE 7.16 Coherent microdiffraction patterns from a thin crystal of $Ti_2Nb_{10}O_{29}$, obtained with a 4-Å-diameter electron probe that is smaller than the lattice spacing of 28 Å. The patterns repeat as the probe is moved from one side of the unit cell to the other in 7.16(a) to 7.16(d). Only the intensity within the region of overlap of the orders varies as the probe is moved. (From Cowley, 1981.)

for HRTEM. However, the loss of contrast in a coherent microdiffraction pattern owing to vibration is a function of the scattering angle and focus setting; and in general, for patterns with overlapping orders (or from defects), the finest detail that can be extracted is expected to be about the same as the detail that would appear in the corresponding STEM lattice image. High-angle scattering is most sensitive to instabilities.

For many problems in materials science, these interferometric effects and their dependence on instrumental parameters that result from the use of the smallest probes constitute an unnecessary complication. A more fruitful approach is then to observe characteristic disturbances in certain nonoverlapping orders diffracted by interrupted or distorted planes of the crystal lattice. Then, in the spirit of $g \cdot b$ analysis in TEM, by noting which of the microdiffraction spots are unaffected, one may be able to determine fault vectors or to classify defect types by using a probe only slightly larger than the lattice period (Zhu and Cowley, 1983). An example of the approach is shown in Figure 7.17. Earlier dynamical calculations and experiments on coherent microdiffraction from the metal-catalysis particles (Cowley and Spence, 1981) had shown that the normal CBED disks frequently appear as

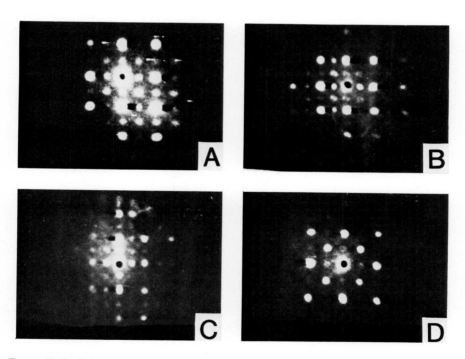

FIGURE 7.17 Subnanometer microdiffraction patterns from antiphase boundaries in Cu_3Au. Particular groups of superlattice reflections are split (appear annular), depending on the type of boundary. (*a*) A "good" boundary. (*b*) A "two good" boundary. (*c*) A "bad 1" boundary. (*d*) A "bad 2" boundary. (From Zhu and Cowley, 1982.)

annular rings, possibly also broken up into segments of arc that may appear simply as small blobs of intensity around the geometrical perimeter of the CBED disk. This effect is only observed when the probe lies near an edge or discontinuity in the lattice and is a coherent interference effect, not due to dynamical scattering in the crystal. The resulting intensity modulation within the CBED disk may make the identification of the geometrical outline of the disk difficult and so complicates the indexing of coherent microdiffraction patterns from small particles. Further complications arise because of additional structure in the disks owing to twinning, which is common in small particles. However, the usefulness of this effect lies in the fact that disks due to diffraction from planes unaffected by a defect will not show this splitting.

The patterns shown in Figure 7.17 were obtained from samples of Cu_3Au containing antiphase-domain boundaries of three possible types, depending on the near-neighbor coordination of atoms in the boundary. The electron-probe size is about 5 Å, smaller than the average domain size. Since the domains represent a discontinuity in the Cu_3Au superlattice, the expected annular splittings are seen in the first-order superlattice reflections only (e.g., those nearest the central beam in Figure 7.16c. By analyzing all the possible boundary arrangements, these authors were able to draw up a classification scheme relating spot splitting to boundary type (Zhu and Cowley, 1983).

This work provides an example of a powerful general approach to a problem that would be insoluble by using the HRTEM technique, since lattice images of Cu_3Au antiphase boundaries have proven extremely difficult to interpret. A similar approach has been applied to stacking faults and twin boundaries in metals, where the determination of fault vectors from spot splittings observed in coherent microdiffraction have also been demonstrated (Zhu and Cowley, 1983).

In all this work, the ability to record the weak-microdiffraction fine structure efficiently by using an axial-parallel-detection system and image intensifier has proved crucial. Many of these effects cannot be observed by using the serial-Grigson-detection scheme. An optical system has also proved useful for on-line dissection of the patterns and to provide a flexible method of varying the shape of the various STEM detectors used (Cowley and Spence, 1979; Cowley, 1985).

7.8 Specimen preparation

This section is intended as a very brief survey of the literature on specimen preparation and may be useful in directing researchers to the many techniques scattered throughout the literature.

The text by Goodhew (1972) covers electrochemical methods and jet thinning at a very practical level. It emphasizes metals, alloys, and some semiconductors and contains a useful review of extraction-replica methods.

These methods may be essential for microdiffraction studies of precipitates. The book by Hirsch et al. (1976) contains an extensive list of techniques indexed according to material. The metallography literature also should not be overlooked (Morris, 1979). An electrothinning apparatus is described in Schoane and Fischione (1966).

For semiconductors, both chemical and ion-beam methods are common. Many chemical methods have been developed for thinning silicon, which are variants of the technique described by Booker and Stickler (1962). For those not wishing to work with hydroflouric acid, alternative procedures are given in Kestel (1982) for thinning silicon, germanium, tantalum, and niobium, using a method that gives more controllable removal of material. Silicon may be conveniently thinned by hand on a grinding wheel if a slice is glued to a thick, optically flat glass disk of known thickness. By measuring the total thickness of the glass block and specimen with a micrometer, one can obtain an estimate of the thickness. With practice, a final thickness of about 30 μm (suitable for ion-beam thinning) is obtained.

A more accurate indication of surface roughness can be obtained by viewing the sample under reflected sodium-lamp illumination in an optical microscope. The optical-transmission properties of silicon allow one to judge thicknesses rather accurately. If the glue is then dissolved by heating in a suitable solvent, small pieces can be retrieved for subsequent thinning by either chemical or ion-beam methods. A very fine, moistened paintbrush is ideal for handling even the most minute, barely visible fragment of silicon that cannot be safely handled by tweezers. Vacuum tweezers are also invaluable for more rugged samples. This general approach can be applied to most brittle materials, such as ceramics and minerals.

Some chemical techniques require a disk sample with a dimpled center; for this purpose an ultrasonic drill for cutting the disks to size for the electron microscope (perhaps 200-μm thick) is ideal. Several grinding devices have recently appeared for performing the dimpling operation (Swann, 1985). In one technique (Alexander et al., 1986), the dimpled silicon disk is held in a teflon holder, which prevents attack by acids around the specimen perimeter; and a drop of thinning solution is placed in the dimple. The thinning is viewed through an optical microscope and stopped at perforation.

For semiconductor devices and interfaces, many groups have perfected techniques for the preparation of cross sections that allow an interface to be viewed with the electron beam in the plane of the interface; a review of these methods can be found in Bravman and Sinclair (1984). For this work, the choice of glues is important; this author's experience has been that superglues such as Eastman 910 are useful for tacking a thinned fragment onto a circular-hole grid (protruding into the central hole), but an additional dab of silver dag is required to retain the specimen under the effects of the electron and ion beams. A variety of epoxies have been tried for glueing layered films together. M-Bond 610 appears to withstand ion-beam thinning and electron-beam effects well. More details and clear diagrams of the cross-sectioning method for interfaces can be found in Fletcher et al. (1980).

For the III–V semiconductors, most labs appear to use a variant of the method of Buiocchi (1967) (bromine and chlorine in methanol). A solution for indium antimonide is given in Holt et al. (1966), which also contains a useful review of chemical methods for many compound semiconductors.

The convenient operation of the modern ion-beam-thinning apparatus has led to a great increase in the popularity of this method of thinning. The unwanted effects of radiation damage and surface roughening and their mechanisms are described in detail by Howitt (1984). The possibility of milling micron-sized regions while viewing the milling process in a type of scanning-ion microscope has recently been demonstrated (Swann, 1985). This precision-ion-milling (PIMS) device is intended to accept electron-microscope holders so that a new way of working becomes possible: The defect or region of interest is sought in the electron microscope, and the sample is transferred to the PIMS device for further thinning of that region alone. A secondary-ion microanalyzer may also be fitted.

Ion milling of compound semiconductors has generally proven difficult, owing to the preferential loss of one species. A solution to this problem has been described by Chew and Cullis (1987), who review the use of other ions to replace the inert argon ions usually used. Some success has also been obtained by using oxygen ions (Liliental, 1985).

For many minerals and ceramics, TEM samples can be made simply by grinding the material in a mortar and pestle under alcohol. Fragments are then picked up on a holey carbon grid. Other ceramics form in smoke; magnesium oxide crystals can be collected by passing a grid through a column of burning magnesium smoke. Humidity controls the size of the crystals and the fraction of platelets formed.

7.9 Real-time image acquisition and videorecording in HRTEM

Since the first real-time videorecordings were made of individual atom movements in STEM (Crewe et al., 1975), many groups around the world have come to realize the importance of real-time recording for the study of atom motion in and on crystals. The spectacular results this technique has provided on surface phase-transitions in semiconductors are described in Chapter 13. Videorecording of HRTEM images in TEM has recently gained popularity for several reasons:

1. The falling cost of TV-rate, digital-image-acquisition systems makes many image-processing techniques available on-line to the TEM user. These techniques include on-line drift correction; on-line, fast-Fourier-transform analysis for diffractogram observation to assist with focusing; astigmation correction; and contrast enhancement.
2. Because signals may be fed back to the microscope, automated focusing, astigmatism correction, and alignment are possible (Saxton et al., 1983).

3. Weak-beam and atomic-resolution "movies" have revealed new aspects of phase transformation and defect interactions in solids, such as the finding that dislocations in semiconductors generally remain dissociated while in motion.
4. Radiation-damage effects can be minimized by examining the first few frames of a recording of a new specimen area before serious damage occurs.
5. For teaching and demonstration purposes, video display is a great asset.

A typical image-intensifier and videorecording system for HRTEM with computer interface is described in Spence et al. (1982) (also see Swann, 1985). In this system, a single-crystal, yttrium-aluminum garnet detector screen has been use to provide higher spatial resolution and speed than that possible with powder-phosphor screens. A more advanced system designed for the quantitative parallel detection of electron diffraction patterns covering a wide dynamic range can be found in Gjønnes, Gjøonnes, Zuo, and Spence (1988). Here a similar detector is coupled by fiber optics to a charged couple device array, held at liquid nitrogen temperature. A general review of the principles of computer interfacing to electron microscopes can be found in Rez and Williams (1982). Using similar systems, recent workers have revealed the growth of small crystals, which can be followed in real time, row by row, from the atomic-resolution images (Wallenberg et al., 1985). Iijima and Ichihashi (1985) have also observed "atomic clouds" outside gold-atom clusters. Surface-profile imaging at atomic resolution has also revealed fascinating new dynamic effects (Sinclair et al., 1981; Smith and Marks, 1985).

Undoubtedly, the direct digital acquisition of images and full computer control of electron microscopes will become the standard practice in the near future. This achievement will be an important step toward truly quantitative work. The problems of the large dynamic range of diffraction patterns, of finding a satisfactory hardcopy format, and of image-detector design at high voltage remain, however. The study of time-dependent phenomena at atomic resolution will probably be one of the major growth areas of HRTEM, but the observation of reproducible effects will require much more highly controlled vacuum conditions than those present on current machines.

REFERENCES

Ahn, C., and Rez, P. (1985). Inner shell edge profiles in ELS. *Ultramicroscopy* **17,** 105.
Alexander, H., Spence, J. C. H., Shindo, D., Gottschalk, P., and Long, N. (1986). Forbidden reflection lattice imaging for the determination of kink densities. *Philos. Mag.* **53,** 627.
Batson, P. E. (1986). High-resolution electron spectrometer for 1-nm spatial analysis. *Rev. Sci. Instrum.* **57,** 43.
———, and Craven, A. J. (1979). Extended fine-structure on the carbon core-ionization edge obtained from nanometer-sized areas with electron-energy loss spectroscopy. *Phys. Rev. Lett.* **42,** 893.

Bianconi, A., Incoccia, L., and Stipcich, S. (1983). *EXAFS and near edge structure.* Springer Tracts in Chemical Physics, Vol. 27. Springer Verlag, New York.

Blake, R. G., Jostsons, A., Kelly, P. M., and Napier, J. G. (1978). The determination of extinction distances by STEM. *Philos. Mag., Ser. A* **37,** 1.

Booker, G. R., and Stickler, R. (1962). Method of preparing Si and Ge specimens for examination by transmission electron microscopy. *Br. J. Appl. Phys.* **13,** 446.

Bravman, J. C., and Sinclair, R. (1984). The preparation of cross-section specimens for transmission electron microscopy. *J. Electron Microsc. Tech.* **1,** 53.

Brown, M. (1984). Personal communication. See also W. Schockley (1953). *Electrons and holes in semiconductors.* Van Nostrand, New York.

———, M. (1985). Personal communication.

———, Peierls, R. E., and Stern, E. A. (1977). White lines in X-ray absorption. *Phys. Rev. B* **15,** 738.

Buiocchi, C. J. (1967). Preparation of GaAs for TEM. *J. Appl. Phys.* **38,** 1980.

Cardona, M., and Ley, L., eds. (1978). *Photoemission in solids, I and II.* Springer Topics in Applied Physics, Vols. 26 and 27. Springer-Verlag, Berlin.

Carpenter, R., and Spence, J. C. H. (1982). Three-dimensional strain information from CBED patterns. *Acta Crystallogr., Sect. A* **38,** 55.

Chan, H. M., Harmer, M. P., Lal, M., and Smyth, D. M. (1984). Calcium site occupancy in barium titentate. In *Materials research society symposium proceedings,* Vol. 31, 345. Elsevier, New York.

Chew, N. G., and Cullis, A. G. (1987). The preparation of transmission electron microscope specimens from compound semiconductors by ion milling. *Ultramicroscopy* **23,** 175.

Chen, C. H., and Silcox, J. (1977). "Direct nonvertical interband transitions at large wave vectors in aluminum. *Phys. Rev. B* **16,** 4246.

Citrin, P. H., Wertheim, G. K., and Schluter, M. (1979). One-electron and many-body effects in X-ray absorption and emission edges of Li, Na, Mg and Al metals. *Phys. Rev. B* **20,** 3067.

Colliex, C. (1984). Electron energy loss spectroscopy. In *Advances in Optical and Electron Microscopy,* Vol. 9, ed. R. Barer and V. E. Cosslet, 65. Academic Press, New York.

———, Manoubi, T., Gasgnier, M., and Brown, L. M. (1984). Near edge fine structures on electron energy loss spectroscopy core loss edges. In *Scanning electron microscopy—1984,* ed. O. Johar, 489. A. M. F. O'Hare, Chicago.

Cowley, J. M. (1981). Coherent interference effects in STEM and CBED. *Ultramicroscopy* **7,** 19.

——— (1985). A new detector system for the HB5 STEM. In *Proceedings of the Electron Microscopy Society of America, 1985,* ed. G. Bailey, 134. San Francisco Press, San Francisco.

———, and Spence, J. C. H. (1979). Innovative imaging and microdiffraction in STEM. *Ultramicroscopy* **3,** 433.

———, and Spence, J. C. H. (1981). Convergent beam electron microdiffraction from small crystals. *Ultramicroscopy* **6,** 359.

Crewe, A. V., Langmore, J. P., and Isaacson, M. (1975). Scanning transmission electron microscopy. In *Physical techniques in electron microscopy,* ed. B. Siegel and D. Beaman, 247. Wiley, New York.

Disko, M. M., Krivanek, O. L., and Rez, P. (1982). Orientation-dependent extended fine structure in electron-energy-loss spectra. *Phys. Rev. B* **25,** 4252.

———, Spence, J. C. H., Sankey, O., and Saldin, D. (1986). The electron energy loss structure of Be_2C. *Phys. Rev. B* **33,** 5642.

Dodd, C. G., and Glen, G. L. (1968). Chemical-bonding studies of silicates and oxides by X-ray K-emission spectroscopy. *J. Appl. Phys.* **39,** 5377.

Durham, P. (1985). Theory of XANES. In *"EXAFS" and "XANES" spectroscopy,* ed. A. Prince and P. Konigsburg. Wiley, New York.

Eades, J. A. (1984). Zone-axis diffraction patterns by the Tanaka method. *J. Electron Microsc. Tech.* **1,** 279.

———, Shannon, M. D., and Buxton, B. F. (1983). Crystal symmetry from electron diffraction. In *Scanning electron microscopy—1983,* Vol. III, ed. O. Johari, 1051. A. M. F. O'Hare, Chicago.

Economou, E. N. (1983). *Green's functions in quantum physics.* Springer-Verlag, New York.

Egerton, R. F. (1986). *Transmission electron energy loss spectroscopy.* Plenum, New York.

———, and Whelan, M. J. (1974). Electron-energy loss spectra of diamond, graphite and amorphous carbon. *J. Electron Spectrosc. Related Phenom.* **3,** 232.

Fletcher, J., Titchmarsh, J. M., and Booker, G. R. (1980). Experimental procedures for preparing cross-section TEM specimens of semiconductors. In *Institute of physics conference series,* No. 52, 153. Institute of Physics, Bristol, England.

Fung, K. K., and Yang, C. Y. (1984). Combined CBED and HREM in a Philips EM400T analytical electron microscopy. *Ultramicroscopy* **13,** 333.

Gjønnes, J., Gjønnes, K., Zuo, J. M., and Spence, J. C. H. (1988). Structure factor refinement by electron diffraction. *Acta Crystallogr.* (forthcoming).

Goodhew, P. F. (1972). *Specimen preparation in materials science.* North-Holland/Elsevier, New York.

Goodman, P. (1975). Practical method of 3-dimensional space-group analysis using convergent-beam electron-diffraction. *Acta Crystallogr., Sect. A* **31,** 804.

———, and Olsen, A. (1981). Combining CBED with HREM. *Ultramicroscopy* **6,** 101.

Graham, R. J., Spence, J. C. H., and Alexander, A. (1987). Infrared cathodoluminescence studies from dislocations in silicon in TEM and ELS/CL coincidence measurements of lifetimes in semiconductors. In *Characterisation of defects in materials,* **82,** p. 235, ed. A. Siegel, R. Sinclair, and J. Weertman. North-Holland, New York.

Grunes, L. A., Leapman, R. D., Wilker, C. N., Hoffmann, R., and Kunz, A. B. (1982). Oxygen K near-edge fine structure: An electron-energy-loss investigation with comparisons to new theory for selected 3d transition-metal oxides. *Phys. Rev. B* **25,** 7157.

Henrich, V. E., and Dresselhaus, G. (1975). Evidence for localized excitations in MgO from electron energy-loss spectroscopy. *Solid State Commun.* **16,** 1117.

Hirsch, P. B., Howie, A., Nicholson, R. B., Pashley, D. W., and Whelan, M. J. (1976). *Electron microscopy of thin crystals.* 2nd ed. Kreiger, Huntington, N.Y.

Hjalmarson, H. P., Buttner, H., and Dow, J. D. (1981). Theory of core excitons. *Phys. Rev. B* **24,** 6010.

Holt, D. B., Porter, R., and Unvala, B. A. (1966). Thinning semiconducting compounds for TEM. *J. Sci. Instrum.* **43,** 371.

Howitt, D. G. (1984). Ion milling of specimens for TEM. *J. Electron Microsc. Tech.* **1,** 405.

Iijima, I., and Ichihashi, T. (1985). Motion of surface atoms on small gold particles. *Jpn. J. Appl. Phys.* **24,** L125.

Isaacson, M. (1979). Electron-beam-induced damage of organic solids—implications for analytical electron microscopy. *Ultramicroscopy* **4,** 193.

Keski-Rahkonen, O., and Krause, M. O. (1974). Total and partial atomic level widths. *At. Nucl. Data* **14,** 139.

Kestel, B. J. (1982). A jet-polishing solution for Si, Ge, Ta, Nb and W–Rh. *Ultramicroscopy* **9,** 379.

Knapp, G. S., Veal, B. W., Pan, H. K., and Klippert, T. (1982). XANES study of 3d oxides: Dependence on crystal structure. *Solid State Commun.* **44,** 1343.

Krishnan, K. (1988). Atom location by chanelling enhanced microanalysis, in "Limits of sub-micron microanalysis," (special issue, ed. J. Hren). *Ultramicroscopy* **24,** 125.

———, and Thomas, G. (1984). A generalization of ALCHEMI. *J. Microsc.* **136,** 97.

———, Rez, P., and Thomas, G. (1985). Crystallographic site occupancy in thin film oxides. *Acta Crystallogr., Sect. B* **41,** 396.

Leapman, R. D., and Cosslett, V. E. (1976). Extended fine-structure above X-ray edge in electron-energy loss spectra. *J. Phys. D* **9,** L29.

———, Fejes, P. L., and Silcox, J. (1983). Orientation dependence of core edges from anisotropic materials determined by inelastic scattering of fast electrons. *Phys. Rev. B* **28,** 2361.

———, Grunes, L. A., and Fejes, P. L. (1982). Study of the L_{23} edges in the 3d transition metals and their oxides by electron-energy-loss spectroscopy with comparisons to theory. *Phys. Rev. B* **26**, 614.
Lee, P. A., Citrin, P., Eisenberger, P., and Kincaid, B. M. (1981). Extended X-ray absorption fine structure—its strengths and limitations as a structural tool. *Rev. Mod. Phys.* **53**, 769.
Lehmpfuhl, G., and Taftø, J. (1980). Reciprocity in EELS. In *Proceedings of the 7th European conference on electron microscopy* Vol. 3, ed. P. Brederoo and V. Cosslet, 62. Electron Microscopy Foundation, Leiden.
Liliental, Z. (1985). Personal communication.
Lindner, T., Sauer, H., Engel, W., andbe, K. (1985). Near edge structure in EELS of MgO. *Phys. Rev. B* **33**, 22.
Marks, L. D. (1985). STEM probe spreading. *Mat. Res. Soc. Symp. Proc.*, Vol. 41, 247.
Mele, E. J., and Ritsko, J. J. (1979). Fermi-level lowering and the core exciton spectrum of intercalated graphite. *Phys. Rev. Lett.* **43**, 68.
Moodie, A. F., and Whitfield, H. J. (1984). Combined CBED and HREM in the electron microscope. *Ultramicroscopy* **13**, 265.
Morris, F. W. (1979). The preparation of difficult samples for transmission electron microscopy by ionbeam thinning. *Practical metallography* **16**, 312.
Mory, C. (1985). Ph.D. diss., University of Paris.
———, Colliex, C., and Cowley, J. M. (1987). Optimum defocus for STEM imaging and microanalysis. *Ultramicroscopy* **21**, 171.
Muller, J. E., Jepsen, O., and Wilkins, J. W. (1982). X-ray absorption spectra: K-edges of 3d transition metals, L-edges of 3d and 4d metals, and M-edges of palladium. *Solid State Commun.* **42**, 365.
———, Jepsen, O., Andersen, O. K., and Wilkins, J. W. (1978). Systematic structure in the K-edge photoabsorption spectra of the 4d transition metals: Theory. *Phys. Rev. Lett.* **40**, 720.
Natoli, C. (1983). Near edge absorption structure. In *EXAFS and near edge structure*. Springer Tracts in Chemical Physics, Vol. 27, ed. A. Bianconi, L. Incoccia, and S. Stipcich, 43. Springer-Verlag, New York.
Neuman, K. D. and Spence, J. C. H. (1987). The electron energy loss near edge structure of MgO. In *Proceedings of the electron microscopy society of America*, 1987, ed. G. W. Bailey, p. 126. San Francisco Press, San Francisco, U.S.A.
Otten, M., and Buseck, P. R. (1987). The determination of site occupancies in garnet by planar and axial ALCHEMI. *Ultramicroscopy* **23**, 151.
Pantelides, S. T. (1985). Electronic excitation energies and the soft–X-ray absorption spectra of alkali halides. *Phys. Rev. B* **11**, 2391.
Pennycook, S. J. (1981). Investigation of the electronic effects of dislocations by STEM. *Ultramicroscopy* **7**, 99.
———, Brown, L. M., and Craven, A. J. (1980). Observation of cathodoluminescence at single dislocation by STEM. *Philos. Mag., Ser., A*, **41**, 589.
——— (1987). Impurity atom location using axial electron channeling. In *Scanning electron microscopy—1987*, ed. O. Johari, A. M. F. O'Hare, Chicago, IL. (forthcoming).
Petroff, P., Logan, R. A., and Savage, A. (1980). Nonradiative recombination at dislocation in III–V. *Phys. Rev. Lett.* **44**, 287.
Porter, A. J., Ecob, R., and Ricks, R. (1983). Coherency strain fields magnitude and symmetry. *J. Microsc.* **129**, 327.
Rehr, J. J., and Stern, E. A. (1983). Role of inelastic effects in EXAFS. In *EXAFS and near edge structure*. Springer Tracts in Chemical Physics, Vol. 27, ed. A. Bianconi, L. Incoccia, and S. Stipcich, 22. Springer-Verlag, New York.
Rez, P., and Leapman, R. (1981). Core loss shape and cross section calculations. In *Analytical electron microscopy—1981*, ed. R. Geiss, 81. San Francisco Press, San Francisco.
———, and Williams, D. B. (1982). Electron microscope/computer interactions: A general introduction. *Ultramicroscopy* **8**, 247.

Ritsko, J. J., Shnatterly, S. E., and Gibbons, P. C. (1974). Simple calculation of LII, III absorption-spectra of NA, AL and Si. *Phys. Rev. Lett.* **32**, 671.

Roberts, S. H. (1981). Cathodoluminescence in STEM. In *Microscopy of semiconductor materials 1981*, ed. A. G. Cullis and D. C. Joy, 377. Institute of Physics, London.

Robertson, J. (1983a). Electronic structure and X-ray near-edge core spectra of Cu_2O. *Phys. Rev. B* **28**, 3378.

────── (1983b). Electronic structure and core exciton of hexagonal boron nitride. *Phys. Rev. B* **28**, 3378.

Rossouw, C. J., and Maslen, V. W. (1987). Localisation and Alchemi for zone axis orientations. *Ultramicroscopy* **21**, 277.

Saxton, O., Smith, D., and Erasmus, S. K. (1983). Procedures for focussing, stigmating and alignment in HREM. *J. Microsc.* **130**, 187.

Schoane, R. D., and Fischione, E. A. (1966). Automatic unit for thinning TEM specimens. *Rev. Sci. Instrum.* **37**, 1351.

Self, P. G., and Buseck, P. R. (1983). Low-energy limit to channeling effects in the inelastic scattering of fast electrons. *Philos. Mag., Ser. A* **48**, L21.

Sinclair, R., Yamashita, T., and Ponce, F. A. (1981). Atomic motion on the surface of a cadmium telluride single crystal. *Nature* **290**, 386.

Smith, D., and Marks, L. D. (1985). Direct atomic imaging of solid surfaces. III. Small particles and extended Au surfaces. *Ultramicroscopy* **16**, 101.

Spence, J. C. H. (1978). Approximations for the calculation of CBED patterns. *Acta Crystallogr., Sect. A* **34**, 112.

────── (1985). The structural sensitivity of ELNES. *Ultramicroscopy* **18**, 165.

──────, and Carpenter, R. (1986). Electron microdiffraction. In *Principles of analytical electron microscopy*, 2nd ed., ed. D. Joy, A. Romig, and H. Goldstein, Chap. 3. Plenum, New York.

──────, and Taftø, J. (1983). ALCHEMI: A new technique for locating atoms in small crystals. *J. Microsc.* **130**, 147.

──────, Krivanek, O. L., Taftø, J., Disko, M. (1981). The crystallographic information in electron energy loss spectra. In *Electron microscopy and analysis*, 1981, Vol. 61, ed. M. J. Goringe, 253. Institute of Physics, London.

──────, Graham, R. J., and Shindo, D. (1986). Cold ALCHEMI. Impurity atom site location and the temperature dependence of dechanneling. In *Materials research society symposium proceedings*, Vol. 62, 153. Elsevier, New York.

──────, Disko, M., Higgs, A., Wheatley, J. and Hashimoto, H. (1982). A digital on-line diffractometer and image processor for HREM. In *Electro-microscopy—1982*, Vol. 1, ed. The Congress Organizing Committee, 5. Deutsche Gesellschaft für Elektronenmikroscopic, Frankfurt.

Srivastara, U. C., and Nigam, H. L. (1972–1973). X-ray absorption edge spectrometry (XAES) as applied to coordination chemistry. *Coord. Chem. Rev.* **9**, 275.

Steeds, J., and Vincent, R. (1983). Use of high-symmetry zone axis in electron diffraction in determining crystal point and space groups. *J. Appl. Crystallogr.* **16**, 317.

Stern, E. A. (1982). Comparison between electrons and X-rays for structure determination. *Optik* **61**, 45.

──────, and Rehr, J. J. (1983). Many-body aspects of the near-edge structure in X-ray adsorption. *Phys. Rev. B* **27**, 3351.

Stohr, J., Sette, R., and Johnson, A. L. (1984). Bond length determination with a ruler. *Phys. Rev. Lett.* **53**, 1684.

Swann, P. (1985). Personal communication (Gatan Corporation, Pleasanton, Calif.).

Taftø, J. (1979). Chanelling effects in electron-induced X-ray emission from diatomic crystals. *Z. Naturforsch.* **34a**, 452.

────── (1982). The cation distribution in a $(Cr, Fe, Al, Mg)_3O_4$ spinel as revealed from the channeling effect in electron-induced X-ray emission. *J. Appl. Crystallogr.* **15**, 378.

────── (1984). Absorption edge fine structure study with subunit cell spatial resolution using electron channeling. *Nucl. Instrum. Methods* **B2**, 733.

———, and Buseck, P. (1982). Quantitative study of Al–Si ordering in an orthoclase feldspar. *Am. Min.* **68,** 944.

———, and Krivanek, O. L. (1982). Site-specific valence determination by electron energy-loss spectroscopy. *Phys. Rev. Lett.* **48,** 560.

———, and Lehmfuhl, G. (1982). Direction dependence in electron energy loss spectroscopy from single crystals. *Ultramicroscopy* **7,** 287.

———, and Liliental, Z. (1982). Studies of the cation distribution in $ZnCr_xFe_{2-x}O_4$ spinels. *J. Appl. Crystallogr.* **15,** 260.

———, and Spence, J. (1982). Crystal site location of iron and trace elements in an Mg–Fe-olivine using a new crystllographic technique. *Science* **218,** 49.

———, and Spence, J. (1983). Crystal site location of dopants in semiconductors. *J. Appl. Phys.* **54,** 5014.

———, and Zhu, J. (1982). Electron energy loss near edge structure (ELNES), a potential technique in the studies of local atomic arrangements. *Ultramicroscopy* **9,** 349.

———, Clarke, D., and Spence, J. (1983). Determination of crystal site occupancy in SYNROC-D. In *Materials research society symposium proceedings,* Vol. 15, 9. Elsevier, New York.

Teo, B. K., and Joy, D. C., eds. (1981). *EXAFS spectroscopy, techniques and applications.* Plenum, New York.

Waddington, W. G., Rez, P., Grant, I. P., and Humphreys, C. J. (1986). White lines in the $L_{2,3}$ electron energy loss spectra of 3d transition metals. *Phys. Rev. B* **34,** 1467.

Wallenberg, L. R., Bovin, J., and Smith, D. J. (1985) *Naturwissenschaften* **72,** 539.

Yamamoto, N., and Ishizuka, K. (1983). Analysis of the incommensurate structure of $Sr_2Nb_2O_7$ by HREM and CBED. *Acta. Crystallogr., Sect. B* **39,** 210.

———, Spence, J. C. H., and Fathy, D. (1984). Cathodoluminescence and polarization studies from individual dislocations in diamond. *Philos. Mag., Ser. B* **49,** 609.

Zhu, J., and Cowley, J. M. (1982). Microdiffraction from antiphase domain boundaries in Cu_3Au. *Acta Crystallogr., Sect. A* **38,** 718.

———, and Cowley, J. M. (1983). Microdiffraction from stacking faults and twin boundaries in f.c.c. crystals. *J. Appl. Crystallogr.* **16,** 171.

8
CALCULATION OF DIFFRACTION PATTERNS AND IMAGES FOR FAST ELECTRONS
PETER G. SELF AND MICHAEL A. O'KEEFE

8.1 Introduction

An unfortunate aspect of the electron-diffraction process is that it is irreversible. That is, given a series of images or a diffraction pattern from a specimen of unknown structure, it is not possible to derive a unique structure for that specimen except under very special circumstances, as for instance when the weak-phase-object (WPO) approximation can be applied. The WPO and similar approximations apply only to the thinnest of specimens (Chapter 1, section 1.3.2). Given the difficulty of preparing large areas of very thin specimen, rarely can approximations such as the WPO approximation be constructively used. Furthermore, images from thin specimens tend to be dominated by extremes in potential such as columns of heavy atoms or large tunnels in the structure. Correspondingly, images from thin crystals are insensitive to small changes in atomic positions and to the replacement of atoms in the structure by atoms of similar atomic number.

Given the irreversible nature of electron diffraction and imaging, structure determination by electron microscopy involves modeling a structure, computing images and diffraction patterns for the model, and attempting to match the computed and experimental results (for example, O'Keefe et al., 1978). If there is a mismatch between the computed and experimental results, then the structural model must be refined and the calculations repeated until a satisfactory match is obtained (O'Keefe and Buseck, 1979). The process is simplified slightly when there are only a few possible models for the structure; but even then, one must be able to perform the calculations quickly and with high accuracy.

The electron optical system used in calculations to approximate the transmission electron microscope is shown in Figure 8.1. The wave incident on the specimen is represented by ψ_i, which characterizes the electron optics of the electron gun and condenser lenses. Postspecimen lenses are represented by a single lens. Because of the small scattering angles in electron diffraction, the characteristics of this single lens are dominated by the behavior of the microscope objective lens, so for most purposes, the single lens is referred to as the objective lens. Figure 8.1 is essentially the same as Figure 1.4, but Figure 8.1 has been drawn so as to purposely accentuate the microscope-alignment possibilities that must be allowed for in any calculation. That is, the specimen normal, the incident-beam direction, and the objective-lens optic axis are not necessarily colinear.

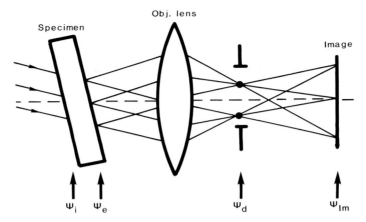

FIGURE 8.1 Electron-optical system used in the calculation of diffracted-electron amplitudes and phases.

For high-energy electrons, it is possible to neglect backscattering and thus, the system shown in Figure 8.1 may be treated sequentially. The computation of diffracted-electron amplitudes and phases is a step-by-step calculation of ψ_i, the incident wave, ψ_e, the exit-surface wave function, ψ_d, the diffraction pattern, and, finally, ψ_{Im} in the image plane. The form of ψ_i can be readily defined from a knowledge of the prespecimen electron optics. Also, the relationship between ψ_e, ψ_d, and ψ_{Im} is well known and, as discussed in previous chapters, simply involves the use of Fourier transforms and appropriate phase factors. The most time-consuming part of any calculation is computing ψ_e for a given ψ_i. There are several methods for performing this section of the calculation, such as the Bloch-wave formulation, the Howie-Whelan equations, the scattering-matrix method, and the multislice method (see Chapter 3 and Self et al., 1983).

Of these methods, the two that have found the most use in calculations are the Bloch-wave formulation (Bethe, 1928; Hirsch et al., 1977; Cowley, 1981) and the multislice method (Cowley and Moodie, 1957, 1958, 1959a, 1959b; Goodman and Moodie, 1974; Cowley, 1981). The former method provides an intuitive feel for the nature of electron diffraction in that, once the Bloch waves have been determined, the wave function of the specimen may be readily calculated for all crystal thicknesses. Furthermore, the wave function in the specimen can usually be described by a few dominant Bloch waves, making qualitative interpretations of diffraction systems straightforward. Thus, the Bloch-wave formulation is very useful in the understanding of diffraction symmetries, beam absorption, electron-current distributions, and diffraction from extended defects. As the Bloch-wave formulation restricts diffraction to only the beams included in a given calculation, the formulation remains self-consistent irrespective of the number of beams in the calculation. Hence, the results of Bloch-wave calculations that use a small number of beams will often be adequate to match the general features of electron diffraction patterns and images. Readers interested in the Bloch-wave formulation are referred to the excellent article on the subject by Metherell (1975).

The multislice method provides very little intuitive sense of the diffraction process in a particular system. With the multislice method, predicting the wave function at a given specimen thickness from the wave function at another thickness is difficult without the use of a computer. Moreover, unlike the Bloch-wave formulation, the multislice method is not self-consistent and requires that scattering to beams not included in the calculation be insignificant. Thus, the multislice method cannot be used to predict electron-diffraction features when only a few beams are included in the calculation.

The major advantage of the multislice method for the accurate calculation of diffracted amplitudes and phases is that, of all the computational techniques in electron diffraction, the multislice method requires the least computer memory and the least computer time to calculate the exit-surface wave function for a given number of diffracted beams. Hence, most calculations for structure determination by transmission electron microscopy use the multislice method. When discussing the calculation of the electron wave function at the exit surface of a specimen, this chapter deals exclusively with the multislice formulation. Of course, other sections of this chapter (in particular, Section 8.4) are generally applicable and are independent of the formulation used to calculate the exit-surface wave function. The aim of this chapter is to provide a reader who has an elementary knowledge of crystallography and computer programming the information necessary to be able to set up and run multislice calculations for various systems. The reader should be aware that, throughout any formulation of electron-diffraction theory, he or she must maintain consistency in the definition of phase relationships for electron wave functions. For a discussion on phase conventions in electron diffraction, see the article by Saxton et al. (1983).

8.2 Calculation of diffracted amplitudes and phases using multislice

The basis of the multislice method (Chapter 4, section 4.3) is to divide the specimen into a number of thin slices perpendicular to the incident-beam direction. In each slice, the effects of Fresnel diffraction (propagation) in the slice and the effects of the specimen potential in the slice (transmission) are treated separately. Following Huygens's principle (see Cowley, 1981, section 2.1.3), the real-space wave function after the nth slice, $\psi_n(xy)$, is given by

$$\psi_n(xy) = [\psi_{n-1}(xy) \ast p_{n-1}(xy)] \cdot q_n(xy), \qquad (8.1)$$

where $q_n(xy)$ is the transmission function of the nth slice and $p_{n-1}(xy)$ is the free-space propagator for the distance between the $(n-1)$th and nth slices. The symbol ✱ represents convolution, and xy represents the real-space coordinates in planes perpendicular to the incident-beam direction. This formulation is exact in the limit that the number of slices becomes large as the slice thicknesses tends to zero. For the purposes of calculation, one may use slice thicknesses such that the difference between the exact solution and the calculated solution is negligible.

For initiation of the iteration of (8.1), the zeroth-slice wave function $\psi_0(xy)$ is set to the incident wave function ψ_i. Equation (8.1) can be used to model any electron-diffraction system provided a mathematical representation suitable for numerical computation can be specified for the system. The simplest system to represent is a plane wave incident on a crystalline material. Although this system applies to conventional transmission electron microscopy (CTEM) and high-resolution transmission electron microscopy (HRTEM) particularly, it also is the basis of computations for scanning TEM (STEM) and convergent-beam electron diffraction (CBED) as well as computations for defect structures and even amorphous materials. For periodic objects, such as crystalline materials, the reciprocal-space formulation of (8.1) provides a more convenient form for numerical evaluation. The reason is that, for periodic objects, the reciprocal-space transmission function exists only at the discrete points of the reciprocal lattice. That is, diffraction is to the reciprocal-lattice points only. The reciprocal-space transmission function is simply the Fourier transform of the corresponding real-space transmission function. Similarly, the form of the reciprocal-space multislice equation is obtained by Fourier-transforming (8.1) to give

$$\Psi_n(uv) = [\Psi_{n-1}(uv) \cdot P_{n-1}(uv)] \ast Q_n(uv), \tag{8.2}$$

where uv represents the reciprocal-space coordinates corresponding to xy. For nonperiodic objects, such as defect structures or amorphous materials, one may still use a discrete-valued, reciprocal-space formulation by applying the method of periodic continuation.

When the incident wave function is a plane wave, $\Psi_0(uv)$ is a delta function defining, in reciprocal space, the direction of the incident beam. Thus, $\Psi_1(uv) = Q_1(uv)$, and consequently, one need calculate $P_n(uv)$ at the reciprocal-lattice points only. Non-plane-wave illumination can be modeled by considering the illumination to be a sum of plane waves. The exact form of this sum will depend on whether the incident non-plane-wave illumination is coherent, partially coherent, or incoherent.

A second advantage of calculating in reciprocal space is that, because elastic scattering falls off with scattering angle, one need only consider a limited number of Fourier coefficients in the calculation. The number of Fourier coefficients considered defines the number of (diffracted) beams in the calculation. As implied previously, if the multislice calculation is to be accurate, the scattering to beams not included in the calculation must be insignificant. If this condition does not hold, more Fourier coefficients must be included when one is evaluating (8.2).

In the following sections, the setting up of the transmission functions and the propagation functions and the method for performing the multislice iteration for crystalline materials with plane-wave illumination will be discussed in detail. For convenience of discussion, a^* and b^* will be used to describe the lengths of the basis vectors (not necessarily orthogonal) of the reciprocal-lattice net in the plane perpendicular to the incident-beam direction; c^* will be used to describe the length of the basis vector parallel to be beam. The notation a, b, and c will be used for the lengths of the real-space lattice vectors corresponding to the reciprocal-space lengths a^*, b^*, and c^*, respectively. The indices hkl will be used to describe the coordinates of reciprocal-lattice vectors. These basis vectors and indices do not necessarily correspond to the unit cell and Miller indices of the specimen. Furthermore, although the multislice method will be described in two-dimensional form with diffraction occurring to a plane in reciprocal space, the arguments used apply equally well to one-dimensional (systematic-row) diffraction for which one of the coordinates x or y can be neglected.

8.2.1 The transmission function

The transmission function $q_n(xy)$ used in the multislice formulation must model, for an electron wave function, the amplitude and phase changes that are induced by passage through the electrostatic potential of each slice of the specimen. A suitable form for the transmission function for multislice calculations is a phase grating such that

$$q_n(xy) = \exp[-i\sigma\phi_n(xy)\,\Delta z_n], \qquad (8.3)$$

where Δz_n is the slice thickness of the nth slice and $\phi_n(xy)$ is the nth-slice projected potential per unit length for a projection direction parallel to the incident-beam direction. The interaction constant for electrons for an accelerating voltage E is defined as

$$\sigma = \frac{2\pi m_0 e \lambda (1 + eE/m_0 c^2)}{h^2}, \qquad (8.4)$$

where m_0 is the electron rest mass, e is the electron charge, h is Planck's constant, c is the speed of light, and λ is the free-space (relativistically corrected) electron wavelength for electrons of energy eE. The projected potential is calculated by integrating the specimen potential over the distance Δz_n. The derivation of the phase grating in equation (8.3) is given for relativistic electrons by Fujiwara (1961). (See also Chapters 1 and 2 for an outline of the derivation and uses of the phase grating.)

CALCULATION OF DIFFRACTION PATTERNS AND IMAGES   249

The iteration in (8.2) is a general form and does not require slices of equal thickness; however, a considerable computational advantage results if the slices can be defined such that $Q_n(uv)$ and $P_n(uv)$ are unchanged from one slice to the next or if slice thickness are chosen so that one has a minimum number of different propagation and transmission functions. For most crystals, it is possible to divide the crystal into slices of equal thickness with equal potential in each slice. The average projected potential per unit length is found by integrating over the crystal-repeat distance parallel to the beam direction; that is,

$$\phi(xy) = \frac{1}{c}\int_0^c \sum_{hkl} \Phi(hkl) \exp\left[-2\pi i\left(\frac{hx}{a} + \frac{ky}{b} + \frac{lz}{c}\right)\right] dz$$

$$= \sum_{hk} \Phi(hk0) \exp\left[-2\pi i\left(\frac{hx}{a} + \frac{ky}{b}\right)\right], \qquad (8.5)$$

where $\Phi(hkl)$ are the Fourier coefficients of the potential. Of course, the slice thickness must be small enough so that the phase-grating approximation gives an accurate representation of diffracted amplitudes and phases. For most cases, a slice thickness of 1 to 3 Å is sufficient. In many cases, the slice thickness will be less than c; and in these cases, a convenient but not essential technique is to define the slice thickness as an integer submultiple of c. Although each slice is assigned the average projected potential, the use of (8.5) for slice thicknesses less than c remains valid as long as there is no significant diffraction to beams other than the $hk0$-beams (i.e., when there is no diffraction to the first or higher-order Laue zones). Hence, the use of (8.5) is applicable to crystals with a repeat distance parallel to the beam of less than 5 to 10 Å. The situations where the crystal-repeat distance is longer than this distance or where there is strong coupling to higher-order Laue zones are discussed later in this chapter.

The Fourier coefficients of the potential can be determined from the electron-structure factors for the crystal, $F(hkl)$, using the relationship

$$\Phi(hkl) = \frac{h^2 F(hkl)}{2\pi m_0 e V_c}, \qquad (8.6)$$

where V_c is the volume of the unit cell (Cowley, 1981, section 4.2.2). Values of $\Phi(000)$ range from approximately $+5$ V for specimens containing only light elements to approximately 30 V for specimens containing mostly heavy elements. The structure factors are given by the standard formula (see, for example, Warren, 1969, Chap. 3; Hirsch et al., 1977, Chap. 4),

$$F(hkl) = \sum_j {}^e f_j(s_{hkl}) \exp(-B_j s_{hkl}^2) \exp[+2\pi i(\mathbf{g}_{hkl} \cdot \mathbf{r}_j)]. \qquad (8.7)$$

The index j in (8.7) indicates a sum over all atoms in the unit cell, with the fractional coordinate of the jth atom being $\mathbf{r}_j$ and $\mathbf{g}_{hkl}$ being the reciprocal-lattice vector with indices hkl.

The electron-scattering factor for the jth atom, ${}^e f_j(s_{hkl})$, may be obtained either from experimental determinations or from tabulated values for free atoms (*International Tables for X-Ray Crystallography*, Vol. IV; Doyle and Turner, 1968). If electron-scattering factors are unavailable for any particular atom or group of atoms, they can be deduced from the corresponding X-ray–scattering factor by using the Mott formula (Mott, 1930). The variable s_{hkl} is specified as $s_{hkl} = \sin\theta_{hkl}/\lambda$, where θ_{hkl} is the Bragg angle for the hkl-reflection. Thus, by Bragg's law, $s_{hkl} = 1/(2d_{hkl})$, where d_{hkl} is the interplanar spacing of the diffracting planes. Finally, B_j is the Debye-Waller factor for the jth atom. Again, the values of B for various atoms can be obtained from experimental determinations or from tabulated values such as those given in volume III of the *International Tables for X-ray Crystallography*. A vector notation is adopted in (8.7) to indicate that the equation is applicable to the case of anisotropic-scattering factors and Debye-Waller factors.

Note, at this point, that (8.5) and (8.6) describe the crystal potential, which is independent of the energy of the incident electron; whereas the electron-scattering factor does vary with the incident-electron energy and must be corrected for the change, with energy, of the mass of the electron. In (8.7), the scattering factors used should not be corrected for this mass change; the change in scattering power with incident-electron energy is taken into account by the interaction constant. The use of free-atom-scattering factors for calculating the structure factors of the crystal reflects the fact that matching computed results to experimental results is, at present, a fairly qualitative process. There are few cases where attempts have been made to quantify bonding effects in crystals (for an example, see Anstis et al., 1973). Similarly, the Debye-Waller factors are usually assigned rather approximate values. This arbitrariness does not depict inaccuracy in the multislice calculation but, rather, the problems of accurately quantifying experimental electron intensities. With the increasing use of electronic-recording systems on TEMs, the measurement of electron intensities is becoming more and more accurate, which will, in turn, require the use of more realistic scattering factors and Debye-Waller factors in multislice calculations.

Once one has computed the projected potential and the real-space transmission function, the reciprocal-space transmission function is found by Fourier transformation. Since $\phi(xy)$ is continuous, its values can only be calculated at discrete points in real space. Therefore, since $q(xy)$ is calculated at the same points, $Q(hk)$ is found by approximating the Fourier integral by a sum. The choice of a real-space sampling interval for the calculation of $Q(hk)$ is specimen-dependent; but in general, an interval of 0.2 Å is sufficient for samples containing only light elements, and an interval of 0.1 Å is adequate for specimens predominantly containing heavy elements (such as gold). If the values of a and b are significantly different, then one must calculate $\phi(xy)$ [and therefore $q(xy)$] with more sampling points

CALCULATION OF DIFFRACTION PATTERNS AND IMAGES

in one direction than the other in order to maintain approximately equal sampling spacings in both the a- and b-directions. Similarly, the reciprocal-lattice net must be calculated with different dimensions in the a^*-direction and in the b^*-direction.

The transmission function given in (8.3) represents a true phase grating in that it only affects the phase of the electron wave and not the amplitude. In real systems, the electrons are not only elastically scattered but also absorbed (inelastically scattered) by the specimen. The effects of absorption are modeled phenomenologically by defining an absorption function $\mu(xy)$ and modifying the transmission function so that

$$q(xy) = \exp[-i\sigma\phi(xy)\,\Delta z]\exp[-\mu(xy)]. \tag{8.8}$$

A more convenient form of (8.8) includes the absorption as an imaginary part of the potential, giving

$$q(xy) = \exp\{-i\sigma[\phi(xy) - i\phi_a(xy)]\,\Delta z\}, \tag{8.9}$$

where $\phi_a(xy)$ equals $\mu(xy)/(\sigma\,\Delta z)$. As a first approximation, the absorption potential can be taken as proportional to the crystal potential with a constant of proportionality of between 0.05 and 0.10. Given that the crystal potential peaks on the atom sites, the use of $\phi_a(xy)$ proportional to $\phi(xy)$ will mean that the absorption also peaks on the atom sites, as expected. A more accurate representation of the absorption potential can be obtained by using formulations such as those given by Humphreys and Hirsch (1968) or Radi (1970). In these representations, the absorption potential peaks more strongly on the atom sites than $\phi(xy)$. The effect of the absorption potential is to remove some electrons from the calculation. In reality, these electrons are not lost; they simply do not contribute to the elastic wave function. These "absorbed" electrons form a diffuse, inelastic background in images and diffraction patterns. In general, the effect of this background is to lower contrast, which can be a severe limitation in HRTEM and STEM of thick crystals.

8.2.2 The propagation function

The reciprocal-space form of the free-space propagation function for propagation through a distance Δz (Allpress et al., 1972) is given by

$$P(hk) = \exp[-2\pi i\zeta(hk)\,\Delta z], \tag{8.10}$$

where $\zeta(hk)$ is the excitation error of the hk-reciprocal-lattice point, as shown in Figure 8.2. The excitation error is defined as negative when the reciprocal-lattice point lies outside the Ewald sphere. The strict definition of the excitation error specifies it to be parallel to the specimen-surface normal. Because the specimen-surface normal varies from specimen to specimen and therefore is an ill-defined direction, multislice calculations use the assumption that the excitation error (and therefore, the specimen-

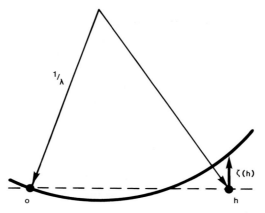

FIGURE 8.2 Geometry defining the excitation error $\zeta(hk)$, as used in the multislice propagation function.

surface normal) is perpendicular to the hk-plane. From geometric arguments, one can show that

$$\zeta(hk) = \left\{ \left(\frac{1}{\lambda}\right)^2 - [(h - u_0)^2 a^{*2} + (k - v_0)^2 b^{*2} \right.$$
$$\left. + 2(h - u_0)(k - v_0) a^* b^* \cos \beta^*] \right\}^{1/2}$$
$$- \left\{ \left(\frac{1}{\lambda}\right)^2 - [u_0^2 a^{*2} + v_0^2 b^{*2} + 2 u_0 v_0 a^* b^* \cos \beta^*] \right\}^{1/2}, \quad (8.11)$$

where β^* is the angle between the reciprocal-lattice basis vectors and $(u_0 v_0)$ is the center of the Laue circle (i.e., the point of perpendicular projection of the center of the Ewald sphere onto the hk-plane).

When equation (8.11) is used to evaluate the excitation error, $\zeta(00)$ will always be assigned the value zero. Thus, the propagator models the relative phase of the beams rather than the absolute phases. (The absolute phases can be calculated by adding a constant phase to all beams, where the value of the additional phase is determined by the propagation of the incident beam through free space by a distance corresponding to the specimen thickness.) Hence, the phase change of an individual beam is meaningless unless considered in the context of the other beams in the calculation. Thus, usually, one considers $\theta(hk) - \theta(00)$ instead of just $\theta(hk)$, the calculated phase of the (hk)-beam.

By the inclusion of a term dependent on the center of the Ewald sphere in (8.11), allowance has been made for changing the direction of the incident illumination. No such allowance has been made in the transmission function. In fact, in deriving (8.5), we stated that the average projected potential is obtained by integrating over the crystal-repeat distance parallel to the beam direction. The existence of a crystal-repeat distance necessarily implies that the incident-beam direction is parallel to a crystal-zone axis. When the beam is not aligned with a zone axis, the representation of $\phi(xy)$ by a sum over $\Phi(hk0)$ will not be accurate. However, the error introduced by using slices projected along a zone axis and a propagation function that uses (8.11) to allow for beam tilt leads to negligible error in calculated diffracted-electron amplitudes and phases for beam tilts of up to 10° away from the zone axis. Similarly, the assumption that the surface normal is perpendicular to the hk-plane (i.e., parallel to a zone axis) causes negligible error for crystals with surfaces inclined by up to 45° from the hk-plane. A rigorous multislice formulation for crystal tilt and inclined surfaces is given by Ishizuka (1982).

As stated previously, equation (8.11) is not a strict definition of $\zeta(hk)$ in the sense that, in the derivation of this equation, we assumed that the normal to the specimen surface is perpendicular to the hk-reciprocal-lattice plane. Consequently, other expressions for $\zeta(hk)$ that are not exactly equivalent to that of equation (8.11) may be found throughout the literature. An often-used alternate method of evaluating $\zeta(hk)$ is to use a paraboidal approximation to the Ewald sphere (Cowley, 1981, section 11.4). Similarly, there are different approaches for including the effects of crystal tilt in multislice calculations. For example, Lynch (1971) includes crystal tilt by successive shearing movement of the crystal slices. In this method, the Fourier coefficients of the transmission function are multiplied by phase factors after each slice. In the formalism described in this chapter, these phase factors are, in effect, incorporated directly into the propagation function by the terms in equation (8.11) dependent on the position of the center of the Laue circle.

8.2.3 Multislice iteration

When the reciprocal-space wave function and transmission functions are defined only at the reciprocal-lattice points, the multislice iteration can be written as

$$\Psi_n(hk) = \sum_{h'k'} \Psi_{n-1}(h'k') P_{n-1}(h'k') Q_n(h-h', k-k'), \quad (8.12)$$

where the sum is over all beams included in the calculation. For plane-wave illumination, $\Psi_1(hk) = Q_1(hk)$. If equal slice thicknesses are used in the way described previously, the subscripts on P and Q need not be included in the formula for the multislice iteration;

$$\Psi_n(hk) = \sum_{h'k'} \Psi_{n-1}(h'k') P(h'k') Q(h-h', k-k'), \quad (8.13)$$

and

$$\Psi_1(hk) = Q(hk).$$

From (8.12) and (8.13), we see that the effect of the convolution in the multislice iteration is such that, at each slice, any given beam is modified by a contribution from all other beams in the calculation. This clearly demonstrates the dynamical (multiple-scattering) nature of electron diffraction. Unfortunately, multiple scattering means that if there is a total of N beams included in the calculation, then evaluation of the wave function at the nth slice from that at the $(n-1)$th slice requires N^2 complex additions and $2N^2$ complex multiplications. The computation time required for this evaluation can be prohibitive, considering that the evaluation must be repeated for every slice until the required crystal thickness is reached.

An alternate method of evaluating a convolution is to use the convolution theorem for Fourier transforms—that is, the convolution of two functions G and R can be expressed as

$$G * R = \mathcal{F}[\mathcal{F}^{-1}(G)\mathcal{F}^{-1}(R)], \tag{8.14}$$

where $\mathcal{F}$ and $\mathcal{F}^{-1}$ denote the forward and inverse Fourier transforms, respectively. That is, the convolution of two functions can be evaluated by inverse-Fourier-transforming both functions into real space, multiplying the two real-space functions obtained in the transformations, and then completing the convolution by a Fourier transform back to reciprocal space. With this theorem, the multislice iteration is

$$\Psi_n(hk) = \mathcal{F}\{\mathcal{F}^{-1}[\Psi_{n-1}(hk)P_{n-1}(hk)]\mathcal{F}^{-1}[Q_n(hk)]\}. \tag{8.15}$$

If the Fourier transforms are calculated by using the conventional Fourier summation, the iteration in (8.15) is slightly slower than the direct summation (8.12). However, with the use of the fast-Fourier-transform (FFT) algorithm (Cooley and Tukey, 1965) and a choice of the number of beams in the calculation that suits this algorithm, the use of (8.15) becomes much quicker than the use of (8.12) (Ishizuka and Uyeda, 1977). The FFT method is extremely fast when programmed in conjunction with an array processor. For a full discussion on the methodology of the FFT, see the monograph by Brigham (1973) (see also Appendix A).

FFTs have found a wide use in all fields of science, and so a large number of subroutines for the computation of complex FFTs are available from the scientific software packages. Not all of these routines are suitable for the particular application of the multislice iteration. The FFT routine used in a multislice calculation must be able to handle forward and inverse transforms for variable numbers of Fourier coefficients in at least a two-dimensional space. Also, during the course of the iteration, the FFT procedure will be used many times. So the routine should allow tables of the sine and cosine values used in the transformation to be set up once and then stored for use in subsequent transforms so that sine and cosine values are not repeatedly calculated throughout the iteration.

The FFT can only be applied to transformations where the number of Fourier coefficients can be highly factorized. Although mixed-radix routines exist for the calculation of FFTs (e.g., Singleton, 1969), usually, a faster and more convenient technique is to use a routine that is based on the number of Fourier coefficients being a power of a single factor. In general, this means choosing the number of Fourier coefficients to be a power of 2 or 3. Although routines for powers of 3 are slightly faster than those using powers of 2, routines based on powers of 2 allow a greater flexibility in the choice of the number of beams to be included in the multislice calculation. The computer time taken to evaluate an FFT based on powers of 2 with M Fourier coefficients is approximately proportional to $M \log_2 M$ (Cooley and Tukey, 1965), and for calculations where the number of beams is greater than 32, the iteration using FFTs is faster than evaluating the summation (8.12).

The use of (8.14) to carry out the convolution in the multislice iteration involves sampling the real-space wave function. When one is using an FFT, the number of real-space sampling points is equal to the maximum number of Fourier coefficients that can be included in the FFT. One can obtain very fine sampling of real space by calculating the FFT with an overly large array of Fourier coefficients, with the outer coefficients set to zero. For the purpose of evaluating the multislice convolution, the real-space sampling will be adequate when the number of beams in the calculation satisfies the usual requirement that scattering to beams not included in the calculation is insignificant. However, because real space is discretely sampled and because of the sampling theorem for Fourier transforms (see for example Arsac, 1966), the reciprocal-space wave function appears periodic.

The nature of this periodicity may be demonstrated by considering the one-dimensional formula for the convolution of two reciprocal-space functions (G and R) that have Fourier coefficients that extend from $-M$ to M. The convolution is given by

$$G(h) * R(h) = \sum_{h'=-M}^{M} G(h')R(h-h'). \qquad (8.16)$$

Since both $G(M)$ and $R(M)$ exist, the term $G(M)R(M)$ is a valid term for the summation; and because $h' = M$ for this term, $h = 2M$. From this argument, the Fourier coefficients for the convolution extend from $-2M$ to $2M$. When (8.12) is used for the multislice iteration, terms that lie outside the range corresponding to $-M$ to M are simply not calculated, and so the number of beams remains constant after each slice. If the Fourier-transform method is used to calculate the convolution of $G(h)$ and $R(h)$, then the Fourier coefficients are constrained to exist from $-M$ to M; and so the reciprocal lattice will appear to repeat every $2M + 1$ points. Thus, a term such as $G(M)R(M)$ will not contribute to the $2M$ Fourier coefficient but, rather, will incorrectly contribute to the -1 Fourier coefficient.

The term for this false periodicity in reciprocal space is aliasing; and so that aliasing is avoided in the FFT multislice formulation, the Fourier coefficients of the wave function (i.e., the diffracted beams) must be embedded into an array such that the diffracted beams extend to, at most, two-thirds of the array size in any direction. Thus, for example, if a 32 × 32 array is used for the Fourier coefficients in an FFT multislice iteration, the maximum number of beams that can be used in the calculation is 21 × 21. Similarly, if a 64 × 32 array is used, the maximum number of beams is 42 × 21. The array elements outside this limit must be set to zero after each iteration. This can be achieved either by setting the corresponding propagation coefficient for these array elements to zero or, preferably (from the point of view of computation speed), by directly assigning the value zero to these elements after each slice in the multislice iteration.

A characteristic of FFTs that is similar to aliasing is that, usually, FFTs use the first element in the array of Fourier coefficients as the zeroth-order coefficient. For multislice calculations, one usually defines the zeroth-order coefficient to be near to the middle of the array and symmetrically surrounded by the diffracted beams. This incompatibility is overcome by applying phase factors to the real-space function after an inverse FFT and the opposite phase factors before a forward FFT. A preferred procedure is to shift the reciprocal-space wave function so that the transform goes from $h = -2^{n-1}$ to $2^{n-1} - 1$ and $k = -2^{m-1}$ to $2^{m-1} - 1$ (for a $2^n \times 2^m$ array), because then the necessary phase factors are $(-1)^{i+j}$ for the (ij)-sampling point in real space. Since the multislice iteration involves an inverse FFT followed by a forward FFT, no changes need be made to the iteration to suit the use of an FFT. The strict form of the FFT multislice iteration is therefore

$$\Psi_n(hk) = F^+\{F^-[\Psi_{n-1}(hk)P_{n-1}(hk)]\mathscr{F}^{-1}[Q_n(hk)]\}, \quad (8.17)$$

where F^+ represents an FFT from real to reciprocal space and F^- represents an FFT from reciprocal to real space. Note that, at intermediate steps in the iteration, the calculation may not give a true representation of the real-space wave function because of phase changes introduced by the shift of origin in reciprocal space.

Since $\mathscr{F}^{-1}[Q(hk)] = q(xy)$, one need not find $Q(hk)$ to evaluate (8.17). However, to avoid aliasing effects, one should use $q(xy)$ with the Fourier coefficients truncated in the way specified previously. Thus, the preferable method of evaluation of $\mathscr{F}^{-1}[Q(hk)]$ is as follows:

1. Calculate $\Phi(hk0)$ for all reciprocal-lattice points in the FFT array [i.e., $\Phi(hk0)$ are not truncated].
2. Calculate the projected potential by using an inverse FFT and applying phase factors as necessary.
3. Evaluate $q(xy)$; apply phase factors, if necessary, and FFT to give $Q(hk)$.
4. Truncate $Q(hk)$ to allow for aliasing effects.
5. Inverse-FFT the truncated $Q(hk)$, applying phase factors after the

transformation, if necessary, to give $\mathcal{F}^{-1}[Q(hk)]$, and store for use in the multislice iteration.

The notation $\mathcal{F}^{-1}[Q(hk)]$—as distinct from $F^{-}[Q(hk)]$—has been used in (8.17) to signify that this term represents the exact real-space function rather than the real-space function that has been modified by phase factors because of the use of an FFT. Note that the FFT is not only of use in the multislice iteration but also in the evaluation of $\phi(xy)$ and $q(xy)$.

8.2.4 Consistency tests

The two major reasons why a multislice calculation can fail to give accurate values for diffracted amplitudes and phases are the use of a slice thickness that is too large and the use of an insufficient number of beams in the calculation. Therefore, a worthwhile procedure is to formalize some methods of determining whether particular values of slice thickness and number of beams will be adequate for use in the multislice iteration. The use of such methods will allow the computation time for obtaining accurate specification of the exit-surface wave function of the crystal to be minimized. The choice of the two parameters, slice thickness and number of beams, is not an independent process for each parameter, because to some extent the effect of having too large a slice thickness can be countered by using overly large numbers of beams; similarly, having too few beams can be countered by using very thin slices. Thus, a given test should not be interpreted as applying to one parameter to the exclusion of the other; however, from the way in which a test fails, one can predict which of the two parameters is more likely to be inadequate for the multislice iteration.

Since the phase grating forms the basis of the multislice iteration, a profitable exercise is to test the accuracy of the phase grating used for the transmission function before the start of each evaluation of the electron wave functions. The unitarity test is suitable for this purpose. This test is based on the fact that $|q(xy)|^2 = 1$, and hence, the Fourier transform of $|q(xy)|^2$ should be 1 for the zeroth-order Fourier coefficient and zero for all other Fourier coefficients. That is,

$$\mathcal{F}[|q(xy)|^2] = \delta(hk), \qquad (8.18)$$

where $\delta(hk)$ is the Kronecker-delta function. This expression is sometimes written in the expanded form

$$Q(hk) * Q^*(-h, -k) = \sum_{h'k'} Q(h'k')Q^*(h-h', k-k') = \delta(hk). \quad (8.19)$$

The philosophy of the unitarity test is to measure how well (8.18) holds when only the Fourier coefficients corresponding to the beams in the calculation are included in the evaluation of $|q(xy)|^2$.

To carry out the unitarity test, $|q(xy)|^2$ is formed from the values of $q(xy)$ {i.e., $\mathcal{F}^{-1}[Q(hk)]$} obtained by using the restricted number of

Fourier coefficients as determined by the criterion for avoiding aliasing. The values thus obtained are fast-Fourier-transformed, and the result is compared with $\delta(hk)$. Because of the truncation, the zeroth-order Fourier coefficient obtained in this transformation will be slightly less than 1, and the other Fourier coefficients will be nonzero. If the coefficients vary significantly from $\delta(hk)$, the phase grating should be recalculated with thinner slices (1- to 3-Å slices are usually adequate). For the calculation of thick-crystal amplitudes and phases, the differences between the results of the unitarity test and $\delta(hk)$ should be no greater than 1 in 10^5. For thinner crystals (around 100 Å or less thick), the criteria on the unitary test may be reduced to 1 in 10^4.

A second test is to calculate, at regular intervals during the iteration, the sum of the intensity in all the beams being used in the calculation and compare this sum with the total intensity incident on the specimen (i.e., the incident-beam intensity). If there is no scattering to beams other than those included in the calculation, then the calculated sum of intensities will remain equal to the incident intensity throughout the calculation. In practice, the sum will fall off slowly with increasing crystal thickness. If the multislice iteration is to be judged as accurate, then the sum of the intensities in all beams at the exit surface of the crystal should not fall below 0.9 of the incident intensity. If the sum does fall below this value, then more beams must be included in the calculation. If, however, the sum of intensities rises above the incident intensity (even by the smallest amount), then thinner slices are required. Of course, if no attempt is made to allow for aliasing in a FFT multislice calculation, there will be no loss of total intensity; and so this test becomes invalid. Note that if the multislice calculation is set up for plane-wave illumination in the way described previously—that is, such that $\Psi_1(hk) = Q_1(hk)$—then the incident intensity is unity. And indeed, usually one normalizes the incident intensity to 1 for all illumination conditions.

Neither of these tests can be applied when absorption is included in the phase grating, because when absorption is included, $|q(xy)|^2 \neq 1$ and the intensity should fall off with increasing crystal thickness. Thus, before including absorption in calculations for a particular crystal system, one must prove the adequacy of the chosen values for slice thickness and number of beams in calculations for the same system without absorption.

Neither the unitarity test nor the total-intensity test are fully reliable in guaranteeing that a given calculation will be accurate. For example, if the number of beams included in a calculation is insufficient, both the unitarity test and the total-intensity test can be made to work by choosing exceedingly thin slices. The most rigorous test is to prove that the calculated diffracted amplitudes and phases are unchanged by decreases in slice thickness and increases in the number of beams in the calculation. If the values of slice thickness and number of beams are suitable for accurate evaluation of the multislice iteration, then the iteration performed with thinner slices and more beams should not significantly change any of the

CALCULATION OF DIFFRACTION PATTERNS AND IMAGES 259

calculated values for amplitude and phase. This test need only be carried out once for each system. The one caution that must be considered while one is carrying out this test is that one can have too many beams in a multislice calculation. Although this possiblity seems contrary to all that has been said previously, it is a consequence of the fact that if beams that scatter to very high angles are included in a calculation, then the phase change introduced by the propagator for some of these beams may be 2π, which will cause the intensity in these beams to be significantly overestimated. This effect is discussed in the following section on higher-order Laue zones.

8.3 Special systems

8.3.1 Higher-order Laue zones

In the previous discussion, where the incident-electron beam is parallel or almost parallel to a zone normal, diffraction of fast electrons has been described in terms of scattering to a single plane of reciprocal-lattice points. This plane of points, which necessarily contains the origin of reciprocal space, (000), is known as the zero-order Laue zone (ZOLZ). Scattering to reflections out of this plane is described as diffraction to higher-order Laue zones (HOLZ). In the notation used above, the HOLZ reflections are for diffraction to (hkl)-reciprocal-lattice points where $l \neq 0$. By use of the Ewald-sphere construction, these HOLZ reflections will appear in a series of rings around the central ZOLZ pattern of reflections (see Figure 3.5). In electron diffraction from crystals where the repeat distance parallel to the beam is very large or crystals that contain planes of heavy atoms perpendicular to the beam direction, HOLZ reflections are sometimes observed among the ZOLZ reflections at positions corresponding to the projection along the zone normal of the HOLZ reciprocal-lattice points. The intensity of HOLZ reflections is extremely sensitive to changes in atomic positions and to the surface structure of a crystal. Thus, HOLZ reflections are finding increasing use in techniques such as CBED, where the HOLZ reflections are used to accurately measure, among other things, crystal symmetry and strains around lattice defects (see Chapter 12).

The formulation given above obviously cannot treat HOLZ reflections, since the expression used for the projected potential, (8.5), contains information from the $(hk0)$-reflections only. This failure to include the effects of HOLZ reflections means that the formulation cannot be used for crystals with repeat distances parallel to the beam of greater than 5 to 10 Å. To see how to include HOLZ reflections in multislice calculations, we must consider exactly what is to be modeled. Diffraction to a HOLZ reflection is an elastic scattering by a reciprocal-lattice vector that has a significant component parallel to the zone normal (i.e., parallel to the incident-beam direction). For an (hkl)-reflection, the length of this component of the reciprocal-lattice vector is lc^*. Remembering that the amplitudes of the diffracted beams correspond to the Fourier coefficients of the electron wave function, the presence of HOLZ reflections means that there is a periodic modulation of the electron wave function parallel to the beam direction. The wavelength of this modulation will be integer fraction of $1/c^*$. Hence, to sample this modulation, it is necessary to use slices thinner than $1/c^*$ and slices for which the correct projected potential for the slice is used rather than the average projected potential given by (8.5) (Lynch, 1971).

For a slice of thickness Δz centered on z_0, the projected potential per unit length can be derived by placing suitable limits on the integral in equation (8.5) and is therefore given by

$$\phi(xy) = \sum_{hkl} \Phi(hkl) \frac{\sin(\pi l \Delta z/c)}{\pi l \Delta z/c} \exp\left(\frac{-2\pi i l z_0}{c}\right) \exp\left[-2\pi i \left(\frac{hx}{a} + \frac{ky}{b}\right)\right]. \quad (8.20)$$

So that the number of different phase gratings that have to be calculated is mimimized, a convenient technique is to slice the crystal in such a way that there is an integral number of slices in the distance $1/c^*$. The multislice calculation proceeds in the way described earlier, except that the crystal is built up by sequencing through the different slices. If the slices are defined to be of equal thickness, then the sum in (8.20) is most easily calculated by setting up a table of $\Phi(hkl) \sin(\pi l \Delta z/c)/(\pi l \Delta z/c)$-values; then, for each different slice in turn, one calculates the two-dimensional Fourier coefficients

$$\Phi(hk) = \sum_{l} \left[\Phi(hkl) \frac{\sin(\pi l \Delta z/c)}{\pi l \Delta z/c}\right] \exp\left(\frac{-2\pi i l z_0}{c}\right) \quad (8.21)$$

by using the table; for example, see Kilaas et al. (1987). Then, $\phi(xy)$ is calculated by using an inverse Fourier transform (preferably an FFT).

Because, in (8.20), the $\Phi(hkl)$-values are multiplied by a $(\sin x)/x$ function, one need not calculate $\Phi(hkl)$ to very high order in l. However, one must calculate not only $\Phi(hkl)$ but also $\Phi(hk\bar{l})$, because an electron scattered to a HOLZ can scatter back to the ZOLZ via a $(hk\bar{l})$-reciprocal-lattice vector; and so the structure factors of the $(hk\bar{l})$-reflections must be included in the calculation. Since the projected potential is a two-dimensional quantity, it only has (hk)-Fourier coefficients. Thus HOLZ reflections appear in the calculation (and also experimentally) as projections onto the hk-plane. Hence no change is made to the propagation function in calculations that include HOLZ reflections; and from the computational point of view, the contributions of HOLZ reflections become indistinguishable from the contributions of ZOLZ reflections.

When one calculates the intensities of HOLZ reflections, one must choose not only a suitable slice thickness but also the suitable positions to slice the specimen. The choice of slice thickness is usually based on the requirement that there be as few as possible phase gratings to be evaluated but that there be adequate sampling of the modulation of the electron wave function in the direction parallel to the beam. The criteria that should be used in choosing slice thickness and slice positions in the calculation of HOLZ diffraction intensities can be demonstrated by the following (albeit nonphysical) example. Consider a two-dimensional crystal where the atoms are arranged on a body-centered lattice, as shown in Figure 8.3a. This structure can be considered as rows of equally spaced atoms stacked so that there is a half unit-cell shift between each row. In this example, the stacking direction is defined as the (001) direction; the direction perpendicular to the stacking direction is defined as the (100) direction. The allowed reflections for this structure are such that $h + l$ is even. Thus, the ZOLZ reflections will be (000), (200), (400), and so on, and the first-order, Laue-zone reflections will be (101), (301), and so on. For this system, the average projected potential will have nonzero Fourier coefficients only for h even. Thus, when the average projected potential is used, there can be no scattering to reflections ($h0$) with h odd, which correspond to the positions for allowed HOLZ reflections. If the crystal is sliced thinner than c, the projected potential for each slice will have Fourier coefficients not only for h even but also for h odd. The sum over one unit cell of the h-odd Fourier coefficients must be zero; but in multislice calculations, because of the effects of propagation between slices, these h-odd Fourier coefficients will not combine in phase, and so intensity will remain in these reflections.

The stacking arrangement of the crystal structure suggests two suitable slicing schemes. The first, labeled type I, is such that the center of any atom row always corresponds to the center of one of the slices. The second, labeled type II, is such that the center of any atom row always corresponds to the beginning of one of the slices. Diagrams of these slicing arrangements are shown in Figure 8.3a for two slices per unit cell. Figures 8.3b,c and d give the results of multislice calculations for up to eight equally spaces slices per unit cell. The results on the left in Figures 8.3b, c and d are for slice type I while those on the right are for slice type II. For these calculations, the incident-beam direction was parallel to the (001) direction, and the accelerating voltage used was 91.5 kV; so the Ewald sphere passes through the (17, 0, 1) reciprocal-lattice point. For enhancement of the scattering to HOLZ reflections, the atomic-scattering factors used in the calculation correspond to the atomic-scattering factors of a heavy element (approximately equal to gold). For demonstration of the slicing requirements, the unit-cell dimensions used were $a = 4.0$ Å and $c = 2.83$ Å. These unit-cell sizes are not particularly large; and so for a corresponding structure consisting of light elements, the HOLZ reflections would be negligibly weak. However, for structures with a large repeat-distance parallel to the beam direction, HOLZ reflections will be very strong even if the structure only contains very light elements.

Figure 8.3b shows that the amplitude of the kinematically allowed ZOLZ reflections are independent of slice thickness and slice position in all the

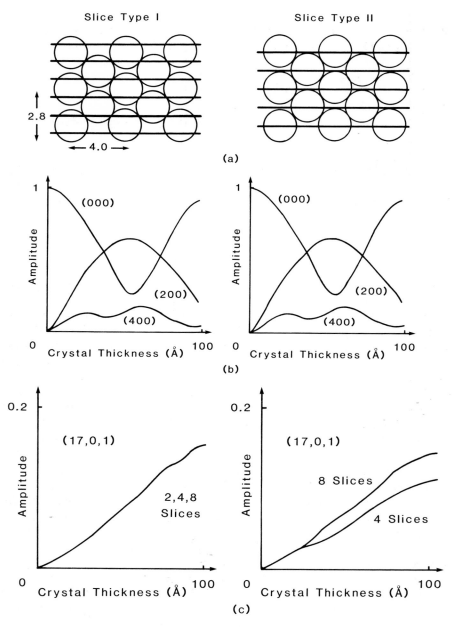

FIGURE 8.3 Results of multislice calculations that include HOLZ reflections for a specimen described in the text. (a) The two types of slicing arrangements used. (b) The results for kinematically allowed ZOLZ reflections [(000), (200), and (400)]. The results for these ZOLZ reflections are independent of slice type and number of slices per unit cell. (c) The results for the Bragg-satisfied (17, 0, 1) reflection. (d) The amplitudes of the (101) reflection. The instrumental parameters used in the calculation are detailed in the text.

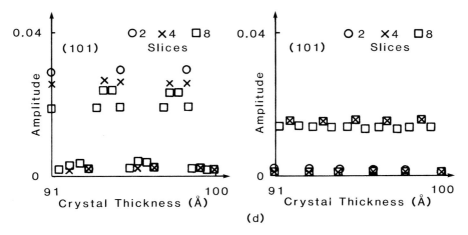

FIGURE 8.3 (*Continued*)

calculated cases. Furthermore, the results for these ZOLZ reflections are indistinguishable from the results calculated by using the average projected potential and one slice per unit cell. For type I slicing, the amplitude of the reflection corresponding to the Bragg-satisfied (17, 0, 1) reflection is unchanged by using two, four, or eight slices per unit cell. The amplitude for this reflection calculated by using type II slicing is in reasonable agreement with the type-I-slicing result only when eight slices are used per unit cell. In fact, type II slicing with two slices per unit cell gives two identical slices and therefore does not show any HOLZ reflections. As a general rule, when one is including HOLZ reflections, one should slice the crystal so that the atomic planes lie at the center of slices and so that there is one slice per atomic plane. Slicing in this way maximizes, for each slice, the Fourier coefficients of the projected potential that correspond to the HOLZ reflections. An often-used alternative method is to slice the crystal so that the atomic planes are at the center of the slices and to include only the potential from atoms within a slice in the evaluation of the projected potential for that slice. This method of evaluation has little effect on the calculated amplitudes and phases but greatly simplies the calculation of the projected potential for each slice.

Given that the intensity of the Bragg-satisfied HOLZ reflections can be evaluated by using type I slicing with two slices per unit cell, the mechanism for intensity transfer to HOLZ reflections can be deduced by studying this system. For type I slicing with two slices per unit cell, the h-odd Fourier coefficients of the projected potential of one slice are equal and opposite to the corresponding Fourier coefficient of the other slice. Kinematically, the net effect is for the h-odd contributions to cancel when summed over the unit cell. However, the reciprocal-space propagator for the $h = 17$ reflection is −1 [since $\Delta z = c/2$ and therefore $2\pi\zeta(17)\,\Delta z = \pi$]. Thus, when one is propagating from one slice to the next, the scattering from a given slice will be in phase with the scattering from the next slice and so will reinforce to give strong scattering in the $h = 17$ reflection.

This argument can be applied to all multislice calculations. For example, the propagator for a Bragg-satisfied ZOLZ reflection is always 1, which means that the phase of the scattering is preserved from one slice to the next, and so, if the reflection is kinematically allowed, the intensity in the reflection will reinforce at each slice. Of course, the intensity of a Bragg-satisfied reflection does not grow indefinitely because of the strong scattering to and from other reflections. The propagator is also 1 when $2\pi\zeta(hz)\,\Delta z$ is any multiple of 2π. For example, if the accelerating voltage used for the two-dimensional crystal is increased to 112 kV, the Ewald sphere passes through the forbidden (18, 0, 1) reflection. When one slice per unit cell is used, $2\pi\zeta(18)\,\Delta z = 2\pi$ (since $\Delta z = c$), and the intensity of the $h = 18$ reflection behaves like an allowed HOLZ reflection. This result is incorrect; and if more than one slice per unit cell is used, the intensity of the $h = 18$ reflection remains very low. To avoid this problem in multislice calculations, one should choose the number of beams used and the slice thickness such that the $2\pi\zeta(hk)\,\Delta z$ is always less than 2π for any reflection in the calculation.

Even for low-order HOLZ reflections, where the propagator does not significantly change the phase of the scattering from a slice, there is some intensity buildup. Figure 8.3d shows the intensity in the (101) reflection over a crystal thickness of two and a half unit cells at a total crystal thickness of around 100 Å. The intensity in this reflection shows a strong modulation in amplitude within the distance of a unit cell. The intensity variation correlates with the atomic layers in the crystal. As expected, at the end of each unit cell, the amplitude does not go exactly to zero as it would kinematically. From the argument used earlier, if the beam is tilted so as to satisfy the Bragg angle for the ZOLZ reflection corresponding to a low-order HOLZ reflection, then the propagator for that reflection becomes 1; and so the phase grating will exactly cancel, and there will be no intensity in the reflection. Of course, there is a second mechanism for diffraction into the reflections corresponding to low-order HOLZ reflections—namely, scattering back from a Bragg-satisfied, high-order HOLZ reflection. This mechanism requires the intensity in the high-order HOLZ reflection to be significant and so is usually seen in thick crystals only (e.g., Steeds and Evans, 1980). For thin crystals, one need not include high-order HOLZ to obtain the intensities of low-order HOLZ reflections.

Figure 8.3d shows that the calculated intensity in the low-order HOLZ reflections is vastly different for the two slicing types and even varies with number of slices for a given slice type. Consideration of the method of slicing reveals that, in each calculation, the top and bottom surface of the crystal has been modeled differently. Because of the rapid modulation in the intensity of these reflections parallel to the beam direction, these reflections are extremely sensitive to the surface structure of the crystal even at crystal thicknesses of more than 100 Å, where surface effects might be expected to be dominated by diffraction from a bulk of the crystal. Cherns (1974) has used low-order HOLZ reflections to study single atomic steps on the (111) gold surface. Van Dyck (1980) showed that, to improve the accuracy in calculated values of HOLZ-diffracted intensities, one must include extra sub-unit-cell-thick slices that allow for crystal-potential eccentricity at the beginning and end of the multislice iteration. Calculations by R. Kilaas (private communication) confirm that Van Dyck's method allows thicker slices to be used in the accurate calculation of intensities for all HOLZ reflections. To accurately model HOLZ reflections (especially low-order HOLZ reflections), one must include the surface structure of the crystal in the multislice calculation by using slices representative of the surface potential at the start and finish of the multislice iteration.

8.3.2 Periodic continuation

The diffracted amplitudes and phases from periodic objects are calculated by using the reciprocal-space formulation of the multislice iteration, because, in reciprocal space, the wave function of a periodic object exists only at discrete points. For nonperiodic objects, such as defects in crystalline materials or amorphous specimens, both the real- and reciprocal-space forms of the electron wave function are continuous. To calculate diffraction patterns or images from nonperiodic objects, one must approximate the electron wave function by a discrete-valued function. This is achieved by defining a supercell that is representative of the nonperiodic object and then laying the supercells together on a regular lattice. The introduction of a periodicity into the system by this method, which is called periodic continuation, allows the reciprocal-space multislice formulation to be used. In the calculation, reciprocal space is sampled at positions corresponding to the Bragg reflections for the supercell lattice (Griton and Cowley, 1971).

An example of the use of periodic continuation is shown in Figure 8.4. The structure is a row parallel to the beam direction of tetragonal tungsten bronze structure in a WO_3 crystal. For the purposes of calculation, the defect was placed in a supercell of 10×10 WO_3 unit cells, as shown in Figure 8.4b (O'Keefe and Iijima, 1978). The dimensions of the supercell are 38×38 Å. With this model, computed images for the defect were matched to an experimental through-focus series.

Several criteria must be considered in defining a supercell:

1. Sampling in both real and reciprocal space must be considered.
2. For defect structures, the strain field associated with the defect should become insignificant within the supercell.

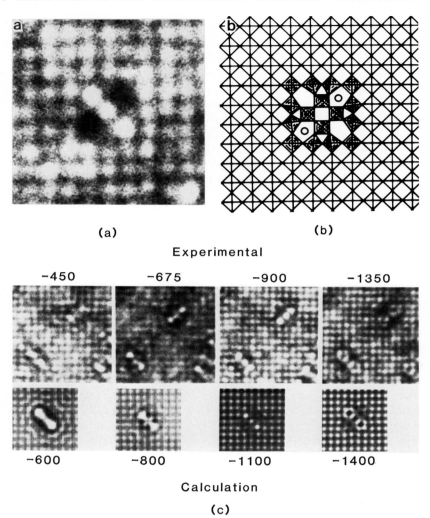

FIGURE 8.4 (a) Structure image of a row parallel to the beam direction of tetragonal tungsten bronze in a WO_3 matrix. (b) The supercell used in the computation of images for the structure. (c) Comparison of an experimental through-focal series with a computed series. Defocus values are marked in angstroms. Because of the difficulty in determining absolute experimental values of defocus these values should be treated as relative only.

3. Defects should be far enough apart that there is no interaction of scattering between neighboring defects in the supercell lattice.

The choice of suitable sampling in real space is, to a certain extent, in opposition with the choice of sampling in reciprocal space. The use of very close sampling of reciprocal space requires a very large supercell and may lead to undersampling in real space; the use of a small, closely sampled supercell may lead to undersampling of reciprocal space. The second and third criteria are consequences of trying to model isolated defects in a structure. To satisfy the second criterion, one needs a knowledge of the defect structure; the third criterion requires a knowledge of specimen thickness and imaging conditions. As a rough approximation, the third criterion can be evaluated by considering the maximum possible scattering distance of an electron in the direction perpendicular to the beam and requiring that this value be less than half the supercell dimension in any direction. The optimum condition is to have as large a supercell as possible and yet maintain diffraction to high-order reflections. The limitation on supercell size is usually determined by the maximum number of beams that can be included in the calculation. In the example shown in Figure 8.4, sampling of real and reciprocal space was for a 128×128 net of points. O'Keefe (1984) states that for calculations using 128×128 sampling, supercell sizes should be less than 40×40 Å.

Periodic continuation may be applied in many systems. Point defects can be modeled by placing a perfect crystal above and below the defect region. Surfaces parallel to the beam direction can be modeled by surrounding a sliver of crystal by free space (Marks and Smith, 1983). Planar defects are also easily modeled; but there is often a bulk displacement of the crystal on either side of the planar fault, and a returning fault must be included to allow a supercell to be defined. Some care is required in the choice of a returning defect, since this defect may dominate the images and diffraction patterns of the system. Figure 8.5 shows computed images for a planar defect (I_2-stacking fault) in CdSe with a returning defect at the edge of the supercell. Since there are now two defects per supercell, and since interaction between the prime and returning defect must be avoided, the size requirements on the supercell are doubled. In Figure 8.5a there is poor matching at the edges of the supercell, so two heavy atoms are lying very close together. The calculated images for this returning defect show an extended modulation around the returning defect that interferes with the region of image around the prime defect. A more suitable returning defect can be formed by extending the supercell a few atomic widths (Figure 8.5b); the images of this system are more satisfactory for the study of the prime defect. Thus, returning defects should be chosen to have as little difference as possible from the surrounding structure.

In systems where a returning defect is required to form a supercell, the calculated diffracted amplitudes are made up of a contribution from both defects in the system and therefore are not representative of the diffraction amplitude of either of the defects. Wilson and Spargo (1982) show that, by adding, with appropriate phase factors, the diffracted-amplitude results of calculations for two sizes of supercell, one can extract the diffraction pattern for a single defect from the mixed pattern. Furthermore, in diffraction patterns of defects calculated by periodic continuation, the calculated

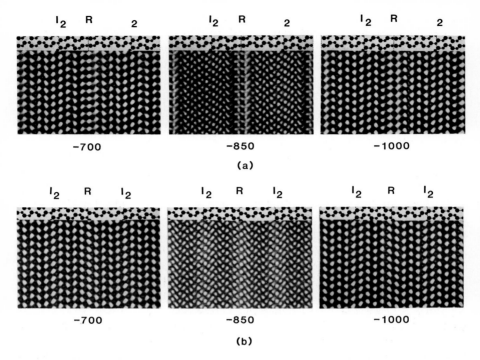

FIGURE 8.5 Through-focal series of images of an I_2-stacking fault (marked I_2) in CdSe, using two different periodically continued cells. Since there is a bulk atomic shift across the I_2-fault, a returning defect is required (marked R). The inserts show the atomic positions for two supercells in the horizontal direction. (a) The supercell is such that two atoms in the returning defect lie unrealistically close together, and consequently, the overly strong contrast of the returning defect makes the contrast of the I_2-fault hard to discern. (b) The returning defect is also an I_2-stacking fault, and the contrast of both defects is readily distinguished. The defocus values are marked in angstroms. The specimen thickness is 200 Å, and the zone is [110]. The instrumental parameters are $C_s = 0.7$ mm, accelerating voltage = 100 kV, chromatic-aberration spread in defocus = 100 Å, and beam-divergence half angle = 1 mrad. (Courtesy of R. W. Glaisher, Ph.D. diss., University of Melbourne.)

intensity of the Bragg reflections that arise from the matrix surrounding defects cannot be regarded as absolute, because increasing the size of a supercell will increase the amount of perfect crystal and thus increase the Bragg scattering.

The computation of diffracted amplitudes and phases for systems such as those in Figures 8.4 and 8.5 is greatly helped by the periodicity of the system parallel to the beam direction and, in the case of Figure 8.5, the short period in one direction perpendicular to the beam. Periodicity in some direction is not a necessary requirement of periodic continuation, and line or planar defects that are highly inclined to the beam may be modeled by using the column approximation and montaging the results by the technique used by Head et al. (1973). In the column approximation, the area of

CALCULATION OF DIFFRACTION PATTERNS AND IMAGES 269

interest of the crystal (which may be micrometers in extent) is divided up into regions (preferably with some overlap) that can be handled in a single, periodic-continuation calculation. Images are computed for each region, and then the total image of the extended defect is formed by placing the individual images together at the correct positions. Often, the slicing of the crystal in the case of highly inclined faults is simplified by the fact that the difference in the atom positions between one slice and the next is given by a uniform shift, and so the phase grating does not have to be recalculated for each slice but simply shifted.

Of course, for amorphous materials, there is no periodicity in any direction. Thus, the diffracted amplitudes and phases from amorphous materials are very hard to calculate. The normal procedure (Bursill et al., 1981) is to computer-generate the positions of the atoms in a given volume of material (using whatever constraints must be satisfied by the amorphous material) and then slicing this volume to give several phase gratings. The large number of phase gratings to be evaluated usually limits the calculation to thin samples (less than 40 Å). Calculations for amorphous material also suffer from the problems of mismatch at the boundaries of the supercell and, more importantly, an unsuitability for defining appropriate slices. In section 8.3.1, we concluded that a preferable technique is to slice a crystal so that rows or planes of atoms lay at the center of slices. Such slicing cannot be done for amorphous materials; hence, the slice thicknesses must be small relative to those used for crystalline materials. Van Dyck's modifications to the multislice iteration are therefore applicable to the calculation of diffracted amplitudes and phases from amorphous materials.

8.3.3 CBED and STEM

The only form of incident-electron beam that has been discussed explicitly so far in this chapter is plane-wave illumination. Although a plane wave may, in many cases, be a reasonable approximation to the form of the illumination in CTEM, it necessarily restricts calculations to a very limited class of microscopical applications. The general case of electron illumination is to have a converging or diverging wave front incident on the specimen. Indeed, for CBED and STEM, the electron beam is focused to a highly converged probe. For the modeling of the general case of electron illumination, the incident-wave function must be represented in reciprocal space by a continuous distribution that maps the amplitude of the incident wave as a function of angle. Usually, the angular range of the incident illumination—and therefore, the extent of the corresponding reciprocal-space distribution—is determined by the final prespecimen (illuminating) aperture (i.e., the final condenser-lens aperture for CBED and CTEM and the final objective-lens aperture for STEM). Normally, this aperture can be considered as uniformly illuminated; so the reciprocal-space distribution is a top-hat function, the shape of which is determined by the shape of the aperture. Thus, for a circular aperture, the distrubution of the incident intensity in reciprocal space is a circular top-hat function, which when diffracted, gives rise to the well-known CBED disk patterns (see Figure 3.6).

The diffracted amplitudes from a specimen illuminated by non-plane-wave radiation can be calculated by using the principle of superposition. That is, the amplitude contribution at each angle in the incident wave can be represented by a plane wave with an appropriate amplitude and phase, and therefore, the total incident wave can be considered as the superposition of these separate plane waves. By consideration of a single plane wave (i.e., a single angle of incidence) at a time, the amplitudes across a diffraction disk can be built up point by point. The simplest situation to model in this way is diffraction from a periodic object when the diffraction disks do not overlap (i.e., when the range of incident angles is less than twice the Bragg angle for the lowest-order allowed reflection). Because the range of angles in the incident beam is continuous, one cannot evaluate the diffracted amplitude corresponding to every angle. Rather, the incident angles are sampled on a suitable grid, and multislice calculations for plane-wave illumination are run with tilts corresponding to each point on this grid. A CBED pattern is constructed by collating the results of the separate multislice runs onto the angular grid, giving what is essentially rocking curves for the specimen. One normally uses a plane wave with unit amplitude and zero phase at the entrance surface of the crystal in all of the multislice calculations required to make up the CBED pattern. The variation with angle in the amplitude and phase of the incident-wave function is then included when one collates the CBED pattern by multiplying the diffracted amplitudes of all the beams in a given multislice run by the amplitude and phase factor of the incident beam in the direction corresponding to the tilt used in the multislice run.

The definition of a suitable grid is somewhat subjective because the sampling of incident angles will depend on the specimen type and thickness. Since multislice calculations must be run many times to build up a single CBED pattern, considerable computation time can be saved by choosing an angle as large as possible between grid-sampling points. However, the sampling should be fine enough to give a proper representation of the variation in diffracted amplitude across the diffraction disks. The variation with angle of the diffracted amplitude increases with increasing crystal thickness, and so the angular sampling should be greater for thick crystals than for thin. The calculation time can be reduced significantly by taking advantage of any known symmetry of the diffracted disks. For example, for centrosymmetric crystals, one needs to calculate the diffracted amplitudes for only one of a complementary pair of incident angles and then invert the calculated amplitudes through the origin of reciprocal space to generate the diffracted amplitudes for the complementary incident angle. Note, also, that the geometry of CBED patterns is such that diffracted amplitudes calculated with a tilt such that the center of the Laue circle is at the reciprocal-lattice point (uv) should, in fact, be placed so that the (000) beam for the calculation lies at the point $(-u, -v)$.

In the previous situation in which diffraction disks to not overlap, calculation of the intensities in a CBED pattern does not require a knowledge of the phase relationships between waves traveling at different angles in the incident beam. When the disks to overlap, these phase relationships must be well known in order to generate the correct CBED patterns. Thus, the electron optics of the probe-forming lenses must be well characterized. A CBED pattern with overlapping diffraction disks is built up in the same way as for the case with no overlap but with the added step of combining the results from the various different incident angles that contribute to areas of disk overlap. To obtain the correct correlation between the different incident angles, one must choose a sampling grid such that the number of sampling points over any Bragg angle in the reciprocal-lattice plane of interest is integral.

For most CBED cases, the probe size is much larger than the unit-cell dimensions of the specimen. In this case, the illuminating aperture can be considered as incoherently filled. That is, the phase relationship between waves traveling at different angles in the incident illumination is not fixed. Therefore, the total intensity at a region of overlap in a CBED pattern is the sum of all intensities from the various angles of incidence contributing to the region. Figure 8.6 shows a calculated CBED pattern of the [001] zone of rutile. This pattern was calculated for incoherent radiation at an accelerating voltage of 100 kV and a crystal thickness of 150 Å. The illuminating aperture for the system was taken as circular, and the range of incident angles in the illumination can be gauged by the size of the central beam [which is the (000) reflection]. {In [001] rutile, the intensity of $\langle 100 \rangle$ reflections is very low, and so the $\langle 100 \rangle$ reflections do not appear in this pattern.} The grid used in the calculation sampled reciprocal space at intervals of $\frac{1}{10}$ of the (100) Bragg angle. Since rutile is centrosymmetric, multislice calculations were run for only half of the grid points inside the incident-beam disk, the other half being generated by inversion symmetry.

For the cases of an incoherently filled illuminating aperture and of nonoverlapping diffraction disks, the probe size must be larger than the unit-cell size of the crystal. For STEM, probes can have diameters of 5 Å or less (Chapter 1, section 1.6). To model the diffracted amplitudes for such a probe, one must consider the illuminating aperture as coherently filled. Furthermore, for crystals with unit-cell dimensions approximately equal to or larger than the probe size, the CBED diffraction disks will necessarily overlap. In these regions of overlap, because of the coherent illumination, one must add the amplitudes of scattering from the various angles of incidence contributing to the region. From Chapter 2, section 2.5, the phase relationship between waves traveling at different angles in the incident illumination is specified in the form (Spence and Cowley, 1978)

$$\exp\left[i\chi(U) + if(uv)\right]. \tag{8.22}$$

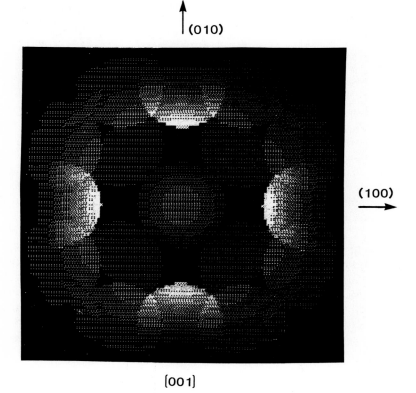

FIGURE 8.6 Calculated CBED pattern of [001] rutile for an incoherently filled illuminating aperture.

The term $\chi(U)$ describes the electron optics of the prespecimen lens and is such that

$$\chi(U) = \pi\lambda \, \Delta U^2 + \tfrac{1}{2}\pi C_s \lambda^3 U^4 \tag{8.23}$$

for a lens-spherical-aberration coefficient C_s and defocus Δ. The angular dependence in this term is specified by $U = 2(\sin \theta)/\lambda$, where 2θ is the angle of the incident illumination relative to the optic axis of the prespecimen lens. The term $f(uv)$ describes the position of the center of the electron probe relative to the origin of a unit cell. For calculations, the origin of the unit cell is defined by the coordinates used in deriving the structure factors of the crystal. The expression for $f(uv)$ is

$$f(uv) = 2\pi(x_0 au + y_0 bv), \tag{8.24}$$

where $(x_0 y_0)$ are the fractional coordinates of the center of the probe relative to the origin of the unit cell.

Figure 8.7 shows a set of CBED patterns of [000] rutile calculated for

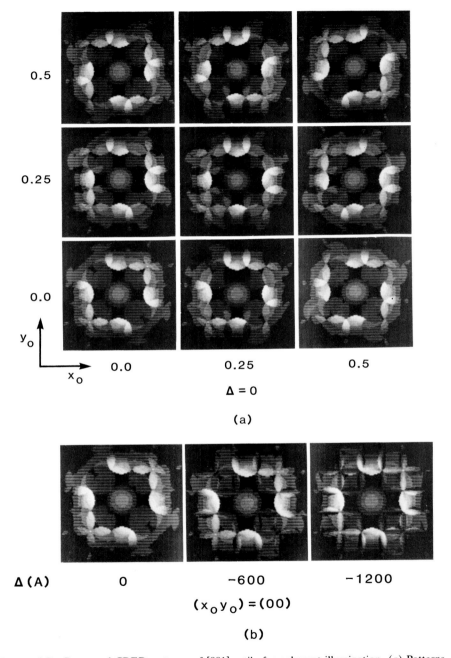

FIGURE 8.7 Computed CBED patterns of [001] rutile for coherent illumination. (a) Patterns for the probe centered on different positions in the unit cell and with the probe defocus constant. (b) Patterns for various values of probe defocus and with the probe centered on the origin of the unit cell (a titanium site).

273

coherent, 100-kV radiation with various values of Δ and $(x_0 y_0)$. Figure 8.7a shows a series of patterns for $\Delta = 0$ with the probe at different positions in the unit cell. In Figure 8.7b, the probe is held fixed at the origin of the unit cell, and the defocus is varied. (The origin of the unit cell is defined to be at a titanium site.) These patterns were calculated from the same set of multislice results used to calculate the pattern shown in Figure 8.6. In regions where the diffraction disks do not overlap, the intensities in the coherent and incoherent patterns are identical; in regions of overlap, the intensities are markedly different. The most notable change for coherent and incoherent illumination is the change in pattern symmetry. The CBED pattern in Figure 8.6 shows 4 mm symmetry; whereas patterns for coherent illumination show the symmetry of the structure at the point corresponding to the center of the incident-electron probe. For example, from Figure 8.7a, we see that, when the probe is positioned at $(x_0 y_0) = (0, 0)$ or $(0.5, 0.5)$, the CBED pattern shows 2 mm symmetry. Since the patterns of Figure 8.7 are generated by the collation of separate multislice runs by using the propagator specified by (8.10) and (8.11), no allowance is made in the propagation function for the phase differences of the waves at various incident angles. Hence defocus values are referred to the exit surface of the specimen (i.e., $\Delta = 0$ corresponds to focusing the electron probe onto the exit surface of the crystal). Also, a negative value of Δ corresponds to weakening (i.e., underfocusing) the probe-forming lens, causing the probe to focus beyond the exit surface of the specimen.

A STEM image can be calculated by taking the intensity at the point corresponding to the position of the STEM detector in a set of CBED patterns calculated with various values of $(x_0 y_0)$. If the detector is situated in a region of no disk overlap, then there will be no contrast in the STEM image. If the detector is situated in a region, of disk overlap, there will be a variation in intensity across a unit cell, giving rise to a lattice-fringe STEM image. As discussed by Spence and Cowley (1978), there is a one-to-one correspondence between lattice imaging in STEM and HRTEM. Beam divergence in HRTEM corresponds to using a finite size of STEM detector; and in both STEM and HRTEM, chromatic aberration can be modeled by averaging patterns or images over a range of defocus values. A detailed discussion on beam divergence and chromatic aberrations in HRTEM is given in the following section.

For periodic objects, there is no strict criterion placed on the reciprocal-space sampling grid used to calculate a CBED pattern. In fact, for periodic objects, for calculation of the intensity at a single point in a CBED pattern, the reciprocal-space sampling need be no finer than the Bragg angle of the lowest-order allowed reflection. The case is different for nonperiodic objects. For nonperiodic objects, the incident wave scatters to all angles in reciprocal space, so every point in a CBED pattern will contain an amplitude contribution from every angle of incidence in the illuminating wave function. Hence, for nonperiodic objects, the reciprocal-space sam-

CALCULATION OF DIFFRACTION PATTERNS AND IMAGES 275

pling of the incident-wave function should be as fine as possible. The conditions that must be satisfied by the reciprocal-space sampling grid are identical to the criteria used in choosing a supercell for the periodic continuation of the nonperiodic object. Thus, to calculate CBED patterns of nonperiodic objects, one need only sample the incident-wave function at the same points in reciprocal space as the periodic-continuation lattice.

This coincidence of the reciprocal-space sampling of the incident illumination and the periodic-continuation lattice suggests an alternative method of calculating CBED patterns. That is, the effects of the angular range in the incident illumination can be combined with periodic continuation by initiating a periodically continued multislice calculation with an incident wave [$\psi_0(xy)$ in (8.1)] equal to the total incident-wave function. This technique can also be applied to periodic objects by forming a supercell consisting of a number of crystalline unit cells. Although this technique removes the need to collate results to form a CBED pattern, it has many disadvantages compared with the technique of running individual, plane-wave, multislice calculations. In particular, the technique is slower when one is calculating CBED patterns of periodic objects, since the multislice calculation for the periodically continued cell allows unnecessarily for scattering between all incident angles. Furthermore, the technique is less flexible for handling CBED patterns in which the disks overlap because, if the overlap is included in a single, periodically continued, multislice calculation, the illumination is treated as coherent; the effects of changes in the phase relationship across the illuminating aperture must be evaluated by running a series of periodically continued multislice calculations. Finally, when one is using this technique, the propagator for the waves at each angle of incidence in the illuminating beam will be different; and so, in this case, defocus values are referred to the entrance surface of the crystal.

8.4 HRTEM imaging

Imaging theory in CTEM and HRTEM deals with the transmission of the electron wave function at the exit surface of the specimen through the objective lens to the diffraction plane and onto the imaged plane. The basis of this theory is covered in Chapters 1 and 2; before discussing the specific case of HRTEM imaging, the results given in the previous chapters will be quickly reviewed.

The diffraction pattern of a specimen is given by the Fourier transform of the exit-surface wave function modified by phase factors introduced by the lens system. In the notation of Figure 8.1,

$$\Psi_d(u'v') = \mathcal{F}[\psi_e(xy)] \exp[i\gamma(uv)] = \Psi_e(uv) \exp[i\gamma(uv)], \quad (8.25)$$

where $\gamma(uv)$ represents the phase changes introduced by the lens. This phase factor is irrelevant when computing diffracted intensities as

$$|\Psi_d(u'v')|^2 = |\Psi_e(uv)|^2. \tag{8.26}$$

The use of coordinates $(u'v')$ in the diffraction plane is to indicate that these cooordinates include the effect of diffraction-camera length (L); so in the small-angle approximation, $u' = u\lambda L$ and a similar expression exists for v'. Because diffraction patterns are not referred to any particular camera length, the term λL is ignored in most applications.

The image-plane wave function is given by applying a further Fourier transform, so

$$\psi_{Im}(Mx, My) = \mathcal{F}^{-1}\{\Psi_e(uv) \exp[i\chi(uv)]\}. \tag{8.27}$$

Since images are usually compared with the original object, the magnification factor (M) and the coordinate inversion caused by the forward Fourier transform are generally ignored. So the expression

$$\psi_{Im}(xy) = \mathcal{F}^{-1}\{\Psi_e(uv) \exp[i\chi(uv)]\} \tag{8.28}$$

is used in computing HRTEM images. The phase factor $\chi(uv)$ describes the phase at the image plane introduced by the lens system. Thus,

$$\chi(uv) = \pi\lambda \, \Delta \mathbf{U}(uv)^2 + \tfrac{1}{2}\pi C_s \lambda^3 \mathbf{U}(uv)^4. \tag{8.29}$$

The equivalence of (8.23) and (8.29) demonstrates the one-to-one correspondence of HRTEM and STEM lattice imaging. In (8.29), the generalized vector notation is adopted for $\mathbf{U}(uv)$, the reciprocal-space coordinate of the point (uv) relative to the optic axis of the objective lens. The magnitude of this vector is $|\mathbf{U}(uv)| = 2[\sin \theta(uv)]/\lambda$, where $2\theta(uv)$ is the angle between the (uv)-reflection and the optic axis. For electron diffraction, the small-angle approximation can be applied, so $|\mathbf{U}(uv)| = 2\theta(uv)/\lambda$. If the incident-electron-beam direction coincides with the optic axis, then for the (hk) Bragg reflection, $\theta(hk) = \theta_B(hk)$ and $|\mathbf{U}(hk)| = 1/d_{hk}$, where d_{hk} is the interplanar spacing of the $(hk0)$ planes. The term $|\mathbf{U}(hk)|$ should not be confused with the parameter s_{hkl} that is used to specify atomic-scattering factors in section 8.2.1 as $s_{hkl} = [\sin \theta_B(hkl)]/\lambda = 1/2d_{hkl}$.

Equation (8.29) is often expressed in terms of the generalized coordinates $\mathbf{v} = \mathbf{U}(uv)(C_s\lambda^3)^{1/4}$ and $\Delta' = \Delta/(C_s\lambda)^{1/2}$, so

$$\chi(uv) = \pi v^2(\Delta' + \tfrac{1}{2}v^2).$$

By using these generalized coordinates, one can compare HRTEM-imaging effects at several, different, microscope-operating conditions and compare the relative performance of different microscopes. Using the generalized coordinates $\mathbf{v}$ and Δ', one can express commonly used imaging conditions as values that are independent of microscope parameters. For example, the condition for optimum defocus is $\Delta' = 1.2$, and the first zero in the imaginary part of the contrast-transfer function at optimum defocus occurs when $v = 1.5$ (see Chapter 1, section 1.3.3).

The intensity of a recorded image $I(xy)$ will be given by the intensity of the image-plane wave function. That is,

$$I(xy) = |\psi_{\text{Im}}(xy)|^2 = \psi_{\text{Im}}(xy)\psi^*_{\text{Im}}(xy), \tag{8.30}$$

or, equivalently, in reciprocal space,

$$I(uv) = \psi_{\text{Im}}(uv) * \psi^*_{\text{Im}}(-u, -v). \tag{8.31}$$

As any computational method samples reciprocal space at discrete points, (8.31) is best written as the summation

$$I(uv) = \sum_{(u'v')} \psi_{\text{Im}}(u + u', v + v')\Psi^*_{\text{Im}}(u'v'). \tag{8.32}$$

The sum in (8.32) represents a sum over all reciprocal-lattice points $(u + u', v + v')$ and $(u'v')$ that contribute to the image. For convenience, these reciprocal-lattice points will be represented by vectors $\mathbf{U} + \mathbf{U}'$ and $\mathbf{U}'$, so

$$I(\mathbf{U}) = \sum_{\mathbf{U}'} \Psi_{\text{Im}}(\mathbf{U} + \mathbf{U}')\Psi^*_{\text{Im}}(\mathbf{U}'). \tag{8.33}$$

The basic form of $\Psi_{\text{Im}}(\mathbf{U})$ is given by (8.28); however, (8.28) does not include the effects of an objective-lens (diffraction) aperture, lens astigmatism, or mechanical instabilities (vibration) of the microscope.

The effect of the diffraction aperture is to truncate the number of beams contributing to the image. The aperature may be modeled by the function

$$A(\mathbf{U}) = \begin{cases} 1, & \text{if } \mathbf{U} \text{ lies inside the diffraction aperture,} \\ 0, & \text{if } \mathbf{U} \text{ lies outside the diffraction aperture.} \end{cases}$$

In practical terms, the summation in (8.33) must be restricted only to those reciprocal-lattice points (beams) that fall within the aperture. These beams can be readily determined from an experimental diffraction pattern. From (8.33), although the vectors $\mathbf{U} + \mathbf{U}'$ and $\mathbf{U}'$ in the summation are restricted to lie within the diffraction aperture, the Fourier coefficients of the image—and therefore, the lattice spacings in the image as determined by the vector $\mathbf{U}$—can, in fact, extend beyond the limit set by the diffraction aperture.

Astigmatism in an image is caused by the objective lens being of unequal strength in different directions, so beams traveling at equal angles to the optic axis but in different directions do not experience the same phase changes. Thus, astigmatism introduces an extra phase factor to the imaged-plane wave function and can be modeled by the expression

$$S(\mathbf{U}) = \exp\left[\tfrac{1}{2}\pi i\lambda C_A U^2 \cos(2\phi)\right], \tag{8.34}$$

where C_A is the defocus difference between the maximum and minimum astigmatism-induced defocus values and ϕ is the angle between the vector $\mathbf{U}$ and the direction of maximum defocus change. We usually assume that astigmatism is fully corrected in all HRTEM images (i.e., $C_A = 0$). Indeed, because the parameters C_A and ϕ are very difficult to determine, image matching in HRTEM relies heavily on this being the case.

Any vibration of the microscope during the exposure time of a micrograph will cause a blurring or smearing or the image. This effect can be modeled by convoluting the calculated image by a function representative of the vibration. More conveniently, the effect can be modeled by multiplying the Fourier coefficients of the image by a suitable function. Assuming that the direction of vibration is random, one such suitable function is the Gaussian

$$V(\mathbf{U}) = \exp(-2\pi^2 e^2 \mathbf{U}^2), \tag{8.35}$$

where e is the mean amplitude of vibration. The mechanical stability of microscopes is usually such that the amplitude of vibration is well below the maximum attainable resolution. Hence, the vibration term may be neglected and is included here for completeness only.

If we combine the previous terms with (8.28) and (8.33), the image-intensity spectrum can be written as

$$I(\mathbf{U}) = \sum_{\mathbf{U}'} \Psi_e(\mathbf{U} + \mathbf{U}')\Psi_e^*(\mathbf{U}') \exp\{i[\chi(\mathbf{U} + \mathbf{U}') - \chi(\mathbf{U}')]\}$$
$$\times A(\mathbf{U} + \mathbf{U}')A(\mathbf{U}')S(\mathbf{U} + \mathbf{U}')S(\mathbf{U}')V(\mathbf{U}). \tag{8.36}$$

This equation gives the Fourier coefficients of an image of the coherent, exit-surface wave function, $\psi_e(xy)$. Because (8.36) deals only with a coherent-wave function, it neglects two very important incoherent effects in image formation, namely, chromatic aberration and beam divergence (O'Keefe and Sanders, 1975). Chromatic aberrations are caused by instabilities in the objective-lens power supply and by the spread of energy in the incident-electron beam (caused by both high-voltage instability and filament-emission effects). Beam divergence is the result of using a converged incident beam in order to increase brightness.

Chromatic aberrations will have two effects on the wave function at the imaged plane. First, the spread of energies means that the incident beam must be considered as having a continuous range of wavelengths. Second, the combined effects of energy spread and lens instability will cause the recorded image to be the average of images from a continuous range of defocus values. For microscopes with accelerating voltages of 100 kV or more, a typical value of the electron-energy spread in the incident beam is 1 part in 10^5. Given that electron wavelength varies as $E^{1/2}$, this value of energy spread will cause only a small change in wavelength; and a reasonable assumption is that the wave function at the exit surface of the specimen is uneffected by the energy spread in the incident-electron beam.

However, the change in wavelength does have a significant effect on the phase change introduced by the objective lens.

The effect of both incident-beam energy spread and lens instability can be modeled by a change of objective-lens defocus. Hence, for formation of the chromatically aberrated image, the images calculated by using (8.36) must be averaged over a range of defocus values; and the distribution of defocus values must describe the defocus spread caused by chromatic aberrations. Thus,

$$I_D(\mathbf{U}) = \int E(d, \Delta)I(\mathbf{U})\, dd, \tag{8.37}$$

where $E(d, \Delta)$ is the distribution of defocus values (d) around the average value (Δ) and $I(\mathbf{U})$ is calculated by using (8.36). Given that the instabilies are random processes,

$$E(d, \Delta) = \left(\frac{1}{(D\sqrt{2\pi})}\right)\exp\left[\frac{-(d-\Delta)^2}{2D^2}\right]. \tag{8.38}$$

The standard deviation in defocus (D) is given by the expression

$$D = C_c\left[\left(\frac{\Delta E}{E}\right)^2 + 4\left(\frac{\Delta J}{J}\right)^2\right]^{1/2}, \tag{8.39}$$

where ΔE is the mean energy spread in the incident-beam energy (E) and ΔJ is the variation in the objective-lens current (J). The distribution of (8.38) is normalized so that the total intensity of the illumination is 1. Typically, $\Delta J/J$ is of the order of 1×10^{-6} to 5×10^{-6}; and for most high-resolution electron microscopes, the chromatic-aberration coefficient (C_c) is such that D is of the order of 100 Å. In practice, $I(\mathbf{U})$ varies fairly slowly with defocus, and so $I_D(\mathbf{U})$ can be evaluated by averaging the values of $E(d, \Delta)I(\mathbf{U})$ calculated at defocus intervals of roughly 50 Å over the defocus range $\Delta - 2D$ to $\Delta + 2D$.

Beam divergence is treated in the same way as calculating CBED patterns (section 8.3.3) but over a much smaller angular range. In HRTEM imaging, beam divergence means not only that the exit-surface wave function is made up of a set of wave functions corresponding to slightly different specimen tilts, but also that the images of these wave functions are formed with beams covering a range of angles relative to the optic axis of the objective lens. The total image, including the effects of beam divergence, is therefore the average of the set of images formed for all angles in the incident beam. Thus,

$$I_{D\alpha}(\mathbf{U}) = \int B(\mathbf{s})I_D(\mathbf{U})\, d^2s = \iint B(\mathbf{s})E(d, \Delta)I(\mathbf{U})\, dd\, d^2s, \tag{8.40}$$

where $B(\mathbf{s})$ is the angular distribution of intensity in the incident illumination. Usually, the incident illumination fills the condenser aperture uni-

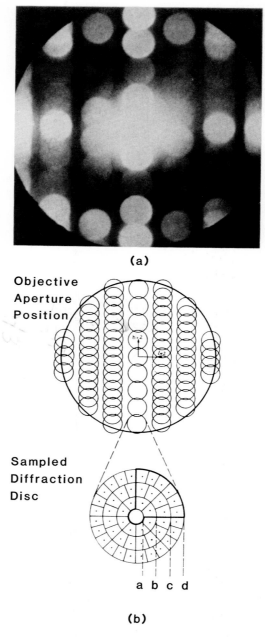

FIGURE 8.8 Example of modeling beam divergence in HRTEM images. (*a*) An experimental diffraction pattern of [010] $Nb_{12}O_{29}$. (*b*) The division of the diffraction disks. (*c*) The computed images for each of the divisions. (*d*) Comparison of an image computed with no allowance for divergence (insert a) and an image that is the average of images from each division of the diffraction disk (insert b) with an experimental image.

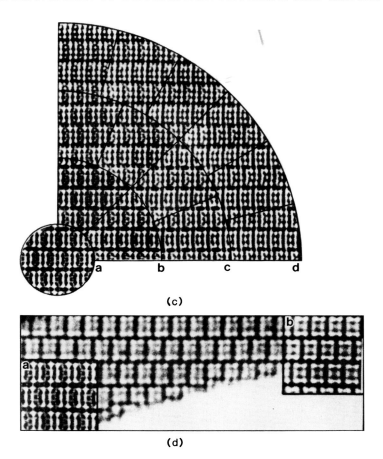

FIGURE 8.8 (Continued)

formly, and so

$$B(\mathbf{s}) = \begin{cases} \dfrac{\lambda^2}{\alpha^2 \pi}, & \text{for } |\mathbf{s}| \leq \dfrac{\alpha}{\lambda}, \\ 0, & \text{for } |\mathbf{s}| > \dfrac{\alpha}{\lambda}, \end{cases} \qquad (8.41)$$

where α is the divergence half angle. An alternative procedure that is often used is to assume a normalized Gaussian distribution, so

$$B(\mathbf{s}) = \frac{1}{\pi s_0^2} \exp\left(\frac{-s^2}{s_0^2}\right), \qquad (8.42)$$

where $s_0 = \alpha/\lambda$. Both (8.40) and (8.41) have been normalized so that the total intensity (in the two-dimensional image space) of the incident beam is

1. In practice, the results from using these two distributions are indistinguishable.

Figure 8.8 shows a practical example of calculating the total image including beam divergence (from O'Keefe and Sanders, 1975). Figure 8.8a shows an experimental diffraction pattern of [010] $Nb_{12}O_{29}$ taken with the same illuminating conditions used to record high-resolution images. Experimental diffraction patterns such as in Figure 8.8a can be used to measure the divergence half angle, and in this case, $\alpha = 1.4$ mrad. For the image simulation, the diffraction disk is divided into 49 segments, as shown in Figure 8.8b, and images are computed for the illumination conditions corresponding to each of these segments (Figure 8.8c). As the structure has fourfold symmetry, one need only calculate the images for one quadrant of a diffraction disk. Finally, the total image is formed by averaging the images from the 49 segments of diffraction disk. In Figure 8.8d, the image corrected for divergence (labeled b) and the image not corrected for divergence (labeled a) are compared with an experimental image. The necessity of including the effects of beam divergence can be clearly seen from this figure. For these images, the accelerating voltage is 100 kV; the specimen thickness was taken as 50 Å, C_s as 1.8 mm, and defocus as -600 Å.

For a matching of computed images to experimental micrographs, accurate values for C_s and C_c are needed. The values of these parameters given by a microscope manufacturer are nominal, and the use of experimentally determined values is desirable. Krivanek (1976) gives an elegant method of determining C_s. This method uses diffractogram analysis of high-resolution images of an amorphous material such as evaporated carbon. A method for determining C_c is described by Spence (1980, Chap. 8). One must remember, when calculating HRTEM images, that C_s and C_c change with the operating conditions of the microscope. In particular, these parameters vary as a function of the specimen height in the objective lens. Thus, to obtain the optimum match of experimental and computed images, one may need to vary the values of C_s and C_c (i.e., defocus spread D) used in the calculations. In fact, suitable values of C_s and D for a particular microscope are often found only after attempts have been made to match images from a number of different specimens.

The use of the above method to allow for beam divergence and chromatic aberrations requires a considerable amount of computation time, because many intermediate images must be calculated to form one final image. However, approximations can be made in HRTEM-imaging theory, and they can be used to derive analytical expressions for the integral in (8.40). The form of these approximations is generally grouped into the two classes, linear and nonlinear imaging.

8.4.1 Linear imaging

The term *linear image* is applied to images in which the majority of the intensity is carried by the zeroth-order beam. Thus, in linear images, the assumption is that the only nonnegligible terms in the summation of (8.33) are those for which either $\mathbf{U}+\mathbf{U}'=0$ or $\mathbf{U}'=0$. Thus, the summation in (8.33) reduces to

$$I(0) = \Psi_{\text{Im}}(0)\Psi_{\text{Im}}^*(0),$$
$$I(\mathbf{U}) = \Psi_{\text{Im}}(\mathbf{U})\Psi_{\text{Im}}^*(0) + \Psi_{\text{Im}}(0)\Psi_{\text{Im}}^*(\mathbf{U}) \quad \text{for} \quad \mathbf{U} \neq 0. \quad (8.43)$$

Combining these values with (8.36) and (8.40) yields

$$I(0) = \iint \Psi_e(0)\Psi_e^*(0)A(0)^2 S(0)^2 B(\mathbf{s})E(d, \Delta) \, dd \, d^2\mathbf{s},$$

$$I(\mathbf{U}) = \iint \Psi_e(\mathbf{U})\Psi_e^*(0)A(\mathbf{U})A(0)S(\mathbf{U})S(0)$$
$$\times \exp\{i[\chi(\mathbf{U}) - \chi(0)]\} B(\mathbf{s})E(d, \Delta) \, dd \, d^2\mathbf{s}$$
$$+ \iint \Psi_e(0)\Psi_e^*(-\mathbf{U})A(0)A(-\mathbf{U})S(0)S(-\mathbf{U})$$
$$\times \exp\{i[\chi(0) - \chi(-\mathbf{U})]\} B(\mathbf{s})E(d, \Delta) \, dd \, d^2\mathbf{s}, \quad \text{for} \quad \mathbf{U} \neq 0, \quad (8.44)$$

where vibration has been ignored. In the evaluation of (8.44), the spread in the incident-electron energies and the beam divergence are assumed to be small enough that $\Psi_e(\mathbf{U})$ does not change over the range of these two parameters. Hence, the Ψ_e-terms can be taken outside the integrals. Similarly, astigmatism effects in an image are assumed to be constant and may also be taken outside the integral. The function $A(\mathbf{U})$ representing the diffraction aperture may not be removed from the integral in such a simple way. In the diffraction pattern shown in Figure 8.8a, because of beam divergence, many of the diffraction disks are truncated by the diffraction aperture. Thus, these truncated disks will not contribute at full strength to the final image. The truncation of diffraction disks by the diffraction aperture can be allowed for by using a modified aperture function,

$$A_\alpha(\mathbf{U}) = \frac{\text{area of diffraction disk inside diffraction aperture}}{\text{total area of diffraction disk}}.$$

With these approximations, (8.43) can be written as

$$I(0) = \Psi_e(0)\Psi_e^*(0)A_\alpha^2(0)S^2(0),$$
$$I(\mathbf{U}) = \Psi_e(\mathbf{U})\Psi_e^*(0)A_\alpha(\mathbf{U})A_\alpha(0)S(\mathbf{U})S(0)$$
$$\times \iint \exp\{i[\chi(\mathbf{U}) - \chi(0)]\} B(\mathbf{s})E(d, \Delta) \, dd \, d^2\mathbf{s}$$
$$+ \Psi_e(0)\Psi_e^*(-\mathbf{U})A_\alpha(0)A_\alpha(-\mathbf{U})S(0)S(-\mathbf{U})$$
$$\times \iint \exp\{i[\chi(0) - \chi(\mathbf{U})]\} B(\mathbf{s})E(d, \Delta) \, dd \, d^2\mathbf{s}, \quad \text{for} \quad \mathbf{U} \neq 0. \quad (8.45)$$

The integrals in (8.45) define the envelope functions $D(\mathbf{U})$ and $D(-\mathbf{U})$. Frank (1973) showed that by expanding $\chi(\mathbf{U})$ as a Taylor series to first-order terms, these integrals could be evaluated analytically with a form

$$D(\mathbf{U}) = C(\mathbf{U})E(\mathbf{U}), \tag{8.46}$$

where $C(\mathbf{U})$ represents the chromatic-aberration envelope function and $E(\mathbf{U})$ represents the divergence envelope function. From the distribution in (8.38) for chromatic aberrations and the distribution in (8.41) for divergence, we obtain

$$C(\mathbf{U}) = \exp\left(-\tfrac{1}{2}\pi^2\lambda^2 D^2 \mathbf{U}^4\right), \tag{8.47}$$

$$E(\mathbf{U}) = \frac{2J_1[2\pi\alpha(\Delta|\mathbf{U}| + C_s\lambda^2|\mathbf{U}|^3)]}{2\pi\alpha(\Delta|\mathbf{U}| + C_s\lambda^2|\mathbf{U}|^3)} \tag{8.48}$$

where J_1 is the first-order Bessel function. If the Gaussian distribution in (8.42) is assumed for divergence, then

$$E(\mathbf{U}) = \exp\left[-\pi^2\alpha^2(\Delta|\mathbf{U}| + C_s\lambda^2|\mathbf{U}|^3)^2\right]. \tag{8.49}$$

Wade and Frank (1977) extended this analysis by expanding the Taylor series for $\chi(\mathbf{U})$ to second order and, assuming the Gaussian distribution of (8.41) for divergence, showed that the second-order terms slightly modify $C(\mathbf{U})$ and $E(\mathbf{U})$ as well as introduce the two extra multiplicative terms that involve both divergence and chromatic aberration. However, these extra terms may be neglected because they have little effect on the overall value of $D(\mathbf{U})$.

The advantage of using linear-imaging theory combined with the use of envelope functions to describe chromatic-aberration and beam-divergence effects is that, for this system, the Fourier coefficients of the image-plane wave function are separable. That is,

$$I(xy) = |\mathcal{F}^{-1}\{\Psi_e(\mathbf{U}) \exp[i\chi(\mathbf{U})]A_\alpha(\mathbf{U})S(\mathbf{U})D(\mathbf{U})\}|^2. \tag{8.50}$$

Thus, an image may be formed by multiplying the Fourier coefficients of the wave function at the exit surface of the specimen by the appropriate value of $\exp[i\chi(\mathbf{U})]A_\alpha(\mathbf{U})S(\mathbf{U})D(\mathbf{U})$, inverse-Fourier-transforming, and taking the modulus squared of the resulting real-space distribution. Thus, linear imaging reduces the calculation of images including the effects of chromatic aberration and beam divergence from one requiring the calculation of many images to a single-image calculation. In practical terms, the Fourier transform in (8.50) is best carried out by using an FFT.

Linear imaging applies particularly to weak phase objects for which

$$\psi_e(xy) = 1 - i\sigma\phi(xy), \tag{8.51}$$

and therefore

$$\Psi_e(\mathbf{U}) = \delta(\mathbf{U}) - i\sigma\Phi(\mathbf{U}). \tag{8.52}$$

Image formation for this system is discussed in detail in Chapter 2. In

Chapter 2, it is shown that only the imaginary part of the phase factor $\exp[i\chi(\mathbf{U})]$ is important in the image formation from weak phase objects. From the arguments used in Chapter 2 and those used previously to allow for beam divergence and chromatic aberrations, we obtain

$$I(\mathbf{U}) = \delta(\mathbf{U}) + \sigma\Phi(\mathbf{U})\sin[\chi(\mathbf{U})]D(\mathbf{U}) \tag{8.53}$$

(assuming no astigmatism or diffraction aperture). Thus, images of weak phase objects will resemble the projected potential of that object if $\sin[\chi(\mathbf{U})]D(\mathbf{U})$ equals $+1$ for all $\mathbf{U}$ or -1 for all $\mathbf{U}$. Hence, as detailed in Chapter 1, section 1.3, the direct-structure resolution limit is given by the value of $\mathbf{U}$ at the first zero in $\sin[\chi(\mathbf{U})]$ at optimum defocus; the minimum point-to-point resolution in an image (i.e., the information limit) is determined by the value of $\mathbf{U}$ at which $D(\mathbf{U})$ goes to zero. Plots of $\sin[\chi(\mathbf{U})]$, $\cos[\chi(\mathbf{U})]$, $E(\mathbf{U})$, and $C(\mathbf{U})$ for various values of defocus are shown in Figure 8.9 (see also Figures 1.8 and 1.9). For Figure 8.9, the electron optical parameters are $C_s = 1.4$ mm, an accelerating voltage of 200 kV, spread of defocus $D = 100$ Å, and divergence half angle $\alpha = 0.9$ mrad. At optimum defocus (Figure 8.9a), $\sin[\chi(\mathbf{U})]$ is -1 for a considerable range of $\mathbf{U}$, and thus, at optimum defocus, an image will show a limited-resolution picture of the projected potential of the weak-phase object. Such images are usually referred to as structure images. Since $\sin[\chi(\mathbf{U})]$ is negative, the image intensity will be such that regions of high projected potential in the specimen (atomic columns) will show low intensity (black) and areas of low projected potential will show high intensity (white) in positive prints.

The values of defocus used in Figure 8.9 have been chosen so that the minimum value of $\chi(\mathbf{U}) \sim -n\pi$ for $n = 1, 2, 3, 4$. At these defocus values, there is a band of $|\mathbf{U}|$-values such that $\sin[\chi(\mathbf{U})] = \pm 1$. These defoci define the so-called broadband-imaging conditions at which there can be (depending on specimen structure) a large number of Fourier coefficients of the image equal to the Fourier coefficients of the projected potential (except for a possible sign change). For those broadband defoci where $\sin[\chi(\mathbf{U})] = +1$, the image is referred to as having reverse contrast.

As might be expected, the divergence envelope function $E(\mathbf{U})$ becomes small when $\sin[\chi(\mathbf{U})]$ varies rapidly with $|\mathbf{U}|$. We can see from Figure 8.9 and from (8.48) and (8.49) that $E(\mathbf{U})$ is defocus-dependent and extends to larger values of $|\mathbf{U}|$ for higher underfocus values. The chromatic-aberration envelope function $C(\mathbf{U})$, in contrast, is independent of defocus. Thus, for a particular microscope, if the maximum observable resolution (i.e., the information limit) is invariant with defocus, then resolution is limited by chromatic aberrations. Similarly, if the maximum observable resolution increases with increasing values of underfocus, then resolution is limited by beam divergence. A suitable way of measuring resolution variations is by using diffractogram analysis of a through-focal series of images from an amorphous material. The plots in Figure 8.9 model the characteristics of the

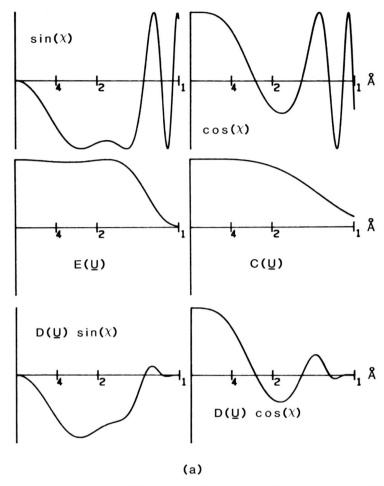

FIGURE 8.9 Plots of $\sin[\chi(\mathbf{U})]$, $\cos[\chi(\mathbf{U})]$, $E(\mathbf{U})$, $C(\mathbf{U})$, $D(\mathbf{U})\sin[\chi(\mathbf{U})]$, and $D(\mathbf{U})\cos[\chi(\mathbf{U})]$ for four values of defocus. (a) −680 Å. (b) −1050 Å. (c) −1360 Å. (d) −1600 Å. The instrumental parameters are an accelerating voltage of 200 kV, $C_s = 1.44$ mm, divergence half angle $\alpha = 0.9$ mrad, chromatic spread in defocus $D = 100$ Å.

ultrahigh-resolution pole piece of a JEOL 200CX electron microscope and show the typical balance between the chromatic-aberration and beam-divergence envelope functions.

8.4.2 Nonlinear imaging

Linear-imaging theory cannot be applied to images in which the intensity in the zero-order beam is comparable to the intensity of the other reflections

CALCULATION OF DIFFRACTION PATTERNS AND IMAGES 287

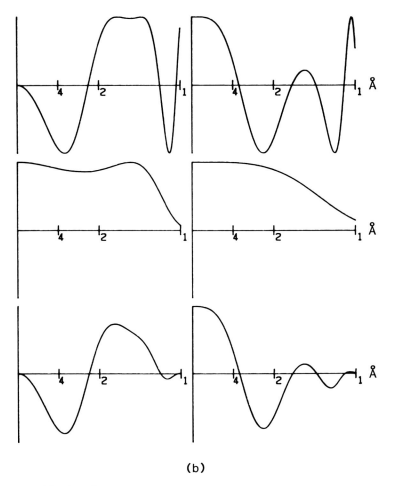

(b)

FIGURE 8.9 (*Continued*)

contributing to the image. Similarly, linear-imaging theory cannot be applied to dark-field images because the zeroth-order beam is excluded by the diffraction aperture. In nonlinear-imaging theory, no assumption is made about the relative intensities of the beams contributing to the image, and the Fourier coefficients of the image are taken to be those given in the full form of (8.40). That is,

$$I(\mathbf{U}) = \iint \sum_{\mathbf{U}'} \Psi_e(\mathbf{U} + \mathbf{U}')\Psi_e^*(\mathbf{U}') \exp\{i[\chi(\mathbf{U} + \mathbf{U}') - \chi(\mathbf{U}')]\}$$
$$\times A(\mathbf{U} + \mathbf{U}')A(\mathbf{U}')S(\mathbf{U} + \mathbf{U}')S(\mathbf{U}')V(\mathbf{U})B(\mathbf{s})E(d, \Delta)\,dd\,d^2\mathbf{s}. \quad (8.54)$$

The same assumptions used in linear-imaging theory—namely, $\Psi_e(\mathbf{U})$ and

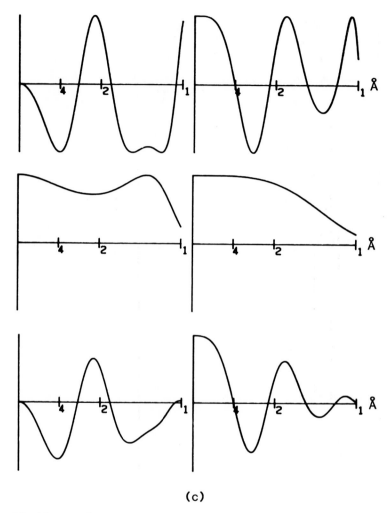

(c)

FIGURE 8.9 (Continued)

$S(\mathbf{U})$ are independent of divergence and chromatic aberration and $A(\mathbf{U})$ can be replaced by a modified aperture function $A_\alpha(\mathbf{U})$—are now applied to (8.54). Thus,

$$I(\mathbf{U}) = \sum_{\mathbf{U}'} \Psi_e(\mathbf{U} + \mathbf{U}')\Psi_e^*(\mathbf{U}')A_\alpha(\mathbf{U} + \mathbf{U}')A_\alpha(\mathbf{U}')S(\mathbf{U} + \mathbf{U}')S(\mathbf{U}')V(\mathbf{U})$$

$$\times \iint \exp\{i[\chi(\mathbf{U} + \mathbf{U}') - \chi(\mathbf{U}')]\}B(\mathbf{s})E(d, \Delta) \, dd \, d^2\mathbf{s}. \qquad (8.55)$$

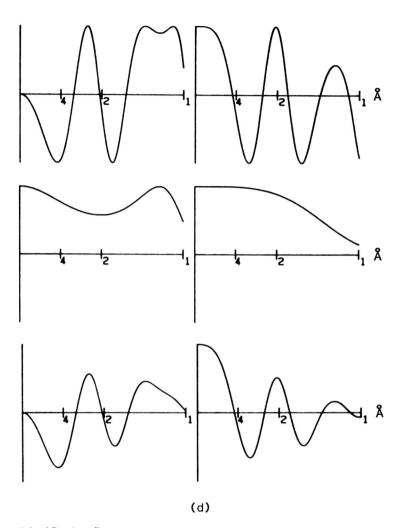

(d)

FIGURE 8.9 (*Continued*)

The integral in (8.55) defines the nonlinear-damping function $D(\mathbf{U}+\mathbf{U}', \mathbf{U}')$.

The integral defining the envelope function may be evaluated by extending the analysis of Wade and Frank (1977) (O'Keefe, 1979; Ishizuka, 1980). So to second order in the Taylor-series expansion of $\chi(\mathbf{U})$ and for the distributions in (8.38) and (8.42), we have

$$D(\mathbf{U}+\mathbf{U}', \mathbf{U}') = C(\mathbf{U}+\mathbf{U}', \mathbf{U}')E(\mathbf{U}+\mathbf{U}', \mathbf{U}')E_x(\mathbf{U}+\mathbf{U}', \mathbf{U}')P(\mathbf{U}+\mathbf{U}', \mathbf{U}'),$$
(8.56)

with

$$C(\mathbf{U}+\mathbf{U}', \mathbf{U}') = \exp\left\{\frac{-\tfrac{1}{2}D^2\pi^2\lambda^2[(\mathbf{U}+\mathbf{U}')^2 - (\mathbf{U}')^2]^2}{1+a\mathbf{U}^2}\right\},$$

$$E(\mathbf{U}+\mathbf{U}', \mathbf{U}') = \exp\left(\frac{-\pi^2\alpha^2\{\Delta\mathbf{U} + C_s\lambda^2[(\mathbf{U}+\mathbf{U}')^2(\mathbf{U}+\mathbf{U}') - (\mathbf{U}')^2\mathbf{U}']\}^2}{1+a\mathbf{U}^2}\right)$$

$$E_x(\mathbf{U}+\mathbf{U}', \mathbf{U}') = (1+a\mathbf{U}^2)^{-1/2}$$

$$\times \exp\left(\frac{-2(\pi^2\alpha^2 DC_s\lambda^2)^2[(\mathbf{U}+\mathbf{U}')^2 - (\mathbf{U}')^2]}{\times\{(\mathbf{U}+\mathbf{U}')^2(\mathbf{U}')^2 - [(\mathbf{U}+\mathbf{U}')\cdot\mathbf{U}']^2\}}{(1+a\mathbf{U}^2)}\right)$$

$$P(\mathbf{U}+\mathbf{U}', \mathbf{U}')$$

$$= \exp\left(\frac{-2\pi i\lambda(\pi\alpha D)^2[(\mathbf{U}+\mathbf{U}')^2 - (\mathbf{U}')^2]}{\times\{\Delta\mathbf{U}^2 + C_s\lambda^2[(\mathbf{U}+\mathbf{U}')^2(\mathbf{U}+\mathbf{U}') - (\mathbf{U}')^2\mathbf{U}']\cdot\mathbf{U}\}}{1+a\mathbf{U}^2}\right),$$

where $a = 2\pi^2\alpha^2 D^2$. In this expression, $\mathbf{U}+\mathbf{U}'$ and $\mathbf{U}'$ must be taken strictly as vector quantities, and the symbol · represents the vector dot product. These expressions have a form similar to that used in linear-imaging theory; and, indeed, the expressions in (8.56) reduce to the linear-image form if $\mathbf{U}' = 0$.

The cross term $E_x(\mathbf{U}+\mathbf{U}', \mathbf{U}')$ and the phase term $P(\mathbf{U}+\mathbf{U}', \mathbf{U}')$ can become quite large, especially for large values of $|\mathbf{U}|$. However, when these terms are large, the terms $E(\mathbf{U}+\mathbf{U}', \mathbf{U}')$ and $C(\mathbf{U}+\mathbf{U}', \mathbf{U}')$ are usually very small; and so, in general, the terms $E_x(\mathbf{U}+\mathbf{U}', \mathbf{U}')$ and $P(\mathbf{U}+\mathbf{U}', \mathbf{U}')$ can be neglected. Similarly, the term $1+a\mathbf{U}^2$ can be replaced by 1 in most cases. The expression corresponding to (8.56) (but without the cross term and phase term) for the divergence distribution of (8.40) can be derived by inspection from the linear-imaging-theory envelope function.

Figure 8.10 shows the envelope functions $E(\mathbf{U}+\mathbf{U}', \mathbf{U}')$ and $C(\mathbf{U}+\mathbf{U}', \mathbf{U}')$ compared with the envelope functions that would be obtained by using linear-imaging theory. Since the envelope functions $E(\mathbf{U}+\mathbf{U}', \mathbf{U}')$ and $C(\mathbf{U}+\mathbf{U}', \mathbf{U}')$ are defined over a four-dimensional space (two coordinates for $\mathbf{U}+\mathbf{U}'$ and two coordinates for $\mathbf{U}'$), we can only show the functions that would be used for an image with Fourier coefficients in one dimension. The plots in Figure 8.10 have the value $|\mathbf{U}'|$ for the horizontal axis and the value $|\mathbf{U}+\mathbf{U}'|$ for the vertical axis. The magnitudes of the envelope functions are shown as an intensity (white = 1, black = 0). There are considerable differences between the linear-imaging and nonlinear-imaging envelope functions, particularly when $\mathbf{U}+\mathbf{U}' = \mathbf{U}'$ for both $E(\mathbf{U}+\mathbf{U}', \mathbf{U}')$ and $C(\mathbf{U}+\mathbf{U}', \mathbf{U}')$ and when $\mathbf{U}+\mathbf{U}' = -\mathbf{U}'$ for $C(\mathbf{U}+\mathbf{U}', \mathbf{U}')$. The reason for this behavior can be seen by noting that, when $\mathbf{U}+\mathbf{U}' = \mathbf{U}'$, the term $\exp\{i[\psi(\mathbf{U}+\mathbf{U}') - \psi(\mathbf{U}')]\}$ in the integral defining $D(\mathbf{U}+\mathbf{U}', \mathbf{U}')$ becomes 1, and so the envelope function also

CALCULATION OF DIFFRACTION PATTERNS AND IMAGES

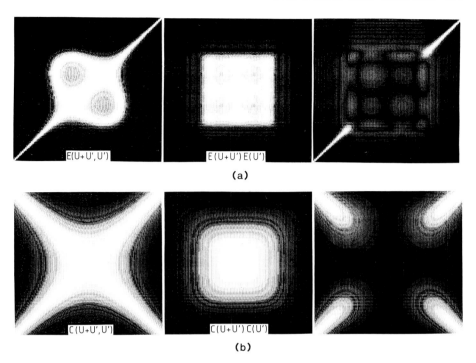

FIGURE 8.10 Envelope functions. (a) For divergence. (b) For chromatic aberrations. The leftmost diagrams show the nonlinear-imaging envelope functions; the central diagrams show the linear-imaging envelope functions. The rightmost diagrams give the difference in the envelope functions. The instrumental parameters are an accelerating voltage of 100 kV, $C_s = 2.2$ mm, $D = 120$ Å, $\Delta = -1105$ Å, and $\alpha = 1.4$ mrad. (See text for explanation.)

becomes 1. In the linear-imaging theory, $\mathbf{U} + \mathbf{U}' = \mathbf{U}'$ only when $\mathbf{U}$ and $\mathbf{U}'$ equal 0. The terms for which $\mathbf{U} + \mathbf{U}' = \pm \mathbf{U}'$ correspond to high-order spacings in the image, and so one can obtain images showing spacings that are much smaller than those expected for the point-to-point resolution of the microscope.

In the calculation of HRTEM images, one does not know, for a general system, under what conditions linear-imaging theory will hold. Hence, calculations should routinely apply nonlinear-imaging theory and only use linear-imaging theory as a special case. Unlike linear-imaging theory, nonlinear-imaging theory is inconvenient in that, the Fourier coefficients of the imaged-plane wave function are not separable. Thus, for calculation of an image, the expression

$$I(\mathbf{U}) = \sum_{\mathbf{U}'} \Psi_e(\mathbf{U} + \mathbf{U}')\Psi_e^*(\mathbf{U}') \exp\{i[\chi(\mathbf{U} + \mathbf{U}') - \chi(\mathbf{U}')]\}$$
$$\times A_\alpha(\mathbf{U} + \mathbf{U}')A_\alpha(\mathbf{U}')S(\mathbf{U} + \mathbf{U}')S(\mathbf{U}')V(\mathbf{U})D(\mathbf{U} + \mathbf{U}', \mathbf{U}') \quad (8.57)$$

must be evaluated and then inverse-Fourier-transformed. Although this procedure is more complicated than that required for linear images, it still requires the calculation of only a single image to allow for the effects of beam divergence and chromatic aberrations.

The envelope functions of the form given in (8.56) are for incident illumination that is spatially and temporarily incoherent. So these functions do not necessarily apply to microscopes fitted with field-emission guns where the condenser-lens aperture may be coherently filled. For spatially coherent illumination, the effects of beam divergence reverts to the linear-imaging case. That is, the divergence envelope function becomes separable, so

$$E(\mathbf{U} + \mathbf{U}', \mathbf{U}') = E_L(\mathbf{U} + \mathbf{U}')E_L(\mathbf{U}'), \qquad (8.58)$$

where the subscript L represents the linear-imaging damping function given by (8.48) or (8.49) (O'Keefe and Saxton, 1983).

8.4.3 Limitations of the envelope functions

The derivation of the envelope function used to model the effects of chromatic aberrations and beam divergence in HRTEM images requires two main approximations. The first of these approximations is that the wave function at the exit surface of the specimen is invariant over the range of beam divergence and chromatic aberrations in the microscope. The second approximation is that the changes in the function $\chi(\mathbf{U})$ caused by beam divergence and chromatic aberrations can be treated by a first- or second-order Taylor-series expansion. These two approximations will set an upper limit on the magnitude of beam divergence and chromatic aberrations for which the envelope functions accurately model these effects.

The validity of the second approximation can be gauged by considering the size of the second-order terms in the Taylor-series expansion of $\chi(\mathbf{U})$. These second-order terms give rise to the terms $E_x(\mathbf{U} + \mathbf{U}', \mathbf{U}')$ and $P(\mathbf{U} + \mathbf{U}', \mathbf{U}')$ in (8.56). As stated earlier, these two terms may be neglected in most cases of HRTEM imaging, and so the use of a first-order Taylor-series expansion is usually more than adequate. In fact, in the absence of beam divergence, a first-order Taylor-series expansion is an exact representation of the variation in $\chi(\mathbf{U})$ caused by chromatic aberrations. Hence, the Taylor-series expansion will be inadequate only when beam divergence is large.

Because electron wavelength varies roughly as $E^{-1/2}$, the effect of chromatic aberrations on the exit-surface wave function of the specimen is insignificant for high-voltage (above 50 kV) microscopes where the energy spread in the incident-electron beam is less than 5 V. Thus, the approximation that the exit-surface wave function of the specimen is invariant over the range of chromatic aberrations is valid for all HRTEM-imaging conditions. For beam divergence, the validity of this approximation will be specimen-dependent, or, more accurately, specimen-thickness-dependent. This approximation requires that there be no variation in amplitude and phase across the diffraction disks of the beams contributing to the image. Hence, a simple method of testing whether the envelope function can be used for HRTEM images of a particular specimen is to examine an experimental diffraction pattern taken under the same illumination conditions and of the same specimen area as the HRTEM images. If there is significant variation in intensity across individual diffraction disks in this diffraction pattern, then the divergence envelope function should not be used. Rather, only the chromatic envelope function should be used and divergence included by averaging images calculated for a series of specimen tilts as described earlier. An alternative method of testing the invariance of the exit-surface wave function is to compare the results of a multislice calculation for a crystal tilt corresponding to the maximum angle in the incident illumination with results obtained from a multislice calculation for a crystal tilt corresponding to the mean illuminating conditions.

The images that are most sensitive to beam-divergence effects are those of specimen zones containing kinematically forbidden reflections caused by glide- or screw-symmetry elements. Under certain illumination conditions, these kinematically forbidden reflections remain dynamically forbidden (Gjønnes and Moodie, 1965). One such illumination condition is when the incident-beam direction is aligned with the zone axis (which is the normal condition used for HRTEM imaging). For small tilts away from the zone axis, these reflections can have significant intensities, especially in thick crystals. Hence, using the envelope functions to calculate the image of such a zone will not include the contributions from these reflections in the image. Often, a more important factor in the images of these zones is the breakdown of the symmetry relationships between nonforbidden reflections when the crystal is tilted slightly away from the zone axis (R. W. Glaisher, private communication). The magnitude of this effect is shown in Figure 8.11. This figure compares images of the [110] zone of hexagonal CdSe computed by averaging several images from a range of specimen tilts and by using the envelope functions. The beam divergence in these images is 1 mrad. Other instrumental parameters are $C_s = 0.7$ mm and an accelerating voltage of 100 kV. The hexagonal CdSe structure contains a c-glide and therefore $(00l)$ reflections are kinematically forbidden for l odd. As shown in Figure 8.11a, images at optimum defocus computed by using the two methods become significantly different at crystal thicknesses above 300 Å. Figure 8.11b shows the same effect in a through-focal series of a thinner specimen. At the crystal thickness used for Figure 8.11b, the intensity of the $(00l)$ reflections is negligible over the range of angles in the incident illumination. However, the phase relationships between the strong reflections such as $\langle 110 \rangle$ and $\langle 111 \rangle$ vary markedly over the range of angles in the incident illumination. For specimens less than 150 Å thick, the envelope functions give an accurate representation of HRTEM images for all imaging conditions.

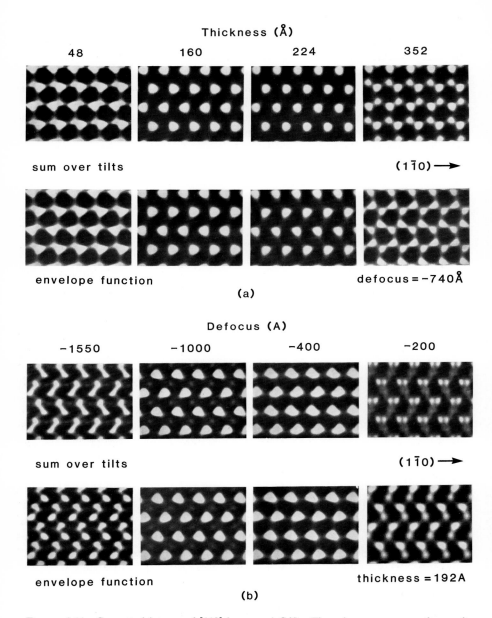

FIGURE 8.11 Computed images of [110] hexagonal CdSe. These images compare the results from allowing for beam divergence by calculating the image as an average of several images at different crystal tilts with the results from using the beam-divergence envelope function. (a) The breakdown of the use of the envelope function for thick crystals. (b) The breakdown of the use of the envelope function with defocus. Since CdSe contains a c-glide and therefore reflections that can be dynamically forbidden, it represents a worst-case situation for the use of the beam-divergence envelope function. (Courtesy of R. W. Glaisher, Ph.D. diss., University of Melbourne.)

8.4.4 Display techniques

After computing an image it is necessary to display this image so that it can be directly compared with experimental images. The calculated intensities can be mapped onto the grey scale of an output device. To give a proper comparison of experimental and calculated images, one must make allowance in the mapping to the grey scale for the contrast of the experimental-recording medium and for the exposure time used to record the experimental image. A suitable form for mapping a calculated intensity I to the grey-scale levels of the output device I_0 is

$$I_0 = M\left[\left(\frac{I - I_{av}}{I_{av}}\right)\text{contr} + \text{mean}\right]. \tag{8.59}$$

Since the grey scale of an output device usually exists at discrete levels only, the value M in (8.59) represents the number of these levels for the output device. (A value of $I_0 = M$ corresponds to the maximum intensity from the output device, and a value of $I_0 = 0$ corresponds to the minimum intensity from the output device.) In (8.59), I_{av} is the average of the calculated intensities. Thus, the term $(I - I_{av})/I_{av}$ reduces the calculated intensities to values in the range $-\frac{1}{2}$ to $+\frac{1}{2}$. The parameter "contr" is used to match the contrast value of the experimental, image-recording system, and the parameter "mean" is used to match the recording (exposure) time of an experimental image. A high value of "mean" will correspond to a long exposure time; and because the calculated intensities are reduced to the range $-\frac{1}{2}$ to $+\frac{1}{2}$, to use values of "mean" outside the range 0 to 1 is usually pointless. For example, for "contr" equal to 1, if "mean" equals $\frac{3}{2}$, then all output levels will be M or greater; and if "mean" equals $-\frac{1}{2}$, then all output levels will be 0 or less. A suitable value for both "contr" and "mean" for high-speed, electron-microscope film is 0.7. We can see from (8.59) that the calculated value of I_0 can exceed M or be less than 0; for such cases, the value of I_0 is simply truncated to M or 0, respectively.

In (8.59), the true value of the contrast is given by (contr/I_{av}). The term I_{av} is introduced to make full use of the dynamic-intensity range of the output device. If a series of experimental images recorded with identical contrast and exposure times is to be compared with a series of computed images, then the values (contr/I_{av}) and (mean − contr) must be kept constant. The value of I_{av} is given by the zero-order Fourier coefficient of the calculated image. From (8.57), we see that the zeroth-order Fourier coefficient of the image, $I(0)$, is constant for a through-focal series of images. Thus, to compare a through-focal series of experimental and computed images, one need only keep the parameters "contr" and "mean" constant. To compare a through-thickness series of experimental and computed images, one may need to change the values of "contr" and "mean" to allow for changes in I_{av}; but in general, the change in I_{av} is small enough that one usually does not need to change "contr" and "mean." In a

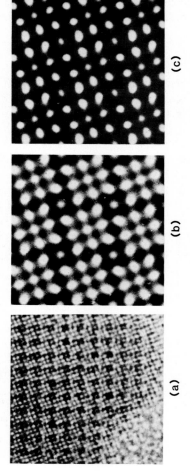

FIGURE 8.12 Example of the processing of an HRTEM image. (a) An image of GeNb$_9$O$_{25}$ taken at roughly optimum defocus on the Cambridge University High Resolution Electron Microscope. (Courtesy of D. J. Smith.) (b) The spatially averaged image (including 4-fold symmetry averaging) of an approximately 20-unit-cell area of the experimental image. (c) Spatially averaged imaged without 4-fold symmetry introduced. Comparison of the three images shows that the introduction of 4-fold symmetry is not justified, and we must conclude that the image was taken with either crystal or electron-optical misalignment.

comparison of images from one specimen with images from another specimen, the change in I_{av} can be quite large; and so in this case, one must change the values of "contr" and "mean."

Over the range of intensities in electron micrographs, the eye can distinguish between 16 and 32 levels of grey. Thus, a device that can reproduce 64 levels of grey is more than adequate for the display of computed electron micrographs. Devices of this type are photowrite systems and high-quality graphics, visual-display units. An example of this type of output is shown in Figures 8.12b–c and 8.13. So that the full range of

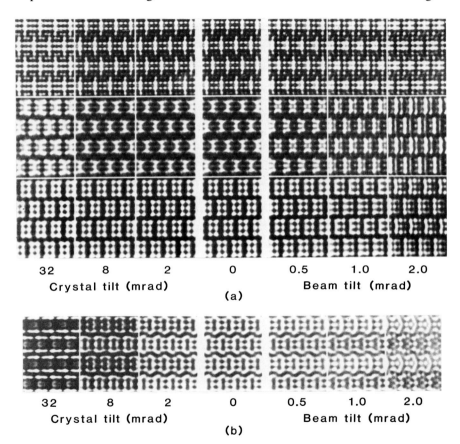

FIGURE 8.13 Simulated images of $Nb_{12}O_{29}$ showing the effect of beam tilt and crystal tilt on HRTEM images. (a) Images for a 26-Å-thick specimen at three values of defocus: −1527 Å (upper), −1000 Å (center), and −625 Å (lower). (b) Images for a 76-Å-thick specimen at −600-Å defocus. The displayed values of beam tilt and crystal tilt used in calculating these images are around an axis lying vertically on the page. Other instrumental parameters are $C_s = 1.8$ mm, an accelerating voltage of 100 kV, divergence half angle = 1.4 mrad, and chromatic-aberration spread of defocus = 140 Å. These images show the dramatic effect small-beam tilts have on HRTEM images for both thin and thick crystals. The effect of crystal tilt is more marked for a thick crystal than for a thin.

intensities from these devices is maintained, the images produced must be recorded on film. Thus, obtaining a hard copy from these devices is usually a slow process. In any image-matching procedure, many images are computed before one decides on the set of images that best matches the experimental micrographs; so an output device that gives a rapid output of images is desirable. In return for this rapid output, a decrease in the quality of the display is usually accepted. Once the set of matching images has been chosen, this matching set can be reproduced on the slower but higher-quality output device.

A printer can produce an adequate 8-level grey scale by using a suitable set of printer characters. Images produced by printers in this way are useful for comparison of a large number of computed images but are not suitable for high-quality reproduction. The major reason is that, given the character set available on most printers, producing regions of complete black is not possible. This problem is overcome by using an overprinting technique (Head et al., 1973). An overprinting technique can produce a 16-level grey scale and has been the traditional method of displaying computed micrographs. A disadvantage of the overprinting technique is that the maximum level of black that can be produced is limited by the space between characters and the space between lines. Thus, images produced in this way have a "blocky" appearance, which can sometimes dominate features in the image. Also, because the font of character sets varies from printer to printer, the set of overprinted characters defining a grey scale on one printer may not define a suitable grey scale on another printer. To create a grey scale for either of the printing techniques, one should print an area of constant intensity for each level of grey and compare these outputs by using a suitable light meter.

Point-addressable, dot-matrix printers can also be used to give high-quality hard copies of images. This technique can also be applied to line-graphics, visual-display units. For these devices, the printed page or visual-display-unit screen can be considered as a rectangular array of dots that can be either on or off. Because the size of the dots and the spacing between the dots are usually such that the dots just overlap, one can produce regions of total black or total white in images. To generate a grey scale, one groups the individual dots into picture elements (PELs) consisting of an $n \times n$ array of dots. The number of intensity levels that can be achieved for an $n \times n$ PEL is $n^2 + 1$. As n increases, the number of grey levels also increases; but the resolution of an image using a fixed total number of dots will decrease. A suitable value of n for electron micrographs is 4, giving 17 levels of grey. A problem with the dot method is that unwanted geometrical patterns in large areas of constant intensity can be generated. This problem can be overcome by careful choice of the dot patterns used in the PELs or by interweaving two different sets of dot patterns (Billington and Kay, 1974). Figures 8.5 and 8.6 were generated by using the dot method with a single set of PEL dot patterns. The results are satisfactory for the images of Figure

8.5; but in Figure 8.6, where there is little variation in intensity over some of the diffraction disks, some unwanted stripping can be seen. Since the intensity in electron micrographs varies continuously, no advantage is gained in using the dithering techniques, which are more suited to images with large, discontinuous changes in intensity (e.g., see Jarvis et al., 1976).

For ease of comparison of calculated and experimental images, the geometry of a displayed image should match the geometry of the corresponding experimental image. That is, the angles and ratio of lengths in the displayed image should be identical to those of the experimental image. The geometry of a calculated image is determined by the modeled structure, and thus, the dimensions of the calculated image are often unequal and, invariably, are nonintegral. For the display of an image, the intervals in real space at which the image is sampled must be made compatible to the output capabilities of the display device. Usually, display devices are such that picture points must be defined on a square or rectangular grid of fixed dimensions. One method of matching the output grid of the display device to the geometry of a calculated image is to sum, for each grid point, the Fourier coefficients of the image-plane wave function with phase factors appropriate to the coordinates of the grid point as determined by the magnification (i.e., scale factor) used to display the calculated image. This method requires a large number of mathematical operations, including repeated calculation of sine and cosine values.

A preferable method is to apply an FFT to the Fourier coefficients of the image-plane wave function and then interpolate the calculated real-space intensity distribution onto the grid required for the display device. The number of elements used in the FFT should be such that sampling in real space will be fine enough that a simple linear interpolation is adequate when one is calculating intensity values for the output grid of the display device. Thus, if possible, the sampling distances of the real-space intensity distribution calculated by the FFT should be approximately equal to the distances fixed by the output grid of the display device. For convenience, one usually aligns one of the principal axes of the modeled structure with either the horizontal or vertical direction of the display device Also, one usually displays more than one unit cell of a calculated image (e.g., 2×2 unit cells) so that contrast effects at the edges of a unit cell may be easily viewed.

8.4.5 HRTEM-image processing

HRTEM-image processing can be divided into two broad classes, off-line processing and on-line processing. In off-line processing, an electron micrograph or series of electron micrographs is recorded and then retained for processing by equipment away from the electron microscope. This approach allows a large number of very complicated operations to be carried out on each micrograph, which can therefore be studied in great detail. On-line processing of an image gives a response directly to the microscope operator. On-line processing therefore requires rapid response from the processing device and can only give a limited amount of information about an image. It is only with the introduction of high-sensitivity, television-imaging systems that on-line processing has become practicable. Before these systems were available, HRTEM-image processing was limited to off-line processing using either optical techniques or the digitization of electron micrographs and subsequent processing by digital methods.

A widely used application of image processing (in particular, off-line processing) is to form a spatially averaged image of a periodic object, thus giving a high-quality image for a single unit cell. With this technique, an area of the image is Fourier-transformed, and the intensity at all but the reciprocal-lattice points of the periodic object is masked out before inverse-Fourier-transforming to give the averaged image. Since only the periodic part of the image is retained, reduction in image noise is enormous. An example of the use of this technique is shown in Figure 8.12, where an area of roughly twenty unit cells of an experimental micrograph of $GeNb_9O_{25}$ (Figure 8.12a), taken on the Cambridge University HVTEM, has been spatially averaged to give the image shown in Figure 8.12b. Because the periodic part of the Fourier transform is 4-fold symmetric, the averaged image shows the known symmetry of the $GeNb_9O_{25}$ structure. This technique has been used with great success in the processing of biological macromolecules. By combining the averages from a set of images taken at different specimen tilts, one can construct a three-dimensional model of these macromolecules (for a review, see Klug, 1978–1979).

For reconstruction of these macromolecules, the image is assumed to be the projection of the molecule. This method neglects lens-transfer effects and so limits the resolution of the reconstructed models to well above the resolution limit of the microscope. The effects of lens transfer can be readily included in image processing provided the linear-imaging (WPO) approximation can be used. From (8.53), we can see that, for reconstruction of the projected potential of an object, the Fourier transform of an image of that object can be divided by $\sin[\chi(\mathbf{U})]D(\mathbf{U})$ to give the Fourier coefficients of the projected potential. This simple approach has severe limitations when the value of $\sin[\chi(\mathbf{U})]D(\mathbf{U})$ is very small—that is, when $\sin[\chi(\mathbf{U})] \simeq 0$ or when $D(\mathbf{U}) \simeq 0$. The criterion that $D(\mathbf{U}) \neq 0$ necessarily limits the resolution of the reconstructed projected potential to the point-to-point resolution of the microscope. The problem of zeros in the term $\sin[\chi(\mathbf{U})]$ can be overcome by reconstructing from a through-focal series of images. For a comprehensive discourse on the reconstruction of the projected potential from a through-focal series of HRTEM images, see the monograph by Saxton (1978). Again, this technique has been used with success for biological molecules (Unwin and Henderson, 1975), where, because of the low atomic density of the molecules, the projection or WPO approximation can be readily applied. This technique can also be applied to very thin regions of inorganic crystals.

Unfortunately, even at resolutions below 2 Å, the projected potential is not a particularly sensitive measure of small atomic displacements. An HRTEM method that potentially can give information about small atomic displacements or changes in atomic-site occupancies is the analysis of images from thick crystals. For thick crystals, because of the dynamical nature of electron diffraction, the projection approximation is not valid. That is, an image that appears like the projected potential of the structure cannot be obtained. However, there are usually values of crystal thickness at which HRTEM images are sensitive to small changes in the projected potential of the specimen. Thus, by processing images of specimens having these values of thickness, one can distinguish between different models of the structure. To initiate this process, it is necessary to analyze images from a thin region of specimen in order to obtain values of defocus and other instrumental parameters to be used in the image processing of the thick-specimen images. To reconstruct the projected potential from images of thick crystals, one must use a method that deconvolutes the summation in (8.57). Saxton (1980) has proposed that a reconstruction can be made from a through-focal series of bright-field/dark-field pairs of images. However, as pointed out in Chapter 1, section 1.5, the practical problems of obtaining dark-field images with a small-central-beam stop makes obtaining suitable dark-field images very difficult.

A problem with image simulation and image matching for thick crystals that is more dire than the experimental problems associated with taking dark-field images is the inclusion of absorption effects in simulated images. The formulation for multislice given previously assumes that inelastically scattered electrons can be neglected or contribute only a uniform background to HRTEM images. For thick crystals, this assumption does not hold. Although there have been attempts of using the Bloch-wave method with a limited number of diffracted beams to model the contrast of images formed by inelastically scattered electrons (e.g., Rossouw and Whelan, 1981), there have been no serious attempts to include inelastic effects in images calculated by the multislice method.

The major limitations on the image analysis of inorganic materials at resolution approaching the point-to-point resolution of the microscope for both thin and thick crystals are the misalignment of the beam from the zone axis of a crystal (crystal tilt) and the misalignment of the beam from the optic axis of the microscope (beam tilt). Figure 8.13 shows computed images of $Nb_{12}O_{29}$ for various values of crystal tilt and beam tilt. For even a small amount of beam tilt (less than the divergence half angle of normal-illumination conditions), the images show marked changes in comparison to images calculated for perfectly aligned conditions. This effect can also be seen in Figure 8.12c, where the image shown has been spatially averaged but not fourfold rotationally averaged. This non–rotationally averaged image does not show the 4-fold symmetry of the $GeNb_9O_{25}$ structures in this [001] projection. The breakdown in the symmetry of the image is caused by a small beam tilt. If the magnitude and the direction of the beam tilt are known, then it is possible to allow for this tilt in an image analysis. However, if the magnitude and the direction of beam tilt are not constrained in any way, these quantities are very difficult to measure to the accuracy required for image analysis. Furthermore, not only beam tilt and crystal tilt but also any residual astigmatism must be evaluated. Thus, every effort should be made to eliminate beam tilt, crystal tilt, and astigmatism from experimental images if a satisfactory image analysis is to be carried out on these images.

The correction of beam tilt and astigmatistm has been the predominant use of on-line image processing. The realization that the useful information in images such as in Figure 8.12a is seriously degraded by beam tilt has led to the development of many algorithms for the correction of beam tilt, astigmatism, and, to some extent, crystal tilt (Smith et al., 1983). The most practical of these algorithms for the correction of beam tilt and astigmatism uses the contrast variation in images of an amorphous material at various values of defocus and beam tilt (Saxton et al., 1983). The advantages of this method are that it does not involve the use of Fourier transforms (whose calculation without the aid of an array processor can be time-consuming) and that accurate beam alignment can be obtained independently of astigmatism correction. Thus, the method involves, first, correcting beam tilt and, then, correcting astigmatism. The correction of crystal tilt is more complicated than the correction of beam tilt because any method of correcting the crystal tilt must rely on a known symmetry of the crystal. Fortunately, as indicated in Figure 8.13, the effects of small crystal tilts are much less severe than the effects of beam tilts. The exceptions to this rule are images of zone axes containing kinematically forbidden reflections caused by glide- or screw-symmetry elements, as discussed in section 8.4.3. For these images, the effect of crystal tilt can be even more severe than the effects of beam tilt.

Since methods for the alignment of the electron beam with the optic axis of the microscope are now well characterized, the next development in on-line image processing will be the display at the console of the microscope of computed images side by side with a real-time image from the microscope. By storing previously calculated exit-surface wave functions for various crystal thicknesses and different structural models, the microscope operator will be able to qualitatively match an observed image with a computed image from a particular structural model. This form of image processing will also enable the operator to rapidly determine the optimal values for parameters such as defocus at which to record images for later off-line processing.

CALCULATION OF DIFFRACTION PATTERNS AND IMAGES

Appendix A: The fast Fourier transform

The Fourier transform lies at the heart of most calculations for electron diffraction and HRTEM imaging. The mathematical form of the forward Fourier transform is

$$F(j) = \sum_{k=0}^{N-1} f(k) \exp\left(2\pi i \frac{jk}{N}\right) \quad \text{for } j = 0, 1, \ldots, N-1 \quad (1)$$

and of the corresponding reverse Fourier transform is

$$f(k) = \frac{1}{N} \sum_{j=0}^{N-1} F(j) \exp\left(-2\pi i \frac{jk}{N}\right) \quad \text{for } k = 0, 1, \ldots, N-1 \quad (2)$$

To evaluate either of these transforms as a direct summation requires N^2 complex multiplications and N^2 complex additions. The cyclic nature of $\exp(2\pi i (jk/N))$ means that all exponential terms can be evaluated from a lookup table with very little overhead in computation time. In the multislice iteration, sums similar to those in equations (1) and (2) must be repeated for every slice. Hence, methods of reducing the number of operations in the calculation of the Fourier transform can considerably reduce the computer time required to compute electron diffraction patterns and images.

One such method is the fast Fourier transform (FFT) (Cooley and Tukey, 1965). The philosophy of the FFT is to factorize the sum in equation (1) into smaller, more readily calculated, Fourier transforms. Consider the discrete Fourier transform

$$F(j) = \sum_{k=0}^{N-1} f(k) W^{jk} \quad (3)$$

with the number of points in the transform equal to 2^m (i.e., $N = 2^m$). In equation (3) $W = \exp(2\pi i/N)$. Because $N = 2^m$, the coefficients j and k can be highly factorized and written in the form

$$j = j_{m-1} 2^{m-1} + \cdots + j_1 2 + j_0$$

and $\quad (4)$

$$k = k_{m-1} 2^{m-1} + \cdots + k_1 2 + j_0$$

where the factors j_i and k_i take values 0 and 1. Thus, equation (3) can be rewritten in the form

$$F(j_{m-1}, \ldots, j_0) = \sum_{k_0=0}^{1} \cdots \sum_{k_{m-1}=0}^{1} f(k_{m-1}, \ldots, k_0) W^{jk_{m-1} 2^{m-1} + \cdots + jk_0}. \quad (5)$$

The innermost sum of equation (5) is

$$\sum_{k_{m-1}=0}^{1} f(k_{m-1}, \ldots, k_0) W^{jk_{m-1} 2^{m-1}}.$$

This sum is simplified by observing that

$$W^{jk_{m-1}2^{m-1}} = W^{(j_{m-1}2^{m-1}+\cdots+j_0)k_{m-1}2^{m-1}}$$
$$= W^{j_0 k_{m-1}2^{m-1}} W^{k_{m-1}2^m (j_{m-1}2^{m-1}+\cdots+j_1)}$$

and, as W^{2^m} is 1 and therefore $W^{k_{m-1}2^m(j_{m-1}2^{m-1}+\cdots+j_1)}$ is also 1,

$$W^{jk_{m-1}2^{m-1}} = W^{j_0 k_{m-1}2^{m-1}}.$$

Thus the inner sum of equation (5) can be rewritten as

$$f_1(j_0, k_{m-2}, \ldots, k_0) = \sum_{k_{m-1}=0}^{1} f(k_{m-1}, \ldots, k_0) W^{j_0 k_{m-1} 2^{m-1}}$$
$$= f(0, k_{m-2}, \ldots, k_0) + f(1, k_{m-2}, \ldots, k_0)(-1)^{j_0}$$

The sum in this expression is a Fourier transform containing only two elements and generates the N element array f_1. Hence, after evaluating the inner sum, equation (5) becomes

$$F(j_{m-1}, \ldots, j_0) = \sum_{k_0=0}^{1} \cdots \sum_{k_{m-2}=0}^{1} f_1(k_{m-2}, \ldots, k_0) W^{jk_{m-2}2^{m-2}+\cdots+jk_0}.$$

The procedure for calculating the inner sum for k_{m-1} can now be applied to the sum over k_{m-2}, and so on. Thus, for the pth sum

$$f_p(j_0, \ldots, j_{p-1}, k_{m-p-1}, \ldots, k_0)$$
$$= \sum_{k_{m-p}=0}^{1} f_{p-1}(j_0, \ldots, j_{p-2}, k_{m-p}, \ldots, k_0) W^{(j_{p-1}2^{p-1}+\cdots+j_0)k_{m-p}2^{m-p}}$$
$$= f_p(j_0, \ldots, j_{p-2}, 0, k_{m-p-1}, \ldots, k_0)$$
$$+ f_{p-1}(j_0, \ldots, j_{p-2}, 1, k_{m-p-1}, \ldots, k_0) W^{(j_{p-2}2^{p-2}+\cdots+j_0)}(-1)^{j_{p-1}}$$

$$\text{for} \quad p = 0 \text{ to } m. \quad (6)$$

Equation (6) applies to each of the n elements in the array f_p and it shows that each element in the array can be generated from elements in the array f_{p-1} by 1 complex multiplication and 1 complex addition. Hence, the whole of the array f_{p-1} can be generated by N complex multiplications and N complex additions. Indeed, by noting that the multiplicative operation $(f_{p-1}W)$ is common to the sums with $j_{p-1} = 0$ and $j_{p-1} = 1$, the array f_{p-1} can be calculated with $N/2$ complex multiplications and N complex additions—that is, a total of $3N/2$ operations. The values of $W^{(j_{p-2}2^{p-2}+\cdots+j_0)}$ are evaluated by using a lookup table containing values of W_j for $j = 0$ to m.

Equation (6) defines a recursive series of operations starting from the array $f(k)$ and ending at the array $f_m(j_0, \ldots, j_{m-1})$ after m steps and therefore requiring a total of $(3N/2)m = (3N/2) \log_2 N$ operations. The required Fourier transform is such that

$$F(j) = F(j_{m-1}, \ldots, j_0) = f_m(j_0, \ldots, j_{m-1}).$$

Hence, the elements of the array $F(j)$ are obtained from f_m by bit reversing the indices of f_m. Similar factorization techniques can be applied to any system where N can be broken down into component factors. Furthermore, it is not necessary to develop independent relations for the inverse Fourier transform because the equality

$$\frac{1}{N} \sum_{j=0}^{N-1} F(j) \exp\left(-2\pi i \frac{jk}{N}\right) = \frac{1}{N} \left[\sum_{j=0}^{N-1} F^*(j) \exp\left(+2\pi i \frac{jk}{N}\right)\right]^*$$

can be used. In this equality the asterisk indicates complex conjugation.

The FFT algorithm reduces the number of operations required to evaluate a Fourier transform of $N = 2^m$ points from $2N^2$ to $(3N/2) \log_2 N$ plus some overhead for bit reversal and recalling W^j. In multislice, using the FFT to calculate convolution is faster than a direct sum when the array used is 32 or greater (i.e., when the number of beams is 21 or greater after allowing for aliasing).

REFERENCES

Allpress, J. G., Hewat, E. A., Moodie, A. F., and Sanders, J. V. (1972). n-Beam lattice images. I. Experimental and computed images of $W_4Nb_{26}O_{77}$, *Acta Crystallogr., Sect. A* **28**, 528.

Anstis, G. R., Lynch, D. F., Moodie, A. F., and O'Keefe, M. A. (1973). n-Beam lattice images. III. Upper limits of ionicity in $W_4Nb_{26}O_{77}$. *Acta Crystallogr., Sect. A* **29**, 138.

Arsac, J. (1966). *Fourier transforms and the theory of distributions.* Prentice-Hall, Englewood Cliffs, N.J.

Bethe, H. A. (1928). Theorie der beugung von elektronen an kristallen. *Ann. Phys. (Leipzig)* **87**, 55.

Billington, C., and Kay, N. R. (1974). Pictorial presentation of two-dimensional calculations. *Aust. J. Phys.* **27**, 73.

Brigham, E. O. (1973). *The fast Fourier transform.* Prentice-Hall, Englewood Cliffs, N.J.

Bursill, L. A., Mallinson, L. G., Elliot, S. R., and Thomas, J. M. (1981). Computer simulation and interpretation of electron microscopic images of amorphous structures. *J. Phys. Chem.* **85**, 3004.

Cherns, D. (1974). Direct resolution of surface atomic steps by transmission electron microscopy. *Philos. Mag.* **30**, 549.

Cooley, J. W., and Tukey, J. W. (1965). An algorithm for the machine computation of complex Fourier transforms. *Math. Computation* **19**, 297.

Cowley, J. M. (1981). *Diffraction physics.* North-Holland, Amsterdam.

———, and Moodie, A. F. (1957). The scattering of electrons by atoms and crystals. I. A new theoretical approach. *Acta Crystallogr.* **10**, 609.

———, and Moodie, A. F. (1958). A new formulation of scalar diffraction theory for a restricted aperture. *Proc. Phys. Soc., London* **71**, 533.

———, and Moodie, A. F. (1959a). The scattering of electrons by atoms and crystals. II. The effects of finite source size. *Acta Crystallogr.* **12**, 353.

———, and Moodie, A. F. (1959b). The scattering of electrons by atoms and crystals. III. Single-crystal diffraction patterns. *Acta Crystallogr.* **12**, 360.

Doyle, P. A., and Turner, P. S. (1968). Relativistic Hartree-Fock X-ray and electron scattering factors. *Acta Crystallogr., Sect. A* **24**, 390.

Frank, J. (1973). The envelope of electron microscopic transfer functions for partially coherent illumination. *Optik* **38**, 519.

Fujiwara, K. (1961). Relativistic dynamical theory of electron diffraction. *J. Phy. Soc. Jpn.* **16**, 2226.

Gjønnes, J., and Moodie, A. F. (1965). Extinction conditions in the dynamic theory of electron diffraction. *Acta Crystallogr.* **19**, 65.

Goodman, P., and Moodie, A. F. (1974). Numerical evaluation of n-beam wave functions in electron scattering by the multislice method. *Acta Crystallogr., Sect. A* **30**, 280.

Griton, G. R., and Cowley, J. M. (1971). Phase and amplitude contrast in electron micrographs of biological material. *Optik* **34**, 221.

Head, A. K., Humble, P., Clarebrough, L. M., Morton, A. J., and Forwood, C. T. (1973). *Computed electron micrographs and defect identification.* Defects in Crystalline Solids, vol. 7. North-Holland, Amsterdam.

Hirsch, P., Howie, A., Nicholson, R. B., Pashley, D. W., and Whelan, M. J. (1977). *Electron microscopy of thin crystals.* Krieger, Huntington, New York.

Humphreys, C. J., and Hirsch, P. B. (1968). Absorption parameters in electron diffraction theory. *Philos. Mag.* **18**, 115.

International tables for X-ray crystallography. Kynoch, Birmingham, England.

Ishizuka, K. (1980). Contrast of crystal images in TEM. *Ultramicroscopy* **5**, 55.

―――, (1982). Multislice formulation for inclined crystals. *Acta Crystallogr., Sect A* **38**, 773.

―――, and Uyeda, N. (1977). A new theoretical and practical approach to the multislice method. *Acta Crystallogr., Sect. A* **33**, 740.

Jarvis, J. F., Judice, C. N., and Ninke, W. H. (1976). A survey of techniques for the display of continuous tone pictures on bilevel displays. *Computer Graphics and Image Processing* **5**, 13.

Kilaas, R., O'Keefe, M. A., and Krishan, K. M. (1987). On the inclusion of upper laue layers in computational methods in high resolution transmission electron microscopy. *Ultramicroscopy* **21**, 47.

Klug, A. (1978–1979). Image analysis and reconstruction in the electron microscopy of biological macromolecules. *Chem. Scripta*, **14**, 245.

Krivanek, O. L. (1976). A method for determining the coefficient of spherical aberration from a single electron micrograph. *Optik* **45**, 97.

Lynch, D. F. (1971). Out-of-zone effects in dynamic electron diffraction intensities from gold. *Acta Crystallogr.* **27**, 399.

Marks, L. D., and Smith, D. J. (1983). Direct surface imaging in small metal particles. *Nature* **303**, 316.

Metherell, A. J. F. (1975). Diffraction of electrons by perfect crystals. In *Electron microscopy in materials science*, Vol. 2, ed. U. Valdre and E. Ruedl, 397. NATO Publications, Luxembourg.

Mott, N. F. (1930), The scattering of electrons by atoms. *Proc. R. Soc. London, Ser. A* **127**, 658.

O'Keefe, M. A. (1979). Resolution-damping functions in nonlinear images, In *37th annual proceedings of the electron microscopy society of America*, ed. G. W. Bailey, 556. Claitor's, Baton Rouge.

――― (1984), *Electron image simulation: A complementary processing technique.* Electron Optical Systems, SEM Inc., Chicago.

―――, and Buseck, P. R. (1979). Computation of high resolution TEM images of minerals. *Trans. Am. Cryst. Assoc.* **15**, 27.

―――, and Iijima, S. (1978). Calculation of structure images of crystalline defects. In *Electron microscopy 1978: Ninth international congress on electron microscopy, Toronto,* Vol. 1, ed. J. M. Strugers, 282. Microscopical Society of Canada, Toronto.

―――, and Sanders, J. V. (1975). n-Beam lattice images. VI. Degradation of image resolution by a combination of incident-beam divergence and spherical aberration. *Acta Crystallogr., Sect. A* **31**, 307.

―――, and Saxton, W. O. (1983). The "well-known" theory of electron image formation. In

41st annual proceedings of the electron microscopy society of America, ed. G. W. Bailey, 288. San Francisco Press, San Francisco.

———, Buseck, P. R., and Iijima, S. (1978). Computed crystal structure images for high resolution electron microscopy. *Nature* **274**, 322.

Radi, G. (1970). Complex lattice potentials in electron diffraction calculated for a number of crystals. *Acta Crystallogr., Sect. A* **26**, 41.

Rossouw, C. J., and Whelan, M. J. (1981). Diffraction contrast retained by plasmon and K-loss electrons. *Ultramicroscopy* **6**, 53.

Saxton, W. O. (1978). Computer techniques for image processing in electron microscopy. In *Advances in electronics and electron physics*, Supp. 10, ed. L. Marton and C. Karton. Academic, New York.

——— (1980). Correction of artifacts in linear and nonlinear high-resolution electron micrographs. *J. Microsc. Spectrosc. Electron.* **5**, 661.

———, O'Keefe, M. A., Cockayne, D. J. H., and Wilkens, M. (1983). Sign conventions in electron diffraction and imaging. *Ultramicroscopy* **12**, 75.

———, Smith, D. J., and Erasmus, J. J. (1983). Procedures for focusing, stigmating and alignment in high-resolution electron microscopy. *J. Microsc.* **130**, 187.

Self, P. G., O'Keefe, M. A., Buseck, P. R., and Spargo, A. E. C. (1983). Practical computation of amplitudes and phases in electron diffraction. *Ultramicroscopy* **11**, 35.

Singleton, R. C. (1969). An algorithm for computing the mixed radix fast Fourier transform. *IEEE Trans. Audio Electroacoust.* **AU17**, 93.

Smith, D. J., Saxton, W. O., O'Keefe, M. A., Wood, G. J., and Stobbs, W. M. (1983). The importance of beam alignment and crystal tilt in high-resolution electron microscopy. *Ultramicroscopy* **11**, 263.

Spence, J. C. H. (1980). *Experimental high-resolution electron microscopy*. Clarendon Press, Oxford, England.

———, and Cowley, J. M. (1978). Lattice imaging in STEM. *Optik* **50**, 129.

Steeds, J. W., and Evans, M. S. (1980). Practical examples of point and space group determination in convergent beam diffraction. In *38th annual proceedings of the electron microscopy society of America*, ed. G. W. Bailey, 188. Claitor's, Baton Rouge.

Unwin, P. N. T., and Henderson, R. (1975). Molecular structure determination by electron microscopy of unstained crystalline specimens. *J. Mol. Biol.* **94**, 425.

Van Dyck, D. (1980). Fast computational procedures for the simulation of structure images in complex or disordered crystals: A new approach. *J. Microsc.* **119**, 141.

Wade, R. H., and Frank, J. (1977). Electron microscope transfer functions for partially coherent axial illuminations and chromatic defocus spread. *Optik*, **49**, 81.

Warren, B. E. (1969). *X-ray diffraction*. Addison-Wesley, Reading, Mass.

Wilson, A. R., and Spargo, A. E. C. (1982). Calculation of the scattering from defects using periodic continuation methods. *Philos. Mag., Ser. A* **46**, 435.

9
MINERALOGY
PETER R. BUSECK AND DAVID R. VEBLEN

9.1 Introduction

The development of the transmission electron microscope as an instrument for high-resolution imaging and chemical analysis has produced a major impact in mineralogical studies in a relatively short time. The imaging, diffraction, and analytical capabilities of the TEM closely match the needs of much mineralogical research, and it is the logical complement and extension of some of the more established mineralogical techniques and instruments. Indeed, some people might argue that the study of minerals has been among the most important contributions of high-resolution transmission electron microscopy (HRTEM). This is because many minerals, unlike most metals and other simple structures, have relatively large unit cells and large-scale defects that can be imaged successfully with TEM instruments of moderate resolution (e.g., ≥ 4 Å point to point). Mineralogical applications of HRTEM have been reviewed recently by Buseck (1983, 1984) and Veblen (1985a).

In this chapter, we assume that the reader has a minimal knowledge of minerals but a substantial interest in them and in the crystal-chemical problems encountered in their study. The goals of mineralogical research coincide with those for many other materials, with the important difference that minerals are potential recorders of geological processes and events, and so they potentially are important tools for reconstructing the past. For mineralogical research, it is the problems rather than the techniques that are specialized. The standard methods of imaging, diffraction, and X-ray–emission analysis that are described in Chapter 12 are used. In the past few years, increasing use also has been made of ALCHEMI (Chapter 7) to obtain information about site-occupancy ordering in small areas of crystals.

High-resolution imaging with the TEM undoubtedly has had its greatest mineralogical impact in the study of localized structural and chemical perturbations in minerals, such as crystal defects, intergrowths of different structure types, twinning, and exsolution lamellae. These features commonly contain much of the information regarding the history of a mineral and are thus of geological interest. Yet, these local, nonperiodic structural and chemical variations are among the most difficult to study with other techniques, especially when the features are small or low in abundance. Similarly, structures having large repeat distances or long-period superstructures may be difficult to study with X-ray diffraction but, in many cases, are amenable to TEM investigation. TEM methods also can be useful for assessing the partitioning of minor elements among different crystallographic sites and for determining the exact distributions of planar faults and the presence of domain structures.

Much of this chapter deals with examples of the detection and interpretation of such small, nonperiodic features, presented in the context of the minerals or mineral groups in which they occur. These features commonly represent metastable transition states. They are of interest not only from the viewpoint of pure crystallography but also from the viewpoint of the important insights into structural and chemical disorder, nonstoichiometry, reaction mechanisms, polymorphic and polytypic transformations, and other processes and deviations from ideality that they can provide.

To some extent, the resolutions of available TEMs have limited the types of structural features that have been studied. It has been both productive and far simpler to study structural intergrowths—where resolutions of 4 or 5 Å may suffice—than to study irregularities on close to the atomic scale. As a consequence, many of the mineralogical features in the literature and in this chapter consist of intergrowth features that can be interpreted with one- or two-dimensional, lattice-fringe images. Although the positions and contrast of such fringes do not necessarily have a simple or even unique relationship to the structure of the specimen, they can reveal its basic periodicity as well as local variations in that periodicity. Fringe images thus have been used extensively for minerals where structural units of different dimensions are intergrown either periodically or in an irregular, disordered fashion. For those cases where more detailed information is required from an image, one must obtain the higher-resolution structure images. The next generation of mineralogical studies with the TEM undoubtedly will include an increasing number of studies emphasizing exact atomic positions in defects and the role of smaller-scale features, such as atom clusters and perhaps even point defects, in mineralogical reactions.

The most important pieces of information required to characterize a mineral are its crystal structure and its chemical composition. The power of the TEM is that it can be used to obtain both structural and chemical information from a crystal over volumes far smaller than is possible with other techniques. Moreover, relationships both within and between the grains of a rock can be studied; information can be obtained on grain boundaries, the geometry of fine-grained mineral intergrowths, alteration interfaces, mineral inclusions and precipitates, and compositional zoning on a scale too fine to be resolved with other techniques.

As shown in Chapters 3 and 8, interpretation of high-resolution images and electron-diffraction data can be difficult and is not always uniquely possible. Thus, the TEM has been used more widely for obtaining information about minerals where some structural information already exists than for totally unknown structures; however, the TEM is finding increasing use for minerals where little or no prior information exists. There are now several examples of minerals whose structures have been proposed in whole or in part from high-resolution data or the combination of high-resolution imaging with analytical TEM data. (Nonetheless, X-ray diffraction is still the technique of choice for structure determination of minerals when crystals of sufficient size and adequate perfection can be obtained.)

The goals of current mineralogical research can be subdivided into two separate yet interrelated aspects: (1) obtaining information about geological events and (2) understanding crystal structures and crystal chemistry, reflecting the intrinsic interest in minerals themselves. Rock-forming minerals such as pyroxenes, amphiboles, micas, and feldspars generally are of greatest use for the former goal, since they are major constituents of many rocks (Deer et al., 1966). Accordingly, a number of examples involving such minerals are included in this chapter. Also included are examples of minerals such as graphite, which, although widespread, are not abundant in most rocks; examples of minerals such as the manganese oxides, which occur only in special geological environments; and a few examples of minerals such as bastnaesite and pinakiolite, which are rather rare but are included because they display features of special crystal-chemical interest.

In this chapter, we describe some of the general mineralogical problems and processes that have been addressed by high-resolution and related TEM techniques. Wherever possible, emphasis is on high-resolution structure imaging. We use various minerals or mineral groups as examples of these problems and processes. Through these examples, we hope to illustate why certain mineralogical problems are of interest, the state of current knowledge about minerals at the electron-microscopic scale, and, where appropriate, possible directions for future research.

9.2 Reaction mechanisms

9.2.1 Introduction

Many important geological processes take place within mineral crystals, and therefore reactions in the solid state are of great interest. The reactions that form minerals in metamorphic rocks commonly fall into this category of solid-state reactions (typically, with the participation of a fluid phase as a transport medium or catalyst), and many igneous and sedimentary rocks also contain minerals that have formed or changed within the solid state. While of great importance, such reactions have proven difficult to study adequately by standard techniques. The slow rates of many silicate reactions also preclude experimental replication on laboratory time scales.

There are, however, many instances where minerals have not reacted completely, so there are metastable remnants of precursor phases that persist within the reaction products. Such remnants are especially likely to occur in minerals that react slowly, and these are just the minerals whose reactions are the most difficult to study by laboratory simulations. HRTEM imaging has found extensive applications in the detection of such precursor or intermediary materials, which, in fortunate cases, can be used for the interpretation of reaction paths and mechanisms. The quality and reliability of the interpretations vary widely, depending on the minerals and the amount of work that has been done on them. Examples from different mineral groups will be given in this section.

9.2.2 Biopyriboles

The biopyribole minerals are an abundant group of silicates in both the Earth's crust and mantle. They contain infinite chains of polymerized silicate tetrahedra (Figure 9.1) or two-dimensionally infinite sheets of similarly linked tetrahedra. The structures of the pyroxene group are based on single chains; the amphiboles contain double chains; jimthompsonite has triple chains; and the mixed-chain silicate chesterite has alternating double and triple chains. The biopyribole sheet silicates include talc, pyrophyllite, the micas, and the brittle micas. In all of the biopyriboles, pairs of chains or sheets are connected to strips or sheets of octahedrally coordinated cations, such as Mg^{2+}, Fe^{2+}, and Al^{3+}, as shown in Figure 9.2. These modules, which are called I-beams for the chain silicates, are then stacked to form the three-dimensionally continuous structures. The term *biopyribole* was derived by Johanssen (1911) from the names *biotite* (a mica), *pyroxene*, and

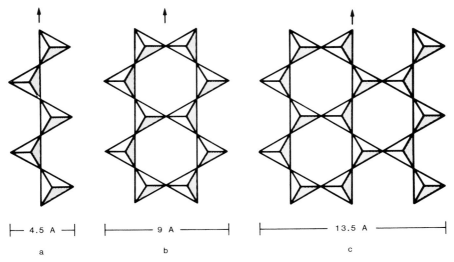

FIGURE 9.1 Different types of silicate chains, shown in polyhedral representation. Silicon atoms are located in the centers of the tetrahedra, with oxygen atoms at the vertices. (*a*) Single silicate chain, as found in pyroxenes. (*b*) Double chain, as in amphiboles. (*c*) Triple chain, as in jimthompsonite. (After Veblen et al., 1977.)

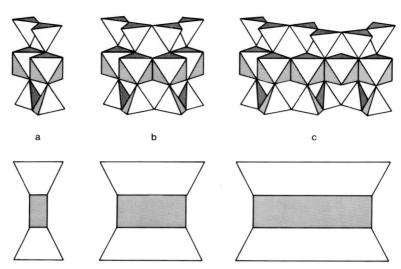

FIGURE 9.2 Chain-silicate modules (I-beams) viewed along the chains, which are articulated to strips of octahedrally coordinated cations. Individual polyhedra are indicated at the top, and simplified representations are shown at the bottom. (*a*) Single chain. (*b*) Double chain. (*c*) Triple chain.

amphibole. The term *pyribole* describes the chain-silicate members of the group (i.e., the biopyriboles excluding the sheet silicates).

Crystallographic shear of pyroxene can produce all the other biopyribole structures (Chisholm, 1973, 1975); in this regard, the biopyriboles are similar to other crystallographic-shear structures, as discussed in Chapter 10. An alternative description is that all of the biopyriboles can be constructed by interleaving two types of slabs, a P-slab consisting of pyroxene structure and an M-slab consisting of mica structure (Thompson, 1970, 1978). Pyroxenes consist purely of P-slabs, and micas and talc are purely M-slabs. As shown in Figure 9.3, other biopyribole structures contain ordered mixtures of M- and P-slabs. For example, amphibole has a 1:1 mixture of M's and P's, or simply (MP). Jimthompsonite has a 2:1 ratio (MMP), and chesterite is (MPMMP). Because they all consist of mixtures of two structurally and stoichiometrically different types of slabs, all the biopyriboles are stoichiometrically collinear (Thompson, 1978), including even those we discuss below that have disordered sequences of M- and P-slabs (Veblen and Buseck, 1979a). Thompson introduced the terms *polysomatism* and *polysomatic series* to describe these types of structural and chemical relationships.

9.2.2.1 Structurally ordered types. Structurally ordered biopyribole types, such as the pyroxenes, amphiboles, and sheet biopyriboles, have been recognized as important mineral groups for well over a century. Likewise,

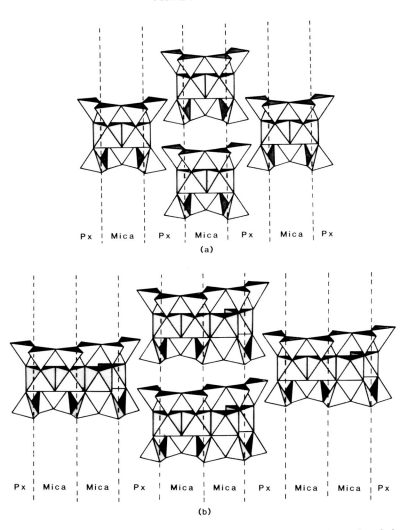

FIGURE 9.3 Schematic diagram of chain-silicate structures viewed along the chains and showing the slabs of mica (M) and pyroxene (P) structure. (*a*) Amphibole structure (MP). (After Thompson, 1978.) (*b*) Triple-chain pyribole structure (MMP).

though more recently discovered, the ordered biopyriboles having triple silicate chains (jimthompsonite and chesterite) can occur in relatively large crystals and were discovered with single-crystal, X-ray–diffraction techniques (Veblen and Burnham, 1978).

In addition to these biopyriboles that occur macroscopically, a number of other ordered biopyribole structures have been disovered with HRTEM. Although they occur in statistically significant amounts, they are too fine-grained to have been recognized by more conventional methods. For example, a structure with a unit-cell chain sequence (2233) can occur intergrown with chesterite (23), even though both structures have the same stoichiometry (Veblen and Buseck, 1979a). Other structures found intergrown with anthophyllite, chesterite, and jimthompsonite include those with unit-cell chain sequences of (233), (2333), (232233), (433323), and (43332343332423). Such unusual structures do not occur in volumetrically large amounts, but the probability that they arose from random combinations of chains of various widths is miniscule [e.g., one in 10^{42} for the structure (2333)].

In spite of only occurring in small amounts, such unusual structures can be important for understanding the growth or reaction processes that formed a mineral crystal. Ordered structures that occur in small amounts intergrown with different ordered structures or disordered material are not restricted to the biopyribole system. They also occur, for example, in the pyroxenoids, the humite/leucophoenicite system, and in various layer minerals. To the extent that such intergrown structures reflect crystal growth and reaction processes, HRTEM can provide a valuable tool for unraveling the history of a given mineral occurrence.

9.2.2.2 Structurally disordered types. Although the thermodynamically stable biopyriboles certainly are the most common ordered types, disordered biopyriboles also can be important for some compositions (e.g., the ferromagnesian biopyriboles). Like ordered structures that occur in small amounts, the detailed microstructures of such disordered material can be crucial for understanding the structural mechanisms of solid-state transformation reactions, and HRTEM is ideally suited for such structure determinations in disordered silicates.

As noted by Veblen and Buseck (1979a, 1979b), the sequences of silicate chains in disordered biopyribole can be random, or nonperiodic sequences can occur nonrandomly. In chesterite (23), defects that consist of one block of (2233) occur more frequently than other perturbations of the ideal chain sequence. Of more importance for understanding detailed reaction mechanisms, however, are the terminations of planar defects that have been arrested during their growth. For example, in amphiboles that have undergone a partial hydration reaction (sometimes called alteration by geologists), defects such as that shown in Figure 9.4 commonly emanate from fractures (Veblen and Buseck, 1980). This defect consists of a slab of sextuple-chain silicate that terminates in the double-chain material of the reactant crystal; the defect implies that at least part of the hydration reaction took place by the nucleation and growth of such features. Diffusion necessary for the reaction presumably took place, at least in part, by relatively fast transport along the channel at the termination of the sextuple-chain slab. As we discuss in the next section, reaction mechanisms of this sort can be important in the replacement of one biopyribole by another.

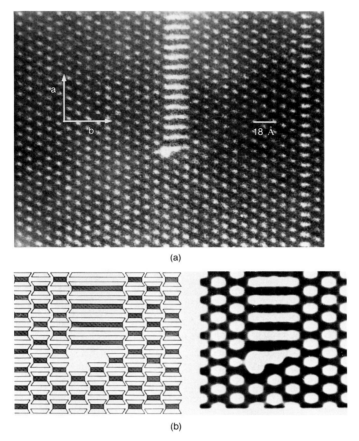

FIGURE 9.4 A slab of sextuple-chain structure terminating in amphibole. (*a*) Experimental HRTEM image. (After Buseck and Veblen, 1978). (*b*) A structural model for the termination, in I-beam representation (left) and a computer-simulated image based on this structural model (right). (After Veblen and Buseck, 1980.)

9.2.2.3 Types of biopyribole reactions. Many types of reaction behavior and reaction-associated microstructures have been observed in the biopyribole system. Rather than providing an exhaustive description of such features, we provide a rough idea of the types of information that have been produced by HRTEM. There is great diversity in biopyribole-replacement reactions occurring under different physical and chemical conditions (Veblen and Buseck, 1979a, 1980, 1981; Nakajima and Ribbe, 1980, 1981; Veblen, 1980). Reviews of replacement reactions in biopyriboles can be found in Buseck et al. (1980) and Veblen (1981).

Although there is great variation in the details of reaction mechanisms by which one biopyribole is replaced by a different ordered or disordered biopyribole, the mechanisms can be separated into two general groups: (1) lamellar reaction mechanisms and (2) bulk reaction mechanisms. An

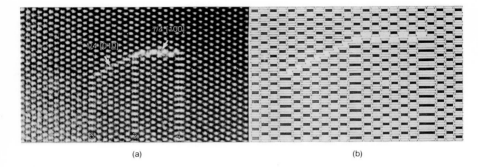

FIGURE 9.5 Cooperative replacement of amphibole by two triple-chain slabs and one quadruple-chain slab. (*a*) The HRTEM image shows that the terminations of these slabs are connected by planar defects having projected displacements of 1/4[010] and 1/4[100]. (*b*) A structural model for this structure, in I-beam representation. The orientation is the same as in Figure 9.4. (After Veblen and Buseck, 1980.)

example of a lamellar reaction was given above, where we indicated that a slab of sextuple-chain silicate could nucleate in amphibole and grow at its termination. There are many other lamellar-reaction mechanisms that also have been observed in biopyriboles, including, for example, the growth of triple-chain slabs or pairs of double-chain slabs into pyroxene. In addition to replacement by simple lamellae, chain silicates can be replaced by the cooperative growth of several different lamellae that are connected by displacive planar faults (Figure 9.5).

The replacement of a pyribole by lamellar mechanisms typically results in a reaction product with a high degree of structural disorder; this may be the origin of many of the biopyribole crystals that have been observed to have apparently random chain sequences. So that ordered crystals are produced from these disordered structures, a second stage of reaction is necessary. In this stage, displacive planar defects may move through the disordered material, replacing it with an ordered chain sequence (Veblen and Buseck, 1980).

The second broad category of replacement reactions in biopyriboles are those that occur by bulk mechanisms. In these reactions a broad reaction front sweeps through the crystal, with the product structure totally replacing the reactant as the reaction proceeds. This bulk mechanism is similar to the massive transformation mechanisms familiar from metallurgy, except that, in the case of biopyribole replacement, the reaction product differs chemically from the reactant. In these reactions, at least part of the chemical transport necessary for the reaction to proceed presumably takes place along the interface between the reactant and product. Bulk-reaction mechanisms have been observed to be important for some cases of pyroxene

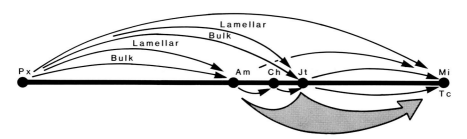

FIGURE 9.6 Some of the paths observed for hydration reactions in biopyriboles. Some reactions take place both by lamellar and bulk-reaction mechanisms. Here, Px = pyroxene, Am = amphibole, Ch = chesterite, Jt = jimthompsonite, Mi = mica, Tc = talc. (After Veblen and Buseck, 1981.)

replacement by amphibole, pyroxene replacement by triple-chain silicate, amphibole replacement by triple-chain silicate, and replacement of any of the pyribole structures by sheet silicates.

HRTEM has shown not only that a given reaction may proceed by more than one structural mechanism but also that a reaction may take place by more than one path; in some cases, several different reaction paths may be followed even within the same hundred-micrometer crystal. Figure 9.6 summarizes some of the reaction paths and mechanisms that have been observed for hydration reactions in a variety of biopyriboles.

The determination of different reaction mechanisms with HRTEM has underlined another important feature of crystalline reactions: The structure of the reactant and the way it fits together with the crystal structures of various possible reaction products can strongly affect the mechanisms of reaction and also which reaction products form. These effects are particularly important for lamellar reactions but may also hold for some bulk reactions. As an example, Figure 9.7 (Veblen, 1981) shows that a slab that is one amphibole chain wide cannot fit coherently into the pyroxene structure. In contrast, a simple triple-chain slab will fit, as will a slab of structure that is two amphibole chains wide. Growth of the latter two types of lamellae are important mechanisms of pyroxene replacement; growth of single amphibole lamellae almost never occurs. This is not a simple thermodynamic effect but results from the geometric ways in which reactant and product crystal structures can fit together. Indeed, effects such as these may help to explain structurally the growth of metastable phases in some geological systems.

The above discussion has been restricted to replacement reactions in biopyriboles. However, both lamellar and bulk reaction mechanisms, as well as reactions proceeding by multiple paths, have been observed with HRTEM in other geologically important mineral groups, such as the pyroxenoids and sheet silicates.

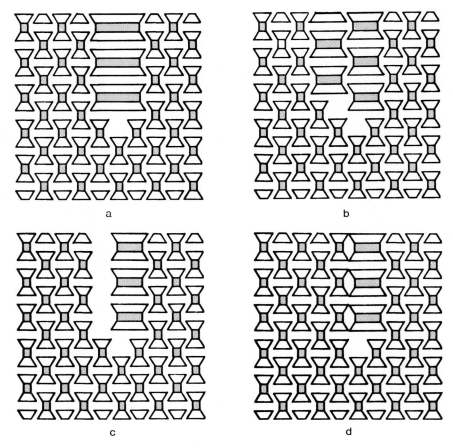

FIGURE 9.7 I-beam diagrams. (*a*) Slab that is one triple-chain wide. (*b*) Slab that is two double-chains wide. These diagrams show that such slabs can fit coherently into the pyroxene structure. (*c*) Slab that is one double-chain wide, resulting in a gap in the structure. (*d*) Slab that is one double-chain wide, resulting in a zone of impossible atomic overlaps.

9.2.3 Graphite crystallization

Many different types of rocks contain elemental carbon or carbon compounds. Carbon occurs, for example, as organic debris in a wide variety of sedimentary rocks, as subcrystalline material in low-grade metamorphic rocks, and as well-crystallized graphite in igneous and high-grade metamorphic rocks. It is also an important constituent in the most primitive and least transformed of the meteorites, the carbonaceous chondrites. The carbon in many sedimentary rocks is of biological origin, and some such rocks are hundreds of millions of years older than the oldest rocks that contain fossils. There is thus the possibility that the study of graphite precursors in such

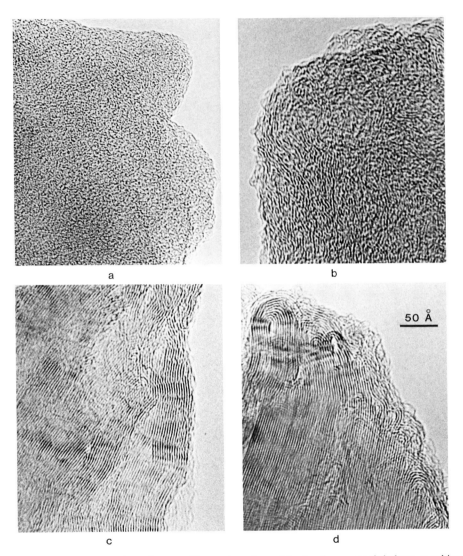

FIGURE 9.8 Images of poorly crystalline carbon formed by heating acenaphthylene, a coking compound, for 1 h. (a) At 500°C. (b) At 1000°C. (c) At 1750°C. (d) At 2000°C. The improved crystallinity as a function of temperature is evident. Few fringes are evident in (a), but regions with well-developed, parallel fringes resembling those of ideal graphite are evident in (d). The material is sufficiently well crystallized so that dislocations can form and are prominent in (c). (After Buseck et al., 1987.)

Precambrian rocks might provide insight into the earliest life forms on earth (Hayes et al., 1983; Buseck, Huang, and Miner, 1988).

Organic molecules can react in innumerable ways to form the planar sheets of pure carbon that are the basis of the graphite structure. Such reaction and polymerization of the carbon is structurally simpler, for example, for aromatic than for aliphatic molecules. Similarly, molecules that are already planar or that contain few or no bonding vacancies (i.e., only six-membered rings) graphitize more readily than molecules that do not display these features. Recent reviews are provided by Oberlin (1984), Lewis (1982), and Buseck, Huang, and Keller (1987).

The details of graphite crystallization depend not only on the development of bonding between carbon atoms but also on the release of other atoms, notably hydrogen and oxygen. These details are not amenable to HRTEM imaging, but the development of graphite crystals from only slightly polymerized sheets of carbon can be observed. Figure 9.8 shows such a sequence for acenaphthylene, one of a set of organic molecules that has been studied by a variety of analytical techniques as a function of annealing temperature in an attempt to understand the development of graphite from noncrystalline, organic precursors. A complete progression can be traced from individual fringes representing short, isolated carbon sheets, through small, partly organized crystals containing a few subparallel stacks of carbon sheets, to fully developed graphite.

9.2.4 *Cordierite transformation*

Cordierite ($Mg_2Al_4Si_5O_{18}$) was one of the first minerals to be imaged successfully by HRTEM (Buseck and Iijima, 1974). It occurs in two polymorphs, with a transformation temperature of about 1450°C (Schreyer and Schairer, 1961). Above 1450°C, the equilibrium form is hexagonal, with space group P6/mcc; and below, it is orthorhombic, Cccm. The structures were refined by Gibbs (1966), Cohen et al. (1977), and Meagher and Gibbs (1977). High-temperature cordierite has the same space group as beryl, and comparison of the HRTEM images of beryl and cordierite shows that, in c-axis projection, their structures are similar (Figure 9.9), even though beryl is a ring silicate and cordierite is a framework silicate. The transformation in cordierite occurs as a result of Al–Si ordering among tetrahedral sites that are equivalent in the hexagonal polymorph, thereby producing the change in symmetry from hexagonal to orthorhombic.

Putnis (1980) and Putnis and Bish (1983) have studied the mechanism and rate of Al–Si ordering through the transition. They found a sequence of modulated structures between the high- and low-temperature forms and found that the transition is first-order. Slow cooling below the transition temperature results in an alternative transformation mechanism that produces metastable states. Localized Al–Si ordering results in modulation waves that permit short-range order while retaining the overall hexagonal

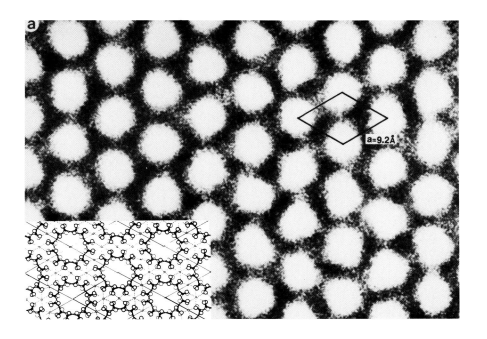

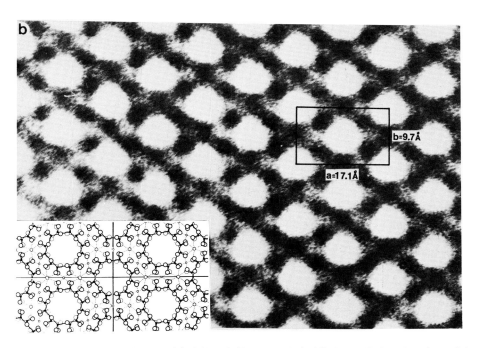

FIGURE 9.9 HRTEM images. (*a*) Of beryl ($Be_3Al_2Si_6O_{18}$). (*b*) Of cordierite viewed parallel to the *c*-axis. The round white areas are channels. Cation ordering produces a change from hexagonal to orthorhombic symmetry in low-temperature cordierite. The arrows in (*b*) mark atom clusters. The insets show the structures as determined by X-ray diffraction. (After Buseck and Iijima, 1974.)

symmetry of high-temperature cordierite, and maxima and minima of the modulations represent Al–Si distributions that are twin-related. With time, the modulation periodicity increases until, eventually, the structural distortions near the boundaries of the locally ordered domains produce sufficient strain that it is energetically advantageous to form twinned orthorhombic cordierite (see the discussion on alkali feldspars in section 9.5.3). During the transformation, the strain energy of the distortions produced by the modulations is replaced by the surface free energy of the sharp twin boundaries. This process in some ways resembles the development of intergrowths that arise from phase separation by a spinodal mechanism.

The cordierite transformation provides an opportunity to compare several methods of studying mineralogical reactions. A variety of techniques have been used to study cordierite, including powder X-ray diffraction—from which comes the widely used distortion index Δ that, in part, reflects Al–Si ordering (Miyashiro, 1957)—polarized-light microscopy (Armbruster and Bloss, 1981), infrared (IR) spectroscopy, and transmission electron microscopy. The study of Putnis and Bish (1983) permits an interesting comparison of these techniques for detecting the short-range ordering that is the first stage in the hexagonal-to-orthorhombic transition. They find that the IR measurements, being highly sensitive to short-range interactions, detect the earliest steps in Al–Si ordering in devitrified cordierite glass after a few minutes annealing at 1200°C. Modulations observable by TEM imaging occur after about 3 h, and coarsening of these modulations to $\simeq 1000$ Å occurs before peak splitting (and a measurable Δ-value) is detectable by powder X-ray diffraction; this occurs after about 100 hours annealing. The 2V measurement (indicating the angle between the two optic axes), made using polarized light in an optical microscope, also records the effects of ordering prior to X-ray diffraction.

9.2.5 Biotite-chlorite reaction

The conversion of biotite to chlorite is a common reaction in slowly cooled igneous and metamorphic rocks that have been allowed to hydrate. Biotite is a mica consisting of layers having two tetrahedral silicate sheets that form a sandwich filled with octahedrally coordinated cations. Monovalent cations, primarily potassium, are situated between these layers. Chlorite possesses a similar sheet structure, but the layers are separated by brucite-like magnesium hydroxide sheets instead of potassium ions (Figure 9.10).

Two simple reaction mechanisms that can produce chlorite from biotite have been described by Veblen and Ferry (1983); other more complicated mechanisms are also possible. One mechanism involves the replacement of a sheet of interlayer cations by a hydroxide sheet (Figure 9.11a). An alternative mechanism involves the dissolution of two sheets of interlayer cations and two tetrahedral sheets, leaving a remnant hydroxide sheet (Figure 9.11b). This second mechanism results in the termination of a mica layer; the first mechanism results in termination of a brucite-like sheet. Since these layers and terminations can be distinguished with HRTEM, observations on incompletely reacted crystals can be used to determine which reaction mechanism operated in any given occurrence.

HRTEM studies on several partially chloritized biotites have determined the mechanisms of reaction. Veblen and Ferry (1983) showed that the mechanism in Figure 9.11b operated in the case of a biotite from a granitic igneous rock. Alternatively, Olives Baños and Amouric (1984) showed that the other mechanism was responsible for chloritization of a metamorphic-biotite specimen, and Olives Baños et al. (1983) suggested that chloritization occurred selectively along layers where slip resulting from deformation had occurred. Eggleton and Banfield (1985) found that both mechanisms operated in two other igneous biotites.

These results are similar in some ways to the observations on pyribole-replacement reactions discussed above. For example, reaction of biotite to

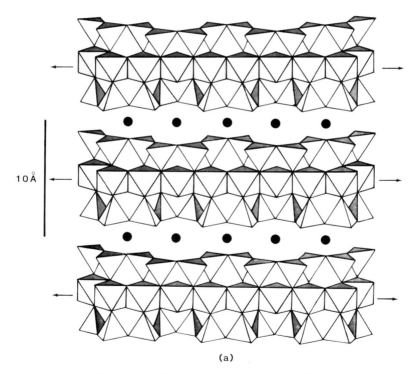

(a)

FIGURE 9.10 Schematic representations of some important sheet-silicate-structure types, all of which contain two-dimensional, infinite silicate sheets. (a) The mica structure, consisting of tetrahedral (T)–octahedral (O)–tetrahedral sandwiches (TOT, or 2:1 layers). Interlayer cations are shown as black circles. (b) The chlorite structure, consisting of TOT layers alternating with hydroxide (brucite-like) sheets. (c) The basic serpentine structure, consisting of TO (or 1:1) layers.

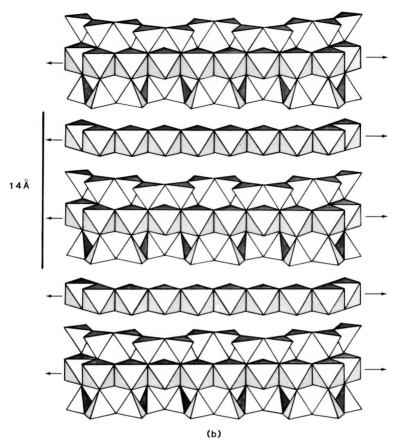

14Å

(b)

FIGURE 9.10—*continued.*

chlorite can proceed by more than one reaction mechanism, even within the same crystal. Reaction to chlorite can also follow different paths, as shown by Olives Baños and Amouric (1984), who observed ordered, intermediate structures that are absent in the other specimens examined.

The reaction mechanism of the biotite-chlorite reaction is intimately related to the detailed reaction chemistry. The mechanisms in Figures 9.11a and b result in a substantial volume increase and decrease, respectively, and the two structural mechanisms have different chemical consequences (Veblen and Ferry, 1983). The first mechanism, for example, implies removal of potassium and introduction of large numbers of octahedrally coordinated cations; the second mechanism requires removal of potassium and large amounts of silica. In addition, the two mechanisms differ drastically in the behavior of H^+, suggesting that the acidity of altering fluids may control the structural mechanism of the reaction. Veblen and

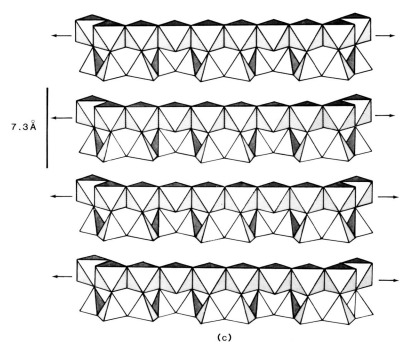

(c)

FIGURE 9.10—continued.

Ferry (1983) showed that the chemical consequences of the reaction mechanism are consistent with the macroscopic, electron-microprobe-scale mineralogical changes in their specimen. Thus, HRTEM-scale reaction mechanisms can be related to the larger-scale chemistry of a solid-state reaction; electron microscopists working with reactions should be aware of the macroscopic chemical implications of their results.

9.3 Stacking disorder and polytypism

9.3.1 Introduction

A large number of minerals have layer structures or have units that can be viewed as layers. An example of the latter are the pyroxenes (section 9.3.4), which are described more conventionally as chain silicates. Where layers occur, there is commonly more than one way to stack them one on top of another, which can result in stacking faults or twinning. If the layers can be stacked in more than one regular sequence, the structures are called polytypes. Both stacking defects and periodic polytypes have been observed in layer minerals studied with the TEM.

For observing stacking faults, twinning, and polytype periods, one

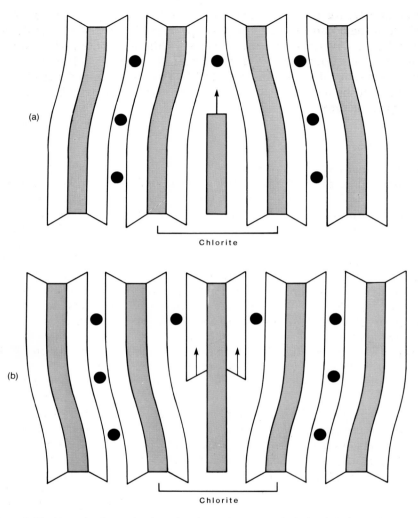

FIGURE 9.11 Two simple mechanisms for the replacement of biotite (mica) by chlorite. (*a*) Hydroxide sheet replacing a sheet of mica interlayer cations. (*b*) Residual hydroxide sheet produced by dissolution of two sheets of interlayer cations. See text for details.

commonly uses one-dimensional fringe images. However, high-resolution structure images or electron-diffraction patterns are required to determine specific polytypes; and even then, additional information may be required. Since HRTEM images are two-dimensional projections of structures, there commonly is an ambiguity in interpreting the direction of a projected stacking vector, thereby precluding a unique interpretation. Thus, either another image or electron diffraction pattern displaying a different orientation for a given crystal is required to obtain correct polytype assignments.

Alternatively, for polytypes that occur in relatively large volumes, X-ray–diffraction data can be used for polytype identification. We shall describe examples of a variety of mineral types. Although considerable work on polytypism has been completed, this area will benefit significantly from the next generation of high-resolution TEMs, because they will permit better definition of the details of stacking sequences and, thus, stacking defect and polytype structures.

9.3.2 Micas

There are many varieties of micas and related minerals, and they can occur in almost all of the major rock types. As described in section 9.2.5, their structures consist of planar tetrahedral (T) and octahedral (O) sheets that are ordered into layers of the type TOT, sometimes called talc-like layers (Figure 9.10). These layers are then arranged one on top of another; in some cases, there are no cations between the layers, as in the structures of talc and pyrophyllite. If the layers are bonded to one another by alkali or alkali-earth cations, the mica and brittle-mica minerals result. When the TOT-layers are separated by Mg, Fe^{2+}, Al hydroxide (brucite- or gibbsite-like) sheets, the chlorite minerals are produced. The symmetry of the (undistorted) T-sheets is hexagonal, and thus there are generally six possible ways in which succeeding layers can be stacked.

There have been many X-ray studies of polytypism in the micas (Smith and Yoder, 1956; Ross et al., 1966; Takeda, 1967; Baronnet, 1975), but these studies define only average polytypic sequences. Because micas can exhibit considerable stacking disorder, there also has been much scope for HTREM imaging in the study of mica polytypism. Iijima and Buseck (1978) studied polytypism in a muscovite and a biotite mica. They showed that HRTEM images could be used to indicate the basic TOT-layers and that the stacking sequences could be observed by noting the relative positions of the white spots that are produced in the positions between adjacent layers (Figure 9.12). Although the images allow one to limit the number of possible stacking sequences, the ambiguity that results because the images are projections was not resolved. Nonetheless, Iijima and Buseck were able to recognize several polytypic sequences having different periodicities and to show that distinct stacking types occur in intimate intergrowths on the scale of tens of angstroms. Some of the questions left unanswered by that study were the definition of the origins of given unit cells and, therefore, of polytypic sequences and the number of repeat units required to define a given polytype.

To understand the significance of polytypism in micas, Baronnet and his colleagues have studied numerous natural and synthetic micas. They used a large range of computed images to interpret and validate their experimental work (Amouric et al., 1981), and they also used HRTEM to study the details of polytypism in micas grown under a variety of experimental

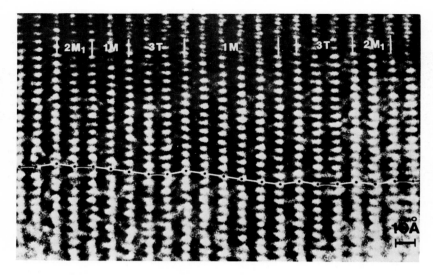

FIGURE 9.12 HRTEM image showing several stacking sequences in biotite (Fe, Mg mica). The vertical rows of white spots define the well-developed cleavage planes of micas and lie along the interlayer cation sites. The segmented line running roughly horizontally defines the projected stacking vectors in this mica. The symbols mark the names of polytypes that could correspond to the stacking sequences in the indicated regions; however, more repetitions are required before such polytype assignments can be considered to be meaningful. (After Iijima and Buseck, 1978.)

conditions. They found that crystal nucleation and growth mechanisms are major factors in determining which polytypes form (Amouric and Baronnet, 1983, 1984). They conclude that environmental factors such as temperature of crystallization and degree of supersaturation of the parent fluid can be important for determining polytypes. However, further work is needed to unravel the details.

9.3.3 Chlorites

The multiplicity of possible stacking configurations found between the layers of micas also can occur between the talc-like layer and the brucite-like sheet in chlorite (Figure 9.10b). However, there is a further complexity, since a similar multiplicity also can occur on the opposite side of the brucite-like sheet, where it comes into contact with another talc-like layer. The result is that the number of distinct stacking sequences and the possibilities for disorder in chlorites are far greater than in micas. In addition, there can be further complexities resulting from extra or missing brucite-like sheets (section 9.4.2).

Although polytypism in chlorites has long been a subject of interest, its complexity has limited the number of comprehensive studies. The basic polytype nomenclature was derived by Bailey and Brown (1962), who

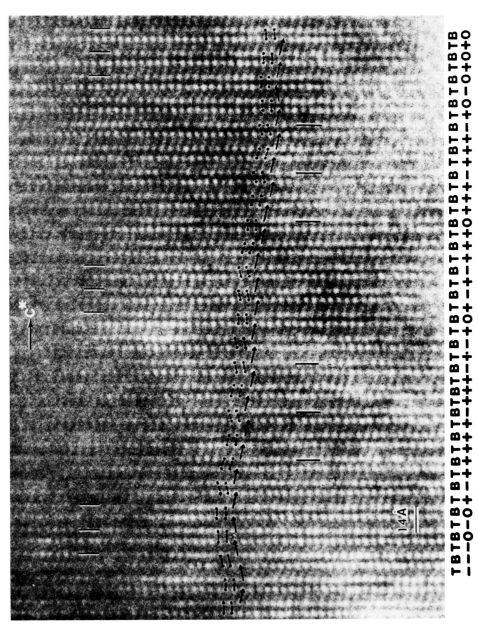

FIGURE 9.13 HRTEM image of chlorite showing regularly alternating layers of talc-like (T) and brucite-like (B) layers. As in Figure 9.12 of mica, the white spots define the cleavage planes parallel to the layering. The arrows and short, roughly horizontal lines indicate the projected stacking sequences, and the vertical lines indicate repeating units. (After Spinnler et al., 1984.)

determined polytypes for numerous chlorites from X-ray–diffraction measurements. Fringe and structure images were first used for determining chlorite polytypes by Iijima and Buseck (1977), and a more detailed study was carried out by Spinnler et al. (1984). They used correlated bright- and dark-field fringe images in order to define the polytypes. Although more detail is available in the dark-field images, including spacings that are two and three times the basic one-layer chlorite structure, they still cannot be interpreted uniquely in terms of exact polytype.

Spinnler et al. (1984) confirmed that HRTEM structure images can be used to distinguish among the one-layer polytypes, to characterize layer-stacking sequences, and to determine projected shift vectors (reflecting the offsets between the tops and bottoms of the respective layers and sheets) across the talc-like layers and the brucite-like sheets. These vectors are then used to determine which of the several polytypes occurs. Spinnler et al. determined that there are five distinct projections of the several polytypes along the major viewing directions ([100], [1$\bar{1}$0], and [110]), and they used calculated images of these to interpret the HRTEM images (Figure 9.13). Similar procedures should be useful for determining polytypes in other mineral systems from HRTEM images, although the details of the calculations and projections are specific to the structures in question. A parallel, theoretical study of chlorite polytypism, but without the use of electron microscopy, also appeared recently (Durovic et al., 1983; Weiss and Durovic, 1983), and it remains for future studies to further correlate and integrate the two approaches.

9.3.4 Pyroxenes

The pyroxene minerals are single-chain silicates and traditionally have not been considered in terms of polytypism. However, adjacent chains parallel to (100) can be viewed as forming slabs of structure (Morimoto and Koto, 1969; Thompson, 1970). The stacking of these slabs can, in turn, be interpreted in terms of polytypism. Iijima and Buseck (1975) used such a model to explain high-resolution images of enstatite ($MgSiO_3$) pyroxene and to show how the three major polymorphs (protoenstatite: orthorhombic, Pbcn; orthoenstatite: orthorhombic, Pbca; and clinoenstatite: monoclinic, $P2_1/c$ at room temperature) can intergrow coherently.

Buseck and Iijima (1975) reported the enstatite polytypes in rocks from a variety of terrestrial occurrences and meteorites and used the crystals to interpret aspects of the geological histories of the host rocks. Based on the stacking sequences, they suggested that one should be able to distinguish among crystals formed by a low-temperature, static-transformation mechanism, thermal quench, or shock. They also reported the occurrence of a few units of structure with larger repeat distances that were thought to represent new polytypes, but subsequent work showed that these structures are not statistically significant (Buseck et al., 1980). The existence of statistically meaningful amounts of long-period polytypes in pyroxene minerals remains to be demonstrated. More recently, stacking variations in enstatite whiskers and platelets from interplanetary dust particles have been used to clarify the thermal history of these particles (Bradley et al., 1983).

9.3.5 Pyrosmalite

The pyrosmalites are layer-silicate minerals that consist of $Si_{12}O_{30}$ tetrahedral sheets that alternate rigorously with brucite-type octahedral sheets to yield a unit layer of composition $(Mn, Fe)_{16}(Si_{12}O_{30})(OH, Cl)_{20}$ (Takéuchi et al., 1983). In the sense of having alternating tetrahedral and octahedral sheets, they resemble the serpentine minerals (section 9.5.2), but some of the tetrahedral apices in pyrosmalites point toward the closest octahedral sheet, while others point away from it.

There are two ways of placing the octahedral sheet onto the tetrahedral one, yielding either the pyrosmalite or the mcGillite structure type. The two possible types of stacking lead to the potential for stacking disorder and, if the two stacking sequences intermix in ordered configurations, to other polytypes. Such polytypism has been studied both theoretically and by HRTEM imaging (Iijima, 1982a, 1982b; Ozawa et al., 1983; Takéuchi et al. 1983). Twinning by layer rotation of 120° around [100] gives rise to a series of polytypes. Iijima reports six mcGillite polytypes. The structures of manganpyrosmalite (which is isostructural with pyrosmalite) and schallerite represent other stacking variations.

9.3.6 Other polytypic minerals

Polytypism has been studied by transmission electron microscopy in a large number of minerals. Like the previous examples, some are routinely viewed as layer structures, and others are commonly considered in terms of chains or other structural units. An infinite variety of stacking sequences theoretically can occur in all of them.

Many of the minerals discussed elsewhere in this chapter display polytypism, and details for some of those minerals are discussed in the cited references. Polytypism has also been studied with HRTEM in wollastonite (Jefferson and Thomas, 1975; Wenk et al., 1976), sursassite (Mellini et al., 1984), zirconolite (White et al., 1984), chloritoid (Jefferson and Thomas, 1979), talc (Veblen and Buseck, 1981), chrysotile and antigorite serpentine (Yada, 1971, 1979; Veblen, 1980), amphibole (Hutchison et al., 1975), cancrinites (Hassan and Grundy, 1984), and sphalerite and wurtzite (Akizuki, 1981, 1983; Fleet, 1983). In all of these cases, the TEM has played a critical role in establishing the presence of stacking irregularities and polytypism. However, as discussed in section 9.3.3, a full description of many polytypes requires tilting capabilities beyond those available on most current TEMs, and complete characterization of stacking variations therefore remains a challenge for the future.

9.4 Intergrowth disorder and nonstoichiometry

9.4.1 Introduction

The preceding discussion has referred to a number of different minerals that consist of chemically and structurally different units that can fit together coherently with one another. Examples include talc-like and brucite-like units, which, when they alternate with one another in a regular sequence, form chlorite. Many other minerals possess analogous structures. When the structural units alternate regularly, and when there are no other types of units that enter the structures, then the minerals can be studied satisfactorily by X-ray or neutron diffraction. However, in numerous instances, the sequence of structural units is irregular or small concentrations of other structure types are intergrown within the host crystals; and it is these cases that are particularly well suited for study by electron microscopy. Depending on the scale of the intergrowths and the dimensions of the relevant structural units, high-resolution imaging may or may not be required. The structural principles are the same regardless of the scale of intergrowth.

Nonstoichiometry in minerals is widespread and can be perplexing, since compositions of mineralogical solid solutions are commonly written in terms of stoichiometric end members. There are a number of ways in which minerals can accommodate small changes in composition without excessive strain—for example, by incorporation of substantial numbers of cation or anion vacancies (Veblen, 1985b). Coherent intergrowth of different but related structural units with distinct chemical compositions is another important mechanism. Intergrowth structures thus have considerable crystal-chemical interest as a means for accommodating small changes in chemical composition. Because intergrowth crystals can consist of mixtures of two or more structure types, they can have bulk compositions intermediate between those of the individual structural units. When the proportions of these structural units are varied, a series that chemically is essentially completely continuous can be generated. Such a system can be called a polysomatic series (section 9.2.2). While conceptually feasible for many structure types, there are thermodynamic constraints arising from structural strain that preclude extensive structural mixing in many minerals.

Some twins can be understood in terms of intergrowth structures. Twinning is common in minerals and can result from a variety of causes, such as crystal growth, deformation, or phase transformations. In some cases, twinning imposes no chemical changes on the host crystal; but in many instances, the chemistry at and immediately adjacent to the twin plane differs from that of the bulk crystal.

Andersson and Hyde (1974) introduced the term *chemical twinning* to explain the phenomenon whereby changes in composition can be accommodated within a crystal by such chemical differences along twin planes. Where such twins are scarce, they can produce subtle changes in chemistry and result in nonstoichiometric compounds. Where they are abundant and perfectly periodic, they result in structures and compositions that differ from those of their parent compounds; in these instances, the chemically twinned derivative structures have specific stoichiometries.

Chemical twinning can also be viewed as a subset of the phenomenon of polysomatism (section 9.2.2) and intergrowth structures. When the chemical twinning is not perfectly periodic, it provides an example of intergrowth disorder. In this instance, the material along the twin plane is the structure that is intergrown with the host; the relationship between nearest host slabs can then be described by a twinning operation. The term and concept of chemical twinning are well ensconced in the solid-state literature, although they are less common in mineralogical papers. We utilize them for the examples in section 9.4.7.

Intergrowth structures can be produced either during initial crystallization and growth or during subsequent reaction and reequilibration in response to changes in temperature, pressure, and chemical environment. In the second case, the intergrown slabs may be viewed as early stages of exsolution or replacement reactions, whereby a structure that is stable under one set of conditions adjusts to chemical or environmental changes by decomposing or reacting to form two or more distinct, coexisting phases.

One of the best examples of a mineral group that exhibits non-stoichiometry resulting from intergrowth is the biopyribole group. This group is discussed extensively in section 9.2.2, and so we will not consider it further. Rather, we will discuss examples of intergrowth in other silicates, rare-earth carbonates, and oxysulfides.

9.4.2 Sheet silicates

The structures and stacking complexities of some of the major rock-forming sheet silicates are discussed in sections 9.2 and 9.3. However, minerals having layer structures clearly are also candidates for intergrowth disorder and that is, in fact, a common feature; clay mineralogists refer to this type of disorder as "mixed layering." Only a few selected examples are considered here.

Micas and chlorites can readily intergrow with other layer silicates and hydroxides, and this is one of the ways in which alteration reactions proceed in the geological environment. The details are best revealed by imaging with the TEM. Veblen (1983) illustrates samples where extra brucite-like and talc-like layers occur in chlorite and brucite-like sheets occur in mica. Extra talc-like layers in chlorite and brucite-like layers in talc have been reported by Buseck et al. (1980) and Veblen and Buseck (1981). In addition, extreme mixed-layering disorder can occur as a result of the biotite-chlorite reaction, as discussed in section 9.2.

Where the alternating structural units are perfectly periodic, new structurally ordered phases result. An example is the new mineral kulkeite, which TEM study suggested consists of units of chlorite-like structure alternating with talc-like layers (Schreyer et al., 1982). Some kulkeite crystals also contain extra layers of chlorite. Olives Baños and Amouric (1984) report a similar structure based on biotite and chlorite. Clearly, there are many possibilities for variations on the structures of chlorite and chlorite-related minerals.

In many of the above examples where disorder occurs, the less abundant intergrown structure results from the early stages of alteration reactions. Had these reactions been able to go to completion, they presumably would have produced a crystal consisting totally of this intergrown structure. There are also examples where just the opposite interpretation is appropriate. Page (1980) gives several examples where the less abundant intergrown layers are residual rather than new. Thus, he has found single layers of smectite (a group of clay minerals) within kaolinite $[Al_2Si_2O_5(OH)_4]$ and units several layers thick of muscovite mica within pyrophyllite $[Al_2Si_4O_{10}(OH)_2]$. Both the smectite and the muscovite were precursor phases. Both of these minerals occur as partial interlayers rather than the complete layers that have been recorded in most other studies.

Kaolinite, smectite, and pyrophyllite, mentioned above, are clay minerals. They typically display a wide range of intergrowth disorder, as do other clay minerals. Many studies have been made by X-ray and spectroscopic methods to characterize this disorder, but they have had only limited success in characterizing the exact nature of intergrowth. TEM imaging obviously would be an ideal method to apply to mixed-layer clays. However, as in the case for the zeolites (section 9.9.3), the problems of radiation damage and sample preparation provide a severe constraint. Nonetheless, some HRTEM studies have been carried out.

Yoshida (1973), in an early investigation of lattice fringes in layer silicates, examined a mixed-layer clay that consists of both nonexpanding (mica-like) and expanding (smectite-like) layers. Expansion was produced by immersing the sample in an organic solvent. Comparison of interlayer spacings of nearby expanded and nonexpanded layers allowed Yoshida to infer surface characteristics of the individual layers.

Clay minerals intimately mixed with other layer silicates have been detected in the natural weathering products of feldspar (Eggleton and Buseck, 1980a) and enstatite (Eggleton and Boland, 1982). They also have been observed in weathered olivine, along with goethite, an iron hydroxide (Eggleton, 1984). In some of these cases, mixed-layer sequences were observed. In the feldspar-weathering study, a variety of clay stacking sequences as well as the progression of stages involved during clay crystallization were observed. These stages include the development of amorphous, ringlike structures, followed by a poorly crystallized material having the 10-Å interlayer spacing of many clays, which in turn formed into crinkled sheets of better-developed clay crystallites. The layer units of different thicknesses presumably developed during growth of the clay crystals.

The alternating mica-like and smectite-like layers of the clay mineral rectorite were imaged by McKee and Buseck (1978), who showed that they do, indeed, repeat regularly in the specimen examined. Clay minerals also have been found in meteorites, although they are rare. HRTEM imaging has revealed complex mixtures of mica and smectite in the Allende carbonaceous chondrite (Tomeoka and Buseck, 1982).

All of the intergrowths described above involve different units parallel to (001), the orientation of the silicate sheets. The serpentine minerals can exhibit this sort of intergrowth with other sheet silicates, but they also show other types of intergrowth as well. In his classic work on chrysotile (serpentine asbestos), Yada (1967, 1971) showed how the curled-layer fibers intergrow with each other. In addition, chrysotile and the other serpentine varieties (lizardite and antigorite) can intergrow in a variety of complex ways. Because these intergrowths bear on the question of an adequate nomenclature for serpentine minerals, they are discussed in more detail in section 9.7 on nomenclature.

9.4.3 Pyroxenoids

In pyroxene minerals, the single silicate chains repeat after every two SiO_4 tetrahedra. The pyroxenoids also contain single silicate chains, but offsets along the chains produce chain periodicities of three or more tetrahedra (Figure 9.14). The various pyroxenoids are distinguished from each other in part by the number (p) of SiO_4 tetrahedra within a repeat unit (Liebau, 1962).

A number of workers have used HRTEM to study faults in pyroxenoids (see the references in Table 9.1). There can be mistakes in the periodicity within individual chains in a crystal. More commonly, entire slabs of

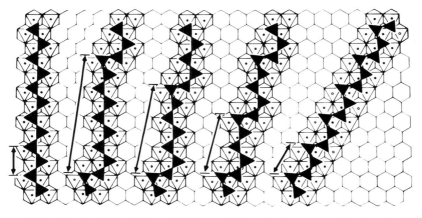

FIGURE 9.14 Sketch of the chains of SiO_4 tetrahedra in the pyroxenoid minerals. (After Czank and Liebau, 1980.)

Table 9.1
Pyroxenoids and Reports of Chain-Periodicity Faults from HRTEM

Mineral	P	Major cations	Detection of chain-periodicity faults	Comments
Wollastonite	3	Ca	Hutchison and McLaren (1976)	Contains stacking faults
Pectolite	3	Ca, Na	Müller (1976)	Contains stacking faults
Serandite	3	Mn, Na	Müller (1976)	Contains stacking faults
Rhodonite	5	Mn, (Ca)	Jefferson et al. (1980)	Contains "inserted" pyroxmangite
Babingtonite	5	Ca, Fe	Czank (1981)	
Pyroxmangite	7	Mn, (Fe, Ca)	Ried and Korekawa (1980)	Contains "inserted" rhodonite
Pyroxferroite	7	Fe (Mg, Mn)	Czank and Liebau (1983)	Contains lamellae with $p = 5$ and $p = 9$
Ferrosilite III	9	Fe	Czank and Simons (1983)	Synthetic; contains slabs with $p = 11$
(Suspected)	11	Fe(?)		Mentioned by Czank and Simons (1983)

material are anomalous in their chain repeats, resulting in planar defects that have been called chain-periodicity faults (Czank and Liebau, 1980). For example, the five-repeat pyroxenoid rhodonite can contain slabs with seven tetrahedra between offsets, and the 7-repeat structure of pyroxmangite can contain slabs with five tetrahedra (Czank and Liebau, 1980; Jefferson et al., 1980; Ried and Korekawa, 1980; Jefferson and Pugh, 1981). Stacking faults also can occur (Hutchison and McLaren, 1976). Rhodonite and pyroxmangite also can form extremely fine-grained intergrowths resolvable only with HRTEM methods (Aikawa, 1984). The 9-repeat pyroxenoid ferrosilite-III can contain a variety of chain periodicity faults, with 11-repeat faults being the most common (Czank and Simons, 1983).

There have been a few reports of ordered structures containing two different chain periodicities. Ried and Korekawa (1980) reported a (555555557) structure in which the unit cell contains eight 5-repeat units followed by one 7-repeat unit, and Veblen (1985c) observed the structures (557) and (57557). Czank and Simons (1983) and Czank (personal communication, 1984) have found ordered structures in several synthetic pyroxenoids. Such ordering of different planar units is analogous to the ordering of different chain widths in the mixed-chain biopyriboles (section 9.2.2.1). In the biopyriboles, structures with complete disorder in the chain widths also occur, and parts of pyroxenoid crystals can similarly have essentially complete disorder in the intergrowth of 5- and 7-repeat slabs (Aikawa, 1984; Veblen, 1985d).

Because the pyroxenes and pyroxenoids have related single-chain structures, they are able to intergrow coherently. The occurrence of chain offsets has been reported from HRTEM studies in both natural (Buseck et al.,

1980) and synthetic (Ried, 1984) pyroxenes. In synthetic pyroxenes, these faults are no doubt introduced during crystal growth. In natural examples, however, they can represent the first stages of a solid-state reaction in which pyroxene is converted to pyroxenoid (Veblen, 1985d).

9.4.4 Bastnaesite-synchysite

One of the earliest TEM studies of intergrowth structures was of an interesting but relatively obscure group of rare-earth carbonate minerals (Van Landuyt and Amelinckx, 1975). In compositional terms, they can be described as mixtures of $LaFCO_3$ (symbolized by B for bastnaesite) and $2LaFCO_3 \cdot CaCO_3$ (symbolized by S for synchysite). The minerals of this series are bastnaesite, B; parisite, BS; roentgenite, BS_2; and synchysite, S.

In previous X-ray studies of bastnaesite, Donnay and Donnay (1953) had noted the disordered nature of some crystals and suggested that this disorder may result from intergrowths of different members of the series; they proposed the term *polycrystal* for such structures. The disordered character of the other members of the bastnaesite-synchysite series has precluded X-ray structure determinations. Using electron microscopy, Van Landuyt and Amelinckx confirmed that this series of minerals could be represented by structural slabs, each of which has unique dimensions parallel to [001]. They found both ordered and disordered members of the mineral series and used the terms *mixed-layer compounds* and *microsyntaxy* to describe a number of structures that could be interpreted as intergrowths of the end-member species in various proportions. They also indicated the presence of three new compounds within this series, as well as regions of crystal where the succession of layers differs from that of the "normal" phase having that composition. BS_4 is a new compound with a composition intermediate between those of roentgenite and synchysite, and it displays complex polytypism.

9.4.5 Humites and leucophoenicite

The humite family of minerals consists of a series of compounds of the type $nMg_2SiO_4 \cdot Mg(OH, F)_2$, where n is known to range from 1 to 4, and reports exist of crystals having higher values. The chemical formulas correspond to mixtures of olivine (an orthosilicate, with SiO_4 tetrahedra that are not linked to one another) and either brucite $[Mg(OH)_2]$ or sellaite (MgF_2). Thompson (1978) suggested that the humite group could more accurately be described as a polysomatic series containing slabs of olivine (Mg_2SiO_4) and norbergite $[Mg_2SiO_4 \cdot Mg(OH, F)_2]$ structure. With the use of these components, the minerals can be represented as follows: norbergite, N; chondrodite, NO; humite, NO_2; and clinohumite, NO_3; where N signifies norbergite and O olivine. This seemed like an ideal mineral series in which to search for structural disorder and new members of the series.

Müller and Wenk (1978) first demonstrated disorder in the sequence of N- and O-slabs in both chondrodite and clinohumite. White and Hyde (1982a, 1982b) used HRTEM to study a wide range of natural and synthetic humite minerals. In addition to the well-known minerals, they found intergrown lamellae with spacings that correspond to species with n always even and corresponding to 6, 8, 10, ..., where n represents the number of O-slabs in the formula as given above. These structures have not been identified as discrete minerals, but their possible discovery in the future is made more likely by the HRTEM results. Another result of interest is the occurrence of superstructures, intergrown within clinohumite, that consist of slabs having a width corresponding to several unit cells of clinohumite followed by one cell of the $n = 6$ material. The largest such superstructure that White and Hyde observed has a spacing of 196 Å.

Leucophoenicite is a manganese-silicate mineral that is related to the humite group. Its idealized composition is $Mn_7[SiO_4]_2(SiO_4)(OH)_2$, which is stoichiometrically equivalent to the humite-type formula of $3Mn_2SiO_4 \cdot Mn(OH)_2$. Leucophoenicite also commonly contains small amounts of other cations, which may be required to stabilize the structure (Dunn et al., 1984). White and Hyde (1983) performed a theoretical analysis and HRTEM study of leucophoenicite, parallel to their study of the humites. They observed a range of ordered and partly ordered sequences, as well as occasional intergrowths of humite structures within leucophoenicite. They propose a series of structures analogous to those of the humite series and interpret lamellar intergrowths within leucophoenicite in terms of those proposed structures. As with their reported lamellae of possible new types of humite minerals, it will be of interest for future workers to determine whether the members of the proposed leucophoenicite series occur as discrete minerals and, if so, what their origin is.

9.4.6 Oxysulfides

Minerals of many structure types and compositions exhibit intergrowth structures. An unusual example is provided by a rare group of minerals, the oxysulfides, of which there are only four known examples: kermesite, sarabauite, versiliaite, and apuanite, the last two of which were described recently by Mellini et al. (1979).

Both versiliaite and apuanite are markedly nonstoichiometric (Mellini et al., 1981). The idealized formula of versiliaite is $Fe_{12}Sb_{12}O_{32}S_2$ (with space group Pbam), but the mineral actually contains 1.33 sulfurs per formula unit rather than 2. The idealized formula of apuanite is $Fe_{20}Sb_{16}O_{48}S_4$ (with space group $P4_2/mbc$), but the mineral contains 3.57 sulfurs per formula unit rather than 4. Mellini et al. describe the two minerals as having structures related to schafarzikite ($Fe_4Sb_8O_{16}$, $P4_2/mbc$) by a complex mechanism based on sulfide insertion, producing Sb^{3+} tetrahedral double chains in versiliaite and sheets in apuanite. The minerals have similar cell

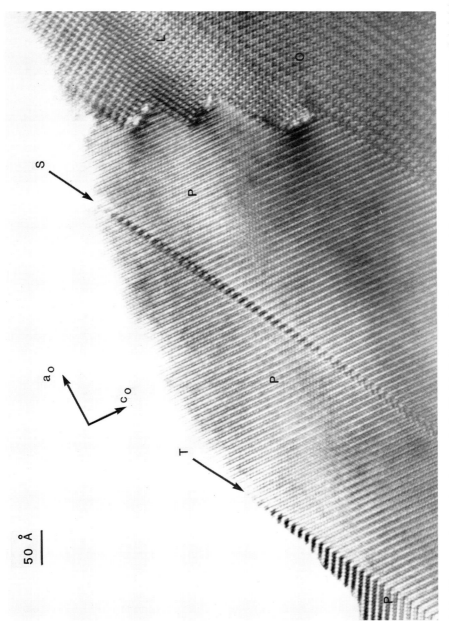

FIGURE 9.15 Electron micrograph of a synthetic $Mg_3Mn_3B_2O_{10}$ crystal showing an intergrowth of the crystal structures of ludwigite (L), orthopinakiolite (O), and pinakiolite (P). Arrows marked S and T indicate a slip boundary and a twin boundary, respectively, in pinakiolite. (Courtesy of J.-O. Bovin and M. O'Keeffe.)

dimensions, except for *c,* which is approximately 6, 12, and 18 Å for schafarżikite, versiliaite, and apuanite, respectively. As sulfur enters the structure, charge balance is maintained by iron oxidation, and optimal coordination is attained by Fe–Sb substitution.

The different *c*-periodicities allow the several minerals to be readily distinguished through the use of lattice-fringe images. Such images show that apuanite commonly contains lamellae of versiliaite. Crystals of versiliaite, in contrast, typically contain large domains of schafarżikite. Mellini et al. (1981) explain the sulfur deficiencies in the two oxysulfides as the consequence of each being intergrown with domains of a phase that contains a lower sulfur content. An unresolved problem is that there is insufficient versiliaite in the apuanite to explain the observed sulfur deficit. The authors therefore hypothesize the existence of a new phase having a *c*-dimension similar to that of apunanite but with a lower sulfur content. They suggest a theoretical formula and space group for this structure, but it remains for future research to determine whether it really exists and, if not, how the apuanite accommodates its unexplained sulfur deficiency. The advent of higher-resolution TEMs may be necessary to solve this and similar problems.

9.4.7 *Oxyborates and chemical twinning*

Electron microscopy has demonstrated several examples of chemical twinning in minerals. Mellini et al. (1984b, 1986) point out that the twin planes of sursassite (section 9.6.4) have the composition and structure of lawsonite [$CaAl_2Si_2O_7(OH)_2 \cdot H_2O$]. The oxyborates (Takéuchi, 1978) also provide a good mineralogical example of chemical twinning.

A family of chemically related oxyborate minerals, having composition M_3BO_5 (M = Mg, Mn, Fe, and, in some cases, minor Al, Sn, Ti), characteristically show structural defects. Many of the compositional relations among the minerals in this group can be explained by chemical twinning. Bovin et al. (1981a, 1981b) and Bovin and O'Keeffe (1981) used electron microscopy to study the borate minerals pinakiolite, ludwigite, orthopinakiolite, and takéuchiite; the last three can be generated by chemical twinning of pinakiolite. All of these minerals plus vonsenite can, in turn, be described as the result of chemical twinning of hulsite, the parent structure of pinakiolite (Figure 9.15). Bovin and colleagues describe missing twin operations as a common defect and ascribe to it some of the observed chemical differences among the various minerals. They also discovered a new, long-period (82.3-Å) structure in one of their synthetic oxyborates.

9.5 Modulated structures and nonstoichiometry

9.5.1 *Introduction*

As is evident from this chapter, there are many mineral structures that cannot be represented adequately by a unit cell that contains all of the basic

TABLE 9.2
Correlations Among Structure Types, Existence of Lattice, and Selected Diffraction Features

	Presence of		Diffraction characteristics			
Structure type	Single sublattice	Super-lattice	Nonintegral spacings	High probability of diffuse spots or streaking	Examples	Reference[c]
Fully ordered substructure	Yes	No	No	No	Halite; quartz; $1M$ mica	
Fully ordered superstructure	Yes	Yes	No	No	All polytypes	
Partially ordered superstructure	Yes	Yes	No	Yes	Intermediate albite; $2M_d$ mica	
Intergrowth structure						
Commensurate	No[a]	Yes	No	Yes	Long-period biopyriboles	(1)
Incommensurate	No[a]	No	Yes	Yes	Pyroxenoids	(2)
Modulated (nonintergrowth) structure						
Commensurate						
Statistical-site occupation	No	Yes	No	Yes	Hollandites	(3)
Exact-site occupation[b]	Yes	Yes	No	No		
Incommensurate (modulation in site occupations or atom positions)	Yes	No	Yes	Yes	Antigorite; e-plagioclase; pyrrhotite; sylvanite and non-stoichiometric calaverite; intermediate tridymite	(4)

Source: After Buseck and Cowley (1983).
[a] Multiple sublattices could be defined to correspond to each of the substructures, but they would lack the essential characteristic of "infinite" periodicity.
[b] Equivalent to the superstructures listed above unless it is one member of a series of mostly incommensurate structures.
[c] (1) Veblen and Buseck (1979b); (2) Veblen, 1985a; (3) Post et al. (1982); (4) Kunze (1956); Smith (1974); Pierce and Buseck (1974); Van Tendeloo et al. (1984).

features of the crystal. Structures that are not rigorously periodic or that are periodic only in a statistical sense fall into this category. Some structures produce complex diffraction patterns with satellite reflections and images that show long-range fluctuations or wavelike structures. Such modulated structures can result from atomic ordering on crystallographically equivalent sites, from atomic positional displacements, from charge-density waves, or from both structural distortions and chemical heterogeneities (Cowley et al., 1979). A recent review of modulated structures applied to mineralogy is given by Buseck and Cowley (1983); a modification of their terminology is given in Table 9.2.

Modulated structures are commonly metastable, resulting from reactions that have not yet gone to completion. As such, they can provide information about precursor phases as well as the reaction mechanisms. In many instances, modulated and intergrowth structures (see section 9.4) give rise to similar diffraction phenomena; thus, both were discussed by Buseck and Cowley (1983).

The diffraction patterns of modulated (and intergrowth) structures can be complex. If the structures contain long-period fluctuations, these superperiodicities can yield superlattice diffraction spots. The fluctuations may or may not be simple multiples of the unit cell of the basic substructure. If they are not simple multiples, then the structures are called incommensurate, and the superlattice diffraction spots are spaced at distances that are not integral or rational with respect to the spacing of the strong spots produced by the subcell. Such complex diffraction patterns are typical of, among others, some feldspars.

In some instances, the modulation direction is not parallel to the fundamental directions of the substructure unit cell; then, the satellite spots are in rows that are aligned in directions that are at an angle to the rows of stronger, main diffraction spots. When the modulation periodicity varies somewhat, or when the intergrown blocks are of varied dimensions, streaking may also occur between the diffraction spots. Clearly, the diffraction patterns from modulated structures can be complex and difficult to interpret.

All of the above comments regarding diffraction patterns from modulated structures apply to patterns obtained with X-rays, neutrons, or electrons; however, electrons interact more strongly with crystals than do the other radiations. Therefore, when the crystal modulations are weak or poorly developed, as is common in minerals, electron diffraction with the TEM can be useful because of the enhancement of weak spots that results from dynamical scattering. In these instances, HRTEM imaging also may be useful for interpreting the structure of the modulations.

9.5.2 Antigorite and pyrrhotite

Several examples of modulated minerals are discussed by Buseck and Cowley (1983). The following discussions of antigorite and pyrrhotite are taken directly from them. Antigorite, a serpentine mineral with composition $Mg_{3-x}Si_2O_5(OH)_{4-2x}$, is an example of a silicate in which a small structural misfit between the octahedral hydroxide sheet and adjacent tetrahedral silicate sheet produces a structure having a corrugated appearance (Figure 9.16). These corrugations, which commonly have periodicities greater than 40 Å (Kunze, 1956), can be imaged by electron microscopy (Yada, 1979;

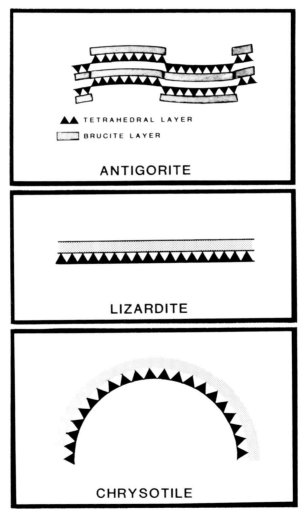

FIGURE 9.16 Sketch of the structures of antigorite, lizardite, and chrysotile, the minerals that make up the serpentine group. There are two types of layers. In one, the black triangles represent cations (normally Si) that are coordinated to four oxygens; the shaded (brucite) layer contains cations (mainly Mg) that are coordinated to six oxygens. The details of the dimensional match or mismatch between these layers determines which structure forms. Chrysotile serpentine is the primary constituent of most asbestos.

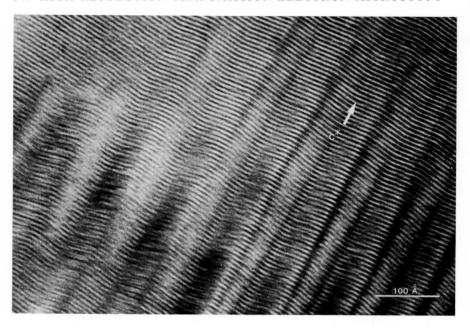

FIGURE 9.17 HRTEM images showing modulation waves in the serpentine mineral antigorite. In a region near the top of the image, the structure loses its waviness, changing to the planar structure that characterizes lizardite, another species of serpentine. (Image courtesy of G. Spinnler, as reported by Buseck, 1984.)

Spinnler et al., 1983). Figure 9.17 shows an image of planar serpentine (lizardite) that is intergrown with poorly modulated serpentine. In many samples, the antigorite modulations are better developed and more continuous.

Pyrrhotite ($Fe_{1-x}S$) is a widespread, nonstoichiometric sulfide that displays a different sort of modulated structure. It typically produces satellite spots having nonintegral spacings in its diffraction patterns; their origin and interpretation has been the subject of controversy. However, high-resolution microscopy has indicated details of some of the structural complexities of the several nonintegral varieties (Pierce and Buseck, 1974, 1976; Nakazawa et al., 1975, 1976; Morimoto, 1978). Imaging with dark-field techniques first indicated the existence of antiphase domains of varied dimensions that alternate parallel to c and that result from supercells produced by ordered iron vacancies. Figure 9.18 shows such a dark-field image, with the antiphase boundaries indicated by arrows. This structure, with an average periodicity along c equal to 5.1 times c, consists of a mixture of domains having dimensions of five and six times the length of c in the pyrrhotite subcell. By analogy, a large number and variety of nonintegral diffraction patterns and incommensurate modulated structures

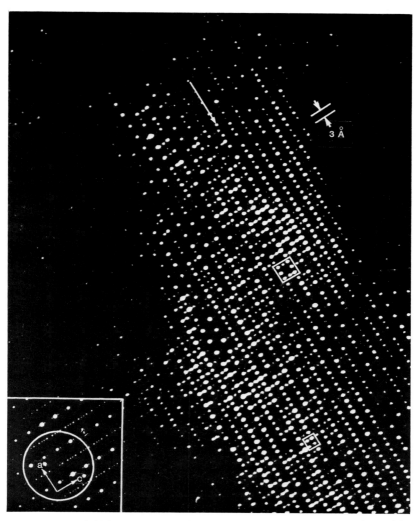

FIGURE 9.18 Dark-field image of pyrrhotite. The white spots correspond to the positions of columns of Fe atoms. Ordering of vacancies on the Fe sites results in superstructures; changes in the ordering pattern produce antiphase boundaries. One such boundary is marked by an arrow. The boxes indicate areas with all Fe sites occupied (small box) and with a central Fe site having few or no Fe atoms (larger box). The inset shows the resulting nonintegral diffraction pattern, with the circle indicating the imaged diffraction spots. (After Pierce and Buseck, 1974.)

can occur in pyrrhotite simply from various mixtures of domains having different dimensions.

9.5.3 Feldspars

Feldspars are framework silicates and comprise the most abundant mineral group in the earth's crust. The important rock-forming feldspars are divided into two chemical series: The plagioclase feldspars range from albite ($NaAlSi_3O_8$; Ab) to anorthite ($CaAl_2Si_2O_8$; An), and the alkali feldspars range from albite to the potassium feldspar composition ($KAlSi_3O_8$). In both series, numerous structural complexities result from the ordering of aluminum and silicon in the tetrahedral framework; from immiscibility related to calcium, sodium, and potassium; and from a combination of these factors (see reviews in Ribbe, 1983b). Among these complexities is the development of modulated structures, which arise in several different ways.

The intermediate plagioclase compositions, from approximately $Ab_{80}An_{20}$ to $Ab_{25}An_{75}$, develop a modulated superstructure at temperatures below about 1100°C. This structure is incommensurate and bears an irrational orientation with respect to the basic feldspar substructure and is thus modulated; both the period and orientation vary as a function of composition. The modulations give rise to satellite diffraction spots that are designated e-reflections, and hence, feldspars possessing this modulated structure commonly are called e-plagioclase. Chao and Taylor (1940) recognized forty-five years ago that the modulations arise from some sort of periodic antiphase structure related to Al–Si ordering; and since that time, numerous, more detailed models for e-plagioclases have been developed (reviewed by Smith, 1974, Chap. 5; 1983). Most of these models recognize that the modulations probably contain albite-like and anorthite-like regions and that local charge balance requires Na–Ca compositional fluctuations that are coupled to the Al–Si pattern. However, various models differ in their details, and a great deal of controversy still surrounds the question of the e-plagioclase structure, even after forty-five years of intensive study.

The modulated structure of e-plagioclases has been imaged in several HRTEM studies. Hashimoto et al. (1976) presented images resembling the structural model of Korekawa and Horst (1974); Nakajima et al. (1977) presented images in support of a different model. Kumao et al. (1981) combined HRTEM with optical-diffraction methods to support a model of the calcium and sodium positions at the $Ab_{50}An_{50}$ composition. However, to date, HRTEM has not provided definitive data that rigorously demonstrates the correct structural principles for the full range of intermediate plagioclase compositions. Indeed, the problem of the e-plagioclase structure may be one that is not resolvable with HRTEM, given the complexity of the problem and the present state of the art.

A prominent feature of the alkali feldspars is a miscibility gap separating sodium- and potassium-rich compositions, and there are probably at least three regions of immiscibility in the plagioclase system, although their details are still only poorly understood. Within the spinodals associated with these gaps, phase separation can take place by the mechanism of spinodal decomposition, in which the unstable, homogeneous crystal develops compositional modulations. These fluctuations may increase in amplitude and eventually develop into larger coherent or incoherent lamellae. However, as a result of kinetic factors, spinodal phase separation may be arrested while the structure is still modulated. In this case, diffraction patterns may display satellite reflections indicating the average modulation periodicity, and the modulated structure may produce images showing diffuse lamellae. A good example of these features is provided by Champness and Lorimer (1976).

Another type of modulated structure in feldspars is related purely to the state of Al–Si ordering and can be illustrated by the case of potassium feldspar. Above approximately 1200°C, K feldspar possesses a completely disordered arrangement of aluminum and silicon, is monoclinic, and is called high sanidine. Below approximately 450°C, the equilibrium state has a highly ordered Al–Si arrangement, is triclinic, and is called microcline. At intermediate temperatures, intermediate states of Al–Si ordering can exist, and the name orthoclase may be applied. X-ray experiments show orthoclase to be monoclinic, but such experiments average the structure over a relatively large volume. As a result, there has been disagreement about the true structure of orthoclase: Is it truly monoclinic at the finest scale, or is it simply a mixture of very small-scale domains of triclinic structure? In a HRTEM study of an orthoclase, Eggleton and Buseck (1980b) imaged wavelike modulations of the structure related to initial Al–Si ordering. Calculations showed that the strain energy arising from the modulations is approximately equal to the energy involved in the monoclinic-triclinic transformation. The modulated structure thus represents a metastable state that, as a result of the strain energy produced by the modulations, is unable to transform to normal microcline. Locally, the modulations are distortions from monoclinic symmetry. Eggleton and Buseck (1980b) suggested that an externally imposed stress may be required for the complete transformation to microcline.

9.5.4 Other minerals

There are many other minerals that display modulated features. The carbonates represent an abundant and widespread group of such minerals. TEM study has resulted in extensive consideration of modulations in calcite ($CaCO_3$) and dolomite [$CaMg(CO_3)_2$] (e.g., Gunderson and Wenk, 1981; Reeder, 1981), but the structural details are complex and the problems are far from resolved. Barber et al. (1983) have proposed that the modulations result from rotations of the CO_3^{2+} units; however, calculations have shown that point-to-point resolution of 1.7 Å is needed to reveal their orientations and 1.3 Å to resolve the carbon and oxygen atoms (O'Keefe and Barber, 1984). Thus, further work on improved TEMs may be required to solve the problem of carbonate modulations.

An interesting example of a modulated mineral occurs in carbonaceous-chondrite meteorites (Tomeoka and Buseck, 1983, 1985). Occurring in extremely primitive material, presumably formed shortly after the solar system, this mineral assumes an importance considerably beyond its interesting structural characteristics. HRTEM imaging has revealed that it is characterized by complex modulations. In many respects, it resembles the serpentine mineral antigorite, in which the modulations result from a slight dimensional misfit between octahedral and tetrahedral sheets (see section 9.5.2). In other places, this meteoritic mineral forms in tubular structures (Figure 9.19) that also resemble the morphology of chrysotile serpentine.

Sulfides, selenides, tellurides, and sulfosalts commonly display modulated structures. Pierce and Buseck (1978) used dark-field imaging to illustrate such features that resulted from Cu–Fe ordering in bornite (Cu_5FeS_4). Barbier et al. (1985) have imaged modulations in a variety of sulfosalts that result from a dimensional mismatch between structural slabs having different compositions; and Van Tendeloo et al. (1983a, 1983b, 1984) have examined modulations in the synthetic precious-metal tellurides calaverite

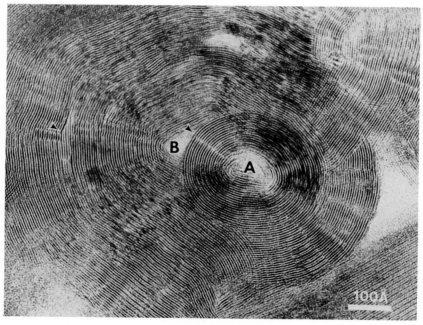

FIGURE 9.19 Tubular structures viewed end-on from an unusual, poorly understood mineral from a carbonaceous-chondrite meteorite. (After Tomeoka and Buseck, 1985.)

($AuTe_2$), sylvanite ($AgAuTe_4$), and krennerite [$(Au, Ag)Te_2$]. It is becoming increasingly evident that modulated structural and chemical variations are important features in a wide variety of minerals.

9.6 Characterization of minerals and structure determination

9.6.1 Introduction

An important component of mineralogy is the identification and characterization of new minerals. Structural characterization is typically done by single-crystal, X-ray–diffraction experiments, from which structure refinements of high quality commonly can be obtained. However, in some instances, insufficient material is available for X-ray–diffraction studies, and in this case, electron diffraction and microscopy may be employed. Unfortunately, as a consequence of dynamical-diffraction effects (Chapters 3 and 4), the intensities of the diffracted electron beams are sensitive to crystal thickness, as well as to many other parameters (variations in composition, orientation) that are generally impossible to evaluate quantitatively. The result is that electron-diffraction intensity data cannot readily be used for structure determinations of totally unknown compounds.

There are instances in which independent information is available about a new mineral, such as a known structure that is likely related to that of the unknown material; then, electron-diffraction data can be of use. Also of considerable use are high-resolution images, especially when they can be compared with computed images (Chapter 8). In such cases, an X-ray structure determination and refinement of a mineral having an unknown structure can be greatly simplified. HRTEM images may lead to initial structural models that can be used as the starting point for X-ray refinement, thus avoiding extensive computations involved in structure solution by direct methods. An example is provided by the new mineral takéuchiite, described below (Bovin et al., 1981b). Finally, although it has not yet been used widely in mineralogy, convergent-beam electron diffraction can provide useful data regarding symmetries of unknown or problematic structures, including the presence or absence of centers of symmetry, screw axes, and glide planes.

The TEM has been used in several different ways to characterize new minerals. Most common is the case where there are intergrowth structures and the lamellar phase has a spacing that is a multiple or a submultiple of the host mineral. If the intergrowth is coherent and free of strain, we can generally assume that the structure of the lamellar material has a simple relationship to that of its host. In some cases, other techniques have limited the number of probable structure types, and in this instance, the TEM data may be used to choose from among the possibilities. An example is provided by the tunnel-structure manganese oxides.

9.6.2 Manganese oxides: fine-grained minerals

The black stains on many rock surfaces, the black nodules on the deep floors of many oceans, the black rocks in many ore deposits, and the black mineral paste in many dry-cell batteries all consist in large part of fine-grained mixtures of manganese oxides. Although they are extremely widespread and abundant minerals, their structures have long been problematical and, in some cases, the sources of vigorous and long-lasting disputes. The difficulty is that these minerals typically occur in extremely small crystals, are intergrown with one another, and display a high degree of structural disorder. As a consequence, X-ray structure determinations have been limited and, in some cases, precluded.

In a series of HRTEM studies, Turner and Buseck (1979, 1981, 1983) and Turner et al. (1982) have used HRTEM images to define the structures, intergrowths, and defects in selected manganese oxide minerals. Some of the structures involved are summarized in Figure 9.20. They consist of edge- and corner-linked octahedral chains, and the structures have roughly square or rectangular cross sections. The classification system results from the HRTEM work; the structures derived from X-ray measurements have their mineral names underneath. Each horizontal row comprises a family of similar structures that can intergrow by sharing a common number of octahedral chains (either 1, 2, or 3). The structural nature of the defects and intergrowths of each group is interesting both intrinsically and for applications, as described below.

The first group of structures is important because nsutite (and its synthetic analogs, known as γ-MnO_2) is a major constituent of the millions of dry-cell batteries that are sold annually. Two of the basic structures forming nsutite were known from X-ray study, but features that made some batches of nsutite suitable for battery use and others unsuitable were not understood. This variability in efficiency of nsutite-type material may be related to defects and intergrowths revealed by HRTEM (Turner and Buseck, 1983). An example of such a defect is shown in the image of Figure 9.21a, and an interpretation of the outlined region is given in Figure 9.21b. Such defects may permit improved diffusion in some nsutites and therefore increase battery efficiency. Additionally, domains of the larger todorokite structures (discussed below) were found intergrown in nsutite, and these domains may also influence battery efficiency.

The second group of structures—hollandite-romanechite—have a larger tunnel size than those of nsutite. These tunnels can accommodate water and large cations such as potassium, sodium, and barium. This feature of the structures has led to consideration of materials having the hollandite structure as minicontainers for storage of actinides produced in radioactive waste (Ringwood et al., 1979; Ringwood, 1982). Structural features could be important to the trapping ability of the structure. Use of HRTEM has shown twinning, defects, and intergrowth of larger structures in hollandite

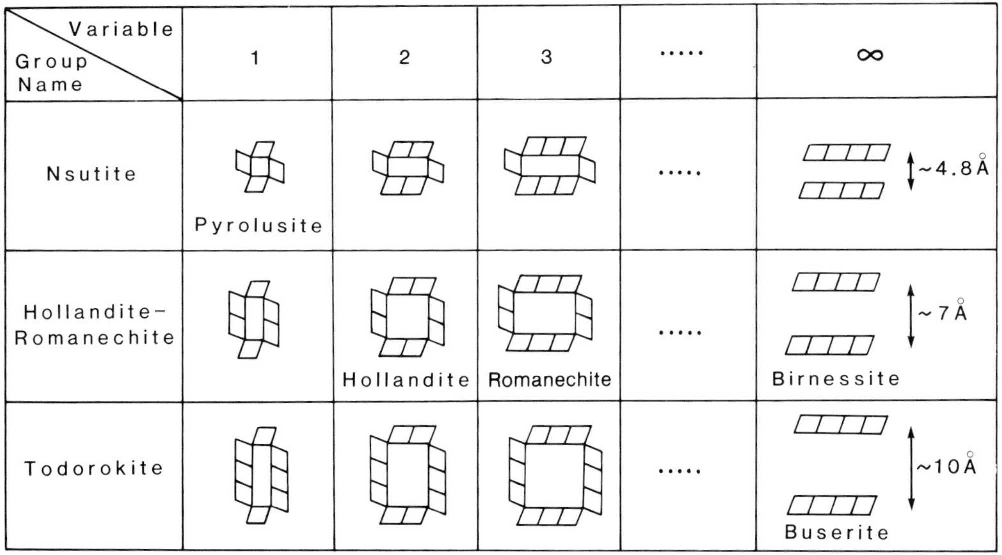

FIGURE 9.20 Sketch showing three groups of Mn oxide tunnel and layer structures. Those that occur as minerals have their names indicated below their structures. The rhombs represent MnO_6^{8-} octahedra, and the squares represent tunnels. The numbers of octahedra in a chain that runs along one side of a tunnel and projects out of the page are indicated along the top of the figure. An infinitely wide chain results in sheet structures. The spacings between sheets are indicated on the right. (Courtesy of S. Turner.)

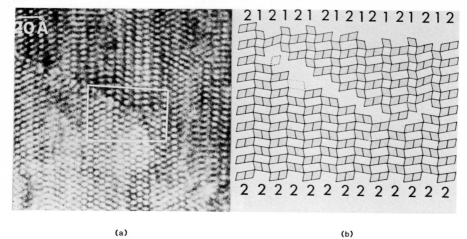

(a)   (b)

FIGURE 9.21 HRTEM image of disordered nsutite (γ-MnO$_2$) and interpretive sketch. (a) Tunnels are shown by the white spots in the image. Note the large number of irregularities (defects). (b) Sketch of the boxed area in (a). The numbers of octahedra in chains are indicated above and below the sketch. The bottom consists of pure double-chain material; alternating single and double chains occur at the top. The area of mismatch defines a large defect. (After Turner and Buseck, 1983.)

(Turner and Buseck, 1979). Such defects, if they occur in material used for radioactive waste, have the potential of allowing easy loss of the radioactive nuclides. Thus, one must understand the structures and their defects on the sub-unit-cell scale in order to evaluate their suitability for storage of radioactive waste.

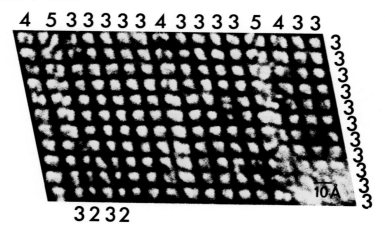

FIGURE 9.22 HRTEM image of todorokite showing the rectangular nature of the tunnels. The numbers of octahedra in the chains lining the tunnels are indicated in the margins. Note the relative uniformity in one direction and the considerable disorder in the other. (After Turner and Buseck, 1981.)

Materials belonging to the third group of structures—the todorokite family—are major constituents of manganese nodules from the seafloors. Todorokite, the principal member of this family, is a potentially important ore mineral and has been the focus of a controversy over its nature and whether it has a tunnel or layer structure (Burns and Burns, 1979; Burns et al., 1985; Giovanoli, 1985). Even the existence of todorokite has been in dispute; in 1970, the International Commission on New Minerals and Mineral Names discredited it as a mineral. However, HRTEM imaging (Turner and Buseck, 1981; Turner et al., 1982) clearly has shown the presence of roughly rectangular channels in both terrestrial and oceanic samples, thereby confirming the tunnel character of todorokite (Figure 9.22). The todorokite structures can be modeled (Turner et al., 1982) by comparison of their HRTEM images to those of hollandite-romanechite. Further confirmation was provided by computed images. An example of disorder in the material is shown in the adjacent rotated domains of Figure 9.23. Todorokite may exist only as an intergrowth material, since pure, well-crystallized material has not been observed to date.

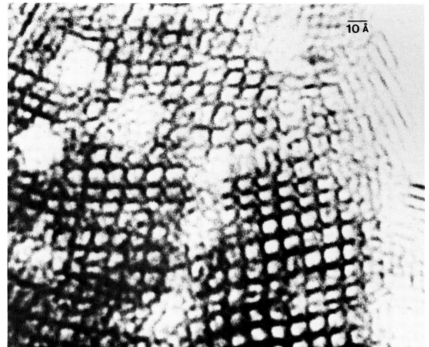

FIGURE 9.23 HRTEM image of well-crystallized todorokite. The small regions of order and the rotated domains explain why single-crystal diffraction methods have been so limited in their usefulness for determining the todorokite structure. (Photograph courtesy of S. Turner.)

9.6.3 Carlosturanite: a new type of chain silicate

Carlosturanite is a newly discovered, rock-forming silicate mineral that grows in fibers up to 20 cm long (Compagnoni et al., 1985; Mellini et al., 1985). Although reasonably abundant, its fibrous habit precluded an X-ray structure determination and even the determination of the unit-cell dimensions, and Mellini et al. had to resort to TEM study to derive a proposed structure. They did so by comparing the new mineral with serpentine, with which it is associated and intergrown. The proposed model retains the basic octahedral sheet of serpentine but introduces ordered vacancies into the tetrahedral sheet. The vacancies are aligned so that triple chains of tetrahedra result, with the $(Si_2O_7)^{6-}$ groups that are eliminated replaced by $[(OH)_6H_2O]^{6-}$ groups. Charge balance is thereby maintained, but the result is a structure with less silicon and more water than serpentine. The infinite chains in this structure are parallel to and define the fiber axis. The proposed model explains the observed physical properties of the new mineral.

An interesting analogy can be drawn between the carlosturanite structure and that of the biopyriboles. The biopyriboles can be viewed as the result of

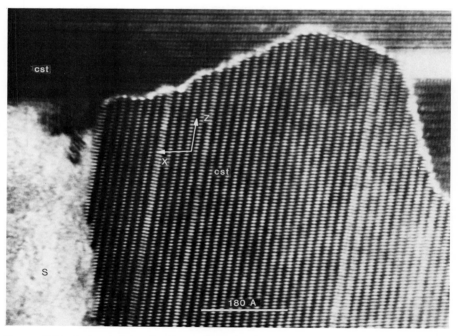

FIGURE 9.24 Isolated slabs of wider a-dimension in carlosturanite (cst). Chain-width-error faults account for these regions of wider structure. Incoherent grain boundaries between adjacent carlosturanite fibers are evident. The neighboring region is the chrysotile variety of serpentine (S). (After Mellini et al., 1985).

MINERALOGY   355

the condensation of individual (pyroxene) chains into ever wider chains until, finally, sheet structures are produced. As Mellini et al. point out, the converse is the case for carlosturanite, for its structure was derived by generating chains from sheets, the reverse of the process for the biopyriboles. However, this difference simply depends on whether a sheet silicate or a chain silicate is viewed as the parent structure, rather than being a fundamental distinction. Indeed, both the biopyriboles and the carlosturanite-serpentine structures are polysomatic series based on the intergrowth of two different types of structural slabs.

The combination of both continuous sheets and infinite silicate chains gives carlosturanite a unique structure. Furthermore, the relation between the sheets in serpentine and the chains in carlosturanite implies that other, related structures could also form with chains having other widths. No such ordered phases were found by Mellini et al. (1985), but the possibility exists that related minerals may in fact occur; isolated planar defects having chain widths other than triple were observed in carlosturanite (Figure 9.24).

9.6.4   Other minerals (sursassite, takéuchiite, etc.)

The TEM has played an important role in the description of a number of new minerals or incompletely characterized materials that may eventually be accepted as new minerals. High-resolution imaging has played a role for some, and electron diffraction has been critical for others. The following discussion provides a few examples.

Sursassite is a rare Mn-Al silicate, $Mn_2Al_3[(OH)_3(SiO_4)(Si_2O_7)]$, that has been the subject of controversy because of its substantial structural disorder. Mellini et al. (1984b) studied sursassite by using X-ray diffraction but had trouble interpreting their data. They resolved the problem through the use of experimental and computed HRTEM images, which showed that the disorder arises in part from its coherent intergrowths with pumpellyite, a Ca-Al silicate mineral that contains the same type of structural layers as sursassite (Figure 9.25). (These features are analogous to some of the intergrowths described in Section 9.4.) The HRTEM measurements allowed Mellini et al. to identify lamellae of pumpellyite ($c = 19$ Å) within the sursassite ($c = 9.8$ Å) matrix and thus to characterize the disorder in detail. They also were able to reinterpret their X-ray data and conclude that sursassite is not isostructural with epidote (Freed, 1964). In addition, the electron microscopy allowed them to identify some polytypic sequences and small regions of ordered intergrowths consisting of multiple repeats of alternating pumpellyite and sursassite lamellae.

Aspects of the oxyborate minerals are described in section 9.4.7. In the course of a study of these materials, a new mineral of composition $Mg_{1.59}Mn_{9.42}^{2+}Mn_{0.78}^{3+}Fe_{0.19}^{3+}Ti_{0.01}^{4+}BO_5$ was discovered by Bovin and O'Keeffe (1980) and named takéuchiite. The structure of takéuchiite was determined by comparison with the features of pinakiolite and other related

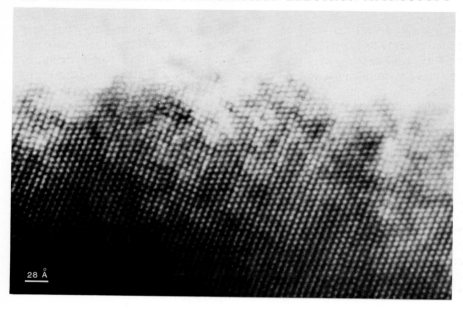

FIGURE 9.25 Highly faulted sequence of sursassitelike (S) and pumpellyitelike (P) domains within a crystal having the bulk characteristics of sursassite. (After Mellini et al., 1984.)

oxyborates, and this structure was tested by comparing experimental HRTEM images with computed ones (Bovin et al., 1981a). The takéuchiite structure is related to that of the other oxyborates and can be derived from that of pinakiolite by periodic chemical twinning.

9.7 Mineral definition and nomenclature

9.7.1 Introduction

The definition and naming of new minerals, controlled by the International Mineralogical Association Commission on New Minerals and Mineral Names, has always been an important aspect of mineralogy. The naming of a new mineral usually requires the demonstration that it is structurally or chemically unique on the basis of X-ray–diffraction data (usually powder diffraction) and chemical analyses, combined with determination of properties such as its refractive indices and density (Donnay and Fleischer, 1970; Fleischer, 1970). Determination of the crystal structure of the new mineral can be an important part of the definition but is not required.

The present international system generally functions well. However, HRTEM and other TEM methods do raise questions for the future about what constitutes a valid mineral species. The problems of structurally disordered minerals, structures that are intergrown on a scale of tens or hundreds of angstroms, and ordered structures that occur only over very limited regions or in crystals too small to study with X-ray methods have not yet been addressed adequately by the formal mineralogical nomenclature. However, nomenclature will probably have to be adapted in the future to take account of such heterogeneities as more and more mineralogists and crystallographers turn to HRTEM as a fundamental tool.

The questions that are raised by transmission electron microscopy for mineral definition and nomenclature are anything but simple and will not be resolved easily. This is demonstrated rather graphically by the fact that the two authors of this chapter have different opinions on the place of TEM data in establishing new named mineral species. One of us (PRB) believes that new mineral descriptions based on TEM observations alone could be sufficient. The other (DRV) believes that, at the present level of development of TEM techniques, it would be unwise to allow naming of new minerals on the basis of electron microscopy alone, since errors undoubtedly would occur and there would be a proliferation of new mineral names of dubious utility. However, both of us believe that TEM methods can play an important role in descriptions of new minerals. For example, the combination of diffraction, imaging, and analytical methods available on modern TEM instruments provides a powerful tool for showing whether a supposed new mineral is truly a homogeneous, single-phase material or whether it is, instead, a submicroscopic mixture of two or more other minerals. In addition, TEM methods can provide powerful supporting data for other, more standard methods for characterizing new minerals.

9.7.2 Structural disorder and intergrowth structures

As indicated in section 9.4, minerals can occur as separate, well-defined crystals, but they also can occur with other structures in extremely fine intergrowths. In the former case, where an essentially infinite, ordered structure is apparent, there is no difficulty in assigning a mineral name. In the latter case, however, should the intimate intergrowth be considered as a mineral species distinct from its structural components? Furthermore, if intimate disordered intergrowths are to be considered distinct types of minerals, then how coarse-grained should the intergrowth be before we consider it to be a mechanical mixture of its two (or more) parent structures?

One example where these questions are important is the new mineral kulkeite (Schreyer et al., 1982). Its ideal structure can be described as alternating chlorite-like and talc-like layers; but in places, it contains slabs of extra chlorite layers. If there are few such layers, they can be viewed as defects. But if they are abundant, then they are better described as a coherent intergrowth of the chlorite mineral within kulkeite. However, the number of layers of chlorite that are required to change it from a defect to a discrete slab of chlorite remains to be defined.

Another example of the problems inherent in these questions is provided by the ferromagnesian biopyribole system and the related chain and sheet minerals of enstatite, anthophyllite, jimthompsonite, and talc. All occur as essentially perfect, single-phase crystals suitable for X-ray–diffraction study. Questions arise with regard to the disorder that occurs in this structural system. It is reasonable to apply the name *anthophyllite* to a crystal that consists predominantly of a double-chain structure with a few intergrown slabs of triple-chain structure. But what are we to call an essentially random intergrowth of double, triple, and even wider chains? Such disordered intergrowths of two or more chain widths are common in some rocks, yet there is no official recognition of their existence in the formal mineralogical nomenclature. In the case of polytypic disorder, where there is still fixed stoichiometry, the letter *d* is commonly affixed to the polytype symbol to indicate that the stacking sequence is disordered. But the mineralogical community has not yet grappled with the cases of nonpolytypic structural disorder.

A related type of disorder occurs in the tunnel manganese oxides, discussed in Section 9.6.2. The disorder in the widths of octahedral chains does not necessarily alter the stoichiometry (MnO_2), but the disorder again is not polytypic. What are we to call the disordered intergrowths of different tunnel types? These structures are extremely important in oceanic manganese nodules and certain manganese-ore deposits and clearly cannot be ignored by mineralogists. Perhaps the name *todorokite* could be applied to these disordered structures (Turner and Buseck, 1981); but in that case, there are a multitude of different structures with varying degrees of order and types of disorder, all called by the same name.

Yet another case of intergrowth disorder occurs in the important and abundant sheet-silicate group of serpentine minerals. From X-ray–diffraction studies, Whittaker and Zussman (1956) separated the serpentines into three distinct mineral species, lizardite (with planar layers), chrysotile (curled layers), and antigorite (corrugated layers, where the curvature reverses periodically). From numerous X-ray investigations, we know that these minerals commonly occur together in the same rock. However, from HRTEM studies, we now know that these three structures can intergrow intimately, with a given layer commonly changing from planar to curved or corrugated shapes (e.g., Figures 9.17, 9.26). Lizardite can contain regions of curved layers, chrysotile can contain planar regions, and antigorite can possess regions of planar structure or places where the curvature fails to reverse. The distinction among these minerals is somewhat blurred in such cases. Thus, the question arises as to whether the division of serpentines into these three species is valid or useful.

Because the three types of serpentine can occur in relatively pure form, it can be argued that the distinctions among them are valid. Several studies have shown that even where the planar and highly curled types are intimately intergrown, there generally is not a continuum of degrees of curvature; the layers tend to be either planar or curled, with layers having

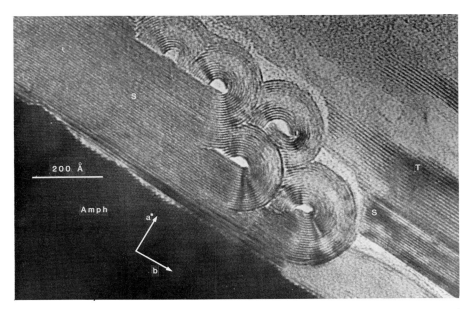

FIGURE 9.26 An intimate intergrowth of planar (lizardite) and curled (chrysotile) varieties of serpentine (S). The serpentine is intergrown with amphibole (Amph) and talc (T). (After Veblen and Buseck, 1979c.)

only slight curvature being relatively rare. Similar observations hold where the modulated variety occurs. Because there generally is not a continuum of structures between lizardite and chrysotile or between them and antigorite, the distinctions among them may be important. Perhaps the serpentine nomenclature is fine the way it is, but it does overlook the transitional materials. However, since intimate structural mixtures do occur, it would be useful if there were a way to denote the degrees of disorder and structural intergrowth in the serpentine system. Presumably, there are other mineral groups that display similar transitional types, as yet unrecognized, thereby providing a fruitful subject for future research.

The increased use of HRTEM in mineralogical studies has resulted in the recognition that structural disorder can be important in many other systems in addition to the biopyriboles, the manganese oxides, and the serpentine minerals. In many cases it is not possible to apply a mineral name as a result of intergrowth disorder. This problem eventually has to be addressed by the international mineralogical community if there is to be a comprehensive and internally consistent mineralogical nomenclature.

9.7.3 Ordered structures

A problem related to those discussed for disordered structures is the determination of when an ordered intergrowth of two or more structure types qualifies as a new compound and therefore mineral. Section 9.4 provides several examples of ordered sequences that have been observed only by electron microscopy. One traditional mineralogical test has been whether the material produces an interpretable X-ray diffraction pattern. However, electron diffraction patterns can be obtained from smaller volumes, and HRTEM imaging can reveal periodicities in still smaller volumes. Which type of observation suffices for a material having a periodic structure to qualify as a mineral?

As noted above, essentially random intergrowths of double, triple, and sometimes wider silicate chains occur in the ferromagnesian biopyribole system. In other cases, slabs of double-chain structure occur intergrown with slabs of triple-chain structure, and there is a continuum of structures ranging from the apparently random mixtures of the two-chain types, through slabs of a given chain type that are a few chains wide, to such slabs that are several hundred micrometers wide. In the first case, no mineral name is currently applied. In the last case, the names *anthophyllite* and *jimthompsonite* obviously should be applied to the double-chain and triple-chain material, respectively. But where is the dividing line? How wide does a slab of structure have to be before we can apply a mineral name to it? In the present case, does the slab have to be two chains wide? Three? Ten? One hundred? Or should the minimum region to which we can apply a mineral name be measured as a dimension in angstroms? Only with the advent of HRTEM techniques has there been a need to ask questions like these, and the answers are not clear.

Another set of problems involves ordered structures that have not been described as macroscopically occurring minerals. One problem involves the question of how many times a structure must be repeated in order to qualify as a "new" compound or structure; small regions of order can occur within any random grouping of two or more structural elements, and we probably should not refer to them as new structures. Several ways have been proposed for distinguishing structures that are significant. Buseck and Iijima (1975) required at least three repeats to qualify as a new ordered structure, Wenk (1978) required ten repeats, and Van Landuyt and Amelinckx (1975) required ten repeats observed in three different crystal fragments. Veblen and Buseck (1979a) used a simple "runs-probability test" on ordered, mixed-chain biopyriboles. Such a statistical test can be applied in all cases where the structure consists of mixtures of two or more structural elements, which includes virtually all of the new structures that have been found with HRTEM. Such a test also can eliminate investigator bias and the arbitrary aspect of deciding whether a given part of a crystal merits consideration as a new compound. We strongly urge that electron microscopists utilize some sort of statistical test before asserting that they have discovered a new compound or an important new ordered structure.

What do we do once a new structure has been found with HRTEM and shown to be statistically significant? There may well be crystalline material that is important and widespread but that never attains the concentrations or crystal sizes needed for conventional characterization. An example is the calcic analog of the triple-chain silicate clinojimthompsonite, which appears to be an extremely widespread, but always submicrometer alteration product of calcic pyroxenes. In the future, the mineralogical community will have to decide whether such minerals can be dignified by proper mineral names or whether they will continue to be referred to by awkward names like "the unnamed calcic analog of clinojimthompsonite."

9.7.4 Phases

As high-resolution microscopists, we are commonly asked by petrologists, mineralogists, and solid-state chemists whether structures in fine-scale intergrowths constitute separate phases. While it is an interesting topic for speculation, we believe that this question is not answerable in terms of the traditional definition of phase. In 1876, Gibbs originally defined *phase* strictly in reference to the phase rule (see Gibbs, 1961, p. 96), and a wise course is to restrict usage of the term to this context of heterogeneous equilibria. According to Ricci (1966, pp. 1–5), "a phase is one of the homogeneous, mechanically separable portions of a heterogeneous system." However, he goes on to say: The term "phase" is not definable on purely physical grounds. The same is true of the expression "number of components," or the order of a system. . . . Both terms may be said to owe not only their existence but their very meaning to the Phase Rule. They are merely aspects of that description of an equilibrium system which is known as phase behavior. There is no general, unequivocal definition of these terms which may be applied universally without reference to phase behavior, which means the relation between β, the number of phases, α, the number of components, and F, the number of degrees of freedom.

Given the above, it makes little sense to examine an electron micrograph and divide it into phases by inspection. It is, perhaps, more interesting to note that there are many reports of materials in which several structures are intimately intergrown, such as the unusual ordered pyribole structures noted in section 9.2.2.1 (see Veblen and Buseck 1979a, Table 1). In most of these cases, if the individual structures are considered to be separate phases, then either the phase rule is violated or some trace constituent is acting as a component in the phase-rule sense. In most cases, the most likely thermodynamic implication is that the coexistence of the various structures is not an equilibrium phenomenon. However, even if a micrograph of an intergrowth cannot be rigorously sliced into a unique set of phases, the differences in structure and coherency strain required to fit the different structures together may have profound effects on thermodynamic properties. Thus, full characterization of a sample is important regardless of whether or not its constituent structures can be defined as phases.

9.8 Experimental techniques

9.8.1 Introduction

Most of the experimental procedures for HRTEM imaging of minerals are standard and similar to those described by Buseck and Iijima (1974) and Spence (1981). The major subsequent changes in mineral-specimen preparation are the use of ultramicrotomy with a diamond knife (Iijima and Buseck, 1978) and, more commonly, ion milling (Barber, 1970; Veblen and Buseck, 1979a). Many experimental details of HRTEM that are specific to mineralogical studies are discussed by Veblen (1985d).

9.8.2 Special imaging to improve resolution (pyrrhotite)

For HRTEM imaging of very thin and weakly scattering specimens, for which the weak-phase-object approximation (WPOA) applies, the best possible point-to-point resolution is normally taken as that given by the Scherzer limit. This resolution is measured by observation of the transfer function of the objective lens by use of the optical-diffraction method or an equivalent digital method (Chapter 1). The smallest spacing that can be imaged correctly corresponds, in reciprocal space, to the outermost diffracted beams that are included in the optimum objective aperture, placed axially, for the Scherzer optimum defocus condition.

Nevertheless, most crystals are either too thick or contain heavy atoms that are strong scatterers, and so the WPOA does not apply. For these cases, interference effects contributing to the image may occur between strongly diffracted beams that are separated by more than the radius of the objective aperture, rather than just from interactions of the primary beam with weakly diffracted beams. Then, detail can be produced in the image on a smaller scale than the Scherzer limit of resolution. Under special circumstances, this detail may produce an image that resembles a projection of the crystal structure. These special cases must be used with great care, however; and they should be verified by comparison with computer simulation of the images, because the resemblance of the images to a projection of the structure can be remote or nonexistent.

One other case where an apparent improvement of resolution can occur is in the use of the tilted-beam (high-resolution), dark-field method. The incident 000 electron beam is tilted by use of deflection coils so that it does not pass through the objective aperture, which transmits only a selection of diffracted beams. In special cases, when a group of diffracted beams that has approximately the same symmetry relationships as the whole diffraction pattern is transmitted in a favorable configuration with respect to the lens transfer function, an image representing a projection of the structure may be produced.

Pyrrhotite has spacings that are close to the theoretical limit for the TEMs that were available during the early years of HRTEM imaging. Pierce and

Buseck (1974, 1976, 1977) used the method of beam tilting to obtain dark-field images of pyrrhotite that have point-to-point resolutions of 1.9 Å. However, this method of extending resolution beyond that normally obtainable is of limited application, because structural features of interest must occur with specific and fortunate spatial frequencies. Furthermore, resolution improvements in modern transmission electron microscopes obviate the use of this technique for most mineralogical problems.

9.8.3 Radiation damage (biopyriboles, serpentines, and zeolites)

Damage to the specimen produced by the incident-electron beam is commonly a limiting factor for HRTEM imaging of minerals. This problem has been considered in detail by Hobbs (1979, 1983) and, for the special case of minerals, by Veblen and Buseck (1983), who discussed the systematic relationships between mineral chemistry and damage rates. Here, we discuss only a few examples of techniques that have been used to overcome experimental difficulties associated with beam damage. The problem of radiation damage, however, is not limited to the minerals discussed here; sheet silicates, in general, are extremely radiation-sensitive, as are many silica-rich minerals (including quartz and its polymorphs) and some minerals rich in water of hydration.

The zeolites are a group of aluminosilicate minerals and synthetic compounds that have open-framework structures that, in theory, could be studied readily by HRTEM imaging. The importance of zeolites as catalysts increases the interest in their study. Nonetheless, because of their great sensitivity to damage in the electron beam, there have been only a few HRTEM studies of zeolites. Limited success has been obtained through extended dehydration in vacuum prior to TEM study, as well as by the incorporation of heavy ions such as UO_2^{2+} into the structure (Bursill et al., 1981). Replacement of structural aluminum by silicon also has modified the rate of radiation damage (Terasaki et al., 1984; Thomas, 1984), but the problem remains severe.

Serpentine minerals and some biopyriboles also can suffer rapid radiation damage. However, much success has been obtained by orienting ion-milled specimens in one area of the crystal, focusing the electron beam, and then moving into regions of previously unexposed material. The sample is then photographed immediately upon exposure to the electron beam. Another approach (Yada, 1979) is to use image-processing methods whereby several successive images are superposed electronically, and optical or electronic filtering is then used to improve the signal-to-noise ratio. Similar image-processing methods are in wide use in fields ranging from medicine to remote sensing from earth-orbiting satellites. Their application to HRTEM imaging of minerals has been limited, in part because there is a risk of introducing artifacts and in part because the method used by Yada is not applicable to nonperiodic features. However, image processing will probably find increased use concurrently with increased TEM resolutions and improved methods for on-line digital recording and processing of TEM signals.

9.8.4 "Controlled" heating by the electron beam (Cu–Fe sulfides)

Minerals commonly undergo changes while they are being examined with the TEM; and under favorable circumstances, one can take advantage of these changes. Many of these effects appear to result from heating, and that is the assumption adopted in the following discussion. However, one difficulty is in distinguishing between the effects of specimen heating and irradiation-induced enhancement of diffusion and reaction processes. There is ample evidence that the electron beam within the TEM can, in the absence of significant increases in temperature, initiate and accelerate reactions that would normally occur at appreciably higher temperatures. It is commonly thought that normal electron-beam irradiation produces a temperature rise of <100°C. Under most circumstances, the heating produced by the electron beam is an undesirable side effect of electron microscopy and is minimized by using appropriate illumination of the specimen. However, in some instances, this heating can be used to advantage by producing phase transformations *in situ* in the TEM. This technique has been used most successfully with sulfide minerals.

Heating of samples in the TEM can be achieved by using a heating stage or, more simply, by using a large condenser aperture and focusing the electron beam on a portion of the specimen, thereby greatly increasing the beam-current density and hence the electron flux through that part of the crystal. Changes in diffraction patterns, including the development of nonintegral spacings and modulations, commonly can be observed during such experiments.

Putnis (1976) was able to track the sulfur loss and subsequent transformation of millerite (NiS) to Ni_7S_6. It, in turn, transforms through a series of polymorphs, between which the reversible transformations were observed by selected-area electron diffraction (SAED). The diffraction effects of superstructure formation could thereby be observed. In a parallel study of copper and Cu–Fe sulfides, Putnis (1977, 1978) observed reversible superstructure formation from SAED patterns.

Pierce and Buseck (1978) studied a family of related structures in several of the Cu–Fe sulfides using both SAED and HRTEM imaging. They observed a wide range of superstructures in the compositional region between digenite (Cu_9S_5) and bornite (Cu_5FeS_4) and were able to transform a number of them in the electron beam. HRTEM imaging showed the modulated character of some of these structures and the separation into vacancy-rich and vacancy-poor clusters, in accord with the model of Morimoto (1964).

Although transformations sometimes can be reversed in the TEM by suitable adjustments of the electron beam, a major difficulty is that the thermal effects cannot be quantified. The temperature of the sample generally cannot be measured; and the thermal effects are, in any case, variable. The dimensions of the irradiated grains, their contact with the supporting substrate, and the conductive character of that substrate (TEM grid or C-film) all can influence temperatures achieved during beam heating. Thus, even for a constant and known beam flux, heating effects will vary from grain to grain of a given mineral, thereby precluding precise calibration.

Heating stages can be used, but they must be designed and manufactured extremely carefully if mechanical stability is to be maintained during heating. Clearly, tolerances are severe, and mechanical drift must be minimized if spacings of a few angstroms are to be imaged. Furthermore, many of the variables affecting temperature calibration and gradients mentioned in the previous paragraph remain. The consequence of all these considerations is that high-resolution experiments at controlled, elevated, and known temperatures are still a goal to be achieved. When such high-temperature techniques are refined, they will find considerable mineralogical applications for the study of phase transformations as well as for certain types of decomposition reactions.

9.8.5 *ALCHEMI and chemical disorder in minerals*

The distribution of chemical elements between coexisting crystals and within individual crystals is of considerable interest. Both can be strongly dependent on temperature. If the dependencies are known, as from controlled laboratory experiments, and if the distributions can be measured for natural minerals, then, in favorable cases, the temperatures of crystallization or equilibration may be determined. Intercrystalline-partitioning determinations are normally performed from bulk-chemical measurements; intracrystalline-site occupancies are usually determined by refinement of X-ray data or bulk-spectroscopic methods. However, the sorts of structural defects described above might, in some cases, severely affect geothermometry and other results based on bulk measurements (Buseck and Veblen, 1978; Veblen and Buseck, 1981).

Site-occupancy disorder, by which we mean the statistical distribution of different types of atoms among given crystallographic sites, occurs within individual crystals and so is not subject to some of the complications connected with structural defects. Moreover, site-occupancy disorder is of interest not only because of temperature determinations, but also because the site distributions affect the thermodynamic state of a mineral and so can be important for accurate thermodynamic calculations. Finally, the element distribution is of interest for an understanding of the crystal chemistry of the mineral.

In some cases, site occupancies can be difficult to determine by standard methods (Hawthorne, 1983). For example, the X-ray–scattering factors of some important and geochemically similar elements such as aluminum and silicon are almost identical; thus, they are difficult to distinguish with X-ray–diffraction techniques, and indirect determinations using bond lengths may be required. Furthermore, crystal-structure refinements can be costly and time-consuming.

Determining site-occupancy distributions with the TEM by using electron channeling has recently become possible (Taftø, 1982; Taftø and Spence, 1982). Since this method can be used to determine occupancies of regions having diameters of 1000 Å or smaller, it has much higher spatial resolution than more standard methods.

ALCHEMI (atom location by channeling-enhanced microanalysis) is based on principles discussed in Chapter 7. It is a relatively new development. To date, there have been only a limited number of mineralogical applications: iron and trace elements in olivine (Smyth and Taftø, 1982; Taftø and Spence, 1982); Al–Si ordering in alkali feldspar (Taftø and Buseck, 1982, 1983); the location of minor phosphorus in olivine (Self and Buseck, 1983a, 1983b); Fe–Mg distributions in pyroxenes (Self et al., 1983); and cation distributions in spinel (Taftø, 1982; Christoffersen et al., 1984) and garnet (Otten and Buseck 1986a, 1986b).

Spinels can be used to illustrate the role of ALCHEMI in mineralogical studies. The distribution of major cations between the tetrahedral and octahedral sites in spinels significantly affects their chemical and physical properties. Unmixing occurs in some spinel solid solutions (Loferski and Lipin, 1983) and may be related to differences in the cation distributions of the end members. Other solid-solution series have nominally ordered cation distributions, although some of them may show significant disorder at high temperature. Both order-disorder and exsolution behavior potentially could provide temperature and cooling-rate information in spinels from igneous and metamorphic environments.

ALCHEMI can be used to study Mg–Al disordering as a function of temperature in heated synthetic and natural spinels that have magnesium, aluminum, chromium, and iron as major cations. Such measurements can help one determine the effects of temperature and Cr and Fe contents on Mg–Al distribution. In $MgAl_2O_4$, the magnesium and aluminum are normally completely ordered between tetrahedral (T) and octahedral sites, respectively. However, Schmocker et al. (1972) and Schmocker and Waldner (1976) report that pure $MgAl_2O_4$ samples quenched from 900° to 1000°C can have up to 30 percent of their T-sites occupied by Al, and recent magic-angle-spinning, nuclear-magnetic-resonance results (Wood et al., 1984; Gobbi et al., 1985) of Al–Mg ordering appear contradictory. O'Neill and Navrotsky (1983) have correlated spinel disorder with variations in heats of solution. The effects of additional major cations on Mg–Al disordering and the extent of disordering above 1000°C are thus of considerable interest, especially for localized regions of crystal.

Christoffersen et al. (1983) used ALCHEMI to study synthetic Mg–Al–Cr spinels crystallized at temperatures up to 1600°C and a natural specimen of Mg–Al–Cr–Fe spinel annealed at 1300°C. Their preliminary results indicated only minor Mg–Al disorder in the samples at all temperatures; but they assumed an immobile chromium reference atom on the octahedral sites, and possibly, some chromium has migrated into the T-sites together with aluminum. Further work is needed to determine the chromium distribution, and this can be done by using electron energy-loss spectroscopy (EELS) ALCHEMI with oxygen as a reference. Such data can provide improved insight into spinel crystal chemistry and can aid in the analysis of thermodynamic data for the geologically important spinel solid solutions.

A new development in ALCHEMI is the use of axial instead of planar electron channeling (Otten and Buseck, 1986). For minerals such as garnets, where some crystallographic sites line up in certain low-index, zone-axis orientations, axial channeling produces much stronger effects and therefore smaller experimental errors on the site-occupancy determinations than does planar channeling.

The work on garnets (Otten and Buseck, 1986, 1987) has shown, however, that the method as proposed by Spence and Taftø (1983) has a serious flaw in that the generation of X-rays commonly is not sufficiently localized for good, quantitative positioning measurements. As a consequence, the calculated site occupancies depend on the strength of the electron channeling, and determining how these relate to the actual site occupancies is difficult. Similar effects have been found for olivine (T. C. McCormick, personal communication, 1986) and spinel (R. Christoffersen, personal communication, 1986), showing that the problem of delocalization of X-ray generation is widespread. Future work on ALCHEMI must solve the delocalization problem before accurate site-occupancy determinations can be achieved in even a semiroutine fashion. A wide area of research remains for the ALCHEMI exploration of crystallographic site ordering in minerals.

9.9 Imaging artifacts and the role of calculations

The discussions in previous chapters clearly indicate that care must be used when one is interpreting high-resolution images. An important aid for minimizing interpretive errors is to compare experimentally obtained images with computed ones (see Chapter 8). Procedures for image simulation with mineralogical examples have been discussed by O'Keefe et al. (1978) and O'Keefe and Buseck (1979). Here, we discuss a few examples of mineralogical studies that have relied on image simulation for correct interpretations.

In an extensive study of several micas and their polytypes, Amouric et al. (1981) computed HRTEM images for a variety of operating conditions. They found several problems with the images. For example, even at optimum imaging conditions, the contrast of unoccupied sites was never fully correct because the calculations did not provide an exact match of either the experimental images or the preconceived notions of what the computed images should look like. The discrepancies made it impossible to obtain intuitive information regarding ordering within the octahedral sites (this might also have been the result of insufficient resolution). They also found that incorrect stacking sequences were obtained at nonoptimum operating conditions, and the apparent thicknesses of the component layers were incorrect, presumably similar to the problems mentioned below for chlorite. Finally, they observed that the brightest spots on the images do not necessarily correspond to tunnels in the interlayer region, in agreement with the results of Veblen and Buseck (1979a) for the similar sites in pyriboles.

As discussed in section 9.2.5, the structure of chlorite can be viewed as regularly alternating layers having many of the characteristics of the layers in talc and brucite. Nonetheless, HRTEM images of chlorite, while displaying the correct overall periodicity of $\simeq 14$ Å, can show talc-like layers that are too thin and brucite-like sheets that are too thick, on the basis of the known structures. Spinnler et al. (1984) discussed this phenomenon in detail and demonstrated that computed images show the same apparent discrepancy. By simulating images obtained from instruments with better resolution (lower values of C_s), they also showed that chlorite images could be obtained with spacings that match X-ray measurements.

The calculations of Spinnler have other intuitively surprising consequences, similar to those noted in other studies. One of these results is related to the white spots that are situated between the talc-like layers and brucite-like sheets in Figure 9.13 and that can be used to interpret images in terms of chlorite polytypes. Although these spots reflect structural aspects of the chlorite structure, the correspondence to that structure is not simple. However, these spots are similar to the spots that have been used to interpret mica polytypes (Iijima and Buseck, 1978) and talc polytypes (Veblen and Buseck, 1980). At increasing resolutions, these spots no longer appear in the same positions as they do in Figure 9.13; and for some experimental conditions, they even disappear.

From the above, it is clear that it is important to use experimental conditions that are appropriate for given purposes—such as, in this instance, determination of chlorite polytypes. A more general statement of this result is that increases in resolution and decreases in C_s, while generally desirable, can produce images that are far more difficult to interpret than results obtained from TEM operating conditions that are less demanding and easier to achieve. Increasing numbers of such images that have high information content but are intuitively confusing can be expected as TEMs with improved resolutions and C_s-values become more widely available. Then, computed images will become still more important for image interpretation.

The previous examples describe interpretation of images of regions of crystals that are periodic. Computation of images of nonperiodic features requires special techniques such as the method of periodic continuation (see section 8.3.2). While this method requires larger computer capacity than calculations for most perfectly regular structures, it allows us to obtain excellent computed matches with experimental images of isolated defects (e.g., Veblen and Buseck, 1980; also, see Figure 9.4).

REFERENCES

Aikawa, N. (1984). Lamellar structure of rhodonite and pyroxmangite intergrowths. *Am. Min.* **69**, 270.

Akizuki, M. (1981). Investigation of phase transition of natural ZnS minerals by high resolution electron microscopy. *Am. Min.* **66**, 1006.

——— (1983). Investigation of phase transition of natural ZnS minerals by high resolution electron microscopy: Reply. *Am. Min.* **68**, 847.

Amouric, M., and Baronnet, A. (1983). Effect of early nucleation conditions on synthetic muscovite polytypism as seen by high resolution transmission electron microscopy. *Phys. Chem. Min.* **9**, 146.

———, and Baronnet, A. (1984). Early first stages of synthetic muscovite polytypism as seen by HRTEM. In *International conference on crystal growth and characterization of polytype structures, Marseilles, France, July 3–6, 1984*. CNRS, Marseilles.

———, Mercuriot, G., and Baronnet, A. (1981). On computed and observed HRTEM images of perfect mica polytypes. *Bull. Min.* **104**, 298.

Andersson, S., and Hyde, B. G. (1974). Twinning on the unit cell level as a structure-building operation in the solid state. *J. Solid State Chem.* **9**, 92.

Armbruster, T., and Bloss, F. D. (1981). Mg-cordierite: Si/Al ordering, optical properties and distortion. *Contrib. Min. Petrol.* **77**, 332.

Bailey, S. W., and Brown, B. E. (1962). Chlorite polytypism. I. Regular and semi-random one-layer structures. *Am. Min.* **47**, 819.

Barber, D. J. (1970). Thin foils of non-metals made for electron microscopy by sputter-etching. *J. Mater. Sci.* **5**, 1.

———, Freeman, L. A., and Smith, D. J. (1983). Analysis of high-voltage, high-resolution images of lattice defects in experimentally deformed dolomite. *Phys. Chem. Min.* **9**, 102.

Barbier, J., Hiraga, K., Otero-Diaz, L. C., White, T. J., Williams, T. B., and Hyde, B. G. (1985). Electron microscope studies of some inorganic and mineral oxide and sulphide systems. *Ultramicroscopy* **18**, 211.

Baronnet, A. (1975). Growth spirals and complex polytypism in micas. I. Polytypic structure generation. *Acta Crystallogr., Sect. A* **31**, 345.

Bovin, J.-O., and O'Keeffe, M. (1980). Two new long-period structures related to beta-alumina. *J. Solid State Chem.* **33**, 37.

———, and O'Keeffe, M. (1981). Electron microscopy of oxyborates. II. Intergrowth and structural defects in synthetic crystals. *Acta Crystallogr., Sect. A* **37**, 35.

———, O'Keeffe, M., and O'Keefe, M. A. (1981a). I. Defect structures in the minerals pinakiolite, ludwigite, orthopinakiolite, and takéuchiite. *Acta Crystallogr., Sect. A* **37**, 28.

———, O'Keeffe, M., and O'Keefe, M. A. (1981b). Electron microscopy of oxyborates. III. On the structure of takéuchiite. *Acta Crystallogr., Sect. A* **37**, 42.

Bradley, J. P., Brownlee, D. E., and Veblen, D. R. (1983). Pyroxene whiskers and platelets in interplanetary dust: Evidence of vapour phase growth. *Nature* **301**, 473.

Burns, R. G., and Burns, V. M. (1979). Manganese oxides. In *Marine minerals*, ed. R. G. Burns, MSA Reviews in Mineralogy, Vol. 6, 1. Mineralogical Society of America, Washington, D.C.

———, Burns, V. M., and Stockman, H. W. (1985). The todorokite-buserite problem: further considerations. *Am. Min.* **70**, 205.

Bursill, L. A., Thomas, J. M., and Rao, K. J. (1981). Stability of zeolites under electron irradiation and imaging of heavy cations in silicates. *Nature* **289**, 157.

Buseck, P. R. (1983). Electron microscopy of minerals. *Am. Sci.* **71,** 175.
―― (1984). Imaging of minerals with the TEM. *Bull. Electron Microsc. Soc. Am.* **14,** 47.
――, and Cowley, J. M. (1983). Modulated and intergrowth structures in minerals and electron microscope methods for their study. *Am. Min.* **68,** 18.
――, and Huang, B.-J. (1985). Conversion of carbonaceous material to graphite during metamorphism. *Geochim. Cosmochim. Acta* **49,** 2003.
――, Huang, B.-J., and Keller, L. P. (1987). Electron microscope investigation of the structures of annealed carbons. *Energy & Fuels* **1,** 105.
――, Huang, B.-J., and Miner, B. (1988). Structural order and disorder in Precambrian kerogens. *Organ. Geochem.* (forthcoming).
――, and Iijima, S. (1974). High resolution electron microscopy of silicates. *Am. Min.* **59,** 1.
――, and Iijima, S. (1975). High resolution electron microscopy of enstatite. II: Geological application. *Am. Min.* **60,** 771.
――, and Veblen, D. R. (1978). Trace elements, crystal defects and high-resolution electron microscopy. *Geochim. Cosmochim. Acta* **42,** 669.
――, Nord, G. L., Jr., and Veblen, D. R. (1980). Subsolidus phenomena in pyroxenes. In *Pyroxenes,* ed. C. T. Prewitt, MSA Reviews in Mineralogy, Vol. 7, 117. Mineralogical Society of America, Washington, D.C.
Champness, P. E., and Lorimer, G. W. (1976). Exsolution in silicates. In *Electron microscopy in mineralogy,* ed. H.-R. Wenk et al., 174. Springer-Verlag, New York.
Chao, S. H., and Taylor, W. H. (1940). Isomorphous replacement and superlattice structures in the plagioclase feldspars. *Proc. R. Soc. London, Ser. A* **176,** 76.
Chisholm, J. E. (1973). Planar defects in fibrous amphiboles. *J. Mater. Sci.* **8,** 475.
―― (1975). Crystallographic shear in silicate structures. In *Surface and defect properties of solids,* Vol. 4, ed. M. W. Roberts and J. M. Thomas, 126. The Chemical Society, London.
Christoffersen, R., Buseck, P. R., and Dickenson, J. (1984). Determination of Mg–Al order-disorder in spinel by electron channeling. *Trans. Am. Geophys. Union* **65,** 289.
Cohen, J. P., Ross, F. K., and Gibbs, G. V. (1977). An X-ray and neutron diffraction study of hydrous low cordierite. *Am. Min.* **62,** 67.
Compagnoni, R., Ferraris, G., and Mellini, M. (1985). Carlosturanite, a new asbestiform rock-forming silicate from Val Varaita, Italy. *Am. Min.* **70,** 767.
Cowley, J. M., Cohen, J. B., Salamon, M. B., and Wuensch, B. J., eds. (1979). *Modulated structures*—1979. American Institute of Physics, New York.
Czank, M. (1981). Chain periodicity faults in babingtonite $Ca_2Fe^{2+}Fe^{3+}H[Si_5O_{15}]$. *Acta Crystallogr., Sect. A* **37,** 617.
―― and Liebau, F. (1980). Periodicity faults in chain silicates: A new type of planar lattice fault observed with high resolution electron microscopy. *Phys. Chem. Min.* **6,** 85.
―― and Liebau, F. (1983). Chain periodicity faults in pyroxferroite from lunar basalt 12021. In *Lunar and planetary science XIV, part I (abstracts),* 144. Lunar and Planetary Science Institute, Houston.
――, and Simons, B. (1983). High resolution electron microscopic studies on ferrosilite III. *Phys. Chem. Min.* **9,** 229.
Deer, W. A., Howie, R. A., and Zussman, J. (1966). *An introduction to the rock-forming minerals.* Wiley, New York.
Donnay, G., and Donnay, J. D. H. (1953). The crystallography of bastnaesite, parisite, roentgenite, and synchisite. *Am. Min.* **38,** 932.
――, and Fleischer, M. (1970). Suggested outline for new mineral description. *Am. Min.* **55,** 1017.
Durovic, S., Dornberger-Schiff, K., and Weiss, Z. (1983). Chlorite polytypism. I. OD interpretation and polytype symbolism of chlorite structures. *Acta Crystallogr., Sect. B* **39,** 547.
Dunn, P. S., Peacor, D. R., Simmons, W. B., and Essene, E. J. (1984). Jerrygibbsite, a new polymorph of $Mn_9(SiO_4)_4(OH)_2$ from Franklin, New Jersey, with new data on leucophoenicite. *Am. Min.* **69,** 546.

Eggleton, R. A. (1984). Formation of iddingsite rims on olivine: A transmission electron microscope study. *Clays and Clay Min.* **32,** 1.

———, and Banfield J. F. (1985). The alteration of granitic biotite to chlorite. *Am. Min.* **70,** 902.

———, and Boland, J. N. (1982). Weathering of enstatite to talc through a sequence of transitional phases. *Clays and Clay Min.* **30,** 11.

———, and Buseck, P. R. (1980a). High resolution microscopy of feldspar weathering. *Clays and Clay Min.* **28,** 173.

———, and Buseck, P. R. (1980b). The orthoclase-microcline inversion: A high-resolution transmission electron microscope study and strain analysis. *Contrib. Min. Petrol.* **74,** 123.

Fleet, M. E. (1983). Investigation of phase transition of natural ZnS minerals by high resolution electron microscopy: Discussion. *Am. Min.* **68,** 845.

Fleischer, M. (1970). Procedure of the international mineralogical association commission on new minerals and mineral names. *Am. Min.* **55,** 1016.

Freed, R. L. (1964). X-ray study of sursassite from New Brunswick. *Am. Min.* **49,** 168.

Gibbs, G. V. (1966). The polymorphism of cordierite. I. The crystal structure of low cordierite. *Am. Min.* **51,** 1068.

Gibbs, J. W. (1961). *The scientific papers of J. Willard Gibbs.* Vol. 1. *Thermodynamics.* Dover, New York.

Giovanoli, R. (1985). A review of the todorokite-buserite problem: Implications to the mineralogy of marine manganese nodules: Discussion. *Am. Min.* **70,** 202.

Gobbi, G. C., Christoffersen, R., Otten, M. T., Miner, B., Buseck, P. R., Kennedy, G. J., and Fyfe, C. A. (1985). Direct determination of cation disorder in $MgAl_2O_4$ spinel by high-resolution ^{27}Al magic-angle spinning NMR spectroscopy. *Chem. Lett.* **6,** 771.

Gunderson, S. H., and Wenk, H.-R. (1981). Heterogeneous microstructures in oolitic carbonates. *Am. Min.* **66,** 789.

Hashimoto, H., Nissen, H.-U., Ono, A., Kumao, A., Endoh, H., and Woensdregt, C. F. (1976). High resolution electron microscopy of labradorite feldspar. In *Electron microscopy in mineralogy,* ed. H.-R. Wenk, 332. Springer-Verlag, New York.

Hassan, I., and Grundy, D. (1984). The character of the cancrinite-vishnevite solid-solution series. *Can. Min.* **22,** 333.

Hawthorne, F. C. (1983). Quantitative characterization of site-occupancies in minerals. *Am. Min.* **68,** 287.

Hayes, J. M., Kaplan, I. R., and Wedeking, K. W. (1983). Precambrian organic geochemistry, preservation of the record. In *Earth's earliest biosphere,* ed. J. W. Schopf, 93. Princeton University Press, Princeton.

Hobbs, L. W. (1979). Radiation effects in analysis of inorganic specimens by TEM. In *Introduction to analytical electron microscopy,* ed. J. J. Hren, J. I. Goldstein, and D. C. Joy, 437. Plenum, New York.

——— (1983). Beam sensitivity in the electron microscope: Strategies for examination. In *41st annual proceedings, Electron Microscopy Society of America,* 346. San Francisco Press, San Francisco.

Hutchison, J. L., and McLaren, A. C. (1976). Two-dimensional lattice images of stacking disorder in wollastonite. *Contrib. Min. Petrol.* **55,** 303.

———, Irusteta, M. C., and Whittaker, E. J. W. (1975). High-resolution electron microscopy and diffraction studies of fibrous amphiboles. *Acta Crystallogr., Sect. A* **31,** 794.

Hyde, B. G. (1979). Some modulation operations and derived structures. In *Modulated structures*—1979, ed. J. M. Cowley, J. B. Cohen, M. B. Salamon, and B. J. Wuensch, AIP Conference Proceedings, Vol. 53, 87. American Instutute of Physics, New York.

Iijima, S. (1982a). High-resolution electron microscopy of mcGillite. I. One-layer monoclinic structure. *Acta Crystallogr., Sect. A* **38,** 685.

——— (1982b). High-resolution electron microscopy of mcGillite. II. Polytypism and disorder. *Acta Crystallogr., Sect. A* **38,** 695.

———, and Buseck, P. R. (1975). High resolution electron microscopy of enstatites. I. Twinning, polymorphism, and polytypism. *Am. Min.* **60**, 758.

———, and Buseck, P. R. (1977). Stacking order and disorder in chlorite and mica. *Trans. Am. Geophys. Union* **58**, 524.

———, and Buseck, P. R. (1978). Experimental study of disordered mica structures by high-resolution electron microscopy. *Acta Crystallogr., Sect. A* **34**, 709.

Jefferson, D. A., and Pugh, N. J. (1981). The ultrastructure of pyroxenoid chain silicates. III Intersecting defects in a synthetic iron-manganese pyroxenoid. *Acta Crystallogr., Sect. A* **37**, 281.

———, and Thomas, J. M. (1975). Resolution of disordered polytypic silicates. *Acta Crystallogr., Sect. A* **31**, S295.

———, and Thomas, J. M. (1979). Topotactical dehydration of chloritoid. *Acta Crystallogr. Sect. A* **35**, 416.

———, Pugh, N. J., Alario-Franco, M., Mallinson, L. G., Millward, G. R., and Thomas, J. M. (1980). The ultrastructure of pyroxenoid chain silicates. I. Variation of the chain configuration in rhodonite. *Acta Crystallogr. Sect. A* **36**, 1058.

Johanssen, A. (1911). Petrographic terms for field use. *J. Geol.* **19**, 317.

Korekawa, M., and Horst, W. (1974). Superstructure of labradorite An52. *Acta Crystallogr., Sect. A* **31**, S90.

Krivanek, O. L., Disko, M. M., Taftø, J., and Spence, J. C. H. (1982). Electron energy loss spectroscopy as a probe of the local atomic environment. *Ultramicroscopy* **9**, 249.

Kumao, A., Hashimoto, H., Nissen, H.-U., and Endoh, H. (1981). Ca and Na positions in labradorite feldspar as derived from high-resolution electron microscopy and optical diffraction. *Acta Crystallogr., Sect. A* **37**, 229.

Kunze, G. (1956). Die gewelte struktur des antigorite. I. *Z. Kristallogr.* **108**, 82.

Lewis, I. C. (1982). Chemistry of carbonization. *Carbon* **20**, 519.

Liebau, F. (1962). Die systematik der silikate. *Naturwissenschaften* **49**, 481.

Loferski, P. J., and Lipin, B. R. (1983). Exolution in metamorphosed chromite from the Red-Lodge district, Montana. *Am. Min.* **68**, 777.

McKee, T. R., and Buseck, P. R. (1978). HRTEM observation of stacking and ordered interstratification in rectorite. In *Electron microscopy* 1978, Vol. 1, ed. J. M. Sturgess, 272. Microscopical Society of Canada,

Meagher, E. P., and Gibbs, G. V. (1977). The polymorphism of cordierite. II. The crystal structure of indialite. *Can. Min.* **15**, 43.

Mellini, M., Ferraris, G., and Compagnoni, R. (1985). Carlosturanite: HRTEM evidence of a polysomatic series including serpentine. *Am. Min.* **70**, 773.

———, Merlino, S., and Orlandi, P. (1979). Versiliaite and apuanite, two new minerals from the Apuan Alps, Italy. *Am. Min.* **64**, 1230.

———, Merlino, S., and Pasero, M. (1984). X-ray and HRTEM study of sursassite: Crystal structure, stacking disorder, and sursassite-pumpellyite intergrowth. *Phys. Chem. Min.* **10**, 99.

———, Merlino, S., and Pasero, M. (1986). X-ray and HRTEM structure analysis of orientite. *Am. Min.* **71**, 176.

———, Amouric, M., Baronnet, A., and Mercuriot, G. (1981). Microstructures and nonstoichiometry in schafarżikite-like minerals. *Am. Min.* **66**, 1073.

Miyashiro, A. (1957). Cordierite-indialite relations. *Am. J. Sci.* **255**, 43.

Morimoto, N. (1964). Structures of two polymorphic forms of Cu_5FeS_4. *Acta Crystallogr.* **17**, 351.

——— (1978). Incommensurate superstructures in transformation of minerals. *Recent Prog. Nat. Sci. Jpn.* **3**, 183.

———, and Koto, K. (1969). The crystal structure of orthoenstatite. *Z. Kristallogr.* **129**, 65.

Müller, W. F. (1976). On stacking disorder and polytypism in pectolite and serandite. *Z. Kristallogr.* **144**, 401.

———, and Wenk, H.-R. (1978). Mixed layer characteristics in real humite structures. *Acta Crystallogr., Sect. A* **34**, 607.

Nakajima, Y., and Ribbe, P. H. (1980). Alteration of pyroxenes from Hokkaido, Japan, to amphibole, clays and other biopyriboles. *Neues Jahrbuch Min. Mh.* **6,** 258.

——, and Ribbe, P. H. (1981). Texture and structural interpretation of the alteration of pyroxene to other biopyriboles. *Contrib. Min. Petrol.* **78,** 230.

——, Morimoto, N., and Kitamura, M. (1977). The superstructure of plagioclase feldspars. *Phys. Chem. Min.* **1,** 213.

Nakazawa, H., Morimoto, N., and Watanabe, E. (1975). Direct observation of metal vacancies by high-resolution electron microscopy: 4C type pyrrhotite (Fe_7S_8). *Am. Min.* **60,** 359.

——, Morimoto, N., and Watanabe, E. (1976). Direct observation of iron vacancies in polytypes of pyrrhotite. In *Electron microscopy in mineralogy,* ed. H.-R. Wenk et al., 304. Springer-Verlag, New York.

Nukui, A., Yamamoto, A., and Nakazawa, H. (1979). Non-integral phase in tridymite. In *Modulated structures*—1979, ed. J. M. Cowley, J. B. Cohen, M. B. Salamon, and B. J. Wuensch, AIP conference proceedings, Vol. 53, 327. American Institute of Physics, New York.

Oberlin, A., (1984). Carbonization and graphitization. *Carbon* **22,** 521.

O'Keefe, M. A., and Barber, D. J. (1984). Interpretation of HREM images of dolomite. *Conf. Ser. Inst. Phys.* **68** (Electron Microsc. Anal., 1983) 177.

—— and Buseck, P. R. (1979). Computation of high resolution TEM images of minerals. *Trans. Am. Cryst. Assoc.* **15,** 27.

——, Buseck, P. R., and Iijima, S. (1978). Computed crystal structure images for high resolution electron microscopy. *Nature* **274,** 322.

Olives Baños, J., and Amouric, M. (1984). Biotite chloritization by interlayer brucitization as seen by HRTEM. *Am. Min.* **69,** 869.

——, Amouric, M., De Fouquet, C., and Baronnet, A. (1983). Interlayering and interlayer slip in biotite as seen by HRTEM. *Am. Min.* **68,** 754.

O'Neill, H. S. C., and Navrotsky, A. (1983). Simple spinels: Crystallographic parameters, cation radii, lattice energies and cation distribution. *Am. Min.* **68,** 181.

Otten, M. T., and Buseck, P. R. (1986). Zone-axis ALCHEMI for the rapid assessment of site occupancies in garnets. In *Proceedings of the 44th annual meeting of the Electron Microscopy Society of America,* 706. San Francisco Press, San Francisco.

——, and Buseck, P. R. (1987). The determination of site occupancies in garnet by planar and axial ALCHEMI. *Ultramicroscopy* **23,** 151.

Ozawa, T., Takéuchi, Y., Takahata, T., Donnay, G., and Donnay, J. D. H. (1983). The pyrosmalite group of minerals. II. The layer structure of mcGillite and friedelite. *Can. Min.* **21,** 7.

Page, R. H. (1980). Partial interlayers in phyllosilicates studied by transmission electron microscopy. *Contrib. Min. Petrol.* **75,** 309.

Pierce, L., and Buseck, P. R. (1974). Electron imaging of pyrrhotite superstructures. *Science* **186,** 1209.

——, and Buseck, P. R. (1976). A comparison of bright field and dark field imaging of pyrrhotite structures. In *Electron microscopy in mineralogy,* ed. H.-R. Wenk et al., 137. Springer-Verlag, New York.

——, and Buseck, P. R. (1977). Experimental high resolution dark field electron microscopy. In *35th annual proceedings, Electron Microscopy Society of America,* 142. San Francisco Press, San Francisco.

——, and Buseck, P. R. (1978). Superstructuring in the bornite-digenite series: A high-resolution electron microscopy study. *Am. Min.* **63,** 1.

Post, J. E., Von Dreele, R. B., and Buseck, P. R. (1982). Symmetry and cation displacements in hollandites: Structure refinements of hollandite, cryptomelane and priderite. *Acta Crystallogr., Sect. B* **38,** 1056.

Putnis, A. (1976). Observations of transformation behavior in Ni_7S_6 by transmission electron microscopy. *Am. Min.* **61,** 322.

—— (1977). Electron diffraction study of phase transformations in copper sulfides. *Am. Min.* **62,** 107.

—— (1978). Talnakhite and mooihoekite: The accessibility of ordered structures in the metal-rich region around chalcopyrite. *Can. Min.* **16,** 23.

—— (1980). The distortion index in anhydrous Mg-cordierite. *Contrib. Min. Petrol.* **44,** 135.

——, and Bish, D. L. (1983). The mechanism and kinetics of Al, Si ordering in Mg-cordierite. *Am. Min.* **68,** 60.

Reeder, R. J. (1981). Electron optical investigation of sedimentary dolomites. *Contrib. Min. Petrol.* **76,** 148.

Ribbe, P. H. (1983a). The chemistry, structure and nomenclature of feldspars. In *Feldspar mineralogy,* ed. P. H. Ribbe, MSA Reviews in Mineralogy, Vol. 2, 1. Mineralogical Society of America, Washington, D.C.

——, ed. (1983b). *Feldspar Mineralogy. MSA Reviews in Mineralogy,* v. 2, 2nd Edition. Mineralogical Society of America, Washington, D.C., 362 p.

Ricci, J. E. (1966). *The phase rule and heterogeneous equilibrium.* Dover, New York.

Ried, H. (1984). Intergrowth of pyroxene and pyroxenoid: Chain periodicity faults in pyroxene. *Phys. Chem. Min.* **10,** 230.

——, and Korekawa, M. (1980). Transmission electron microscopy of synthetic and natural Funferketten and Siebenerketten pyroxenoids. *Phys. Chem. Min.* **5,** 351.

Ringwood, A. E. (1982). Immobilization of radioactive waste in SYNROC. *Am. Sci.* **70,** 201.

——, Kesson, S. E., Ware, N. G., Hibberson, W., and Major, A. (1979). Immobilization of high level nuclear reactor wastes in SYNROC. *Nature* **278,** 219.

Ross, M., Takeda, H., and Wones, D. R. (1966). Mica polytypes: Systematic description and identification. *Science* **151,** 191.

Schmocker, U., and Waldner, F. (1976). The inversion parameter with respect to space group of $MgAl_2O_4$ spinels. *J. Phys. C* **9,** 1235.

——, Boesch, H. R., and Waldner, F. (1972). A direct determination of cation disorder in $MgAl_2O_4$ spinel by E.S.R. *Phys. Lett. A* **40,** 237.

Schreyer, W., and Schairer, J. F. (1961). Compositions and structural states of anhydrous Mg-cordierites: A reinvestigation of the central part of the system $MgO-Al_2O_3-SiO_2$. *J. Petrol.* **2,** 324.

——, Medenbach, O., Abraham, K., Gebert, W., and Müller, W. F. (1982). Kulkeite, a new metamorphic phyllosilicate mineral: Ordered 1:1 chlorite/talc mixed-layer. *Contrib. Min. Petrol.* **80,** 103.

Self, P. G., and Buseck, P. R. (1983a). High-resolution structure determination by ALCHEMI. In *Proceedings of the 41st annual Electron Microscopy Society of America,* 178. San Francisco Press, San Francisco.

——, and Buseck, P. R. (1983b). Low-energy limit to channeling effects in the inelastic scattering of fast electrons. *Philos. Mag. Lett., Ser. A* **48,** L21.

——, Spinnler, G. E. and Buseck, P. R. (1983). Pyroxenes—a novel case for atomic site occupancy determination by ALCHEMI. *Geol. Soc. Am. Abstr. Programs* **15,** 683.

Smith, J. V. (1974). *Feldspar minerals.* Vol. I. Springer-Verlag, New York.

—— (1983). Phase equilibria of plagioclase. In *Feldspar mineralogy,* ed. P. H. Ribbe, MSA Reviews in Mineralogy, Vol. 2, 223. Mineralogical Society of America, Washington, D.C.

——, and Yoder, H. S. (1956). Experimental and theoretical studies of the mica polymorphs. *Min. Mag.* **31,** 209.

Smyth, J. R., and Taftø, J. (1982). Major and minor element site occupancies in heated natural forsterite. *Geophys. Res. Lett.* **9,** 1113.

Spence, J. C. H. (1981). *Experimental high-resolution electron microscopy.* Clarendon Press, Oxford.

——, and Taftø, J. (1983). ALCHEMI: A new technique for locating atoms in small crystals. *J. Microscopy* **130,** 147.

Spinnler, G. E., Veblen, D. R., and Buseck, P. R. (1983). Microstructure and defects of antigorite. In *41st annual proceedings of the Electron Microscopy Society of America,* 190. San Francisco Press, San Francisco, CA.

———, Self, P. G., Iijima, S., and Buseck, P. R. (1984). Stacking disorder in clinochlore chlorite. *Am. Min.* **69,** 252.

Taftø, J. (1982). The cation atom distribution in a (Cr, Fe, Al, Mg)$_3$O$_4$ spinel as revealed from the channeling effect in electron-induced X-ray emissions. *J. Appl. Crystallogr.* **15,** 378.

———, and Buseck, P. R. (1982). Electron channeling: A new and direct method for determining Al–Si ordering in feldspar. *Trans. Am. Geophys. Union* **63,** 1136.

———, and Buseck, P. R. (1983). Quantitative study of Al–Si ordering in an orthoclase feldspar using an analytical transmission electron microscope. *Am. Min.* **68,** 944.

———, and Krivanek, O. L. (1982). Characteristic energy-losses from channeled 100-keV electrons. *Nucl. Instrum. Methods* **194,** 153.

———, and Spence, J. C. H. (1982). Crystal site location of iron and trace elements in a magnesium-iron olivine using a new crystallographic technique. *Science* **218,** 49.

Takeda, H. (1967). Determination of the layer-stacking sequence of a new complex mica polytype: A 4-layer lithium fluorophlogopite. *Acta Crystallogr.* **22,** 845.

Takéuchi, Y. (1978). Tropochemical twinning: Mechanism of building complex structures. *Recent Prog. Nat. Sci. Jpn.* **3,** 153.

———, Ozawa, T., and Takahata, T. (1983). The pyrosmalite group of minerals. 3. Derivation of polytypes. *Can. Min.* **21,** 19.

Terasaki, O., Thomas, J. M., and Millward, G. R. (1984). Imaging the structures of zeolite L and synthetic mazzite. *Proc. R. Soc. London, Ser. A* **395,** 153.

Thomas, J. M. (1984). Dealuminated zeolites. *J. Mol. Catal.* **27,** 59.

Thompson, J. B., Jr. (1970). Geometrical possibilities for amphibole structures: Model biopyriboles. *Am. Min.* **55,** 292.

——— (1978). Biopyriboles and polysomatic series. *Am. Min.* **63,** 239.

Tomeoka, K., and Buseck, P. R. (1982). Intergrown mica and montmorillonite in the Allende carbonaceous chondrite. *Nature* **299,** 326.

———, and Buseck, P. R. (1983). A new layered mineral from the Mighei carbonaceous chondrite. *Nature* **306,** 354.

———, and Buseck, P. R. (1985). Indicators of aqueous alteration in CM carbonaceous chondrites: Microtextures of a layered mineral containing Fe, S, O, and Ni. *Geochim. Cosmochim. Acta* **49,** 2149.

Turner, S., and Buseck, P. R. (1979). Manganese oxide tunnel structures and their intergrowths. *Science* **203,** 456.

———, and Buseck, P. R. (1981). Todorokites: A new family of naturally occurring manganese oxides. *Science* **212,** 1024.

———, and Buseck, P. R. (1983). Defects in nsutite (γ–MnO$_2$) and dry-cell battery efficiency. *Nature* **304,** 143.

———, Siegel, M. D., and Buseck, P. R. (1982). Structural features of todorokite intergrowths in manganese nodules. *Nature* **296,** 841.

Van Landuyt, J., and Amelinckx, S. (1975). Multiple-beam direct lattice imaging of new mixed-layer compounds of the bastnaesite-synchisite series. *Am. Min.* **60,** 351.

Van Tendeloo, G., Amelinckx, S., and Gregoriades, P. (1984). Electron microscopic studies of modulated structures in (Au, Ag)Te$_2$. Part III. Krennerite. *J. Solid State Chem.* **53,** 281.

———, Gregoriades, P., and Amelinckx, S. (1983a). Electron microscopy studies of modulated structures in (Au, Ag)Te$_2$. Part I. Calaverite AuTe$_2$. *J. Solid State Chem.* **50,** 321.

———, Gregoriades, P., and Amelinckx, S. (1983b). Electron microscopic studies of modulated structures in (Au, Ag)Te$_2$. Part II. Sylvanite AgAuTe$_4$. *J. Solid State Chem.* **50,** 335.

Veblen, D. R. (1980). Anthophyllite asbestos: Microstructures, intergrown sheet silicates, and mechanisms of fiber formation. *Am. Min.* **65,** 1075.

——— (1981). Nonclassical pyriboles and polysomatic reactions in biopyriboles. In *Amphiboles and other hydrous pyriboles—mineralogy,* ed. D. R. Veblen, MSA Reviews in Mineralogy, Vol. 9A, 189. Mineralogical Society of America, Washington, D.C.

——— (1983). Microstructures and mixed layering in intergrown wonesite, chlorite, talc, biotite, and kaolinite. *Am. Min.* **68**, 566.

——— (1985a). Direct imaging of complex structures and defects in silicates. *Annu. Rev. Earth Planet. Sci.* **13**, 119.

——— (1985b). Extended defects and vacancy nonstoichiometry in rock-forming minerals. In *Point defects in minerals*, ed. R. N. Schock, AGU Monograph 31, 122. American Geophysical Union, Washington, D.C.

——— (1985c). TEM study of a pyroxene-to-pyroxenoid reation. *Am. Min.* **70**, 885.

——— (1985d). High-resolution transmission electron microscopy. In *Applications of electron microscopy in the earth sciences*, ed. J. C. White, MAC Short Course Handbook, Vol. 11, 63. Mineralogical Association of Canada, Fredericton, New Brunswick.

———, and Burnham, C. W. (1978). New biopyriboles from Chester, Vermont. II. Crystal chemistry of jimthompsonite, clinojimthompsonite, and chesterite, and the amphibole-mica reaction. *Am. Min.* **63**, 1053.

———, and Buseck, P. R. (1979a). Chain-width order and disorder in biopyriboles. *Am. Min.* **64**, 687.

———, and Buseck, P. R. (1979b). New ordering schemes in mixed-chain silicates. In *Modulated structures—1979*, ed. J. M. Cowley, M. B. Cohen, M. B. Salamon, and B. J. Wuensch), AIP Conference Proceedings, Vol. 53, 321. American Institute of Physics, New York.

———, and Buseck, P. R. (1979c). Serpentine minerals: Intergrowths and new combination structures. *Science* **206**, 1398.

———, and Buseck, P. R. (1980). Microstructures and reaction mechanisms in biopyriboles. *Am. Min.* **65**, 599.

———, and Buseck, P. R. (1981). Hydrous pyriboles and sheet silicates in pyroxenes and uralites: Intergrowth microstructures and reaction mechanisms. *Am. Min.* **66**, 1107.

———, and Buseck, P. R. (1983). Radiation effects on minerals in the electron microscope. In *41st annual proceedings, Electron Microscopy Society of America*, 350. San Francisco Press, San Francisco.

———, and Ferry, J. M. (1983). A TEM study of the biotite-chlorite reaction and comparison with petrological observations. *Am. Min.* **68**, 1160.

———, Buseck, P. R., and Burnham, C. W. (1977). Asbestiform chain silicates: New minerals and structural groups. *Science* **198**, 359.

Weiss, Z., and Durovic, S. (1983). Chlorite polytypism. II. Classification and X-ray identification of trioctahedral polytypes. *Acta Crystallogr., Sect. B* **39**, 552.

Wenk, H.-R. (1978). The electron microscope in earth sciences. In *Proceedings of the ninth international congress on electron microscopy*, Vol. 3, 404. Microscopical Society of Canada, Toronto.

———, Müller, W. F., Liddell, N. A., and Phakey, P. P. (1976). Polytypism in wollastonite. In *Electron microscopy in mineralogy*, ed. H.-R. Wenk et al., 324. Springer-Verlag, New York.

White, T. J., and Hyde, B. G. (1982a). Electron microscope study of the humite minerals. I. Mg-rich specimens. *Phys. Chem. Min.* **8**, 55.

———, and Hyde, B. G. (1982b). Electron microscope study of the humite minerals. II. Mn-rich specimens. *Phys. Chem. Min.* **8**, 167.

———, and Hyde, B. G. (1983). An electron microscope study of leucophoenicite. *Am. Min.* **68**, 1009.

———, Segall, R. L., Hutchison, J. L., and Barry, J. C. (1984). Polytypic behaviour of zirconolite. *Proc. R. Soc. London, Ser. A* **392**, 343.

Whittaker, E. J. W., and Zussman, J. (1956). The characterization of serpentine minerals by X-ray diffraction. *Min. Mag.* **31**, 107.

Wood, B. J., Kirkpatrick, R. J., and Montez, B. (1984). Order-disorder phenomena in Mg–Al spinel. *Trans. Am. Geophys. Union* **65**, 1143.

Yada, K. (1967). Study of chrysotile asbestos by a high resolution electron microscope. *Acta Crystallogr.* **23,** 704.
────── (1971). Study of microstructure of chrysotile asbestos by high resolution electron microscopy. *Acta Crystallogr., Sect. A* **27,** 659.
────── (1979). Microstructures of chrysotile and antigorite by high-resolution electron microscopy. *Can. Min.* **17,** 679.
Yoshida, T. (1973). Elementary layers in the interstratified clay minerals as revealed by electron microscopy. *Clays and Clay Min.* **21,** 413.

10
SOLID-STATE CHEMISTRY
LEROY EYRING

10.1 Introduction

10.1.1 Solid-state chemistry

Solid-state chemistry includes the synthesis, analysis, thermodynamics, structure and bonding, and dynamic studies of solids. Although these divisions are common to all chemistry, solids require a distinctive array of techniques and instrumental methods for their study. Solid-state chemistry derives uniquely from the state of aggregation of the atoms of a substance, including their anisotropy and defects. We shall emphasize the crystalline solid state in this account. It may be remembered that all but 13 of the chemical elements are solids at room temperature, as are most of the innumerable compounds. Therefore, solid-state chemistry must occupy a central position in the science of matter.

Since longer-range atomic configurations are relatively enduring, the chemical properties of solids are dependent upon their mode of preparation and history of treatment in a way not associated with the chemistry of liquids and gases. This introduces a new element of concern since equilibrium may be difficult or impossible to achieve. Pitfalls in the interpretation of experimental results are inherent in a world where real solids, in contradistinction to perfect solids, must be treated. This underscores the importance of high-resolution transmission electron microscopy (HRTEM), which is uniquely capable of providing direct visualization of local structure. The justly heralded structural techniques of X-ray and neutron diffraction have provided most of the structural information extant; however, these techniques give accurate information only on the average structure of rather large diffraction volumes. It is apparent that knowledge of the defect nature of real solids, provided by HRTEM, is required for the study of solid-state chemical reactions.

Our preoccupation with chemical equilibrium has been well placed in that we have learned a great deal about the direction and end of spontaneous change, but much greater attention is now being paid to the dynamics of the drift toward equilibrium and of fluctuations and processes far from equilibrium. This attention extends to processes occurring during change and the metastable way stations encountered. To the degree that these mechanisms can be understood, we can exercise greater control by being able not only to engineer materials to our own specifications but also to exercise some control over the path that is taken toward equilibrium. It is possible and proper to study metastable states; indeed, equilibrium observations are frequently made on materials that are not at their lowest free-energy state.

In the chemistry of molecular fluids, lateral resolution, beyond molecular size, is not important since intermolecular relationships are so transient. Molecular spectroscopy of various sorts can be interpreted to yield bond distances and angles with great accuracy. In the simplest cases, these characteristics can be calculated even more accurately than they can be measured. However, for very large molecules or in the limit of quasi-infinite molecules—that is, solids—the methods of spectroscopy often flounder on the uncertainty of convolutional or defect structures where lateral resolution is essential to settle chemical questions. These defect structures are at the very heart of chemical identity and reactivity, whether it be the first events following the absorption of light by rhodopsin and hence the basis of vision, or the rapid motion of charged species in a solid electrolyte, or the performance of the active sites of a functioning catalyst.

It is in its ability to give information about solids at or near atomic resolution that HRTEM has its most lasting impact on solid-state chemistry:

1. It sheds light on the juxtaposition of atoms and atomic groupings that aids the conventional methods of structure determination by suggesting the details of extremely complicated structures.
2. In some cases, it provides approximate information about structures that cannot be obtained by other methods.
3. Of central importance is that it gives high-resolution images of defects in structures that may only be hinted at by other methods. This includes information, down to the atomic level, on planar defects, cluster defects, or even point defects, and the experimental techniques to resolve such defects are steadily improving.
4. Perhaps most important of all, HRTEM can resolve these features in time as well as space. Macroscopic chemical kinetics yields information primarily on the rate-controlling steps of change.

One must utilize this information as a guide through the woods of the more fully resolved information obtained by HRTEM so as not to be totally distracted by the trees.

Because of its capability of high spatial and time resolution, the role of HRTEM is increasing in the study of small particles and thin films. These are one-, two-, or three-dimensionally limited materials that are of increasing importance in the devices of modern technology and the science of surfaces and small systems. HRTEM will play a pivotal role in this development.

There have been a number of reviews and general statements of varying comprehensiveness on the role of HRTEM in the development of solid-state chemistry during the past five or six years. Several have been found particularly useful in the preparation of this chapter. Specific mention should be made of two volumes of reviews: Kihlborg (1978–1979) (see especially Anderson, 1978–1979a and b; Cowley, 1978–1979); Eyring (1982) (see especially Cowley, 1982; Horiuchi, 1982; Thomas, 1982); Smith et al., 1981; and Hutchison, 1984a.

10.1.2 Historical aside

Our purpose here is to provide a brief historical account of some lines of inquiry that brought solid-state chemists to a realization that the techniques of HRTEM could provide information of importance to them. Apologies are offered in advance to those offended by this myopic account.

X-ray and neutron-diffraction analysis of crystalline substances have played a central role in the development of modern chemistry. It has enabled the determination of the precise distances and angles of atoms in molecules and crystals and, hence, aided in the comprehension of the nature of the chemical bond. Beyond this, X-ray and neutron-diffraction analyses have found universal application in the identification and analysis of pure and mixed crystalline substances. From the perspective of solid-state chemistry, the application of HRTEM techniques came out of the visionary belief that it should be possible to observe the relationships of atoms in a two-dimensional array at atomic resolution; such was the view of Menter (1956). HRTEM provides direct information on the structure of substances not adequately handled by X-ray analysis or, more important, on local structure rather than the averaged structure determined by X-ray or neutron analysis.

Wadsley has secured his place among those responsible for the successful application of conventional transmission electron microscopy (CTEM) instrumentation to the study of complex structures. His research into the nature of nonstoichiometry and complex structures in oxides such as $W_4Nb_{26}O_{77}$, which he and his co-workers found to be an ordered intergrowth of two different block mosaics, $WNb_{12}O_{33}$ and $W_3Nb_{14}O_{44}$, is a case in point. In collaboration with Allpress and Sanders (Allpress et al., 1969), Wadsley examined the Nb/W oxides at about 6-Å point resolution. Lattice-fringe images revealed the intergrowth predicted by Wadsley's ingenious, earlier X-ray–structure analysis but also revealed the even more significant feature, that of planar faults and partial intergrowths in these structures. The faults could be as small as one-half a unit cell in width or could involve several unit cells. This irregularity would lead naturally to a composition different from that of the host lattice and, hence, to nonstoichiometry. This work confirmed directly an understanding reached by Wadsley much earlier and suggested a wide application of this technique to solid-state chemical problems. These studies prompted Andersson (1970) to suggest, only a few months after Wadsley's untimely death, that this irregular planar-structural defect be called the *Wadsley defect*. In a profound way, this TEM work foreshadowed much that has been done in solid-state chemistry using HRTEM since that time.

This breakthrough in structural and defect analysis of complex oxide structures was extended at this level of resolution by Bursill (1969), Anderson and Tilley (1970), Tilley and Hyde (1970), Allpress and Sanders (1971), Allpress et al. (1972), Bursill and Hyde (1972), and others.

In 1970, a decision was made by Cowley to pursue the development of HRTEM techniques, and the further interpretation of the diffraction patterns and images obtained. Iijima, working with him in his laboratory, observed the first transmission-electron-microscopic images showing the relative positions of particular atoms in a crystal structure (Iijima, 1971). This was the birth of HRTEM in the sense of this chapter. Iijima, in collaboration with many who had applied lower-resolution techniques (particularly Allpress) to complex structures, examined many different important systems and was the first, for example, to image a Wadsley defect, an isolated $\{102\}$ crystallographic shear (CS) in reduced WO_3 (Iijima, 1975), and point defects in block structures (Iijima et al., 1976). His block-structure images showed in detail the defect intergrowths responsible for nonstoichiometry, observed at lower resolution by Wadsley, Allpress, and Sanders.

The success of Iijima and Cowley in advancing to a resolution permitting a two-dimensional indication of atomic positions further stimulated others already geared up for such work, such as Anderson and Hutchison at Oxford (and later at Aberystwyth), and stimulated the establishment of new centers of high-resolution studies by Tilley at Bradford, Kihlborg at Stockholm, Andersson at Lund, Thomas at Aberystwyth and Cambridge, Gruehn at Giessen, and Caro at Bellevue. In Japan, Uyeda has obtained atomic-level processed images of chlorinated copper phthalocyanine, and Horiuchi and Hirabayashi have developed the high-voltage, high-resolution electron microscopes capable of 2-Å point resolution. Smith and his associates at Cambridge, operating a microscope at 650 keV, have also produced superb high-resolution images, heralding an even brighter future in the visualization of intimate structural details.

In this same period, computational techniques were improved in theory and in the development of computer packages, enabling multislice-image calculations to be made and compared with those observed (O'Keefe et al., 1978; O'Keefe and Buseck, 1979; Skarnulis, 1979; Rae Smith and Eyring, 1982; Saxton and Koch, 1982). The state of high-resolution microscopy in these and other laboratories in 1979 was reviewed in the proceedings of the forty-seventh Nobel Symposium (Kihlborg 1978–1979). At the time of this conference, microscopes capable of resolutions between 2 and 3.5 Å were in use, and most of the results described in this chapter were obtained on them. However, now another threshold has been reached. The newest high-resolution instruments with 1.6- to 1.8 Å resolution will make possible the study of close-packed structures and, hence, virtually any solid material that is stable in the electron beam.

Amelinckx and his co-workers influenced these early developments in solid-state chemistry and continue to this day to make contributions at the vanguard of the application of HRTEM techniques. For example, in the same conference at which the name *Wadsley defects* was given to planar-shear faults, Amelinckx and Van Landuyt (1970) described fine-scale, laminar twin growth in NiMn that resulted in a regular superstructure. Similarly, antiferroelectric domain walls that give rise to a superstructure in WO_3 were imaged. Further, they presented images of elaborate antiphase boundaries in rutile and regularly and irregularly spaced dislocations in WO_3. Parallel to the work of the Belgian scientists, a Japanese group working with Hirabayashi in Sendai, Japan, has developed electron-microscopic examination of alloy systems leading up to their present high-voltage, high-resolution studies. For example, in a recent review article, Hirabayashi (1983) sketches the development of TEM observations from simple lattice-fringe images of alloy structures to atomic-resolution, many-beam images. He alludes to a large number of studies of the overall arrangements of antiphase boundaries, periodic stacking arrays, partially ordered structures, spinodal decomposition, and Guinier–Preston zones. He also shows striking many-beam images of Au–Cd, Au–Mn, and V_3Si obtained at 1 MeV. A one-to-one correspondence exists between the observed images and those calculated from structural models utilizing dynamical-scattering theory. Although these developments have either preceded or paralleled the applications to nonmetallic solids, more conventionally considered to be in the realm of solid-state chemistry, and are of the same type—revealing structure, including defect structure, and being concerned with reactions and transformations—we shall not discuss these results further here.

10.2 Application of HRTEM to solid-state chemistry

10.2.1 The role of HRTEM in solid-state synthesis

The rates of terrestrial chemical reactions involving solids vary from the detonation of high explosives to the disintegration of the most durable rocks. The rate of a reaction cannot be greater than the fastest mass-transport mechanism. Diffusion is the most common form of mass transport in solids, and measured diffusion constants at temperatures common on the earth's surface vary over 13 orders of magnitude in the range of 10^{-3} to 10^{-16} cm^2/s. This means that most synthetic procedures must be carried out at elevated temperatures, since diffusion is an activated process. The temperatures required will vary enormously according to the activation energies. Because of these extremes, most HRTEM studies must be made on specimens quenched in time or temperature.

Following the course of a chemical synthesis will be useful both as a measure of success and as guide to the improvement of the methods. (Serendipitous detection of new configurations and compounds during synthesis could also be very valuable.) Structure and microanalysis are among the best indicators of the course of reaction. The products formed, even if crystalline, frequently have very small particle size—too small for single-crystal methods of analysis. X-ray–powder methods, which also must be used, are instructive but lack essential resolution. The volume of a specimen required for ordinary electron-diffraction analysis is of the order of 10^4 Å in diameter and 10^2 Å thick, or about 10^{10} Å^3. This volume contains about 10^9 atoms. Actually, it is possible at present, using a microfocus STEM instrument, to obtain diffraction information from an area about 10 Å in diameter, if the specimen were 10^2 Å thick, it would contain about 300 atoms. In short, the progress and success of a synthetic method can be followed whether or not crystals of large size are produced.

This same capability of observing the structure of very small regions of space enables observation of the juxtaposition and epitaxial relationship of different structural entities with very high lateral-spatial resolution. Such detailed information could indicate the reaction sequence or the structural principle relating new phases and, hence, suggest new paths of reaction. This capability is of prime importance in cases where the rate of reaction is slow and the time necessary to reach equilibrium, at the temperature required to have a stable product, is very long. Nonequilibrium is a common and informative condition in solid-state chemistry, and HRTEM provides a unique ability to study it in detail.

10.2.2 High-resolution microscopical analysis

Progress in nanometer, analytical electron microscopy (AEM) has gone apace with HRTEM. Every interaction of the electron beam with a specimen that can be observed carries with it information about the chemical identities of the scattering atoms. Hashimoto (1983) has suggested that the identities of the atoms themselves can be determined in the image formed in transmission. Isolated atoms can be observed and tracked as they move on the surface of a low-molecular-weight film in both high-resolution scanning transmission electron microscopy (HRSTEM) and HRTEM (Crewe et al., 1970; Iijima, 1977). Secondary electrons, electron-beam-induced current (EBIC), cathodoluminescence, and atom location by channeling-enhanced microscopy (ALCHEMI) are other techniques that aid in chemical analysis at high lateral resolution (see Chapters 7, 9).

The most important analytical techniques at present are energy-dispersive, X-ray–emission spectroscopy (EDS) and electron-energy-loss spectroscopy (EELS). In the former, an electron probe that is typically 100 to 200 Å in diameter (but may be as small as 50 Å in diameter in a field-emission microscope) excites the several characteristic X-ray emissions from the atoms struck by the electron beam. In the latter, the characteristic absorption edge for electronic transitions are observed, including near-edge losses that reflect the chemical state and environment of the atom. Spectrographs for EDS and EELS are commercially available, and techniques are well developed. A discussion of these several techniques is given in Chapter 7.

10.2.3 Nonstoichiometry and solid-state reactions

Nearly two centuries ago, the question of whether or not compounds consisted of elements in definite proportions of small whole numbers was settled in the affirmative even in the presence of much evidence to the contrary. It is clear now that this generalization, with the exception of molecular solids and the elements, does not apply to the solid state. That is, at any temperature, all solids with two or more components are susceptible to adjustment of their composition as required by the ambient chemical potential of their constituents. The chemical state of a solid cannot be defined without specifying the activities of the components.

The definition of activity in terms of the chemical potential of the components in MX_y is written as

$$\mu_M = \mu_M^\circ(g) + RT \ln a_M,$$
$$\mu_X = \mu_X^\circ(g) + RT \ln a_X,$$

where a_M and a_X are the activities of the components and μ° is the chemical potential of the elements in their standard states. It is obvious that the composition depends entirely on the activity of the components. In cases such as NaCl, the compositional sensitivity to the partial pressure of Na(g) or Cl_2(g) in the environment is small. However, to ignore the small compositional changes would be to fail to account for the profound changes in optical properties as the various color centers are formed accompanying the small compositional variation. In contrast, if praseodymium oxide is equilibrated with oxygen at temperatures above about 650°C, the composition can be varied continuously from near PrO_2 to $PrO_{1.72}$ by changing the oxygen pressure over a range of a few atmospheres. This behavior is illustrated in Figure 10.1. The minimum of free energy is sharply defined for phases A and B but is slowly varying for AB_x. The compositional width of the phase AB_x is determined by the common tangents, as illustrated. It may be very narrow if the minimum in the AB_x curve is sharp or may be wide, as in the case shown.

There are many instances where the situation is quite different. The region between A and B in Figure 10.1 may instead be occupied by a large number of sharp, free-energy minima, as shown in Figure 10.2. In this case, one would have many phases of narrow composition range occupying the intermediate region rather than a nonstoichiometric phase of wide range.

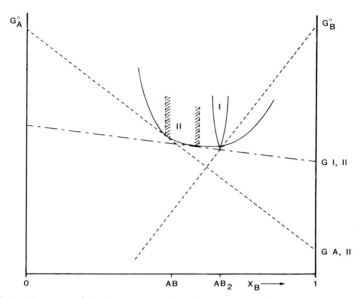

FIGURE 10.1 The composition dependence of the free energy of hypothetical phases occurring in the binary system AB. The two-phase regions are defined by the tangents drawn to the curves. Notice that the composition of the nonstoichiometric phase AB_x depends on the phase in equilibrium with it.

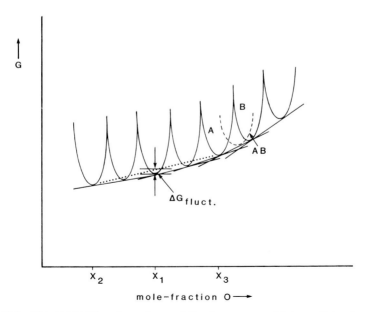

FIGURE 10.2 The composition dependence of the free energy of intermediate phases in the binary system AB, in which nonstoichiometry in each phase is small.

The history of phase analysis is that many regions where wide-range nonstoichiometry was thought to exist (e.g., in TiO_{2-x}) were later found to consist of a series of intermediate phases of very narrow composition range (Ti_nO_{2n-1}). There are, however, still several dramatically wide-composition nonstoichiometric materials, such as PrO_{2-x} in which no ordered intermediate phases have been established. In some of these cases, ordered phases are observed to form at lower temperatures (Pr_nO_{2n-2}).

It is apparent from this discussion that, although perfect crystals in which every atom is in its place (except for the small concentration of defects necessarily present above the absolute zero due to energetic and entropic considerations) could exist, additional defects must be present due to the varied activities of the components. This leads to the conclusion that there must be defects in most solids either in the form of partial occupancy of lattice sites or actual disturbances of the original structure in which sites not normally occupied have been filled. There is, of course, the alternative possibility that the structure itself can adjust to give perfect order (unit cells) within a certain limited composition range. Such a possibility has been termed *infinitely adaptive structure* by Anderson (1973). Exploration of these various alternatives is possible with HRTEM.

This nonstoichiometry that leads to defect formation is, incidentally, largely responsible for the chemical and physical properties of a specimen. Indeed, the defects are the agents by which chemical and physical change takes place. The defects are features that, when regularly incorporated within the specimen, could become elements of a new structure. Furthermore, defects are frequently regions of the structure of an adjacent stable phase.

Every finite crystal has defects at least in the form of its surfaces or interfaces. The surface structure and composition are necessarily different from that of the bulk. There may be internal surfaces, as well, in the form, for example, of grain or twin boundaries. Unless great care is exercised in synthesis, there will also be gross defects such as dislocations or other massive faults. These defects will strongly affect the mechanical- and crystal-growth properties. At nearer atomic level, there may be planar faults or clusters resulting from, for example, stacking disorder or the formation of intergrowths of some other structure along particular crystallographic directions. In addition, there are undoubtedly true point defects, such as single-atom interstitials or vacancies, but they are not easily studied by HRTEM.

The types of defects observed in a material depend entirely on its structure and chemical properties. In the transition-metal oxides, the phenomenon of crystallographic shear (CS) is most common. In the TiO_{2-x} system, for example, CS is observed over the entire range of compositions $0 \leq x \leq 0.5$, where the so-called homologous series exist. In contrast, CS gives way to pentagonal columns (PC) as oxygen is lost in WO_{3-x} (CS, $0 \leq x \leq 0.12$; PC, $0.12 \leq x \leq 0.30$). In the widely nonstoichiometric TiO_{1+x}, the composition variation is achieved by both titanium and oxygen vacancies on every third (011) plane of the NaCl-type structure. When $x = 0$, the vacancies amounting to about 25 percent of each type occur; in the titanium-rich side, the oxygen vacancies predominate, and vice versa for oxygen-rich TiO_{1+x}. In one series of the anion-deficient fluorite structures of the rare-earth oxides, the defect slab consists of oxygen vacancies paired across the body diagonal of all the otherwise cubic-coordinated metal atoms in certain (135) planes. When this feature is regular, a new structure is attained. The phase rule must be modified in these cases to include independent variables such as elastic strain to deal with crystals with irregular defects.

The lack of reversibility in many instances makes the properties of specimens history-dependent, so any particular crystal might behave as though true equilibrium were achieved with respect to a variation of one of the components but not the other. A case in point is the ternary, fluorite-related, oxygen-deficient phases such as calcia-stabilized zirconia, in which the oxygen substructure is mobile and can be in equilibrium with oxygen gas but the cation substructure exhibits very small diffusion coefficients and may be very far from equilibrium. However, in the rare-earth oxides of the same substructure, the oxygen substructure is again very labile; and in this case, the electron mobility is sufficiently high that ordering of the cation substructure can be achieved quickly without cation diffusion and true equilibrium is easy to attain (Bevan and Summerville, 1979).

10.2.4 Electron-beam-induced chemical change

Anyone utilizing HRTEM as a tool in the study of materials must be aware of the effects of the electron beam on the specimen. Frequently, the electron beam will be deliberately used to heat or modify a specimen, but changes will result from HRTEM studies willy-nilly. The effect varies between being catastrophic in some organic materials to nearly negligible in some refractory substances. At present, it is impossible to deal with this problem theoretically, so empirical observations must be relied on (Klug, 1978–1979; Cosslett, 1979; Isaacson, 1984; see expecially Hobbs et al., 1978; Hobbs, 1984). It is necessary to treat each category of material independently, since factors affecting radiation damage (electron energy and intensity, chemical bonding in the specimen, temperature, atomic masses, etc.) will differ in their relative contribution.

Aside from the study of radiation effects themselves, a legitimate concern and interest is in the evaluation of the specific effect of the electron beam on the chemistry being studied. Solid-state chemistry is carried out normally in the absence of an electron beam, although usually at higher-than-normal temperatures. It is therefore necessary to determine the effect of the electron beam beyond ordinary thermalization effects. Some questions must be answered. Are mechanisms, defect species, or pathways the same (or even related) between materials studied in the specimen chamber of the microscope or in an ordinary vacuum line? It is essential, if the information obtained from HRTEM studies is to be combined with that from more conventional observations, to differentiate between direct electron-beam effects and those that result from temperature changes. Environmental cells that include wide-range temperature control must be made compatible with double tilt and high resolution. Beyond the contribution HRTEM can make to the expansion of knowledge concerning solids gleaned from conventional means, the chemical changes wrought by the electron beam itself are of fundamental chemical interest.

For the study of materials using HRTEM, especially for the recording of chemical events in real time utilizing video techniques, the radiation rate must be very high. Perhaps the lower limit of this rate is 10^6 electrons per square nanometer in order to obtain statistically significant information. Since inelastic scattering of electrons can be of the same order of magnitude as elastic scattering, this means that every atom in a 100-Å-thick sample inelastically scatters 500 or more electrons per second. It is amazing that there are as many materials that can be profitably studied in HRTEM as there are.

Substances can be broadly classified as compact (e.g., metals and close-packed solids), network (e.g., covalent materials), or molecular crystals (e.g., carbonates) with decreasing resistance to radiolysis. Interactions that result in radiation-induced alteration are classified as either knock-on or radiolytic The ratio of cross sections for electronic or nuclear interactions is about 40. At 100 kV, direct momentum transfer is of concern for hydrogen, lithium, beryllium, and carbon only; at 200 kV, sodium, magnesium, and aluminum must be added to the list, as well as gallium and arsenic in sulfides and selenides. At 300 kV, silicon is displaced, as are the elements in most III/V and II/VI semiconductors. Between 300 and 400 kV, oxygen in oxides is displaced. Knock-on is also accompanied by sputtering and radiation-enhanced diffusion.

Radiolysis includes many different processes. Ionization of core electrons is followed by X-ray emission and many secondary phenomena. Valence electrons may also be removed, resulting in bond breakage, especially in covalent materials. Another common event is the elevation of valence electrons to locally bound, electron-hole-pair excitonic states. Collective excitation of weakly bound valence electrons into long-wavelength, semi-localized oscillations (plasmons) is also to be expected.

Radiolysis can also cause atomic displacements under certain situations. The electronic excitation must be localized for displacement to occur. This requirement exempts metals and reduces displacive events in semiconductors. The excited lifetime must also be of the order of the period of vibration, or about 1 ps. The potential energy must be greater than the displacement energy, which is about 20 eV but is notoriously variable.

Sensitization by residual gases such as H_2O includes secondary-energy transfer.

A consequence of all this is that the energy required for the radiolytic processes that occur is much greater than the stored energy that remains (e.g., in broken bonds), and the excess heats up the specimen. The temperature increase depends mainly on thermal contact and thermal conductivity and varies from about one degree for metals to a few hundred for insulators under normal observation.

Correlated recombinations are very significant in limiting the extent of the observed radiation effects and in preserving long-range order. As mentioned before, however, there is a great deal of chemistry to be learned from the nature and rate of radiolysis. This is especially true since it exposes a regime of localized energy distribution not reached in normal-heating experiments.

The following chronology of events should be considered relevant to knock-on or radiolytic events [adapted from those suggested for nuclear-electron-capture processes by Sano and Gütlich (1984)]:

t (s)	
0	Knock-on event/radiolytic event Dilatation of the electron shell; emission of Auger electrons and X-rays • Highly charged species → Coulomb explosion • Radiolysis of ligands → electron deficiency of ligand radical • Local heating • Electronic excitation
$\simeq 10^{-15}$	Electron recombination • Normal- and aliovalent-charge states • Excited-crystal field states • Local pressure (size effect) • Vibrational excitation
$\simeq 10^{-13}$	Lattice vibrational relaxation • Fast relaxation of local heat and pressure • Geometrical rearrangement of the ligand sphere • New bond formation
$\simeq 10^{-12}$	Allowed optical transitions • Fast relaxation of excited states • Fast intramolecular-electron transfer • Stabilization of aliovalent-charge states • Fast redox reactions
$\simeq 10^{-11}$	• Slow relaxation of local heat • Cooling down of "thermal spikes"

$\simeq 10^{-9}$ to 10^{-7} Aftereffects
- Change of charge state
  Incomplete electron recombination
  Redox reaction following autoradiolysis and change of redox potentials
- Geometrical ligand rearrangement; "linkage isomerism"
- Change of spin state
  Spin-forbidden de-excitation in unchanged chemical environment; "vertical relaxation"
  Change of chemical environment and ligand-field potential; "horizontal relaxation"
- Long-lived, low-energy excitations with the manifold of the ligand-field ground state (slow relaxation of nonthermalized populations of spin-orbit and Zeeman levels owing to hindered spin-lattice interaction)

10.2.5 *Image processing and structure determination*

Images obtained in the electron microscope are seldom intuitively interpretable with a high level of confidence. Although, up to the present, the systems studied have frequently been interpreted simply by direct visual comparison with the projected structures, only for the very thinnest crystals will this be considered satisfactory. More and more often, the specimens will fail to meet the assumed characteristics on which the theory is based. For these, corrections of various kinds will have to be applied that make reliable interpretation possible even when there is no obvious similarity between the image and the projected model. This means that it is necessary to compare the images obtained with those calculated from specimen specifications and microscope characteristics and settings. To do this, the microscope settings must be known, the microscope characteristics must be taken into account, the crystal must be accurately aligned, and its thickness must be known. In Chapter 12, these requirements are discussed in detail. At one level, however, inspection and intuitive interpretation of images is all that is required. For example, if one wishes only to identify regions of the crystal in terms of the phases present (after they have already been studied by HRTEM in detail), the fringe or spot spacings and angles suffice. For these studies, of great importance in chemistry, the highest resolution or knowledge of the exact structure is not required.

Image calculations based on an assumed structural model are made by computer either in a batch or interactive mode. With the increase of minicomputer capacity, it is becoming more common to make the calculations interactively to reduce the time and trouble of batch-mode operation. Even computer programs taking into account every tractable consideration including thicker crystals can be run in a reasonably short time on a minicomputer.

Once the image has been simulated, it remains to compare it with that observed. This has usually been done visually with selected regions of the image. This comparison is entirely subjective and therefore limited in its reliability. Programs permitting objective, interactive comparisons utilizing statistical, pixel-by-pixel calculations have been implemented on a dedicated minicomputer (Rae Smith and Eyring, 1982). The statistical measures of agreement enable one to obtain an objective comparison as different models are proposed. More needs to be done to improve objective matching of simulated and observed images.

Methods of structure determination using HRTEM are being practiced to some degree by nearly all those utilizing the technique. A combination of very high resolution provided by a 1-MeV electron microscope with 2-Å point resolution and thin crystals and the use of convergent-beam electron diffraction (CBED) has been described by Bando et al. (1981) in the structure analysis of $Cr_4YFe_5O_{13}$. In this technique, high-resolution images taken along two or more principal axes in which each cation site is well resolved are used for the analysis. Each of these reveal the projection of the atoms in only two dimensions; therefore, this is not always a unique method of determining a three-dimensional space group. In general, it is necessary to make this determination by a separate CBED measurement, which yields the point group of the crystal. A unique determination of the space group results from a combination of the two types of microscopic observations. Bando gives several references to structures determined in this way, which provide lattice parameters to 0.1 Å.

The combination of X-ray analysis with HRTEM is illustrated in the case of $Na_{11}Nb_7O_{18}$ by Marinder and Sundberg (1984). High-resolution images were compared visually with those calculated by using the multislice method, and X-ray–powder diffraction was used to refine the structure. Thin crystals are essential in this intuitive interpretation of the images. It was possible by using this method to propose a space group and give lattice parameters to 10^{-3} Å.

Recently, Hovmöller et al. (1984) have applied image-processing methods to calculate accurate *average* atomic positions from electron microscopy alone. This technique relies on electron micrographs of thin crystals from instruments having resolutions of at least 2.5 Å to obtain atomic resolution. In this technique, HRTEM imaging is coupled with computerized image processing to yield atom positions to an accuracy of 0.1 Å. Averaging over many identical unit cells makes this accuracy possible. Their results on $K_{8-x}Nb_{16-x}W_{12+x}O_{80}$, where x is approximately one, compare favorably with an X-ray determination of the structure of the isostructural sodium compound.

10.3 Structure and structural defects in the binary-tungsten oxides

X-ray–structure determination of the tungsten oxides and related systems was a crucial element leading to an understanding of homologous series of intermediate phases, the Magnéli phases (Magnéli, 1953). X-ray structural analysis continues to provide detailed information about averaged structures. Because of the extraordinary flexibility of the tungsten oxides, it is necessary to obtain structural information with much greater resolution. HRTEM is perfectly suited to this task. Many of the most competent chemical microscopists have contributed to the HRTEM study of the tungsten oxides. Because of the central importance of these substances in the development of modern solid-state chemistry, they will serve here as a major illustration of the application of HRTEM techniques.

Two recent reviews of the thorough investigations into the details of the $WO_{3-\delta}$ system from different modes of preparation contain extensive lists of references to the original literature in this field. These two papers afford us the overview we require, and they shall serve as the basis of our discussion. The first of these by Booth et al. (1982) describes the preparation and characterization of the reduced oxides, $WO_{3-\delta}$ ($0 \le \delta \le 1$). Reduction was accomplished by heating sealed quartz capsules containing mixtures of WO_3 and W with the desired overall composition at temperatures between 873 and 1373 K for periods of up to five months. In some cases, the capsules were charged with a mineralizer such as HCl or Cl_2. Equilibrium was approached from only one direction. The resulting specimens were examined by utilizing both HRTEM (JEM 100B) and X-ray–powder diffraction (Guinier-Hägg). The phase diagram resulting from these studies and previous information is shown in Figure 10.3. This simple sketch asserts

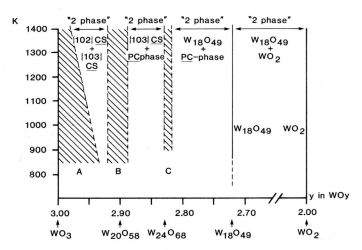

FIGURE 10.3 Schematic representation of the phases occurring in the WO_3–WO_2 region of the binary W–O system. Most phase regions have uncertain boundaries, and equilibrium is not usually achieved, the structures being disordered. Thus, region A contains disordered/quasi-ordered {102} CS phases, region B contains more or less well-ordered {103} CS phases, and region C contains disordered PC phases of the $W_{24}O_{68}$ type in contrast. Compound $W_{18}O_{49}$ appears to be a strictly stoichiometric equilibrium phase. (After Booth et al., 1982.)

that, between $WO_{3-\delta}$ and WO_2, only three additional phases exist: two at approximately $W_{20}O_{58}$ and $W_{24}O_{68}$, with relatively wide composition range, and one at $W_{18}O_{49}$, which can be considered a line phase. Only the $WO_{3-\delta}$ phase shows a definite composition dependence with temperature, widening at lower temperatures.

The second of these reports, Sahle (1983), describes alternative methods of reduction of WO_3, including reduction by dry hydrogen, with or without a carrier gas, and by buffer systems such as H_2-H_2O or $CO-CO_2$ with well-defined oxygen activities. For the experiments described by Sahle, large, single crystals of stoichiometric WO_3 were grown by chemical transport with water vapor and the stoichiometry adjusted by equilibration with an appropriate gas buffer. This technique would avoid the danger of having two types of crystal, one transported and the other not, as when reduction takes place in a sealed ampoule. Furthermore, the gas-equilibration technique allows a reliable method of obtaining information that could shed light on the mechanism of reduction. The results of these two methods, as far as they are comparable, agree quite well. For present purposes, the differences are minor; and only where necessary will discrepancies be commented upon.

X-ray–powder diagrams are essential for monitoring preparations and as an aid to their characterization; but in these systems, inhomogeneities due to disorder or the presence of fine-scaled occurrence of multiphases lessen their value. It is just in such circumstances, where the local structure must be observed, that HRTEM methods are the most powerful and necessary. Both imaging and selected-area electron diffraction were used.

The structure of WO_3 is simply related to the cubic ReO_3 type from which all the intermediate phases may be considered derived. The WO_3 structure is illustrated in a [100] projection in Figure 10.4a. Hatched squares represent octahedra sharing corners in all three directions. The short unit-cell dimension provides an axis along which the structure may be

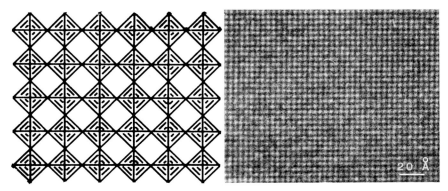

FIGURE 10.4 The structure and lattice image of stoichiometric WO_3. (a) Idealized representation of the WO_3 structure. (b) Corresponding HRTEM image. (After Sahle, 1983.)

profitably imaged with possible intuitive interpretation of the contrast. Figure 10.4b shows an image of WO_3. The white dots correspond to the square tunnels in the structure.

10.3.1 The {102} CS phase

Removal of oxygen from WO_3 down to a composition of about $WO_{2.94}$ at 873 K results in disordered crystallographic shear (CS) along (102) planes of the parent structure. The crystals are chunky like the cubic WO_3 crystals from which they are derived. Although CS along other directions was occasionally observed in these preparations, it was negligible when δ (in $WO_{3-\delta}$) was less than 0.005 and increasingly appreciable up to a δ-value of 0.075.

Crystallographic shear is the name given to the planar, structural boundary separating identical regions in transition-metal oxides when a compositional change results. (A relationship to polysomatism is drawn in section 9.2.2.) Crystallographic shear may be imagined to occur if all of the oxygen atoms along some plane in an oxide are removed and the structure is collapsed in such a way as to close the gap and restore the original anion substructure. Figure 10.5a illustrates a projection of the WO_3 structure after {102} CS. This loss of oxygen with the concomitant change in composition brings metal atoms closer together along the shear plane. There are four ways that {102} CS can occur in the cubic WO_3 structure if the crystal is stressed by reducing the oxygen activity in the ambient gas phase [an isolated element of shear has been named a "Wadsley defect" by Andersson (1970), as noted above]. Crystallographic-shear planes along {102} are the first to form as reduction begins, and they are observed in all the possible directions. The first HRTEM image of a Wadsley defect was of {102} CS in $WO_{3-\delta}$ by Iijima (1975).

When these Wadsley defects are parallel and regularly spaced, they become an element of a new structure that has the composition W_nO_{3n-1}, where n is a positive integer expressed as the number of WO_3 octahedra separating the CS planes. Notice that the phase field for {102} in Figure 10.3 is wide, indicating a continuous range of composition. HRTEM examination of many crystals in this region shows them to have a very high degree of disorder. As the concentration of Wadsley defects increases, they tend to be aligned into domains of the various W_nO_{3n-1} {102} CS series. It is not unusual for the spacing to be irregular, which results in slabs of several members of the homologous series, as shown in Figure 10.5b.

In principle, oxides with {102} CS could order into the entire series of W_nO_{3n-1} covering the whole composition range from WO_3 ($n = \infty$) to WO_2 ($n = 1$), with two-phase regions between each stoichiometric phase, although the stoichiometry could be rather strange. In fact, only a few instances of ordered domains have been seen. These include members with n equal 26, 14, or 12 observed either from the images or from sharp electron

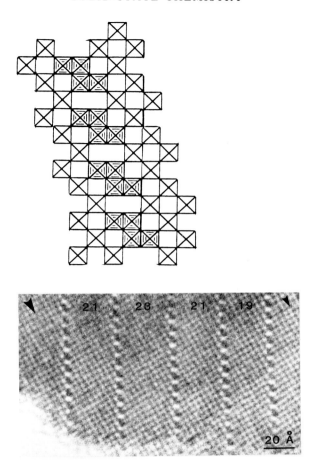

FIGURE 10.5 The structure and image of nonstoichiometric $WO_{3-\delta}$. (a) Idealized illustration of the octahedral coordination along a {102} CS plane. (b) HRTEM image of a {102} CS structure. The numbers of octahedra between the CS planes in the direction indicated by the arrow are specified. This number corresponds to the n-value in the formula for the ordered phases of the homologous series. (After Sahle, 1983.)

diffraction patterns. Usually, extensive disorder indicated by continuously streaked diffraction spots in the $\langle 102 \rangle$ direction is observed. Calculations based upon minimum-elastic-strain energies have predicted that structures with $m = 26$, 14, and 12 would be particularly favorable (Iguchi and Tilley, 1977).

As reduction approaches the lower limit of this phase, isolated {103} CS Wadsley defects begin to appear in small but increasing numbers, until finally a two-phase region is reached, with fragments either of {102} CS or {103} CS or an intergrowth of the two. The preparations clearly reflect this region as having crystals with two markedly different habits. The chunky crystals are replaced by those having needle shapes as {102} CS is replaced by {103} CS.

10.3.2 The {103} CS phase

In the composition range $WO_{2.920-2.985}$, fragments from the needlelike crystals show a much more ordered array of shear planes of the {103} type. In this case, condensed, edge-sharing WO_6 octahedra containing six members—rather than four as in {102} CS—are deployed along the shear plane. There is much less gross disorder observed in these crystals, although fine-textured intergrowths of different widths of the WO_3-like slabs is common. Crystals with regular {103} shear would have the composition W_nO_{3n-2}, and in the observed composition range, n would vary from 25 to 19. Fragments of crystals with domains between 25 and 15 have been reported.

10.3.3 The $W_{24}O_{68}$ phase

The compositional range of stability of this phase is much narrower, from about $WO_{2.83}$ to $WO_{2.82}$. The structure has been described as consisting of an intergrowth of slabs of a tetragonal tungsten bronze (TTB) type and slabs of WO_3. An element of the TTB structure results when a column of four WO_6 octahedra in cross section rotates by 45°, within the WO_3 structure, and maintains all corner sharing of oxygens but creates four pentagonal and four triangular tunnels in the process. In the $W_{24}O_{68}$ structure, one-half of the pentagonal tunnels are filled with –W–O–W–O– strings, forming what are called pentagonal columns (PC). Figure 10.6a illustrates the structure of $W_{24}O_{68}$ by a drawing, and Figure 10.6b shows a micrograph. Fragments of this phase are frequently twinned and faulted by having variable widths of the WO_3-like structure, shown by the light zigzag band in Figure 10.6a. At the most oxidized side, regions of {103} CS were found to coexist, suggesting a mechanistic relationship between these two structures. Both the images and the diffraction patterns confirm a high level of disorder.

The three nonstoichiometric $WO_{3-\delta}$ phases described so far illustrate a very common problem in solid-state chemistry that HRTEM was able to solve. Structures based on a common substructure such as in {102} CS, {103} CS, and the pentagonal column structure of $W_{24}O_{68}$ can easily intergrow (or possess defects related to the various series) with little increase in energy or decrease in entropy. This results in phases that may or may not actually be at equilibrium with respect to the ambient oxygen or tungsten activity. Furthermore, the elastic stress within these crystals can change the atomic potentials in a way that will stabilize the disordered phase. The result is that rather wide ranges of stoichiometry exist for each phase.

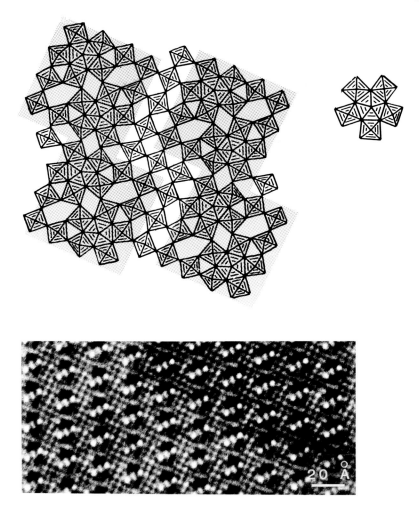

FIGURE 10.6 The structure and lattice image of $W_{24}O_{68}$. (a) The idealized representation of $W_{24}O_{68}$, projected onto the (010) plane. The square TTB-type units are shadowed. A pentagonal column (PC), which is a building unit in the structure, is shown to the right. It consists of a WO_7 pentagonal bipyramid sharing all of its equatorial edges with WO_6 octahedra and is linked by corner sharing to identical polyhedra above and below, along the line of projection. (b) Lattice image of the $W_{24}O_{68}$ phase. (After Sahle, 1983.)

10.3.4 The $W_{18}O_{49}$ phase

In contrast to the other intermediate phases between WO_3 and WO_2, the $W_{18}O_{49}$ phase has no observed range of composition. Fragments examined by HRTEM are rarely twinned and show no other defects. The structure consists of pairwise-linked pentagonal columns possessing a very short W–W distance characteristic of metal–metal bonding. The pentagonal

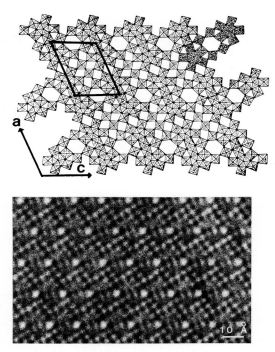

FIGURE 10.7 The structure and lattice image of $W_{18}O_{49}$. (*a*) The idealized structure model of $W_{18}O_{49}$ projected onto (010). One PC–HT–PC unit is marked. (*b*) HRTEM image of a $W_{18}O_{49}$ fragment. (After Sahle, 1983.)

columns are edge-connected to form pairs that are, in turn, corner-linked to give pseudohexagonal tunnels (PC–HT–PC). A PC–HT–PC unit clearly displayed in the drawing in Figure 10.7*a* can be visualized in the micrograph of Figure 10.7*b*. This element of the $W_{18}O_{49}$ structure occurs as a defect in other structures, as will be described later.

10.3.5 Defect structures

The four intermediate $WO_{3-\delta}$ phases described previously each have a characteristic defect that, when ordered, gives a stoichiometric member of a series within the phase field. Each of the defects in the three most oxygen-rich phases intergrow coherently with the parent WO_3 structure; hence, it is not surprising that they not only are capable of disorder in the WO_3 matrix but may be found together in the same crystal. In some cases, this juxtaposition of defect types suggests a mechanistic connection with the processes of oxidation or reduction and a metamorphosis of one type into another.

Disorder in the {102} CS structures is the rule. The {103} CS structures, although more ordered, also possess an abundance of disorder. In the two-phase region or during reaction, intergrowths of the two are frequently observed in a reduced $WO_{3-\delta}$ crystal.

There have been several models suggested for the generation, movement, and transformation of CS planes. In this case, it is suggested that groups of four edge-sharing octahedra of the {102} type acquire the grouping of six edge-sharing octahedra characteristic of the {103} CS structure at the crystal edge and are transported inward by successive shifts of $-W-O-$ strings along the CS planes, ultimately effecting the transition. This model is consistent with one proposed by Andersson and Wadsley (1966) and later modified by Allpress (1972).

Of course, the basic similarities of the connectivities of the various defect types with WO_3 always provide a coherent boundary, as exemplified by the termination of Wadsley defects or the interface of intergrown structures. One particularly interesting feature is that sometimes a Wadsley defect in a shear structure is terminated by a PC–HT–PC defect, or sometimes simply PC–HT.

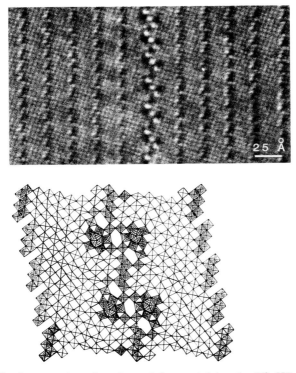

FIGURE 10.8 The incorporation of a planar defect containing the PC–HT element. (*a*) A lattice image of a {103} CS crystal with a special type of planar defect. (*b*) An interpretation of (*a*). Every second group of six edge-sharing octahedra along the CS plane has been replaced by a PC–HT–PC unit. (After Sahle, 1983.)

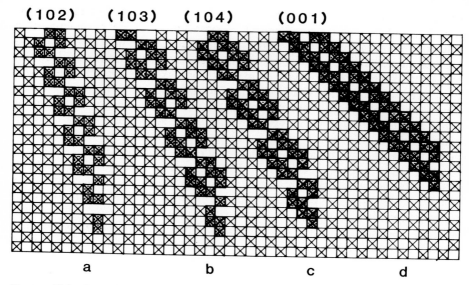

FIGURE 10.9 Octahedral arrangements for [010] projections through pairs of (102), (103), (104), and (001) CSP in a WO_3 matrix. The M^{5+} cation sites (heavily lined) are edge-shared octahedral sites. (After Bursill, 1983.)

Figure 10.8 illustrates the incorporation of an unusual type of planar defect containing the PC–HT element alternating with the cluster of six edge-sharing octahedra along a (103) shear plane. This represents a local compositional heterogeneity with a smaller oxygen content along this defect.

Bursill (1983) has turned his attention to the structure of small defects in WO_{3-x} and, on the basis of the wide background of HRTEM information, proposes structures and mechanisms of aggregation. These models include CSP and PC–HT groups. Linear defects, consisting of two pairs of edge-shared octahedra, which readily explain a large fraction of the HRTEM observations of precipitation and dissolution phenomena in pure and doped $WO_{3-\delta}$, are shown in Figure 10.9. These linear pairs of edge-shared octahedra may be identified by focusing on the vertical rows of shaded octahedra. These, when condensed as shown here, form pairs of crystallographic shear planes (CSP). Pairs of CSP can form without shear of the WO_3 substructure.

10.3.6 Mechanisms of solid-state reactions

One of the principal objectives of solid-state chemistry is to delineate the sequence of atomic shifts that transform reactants into products. Such a stated sequence of events is termed the mechanism of the reaction, and the overall rate is controlled by the slow step in the sequence. The rate law for the reaction is normally determined by observing the overall macroscopic rate of reaction under a variety of controlled conditions of starting concentrations, temperature, and so on. The rate law is determined by the rate-controlling step and points to possible pathways and bottlenecks. The existence of a technique that can image local structure at the atomic level raises the possibility of observing directly the initial steps in the formation of a reaction nucleus and its subsequent accretion. This prospect is especially inviting since it is now possible to record high-resolution images on videotape in real time, as will be discussed in some detail later.

Certain reactions do not proceed at the temperatures available to existing microscope specimen holders capable of allowing high-resolution observation. It is necessary therefore to treat the specimen outside the microscope, followed by a microscopical study. If one reduces a $\{103\}$ CS WO_{3-x} phase with a gaseous buffer, ribbons of a linked TTB-like material grow into the structure as reduction proceeds. The $W_{18}O_{49}$ appears to nucleate and grow from regions of $W_{24}O_{68}$ that previously had become amorphous. No observation of nucleation within the crystalline $W_{24}O_{68}$ has been observed, suggesting a completely reconstructive nucleation. When $W_{18}O_{49}$ is oxidized at lower temperatures ($\simeq 500$ K) than required for reduction ($\simeq 1170$ K), a solid-state reaction is observed in which the features PC and HT disappear in favor of a WO_3-type structure. The other phases could then nucleate and grow in this newly formed WO_3 structure. A mechanism for this process has been proposed (Sahle 1982) from extensive HRTEM observations. Sahle and Kljavens (1985) oxidized a $\{103\}$ CS structure at equilibrium with a controlled O_2 activity; then, they observed the sequence of reactions back to WO_3. They concluded that $\{103\}$ CS oxidizes to a WO_3-like structure that subsequently forms $\{102\}$ CS in a twinned crystal and then oxidizes to WO_3.

Booth et al. (1982) rationalize the collective experience of the large number of studies of the $WO_{3-\delta}-O_2$ system and, in particular, those mentioned by Sahle (1983). They take into account the lower formation energy of $\{102\}$ CS but higher strain energy to explain the differences observed in the $\{102\}$ CS and $\{103\}$ CS regions, considering the methods used for specimen preparation. In the ampoule method used by Booth and co-workers, there is a possibility of growth of new phases from the vapor not available in the buffer-gas reduction used by Sahle. Although it is concluded that equilibrium is probably never reached in any of the ampoule preparations—or if it were, not maintained during the cooling required for microscopical investigation—the phase behavior discussed previously utilizing gas-buffer preparation probably does reflect the equilibrium situation. The strain energy of structures based on PC is considerably lower than for shear structures and, hence, pentagonal columns are favored at lower temperatures or higher degrees of reduction, as observed. Furthermore, they are sometimes seen as a terminus to CS defects, suggesting the dissipation of the relatively high strain energy of crystallographic shear.

10.4 Doped WO_3 to give $(W, Ta, Nb)O_{3-\delta}$

10.4.1 The ternary systems $(W, Nb)O_{3-\delta}$ and $(W, Ta)O_{3-\delta}$

The metal-to-oxygen ratios can be varied through the range characteristic of the $WO_{3-\delta}$ phases without a reduction in the oxidation state of the tungsten. This virtual reduction is accomplished by substituting an altervalent metal atom of lower oxidation number for tungsten in varying concentrations. To accomplish this, Ta_2O_5 and Nb_2O_5 have been added in varying amounts to WO_3, both separately, to give ternary phases, and together, to give quaternary phases. Many interesting solid-state chemistry questions may be answered by such a study.

The consequence of adding either tantalum or niobium is to reduce the amount of oxygen required to produce a stoichiometric compound. However, it would be nonstoichiometric if the reference is to the original WO_3 structure. Under these circumstances, will the structural accommodation to the oxygen deficiency in the WO_3 lattice be the same as or different from the binary system? Can transformations between species be observed? Furthermore, will the structures observed be different depending on whether tantalum or niobium is substituted? The outer electronic structures of Ta^{5+} and Nb^{5+} are the same, but the Nb^{5+} lacks filled $4f$- and $4d$-orbitals underlying the valence shell that are present in Ta^{5+}. A very large mass difference also exists between niobium $(Z=41)$ and tantalum $(Z=73)$. Tilley and his collaborators (England et al., 1982; England and Tilley, 1982, 1983, 1984), following extensive investigations of others utilizing X-ray-diffraction and lower-resolution electron microscopy (see these papers for references), have provided many of the answers to these questions.

High-resolution, electron-microscopical studies supported by X-ray-diffraction analysis were made on Nb_2O_5-WO_3 and Ta_2O_5-WO_3 systems of composition near $(MM')O_3$. In the Nb_2O_5-WO_3 system, the mixtures were fired at 1600 K for up to 20 days. The resulting material exhibited two types of crystal habits, rather reminiscent of the $WO_{3-\delta}$ system, one chunky and one needlelike. The chunky material was characterized by CSP and the needles by PC. As in the $WO_{3-\delta}$ studies, it is likely that true equilibrium was never achieved. Phases containing {001} CS were ubiquitous; however, ordered phases were observed in the range $(NbW)O_{2.9375}$ to $(NbW)O_{2.875}$, forming regions of $(NbW)_nO_{3n-1}$, with n varying from 16 to 8, respectively. The {001} CS phase field is assigned the region $MO_{2.90}$–$MO_{2.93}$. The advantage of HRTEM methods is demonstrated here, where the local structure and its boundary regions are clearly delineated rather than the average over the whole crystal, as obtained by X-ray analysis.

The other prominent mode of crystallographic shear in the $(Nb, W)O_{3-\delta}$ series was along {104}. These CS planes yield a series of $(Nb, W)_nO_{3n-3}$ when they are ordered. The values of n found varied from 52 to 65, corresponding to compositions $(Nb, W)O_{2.942}$ to $(Nb, W)O_{2.954}$; but not only was there substantial uncertainty about the regularity of the ultrafine spacing, there was also variability in the number of edge-shared octahedra grouped along the shear plane. Notice that this phase field lies outside the range of that of {001} CS and nearer to MO_3.

Crystal fragments in the range between {104} CS and {101} CS revealed complex wavy patterns of shear planes that could be interpreted as being made up of attached short segments of {104} CS and {101} CS in most cases but also occasionally {103}. No phases with extensive {103} CS occurring alone were observed. However, {102} CS planes were frequently seen to occur in a disordered array in regions of some crystal fragments. This disorder is entirely analogous to the {102} CS regions of $WO_{3-\delta}$.

When Ta_2O_5 is added to WO_3 to form compositions near MO_3 and fired at 1623 or 1673 K, {103} CS phases were found at compositions near $MO_{2.95}$. Regions of $(Ta, W)_nO_{3n-2}$ varied between $n = 43$ and 50 in one preparation. The {103} CSP, although of wider spacing, was very similar to that observed in the $WO_{3-\delta}$ {103} CS phases. These shear planes are also observed as isolated Wadsley defects. Phases with {104} shear were observed for the first time in Ta–W–O oxide in specimens prepared at 1673 K but not at lower temperatures. This suggested a possible niobium contamination, but none was found. The {102} CS phases were disordered, as they are in $WO_{3-\delta}$ or $(NbW)O_{3-\delta}$, as previously discussed.

As might be expected, the CS behavior of the Ta_2O_5–WO_3 system is much more like the $WO_{3-\delta}$ system than is Nb_2O_5–WO_3. In the first, the predominant shear is along {103} or {104}, with {102} CS also evident; in the last, {001} and {104} CS dominate, with disordered {102} also observed. Shear structures in the composition range where {001} CS phases are found in the Nb–W–O system are not observed in the Ta–W–O system. There appears to be a two-phase region between WO_3 and the nearest shear-structure phases, with lower oxygen content in both the tantalum- and niobium-doped regions. The upper limit of the sheared phases seems to be $(Nb, W)_{65}O_{192}$ and $(Ta, W)_{50}O_{148}$ in the two systems studied. This suggests that the dopant atoms are not randomly replacing tungsten atoms but are concentrated along the shear planes. It would be interesting to see if this segregation could be observed by an EDS scan across the planes utilizing the small probe of the STEM or by ion-probe-depth analysis perpendicular to the shear planes. In the two-phase region, isolated shear planes exist. England et al. (1982) conclude from their experimental observations that, in the two-phase region, all the pentavalent atoms are consumed in the formation of Wadsley defects.

These systems are not infinitely adaptable, as might be expected from, say, the Cr_2O_3–TiO_2 oxides, where, by virtue of "swinging shear planes," a perfect structure may be observed anywhere within the composition range by the combination of antiphase-boundary and shear-plane combinations. [Refer to Bursill and Hyde (1972) for a detailed discussion of "swinging shear planes" proposed from their CTEM studies.] Intergrowths and wavy shear planes are seen but not the regular, long-range order required.

Another interesting observation is that the spacing between shear planes is virtually the same among the possible systems. For example, the spacing in $\{104\}$ CS in $(Nb, W)_{52}O_{153}$ corresponds exactly to that of the $\{001\}$ CS in $(Nb, W)_{13}O_{38}$. It appears that transformation between these CS systems occurs when the spacings between the top of one series coincides with the bottom of another. Similarly, in regions between different CS structures, regions of wavy CS occur that appear to minimize the strain that might exist without this transition.

10.4.2 The quaternary (Nb, Ta, W)$O_{3-\delta}$ systems

England and Tilley (1983), after studying the separate $(Nb, W)O_{3-\delta}$ and $(Ta, W)O_{3-\delta}$ systems discussed previously, turned their attention to the system $(Nb, Ta, W)O_{3-\delta}$. Specimens were prepared by heating the oxide mixtures in sealed platinum ampoules at 1623 K. A CS phase field was observed between the compositions limits $(Nb, W)O_{2.965}$ and $(Nb, W)O_{2.875}$ on the niobium-rich side of the phase interval and between $(Ta, W)O_{2.96}$ and $(Ta, W)O_{2.943}$ on the tantalum-rich side. Regions of CS phases along $\{001\}$, $\{104\}$, and $\{103\}$ varied consistently across the region. Three morphologies were observed in the mixed crystals: chunky, smooth and rounded, and fine needles. The chunky crystals were found to contain CS phases. The rounded crystals were WO_3 with isolated $\{102\}$ Wadsley defects, and the needles were tetragonal tungsten bronze type of nominal composition $2M_2O_5:7WO_3$ (2:7). The defect types found are grouped in a reasonable way, considering the corresponding phases for the ternary systems.

Segmented shear characterizes many of the preparations. These are jerky intergrowths of $\{104\}$ and $\{001\}$ CS. The waves are more or less parallel in these specimens. One of the more interesting features of this study is the occurrence of PC structures together with CS planes. In the crystals where this was observed, the PCs lie along the $\{102\}$ planes (Figure 10.10) and appear to occur in the same configuration as in $WO_{2.83}$.

The structural consequences of reducing the M/O ratio from 3.0 to about 2.86 in the $WO_{3-\delta}$, $(Nb, W)O_{3-\delta}$, and $(Ta, W)O_{3-\delta}$ systems have been studied. This region is dominated by crystallographic-shear structures accommodating oxygen loss either by reduction of WO_3 or, in fully oxidized systems of similar oxygen composition, by the substitution of niobium or tantalum for tungsten. These studies confirm the difference between niobium and tantalum in their anomalous replacement of tungsten in these ReO_3-related structures. Tantalum is electronically much more like tungsten, and the shear types remain essentially the same. But niobium promotes a different shear structure covering a greater range of composition. A particularly important point is that nowhere in these phases has the phenomenon of "swinging shear" been observed.

All evidence points to an inhomogeneity across a crystal owing to the

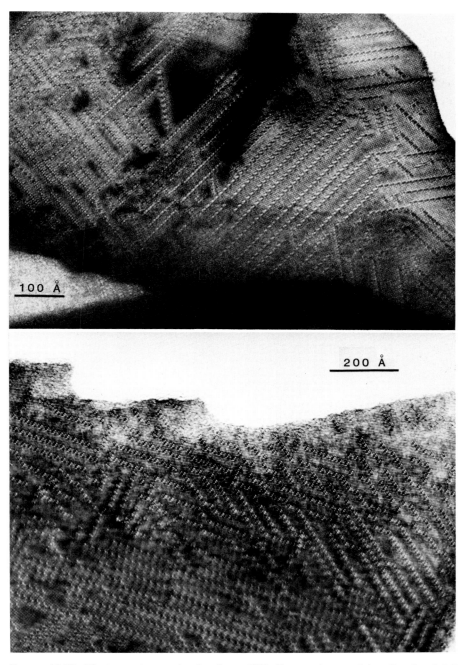

FIGURE 10.10 Electron micrographs showing a WO_3-like matrix containing disordered PC units. (a) A low magnification image revealing a considerable tendency toward ordering of the {102} chains of PC units. (b) A higher magnification image of the edge of the crystal in (a), revealing that the features are indeed PC units. (After England et al., 1984.)

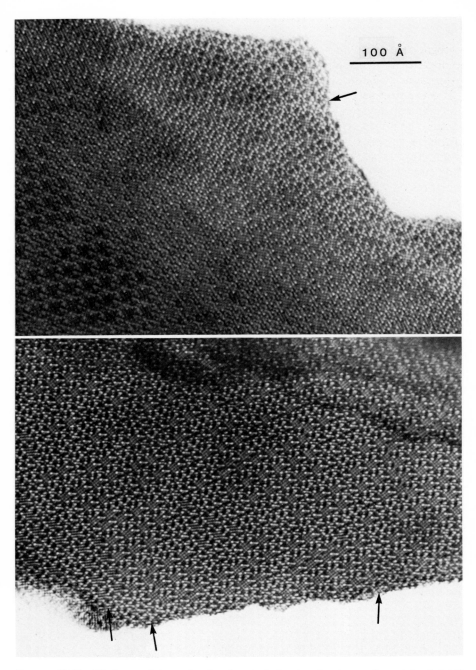

FIGURE 10.11 Electron micrographs showing typical microstructures of crystals. (a) Of $2Nb_2O_5:7WO_3$. (b) Of $2Ta_2O_5:7WO_3$. The PC units are imaged as dark blobs of contrast, which occur in pairs within the WO_2 matrix, imaged as a square array of grey dots. Some boundaries between ordered domains of structure are marked, including one in (a) that consists of a double row of PC units instead of a single row. (After England et al., 1984.)

segregation of niobium into the shear regions; tantalum appears more homogeneous. This may account for the chemical sluggishness of these systems and the difficulty in obtaining equilibrium.

The occurrence of PC defects and structures in the ternary and quaternary phases has been studied by England and Tilley (1982, 1984). In these studies, the mixed crystals were annealed at temperatures between 1073 and 1573 K, a region where the development of PC will not be the result of vapor growth, at least at the lower end of the temperature region. Even after annealing 12 days, the CS phases do not change at all; however, PC structures develop in the WO_3-like slabs between the CS planes. This points again to the difficulty in reaching equilibrium, owing in part to the absence of reasonably rapid cation diffusion. The PCs form at random in the WO_3-like matrix and always in pairs, as modeled by Anderson and Hyde (1967), Bursill and Hyde (1972), and Hyde and O'Keefe (1973). There is speculation about the mechanism, but little evidence exists in the micrographs obtained to support these ideas. Ingenious experiments with the higher-resolution instruments becoming available are needed to settle the question.

England and Tilley (1984) have elaborated on their observations of PC formation in the ternary and quaternary oxides. When oxides in the composition range $MO_{2.95}$ and $MO_{2.82}$ are annealed at temperatures close to 1600 K, PC TTB-type regions were found in all specimens. None were perfectly ordered, and a wide variety of microstructures were found in solid solution. The 2:7 compounds ($2Nb_2O_5:7WO_3$, $2Ta_2O_5:7WO_3$) are found most often and most ordered in regions nearest WO_3 in composition. Figure 10.11 illustrates these results.

10.4.3 Pentagonal-tunnel structures and structural defects in the W–Nb–O system

By means of HRTEM studies, the structures of compounds of the formula $(M, M')O_{3-\delta}$, where δ is greater than about 0.17 and less than about 0.30, M is tungsten, and M' is either niobium or tantalum, have been sorted out. If δ is less than 0.17, the stable phases are CS structures, as described earlier; if δ is greater than 0.30, they are block structures, as will be discussed later. In between, the structural element that dominates the ordered or disordered materials is PC. This element was first recognized in this way by Lundberg and has been expanded to show a general applicability to structures in this composition region by Lundberg et al. (1982). They clearly describe the connectivities of PC for a wide range of structures. In the binary $WO_{3-\delta}$ system, PC structures with a composition near $W_{24}O_{68}$ ($WO_{2.83}$) and $W_{18}O_{49}$ ($WO_{2.72}$) are established, as indicated previously. In the (Nb, W)$O_{3-\delta}$ phases, PC structures with compositions $2Nb_2O_5·7WO_3$ (Nb, W)$O_{2.82}$ and $4Nb_2O_5·9WO_3$ (Nb, W)$O_{2.76}$ have been extensively studied in HRTEM, first by Iijima and Allpress (1974a, 1974b), who

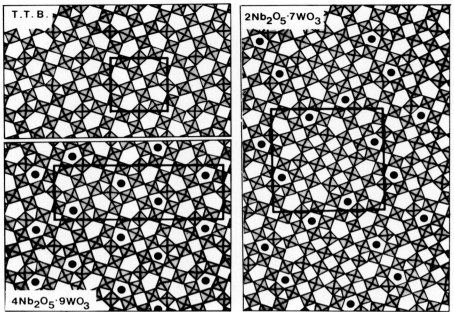

FIGURE 10.12 Structural relationship among structures. (a) Of tetragonal tungsten bronze. (b) Of $4Nb_2O_5 \cdot 9WO_3$. (c) Of $2Nb_2O_5 \cdot 7WO_3$. (After Eyring and Tai, 1976.)

clarified the defect structures, and later by Horiuchi (1982), who formulated mechanisms of defect formation.

The early observations of Iijima were summarized by Eyring and Tai (1976). These HRTEM studies indicated the nature of intergrowth in crystals in this composition region. Figure 10.12 illustrates the projection of the TTB structure of the alkali tungsten bronze and the closely related structures of the 4:9 and 2:7 materials. The departure from the WO_3 corner-sharing square net is the effective rotation of a 2×2 or a 4×4 block of MO_6 octahedra without losing the corner-sharing linkages. This provides new tunnels in the structure that are either pentagonal or triangular in projection. Some of the tunnels are occupied by $-Nb-O-Nb-O-$ chains; and when ordered as shown in Figure 10.12, they have the compositions indicated. However, complete order in these structures is unusual. Some types of defects observed are illustrated in Figure 10.13, which is a high-magnification image of a fragment nominally $4Nb_2O_5 \cdot 9WO_3$. The region marked A is shown in a high-resolution image in Figure 10.14, with the projected network of corner-sharing octahedra shown at the bottom. It is apparent that the projected unit cell of 4:9 contains three TTB subcells, where 4 of the 12 pentagonal tunnels (PT) are filled (PC) and appear as the center of dark spots and where the 8 unfilled PT appear as the larger white

FIGURE 10.13 High-magnification, two-dimensional lattice image from a faulted crystal of $4Nb_2O_5 \cdot 9WO_3$, showing several kinds of defect boundaries. (See Eyring and Tai, 1976.)

spots. The square tunnels appear mainly as faint white spots. This region, then, consists of bands of 4:9, with a wider band of another composition having 5 × 2 TTB subunits with a different filling of the PTs. Of the 40 PTs available, only 14 are PCs, leading to a smaller O/M ratio in this band. The circled region is the configuration of TTB without tunnel filling. This same type of defect has been observed to form at twin boundaries of the 4:9 compound (Eyring and Tai, 1976). Bands with different configurations are observed to separate twinned regions in a stepped fashion as well. These have been observed stepped along {130} and {110}, for example. The TTB substructure is frequently continuous through the defect regions, and the defect bands themselves are consequently integral multiples of the basic TTB unit. Analogous defect structures are observed in $2Nb_2O_5 \cdot 7WO_3$.

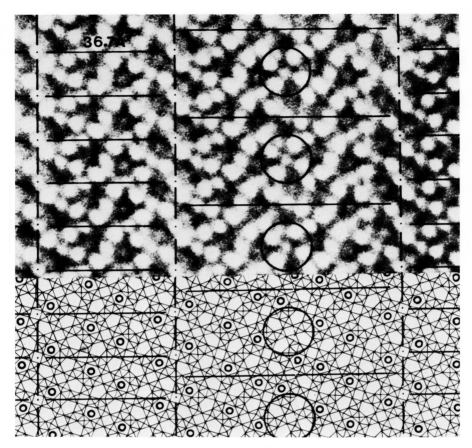

FIGURE 10.14 High-resolution image of a defect boundary $\delta[100]$ in $4Nb_2O_5 \cdot 9WO_3$. (See Eyring and Tai, 1976.)

Horiuchi (1982), using a high-voltage, high-resolution electron microscope operating at 1 MeV and a resolution of 2 Å, has studied stages of reaction in $4Nb_2O_5 \cdot 9WO_3$. Horiuchi suggests that crystals that have been slightly reduced have oxygens missing in the –M–O–M–O– chains in the PCs of the structure. These vacancies loosen adjacent M atoms that are observed to be knocked-on in the 1-MeV electron beam into adjacent, empty pentagonal tunnels. Atomic-level structural images show the contrast change expected from this model and from image simulations. Figure 10.15 illustrates the contrast variation that is interpreted successfully by taking the composition into account. This is a prime example of the level of understanding that can be reached with careful preparation of specimens, skillful use of the best resolution available, the coupling of these steps with

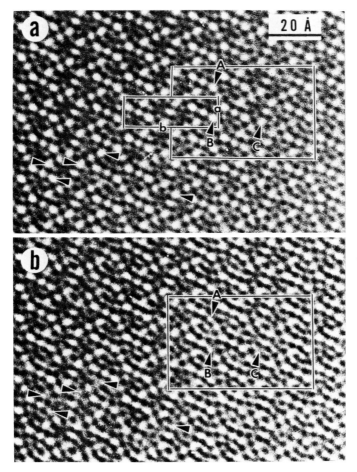

FIGURE 10.15 High-voltage HRTEM image of $4Nb_2O_5 \cdot 9WO_3$. (a) Before strong irradiation. (b) After strong irradiation. (After Horiuchi, 1982.)

chemical information about defect concentration, and then followup by comparisons with calculated images.

In another experiment, a crystal of $2Nb_2O_5 \cdot 7WO_3$ was reduced at high temperature to produce domains of $4Nb_2O_5 \cdot 9WO_3$, and images recording the transition were taken. The concentration of PCs is higher in the reduced material as WO_3 regions are transformed to TTB-type structures. The contrast change can be accounted for on the basis of loosened cations that are knocked into a previously empty pentagonal tunnel. Later, when the transformation is nearly complete but many localized defects remain, they are always found within the disordered interdomain regions. Even an isolated domain of 4:9 is stable enough to resist radiation damage.

Finally, the oxidation of the block structure $Nb_{22}O_{54}$ to Nb_2O_5 at 473 K was observed. Defects that are atomic size in projection have been modeled, and these models have been tested to reach a satisfactory description of the role of interstitial oxygen in the oxidation reaction. Explained also is the superstructure formed as the defect concentration is increased. After 600 h of heating, the c-axis is doubled, with no change in a or b. The doubling results from a regular sequence of layers of (3×3) blocks and $(3 \times 3) + (3 \times 4)$ blocks.

10.5 The tungsten bronzes and related compounds

10.5.1 The tungsten and vanadium bronzes

When appropriate amounts of WO_3, WO_2, and an alkali or thallium tungstate are mixed and heated in sealed, evacuated platinum capsules at about 1100 K, a remarkable series of compounds is formed. They have the formula A_xWO_3, where A is an alkali metal or thallium and x has a value between 0.05 and 0.54. Their intense color and metallic luster and frequently metallic conduction invited the name *bronzes* to be applied. Brief reviews of HRTEM studies of some of these compounds have recently been written (Kihlborg, 1982; Kihlborg and Sharma, 1982); from which we shall draw our even briefer summary. These papers serve as a guide to the large body of research concerning their structures and compositions.

Recall from the discussion of the $WO_{3-\delta}$ phases that WO_3 is a three-dimensional, chessboardlike framework of corner-sharing WO_6 octahedra that have square tunnels when viewed along {100} directions. In the traditional bronzes, the addition of alkali-metal atoms promotes the creation of new types of lattice sites without removing the corner sharing of the WO_6 octahedra. The new linkages consist in either a mixture of pentagonal and triangular tunnels, as found in the tetragonal tungsten bronze (TTB) structures (Figure 10.12), or hexagonal and triangular tunnels, as found in the hexagonal tungsten bronze (HTB) structures (Figure 10.16). The bronze-forming atoms are then to be found in either the pentagonal or hexagonal tunnels. These tunnels can be considered as intergrowth elements in the WO_3 parent structure. If the structure consists only of these elements, the HTB or TTB structures result. Figure 10.17 is a schematic phase diagram showing the existence range of these bronzes. In the formula A_xWO_3 of HTB, x can vary from approximately 0.19 to 0.33 as alkali atoms half fill and completely fill the available sites in the hexagonal tunnels. The c-axis extends over two octahedral layers owing to a slight displacement of tungsten atoms in the HTB structure. There is no X-ray–diffraction evidence of ordering of the partly occupied metal atoms.

Incommensurate superstructures have been found in K_xWO_3 ($0.24 \leq x \leq 0.26$) by Bando and Iijima (1980) and Iijima and Bando (1981), utilizing HRTEM techniques and convergent-beam electron diffraction. Their inter-

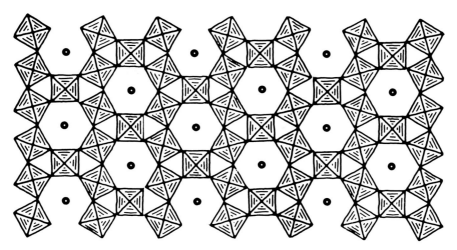

FIGURE 10.16 The HTB structure depicted as WO_6 octahedra-sharing corners. The (ideal) positions of the alkali ions in the hexagonal tunnels are indicated by dots. (After Kihlborg, 1982.)

pretation of the results, particularly the HRTEM images, is that layers parallel to {001} occur, with alkali sites completely filled or completely empty. An empty layer every four or five filled layers leads to the nonintegral value of superstructure spacing along the c-axis (i.e., $2.06c$ for $x = 0.24$ and $2.20c$ for $x = 0.26$). This illustrates the power of electron optical methods to reveal features of a real structure that can be missed even by careful X-ray examination.

Less well known are the alkali and thallium vanadium fluoride bronzes, $A_x VF_3$ (Hong et al. 1979). They, too, form structures analogous to the HTB and TTB compounds in the same range of x-values. The HTB-type

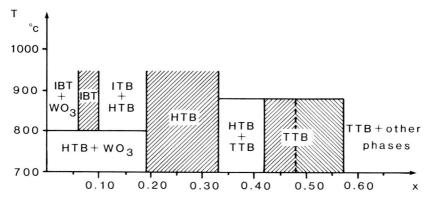

FIGURE 10.17 Phase diagram for $K_x WO_3$. The Rb and Cs do not form a TTB phase; but except for that, the phase diagrams are similar. (After Kihlborg, 1982.)

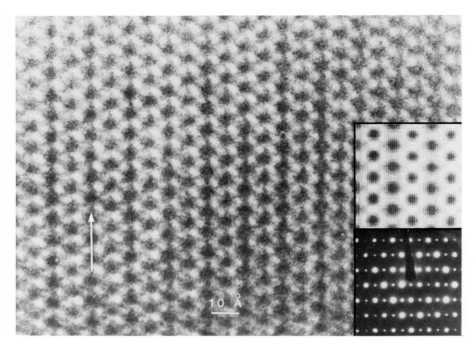

FIGURE 10.18 Electron micrograph of $Rb_{0.25}VF_3$ showing the doubling of the unit cell along a^*. This is a result of ordered partial filling of the hexagonal tunnels by Rb atoms.

structures have been studied by Rieck et al. (1982), utilizing HRTEM. In these, the ordered, fractional partial filling of the hexagonal tunnels with A atoms leads to superstructures where $a^* = 2a^*_{HTB}$, $b^* = b^*_{HTB}$, $c^* = c^*_{HTB}$; $a^* = 2a^*_{HTB}$, $b^* = 2b^*_{HTB}$, $c^* = c^*_{HTB}$; and $a^* = a^*_{HTB}$, $b^* = b^*_{HTB}$, and $c^* = 3c^*_{HTB}$. Images obtained from microscopic examination were matched with calculated images based on a regularly arranged, partially filled tunnel structure. Figure 10.18 illustrates the structure where $a^* = 2a^*_{HTB}$ and the filling corresponds to a pattern of 0.25, 0.75 in alternate tunnels. The defect structures of the A_xVF_3 were also studied by Rieck et al. (1983).

10.5.2 Intergrowth tungsten bronzes (ITB)

At temperatures above about 1170 K and values of x in the range 0.06 to 0.10, compounds are formed that may be regarded as a stacking of layers of the HTB and WO_3 structures. Figure 10.19 shows a high-resolution micrograph and the corresponding interpretation of one such compound formed. The slabs of HTB consist of two rows of hexagonal tunnels separated by an undulating slice of corner-sharing octahedra, labeled 1, and a slab of WO_3 consisting of seven layers of corner-sharing octahedra. This

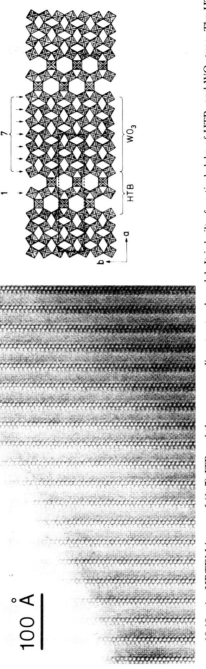

FIGURE 10.19 An HRTEM image of (1, 7) ITB and the corresponding structural model. It is built of vertical slabs of HTB- and WO_3-type. The HTB slabs comprise two rows of hexagonal tunnels, and the WO_3 slabs comprise seven rows of octahedra. The unit cell is indicated. (After Kihlborg, 1982.)

TABLE 10.1
Observed Ordered ITB Structures

Bronzes	Bronzoids, A = K, Cs; M = Nb, Ta
(7)–(11)	(1, 2)–(1, 9)
(1, 4)–(1, 14)	(1, 1, 3)–(1, 1, 7)
	(1, 1, 1, 3)–(1, 1, 1, 7)
	(1, 1, 1, 1, 4)

structure is then indicated as a (1, 7) ITB. The low concentration of alkali metal atoms stabilizes only a fraction of the WO_3 specimen into the HTB structure, which then segregates into slabs, minimizing the free energy. This gives an inhomogeneous structure where the alkali atoms are concentrated into bands—undoubtedly, a common phenomenon that had been encountered earlier for layers of a niobium-rich HTB structure, which were observed to segregate within a WO_3 matrix.

One would expect that an almost unlimited number of ordered structures would be possible simply by adjusting the width of the HTB and WO_3 slabs independently. Table 10.1 records those that have been observed so far. The slab description is given for each in parentheses. Ones to the left of a comma indicate the number of corner-shared layers one octahedron wide of WO_3 separating hexagonal layers, and the last number gives the width (number of WO_6 layers) of the WO_3 slab. In addition, these slabs may be intergrown in virtually any combination. The extraordinary flexibility of the WO_3 structure should be beginning to make an impression. Although this is one of the best examples of flexibility in variable-valence compounds, it is a quite general phenomenon. It is doubtful that the rich variation displayed in the transition-metal and rare-earth compounds could ever have been established without the use of HRTEM techniques.

10.5.3 The ITB bronzoids

As outlined previously, CS and PC structures analogous to those found in the reduced $WO_{3-\delta}$ structures can be made in ternary or quaternary systems utilizing niobium or tantalum to effect a virtual reduction of tungsten. This same behavior is exhibited in the ITB structures with the formula $A_xM_xW_{1-x}O_3$. If specimens are heated to high enough temperatures (1170 K), ITB phases containing vanadium, niobium, or tantalum can be produced with a nominal composition $x = 0.10$. These fully oxidized phases (called bronzoids) have analogous ITB structures but entirely different electronic structures and physical properties. Bronzoid structures so far reported are also listed in Table 10.1. In these phases, the HTB slabs tend to dominate the structures with WO_3 slabs only two or three octahedra wide. Disorder is also frequently observed in the bronzoids, as in the authentic ITB phases, including isolated layers of hexagonal tunnels in an otherwise WO_3 structure and vice versa.

10.5.4 Superstructures in ITB

Occasionally, long-period superstructures are observed in A_xWO_3 structures, but they are much further elaborated in the $A_xM_xW_{1-x}O$ systems. Figure 10.20, from the work of Kihlborg and Sharma (1982), shows a repeat

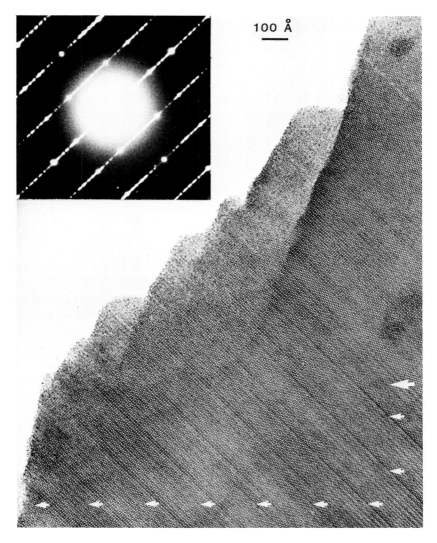

FIGURE 10.20 HRTEM images and the corresponding diffraction pattern of a crystal nominally $Cs_{0.08}Nb_{0.08}W_{0.92}O_3$. Notice the long-range order with sequence 7 of Table 10.2 up to the arrow. The triple-tunnel row is marked with an arrow. The crystal becomes pure HTB at the top. (After Kihlborg and Sharma 1982.)

TABLE 10.2
Observed ITB Sequences Forming Superstructures

No.	Sequence	Multiplicity of tunnel rows	Repeat length (Å)
1	(1, 1, 3, 1, 1, 4)	3	57
2	(1, 1, 6, 1, 1, 8)	3	83
3	(1, 3, 1, 1, 4, 1, 1, 1, 5)	2, 3, 4	91
4	(1, 3, 1, 7, 1, 1, 6)	2, 3	93
5	(1, 1, 1, 1, 5, 1, 1, 1, 1, 7)	5	101
6	[1, 1, 4, (1, 1, 5)$_3$]	3	132
7	[1, 1, (2)$_5$, (1, 2)$_2$, (2, 1)$_2$, 3]	1, 2, 3	143
8	[1, 1, 1, 3, (1, 1, 4)$_2$, (1, 1, 1, 1, 5)$_2$]	3, 4, 5	187
9	[1, 1, 3, (1, 1, 4)$_2$, 1, 1, 6, (1, 1, 4)$_3$]	3	216

of structure every 143 Å in a crystal of composition $Cs_{0.08}Nb_{0.08}W_{0.92}O_3$. The repeat sequence is listed among those so far observed as sequence 7 in Table 10.2. These repeat sequences are unique. This suggests that they are not a manifestation of long-period structures with thermodynamic stability but, rather, are relics of their formation by some chance replication by a template such as a screw dislocation. This mechanism has been previously invoked to explain the long-period polytypic phases.

10.5.5 ITB structures in other systems

The studies of alkali-metal ITB structures have provided a framework for visualizing a number of compounds where tin, lead, or barium are incorporated into WO_3. This work, some of which antedates the recognition of the alkali ITB phases, has been carried forward principally by Tilley and his co-workers. References contained in the review by Kihlborg (1982) should be consulted for an introduction to this work.

Just as tungsten bronze structures are not limited to the alkali metals in WO_3 but are known when the dopant is H^+ or NH_4^+, they also occur when one has neither alkali nor WO_3. For example, a phase of composition $Sb_{0.2}MoO_{3.1}$ has a structure of (2) ITB type (Parmentier et al., 1979, 1980).

Furthermore, there has been a new type of ITB-like structure described in which WO_3 slabs are found separated by layers consisting of P_2O_7 groups and tunnels containing alkali atoms. A compound $Rb_{0.4}P_2W_8O_{28}$ has recently been described by Giroult et al. (1980) and briefly by Kihlborg (1982), who has named them "phosphate tungsten bronzes." Figure 10.21a is an image of a crystal of $A_xP_2O_5W_nO_{3n-1}$, where $n = 10$ and in which there are intergrowths of $n = 8$ and 9 (Labbé et al., 1983). In Figure 10.21b, the structure of $n = 8$ is represented, showing the slabs of corner-shared WO_3, the P_2O_7 groups, and the large tunnels they border in the phosphate groups. In this case, single-crystal, X-ray–structure determination has been accomplished so that the exact tilting of the WO_6 octahedra is known. It is

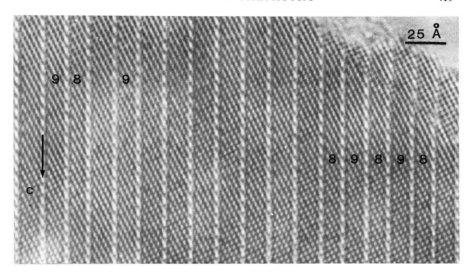

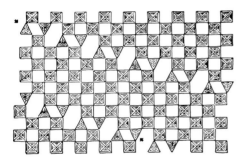

FIGURE 10.21 An image and structural representation of a phosphate tungsten bronze. (a) Micrograph, viewed along b, of a crystal fragment with nominal composition $m = 10$ showing disorder; $m = 8$ and $m = 9$ members can be seen in the matrix. (b) The structure of $Rb_{0.4}P_2W_8O_{28}$ depicted as linked WO_6 octahedra and P_2O_7 groups (triangles). The Rb atoms (not indicated) are located in the tunnels formed by four octahedra and two phosphate groups. The width of the WO_3 slab may be specified by the number of octahedra in the direction indicated by arrows; this number gives the n-value for the homologue, just as for the shear structures. Here, $n = 8$. (After Kihlborg, 1982.)

clear that, where feasible, a full structure determination by X-ray methods is desired to give the average structure, but it cannot give the nature of the intergrowth illustrated in Figure 10.21a, characteristic of real crystals, that can lead to an understanding of their properties. These phosphate bronzes have bronzelike optical and electrical properties and hence are true bronzes.

Kihlborg points out that the P_2O_7-tunnel layer is along {102} and functions to separate the WO_3 slabs in a way completely analogous to the edge sharing that follows along {102} in the W_nO_{2n-1} {102} CS series.

Hence, this new type of compound represents a link between the CS structures and ITB structures.

These features are underscored and developed in recent publications by Hervieu and Raveau (1983a, 1983b), in which HRTEM studies of highly imperfect crystals of high n-values are reported. This work confirms that $\{102\}$ CS is common but that $\{001\}/\{103\}$ CS is also observed, as discussed before for the $WO_{3-\delta}$ and $(M,W)O_{3-\delta}$ phases. A striking observation of intergrowth of W_nO_{2n-1} (102 CS) and $A_xP_2O_5W_nO_{3n-1}$ (shear along $\{102\}$) was observed, including junctions of the pyrophosphate layer with a $\{102\}$CS plane.

10.6 The $TiO_{2-\delta}$ phases

Among those first to utilize CTEM in the study of the structural relationships of intermediate oxide phases in the TiO_x system were Hyde and Bursill, who applied the technique with remarkable success. This system had been studied widely for many years by using X-ray diffraction to determine the structure of Ti_5O_9 and others of the intermediate phases. Bursill and Hyde (1972), using a microscope with only about 6-Å resolution, worked out the structural principles involved in the formation of the intermediate phases.

In the composition range between $TiO_{1.750}$ and $TiO_{1.889}$, a sequence of structures were confirmed belonging to the homologous series Ti_nO_{2n-1} ($4 \leq n \leq 9$). These phases are formed when CS occurs on $(121)_R$ at regular intervals with the required spacing. At higher oxygen content, from $TiO_{1.938}$ to $TiO_{1.98}$ ($16 < n <$ about 40), they found phases with shear along $(132)_R$. Between these two regions, they found phases of composition Ti_nO_{2n-p}, where n and p are integers with $p > 1$ and crystallographic shear changed smoothly between $(121)_R$ and $(131)_R$. The displacement vector was found to be $R \simeq \frac{1}{2}[011]_R$.

The study of the $TiO_{2-\delta}$ system has now been resumed by utilizing high-resolution microscopes capable of displaying much more detail in the structure. These HRTEM studies established that conclusions made on the basis of lower resolution must be extensively modified.

When the oxygen content of TiO_x was greater than $TiO_{1.97}$, Baumard et al. (1977) found a homogeneous solid solution of defects without CSP in rapidly quenched samples. Blanchin et al. (1980) examined the defect structure of slightly reduced rutile to resolve discrepancies in earlier reports as to whether an appreciable point-defect concentration exists in the $TiO_{2-\delta}$ phase or whether all point defects are aggregated into Wadsley defects. Crystals of $TiO_{1.997}$ to $TiO_{2.000}$ were reduced in a CO/CO_2 atmosphere at high temperatures, and specimens were quenched according to varied temperature patterns and then examined in the electron microscope. They found that specimens cooled in vacuum from 1323 K had no CSP and only a low density of dislocations, as would be expected in TiO_2. Furthermore, a

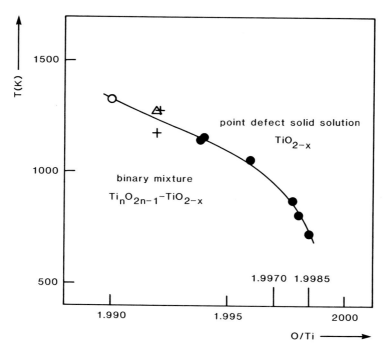

FIGURE 10.22 Phase diagram for the Ti–O system near TiO_2. (After Blanchin et al., 1980.)

specimen of $TiO_{1.9985}$ cooled to 893 K and annealed 12 h, then quenched from 773 K, had no CS. However, a specimen of $TiO_{1.9970}$ cooled slowly to 973 K, then quenched, possessed CS on both $\{121\}_R$ and $\{132\}_R$, with stepped faults between. They also observed that if the crystals were stressed (giving them a high dislocation density), reduced, and quenched from 1323 K, they had both $\{132\}_R$ and $\{121\}_R$ CS. Figure 10.22 shows the temperature-composition range of the $TiO_{2-\delta}$ and the CS regions as revealed in these HRTEM studies.

Closely spaced pairs of CSP have been observed by Bursill et al. (1982). The CSP, rather than being planar, are corrugated, with short segments of varied CSP connected along some average directions, as illustrated in Figure 10.23. The spacing between the corrugated shear planes is varied also, being thicker the more closely the planes are, on average, $(121)_R$ CS. The image of Figure 10.23 is good enough to see the CS segments and even to suggest the arrangement of the cation columns in the gap in some cases. Figure 10.24 suggests the analysis of the laterally spaced pairs and accounts for the contrast reduction in the gap. Intergrowth structures of TiO_2, Ti_nO_{2n-1}, and Ti_2O_3 have been suggested. The CSP have now become slabs.

Otero-Diaz and Hyde (1984) have recently made a study of the form of shear planes in doped and undoped reduced rutile. The slightly reduced

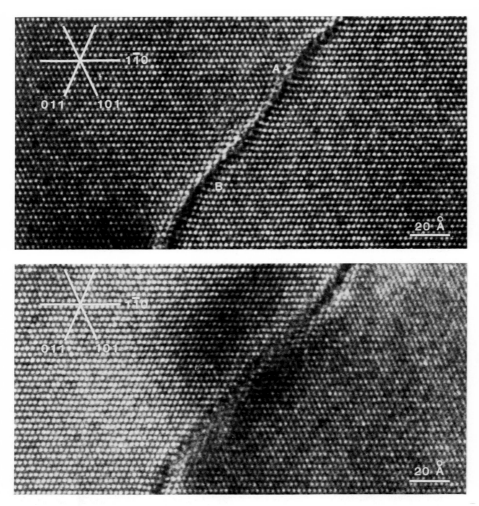

FIGURE 10.23 Enlargement showing contrast features (A and B) for ordered and disordered (132) CS surfaces. (After Bursill et al., 1982.)

rutile contained a few areas with CSP roughly parallel to $(132)_R$, which were quite thick in projection along $[111]_R$ (Figure 10.25). Vanadium-doped rutile of composition $(V, Ti)O_{1.875}$, quenched from 1600°C, exhibited a $(253)_R$ CS structure with $n \simeq 23$. The CSP were observed to change direction in a jerky fashion, presenting a wavy appearance. The waviness was tentatively proposed to arise from the condensation of Ti^{3+} units found in slightly reduced rutile at high temperatures (Blanchin et al., 1980) and proposed to exist in the rutilelike slabs between shear planes. Short diffusion paths for the Cr^{3+} to reach the CSP result in the frequent change of direction.

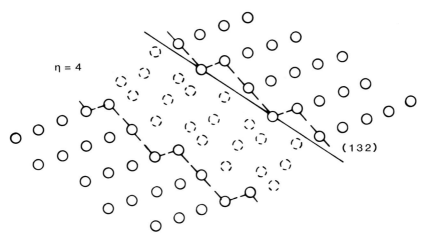

FIGURE 10.24 Arrangement of white dots expected for pairs of (132) CSP corresponding to intergrowth of Ti_4O_7 within a rutile matrix. (After Bursill et al., 1982.)

The precipitation of {100} platelet defects (Bursill et al., 1984) and/or CSP in slightly reduced rutile, depending on the controlled preparation or treatment, presents a well-developed example of the application of HRTEM to suggest an atomistic mechanism for their generation. The platelets frequently decorate CSP as well, and they generally form at a lower temperature than does CS. Frequently, defects intermediate in size between

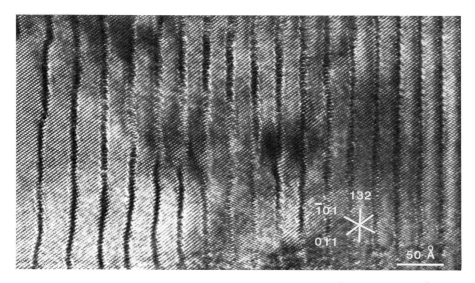

FIGURE 10.25 A $(132)_R$ CSP region with d(CSP) varying from $\simeq$52 Å at the left to 23 Å at the right. (After Otero-Diaz and Hyde, 1984.)

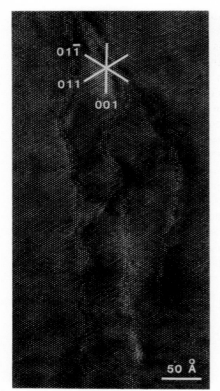
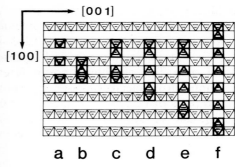

a   b   c   d   e   f

FIGURE 10.26 An image of platelet formation in $TiO_{2-\delta}$ and the probable point defect structure in oxygen deficient rutile. (a) HRTEM images (500 kV) of areas showing zigzag-type platelets. Defect trace varies from $(20\bar{1})$ to $(20\bar{1})$. Note disorder, changes in platelet structure from zero shift to half-lattice shift segments, and strain contrast. (After Bursill et al., 1984.) (b) Bounded projection (along [200], for $-\frac{1}{4} \leq y \leq \frac{1}{4}$) of the structure of rutile, containing a) no defect; b) traditional interstitial defect, that is, additional Ti^{3+} cations; c) two pairs of octahedrally coordinated Ti^{3+} cations sharing faces. d), e), and f) show linear extensions of c) to produce more widely separated pairs of face-shared octahedra. (After Bursill et al., 1983.)

point and fully developed Wadsley defects are observed. In Figure 10.26a, the complex defect structure is shown in an HRTEM image taken along [100]. The specimen is $TiO_{1.9965}$ quenched from 1323 K. The fringe contrast characteristic of CSP is absent in this case, but a complex platelet contrast is evident, as are variations in thickness and deviations in orientation.

Observations of this complex defect pattern in $TiO_{2-\delta}$ necessitated the development of small-defect models that provide satisfactory accommodation of M^{3+} ions in the structure and account for the absence of lattice displacement. The underlying structural theme for both linear and extended defect structures appears to be pairs of face-sharing octahedra as the electrostatically favored environment for Ti^{3+} ions (Figure 10.26a). This could also represent the small defects that exist at high temperatures in nonstoichiometric TiO_{2-x}. These precipitate either as platelets or CS pairs at lower temperatures. Figure 10.26b is a bounded projection of rutile ($-\frac{1}{4} < y \leq \frac{1}{4}$), using an octahedral representation. The traditional interstitial model of a Ti^{3+} sharing opposite faces with adjacent Ti^{4+} octahedra is discounted from electrostatic-energy considerations. The elements shown as c form pairs of $(121)_R$ shear planes, while d, e, and f form pairs of $(253)_R$, $(132)_R$, and $(143)_R$ CS, respectively. The CS in pairs can occur without displacement of the basic rutile structure, whereas one CS plane requires the creation of a dislocation loop. The interaction of a Wadsley defect pair with a dislocation will cause dissociation of the pair consistent with independent observation of deformed specimens that show both CS and platelet formation, even when cooled at a rapid rate.

The change from CSP to a platelet structure is a function of temperature and composition. For example, a specimen of $TiO_{1.9966}$ cooled over a period of seven days showed only very well ordered CSP, but at lower temperatures, platelets formed. *In situ* studies of TiO_{2-x} ($0 \leq x \leq 0.02$) have confirmed precipitation of platelets at 173 to 452°C but the formation of CS at higher temperatures. The temperature dependence of the morphology of defects is striking and, to be studied properly, demands the improvement of double-tilt-heating stages compatible with HRTEM.

Recently Bursill et al. (1985) have obtained bright-field, phase-contrast images of chromia-doped rutiles and have made comparisons of observed spot densities and contrast calculations. They conclude that the smallest clusters contain about 32 Cr^{3+} ions. Larger clusters occurred with increasing frequency for higher-dopant levels.

At the moment, the mechanism of the accommodation of oxygen loss in rutile as supported by HRTEM studies proceeds approximately as follows:

1. Oxygen loss is accompanied by the establishment of a Ti^{3+} ion that produces two face-shared octahedra along [100] (Figure 10.26b).
2. This vacancy pair is mobile and can diffuse further along [100] to produce defects with two face-shared octahedra along [100] and separated by regular-lattice single octahedra of varying numbers.
3. These defects of various sizes can condense to form platelets or CSP in pairs in various orientations.
4. Whether platelets or CS predominates depends on the temperature (e.g., CSP precipitate at higher temperatures than platelets).
5. CSP occur in disordered, short segments giving segmented, curved surfaces rather than planes, and these occur in pairs. They generally have variable spacing between them, filled with material that, when ordered, may be slabs of known reduced material such as Ti_4O_7, Ti_6O_{11}, Ti_2O_3.
6. When the pairs of CSP encounter a dislocation, they may dissociate.

Hirotsu et al. (1982) have made HRTEM studies on the rutile-related

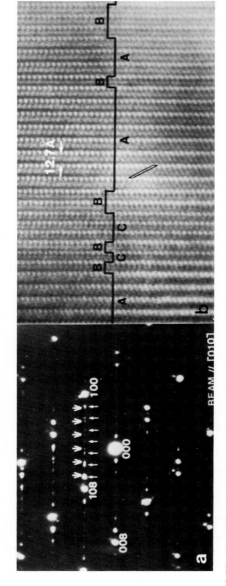

FIGURE 10.27 The diffraction pattern and image of a microsyntactic intergrowth of V_nO_{2n-1} with mixed n. (a) Selected area-diffraction pattern (010)* from an area of microsyntactic intergrowth. A series of diffuse 10l spots from an intergrown V_7O_{13} structure (down arrows) is seen in the matrix V_8O_{15} diffraction pattern. (b) Multiple-beam lattice image of the area by the use of the diffraction spots shown in (a). Diffracted beams within the scattering angle of 2.1×10^{-2} rad are contributing to the imaging. Rows of bright spots are parallel to the CS planes. The regions A, B, and C correspond to V_8O_{15}, V_7O_{13}, and V_6O_{11} structural regions, respectively. The unit cell of V_8O_{15} in A is shown by the parallelogram. The picture was taken at about 1000-Å under focus. (After Hirotsu et al., 1982.)

magnetic phases of vanadium oxide V_nO_{2n-1}. The behavior observed is very similar to that expected of Ti_nO_{2n-1} without the complications just enumerated. Arc-melted specimens gave a mixed microsyntactic intergrowth (with {121} CS) of phases with $n = 6, 7$, and 8, as shown in Figure 10.27. The absence of pairing, segmenting, and so on, appears to be shown at this resolution in contrast to Ti_nO_{2n-1} intergrowths. Another interesting observation was an HRTEM image of periodic microsyntactic intergrowths with $n = 8, 7, 8, 8, 7, 8, 8, 7, 8, \ldots$, in both cases suggesting a facile accommodation by the CS structure to compositional fluctuations. There was, however, an occasional image that was interpreted to have some variation of CSP spacing. One example was shown of a periodic microsyntactic intergrowth of $n = 6, 7$ in which the CSP made steps along their length so that $n = 6, 7$ became $m = 7, 6$, and so on.

10.7 Compounds containing two-dimensional crystallographic shear

Reduced ReO_3-type oxides preserve octahedral-oxygen coordination of the metal atoms by adjusting their corner-sharing structure, which creates planar features (CSP) that contain edge-shared octahedra. In such phases, the O/M ratio has been reduced. In the $WO_{3-\delta}$ system, it was found that this structural adjustment was made in the composition range $ReO_{3-\delta}$, $0 < \delta < 0.10$.

There are a number of oxides with a basic $ReO_{3-\delta}$-type structure where δ is larger and where the MO_6 octahedra are preserved in columns (blocks) of ReO_3 type but are sheared in two dimensions. These structures are the so-called block structures common to NbO_2F and Nb_2O_5. Crystallographic shear occurs in these compounds on (100) and (010) planes. The blocks are joined by edge-sharing and the structures differ according to their block size and the architecture of their corner junctions. These block structures have probably been more generally studied by HRTEM than any other class of compounds. Indeed, it was to shed light on intergrowths of these complicated structures that Allpress et al. (1969) turned to the electron microscope. These successes stimulated solid-state chemists to involve HRTEM in their studies of the structure of complex solids. Reviews of these structures, revealed largely through HRTEM, may be found in Wadsley and Andersson (1970), Tilley (1972), and Eyring and Tai (1976).

These niobium-based oxides and oxyfluorides have unit cells with a periodicity of $\simeq 3.8$ Å along [001], corresponding to the length of the coordination octahedron. Viewed down this short axis, the columns project as blocks at two levels and hence are offset $\simeq 1.9$ Å with respect to their neighboring columns. The oxygen atoms in the shared edges are connected to three metal atoms rather than to two as found in the interior of the columns. The variety of structures observed and defects within these structures arise from the almost unlimited number of ways columns of different cross sections may be joined. The linkages at the block corners

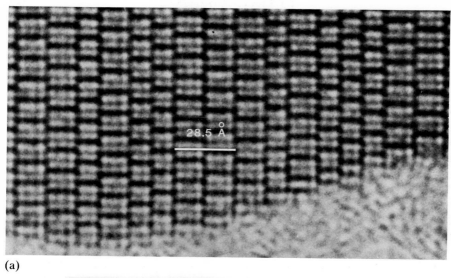

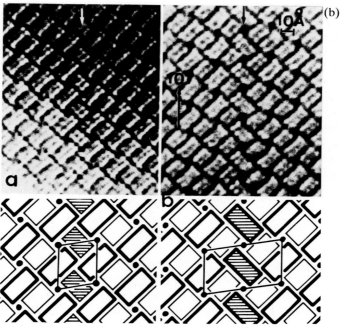

FIGURE 10.28 Structure and structural defects in titanium-niobium-oxides. (a) Electron micrograph of a wedge-shaped crystal of $Ti_2Nb_{10}O_{29}$ ($a = 28.5$ Å, $c = 20.5$ Å) viewed down the b-axis. The thickness varies from about 10 Å at the bottom to 100 Å or more at the top. In the thin region, each grey spot is a row of metal atoms parallel to the beam. The black lines are regions where the metal atoms are close together and unresolved. There are some faults in the a-direction, probably associated with composition fluctuations. (After Iijima, 1971.) (b) Two-dimensional lattice images showing two types of defects of type B, $\delta[101]$ composed of (A) $(4 \times 3)_1$ blocks and (B) $(5 \times 3)_\infty$ blocks. (After Iijima, 1973.)

sometimes involve a string of tetrahedral sites created along the short-axis direction that may be filled with cations.

The first two-dimensional images of a crystal were of $Ti_2Nb_{10}O_{29}$ obtained by Iijima (1971). These images were of sufficient resolution to show, with reasonable clarity, the positions of strings of atoms in the crystal structure. Figure 10.28a is a micrograph of a wedge-shaped crystal with the thin edge at the bottom. In the thin part, the grey spots correspond to rows of octahedrally coordinated metal atoms. The light spots are the square tunnels in the ReO_3 structure, and the dark rectangles are the rows of metal atoms where they approach each other more closely owing to the edge-sharing of their coordinated oxygens. Observe the irregularity of the block pattern with (3×3) rows of block occasionally separating the 3×4 blocks of the $Ti_2Nb_{10}O_{29}$ ideal structure. Oxygen loss and nonstoichiometry is accommodated by this increased edge sharing.

During the next few years, Iijima (1973) occasionally turned his attention to defects observed in these block structures. As an example, crystals of $H-Nb_2O_5$ were observed to possess defects of several different types. The $H-Nb_2O_5$ may be viewed as interleaved layers of (4×3) and (5×3) blocks. If the columns are labeled **A** and B, respectively, the $H-Nb_2O_5$ structure may be represented as **ABABAB**, whereas occasionally the defect **ABABAABAB** occurs [where the regular and bold type represent the columns shifted $\frac{1}{2}$ (3.8 Å) with respect to each other owing to CS]. Another defect (Figure 10.28b) consists of a row of (3×4) or (3×5) blocks parallel to [101], which is rotated by 90° with respect to the host. Double and triple rows of these columns have also been observed.

Two important chemical effects are apparent in this early work. In the block structures, the mode of accommodating a substantial oxygen nonstoichiometry is by insertion of a different block arrangement rather than by random point defects. Incidentally, the columns (blocks) are well ordered down the short axis of the crystal. The defect block structures are frequently elements of another structure. There are many polymorphs of Nb_2O_5, and elements of more than one type may appear intergrown without appreciable change in composition.

After almost a decade of HRTEM observation of various block structures, Anderson (1978–1979a) recorded his impressions of the mechanistic aspects of these studies. In this, he emphasized the importance for solid-state chemistry of the observation of "real" structures as opposed to average structures. The pattern of defects suggest atomistic mechanisms to the solid-state chemist. The block structures are among the most ideal for detecting subtle changes as reaction proceeds, since they are so ideally suited to study by HRTEM. However, it is difficult (if not impossible) to be sure one is approaching equilibrium in these systems. The examples of reactions chosen by Anderson are quasi-homogeneous (essentially nonreconstructive). Most solid-state reactions, however, are reconstructive, and some of these reactions will be illustrated later.

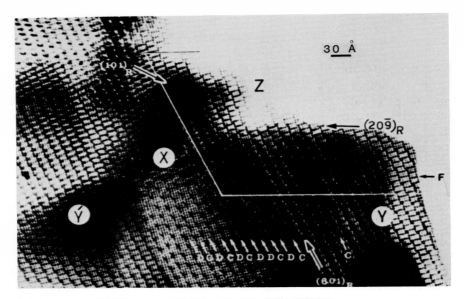

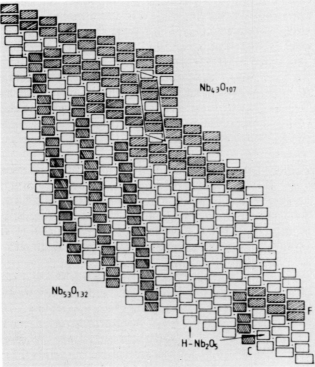

FIGURE 10.29 Structural intergrowth in H–Nb$_2$O$_5$ and their interpretation. (*a*) A domain of a new intergrowth superstructure, Nb$_{43}$O$_{107}$(Z), in reduced H–Nb$_2$O$_5$ abuts on a domain (X) of Nb$_{54}$O$_{132}$, shown by the regular CDCD . . . stacking of rows of Nb$_{25}$O$_{62}$ and rows of H–Nb$_2$O$_5$. At Y are regions of good H–Nb$_2$O$_5$ structure. A double ribbon of (5 × 3) blocks F terminates within the crystal, where it meets a Wadsley defect C of Nb$_{25}$O$_{62}$. (*b*) Structure map of the domains of Nb$_{53}$O$_{132}$ and Nb$_{43}$O$_{107}$ in showing how the double fault C–F is inserted between areas of mutually displaced H–Nb$_2$O$_5$ structure. (After Anderson, 1978–1979a.)

As an illustration of a quasi-homogeneous reaction, consider the reduction of H–Nb_2O_5 (Figure 10.29a). Clearly discernible at Y, the structure is that of well-ordered H–Nb_2O_5. Within the H–Nb_2O_5 host, a pair of intersecting Wadsley defects can be detected at C and F in the drawing of Figure 10.29b. The file of columns marked C has the composition $Nb_{25}O_{62}$ and terminates where it meets the double row of $(5 \times 3)_\infty$ blocks proceeding from the surface at F. The resulting configuration, (4×3)–$2(5 \times 3)_\infty$, appears as a regular insertion at Z. Notice that, across this chevron, the H–Nb_2O_5 structure is shifted by one block. A model for the reaction is suggested and illustrated in the text of Anderson's paper. The scheme provides a possible mechanism for the development of an intergrowth of Wadsley defects of $Nb_{25}O_{62}$ in H–Nb_2O_5 starting with a loss of oxygen at the surface. One new element of $Nb_{25}O_{62}$ is formed with each electron transferred. The reduction proceeds by a cooperative climb suggested by Andersson and Wadsley (1966). If the sequence of cooperative processes were initiated by reductive attack at every (4×3) block in a file, the product would be an intergrown domain of $Nb_{53}O_{132}$, as appears to be the case.

Anderson further points out that intergrowth elements such as these are frequently stabilized in this configuration and do not have an independent existence as a stable phase. Another rule suggested from his observation is that, at the lowest temperatures, the reaction involves the minimum topological change. He also observes that disorder along the short axis of these block structures is rare and that cooperative atom movements such as $\frac{1}{2}a_R \langle 110 \rangle_R$, characteristic of the Wadsley defect involved above, can generate displaced CSP that usually run completely through the crystal along [010].

The work of Iijima and of Anderson and their collaborators on block structures was joined by Horiuchi et al. (1976) utilizing his high-voltage HRTEM (1-MeV) microscope with 2-Å resolution. They produced images of many block structures with the obvious advantage of greater resolution, since the regions of close cation distances and the configurations at the corners could be imaged. Figure 10.30 illustrates the improved images of $Nb_{12}O_{29}$ obtained by high-voltage HRTEM at 2-Å resolution.

An essay (Gruehn, 1982) on the use of HRTEM in preparative solid-state chemistry is illustrated throughout by the preparation and characterization of new block structures as well as references to the original literature. Samples of $(V, Nb)_{12}O_{29}$ were prepared by heating (980°C) mixtures of the binary oxides Nb_2O_5, V_2O_3, and NbO_2 in closed quartz ampoules. When the composition approached $V_3Nb_9O_{29}$, a new and complex Guinier-powder diffraction pattern appeared. The absence of crystals of sufficient size and even of microcrystals free of defects required that HRTEM be used for identification. The structure consisted of (3×4) blocks, as are found in $Nb_{12}O_{29}$ with a different connectivity, predicted by Roth and Wadsley (1965) but not previously observed. In this case, the (3×4) blocks are joined at their long edges. An image shows not only the regular structure of

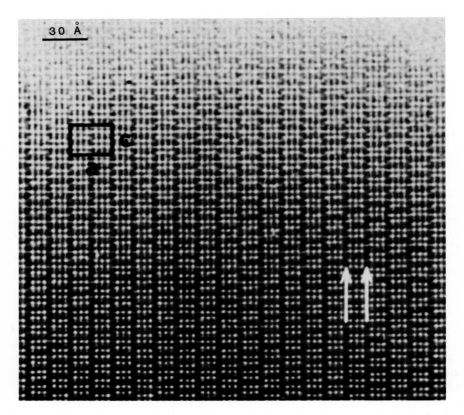

FIGURE 10.30 Lattice image of $Nb_{12}O_{29}$ taken with the incident beam parallel to the b-axis. The accelerating voltage is 1000 keV. Notice that the site of each interstitial atom is resolved. (After Horinchi et al., 1976.)

(3×4) blocks but also a faulted region of (3×3) blocks two columns wide. Each (3×3) block is surrounded by four tetrahedral positions, forming an element of the VNb_9O_{25} structure.

When Nb_2O_5, WO_3, and NbO_2 are heated in closed vessels, a series of solid solutions is obtained in which the W/Nb ratio is increased from 0.444 (the 9/8 phase) to a maximum value of 1.364. Metastable oxidation products of Nb_2O_5/WO_3 with ratios of 6/1, 7/3, 8/5, and 9/8 were studied. These oxidized products have poor crystallinity, but the building principles are revealed in their rather disordered makeup. These oxidized materials might be preserved in the block-type structures into the composition regions where the TTB structure would be thermodynamically stable.

Oxidation should lead to an increase in block size to reduce the number of edge-shared octahedra. In Figure 10.31, it is apparent that the width of the original 6:1 structure is maintained—that is, (3×4)—but when the compound is oxidized from $Nb_{10}W_3O_{33}$ to $Nb_{10}W_3O_{34}$, the length of some of

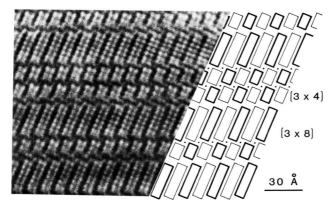

FIGURE 10.31 Metastable oxidation product $Nb_{10}W_3O_{34}$ prepared from a sample of $W_{4/4}[Nb_{10}W_2O_{33}]$, which has the (3×4) block structure. The electron micrograph taken with HRTEM is schematically interpreted at the right. Tetrahedral positions are marked by dots. The larger distances without dots between some rows of blocks can be explained by special tunnels within the structure. (After Gruehn, 1982.)

the blocks are doubled (3×8). Remnants of the original structure are preserved. Some of the rows of tetrahedral positions are maintained, and other types of tunnel structures are introduced. This demonstrates that metastable phases can be synthesized in which new structural principles are exhibited. In this case, the block structure is preserved with an increase of the size of the block of the same width rather than a transition to a new stable form, say, TTB.

Gentle oxidation of $Nb_{12}O_{29}$, in either of its two forms, by different paths yielded three metastable forms of Nb_2O_5 for each of the two starting materials. Upon annealing, these six metastable materials transformed to the equilibrium polymorph, $H-Nb_2O_5$. The almost infinite variability of these block structures will give many new insights into solid-state chemistry.

10.8 The structural characterization of zeolites

The recent success of Thomas and his collaborators (1983) in elucidating the structure and structural defects of some zeolitic-catalyst materials, particularly ZSM-5 and ZSM-11, provides an example of the application of HRTEM to the study of beam-sensitive complex materials. There is a need to understand the intimate details of the structure of catalytic materials to divine the basis of their activity and any loss thereof. Aside from the inherent chemical interest of these studies, they are made with a view to the enhancement of the catalysts for industrial application.

Only a few years ago, Thomas (1982) cryptically described the beginnings of this study by illustrating an HRTEM image of Zeolite-A showing fleeting rafts of crystalline regions along [010] surrounded by a growing region of amorphous aluminosilicate material (Bursill et al., 1980). This image was considered a coup and was credited to improved techniques in handling beam-sensitive materials. At that stage, one could made out the array of channels responsible for the shape selectivity and catalytic activity, with, of course, a promise of improvement as the microscope procedures and data handling were refined.

Since then, however, the studies have taken a different direction, with dramatic success. These microporous materials can be given a greater resistance to radiation damage if the aluminum in the structure is replaced with silicon while preserving essentially the same framework. This dealumination is accomplished either by hydrothermal treatment or by exposure to $SiCl_4$ (Thomas 1983). The resulting product has been studied in detail by HRTEM. The dealumination procedure also expels the exchangable cation in the structure, facilitating comparison between observed and calculated images. This makes possible exploration of structural intergrowths and other defects that would affect the chemical nature of the original material.

A rather full review of these suggested developments can be found in a recent article by Thomas et al. (1983). The catalysts ZSM–5 and ZSM–11 are closely related in structure to zeolite-Y, being made up of sheets related by inversion and mirror reflection, respectively. It is believed that this type of intergrowth occurs in almost unlimited variation. In Figure 10.32, projections of these structures are shown together with an intergrowth consisting of a strip of ZSM–11 two units wide incorporated within a ZSM–5 structure. In Figure 10.33a, the zone axes in A and B are [010] and [100], respectively, for ZSM–5. The images are usually not good enough to ensure correct identification in view of the relative instability of the zeolite in the electron beam. Fortunately, however, the diffraction patterns are quite distinctive, as illustrated in Figure 10.33a. The inserted diffraction patterns are optical diffractograms obtained from the images themselves. By this technique, it is possible to get diffraction information from very small regions of crystal space, and since the specimens are heterogeneous, the technique is particularly important. This method can at least distinguish the zones from the images obtained; and if sufficient care is exercised in exposing the crystal to the electron beam, enough resolution can be obtained to identify the zone directly from the image.

It is possible to observe planar defects of the type illustrated in Figure 10.32d. If the sheets characterized by an inversion axis are called i and those with a mirror plane called σ, the [010] zone of ZSM–5 would be written as i i i i i, and that of ZSM–11 would be $\sigma\ \sigma\ \sigma\ \sigma\ \sigma$; with an intergrowth, it might be i $\sigma\ \sigma$ i i, as shown in Figures 10.32a, c, and d, respecively. Notice, for example, the crystal of ZSM–5 imaged in Figure 10.33b, which has two defect planes exhibiting mirror (σ) symmetry inserted into a crystal with inversion-plane symmetry, to give a planar sequence

$$i\ i\ i\ i\ i\ i\ \sigma\ i\ i\ i\ i\ i\ i\ i\ \sigma\ i\ i\ i\ i\ i\ i\ i\ i\ i\ i.$$

Compare the weak-spot pattern of the drawing and the image below.

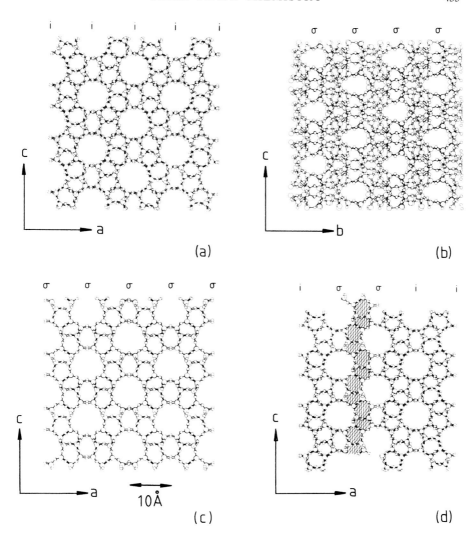

FIGURE 10.32 Structure projections. (*a*) ZSM–5 along [010]. (*b*) ZSM–5 along [100]. (*c*) ZSM–11 along [010]. (*d*) ZSM–5/ZSM–11 intergrowth along [010]. Planes of mirror symmetry (σ) and planes containing centers of inversion (i) are indicated. The ZSM–11 [100] is identical to ZSM–11 [010]. (After Thomas et al., 1983.)

In some cases, electron diffraction patterns show streaking along *a* believed to be due to irregular spacing of σ-sheets in a ZSM–5 host. Optical diffractograms were obtained from model drawings of [010] projections of ZSM–5 with σ-sheets inserted at regular intervals along *a* that showed superlattice spots where streaking is observed in the electron diffraction patterns. This suggests that the planar defects are very irregularly spaced in the real crystals.

Terasaki et al. (1984) have observed a tendency in zeolite-L to form coincidence boundaries resulting from a rotation of part of a crystal 32.2° with respect to the other. The generated interface is a $\sqrt{13} \cdot \sqrt{13}$ R 32.2° superstructure boundary. This coincidence structure becomes apparent in high-resolution images in projection along [001]. This type of defect would grossly diminish the diffusion rate of guest atoms and hence impair the catalytic properties.

10.9 The anion-deficient, fluorite-related, rare-earth oxides

10.9.1 The system

The application of HRTEM to the study of the higher oxides of the rare earths is important for many reasons:

1. These oxides are structurally related to compounds formed by at least one-third of the chemical elements, including many of great technological importance.
2. They are representative of common structure types not ideally suited to HRTEM techniques.
3. The structural principle believed to account for their homologous series of intermediate phases is not crystallographic-shear or polygonal-tunnel-related, although it may be considered as a planar-defect structure that can order with different spacing to form the members of a series. Therefore, new structural principles including defect formation and

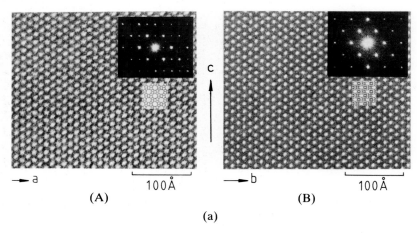

(a)

FIGURE 10.33 Successful techniques utilized to obtain structural detail on beam-sensitive zeolites. (a) Electron micrographs of ZSM-5 projected along (A) [010] and (B) [100]. Optical diffraction patterns (insets) exhibit the rectangular and triangular arrays of spots expected for the respective zone axes. This distinction is not directly apparent by mere inspection of the images. (b) High-resolution image of a region of ZSM-5 ([010] projection) containing two σ-defect planes. The explanatory schematic is shown at the top. (After Thomas et al., 1983.)

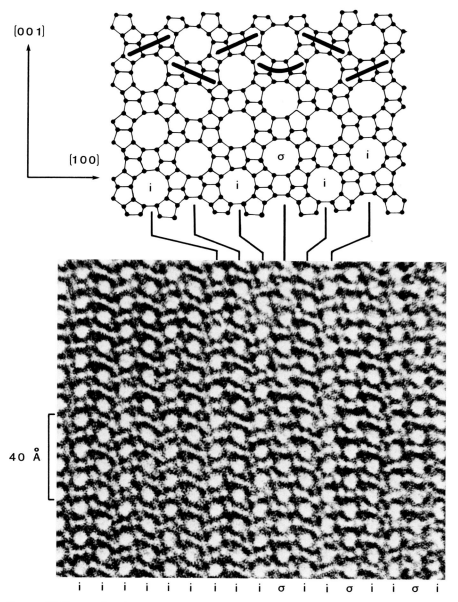

FIGURE 10.33—*continued*.

(b)

438 HIGH-RESOLUTION TRANSMISSION ELECTRON MICROSCOPY

order-disorder can be studied in these systems that are dynamic under the electron beam.
4. Transformations and reactions can be conveniently studied during electron-beam irradiation.
5. Finally, it is not possible to obtain the structural information necessary to understand the chemistry of these materials by conventional means.

Although it had not been possible even to determine the unit cells of the intermediate phases [many of them known to belong compositionally to a series that could be represented as R_nO_{2n-2} ($4 \leq n \leq \infty$, n an integer)] by using X-ray–diffraction methods, it was possible for Kunzmann and Eyring (1975) to determine the unit cells and the relationships between them by utilizing selected area electron diffraction (SAED). Figure 10.34a illustrates the relationships for the series R_nO_{2n-2}. In the [100] projection (down the

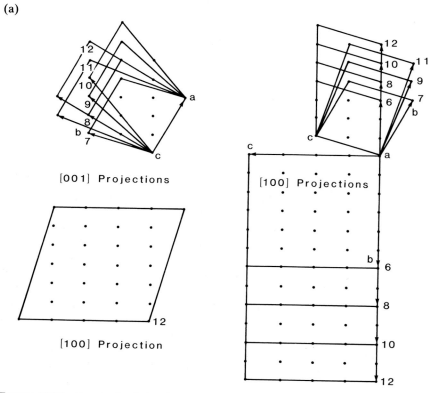

FIGURE 10.34 Structural relationships and the prototype structure in the higher rare earth oxides. (a) Projections of the unit cells of the series R_nO_{2n-2} in the $\langle 211 \rangle$ fluorite zone. (b) Projection of the R_7O_{12} structure onto the $(21\bar{1})$ fluorite plane. The large and intermediate circles represent oxygens at different levels; the small circles are metal atoms. The dark lines outline the six seven-coordinated and the one six-coordinated metal atoms about the two oxygen vacancies represented as hatched circles.

(b)

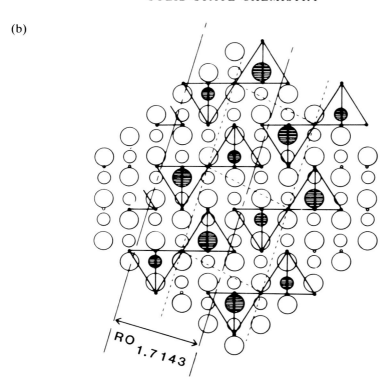

FIGURE 10.34—continued.

a-axis common to all members of this series), the oxides are observed to be of two types. There is a primitive set at the upper right possessing a common ac-plane along {135} and consisting of both odd and even members. This subset is shown along [001] at the upper left. The ac-planes are separated by a distance determined by the b-axis length to form the homologous series. The ac-plane may be considered to be one containing all the six-coordinated metal atoms (metal atoms are eight-coordinated in the fluorite structure), with the two vacant oxygen positions deployed along the $[111]_F$ direction, canted about 70° to the [100] plane. A projection of the structure of R_7O_{12} along [100] is shown in Figure 10.34b. The slab, marked $RO_{1.7143}$ between the broken lines, is the end-on ac-plane with its associated six- and seven-coordinated cations (small circles) and its filled and vacant oxygen positions (two-sized larger circles). As b increases with increasing n, regions of fluorite structure separate these slabs.

The odd members of the homologous series are believed to possess structures that can be described in this way, but only the structure of Pr_7O_{12} among the intermediate phases has so far been determined (Von Dreele et al., 1975) by neutron diffraction. The even members of the series are viturally never observed in the primitive structure but are twinned at the unit-cell level to give the monoclinic projections shown at the bottom of Figure 10.34a. Here, the ac-plane is still common to all these monoclinic cells, but it is different from that considered previously (it could possibly be considered folded {135} segments). It is unknown whether the vacancy orientation about alternate cations is different, requiring the observed doubling of the unit-cell dimension. It is likely that if an answer is obtained to this question, it will come from HRTEM studies. Although single crystals of these oxides of sufficient size for X-ray studies can be grown hydrothermally, the strong absorption and the disparity in the atomic number of rare earths and oxygen, in addition to their unpredictably multitwinned character, make it unlikely that the structure could be refined. The number of atoms in the unit cell increases very quickly to an unmanageable size for structure determination by neutron diffraction with present capabilities.

10.9.2 Structural models for $Ce_{11}O_{20}$ and $Pr_{24}O_{44}$

To illustrate the level of success obtained in matching observed images of the intermediate phases with microscopes of worse than 2.5-Å resolution with those calculated from the models just proposed, we take two examples, $Ce_{11}O_{20}$ ($n = 11$) (Knappe and Eyring, 1985) and $Pr_{24}O_{44}$ ($n = 12$) (Rae Smith et al., unpublished). A model is constructed by assuming the same shifts of the tetrahedrally coordinated cations about the proposed oxygen vacancies in the unknown compound as is known to occur in Pr_7O_{12} from the determination of its structure (Von Dreele et al., 1975) from neutron-powder diffraction using the Rietfeld method. The other cation, oxygen, and vacancy positions are left unchanged from their positions in the fluorite subcell. From this structural model, the conditions of the experiment, and the characteristics of the microscope, the image is calculated by employing the multislice method. Rae Smith and Eyring (1982) developed interactive software to match processed, digitized, observed images with those calculated to give a quantitative comparison. The program takes 64×64-pixel intensity maps of both unit cells and displays them together on a monitor. When the points of coincidence are chosen by the graphics cursor, the program calculates a set of residuals by varying the translation vector between the two images within small limits. It then displays the map of the residuals around the optimum point together with the various statistics of a pixel-by-pixel match of the two images. The difference image, the simulated image, or the observed image can be juxtaposed as desired. This process can be repeated with alternative models. The whole process of calculation and match to a raw image is accomplished in about 20 min on an Eclipse S/140 minicomputer.

Figure 10.35 (Knappe and Eyring, 1985) shows the raw image from a thin region of a crystal of $Ce_{11}O_{20}$ taken on the JEOL 200CX. In the lower right corner, the (111) fringes of the basic fluorite structure fade away at the edge. The rectangle marks the region processed for the comparison, shown

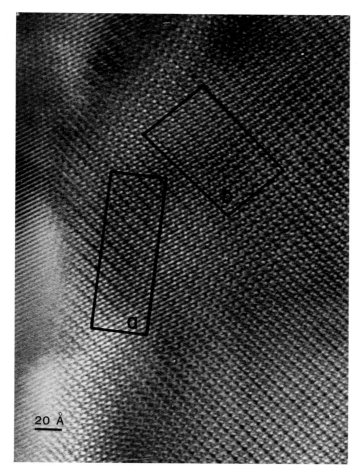

FIGURE 10.35 High-resolution, electron-microscope image of $Ce_{11}O_{20}$.

in Figure 10.36. The agreement factor (FMAD, fractional mean absolute deviation) is 8 percent, using a measure analogous to the R-value in X-ray refinement. In this method, comparisons with the best images of the most favorable specimens where the structures are known by other methods (O'Keefe et al., 1978) give no better than 7 percent.

Rae Smith et al. (unpublished) have made a comparison of the application of the image-match program for two different models of $Pr_{24}O_{44}$ with the observed image. Figure 10.37 shows the results. The model for the calculation on the left is a simple twinning of the primitive $n = 12$ cell of Figure 10.34a, shown as the rectangle with $b = 6$ in the monoclinic projection. This assumed structure projected on [100] is shown in Figure 10.38. The model for the match on the right simply assumes a 50 percent

Calculated Image　　　　　　Experimental Image

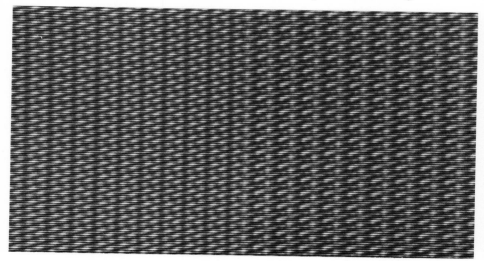

FIGURE 10.36 Calculated and observed image match for $Ce_{11}O_{20}$ from region (a) of Figure 10.35. (After Knappe and Eyring, 1985.)

β(1) $Pr_{24}O_{44}$

Ideal Model	Rotated Vacancy Model
Averaged　Calculated	Averaged　Calculated

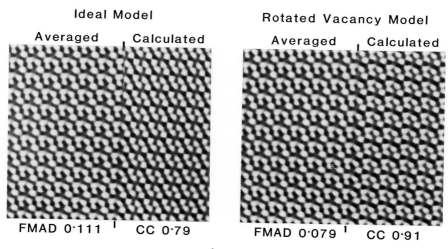

FMAD 0·111　　CC 0·79　　　　FMAD 0·079　　CC 0·91

Defocus −600 Å　Thickness 40 Å

FIGURE 10.37 Comparison of the averaged images of $Pr_{24}O_{44}$ with the model of Figure 10.38 (on the left). The model on the right assumes a partial occupancy of the vacant positions at the middle of the unit cell. (After Rae Smith et al., unpublished.)

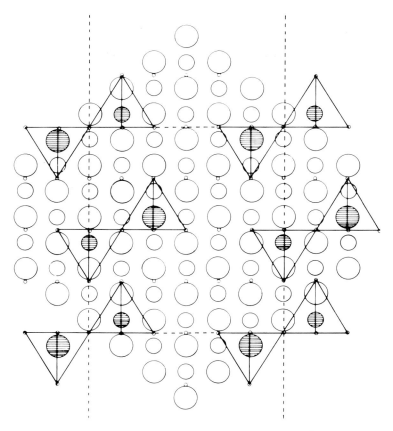

FIGURE 10.38 A projection of the $R_{12}O_{22}$ structure onto the $(21\bar{1})$ fluorite plane, as in Figure 10.34b.

random rotation of the horizontal central-vacancy pairs of Figure 10.38. The improved match suggests that the structure does require partial filling of alternative body-diagonal vacancies of one of the six-coordinated cations in the unit cell. Actual comparison must await an accurate structure determination. The resolution of powder-neutron-diffraction profile analysis is approaching that level necessary to test this suggestion. In the meantime, real-space, crystal-structure elucidation utilizing HRTEM methods may be the best available for certain difficult structures.

The study of the CeO_{2-x} system by Knappe and Eyring (1985) turned up another interesting advantage of HRTEM even if only the electron diffraction patterns (SAED) are used. From X-ray– and neutron-diffraction studies of materials prepared in the composition range $CeO_{1.71}$ to $CeO_{1.82}$, it was suggested that there were two phases in the interval, Ce_9O_{16} and $Ce_{10}O_{18}$, that were members of the homologous series (Ray and Cox, 1975;

Height and Bevan, unpublished). The HRTEM study revealed two quite stable intermediate phases whose unit cells could be determined and were almost certainly $Ce_{19}O_{34}$ and $Ce_{62}O_{112}$ but did not reveal Ce_9O_{16} or $Ce_{10}O_{18}$. The $Tb_{62}O_{112}$ with the same unit cell is known, but $R_{19}O_{34}$ has not otherwise been observed. Microscopes with ultrahigh resolution now available will make the study of these close-packed structures more certain.

The fluorite-related, intermediate, rare-earth oxides are oxygen-deficient. Composition adjustment is made by the incorporation of oxygen vacancies in the fluorite structure, resulting in high oxygen mobility. When these oxides are studied in a 100-keV electron microscope, they must be exposed to minimal radiation if they are not to be reduced in the vacuum of the microscope. It is commonplace to observe compositional change as oxygen moves into or out of the region being imaged. In the rare-earth-oxide system, intergrowth along the {135} planes that bound the R_7O_{12} slab is

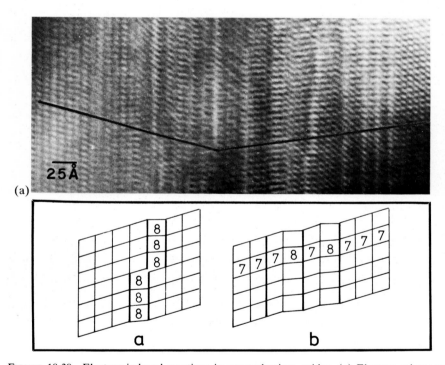

FIGURE 10.39 Electron-induced reactions in praseodymium oxides. (a) Electron micrograph ([21$\bar{1}$] zone) of the phase reaction $Pr_7O_{12} \rightleftarrows Pr_9O_{16}$. The Pr_9O_{16} is the phase on the left, and Pr_7O_{12} is on the right. The disordered central region is an intergrowth along the common a,c-plane. The b-axis projections are drawn as dark lines. The marked regions are modeled at the bottom. (After Eyring, 1980.) (b) Image of a region of a crystal that, a moment before, was ordered $Pr_{40}O_{72}$ undergoing disproportionation. Notice the appearance of {135} fringes. (c) Image of the same region as in (b), in which the disproportionation is further advanced. The $Pr_{40}O_{72}$ remains at the bottom, primitive $Pr_{12}O_{22}$ is in the central region, and a 1:1 intergrowth of Pr_9O_{16} and Pr_8O_{14} surrounds the $Pr_{12}O_{22}$.

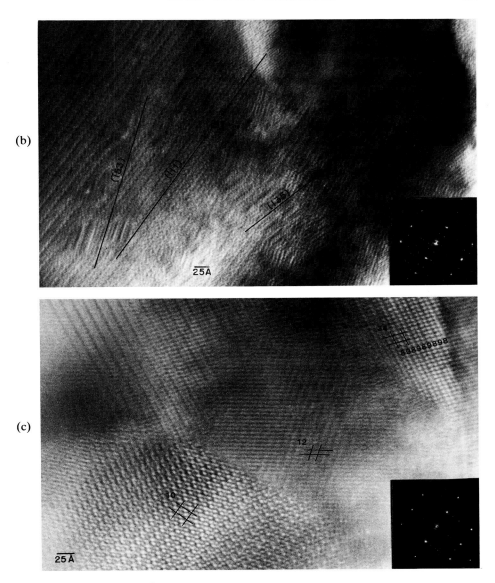

FIGURE 10.39—continued.

coherent. An example of such an intergrowth with stacking variation is shown in Figure 10.39a. In this case, the system consists of an intergrowth of Pr_9O_{16} and Pr_7O_{12}. More or less ordered regions of these phases exist, as shown on the left and right, respectively. The lines are drawn along the b-axis in each case. Between these two ordered regions is an area with a quite-faulted stacking arrangement, where slabs of $n = 7$ (mainly), $n = 8$, and $n = 9$, one or a few unit-cells wide, occur coherently intergrown. An offset intergrowth of $n = 8$ is illustrated. A disproportionation reaction is illustrated in Figures 10.39b and c (Eyring, 1980), where a region of a crystal that, a moment before, had been well-ordered $Pr_{40}O_{72}$ is well on its way to a reaction that will result in part of the region viewed being oxidized to $Pr_{12}O_{22}$; it is a primitive polymorphic form rarely observed. The region around it remains unchanged or is reduced to a 1:1 intergrowth of Pr_8O_{14} and Pr_9O_{16}. This transformation is heralded by the appearance of fringes along [135] characteristic of the primitive $Pr_{12}O_{22}$, Pr_8O_{14}, and Pr_9O_{16}. Figure 10.39c shows the development of these three regions a few moments later, as marked. Regions of $Pr_{40}O_{72}$ have disproportionated to $Pr_{12}O_{22}$ and $PrO_{1.76}$. (This latter is a regular 1:1 intergrowth of Pr_8O_{14} and Pr_9O_{16}, which, although not unknown, is very unusual over such a large area.)

An astonishing observation has just recently been made on the rare-earth-oxide systems (Gasgnier et al., 1986). Thin, uniform films of the sesquioxides of Pr and Tb have been observed to oxidize up to $Pr_{12}O_{22}$ and $Tb_{12}O_{22}$, as well as to the intervening intermediate structures in the electron beam, suggesting a far-from-equilibrium oxidation probably induced by the high-energy electrons. In any case, it is apparent that electron-induced chemical changes must be more thoroughly studied.

10.10 The study of chemical reactions in thin films using HRTEM

10.10.1 Introduction

Specimens prepared for high-resolution, electron-microscopic observation must be thin in the electron-beam direction. A bulk crystal must therefore be shattered, ground, ion-beam-thinned, or in some other way brought to a thickness of no more than a few hundred angstroms. One can speculate on which features of the final specimen are a result of its treatment and which are characteristic of the original material. The specimen is then irradiated by the intense beam of electrons required to form a diffraction pattern, or with even more intensity, an image. This irradiation frequently causes marked changes in the specimen under observation. Certain classes of materials are excluded from observation altogether because of this radiolysis damage, and others must be handled gingerly if observations are to be made and reasonably interpreted. When changes occur in such specimens, they are affected by the dimensions of the sample and by the intense beam of the electrons. If the results of the HRTEM investigations are to be understood in terms of what could happen in a bulk sample, these features of the experiment must be taken into account.

Although we discussed observation of change and even mechanisms of change earlier, our focus here is to emphasize studies in HRTEM where chemical reaction is the main purpose of the study. Change is brought about by chemical potential differences existing within the specimen or between the specimen and its environment, radiation effects (including heating, radiolysis, and knock-on), the small-systems effects resulting from the existing physical dimensions of the sample or from a domain structure stabilized by intergrowth and coherence, and no doubt more subtle causes. It will be difficult to know the weighting factors for the causes that brought it about. Alteration of the specimen is frequently unwanted but unavoidable.

Change is sometimes induced deliberately by treatment outside the microscope, followed by reintroduction for sequential observation; or sometimes, the specimen is simply observed carefully as change is brought about by the ambience. The use of the electron microscope as a tool to investigate the mechanism of chemical reactions at the atomic level will certainly be one of its most valuable contributions to solid-state chemistry. Already, a large body of knowledge has accumulated suggesting atomistic mechanisms of reaction. Some reactions are serendipitous, others deliberately induced in the microscope, and still others contrived by the juxtaposition of reactive species before introduction into the microscope.

Some comments about the electron-beam effects have been made in section 10.2.4. A much more detailed account may be found in Chapters 5, 6, and 7.

10.10.2 Chemical reactions studied by means of high-resolution diffraction methods

The growth of a protective oxide coating on thin films of chromium have been studied by Watari and Cowley (1981), using selected-area, electron-diffraction (SAED) and microdiffraction (MD) techniques. Thin films of chromium were prepared in a number of orientations by evaporation and then oxidized by heating at 600 to 1100°C in a dynamic vacuum of 10^{-5} to 10^{-6} torr. The microdiffraction patterns, observed by use of an optical-image-analyzer attachment to the HB5 microscope, permit the identification of oxide phases as well as their epitaxial relationship with the chromium substrate. Figure 10.40 shows the resolution of an observed diffraction pattern into its components by using MD.

First, a metastable, spinel-like oxide grows in a thin, uniform film; it then disappears on further oxidation to the rhombohedral Cr_2O_3 phase. The spinel is derived from a simple extrapolation of the adsorbed-oxygen layer on the body centred cubic (bcc) metal. It forms quickly before being consumed as the stable Cr_2O_3 is nucleated at preferred sites and grows to form the protective layer. The spinel structure is most often in the (111) orientation on a Cr (011) face.

Rhombohedral Cr_2O_3 grows on the various chromium crystal faces in preferred epitaxial relationships. These were determined by a combination of SAED and MD techniques that enabled sorting out the SAED patterns from heterogeneous regions by judicious application of MD on the small

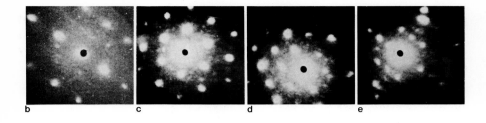

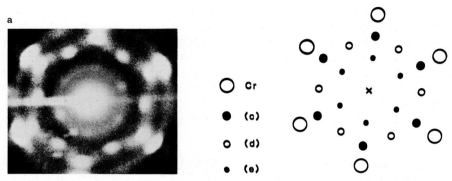

FIGURE 10.40 Resolution of the SAED pattern of oxidized Cr(011) into its component contributions by use of MD. (*a*) Schematic representation of the diffraction pattern. The MD patterns show the following: (*b*) Spots from one of the three Cr(011) orientations. (*c*) Rhombohedral Cr$_2$O$_3$(001) pattern. (*d*) A spinel (111) pattern. (*e*) A spinel (1$\bar{1}$2) pattern. (After Watari and Cowley, 1981.)

TABLE 10.3
Epitaxial Relationships for the Growth of the Spinel and Rhombohedral Oxides on Cr Surfaces and the Number of Twinned Orientations Possible

Cr	Spinel	Cr$_2$O$_3$	Twins of Cr$_2$O$_3$
(000)[100]*	(001)[1$\bar{1}$0]*	(1$\bar{2}$5)[100]*	4 × 2 = 8
		(001)^c[100]*	2 × 2 = 4
(011)[100]*	(111)[1$\bar{1}$0]*	(001)[100]*	1 × 2 = 2
	(112)^a[1$\bar{1}$0]*		
(111)[0$\bar{1}$1]*	(011)^b[100]*	(105)[1$\bar{2}$0]*	3 × 2 = 6
		(010)^a[001]*	3 × 2 = 6
(113)[1$\bar{1}$0]*		(1$\bar{1}$5)[2$\bar{1}$$\bar{3}$]*	2 × 2 = 4

^{a} Less common contributions
^{b} Three twin components are formed.
^{c} Epitaxy at higher temperature.

patches or crystallites present. Table 10.3 summarizes these epitaxial relationships and in it we see demonstrated the power of high-resolution diffraction in the elucidation of complicated processes not easily tractable by other means.

10.10.3 Decomposition reactions

The general class of solid-state reactions in which a compound decomposes into two or more compounds has been under study for a long time. From microscopic studies, it is possible to learn the state of the decomposed mixture and the mechanism of unmixing. Tilley and Wright (1982) have investigated the reaction in which orthorhombic $Pb_{24}Bi_8S_{36}$ or $Pb_{12}Bi_8S_{24}$ decomposes to form cubic PbS. The structures of the lead bismuth sulfides consist of regularly twinned slabs of PbS structure united along (020) planes of the ternary phase or (311) of the PbS structure. The twin planes contain metal atoms, presumably mostly bismuth, in trigonal prismatic coordination. The final product of decomposition is untwinned PbS oriented in the

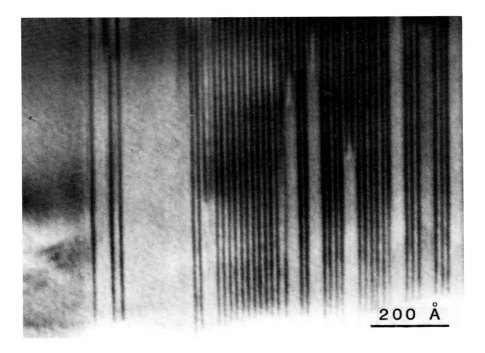

FIGURE 10.41 The edge of a crystal of heyrovskyite after some minutes of observation in the electron microscope, showing the loss of some twin planes and the conversion into strips of PbS. Notice the wide or narrow strips of PbS between ordered twin planes, which extend to the edge of the crystal. This micrograph also reveals the tendency of the twin planes to be lost as pairs from the crystal. (After Tilley and Wright, 1982.)

same way as in the slabs of the original twinned structure. Apparently, the bismuth is lost by evaporation of Bi_2S_3, leaving essentially pure PbS.

The reaction occurs very rapidly, and since it is over in a matter of a few seconds, it has not been studied directly at high resolution. Rather the story has been pieced together from an accumulation of observations on several crystals. The electron beam is used to cause decomposition primarily by heating the specimen. It is striking that the twin planes are lost in pairs with a slight reduction in dimension. A partially decomposed crystal is shown in Figure 10.41. The reaction, which is rapid and topotactic, has been modeled in terms of atomic shuffles of short distances. In this reaction, as in the precipitation of platelets in TiO_{2-x} discussed earlier, coherence is preserved in the matrix when planar defects are eliminated (or formed) in pairs. This example emphasizes the facility with which some complex reactions can occur rapidly at low temperatures.

10.10.4   The chemistry of thin-film reactions

There is, in science, an expanding interest in the chemical behavior of thin films beyond the practical need to make large-scale semiconductor devices. This interest is driven by a curiosity about the chemical behavior of near-surface regions—that is, the effects owing to the condition of thinness on chemical behavior. The differences to be expected would originate in the large surface-to-bulk ratio, with a substantial fraction of the specimen either in or near the surface. Transmission electron microscopy (TEM) can only be carried out on thin specimens, and if high-resolution studies are to be done, the thickness is limited to a very few hundred angstroms. Beyond such thicknesses, there would be little hope of interpreting the images observed.

By employing thin-film techniques, one can cause reactions to occur within the thinned region *in situ* during observation in the microscope. An obvious technique involves the lateral juxtaposition of the reactants on a scale allowing observation in the microscope as the reaction accompanies the material transport of one or more of the reactants. This is most commonly achieved for solid-solid reactions because of the limitations of the vacuum of the microscope column. However, reactions are observed between the solid specimen and the residual gases of the specimen chamber at times and could certainly be effected by a controlled leak. Unfortunately, full-fledged environmental chambers allowing control of the temperature and the gaseous atmosphere have not yet been perfected for use in high-resolution work.

The study of dynamic processes *in situ* in the TEM means the study of reactions in thin films. The change in contrast observed is frequently the result of material transport within the specimen and might be classified as a phase transition, radiolytic effect, or other chemical reaction. The chemical reaction might be a decomposition reaction in which an initially pure compound separates into some components or the combination, by mass transport, of reactants to form products. The cause of the reaction will always have a radiation-effect component, which itself has at least the elements of radiolysis and heating. It will be necessary to sort out the contribution of each in any particular case. The other cause of change is the existence of chemical potential differences within the specimen region. This results in transport, primarily by diffusion, of the components of the reaction to a condition of lower potential. This transport can be through the bulk of the film, on its surface, or both. There will always be complications that arise from contamination from the gas phase of the specimen chamber, the presence of grain boundaries or other interfaces, inhomogeneities in the physical conditions in the region of the change, and so forth. Each can be dealt with to some extent—an ultrahigh-vacuum microscope can be utilized, large grains can be produced, the uniformity of physical conditions can be controlled, and so on.

Another set of problems arises when one attempts to make quantitative and reproducible experiments: The initial conditions of the reaction configuration must be reproduced. This means that contacts between the reactants, temperature, and radiation ambience must be established and maintained in an equivalent manner between experiments. This is difficult to achieve, but the value of direct observation of reactions at the atomic level is more than enough compensation for the effort.

It will require a great deal of experience and effort to assemble the experimental skill and body of knowledge to deal quantitatively with thin-film reactions without ambiguity. The state of knowledge is, at present, sketchy and tends to be anecdotal. In the following subsections, we relate a number of studies that have sketched out the field and indicated its potential. Poppa and Heinemann (1980) have advanced this work by the development of an ultrahigh-vacuum TEM for the study of catalyst surfaces. They also review the relevant literature in this development.

10.10.4.1 Thin-film reactions involving chalcogenide formation. For several years, serious efforts to utilize HRTEM in the study of reactions in crystalline and amorphous thin films, smoke particles from various reactions, monoatomic adsorption, and epitaxial growth, and so forth, have been carried out by Shiojiri and his co-workers. In the interpretation of their results, they have emphasized the visual comparison between the images they obtained and those calculated from reasonable models. They have illustrated their results on thin-film reactions of chalcogenide crystals and given reference to the recent literature (Shiojiri and Kaito, 1983).

A technique has been developed in which a holey amorphous film on a microscope grid receives a deposit of metal (copper, silver, etc.) some thousands of angstroms thick to become a holey grid of the deposited substance. This lacey platform is overlaid by a thin film of a reacting material about 200 Å thick, formed also by evaporation onto some substrate and subsequently stripped before it is applied to the holey grid. The progress of the reaction is observed as the thin film is transformed into reaction product (Shiojiri et al., 1981).

In one experiment, a holey copper grid was covered by a continuous selenium film 200 to 300 Å thick. CuSe was formed almost at once, as revealed by the spacing of the micrograph fringes. Cu_3Se_2 appeared after 20 days, and $Cu_{1.8}Se$ appeared after 50 days as a result of essentially room-temperature diffusion of copper into selenium. The reaction was faster after a "cleaning" procedure utilizing nitric acid was performed on the copper grid before the selenium film was applied. (So far, no experiments have been reported in which the specimen has not at some stage been exposed to a contaminating atmosphere.)

Another experiment, described by this same group, demonstrates the application of the HRTEM technique to describe the structures of crystals formed at such low temperatures that it is impractical to obtain single crystals or single-phase powders allowing conventional structure-determination methods to be applied. In this example, a silver lacey film is overlaid by a thin selenium film (Kaito et al., 1982), which yielded a film of Ag_2Se said to be accurately measured to be 216 Å thick. HRTEM images were taken at different focus settings of the microscope along different directions [e.g., perpendicular to the (010) and (001) surfaces of orthorhombic Ag_2Se] by using axial illumination. Image simulations and diffraction patterns were calculated from assumed structural models over the range of defocus, and the observed and calculated images and diffraction patterns were compared visually. Figure 10.42 illustrates the results for a triclinic crystal. The Ag_2Se crystal is imaged along [101]. The 7.8- and 2.7-Å fringe spacings correspond to the (010) and ($1\bar{2}1$) planes. The images are simulated for the [101] beam direction and for a beam direction tilted 0.37° about the b^*-axis because of the bending of the film, as shown in Figures 10.42c and e. These images correspond to the regions marked by circles at A and B. The calculated images shown in Figures 10.42b and d, based on the triclinic unit cell, show reasonable agreement with those regions. This is an example of real-space crystallographic determination using HRTEM where it is not practicable to carry out the analysis by X-ray or neutron diffraction. Images calculated for structures with atomic parameters differing by more than 0.2 Å from those used in the calculated images shown were appreciably different, indicating their considered range of confidence. They also proposed structures for tetragonal and orthorhombic polymorphs.

Shiojiri and Kaito (1983) include, in addition to the studies just outlined, the observation of defects in II/VI crystals of CdS, ZnSe, and ZnTe having a noncentrosymmetric zinc blende structure and studies of gold adsorption and epitaxial growth on ZnTe surfaces. In these studies, also, comparisons are made between the calculated and observed images to justify the models suggested. In the latter study, they conclude that it is possible to obtain, not only the microtopographical structure and the epitaxial relationships but

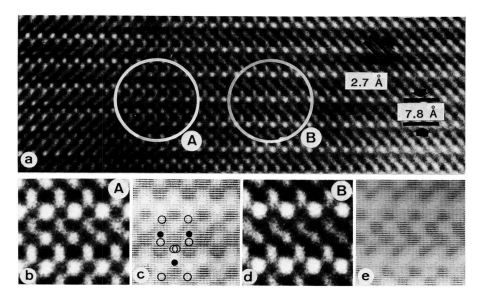

FIGURE 10.42 (a) An electron micrograph of the triclinic Ag_2Se crystal with the [101] direction parallel to the film normal. Fringes with 78- and 27-Å spacings correspond to the (010) and (12$\bar{1}$) planes. (b), (d) Enlarged photographs near A and B in (a). (c), (e) The corresponding simulated images ($\Delta f = 450$ Å). The incident beam was parallel to the [101] direction near A; it was relatively slightly tilted near B. The projections of Ag (open circles) and Se atoms (full circles) are shown. (After Kaito et al., 1982.)

also the actual positions of the adsorbed or overgrown atoms with respect to the substrate lattice.

10.10.4.2 New structures formed from thin-film crystallization or precipitation. Amorphous films of WO_3 about 100 Å thick were obtained by evaporation in vacuum (Miyano et al., 1983) and then transformed by beam heating to monoclinic crystals. With further irradiation, needles grew at the expense of the original WO_3 along the *b*-axis with (001) parallel to the surface. Their composition was W_4O_{11} ($n = 4$ of W_nO_{3n-1}). Crystallographic shear (CS) along (100) was observed for $n = 2$ to 12. These CS phases have not previously been reported in the binary $WO_{3-\delta}$ system; rather, CS occurs along $(102)_R$ and $(103)_R$. They suggest that this difference may result from a minimization of the strain energy for the $(100)_R$ CS plane. This is justified by elastic-strain-energy calculations according to Iguchi and Tilley (Iguchi and Tilley, 1977; Tilley, 1979). These crystals with CS were oriented with the (001) surface and the shear vector not [110]$a/4$ but [101]$a/4$ oblique to the surface. A mechanism of shear is suggested in which a vacancy wall is formed parallel to (100) and a gradual shear by [101]$a/4$ is finally achieved, giving the observed (100) shear structure. The pentagonal columnar structures were not observed.

Compound $TiO_{2-\delta}$, extensively studied previously by thinning bulk-phase crystals, was observed as just outlined for $WO_{3-\delta}$. Vacuum-deposited amorphous films were electron-beam-heated, and images of the resulting crystals showed (110) CS planes not previously reported (Kaito et al., 1983). These structures form by $\frac{1}{2}[0\bar{1}1]$ CS to give members of the Ti_nO_{2n-1} series with $n = 2$ and 4. A blocklike structure with crossed {110} CS planes was also found. A new form of TiO was observed, derived from rutile by successive, partial CS operations of $[0\bar{1}1]/3 \cdot (110)$ and $[0\bar{1}1]/5 \cdot (1\bar{1}0)$. This is quite different from the structure of TiO derived from the NaCl structure. Is this behavior caused by the thin-film material, impurities, electron-induced processes, or some other cause?

When thin, evaporated $BaTiO_3$ films, found to be amorphous, were heat-treated, crystallization occurred to give several structures not previously known in the bulk $BaTiO_3$ material (Shibahara and Hashimoto, 1984). Agreement between electron diffraction patterns and HRTEM images confirmed the existence of an unknown phase and a structure with long periodicity along the [111] direction of the cubic phase. In addition, two new arrangements described as two- and four-layer structures, referring

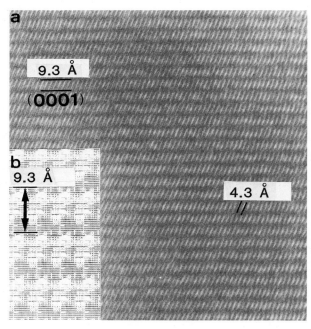

FIGURE 10.43 Lattice image of a four-layer hexagonal phase of $BaTiO_5$. The theoretical image contrast projected along the $[10\bar{1}0]$ direction with a thickness of 200 Å was calculated using the multislice method. The images show the fringes parallel to $(10\bar{1}1)$ with a spacing of 43 Å and the fringes parallel to (0001) with a spacing of 93 Å, which characterize the four-layer structure. (After Shibahara and Hashimoto, 1984.)

FIGURE 10.44 High-resolution image of a grain-boundary-nucleated precipitate (bottom)/matrix (top) interface. Atomic detail of interfacial structure is highlighted along with the location of the interface determined from an extended through-focus series. (After Penisson and Gronsky, 1981.)

to the stacking arrangements of TiO_6 octahedra in the hexagonal array, were observed. Figure 10.43 is an image of this new phase, with the calculated image inset.

Incipient stages of precipitation at grain boundaries in Al–Zn alloys have been studied by utilizing HRTEM (Penisson and Gronsky, 1982). The atomic positions delineated by intersecting (111) planes are within the resolution limits of the microscope. Monolayer steps on interfacial facets appear in early stages to reveal the operation of a ledge mechanism from the onset of growth. Figure 10.44 illustrates the parallel ledges formed in well-defined orientation at the atomic scale. This shows the dominant influence of crystallography on interface formation and propagation.

10.10.4.3 Thin-film reactions in the Au–Sn and Au–In systems. The Au–Sn diffusion couple was chosen by Dufner and Eyring (1986) for the study of reactions in thin films because diffusion and reaction occur at room temperature and they have been extensively studied by other means. It should, therefore, be possible to infer which unusual observations in the thin-film regime are a consequence of the geometry and which are from the radiation field.

Gold diffuses very easily in tin but less readily in the Au–Sn alloys. In fact, bimetallic, thin-film couples have been found by X-ray–diffraction studies to interdiffuse readily at room temperature to form intermetallic compounds. See Dufner and Eyring (1986) for references to previous work utilizing other techniques.

Two methods of specimen preparation were used in the HRTEM study. A two-film method, introduced by Shiojiri and his co-workers as described before, provides an unlimited supply of gold for diffusion into the thin tin film. The other technique consisted of the deposition of tin film on KBr, followed by a deposition of gold islands onto the tin without breaking the vacuum, then application of the wet-stripped composite film onto a holey carbon grid for observation.

The experience in the Au–Sn two-film method was that reaction did not occur at all at room temperature, as expected, but it did occur when the films were heated to 200°C for 10 min. The reason for this is that a contamination layer between the two films was always formed and was a barrier to room-temperature diffusion. This contamination layer consisted of a carbonaceous deposit from the oil-pumped, residual-gas atmosphere, from impurities in the metal samples being evaporated, from oxide formation during the preparation procedures, and from slow oxidation by residual gases in the vacuum system during observation. This barrier was turned to advantage by using it as a throttle to control diffusion and, hence, the reaction rate. When the gold islands were deposited on the tin film without removing it from the vacuum system, the rate of reaction could be controlled by varying the time and system conditions of the interval between gold and tin deposition.

Without heating, no alloy was formed, and electron-microscopic observation always showed only (white) β-tin encapsulated in a contamination envelope. The films were found oriented along [100], [001], [110], [101], and [133]. Images of the films could be obtained on the JEM 200CX microscope, processed, and matched with calculated images from the known structure by using the system described by Rae Smith and Eyring (1982). Figure 10.45a shows the high-resolution image of the [100] zone of β-tin. The contrast difference is due to a discontinuity in thickness between the two regions. Figure 10.45b is a matched image of the low-contrast region. Comparison of the observed images with a calculated through-focus series established the thickness of the light region at about 17 Å and the darker region at about 150 Å. The mottling on the surface is due to SnO_2 particles in the amorphous layer as well as oxide formation on the tin surface itself.

As an example of the single-film-diffusion arrangement, gold was deposited as 25- to 50-Å islands on a tin film that had been stored for five months in a desiccator. The electron beam was condensed on the film for about 5 min. A high-resolution image of part of the film is shown in Figure 10.46. The image shows fringes from the (002) and (1$\bar{1}$1) planes of $AuSn_4$ (5.8- and 4.2-Å spacing, respectively). The dark spots represent the locations of the

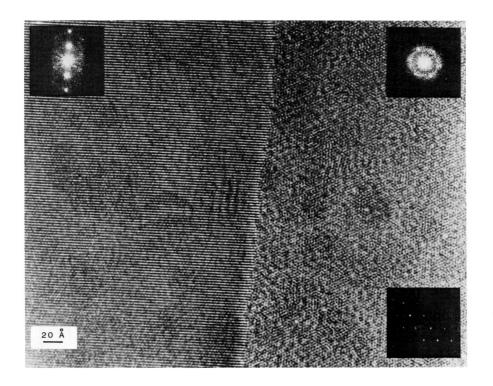

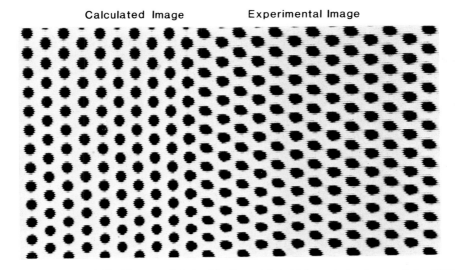

FIGURE 10.45 HRTEM images of β-tin thin films. (a) A high-resolution image of the [100] zone of β-Sn. The boundary separates a thin area at the right from a thicker area at the left. (b) A comparison of the averaged experimental image of the thin region of part (a) with the calculated image at 17-Å thickness and -625-Å defocus.

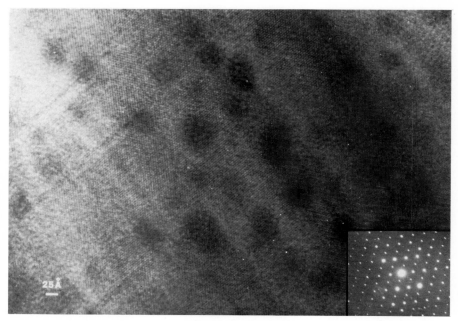

FIGURE 10.46 A high-resolution image of [110] AuSn$_4$. Dark specks represent regions formerly occupied by the gold islands.

original gold islands. The same reaction was observed without intense beam heating when the gold islands were deposited without breaking the vacuum after tin-film deposition on KBr.

The Philips FEG 400 was used to analyze the phases formed in the two-film method on a freshly prepared specimen. The grid had been heated at 200° in vacuum for 10 min before observation. The alloy film is shown in Figure 10.47, where the numbers mark the positions of EDS analyses as listed in Table 10.4. The immediate reaction produced a region of AuSn near the gold supply, with an AuSn$_2$ phase further away. Without additional heating, it took on the order of 130 days to show appreciable further reaction. After this time, the AuSn electron diffraction pattern became dominant in the former AuSn$_2$ region.

The AuSn$_2$ formed in this and many other experiments is not the stable, bulk-phase AuSn$_2$. A high-resolution image of a crystal thought to have this composition is shown in Figure 10.48. This pattern of 8×13-Å blocks of this unknown structure is frequently suggested in these studies. Notice the stacking fault in the top of the figure and the sharp boundary at the bottom. A large number of diffraction patterns were observed that could not come from any of the known bulk phases.

10.10.4.4 The gold-indium reaction system. Gold diffuses rapidly into

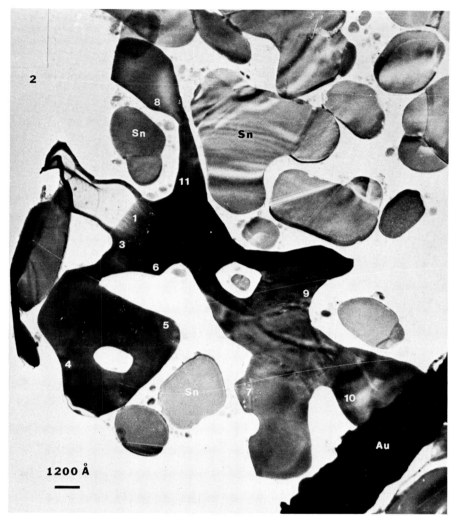

FIGURE 10.47 An image of thin-alloy film, with the microanalysis sampling regions marked. Regions 7, 9, and 10 have the AuSn composition. The remaining regions have a nominal composition of $AuSn_2$. The tin islands are suspended over the hole by an amorphous contamination layer.

indium at room temperature. This means that the techniques employed in the Au–Sn system sketched previously are also applicable in this system. The majority of differences between the two systems arises from the greater oxidizability of indium. The approach is to produce configurations that place the elements apart initially in such a way that diffusion can occur at a rate that can be followed in the HRTEM study—that is, with half-times of

TABLE 10.4
X-ray Microanalysis Results from the Gold–Tin Crystal No. 1 in Figure 10.47.

Region	Composition (by days)					
	1	2	3	4[a]	50	130
1	$AuSn_{2.3}$	$AuSn_{2.0}$	$AuSn_{1.8}$	$AuSn_{2.1}$	$AuSn_{2.5}$	$AuSn_{1.6}$
2			Hole count			
3	$AuSn_{2.2}$	$AuSn_{1.9}$	—	$AuSn_{2.4}$	$AuSn_{2.2}$	$AuSn_{1.3}$
4	$AuSn_{2.2}$	$AuSn_{2.2}$	—	$AuSn_{2.3}$	$AuSn_{2.6}$	$AuSn_{1.4}$
5	$AuSn_{2.5}$	$AuSn_{2.2}$	—	$AuSn_{2.3}$	$AuSn_{2.8}$	$AuSn_{2.0}$
6	$AuSn_{2.1}$	$AuSn_{2.0}$	—	$AuSn_{2.3}$	$AuSn_{2.3}$	$AuSn_{1.1}$
7	AuSn	$Au_{1.1}Sn$	AuSn	$AuSn_{1.3}$	$AuSn_{1.9}$	$Au_{1.1}Sn$
8	$AuSn_{2.3}$	$AuSn_{2.0}$	—	$AuSn_{2.6}$	—	$AuSn_{2.0}$
9	AuSn	$AuSn_{1.1}$	—	—	—	AuSn
10	$AuSn_{1.2}$	$AuSn_{1.3}$	—	—	—	$Au_{1.1}Sn$
11	$AuSn_{2.0}$	$AuSn_{2.0}$	—	—	—	$AuSn_{1.5}$

[a] Data taken at $T = 22°C$.

Note: Experimental conditions are 300-s counts; ≃50- to 100-Å probe size; tilt: $x = 10°$, $y = 0°$; $T = -184°C$.

seconds to hours. Longer reaction times can be tolerated by sequential observation using the conventional HRTEM techniques.

Indium films were produced by evaporation of melted indium shot (99.9999 per cent pure) in a vacuum of less than 10^{-6} torr (Goral and Eyring, 1986). The deposition rate was varied between 10^{-1} and 10^4 Å/s, with the most useful films obtained at a deposition rate of ≃10 Å/s. Either NaCl or KBr, air-cleaved along (100) and maintained at 23 to 40°C, served as the substrate on which the In film was formed. The films were floated off in water and picked up on holey carbon grids.

An island morphology resulted from depositions with thicknesses less than 300 Å, a thickness required for the HRTEM studies. The electron diffraction patterns of individual islands had multiple spots and/or streaking, indicating imperfect crystalline growth. Particles with the [100] direction perpendicular to the plane of the film were observed most often, although islands with the orientations [001], [110], [111], [211], and [311] perpendicular to the film plane were also observed.

A close examination of the diffraction patterns disclosed that almost all had had some trace of In_2O_3. High-resolution images exhibit irregular, wavy fringes of 20- to 40-Å spacing. These have been interpreted as moire fringes arising from a growth of In_2O_3 on the indium film. The development of these wavy fringes progresses during observation in the microscope, suggesting electron-beam-induced oxidation of the indium film by the residual gases. The oxide has been observed to grow epitaxially on an indium particle in the [100] orientation such that In [010] ∥ In_2O_3 [0$\bar{1}$1] and

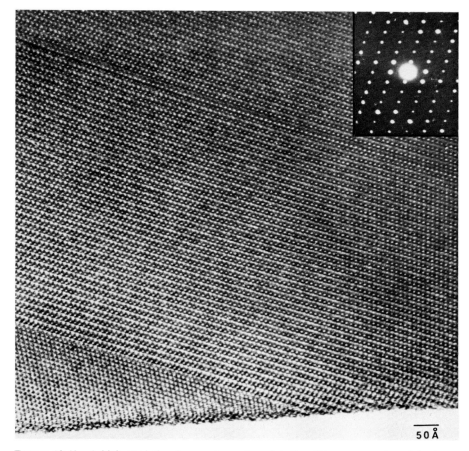

FIGURE 10.48 A high-resolution image of a region of $AuSn_2$. The upper region of the image consists of 8-by-13-Å blocks, as indicated by the accompanying diffraction pattern. In the bottom portion, a region of a hexagonallike array of 7-Å fringes is defined by a sharp boundary.

In [001] || In_2O_3 [100]. The a-axis of In_2O_3 is expanded $\simeq 5$ percent to accommodate epitaxial growth on the underlying indium film.

The two-film, diffusion-couple technique with gold as the support was also applied here but to no avail, owing to effective diffusion barriers. To reduce the thickness of the contamination layer and ensure more intimate contact, the two metals were evaporated in succession without breaking the evaporator vacuum. In all the samples prepared in this way, where the film thickness had to be limited to about 200 Å, the island diameters were not larger than about 10^3 Å, which is much smaller than desired. Since the gold-indium interface is so ill defined, depending as it does on gold-island size and intimacy of contact, it will be difficult to treat the reactions

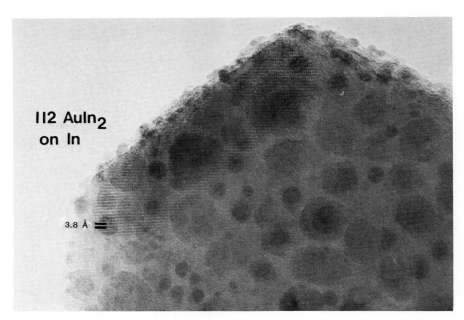

FIGURE 10.49 Growth of [112] AuIn$_2$ on indium.

quantitatively. It will be necessary to develop better specimen configurations so that the boundary conditions for reaction-initiation are more precisely known. Nevertheless, this technique was productive in illustrating the final products of reaction. In one case, a region of AuIn$_2$ was observed to increase in size during the collection of a through-focal series of images (Figure 10.49). It is assumed that the reaction was induced to its conclusion by the electron beam. This observation suggested the alteration of the procedure responsible for observation of the reaction *in situ* that is described in some detail later.

All of the low-temperature compounds identified by Hiscocks and Hume-Rothery (1964) except for the fcc solid solutions were observed by utilizing these techniques. In addition, several new phases, with hexagonal, cubic, or orthorhombic symmetry, were found. It is not known whether these phases could be observed at equilibrium in bulk preparations or whether they are metastable or stabilized by the thin-film configuration. All assignments of compositions were made on the basis of spacings and symmetries observed in the electron diffraction patterns or images. Therefore, the compositions of the new phases are not known. It is necessary to couple these studies with analytical microscopic determinations.

As an example, consider the results on the Au$_4$In phase (hexagonal, $a = 2.94$, $c = 4.94$). Figure 10.50a is a high-resolution image of Au$_4$In in the [011] projection. Multislice calculations of this image have been matched to

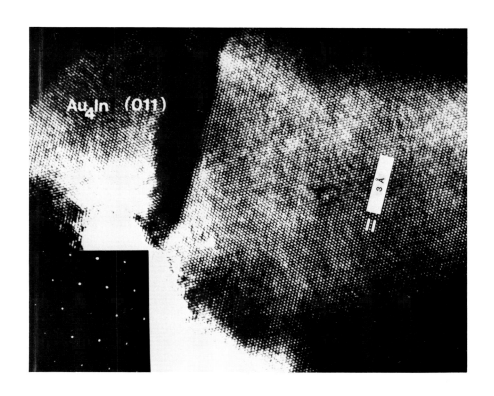

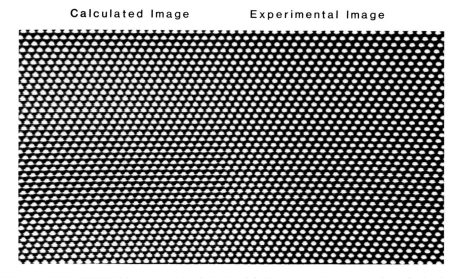

FIGURE 10.50 HRTEM image match of Au_4In. (a) High resolution image of Au_4In in the [011] zone. (b) Match of the averaged image of (a) and the calculated image of Au_4In.

the averaged, observed, digitized image, as shown in Figure 10.50b. The match indicates quantitative agreement well within the range obtained from well-known structures. In addition, a number of superstructures have been observed. A superstructure of Au$_4$In with $a_f^* = \sqrt{13}\, a^*$ has previously been reported (Bhan and Schubert, 1960). Superstructures corresponding to $a_f^* = \sqrt{7}\, a^*$ and $a_f^* = \sqrt{9}\, a^*$ were found. Furthermore, a [001] orientation of Au$_4$In exhibiting a regular array of elongated superstructure spots was recorded. Attempts to suggest the structures of these superstructures by image match has not been successful, probably because of inadequate theory to treat samples of the thickness prepared.

10.11 *In-situ* HRTEM studies of chemical reactions in thin films

10.11.1 The demands of technique development

The commensuration of time and resolution within human scale requires that images evolve at rates that are agreeably perceptible. For this reason, time-lapse and slow-motion photography has been utilized where change is either too slow or too fast for convenient observation. If one is interested in the dynamics and mechanism of chemical changes observed in the electron microscope at high resolution, analogous techniques must be employed. If the change to be observed is slow—of the order of hours, days, or months—then the methods developed for time-independent studies may be employed with time-lapsed image displays. Until recently, this is the way in which dynamic processes have been recorded.

In the other extreme, now that atomic resolution has been achieved in HRTEM, it would be marvelous if the actual atomic migrations could be recorded and played back in slow motion to accommodate our sluggish senses, but at present, no such direct capability is imaginable. Present technology does, however, afford us the possibility of processing and recording images at TV rates. This rate is euphoniously called real time. In the present TV technology, an image frame is produced each $\frac{1}{30}$ s. This image consists of two interleaved fields composing a frame. Played back at this rate, the apparently continuous movement we are accustomed to seeing on TV sets is achieved. Obviously, with this technique, one is limited to the study of atomic events that are of the order of $\frac{1}{30}$ s in duration. It remains now to develop the techniques necessary to (1) record and process images on-line at TV rates, (2) process individual frames off-line, (3) enable the full array of analysis techniques accorded conventionally obtained images to be applied to single frames, (4) recreate edited videorecordings of processes and enhanced sequences that clarify dynamic events more clearly.

We center our attention here on the manipulation of images obtained from the conventional high-resolution electron microscope rather than on spectroscopic information because of the emphasis of this chapter and also because of the requirements for manipulating massive quantities of data in TV time. The scale of the problem has been drawn succinctly in a brief review by Smith (1981). The advantages of on-line operation are so great that it is well worth the expense and effort expended into it. Furthermore, the specimen is being continuously degraded by the electron beam and frequently not in obvious ways. To obtain a processed image from a photograph is a very slow process and must be done off-line and at another time. On-line processing and recording reduce or eliminate all these disagreeable limitations. For example, the amount of radiation received by the specimen can be reduced drastically since an image is recorded at $\frac{1}{30}$-s intervals rather than in the few seconds needed conventionally. Furthermore, various types of averaging can be used to improve the image quality. If the images are recorded on videotape (VT), they can be stored most economically and are easily retrieved. One should be aware, however, that the area of the specimen that is monitored on the TV screen is very small compared with that of a photographic plate; hence, one frequently must record and store photographic negatives so that concurrent events in all parts of the crystal that is photographed can be correlated with the more limited region recorded on VT. Another important advantage of TV recording is that one can go directly to digital recording on a framestore, providing a more faithful representation than that obtained through microdensitometry from the photographic negative; and this information can be treated more accurately. However, the resolution on the TV image is inherently less than can be obtained photographically. This approach to acquisition and management of HRTEM data is growing (Boyes et al., 1981; Erasmus and Smith, 1981; Strahm and Butler, 1981; Skarnulis, 1982),

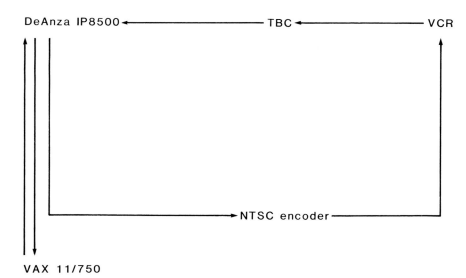

FIGURE 10.51 Schematic drawing of image-processing-facility hardware for videotape image analysis. (Courtesy of G. J. Wood.)

as is automatic alignment and alignment maintenance during the recording of microscopic data. These matters are described briefly by Eyring et al. (1985) and in Chapter 12.

Complete interactive off-line systems for image calculation and simulation, experimental-image digitization, processing, matching, and display have been in use for some time (Rae Smith and Eyring, 1982), as have advanced program packages for interactive-image-processing off-line (Saxton and Koch, 1982).

A multipurpose, multiuser laboratory for the processing and interpretation of electron-microscopic data in both micrograph and real-time video format proposed by Graeme Wood and Andrea Holladay (see Eyring et al., 1985) is diagrammed in Figure 10.51. The installation provides for control of interactive and batch-mode digital processing and storage of digital data.

10.11.2 The recording of the formation of Au_3In in situ

A fluorescent screen can be installed at the base of the microscope column to receive the focused electron beam carrying the image information. A video camera receives the image, and its output is sent alternatively or to both an image processor and comparitor for on-line digitization and processing in real time, or to a videorecorder for off-line processing and analysis (Spence et al., 1982). An experiment utilizing such a system was the $3Au + In \rightarrow Au_3In$ reaction recorded on videotape (Goral et al., 1985).

The reaction pair was prepared by first depositing an indium film by evaporation onto a KBr substrate and then waiting a calibrated period while oxidation and contamination created the necessary diffusion barrier so that the reaction would advance at a manageable rate when gold islands were deposited upon the contaminated film. The composite film was wet-stripped and picked up on a holey carbon grid and immediately placed into the microscope for study. By the time the specimen could be imaged, the relatively gold-rich alloy Au_3In in the $[1\bar{1}1]$ orientation had already formed on the surface of the indium film. The lateral growth of this alloy particle was recorded on videotape. Figure 10.52 shows four images taken from the real-time videotape at 2-s intervals. The averaged experimental image is matched with the calculated image in Figure 10.53. Each of these is an average of eight single frames (taken at $\frac{1}{30}$-s intervals). Thus, each image is averaged over a time period of 0.23 s. In each of the four sequential photographs in Figure 10.52, Au_3In is imaged at about 3-Å resolution in the upper half of the field. The underlying indium film is the lighter region in the lower half. The alloy is seen to grow outward in the $[\bar{1}01]$ and $[110]$ directions as the thin indium grain is consumed. The fringe spacings are 3.9 and 3.7 Å, respectively. Observed in real time, the reaction front proceeds discontinuously, adding a layer one unit cell thick in each jerky step. Apparently, the crystallization nucleates outside the field of view, and the step, one unit cell thick, grows extremely rapidly parallel to $[\bar{1}01]$ and $[110]$ but erratically perpendicular to these planes. The apparently random

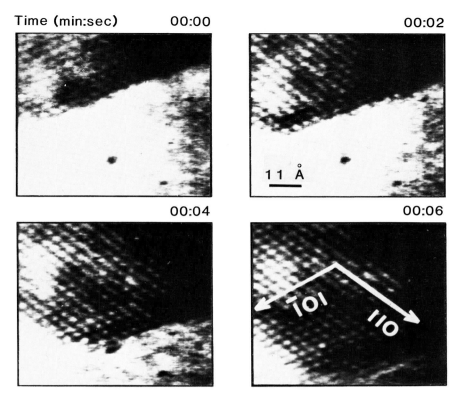

FIGURE 10.52 Reaction sequence of $3Au + In \rightarrow Au_3In$. These eight-frame averages were taken 2 s apart. Growth is along (110) and (101).

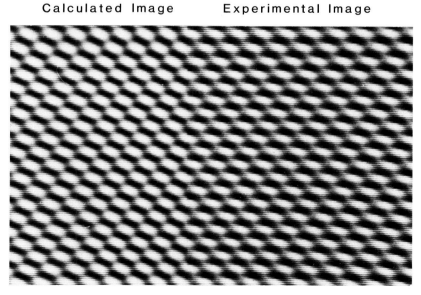

FIGURE 10.53 Digital comparison of calculated and experimental (from videotape) electron-microscope images of Au_3In from the experiment shown in Figure 10.52. The statistical measures of the agreement between the images is satisfactory.

nucleation rate may be perceived by the irregular advance of the front and appears to be the rate-controlling step in the crystallization rate. A slab of at least 30 unit cells long (120 Å) appears in less than $\frac{1}{30}$ s. The advance of the crystal perpendicular to this slab is much slower. The latter motion proceeds outward at about 10 Å s.

10.12 Current and near-term extensions of HRTEM techniques applied to solid-state chemistry

In this chapter, the contribution of HRTEM to the understanding of structures, structural defects, and structural changes reflecting the condition of solid-state chemical systems has been illustrated by several examples. These studies (nearly all between 1971 and 1986) were made with microscopes of 2.0- to 3.5-Å resolution. This limited resolution restricted observation almost entirely to long-period superstructures and changes occurring therein. Although this covers a large number of important chemical systems, it falls short in its application to solid-state chemistry in general.

Some relatively new techniques and instrumentation are beginning to yield results of importance to solid-state chemistry. Furthermore, instrumentation now being developed promises to have even more profound implications. These developments come in improved resolution in the intermediate and high-voltage atomic-resolution microscopes and in the improvement of the vacuum and environmental variability in the specimen chamber accompanied by ultrahigh-vacuum (UHV) preparative and surface-analytical capabilities. The greater control gained of the variables that influence chemical stasis and change, such as temperature and component thermodynamic activity, the better will be the chemistry derived.

10.12.1 Structure elucidation in the bulk and in surfaces and interfaces

Perhaps the most important contribution of the atomic-resolution microscopes (1.5- to 1.7-Å resolution) to chemistry will be the extension of the many methods described in this chapter to the vast class of close-packed structures accompanied by an improved capability in analyzing both structure and defect structure. Improved image-computation packages will complete the set of tools that should bring all substances within range of detailed structural study by HRTEM, provided they are stable enough in the electron beam.

Other high-resolution techniques are already supplying phase identification, crystal-symmetry determination, variation of lattice parameters, atomic potential, crystal-structure information, and orientational relationships from thin films or volumes of only a few cubic nanometers. These methods include microdiffraction and convergent-beam microdiffraction from instruments with field-emission electron guns.

Structural information from regions of thin films only 10 Å or less in diameter can be obtained. This is important for very small particles, point defects, or even different regions of large unit cells. For example, clusters and small particles supported on a substrate can be characterized, including possible chemical reaction. A discussion of these techniques together with a selection of applications has been outlined by Cowley (1984, 1985).

A specialized application of the highest resolution is surface-profile imaging. In the thinnest edge, the contrast is usually low, and great care in astigmatism correction and objective-lens focusing is demanded. In this mode, the short columns of atoms that define the surface edge are observed in sharp profile to give information about the relative frequency and degree of development of various faces, special surface structures, or changes in the spacing of the outermost rows of atoms. Smith (1985) has reviewed this informative extension of HRTEM techniques and their application When this technique is coupled with videorecording methods, the dynamics of surface modification of the various surfaces and other physical and chemical changes can be observed. Kang et al. (1986) have studied TbO_x ($1.5 \leq x \leq 2.0$) by surface-profile imaging in real time and found that the low-index surfaces vary in atomic mobility with $(111) < (100) < (110)$. Other dynamic events were observed, including the untwinning of a crystal and the accretion of a reaction product to the substrate.

For another example, the surface of a crystal of WO_3 has been observed to be reduced to tungsten during observation in the electron beam (Smith and Bursill, 1985; Petford et al., 1986). Surface-profile imaging will continue to provide information about heterogeneous chemical reactions and surface structures and their dynamic changes.

New UHV microscopes are being designed that will combine medium-voltage, high-resolution capability, specimen preparation, and treatment capacity both outside and inside the column; and surface and thin-specimen analytical instrumentation promises to provide an abrupt improvement in HRTEM applications to chemistry. With such tools, virtually any substance compatible with HRTEM can be prepared without serious contamination, treated appropriately, and introduced to the specimen chamber for further treatment during characterization in a controlled atmosphere, including high temperatures and UHV. Videorecording systems will make possible the observation of physical and chemical changes of such especially prepared specimens or samples being reacted by, for example, molecular or atomic beams.

The development of microscopes specifically for surface studies with resolutions in the range of 10 Å are being designed. These UHV instruments will be particularly valuable in the chemical study of heterogeneous catalysts and in clarifying the nature of chemical reactions on surfaces.

10.12.2 Nanometer analytical electron microscopy
Little has been said here about the importance to solid-state chemistry of *in situ* analysis of substances at high lateral resolution. This is of particular

importance when the elemental constitution is unknown, as will happen during chemical change in the microscope, and for the analysis of the oxidation state and environment of atoms in specific sites. The techniques of EDS—with field-emission guns and parallel detection—and EELS are well established and, especially for EELS, are being improved rapidly. The ability to observe at high lateral resolution the occurrence of a particular type of atom by its absorption edge and to determine its oxidation state and local environment from the near-edge (ELNES) and other fine structure (EXELFS) is essential to the full chemical characterization of materials. Colliex (1984) has reviewed EELS methods and their implications.

REFERENCES

Allpress, J. G. (1972). Crystallographic shear in $WO_3 \cdot xNb_2O_5$ ($x = 0.03$–0.09). *J. Solid State Chem.* **4**, 173.

———, and Sanders, J. V. (1971). *n*-Beam lattice images of complex oxides. In *Electron microscopy and structure of materials,* ed. G. Thomas, 134. University of California Press, Berkeley.

———, Sanders, J. V., and Wadsley, A. D. (1969). Multiple phase formation in the binary system $Nb_2O_5WO_3$. VI. Electron microscopic observation and evaluation of nonperiodic shear structures. *Acta Crystallogr., Sect. B* **25**, 1156.

———, Hewat, E. A., Moody, A. F., and Sanders, J. V. (1972). *n*-Beam lattice images. I. Experimental and computed images from $W_4Nb_{26}O_{77}$. *Acta Crystallogr., Sect. A* **28**, 528.

Amelinckx, S., and Van Landuyt, J. (1970). The use of electron microscopy in the study of extended defects related to nonstoichiometry. In *The chemistry of extended defects in nonmetallic solids,* ed. L. Eyring and M. O'Keeffe, 296. North-Holland, Amsterdam.

Anderson, J. S. (1973). On infinitely adaptive structures. *J. Chem. Soc. Lond. Dalton Trans.* **10**, 1107.

——— (1978–1979a). Electron microscopy in the study of solid state reactions. *Chem. Scripta* **14**, 129.

——— (1978–1979b). Direct imaging of atoms in crystals and molecules. Status and prospects for chemistry. *Chem. Scripta* **14**, 287.

———, and Hyde, B. G. (1967). On the possible role of dislocations in generating ordered and disordered shear structures. *J. Phys. Chem. Solids* **28**, 1393.

———, and Tilley, R. J. D. (1970). Crystallographic shear in oxygen-deficient rutile: An electron microscope study. *J. Solid State Chem.* **2**, 472.

Andersson, S. (1970). Arthur David Wadsley. An addendum to the preface in *The chemistry of extended defects in non-metallic solids,* ed. L. Eyring and M. O'Keefe. North-Holland, Amsterdam.

———, and Wadsley, A. D. (1966). Crystallographic shear and diffusion paths in certain higher oxides of niobium, tungsten, molybdenum and titanium. *Nature* **211**, 581.

Bando, Y., and Iijima, S. (1980). An incommensurate superstructure of hexagonal tungsten bronze. In *Proceedings, Electron Microscopy Society of America,* ed. G. W. Bailey, *38th Annual Meeting, Las Vegas, Nevada,* 166. Claitor's Publishing Div., Baton Rouge.

———, Sekikawa, Y., Yamamura, H., and Matsui, Y. (1981). Crystal structure analysis of $Ca_4YFe_5O_{13}$ by combining 1 MeV high-resolution electron microscopy with convergent-beam electron diffraction. *Acta Crystallogr., Sect. A* **37**, 723.

Baumard, J. F., Panis, D., and Anthony, A. M. (1977). A study of Ti–O system between Ti_3O_5 and TiO_2 at high temperature by means of electrical resistivity. *J. Solid State Chem.* **20**, 43.

Bevan, D. J. M., and Summerville, E. (1979). Mixed rare earth oxides. In *Handbook on the physics and chemistry of rare earths,* ed. K. A. Gschneidner, Jr., and L. Eyring, 401. North-Holland, Amsterdam.

Bhan, S., and Schubert, K. (1960). Constitution of the systems cobalt-germanium, rhodium-silicon, and some related alloys. *Z. Metallk.* **51,** 327.

Blanchin, M. G., Faisant, P., Picard, C., Ezzo, M., and Fontaine, G. (1980). Transmission electron microscope observations of slightly reduced rutile. *Phys. Status Solidi a* **60,** 357.

Booth, J., Ekstrom, T., Iguchi, E., and Tilley, R. J. D. (1982). Notes on phases occurring in the binary tungsten-oxygen system. *J. Solid State Chem.* **41,** 293.

Boyes, E. D., Muggridge, B. J., Coringe, M. J., Hutchison, J. L., and Catlow, G. (1981). On-line image analysis and viewing for very high-resolution electron microscopy. Paper presented at EMAG, Cambridge, September 1981. In *Institute of Physics Conference Series,* ed. M. J. Goringe, No. 61, Chap. 3, 119. Institute of Physics, Bristol.

Bursill, L. A. (1969). Crystallographic shear in molybdenum trioxide. *Proc. R. Soc. London, Ser. A* **311,** 267.

────── (1983). Structure of small defects in nonstoichiometric WO_{3-x}. *J. Solid State Chem.* **48,** 256.

──────, and Hyde, B. G. (1972). Crystallographic shear in the higher titanium oxides: Structure, textures, mechanisms and thermodynamics. In *Progress in solid state chemistry,* ed. G. M. Rosenblatt and W. L. Worrell, 117. Pergamon Press, Oxford.

──────, Blanchin, G., and Smith, D. J. (1982). The nature and extent of disorder within rapidly cooled $TiO_{1.9985}$. *Proc. R. Soc. London, Ser. A* **384,** 135.

──────, Blanchin, M. A., and Smith, D. J. (1984). Precipitation phenomena in nonstoichiometric oxides. II. {100} platelet defects in reduced rutiles. *Proc. R. Soc. London, Ser. A* **391,** 373.

──────, Lodge, E. A., and Thomas, J. M. (1980). Zeolitic structures as revealed by high-resolution electron microscopy. *Nature* **286,** 111.

──────, Smith, D. J., and Lin, P. J. (1985). Small defect clusters in chromia-doped rutiles. *J. Solid State Chem.* **56,** 203.

──────, Blanchin, M. G., Mebarek, A., and Smith, D. J., (1983). Point, linear and extended defect structure in nonstoichiometric rutile. *Radiation Effects,* **74,** 253.

Colliex, C. (1984). Electron energy loss spectroscopy in the electron microscope. In *Advances in optical and electron microscopy,* Vol. 9, ed. R. Barer and V. E. Cosslett, 65. Academic Press, London.

Coslett, V. E. (1979). Radiation damage: Experimental work. In *Advances in structure research by diffraction methods,* eds. W. Hoppe and R. Mason, **7,** 81. Friedr. Vieweg and Sohn, Braunschweig Wiesbaden.

Cowley, J. M. (1978–1979). Direct imaging of atoms in crystals and molecules. Status and prospects for physics. *Chem. Scripta* **14,** 279.

────── (1982). The accomplishments and prospects of high-resolution imaging methods. *Ultramicroscopy* **8,** 1.

────── (1984). Scanning transmission electron microscopy and microdiffraction techniques. *Bull. Mater. Sci.* **6,** 477.

────── (1985). Electron microscopy and diffraction techniques for the study of small particles. *ACS Symposium Series* **288,** 329.

Crewe, A. V., Wall, J., and Langmore, J. (1970). Visibility of single atoms. *Science* **168,** 1338.

Dufner, C., and Eyring, L. (1986). High-resolution transmission electron microscopy and X-ray microanalysis of chemical reactions in the gold-tin thin film system. *J. Solid State Chem.* **62,** 112.

England, P. J., and Tilley, R. J. D. (1982). The microstructures occurring in some Nb_2O_5–WO_3 and Ta_2O_5–WO_3 crystallographic shear phases on annealing at 1173 K. *Chem. Scripta* **20,** 102.

──────, and Tilley, R. J. D. (1983). Crystallographic shear in the Nb_2O_5–Ta_2O_5–WO_3 quaternary system at 1623 K. *Chem. Scripta* **22,** 108.

──────, and Tilley, R. J. D. (1984). An electron microscope study of some tetragonal tungsten bronze related phases in the Nb_2O_5–WO_3, Ta_2O_5–WO_3 and Nb_2O_5–Ta_2O_5–WO_3 systems. *Chem. Scripta* **23**, 15.

———, Booth, J., Tilley, R. J. D., and Ekstrom, T. (1982). Crystallographic shear in the Nb_2O_5–WO_3 and Ta_2O_5–WO_3 systems. *J. Solid State Chem.* **44**, 60.

Erasmus, S. J., and Smith, K. C. A. (1981). Real-time digital image processing in electron microscopy. Paper presented at EMAG, Cambridge, September 1981. In *Institute of Physics Conference Series*, No. 61, Chap. 3, 115. Institute of Physics, Bristol.

Eyring, L. (1980). Phase relationships, reaction mechanisms, and defect structures in rare earth oxide-oxygen systems. In *Science and technology of rare earth materials*, ed. E. C. Subbarao and W. E. Wallace, 99. Academic Press, New York.

———, guest ed. (1982). Proceedings of the American Chemical Society symposium on high-resolution electron microscopy applied to chemical problems. *Ultramicroscopy* **8**, 1.

———, and Tai, L.-T. (1976). The structural chemistry of some complex oxides: Ordered and disordered extended defects. In *Treatise on solid state chemistry*, Vol. 3, ed. N. B. Hannay, Chap. 3, 216. Plenum, New York.

———, Dufner, C., Goral, J. P., and Holladay, A. (1985). High-resolution electron microscopic studies of chemical reactions in thin films. *Ultramicroscopy* **18**, 253.

Gasgnier, M., Schiffmacher, G., Caro, P., and Eyring, L. (1986). The formation of rare earth oxides far from equilibrium. *J. Less-Common Metals* **116**, 31.

Giroult, J. P., Goreaud, M., Labré, Ph., Provost, J., and Raveau, B. (1980). $Rb_xP_8W_{32}O_{112}$: A tunnel structure built up from ReO_3-type blocks and P_2O_7 groups. *Acta Crystallogr., Sect. B* **36**, 2570.

Goral, J. P., and Eyring, L. (1986). The gold-indium thin film system: A high-resolution electron microscopy study. *J. Less-Common Metals* **116**, 63.

———, Holladay, A., and Eyring, L. (1985). Time-resolved analysis of high-resolution electron microscope images. *Ultramicroscopy* **18**, 275.

Gruehn, R. (1982). Can electron microscopy be a help in preparative solid-state chemistry? In *Studies in inorganic chemistry*, Vol. 3, ed. R. Metselaar, H. J. M. Hijligers, and J. J. Schoonman, 33. Solid State Chemistry 1982. Proceedings of 2nd European conference, Veldhoven, The Netherlands.

Hashimoto, H. (1983). Direct imaging of atomic processes in crystals: Some personal steps towards this goal. *Roy. Micros. Soc. Proc.*, **18**, 298.

Height, T. M., and Bevan, D. J. M. Unpublished results.

Hervieu, M., and Raveau, B. (1983a). High-resolution electron microscopy study of phosphate tungsten bronzes $KP_4O_8(WO_3)_{2m}$. *Chem. Scripta* **22**, 117.

———, and Raveau, B. (1983b). High-resolution electron microscopy study of phosphate tungsten bronzes $KP_4O_8(WO_3)_{2m}$. *Chem. Scripta* **22**, 123.

Hirabayashi, M. (1983). Many-beam imaging studies of crystal structure of ordered alloys. *Trans. Jpn. Inst. Metals* **24**, 317.

Hirotsu, Y., Tsunashima, Y., and Nagakura, S. (1982). High-resolution electron microscopy of microsyntactic intergrowth in V_nO_{2n-1}. *J. Solid State Chem.* **43**, 33.

Hiscocks, S. E. R., and Hume-Rothery, W. (1964). The equilibrium diagram of the system gold-indium. *Proc. R. Soc. London, Ser. A* **282**, 318.

Hobbs, L. W. (1984). Radiation effects in analysis by transmission electron microscopy. In *Quantitative microscopy*, ed. J. N. Chapman and A. J. Craven, 399. Scottish Universities Summer Schools in Physics, Edinburgh.

———, Howitt, D. G., and Mitchell, T. E. (1978). Electron microscopy and electron diffraction of electron-sensitive materials. In *Institute of Physics Conference Series*, No. 41, Chap. 7, 402. Institute of Physics, Bristol.

Hong, Y. S., Williamson, R. F., and Boo, W. D. J. (1979). Lower-valence fluorides of vanadium. 3. Structures of the pseudohexagonal A_xVF_3 phases (where A = K, Rb, Tl or Cs). *Inorg. Chem.* **18**, 2123.

Horiuchi, S. (1982). Detection of point defects accommodating nonstoichiometry in inorganic compounds. *Ultramicroscopy* **8**, 27.

———, Matsui, Y., and Bando, Y. (1976). A high-resolution lattice image of $Nb_{12}O_{29}$ by means of a high-voltage electron microscope newly constructed. *Jpn. J. Appl. Phys.* **15,** 2483.
Hovmöller, S., Sjögren, A., Garrents, G., Sundberg, M., and Marinder, B.-O. (1984). Accurate atomic positions from electron microscopy. *Nature* **311,** 238.
Hutchison, J. L. (1984a). Where is high-resolution electron microscopy going? In *Institute of Physics Conference Series*, ed. P. Doig, No. 68, Chap. 5, 159. Institute of Physics, Bristol.
——— (1984b). Applications of high-resolution electron microscopy in the study of complex oxides. 8th European Conference on Electron Microscopy, Budapest, Hungary, August 13–18, 1984.
——— (1984c). Recent advances and applications of high-resolution electron microscopy. *J. Microsc.* **136,** 127.
Hyde, B. G., and O'Keeffe, M. (1973). Relations between the $DO_9(ReO_3)$ structure type and some "bronze" and "tunnel" structures. *Acta Crystallogr., Sect. A* **29,** 243.
Iguchi, E., and Tilley, R. J. D. (1977). The elastic strain energy of crystallographic shear planes in reduced tungsten trioxide. *Philos. Trans. R. Soc. London, Ser. A* **286,** 55.
Iijima, S. (1971). High-resolution electron microscopy of crystal lattice of titanium-niobium oxide. *J. Appl. Phys.* **42,** 5891.
——— (1973). Direct observation of lattice defects in $H-Nb_2O_5$ by high-resolution electron microscopy. *Acta Crystallogr., Sect. A* **29,** 18.
——— (1975). High-resolution electron microscopy of crystallographic shear structures in tungsten oxides. *J. Solid State Chem.* **14,** 52.
——— (1977). Observation of single and clusters of atoms in bright field electron microscopy. *Optik (Stuttgart)* **48** (2), 193.
———, and Allpress, J. G. (1974a). Structural studies by high-resolution electron microscopy tetragonal tungsten bronze-type structures in the system $Nb_2O_5-WO_3$. *Acta Crystallogr., Sect. A* **30,** 22.
———, and Allpress, J. G. (1974b). Structural studies by electron microscopy: Coherent intergrowth of the ReO_3 and tetragonal tungsten bronze structure types in the system $Nb_2O_5-WO_3$. *Acta Crystallogr., Sect. A* **30,** 29.
———, and Bando, Y. (1981). High-resolution microscopy of incommensurate superlattices. In *Electron microscopy society of America, 39th annual meeting*, 102. ed. G. W. Bailey. Claitor's Publ. Div., Baton Rouge.
———, Kimura, S., and Goto, M. (1976). Direct observation of point defects in $Nb_{12}O_{29}$ by high-resolution electron microscopy. *Acta Crystallogr., Sect. A* **29,** 632.
Isaacson, M. (1984). Radiation damage in the electron microscope: Bane or boon? In *Institute of Physics Conference Series* ed. P. Doig No. 68, Chap. 1, 1. Institute of Physics, Bristol.
Kaito, C. Iwanishi, M., Harada, T., Miyano, T., and Shiojiri, M. (1983). High-resolution electron microscopic studies of crystallographic shear structures in reduced rutile crystals. *Trans. Jpn. Inst. Metals* **24,** 450.
———, Nakamura, J., Teranishi, K., Sekimoto, S., and Shiojiri, M. (1982). High-resolution electron microscopic studies of the polymorphic transformation and crystal structures of low-temperature Ag_2Se phases. *Phys. Status Solidi a* **71,** 109.
Kang, Z. C., Eyring, L., and Smith, D. J. (1986). Dynamic edge and surface processes in terbium oxides. *Ultramicroscopy* **22,** 71.
Kihlborg, L., ed. (1978–1979). Proceedings of the forty-seventh Nobel symposium on direct imaging of atoms and crystals and molecules. *Chem. Scripta* **14,** 1.
———, (1982). Tungsten bronzes and related compounds. In *Studies in inorganic chemistry*, Vol. 3, eds. R. Metselaar, H. J. M. Heijligers, and J. Schoonman, 143. Proceedings of the 2nd European Conference, Veldhoven, The Netherlands.
———, and Sharma, R. (1982). Order and disorder in compounds with tungsten bronze structures. *J. Microsc. Spectrosc. Electron.* **7,** 387.
Klug, A. (1978–1979). Direct imaging of atoms in crystals and molecules. Status and prospects for biological sciences. *Chem. Scripta* **14,** 291.

Knappe, P., and Eyring, L. (1985). Preparation and electron microscopy of intermediate phases in the interval Ce_7O_{12}–$Ce_{11}O_{20}$. *Solid State Chem.* **58**, 312.

Kunzmann, P., and Eyring, L. (1975). On the crystal structures of the fluorite-related intermediate rare-earth oxides. *J. Solid State Chem.* **14**, 229.

Labbé, Ph., Quachee, D., Goreaud, M., and Raveau, B. (1983). Bronzes with a tunnel structure $K_xP_4O_8(WO_3)_{2m}$: The tenth member of the series $KP_8W_{40}O_{136}$. *J. Solid State Chem.* **50**, 163.

Lundberg, M., Sundberg, M., and Magnéli, A. (1982). The "pentagonal column" as a building unit in crystal and defect structures of some groups of transition metal compounds. *J. Solid State Chem.* **44**, 32.

Magnéli, A. (1953). Structures of the ReO_3-type with recurrent dislocations of atoms: "Homologous series" of molybdenum and tungsten oxides. *Acta Crystallogr.* **6**, 495.

Marinder, B.-O., and Sundberg, M. (1984). The structure of $Na_{11}Nb_7O_{18}$ as deduced from HREM images and X-ray powder diffraction data. *Acta Crystallogr., Sect. B* **40**, 82.

Menter, J. W. (1956). The direct study by electron microscopy of crystal lattices and their imperfections. *Proc. R. Soc. London, Ser. A* **236**, 119.

Miyano, T., Iwanishi, M., Kaito, C., and Shiojiri, M. (1983). High-resolution electron microscopic studies of CS structure in reduced WO_3 twin crystals. *Jpn. J. Appl. Phys.* 22, 863.

O'Keefe, M. A., and Buseck, P. R. (1979). Computations of high-resolution TEM images of minerals. *Trans. Am. Cryst. Assoc.*, **15**, 27.

———, Buseck, P. R., and Iijima, S. (1978). Computed crystal structure images for high-resolution electron microscopy. *Nature* **274**, 322.

Otero-Diaz, L. C., and Hyde, B. G. (1984). A HREM study of disorder in two types of rutile-related CS phases. *Acta Crystallogr., Sect. B* **40**, 237.

Parmentier, M., Gleitzer, C., and Tilley, R. J. D. (1980). Etude du systéme Sb–Mo–O a 500°C: Mise en évidence de deux nouveaux oxydes de molybdéne-antimoine. *J. Solid State Chem.* **31**, 305.

———, Gleitzer, C., Courtois, A., and Protac, J. (1979). Structure cristalline de $Sb_2Mo_{10}O_{31}$. *Acta Crystallogr., Sect. B* **35**, 1963.

Penisson, J. M., and Gronsky, R. (1982). Experimental studies on the atomistics of grain boundary precipitation. Paper presented at TMS/AIME meeting on solid-solid phase transformations, Pittsburgh, Pa., August 1981. Lawrence Berkeley Laboratory, Berkeley, CA LBL–138521, March 1982.

Petford, A. K., Marks, L. D., and O'Keeffe, M. (1986). Atomic imaging of oxygen desorption from tungsten trioxide. *Surf. Sci.* **172**, 496.

Poppa, H., and Heinemann, K. (1980). Basic studies in catalysis by electron microscopy *Optik*, (*Stuttgart*) **56**, 183.

Rae Smith, A., and Eyring, L. (1982). Calculation, display and comparison of electron microscope images modelled and observed. *Ultramicroscopy* **8**, 65.

———, Tuenge, R. T., and Eyring, L., Unpublished results.

Ray, S. P., and Cox, D. E. (1975). Neutron diffraction determination of the crystal structure of Ce_7O_{12}. *J. Solid State Chem.* **15**, 333.

Rieck, D., Langley, R., and Eyring, L. (1982). Structures of hexagonal vanadium fluoride bronzes: A high-resolution electron microscopic study. *J. Solid State Chem.* **45**, 259.

———, Langley, R., and Eyring, L. (1983). Defects in hexagonal vanadium fluoride bronzes: A high-resolution electron microscope study. *J. Solid State Chem.* **48**, 100.

Roth, R. S., and Wadsley, A. D. (1965). Mixed oxides of titanium and niobium: the crystal structure of $TiNb_{24}O_{62}$ (TiO_2, $12Nb_2O_5$). *Acta Crystallogr.* **18**, 724.

Sahle, W. (1982). Electron microscopy studies of $W_{18}O_{49}$. 2. Defects and disorder introduced by partial oxidation. *J. Solid State Chem.* **45**, 334.

——— (1983). Electron microscopy studies of tungsten oxides in the range WO_3–$WO_{2.72}$. Phase relations, defects structures, structural transformations and electrical conductivity. *Chem. Commun. Univ. Stockholm*, **4**.

——, and Kljavins, J. (1985). On the mechanisms of oxidation of {103}–CS structure in the W–O system. *J. Solid State Chem.* **56**, 255.
——, and Sundberg, M. (1980). A new type of defect in reduced tungsten trioxide as revealed by high-resolution electron microscopy. *Chem. Scripta* **16**, 163.
Sano, H., and Gütlich, P. (1984). Hot atom chemistry in relation to Mössbaur emission spectroscopy. In *Hot atom chemistry,* ed. T. Matsuura, 265. Elsevier, Amsterdam.
Saxton, W. O., and Koch, T. L. (1982). Interactive image processing with an off-line minicomputer: Organization, performance and applications. *J. Microsc.* **127**, 69.
Shibahara, H., and Hashimoto, H. (1984). Electron microscope study of the crystallization from amorphous phase in $BaTiO_3$ thin films. *J. Cryst. Growth* **67**, 227.
Shiojiri, M., and Kaito, C. (1983). The high-resolution transmission electron microscopic observation of chalcogen crystals. *JEOL News* **21E** (1) 2.
——, Kaito, C., Saito, Y., Teranishi, K., and Sekimoto, S. (1981). High-resolution electron microscopic study of the growth of Cu–Se crystals by a solid–solid reaction. *J. Cryst. Growth* **52**, 883.
Skarnulis, A. J. (1979). A system for interactive electron image calculations. *J. Appl. Cryst.* **12**, 636.
——, (1982). A computer system for on-line image capture and analysis. *J. Microsc.* **127**, 39.
Smith, D. J. (1985). Atomic-resolution studies of surface dynamics by electron microscopy. *J. Vac. Sci. Technol. B* **3**, 1563.
——, and Bursill, L. A. (1985). "Metallisation" of oxide surfaces observed by *in situ* high-resolution electron microscopy. *Ultramicroscopy* **17**, 387.
——, Cosslett, V. E., and Stobbs, W. M. (1981). Atomic resolution with the electron microscope. *Interdisciplinary Sci. Rev.* **6**, 155.
Smith, K. C. A. (1981). On-line digital computer techniques in electron microscopy. Paper presented at EMAG Conference, Cambridge, England, September 1981. In *Institute of Physics Conference Series,* No. 61, Chap. 3, 109. Institute of Physics, Bristol.
Spence, J. C. H., Disko, M., Higgs, A., Wheatley, J., and Hashimoto, H. (1982). A digital on-line diffractometer and image processor for HREM. 10th International Congress Electron Microscopy, Hamburg. In *Electron microscopy 1982,* Vol. 1, Ed. Congress Organizing Committee, 519. Deutsche Gesellschaft für Elektronen-mikroskopie e.V., Frankfurt.
Strahm, M., and Butler, J. H. (1981). Fast digital data acquisition and on-line processing system for an HB5 scanning transmission electron microscope. *Rev. Sci. Instrum.* **52**, 840.
Terasaki, O., Thomas, J. M., and Ramadas, S. (1984). A new type of stacking fault in zeolites: Presence of a coincidence boundary ($\sqrt{13}$, $\sqrt{13}$ R 32.2° superstructure) perpendicular to the tunnel direction in zeolite L. *J. Chem. Soc. Chem. Commun.* **4**, 216.
Thomas, J. M. (1982). Placing the applications of high-resolution electron microscopy to chemical problems into wider perspective. *Ultramicroscopy* **8**, 13.
—— (1983). Dealuminated zeolites. *J. of Mol. Catal.* **27**, 59.
——, Millward, G. R., and Ramadas, S. (1983). New approaches to the structural characterization of zeolites: High-resolution electron microscopy and optical diffractometry. In *Intrazeolite chemistry,* ed. E. D. Stucky and F. G. Dwyer, Chap. 11, 181. American Chemical Society, Washington, D.C.
Tilley, R. J. D. (1972). Crystallographic shear in inorganic oxides. *Int. Rev. Sci.: Inorg. Chem., Ser. One* **10**, 279.
—— (1979). The crystal chemistry of some tungsten oxides containing crystallographic shear planes. *Chem. Scripta* **14**, 147.
——, and Hyde, B. G. (1970). An electron microscopic investigation of the decomposition of V_2O_5. *J. Phys. Chem. Solids* **31**, 1613.
——, and Wright, A. C. (1982). Electron microscope observation of the decomposition of $Pb_{24}Bi_8S_{36}$ and $Pb_{12}Bi_8O_{24}$. *Chem. Scripta* **19**, 68.
Von Dreele, R. B., Eyring, L., Bowman, A. L., and Yarnell, J. L. (1975). Refinement of the

crystal structure of Pr_7O_{12} by powder neutron diffraction. *Acta Crystallogr., Sect. B* **31,** 971.

Wadsley, A. D., and Andersson, S. (1970). Crystallographic shear and the niobium oxides and oxide fluorides in the composition region MX_x, $2.4 < x < 2.7$. In *Perspectives in structural chemistry* Vol. 3, ed. J. D. Dunitz and J. A. Ebers, 1. Wiley, New York.

Watari, F., and Cowley, J. M. (1981). The study of oxide formation on (001), (011), (111) and (113) surfaces of Cr thin films using STEM-microdiffraction methods. *Surf. Sci.* **105,** 240.

11
MATERIALS SCIENCE: METALS, CERAMICS, AND SEMICONDUCTORS

DAVID J. SMITH AND JOHN C. BARRY

11.1 Introduction

11.1.1 Materials

The application of high-resolution electron microscopy at the atomic level to the characterization of defects in metals, ceramics, and semiconductors is comparatively recent, although studies involving lattice-fringe imaging, particularly with tilted illumination, date from the late 1960s. The reason is simple: the typical spacings between projections of atomic columns in these materials, even for major low-index zones, is substantially less than 3 Å, and microscopes with adequate resolving power, when operated with axial illumination, were previously unavailable. Dedicated high-resolution, high-voltage microscopes have since achieved the required performance levels (for details, see Smith et al., 1982); and even 200-kV instruments in ultrahigh-resolution configurations are nearly reaching the 2 Å level, so many ceramics and semiconductors have become accessible to study. Finally, the so-called medium-voltage (300- to 400-kV) machines are becoming widespread and are starting to demonstrate atomic-resolution capabilities even for close-packed metallic systems.

In the broadest sense, materials science includes the study of metals, ceramics, and semiconductors, but it also crosses the traditional boundaries into solid-state chemistry and mineralogy. Since these latter topics have already been covered in depth in earlier chapters, they are generally omitted from consideration here, although passing mention is made in places because of the fuzziness of dividing lines and the universal interest in some materials. For example, ceramics based on the Si–Al–O–N composites are of prime technological importance because of their valuable high-temperature properties, but they are also of great interest to the chemist and the crystallographer. Our emphasis will be on the use of the high-resolution electron microscope to characterize, at the atomic level, the defects present in small-unit-cell materials, with the objective of understanding their crucial role in influencing physical and chemical properties. The examples described below are intended to provide typical applications of the technique rather than represent an exhaustive summary, since papers describing different materials or defects are appearing almost daily.

11.1.2 Types of defects and processes

The attraction of the high-resolution electron microscope for the materials scientist stems from its ability to provide direct information about local inhomogeneities at an unprecedented atomic level. The locations of atomic columns in the vicinity of certain structural defects can be determined to a high degree of accuracy by using proper imaging conditions, provided that the region of specimen being imaged is sufficiently thin. Planar defects such as twin boundaries, platelets, and crystallographic shear planes have all been studied, and detailed computer simulations have been used to confirm the interpretation of experimental micrographs. However, ordering along the beam direction is not always perfect. Moreover, the electron-beam irradiation can result in atomic displacements in many materials, particularly in the vicinity of the planar defects where the strengths of the interatomic forces are usually less than in the surrounding bulk material. Line defects such as dislocations are difficult to image because one must tilt the sample in such a way that the defect lies precisely along the beam direction; and while it is generally assumed that the ordering is perfect along the length of the defect, this is not always the case: In metals, for example, there is usually some relaxation at the place where the defect intersects with the surface. Nevertheless, it should be clear from the examples given here and elsewhere in this book that high-resolution transmission electron microscopy (HRTEM) has been singularly successful as a technique for characterizing the atomic structure of many crystal imperfections. Finally, it should be appreciated that our emphasis is deliberately upon the information achievable by atomic-resolution electron microscopy. Valuable structural details have been obtained about many materials without such extreme resolution. In many such cases, the technique of HRTEM will continue to make valuable contributions to an understanding of materials' properties.

11.2 Imaging requirements

11.2.1 Limitations and applications of lattice-fringe imaging

The initial observation of lattice-fringe images from thin metal foils by Komoda (1964) involved the use of tilted illumination, as illustrated in Figure 11.1a. By tilting the incident beam so that the directly transmitted beam and the diffracted beams were equidistant from the optic axis, the effects of chromatic aberration on the off-axis beams were minimized. This minimization was an important consideration at a time when the stabilities of high-voltage power supplies were comparatively poor. (Such fringes also provided manufacturers with an opportunity to make misleading claims about microscope performance.)

A general assumption was made, though not actually proven, that there was a one-to-one correspondence between the tilted-illumination, lattice-fringe patterns and the atomic or lattice plane distributions within the

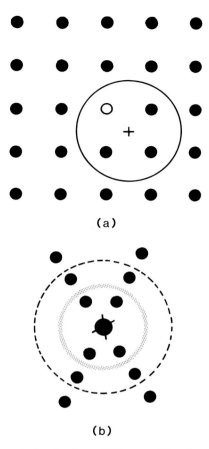

FIGURE 11.1 (a) A schematic drawing for a ⟨001⟩ zone diffraction pattern for a face-centered cubic metal. The circle represents the typical aperture position for tilted-illumination, lattice-fringe imaging; the cross and the small open circle indicate, respectively, the position of the optic axis and the direction of the incident (unscattered) beam. (b) A schematic drawing for the electron diffraction pattern corresponding to one-dimensional, lattice-fringe imaging from two adjacent crystals simultaneously. The circle represents the typical objective aperture, and the hatched disk represents the scattering originating from the amorphous material at the interface.

sample. Detailed studies of dislocations in deformed germanium by Cockayne et al. (1971) later indicated that such a straightforward interpretation was invalid: For example, the number and positions of bent and terminating fringes could not be simply related to crystal defects. In the particular case of two-beam, lattice-fringe images from spinodally decomposed alloys, it was demonstrated that the experimental fringe spacings could be considerably greater or less than the true interplanar spacings (Cockayne and Gronsky, 1981), primarily because of the rapid variations in interplanar spacings. In general, as first shown by Allpress et al. (1972), lattice-fringe imaging can be applied to the study of crystal structure, but axial illumination is preferable; and image simulations should be considered essential for the interpretation of structural information.

Despite these restrictions on the direct interpretability of images recorded with tilted illumination, the technique has been successfully applied in a number of materials studies, mostly involving small-unit-cell alloy phases (for details, see for example, Sinclair, 1977; Clarke, 1979). Moreover, under some circumstances, as described in section 11.2.5, lattice-fringe images can be related with a high degree of accuracy (1 part in 10,000) to the real lattice parameter (see also Self et al., 1981).

Another fruitful application of the lattice-fringe technique has been to the characterization of the amorphous films that occur between crystalline grains in ceramic materials. As represented schematically in Figure 11.1b, the specimen is maneuvered so that one set of lattice planes in each of the adjoining grains is oriented into a diffracting condition. The width of the thin intergranular phase can then be determined. Axial illumination is preferable for the reasons mentioned above, but it may not be possible in the case of very fine lattice spacings and tilted illumination becomes necessary.

11.2.2 Structure images and Fourier images

The correct electron-optical imaging conditions for direct interpretation of high-resolution electron micrographs are well established and have been explained at length in earlier chapters. For large-unit-cell materials, there are many diffracted beams, well dispersed throughout reciprocal space, which are recombined at the optimum defocus to form the structure image, after convolution by the effects of the objective-lens phase transfer function. A serious complication for images of small-unit-cell materials, like those being considered in this chapter, is the periodic recurrence of Fourier or self-images of the crystal lattice with objective-lens defocus (Cowley and Moodie, 1957; Iijima and O'Keefe, 1979). In contrast to the situation for the large-unit-cell materials, these self-images make it difficult to recognize the correct defocus for intuitive structural interpretation, and one must then note the characteristic appearance of a defect or the Fresnel fringe along a nearby crystal edge in order to choose the focus properly. In some cases, an experimental through-focal series is recorded, and the "best" image is chosen only after the negatives have been developed.

The nature of the problem is illustrated in Figures 11.2 and 11.3. Figure 11.2 shows a simulated through-focal series of images for a thin crystal of tin dioxide, SnO_2, in the [100] projection (Smith, Bursill, et al., 1983). The simulations are for a crystal of 28.4-Å thickness, with 500 keV electrons, a C_s-value of 3.5 mm, and focal steps of +100 Å starting from −1400 Å at top left. Note the complete reversals in phase of the image contrast, which

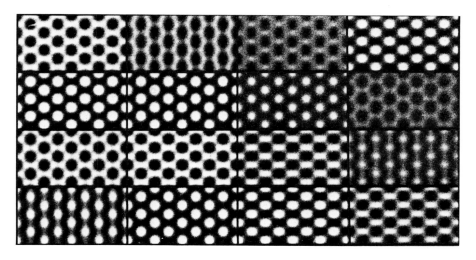

FIGURE 11.2 Through-focal series of simulated images for SnO_2 at 500 kV with $C_s = 3.5$ mm, [100] projection, and crystal thickness of 28.4 Å. Series starts at -1400 Å (top left) with $+100$-Å steps (from left to right), finishing at $+100$ Å (bottom right) (From Smith, Bursill, et al., 1983.)

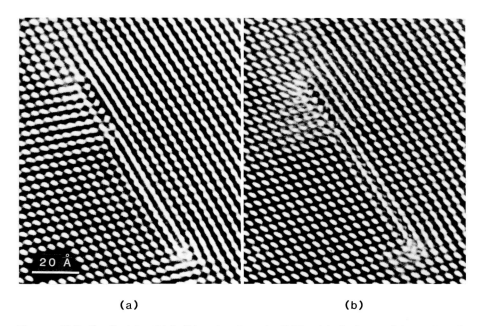

(a)   (b)

FIGURE 11.3 Small, interstitial dislocation loop in CdTe. (a) Optimum defocus (atomic columns are black). (b) Reverse contrast (white atomic columns). (From Smith, Ponce, et al., 1983).

occur with 400-Å increments of focus (this is expected from the half-Fourier period given by d^2/λ, where $d_{020} = 2.37$ Å and λ is the electron wavelength). Figure 11.3 shows an experimental pair of 500-kV micrographs of a short stacking fault in cadmium telluride, CdTe, in the [110] projection (Smith, Ponce, et al., 1983). Since the detailed atomic configurations around the stacking fault were unknown, the imaging conditions could not be determined from the image of the defect alone. The identification of the correct defocus was made possible only by reference to the neighboring crystal edge. Conversely, one can distinguish between a self-image of the lattice and an image with reverse contrast from a knowledge of the point-group symmetry of the stacking fault if the nature of the stacking fault is known (extrinsic or intrinsic) (Olsen and Spence, 1981). This necessity to distinguish between the structure image and other self-images of the lattice, or other images of reverse contrast, always needs to be remembered whenever small-unit-cell materials are studied at high resolution. For structural defects, there is only a small range of focus over which direct interpretation is possible.

11.2.3 Defects and crystal alignment

A prerequisite for successful high-resolution imaging of any crystalline defect is that the crystal lattice itself must be closely aligned with the incident-beam direction. For example, Barry et al. (1983) found that crystal tilts in thick crystals reduced the apparent lattice shift across platelet defects in diamond. On a simple geometrical-projection basis, crystal tilt can be regarded as equivalent to a sideways movement of successive atomic layers. If the greatest allowable phase change through the thickness t of the material is $\pi/4$—which corresponds to $d/8$ sideways shift, where d is the projected atomic-column separation—then the crystal-tilt angle is limited to $\alpha_{max} \le d/8t$. This expression gives 20 mrad for $d = 4$ Å, $t = 25$ Å but only 1 mrad for $d = 2$ Å, $t = 250$ Å. Thus, these simple physical limits on crystal alignment become far more stringent for thick crystals and fine lattice spacings, and rapidly approach the mechanical limits of specimen-tilting holders.

The attainment of sufficient accuracy in *local* crystal alignment is further exacerbated by the fact that many crystals have considerable lattice strain in the vicinity of defects such as dislocations and stacking faults. The traditional alignment methods, which use the selected-area electron diffraction (SAED) patterns from areas of perhaps 5000 Å in diameter, are then not adequate. In instruments with convergent-beam, or microdiffraction, capabilities, one can, in principle, use these techniques to accomplish greater accuracy in the local region of interest. In practice, however, switching back to the imaging mode often results in some deviations of the incident-beam direction. Therefore, the best technique is to carry out preliminary alignment by reference to the SAED or convergent-beam pattern and then to check the appearance of the perfect crystal not far from the defect.

11.2.4 Beam alignment

A serious difficulty in the interpretation of high-resolution images of small-unit-cell materials, which still remains unrecognized by many microscopists, is the presence of imaging artifacts owing to misalignment of the incident beam with respect to the optic axis of the objective lens. Off-axis misalignment results in the introduction of an antisymmetric phase shift in the contrast transfer function of the objective lens, which can be expressed (Smith, Saxton, et al. 1983) by

$$\phi(\mathbf{k}) = -2\pi(k^2 - D)\mathbf{k} \cdot \mathbf{k}_0$$

for spatial frequency $\mathbf{k}$, beam tilt $\mathbf{k}_0$, and underfocus D. These phase shifts, when small (<1 to 2 mrad), are not visible in the optical diffractogram from an image of an amorphous specimen, which therefore cannot be used as a diagnostic guide to their presence. Nevertheless, they result in lateral displacements of finer image detail and remove any centrosymmetry that might otherwise have been expected in the image.

The problem for the microscopist is that the effects of these phase shifts on the images of perfect-crystal regions is not pronounced for materials with small unit cells; they can be easily confused with astigmatism, which is not the case for materials with large unit cells having many diffracted beams well spread out over reciprocal space. These differences are illustrated in Figure 11.4, which shows simulated 500-keV images of SnO_2 in the [111] projection and titanium niobate, $Ti_2Nb_{10}O_{29}$, in the [001] projection, with, from left to right, increasing beam tilts. Tilts of only 0.5 mrad cause reduction of image symmetry in both cases, and yet, even with 2.0-mrad tilt, the cation positions in SnO_2 are still recognizable. The image of the niobate is far more susceptible to beam tilt—a result that is highly significant in our quest for information about atomic distributions around crystal defects, since such faults also give rise to elastic diffuse scattering throughout reciprocal space. The structural information in these beams will be seriously scrambled by the phase shifts owing to beam misalignment, leading to erroneous image interpretation. Therefore, one must resort to alignment techniques more refined than the traditional current- and voltage-center methods, since they do not locate the optic axis of the objective lens with sufficient accuracy under the high-lens excitation conditions normally used for atomic-resolution imaging.

There are a number of possibilities available (Zemlin et al., 1982; Smith, Saxton, et al., 1983). The most straightforward involves successively switching the beam-tilt controls backward and forward in fixed amounts while observing the texture of an amorphous region and adjusting the mean tilt position. With practice, an accuracy approaching 0.1 mrad can be

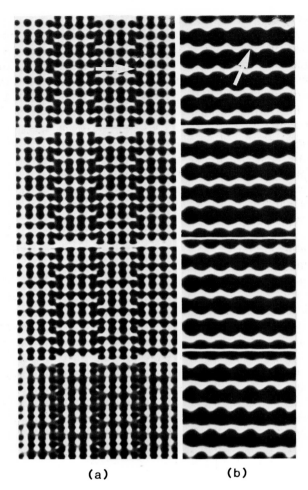

FIGURE 11.4 Simulated 500-kV images showing the effect of beam tilt. (a) $Ti_2Nb_{10}O_{29}$, [001] projection, 26 Å thick, defocus -650 Å. (b) SnO_2, [111] projection, 29.6 Å thick, defocus -550 Å. Common parameters are $C_s = 3.5$ mm, $\alpha = 0.15$ mrad; $\sigma = 200$ Å. Beam tilts, from left to right, are 0, 0.5, 1.0, and 2.0 mrad; the respective tilt axes are shown by arrow. (From Smith et al., 1985).

achieved. Computer control, with a precision beyond that even of highly experienced operators, has also been demonstrated to be feasible (Saxton et al., 1983), but it is not yet being used routinely.

11.2.5 Information finer than the Scherzer limit

Many high-resolution electron microscopes provide information beyond the resolution limit for the structure image, as defined by the first crossover of the phase-contrast transfer function of the objective lens. Since this information is passed through the objective lens with opposite or unspecified phase, its contribution to the image may be interpreted incorrectly. It is thus often more convenient to exclude this higher-resolution detail from the image by inserting an appropriate objective aperture into the back focal plane of the objective lens. In some cases, such as the so-called atom-pair or "dumbbell" images of silicon and other semiconductors (Izui et al., 1978), the image details do not correspond directly to actual specimen features; it has been suggested that these images should be termed interference lattice images (Smith et al., 1982). Calculations (Hutchison et al., 1982) indicate that these silicon images result from interference between {002}- and {111}-type reflections: They are normally only seen in thicker regions, and the image spacings are markedly different from the true lattice spacings (e.g., spot separations of 1.8 Å rather than 1.34 Å). Analysis of weak-phase-object images (Smith and O'Keefe, 1983) suggests that diffracted beams extending out to almost 1 Å will typically be required to contribute to the image in correct phase before atom-pair images will split into two separate spots *at* the atomic sites.

There are relatively few examples where it has proven possible to utilize the extra information available beyond the Scherzer limit. All have necessarily involved lengthy computer simulations to confirm the image interpretation. Bursill and Wood (1978) were able to obtain an acceptable likeness between their experimental and calculated images of Ti_6O_{11} corresponding to an instrumental resolution of 1.6 Å. In imaging thin gold foils, Hashimoto (1978–1979) carefully selected the objective-lens defocus setting so that, despite several oscillations in the transfer function, all the diffracted beams were passed with the same phase through the objective lens. This operating mode was termed the *aberration-free focus* (AFF) setting. However, this particular focus is unique to gold, and it may not be possible to find similar AFF settings for other materials. Moreover, this focus would not be useful for imaging defects, when structural information is contained in beams scattered all over reciprocal space. Finally, it is relevant to note the detailed studies of diamond platelets described in section 11.4.1 (see also Barry et al., 1985). These authors successfully utilized an instrumental resolution well beyond the structure image limit of their particular microscope to discriminate between different models for these platelets, even though the model-sensitive images, recorded at a rather extreme defocus of −2250 Å and a crystal thickness of 300 Å, did not correspond in any simple way to the projected crystal structure.

Meaningful information about *relative* atomic locations can be extracted from small-unit-cell materials at almost an order of magnitude (or more) finer than the Scherzer limit, under very specific conditions. Aided by extensive image simulations and careful comparisons with digitized experimental micrographs, Marks (1984) was able to establish that the positions of atomic columns along a reconstructed 2 × 1 {110} surface of a small gold particle were relaxed by about 0.4 Å relative to the bulk. Methods for the measurement of rigid-body displacements at edge-on boundaries have been analyzed in detail (Stobbs et al., 1984; Wood et al., 1984). An accuracy of about 2 to 3 percent of the lattice-fringe spacing can be achieved, at least in the case of simple $\Sigma = 3$ twin boundaries, provided that the measurements are made sufficiently distant from the boundary that they are not perturbed by the locally variable Fresnel-like contrast, and that the material is uniform and strain-free. When there is rapid, local variation in composition or spacing, one *may* be able by compensating for the imaging conditions by means of a digital-focal-series restoration, to determine the location of atomic columns to within about one-twentieth of the structure-resolution limit (Saxton and Smith, 1985).

11.3 Interfaces and grain boundaries

For materials containing ideal interfaces, one should be able to determine the atomic arrangements in the vicinity of the interface, either from single micrographs or, preferably, from a through-focal series. The interface should have the following properties:

1. The crystals on either side of the interface should both be aligned to an orientation suitable for structure imaging.
2. The interface should present an edge-on configuration to the incident-beam direction.
3. The interface should be flat, with no interfacial steps within the crystal normal to the beam direction.

An interface usually does not have all these properties, although many of the structurally well-defined planar defects described in section 11.4 come close to this ideal. Nevertheless, much structural information can be obtained from high-resolution images of real (nonideal) interfaces, even when the sample is tilted to obtain only systematic reflections on either side of the interface. High-resolution imaging will not be useful when the interface is not edge-on. Finally, when the interface is edge-on but contains interfacial steps, structure analysis is usually possible provided that the density of steps is sufficiently low and that the steps project through the entire crystal.

11.3.1 Semiconductors

The most widespread application of HRTEM has probably been to the characterization of semiconductor devices, defects and interfaces. We restrict ourselves here to a rather selective survey of interfaces viewed in cross section; the interested reader is referred to several recent reviews for more details (e.g., Cherns, 1984; Gibson, 1984; Hutchison, 1984). Note that the various types of dislocations that occur at stacking faults are considered further in section 11.5.

11.3.1.1 Grain boundaries in germanium.
The pioneering studies of semiconductors at atomic resolution were those of Krivanek et al. (1977). By mapping out atomic positions directly from a 500-kV structure image of a high-angle-tilt grain boundary in germanium, these workers were able to construct an elegant model for the boundary. As shown in Figure 11.5, the boundary consists of five- and seven-membered rings of germanium atoms running along the [011] direction without any dangling bonds. Bourret and Desseaux (1979) were subsequently able to identify four different dislocation structures in images of low-angle grain boundaries in germanium.

11.3.1.2 The Si(100)/SiO$_x$ interface.
The single-crystal-silicon/thin-oxide/polycrystalline-silicon configuration is an important component in

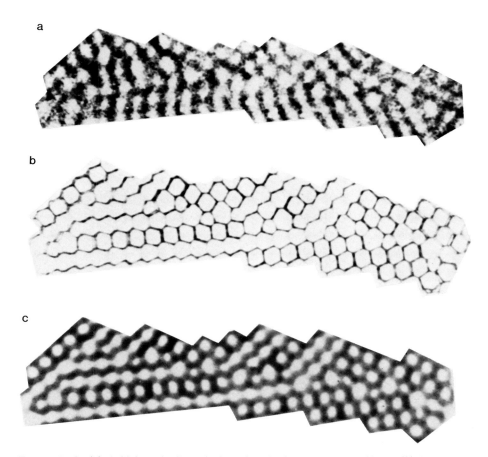

FIGURE 11.5 (a) A high-angle-tilt grain boundary in Ge recorded at 500 kV. (b) Structural model for (a) consisting of alternating five- and seven-membered rings of Ge atoms, with no dangling bonds. (c) Blurred image of (b) showing resemblance to (a). (From Krivanek et al., 1977.)

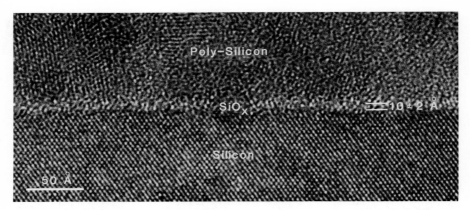

FIGURE 11.6 Cross-section, 200-kV image of an as-grown sample of silicon/oxide/polysilicon. Note the width of the oxide layer. (From Albu-Yaron et al., 1984.)

metal-on-semiconductor (MOS) devices and in high-speed bipolar transistors. The electronic properties of MOS devices are sensitive to the roughness of the Si/SiO_x interface, whereas the electronic properties of the bipolar devices depend on the thickness and contiguity of the thin-oxide layer. This interface was first studied by Krivanek et al. (1978), and many other investigations have since been reported. It has been shown that, for good device material, the interface typically undulates by up to 8 Å in height with a periodicity of around 300 Å. As shown in Figure 11.6, oxide-layer widths (nominal $\simeq 10$ Å) can be measured to within ± 2 Å (Albu-Yaron et al., 1984).

11.3.1.3 Silicon on insulator (SOI). The SOI interface that has received the most attention has been silicon on sapphire (SOS). The active layer in devices based on SOS is located away from the interface so that the interface structure does not affect the device performance directly. However, twins and stacking faults are nucleated at the Si/Al_2O_3 interface, and they propagate through the silicon layer, adversely affecting the device characteristics and effectively placing a lower limit on the useful thickness of the silicon layer. Early studies at high resolution revealed the absence of any amorphous layer at the interface and showed that the defect density dropped off rapidly away from the interface (see Figure 11.7). Recent studies (Paus et al., 1985) have been directed toward discovering the cause of the growth faults in the epitaxial silicon. Twinning occurs within the majority of the initial silicon islands, which are often less than 200 Å across, because the silicon is apparently deposited in two different epitaxial relationships with the support, rather than because of the need to alleviate mismatch strain, as initially proposed (Hutchison et al., 1981).

FIGURE 11.7 High-resolution, 500-kV, lattice-fringe image showing transverse cross-sectional view of as-grown silicon-on-sapphire (SOS). Note the complex twinning and stacking defects in Si leading away from the interface and the absence of any amorphous layer at the interface itself. Electron diffraction pattern (inset) from the region of an SOS interface showing the array of spots corresponding to the (110) zone axis of Si and demonstrating the orientation relationship between Si and sapphire (diffraction spots from the sapphire lie on the prominent arc curving through the Si pattern). (From Smith, Freeman, et al., 1984.)

11.3.1.4 Metal silicide/silicon interfaces. Various metal silicides (e.g., $CoSi_2$, $NiSi_2$, and $PdSi_2$) grow epitaxially on silicon and thus have potential for widespread use as Schottky diodes in device applications. Several high-resolution studies of silicon/silicides have been carried out because it was believed that the barrier height principally depended on the atomic structure at the interface. In the case of $NiSi_2$, there was the possibility of double positioning, (i.e., two twin-related structures, called A- or B-type, were equally likely to develop); but by dint of very careful analysis, Cherns et al. (1982) were able to establish unequivocal models for both alternatives. Subsequently, Tung et al. (1985) used high-resolution imaging to show that predominant A- *or* B-type orientations could, respectively, be grown at specific silicide thicknesses of around 4 and 16 Å. These well-defined layers could then be used as templates to grow silicides of a chosen orientation, and it was highly significant that a difference of 0.2 eV in the Schottky barrier heights for the two orientations has been measured.

11.3.1.5 $GaAs/Ga_{1-x}Al_xAs$ multilayers. Variation in interface structure between GaAs and $Ga_{1-x}Al_xAs$ layers significantly affects the electronic and optical properties of the multiple-quantum-well (MQW) devices based on these materials. An important task, therefore, is to ascertain whether the

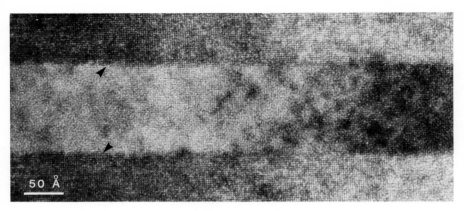

FIGURE 11.8 Cross-section image showing successive interfaces in a sample of GaAs/Ga$_{1-x}$Al$_x$As imaged in the [100] zone at 200 kV. (From Hetherington et al., 1985.)

interfaces are sharp or diffuse and whether or not they are stepped. High-resolution images recorded in the [110] zone, which is normally the preferred orientation for imaging semiconductors, reveal little contrast differences between the layers. This lack of contrast occurs because the {111} beams, which correspond to reflections where structure factors are summed, are relatively insensitive to composition. However, the layers stand out with high contrast (Hetherington et al., 1985) when the material is oriented into a [100] zone, with an objective aperture suitably chosen to include only the {200} beams, which are structure-factor-difference reflections. The high contrast of a GaAs/Ga$_{1-x}$Al$_x$As interface in the latter projection is shown in Figure 11.8. The sharpness of the interface is apparent, and surface steps along the interface are easily seen (arrow).

11.3.1.6 Alkaline-earth fluorides/semiconductors. There is increasing interest in the possible epitaxial growth of mixed and unmixed alkaline-earth fluorides (e.g., CaF$_2$, SrF$_2$, BaF$_2$) on various semiconductors (e.g., silicon, InP, germanium), mainly because these fluorides show promise as insulating layers in heterostructures. HRTEM studies indicate that two types of interface structure occur in these materials, depending on the lattice-mismatch parameter (Gibson and Phillips, 1983). One further complication of studying interfaces in these materials is their sensitivity to electron-beam damage, particularly at the interfaces.

11.3.2 Ceramics

The mechanical properties of ceramics are strongly dependent on grain-boundary structure and composition, particularly when the composite materials originate from fine-grained powders that have been sintered. For example, silicon nitride (Si_3N_4) is a high-temperature engineering ceramic that exhibits a loss of strength at high temperature attributable to the presence of a thin, intergranular, glassy film (Clarke, 1979). These intergranular films can be observed directly by using the imaging conditions drawn in Figure 11.1b. In many other ceramics, well-defined orientation relationships exist across the grain boundaries; some examples are described next.

Low-angle grain boundaries have been observed in hydroxyapatite (HAP) from human tooth enamel (Bres et al., 1984), and it appears that dislocations at the grain boundaries act as sites for preferential attack of the HAP crystals in the early stages of tooth decay. Synthetic apatite crystals, similar in appearance to biological apatites from teeth and bone, also contain grain boundaries (Nelson et al., 1986). Slabs of octacalcium phosphate (a structure closely related to HAP) have been found at grain boundaries in the synthetic material, with thicknesses up to two unit cells in width (see Figure 11.9). It seems likely that similar octacalcium phosphate slabs will occur in biological apatites.

In a dispersion-toughened ceramic, such as ZrO_2 in Al_2O_3, small amounts of ZrO_2 are dispersed in fine-grained Al_2O_3, and the ZrO_2 exists in both intergranular and intragranular incoherent particles. The ceramic toughness arises from the blocking of intergranular cracks by a stress-induced phase transformation of tetragonal ZrO_2 to monoclinic ZrO_2. The ZrO_2/Al_2O_3 interfaces in this material have been successfully imaged, and it was found that the lattice mismatch between ZrO_2 and Al_2O_3 was accommodated by a combination of misfit dislocations and ledges (Heuer et al., 1985).

The development of better preparation conditions for high-temperature ceramics is an important step toward the fullest possible utilization of their valuable properties. With this objective in mind, Hiraga and co-workers

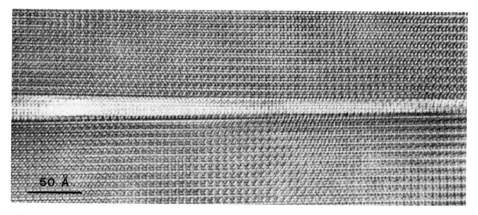

FIGURE 11.9 Low-angle grain boundary in synthetic apatite with interleaving slab of octacalcium phosphate, recorded at 400 kV.

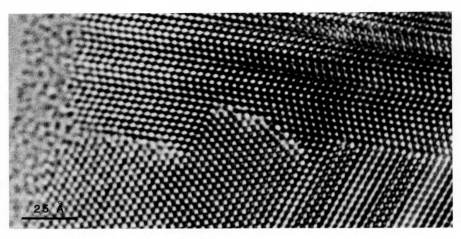

FIGURE 11.10 3C/6H transformation interface in SiC (at 500 kV). Black spots correspond to (unresolved) SiC atom pairs.

have carried out detailed studies of silicon nitride and silicon carbide, utilizing HRTEM (for more details, see Hiraga et al., 1983; Hiraga, 1984). Of particular interest here is the well-defined orientation relationship that was observed between β-Si_3N_4 and TiN precipitates that were dispersed throughout the matrix during chemical-vapor deposition (CVD). Related studies of silicon carbide, grown by CVD, indicate that the grain boundaries have limited disorder and are effectively free of amorphous material (Hiraga, 1984).

The phase transformation between the β- or cubic form of silicon carbide to the various α-polytypes is another important aspect, and HRTEM can be utilized to provide unique information. Figure 11.10 shows an example of a mixed coherent/incoherent transformation interface in a crystal of silicon carbide in the 3C (110) and 6H (11$\bar{2}$0) orientations. Comparison with image simulations (Smith and O'Keefe, 1983) established the imaging conditions and confirmed the direct interpretability of the image features. From structural models, the structure of the disordered central region was eventually deduced (Ness, unpublished).

Structure images of $\langle 001 \rangle$ tilt grain boundaries in nickel oxide have been obtained from bicrystals with several different misorientations (Merkle et al., 1986). Using images such as the one shown in Figure 11.11, it was found that crystallinity continued right up to the grain boundary, with faceting present for both symmetrical and asymmetrical orientations. Experimental micrographs are currently being compared with image simulations based on structural models in order to determine the atomic configuration along the boundaries.

FIGURE 11.11   Structure image of ⟨100⟩ tilt grain boundary in nickel oxide. (From Merkle et al., 1986.)

11.3.3   Metals

Interfacial structure is of fundamental importance in influencing the mechanical properties of metals. However, there have so far been comparatively few high-resolution studies of metal interfaces; they are briefly summarized in the following paragraph. Microscopes have only recently been able to attain structure-image resolution for metals. In addition, many metal boundaries are stepped or curved, so a well-defined atomic projection along the beam direction cannot be located.

One of the earliest studies of grain boundaries in metals was by Penisson and Bourret (1979), who used tilted illumination at 100 kV to characterize details of the structure along [011] low-angle-tilt boundaries in aluminum. Martensite interfaces in TiNi and TiMn have been studied by Knowles (1982a, 1982b) at 200 kV and 2.5-Å structural resolution. This work provided valuable insights into the structure of the interfaces, for example, revealing interfacial steps that corresponded to twinning dislocations. However, it was pointed out that even with an improvement in resolution to better than 1.8 Å, it may still not be possible to deduce details of all the atomic positions along the interfaces; in general, the boundaries are irrational with finite-projection width for all low-index zone directions. The interpretability of 200 kV lattice images of an aluminum/aluminum oxide interface (Timsit et al., 1986) likewise is not necessarily improved by higher-resolution images. However, direct image interpretation of the grain

494 HIGH-RESOLUTION TRANSMISSION ELECTRON MICROSCOPY

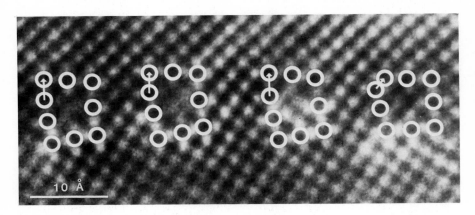

FIGURE 11.12 High-resolution image of a Σ41 grain boundary in a molybdenum bicrystal, with Burgers vectors of $\mathbf{b}_1 = \langle 100 \rangle$ and $\mathbf{b}_2 = 1/2 \langle 111 \rangle$. (From Penisson et al., 1982).

boundary in molybdenum recorded at 200 kV (shown in Figure 11.12) was possible because of the comparatively large spacings between atomic columns and the well-defined atomic configurations along the boundary. Burgers circuits can be drawn around the dislocations, and it is possible to identify two different vectors: $b_1 = \langle 100 \rangle$ and $b_2 = 1/2$ [111] or $b_2 = 1/2$ [11$\bar{1}$] (see Penisson et al., 1982). On the basis of image simulations, Krakow et al. (1985) have provided an interesting discussion of the possibilities for eliciting structural information from both 200- and 400-kV instruments.

Further examples of grain boundaries in metals imaged with atomic resolution and axial illumination have been limited to instruments with voltages of 500 kV or more. Several coherent and incoherent $\langle 110 \rangle$ tilt boundaries in gold have been studied by Ishida et al. (1983), who used image simulations to confirm their structural interpretation. Cook et al. (1983) were able to positively identify stacking defects observed in a Cu–Zn–Al martensitic memory alloy, again on the basis of careful image matching.

11.4 Planar defects

In this section, we briefly consider planar defects with the characteristics of an ideal interface as specified in section 11.3.

11.4.1 Platelets in diamond

The characterization of {100} platelet defects in diamond by Barry et al. (1985) was mentioned earlier with respect to information beyond the resolution limit. Even though there was no one-to-one relationship between the image detail and the crystal structure, these authors showed that their

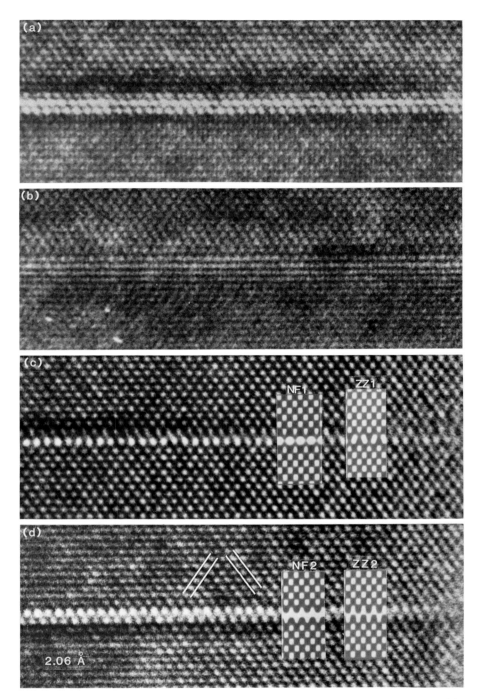

FIGURE 11.13 Experimental images of platelets in diamond, with computed images inset. (a), (b) At 200 kV. (c), (d) At 400 kV. Platelet orientation is [011] in (a) and (c). Platelet orientation is [01$\bar{1}$] in (b) and (d). At 200 kV the zig-zag model (denoted ZZ) provides the best match to the images. At 400 kV the zig-zag model was shown to be inadequate, and a new model (the nitrogen fretwork model denoted NF) was proposed.

images were consistent with only one of the proposed structures. This work was carried out with a 200-kV instrument having a structure-resolution limit of $\simeq 2.5$ Å and an information limit of $\simeq 1.8$ Å. (The largest lattice spacing in diamond is the {111} spacing of 2.06 Å). The platelet images that were recorded close to the optimum defocus were insensitive to defect structure, unlike those taken near -2250-Å underfocus, which proved to be structure-sensitive. The so-called zigzag model clearly produced closer matches to the experimental images than did any of the other models.

These observations of diamond platelets have recently been repeated (Barry, 1988a) by using a 400-keV high-resolution electron microscope, which has a structure resolution of $\simeq 1.7$ Å; so the diamond {111} spacings are now within this limit. Comparison of the platelet images taken at 200 kV (Figures 11.13a and b) with those recorded at 400 kV (Figures 11.13c and d) shows that the latter clearly provide better image detail. Image simulations revealed that the zigzag model did not provide an acceptable match to these 400-kV images, forcing the development of a similar but more refined model, the nitrogen-fretwork (NF) model (Barry, 1988a).

11.4.2 Guinier-Preston zones

Heat treatment of certain Al–Cu, Cu–Be, and many other alloys results in the precipitation of platelike disks, known as Guinier–Preston (GP) zones, on (100) planes of the matrix material. The GP zones have been observed at 100 kV by using both axial and tilted illumination in the AFF mode (Yoshida et al., 1984), and structure images have been recorded at 500 kV under axial-illumination conditions (Dorignac et al., 1980). Although computer simulations of the 500-kV, GP-zone images have been reported (Casanove-Lahana et al., 1982), it appears that the match with the experimental images is still rather qualitative, probably because the extent and composition of the real defects are not as well defined as the proposed structural models.

11.4.3 Twin boundaries

Twinning is the most common crystal-growth defect, and many twinned crystals have been imaged by high-resolution electron microscopy. In structural terms, the $\Sigma = 3$ grain boundary in face-centered cubic metals is the simplest possible twin (since the atomic species on both sides of the interface are identical); but until recently, the available microscope resolution prevented structure images from being obtained [although twin boundaries in gold were imaged under AFF conditions with tilted illumination by Hashimoto et al. (1980)]. Micrographs from gold (Ishida et al., 1983; Stobbs et al., 1984) and copper (Wood et al., 1984) have been

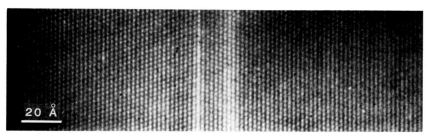

FIGURE 11.14 Σ = 3 (111) twin-boundary image of a gold crystal after image digitization. (Note that the contrast asymmetry of the reflex image has been exaggerated by the digitization process.) (From Stobbs et al., 1984.)

reported that closely match computer simulations, and one is shown in Figure 11.14. From a regressional-analysis technique, it was concluded that there is a rigid-body expansion of 0.09 ± 0.03 Å at this particular twin boundary. However, the appearance alone of a boundary image is not generally an accurate guide to the presence of any relaxation at the interface (Wood et al., 1984).

Twin boundaries are necessarily far more complicated in compound materials than in elemental materials because the number of possibilities for the composition and structure of the twin plane increases markedly. A (011)-type twin interface in tin dioxide was studied in detail by Smith, Bursill, et al. (1983), who eventually showed that this boundary resulted from a simple glide operation. However, an exhaustive analysis of several plausible models, based on computer simulations, was required.

11.4.4 Extended crystallographic-shear defects

Crystallographic shear (CS) planes, described in the previous chapter, are a means whereby many slightly reduced oxides accommodate their loss of stoichiometry. From lower-resolution and electron-diffraction studies, it appears that these extended CS defects consist of well-ordered arrangements of edge- and corner-sharing octahedra (usually denoted C and A steps, respectively). An atomic-resolution study of very slightly reduced rutile (Bursill et al., 1982) has since indicated extensive disorder on the atomic scale both parallel with and normal to the projection axis of the defects. Studies of chromia-doped rutiles likewise indicate considerable variations in the local distributions of C and A steps along the defects, even though close agreement with the fine details of experimental micrographs could be achieved by computer image simulations (Wood et al., 1983), as shown in Figure 11.15. An important conclusion of these studies is that kinetic factors strongly influenced the ordering of the extended defects.

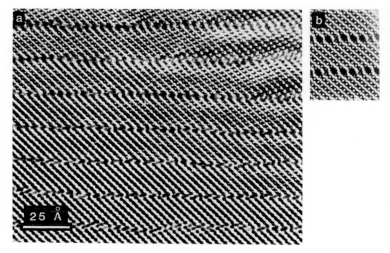

FIGURE 11.15 (a) Experimental micrograph (500 kV, $C_s \approx 2.7$ mm). (b) Computer simulation of ordered (374) extended crystallographic-shear structure ($H = 85.7$ Å, $\Delta f = -600$ Å). The comparison shows a good match of fine detail but confirms disorder of the so-called A and C steps along the boundary (these steps correspond to corner- and edge-sharing octahedra, respectively). (From Wood et al., 1983.)

11.5 Line defects

11.5.1 Dislocations

When planar defects such as twin boundaries, stacking faults, or even CS defects terminate within a material, they do so at some sort of line defect or dislocation. Arrays of dislocations are also common at grain boundaries (see Figure 11.12). In ceramics and semiconductors, crystal structures in the vicinity of dislocations invariably remain rigid, even when the material is thin enough for direct structure imaging. Moreover, atomic ordering along the length of the defects is also apparently maintained. These properties have been significant in the characterization of various line defects and, in particular, of dislocation core structures in semiconductors.

Several groups have studied the 30° and 90° partial dislocations in germanium and silicon (Anstis et al., 1981; Bourret et al., 1981; Olsen and Spence, 1981); and with the benefit of extensive image simulations and prior information about possible defect structures, they were able to distinguish between the glide and shuffle type of dislocation (all workers chose the former). However, it has since been shown that the image of the dislocation core is usually insensitive to the presence or absence of atomic columns at the core unless the specimen thickness and defocus values have been chosen carefully (Bourret et al., 1983). It has been possible to locate all the atomic-column positions at a Lomer dislocation in germanium (Bourret et al., 1982), again after several reiterations of matching simulated images with experimental micrographs, alternated with slight structural refinements.

Observations of various lattice defects in III–V and II–VI compound semiconductors have been reported by several groups (e.g., Echigoya et al., 1982; Sinclair et al., 1983; Suzuki et al., 1983), and Tanaka and Jouffrey (1984) have obtained images of partials from dissociated 60° dislocations in GaAs. The terminating partial dislocation can be identified as being of either Frank- or Shockley-type, and atomistic models for these partials can be drawn by determining the extrinsic or intrinsic nature of the stacking faults commonly observed in these materials. Detailed image simulations that confirm whether or not these models are acceptable (or unique) have not yet been published. Furthermore, while it should be possible to distinguish between glide and shuffle sets in these compounds, it is not clear whether dislocations that terminate on atom A (e.g., gallium) will be distinguishable from those terminating on atom B (e.g., arsenic). Detailed analysis and observations seem warranted, since it is virtually certain that the terminating species will influence electrical properties in these materials.

Compared to the interest shown in semiconductors, terminating faults in ceramic materials have been virtually neglected. Atomic resolution images showing dislocations in SiC have been published by Hiraga (1984). An analysis of defect terminations in reduced tungsten trioxide indicate that they are of intrinsic type, which implies an anion-vacancy nature for the small defects initially responsible for the WO_{3-x} phase (Bursill and Smith, 1984). Observations of extended CS defects in reduced rutile have established the extrinsic nature of the terminating dislocation, thereby providing indirect evidence for an interstitial cation defect (Smith, Bursill, et al., 1984).

11.5.2 Rodlike defects

A common line defect in silicon and germanium, which comes from electron irradiation or ion implantation, is the so-called {311} rodlike defect. The example shown in Figure 11.16 was formed by 1-MeV electron irradiation of a thin silicon foil held at a temperature of 500°C. Various models for these defects have been proposed, based primarily on the precipitation of self-interstitial defects (Salisbury and Loretto, 1979; Tan et al., 1981), and some attempts at image matching have been made (e.g., Hirabayashi et al., 1982). However, a detailed analysis to establish the uniqueness of any particular structural model has not been reported. Some of the problems with solving the structure of these defects are that they are not usually located in the thinnest regions of the foil (the surfaces act as sinks for the small mobile defects) and they do not project through the foil. The likelihood of obtaining a satisfactory image, under well-defined operating conditions, for the purposes of structure refinements should be substantially improved by chemical thinning after irradiation.

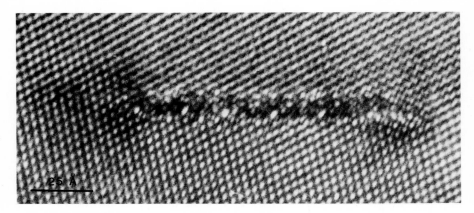

FIGURE 11.16 Micrograph of silicon in the [110] projection at 200 kV, showing a {311} rodlike defect resulting from electron-irradiation damage.

11.6 Point defects

The structure and properties of substitutional or intrinsic small defects, rather than those caused by different types of incident irradiation, continue to be the subjects of many theoretical and experimental investigations. In this section, we briefly review the small number of high-resolution, electron-microscopy studies, postponing discussion of radiation-induced effects to section 11.9. Note that lattice and structure imaging do not seem to be the optimum methods for imaging small defects and differentiating between possible models; scattering from the defects is comparatively weak, and any signal is liable to be swamped by the lattice-fringe image. It is preferable to use different imaging modes, as described in section 11.10, in order to achieve adequate image contrast.

11.6.1 Semiconductors

The possibilities for imaging the split $\langle 100 \rangle$ self-interstitial in silicon have been considered in an image-simulation study reported by Glaisher and Spargo (1984). These workers found that the highest contrast levels from the defect occurred at thicknesses close to the first-extinction distance for {111}-diffracted beams, with an objective aperture set to exclude all diffracted beams so that the image would effectively be formed by the combination of the transmitted (000) beam and the defect-diffuse scattering within the aperture. This result is significant since the usual specification for high-resolution imaging is that the specimen be thin (typically <100 Å).

Experimental dark-field lattice images of silicon samples implanted with gold (and others that had been irradiated by 1-MeV electrons) have been reported by Zakharov et al. (1982). The published micrographs show random spots that are either abnormally bright or abnormally dark, which was attributed to the presence of the gold atoms or lattice vacancies, respectively. Since similar effects were not seen under identical imaging conditions for silicon samples prepared in the same manner but that had not been irradiated, these latter observations thus suggested that surface roughness or contamination were not the cause of the high-contrast spots.

While these results seem promising, it still remains to be seen how closely the spot densities of experimental micrographs can be related to the numbers of atoms or vacancies present. More important, it remains unclear whether the defect images can be uniquely related to a particular defect structure.

11.6.2 Ceramics and metals

The lack of uniqueness in experimental images from ceramics and metals owing to practical factors unrelated to the microscopy might also prove to limit any structure determination of small defects in these materials. For example, as shown in Figure 11.17, extensive calculations for several different types of small defects in rutile (Bursill and Shen, 1984) indicate that isolated single defects could give sufficient image contrast to be visible in bright-field, phase-contrast images (meaning without diffracted beams), for specimen thicknesses close to that of the first-thickness-extinction contour. However, intensive efforts to image small defects in slightly reduced rutile (Bursill et al., 1985) revealed that even very thin ($\approx$5 to

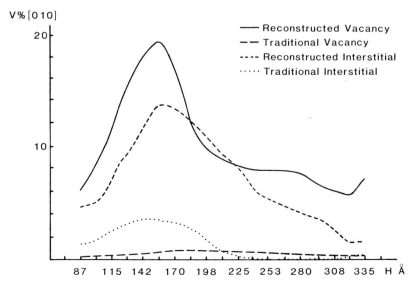

FIGURE 11.17 Plot of visibility (V) versus thickness (H) for reconstructed and traditional vacancy and interstitial models for [010] projection of rutile. (From Bursill et al., 1985.)

10 Å) surface-contamination layers made small defects difficult to locate with certainty. The mobility of the defects also appeared to lead to their aggregation. Extensive image simulations for gold crystals have been reported (Fields and Cowley, 1978), and the self-interstitial defect should give enough contrast to be visible under specific conditions, even in a structure image. However, defect mobility is again highly likely, particularly with the electron-beam current densities required for high-resolution imaging: It might be possible to immobilize the defects by cooling the sample down to liquid-helium temperature.

11.7 Small particles and surface-profile imaging

The most important interfaces in many materials are their external surfaces, since they govern so many of the physical and chemical processes relating to the material. The different methods for studying surfaces within the electron microscope are described in Chapter 13. We restrict ourselves here to a brief survey of surface-profile imaging at atomic resolution, with particular reference to small particles and the surfaces of metals and oxide ceramics.

The initial observations of small gold and silver particles by Marks and Smith (1983) confirmed that it was indeed possible to obtain atomic-resolution images that showed the arrangement of individual column projections along the particle edges, in effect providing an image of the surface profile. However, the shapes of small particles are obscured by the presence of a support film even when the support is exceptionally thin (Gai et al., 1985). Preferably, the particle profile should be clear of the substrate and free of surface contamination. Studies of small metal particles and extended gold surfaces, summarized in Smith and Marks (1985), demonstrate that the technique could provide a wealth of information about surface topography, such as details of steps, ledges, and rearrangements. Similar information about the surfaces of a range of oxides has recently been observed (Smith, Bursill, et al., 1986). Figures 11.18a and b show the changes observed in the surface-profile image from a uranium oxide crystal. Exhaustive image calculations (Marks, 1984) have provided the justification essential for the direct interpretation of various features of such profile images. Although the technique is a recent innovation, it has already been used to study a considerable variety of materials (see Smith, 1986, Table V).

Observations of surface-profile images over extended time periods have made it possible to follow modifications induced by the electron beam. Briscoe and Hutchison (1984) have reported the development of a jagged zigzag structure on the surface of a spinel catalyst, and Kang et al. (1986) have followed the differences in behavior initiated on various surfaces of terbium oxide crystals. Surface modification following *in situ* use in a catalytic reaction chamber has been monitored by Hutchison and Briscoe (1985), who observed small rafts, believed to be ZnO, which were not present in the as-received material.

MATERIALS SCIENCE 503

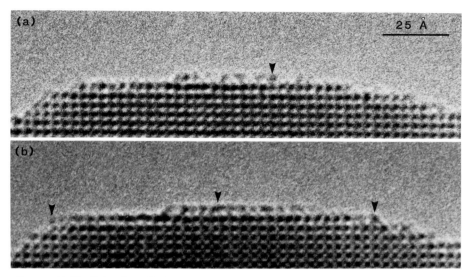

FIGURE 11.18 Pair of profile images taken from a 350 kV through-focal series showing the changes in surface morphology (arrowed) of a uranium oxide crystal.

One striking aspect of surface-profile imaging is that it becomes possible to study dynamic-surface processes in real time by attaching a suitable image-viewing/recording system to the HRTEM (Smith, 1985). Observations have been restricted almost entirely to events initiated by the influence of the electron beam, with the sample held at room temperature. However, there seems to be no fundamental reason why dynamic studies cannot be carried out while the sample is undergoing special treatment such as *in situ* evaporation or annealing in a suitable specimen holder.

The most fascinating dynamic phenomenon observed has undoubtedly been the structural rearrangements of small metal particles, primarily of gold (Bovin et al. 1985; Iijima and Ichihashi, 1986) but also of platinum and rhodium. As shown by the single TV-frame images in Figure 11.19, small metal crystals, which are typically 20 to 50 Å in diameter and clinging to the edges of amorphous support films, are observed, under high-beam-current-density conditions, to be in a state of continual motion. They rotate on the substrate and fluctuate from one crystal structure to another. Several factors strongly influence the particle motion (Smith, Petford-Long, et al., 1986), including the substrate material, the beam current, the particle contact with the substrate, and the metal element itself. Full explanations for the fluctuations are lacking, although it seems reasonable to propose that similar events may be initiated spontaneously if the particular material is heated to sufficiently high temperatures.

A final caveat about the results so far obtained by using profile imaging is that the vacuum in the vicinity of the sample has typically been around 10^{-6}

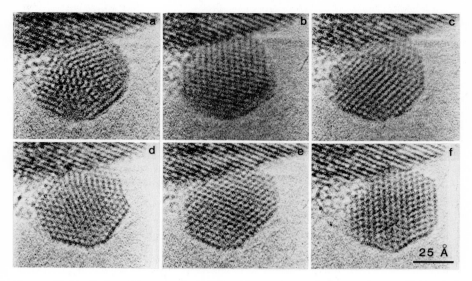

FIGURE 11.19 Series of images of a small gold crystal recorded over about a 20-s period, showing the rapid structural rearrangements that occur as a result of electron irradiation. (From Smith, Petford-Long et al., 1986.)

to 10^{-7} torr, which is inadequate for maintaining clean and well-defined surfaces of most metals and semiconductors. Gibson et al. (1985) have recently reported profile images of reconstructed silicon surfaces under ultrahigh-vacuum (UHV) conditions with the sample heated up to temperatures of 600°C. This study represents the first observations of a clean elemental silicon surface in profile, and it augurs well for the future application of the technique to other materials when a UHV environment will be essential if the intrinsic surface structure is to be determined.

11.8 Metallic-alloy systems

Binary metallic alloys, of the type $A_x B_{1-x}$, form a rich panoply of superlattice structures, many of which have been studied by using HRTEM (for details and references, see Hirabayashi, 1980; Watanabe and Terasaki, 1984; Amelinckx et al., 1985). Since the typical interatomic separations in projection along any major zone axis are 2 Å or less, most studies have not involved direct atomic resolution. In fact, for investigating structural problems such as antiphase domains and long-range order, one need not resolve the fundamental lattice structure, and such levels of resolution are not generally required. The alloys are usually well ordered along the projection axis, and as confirmed by image simulations (e.g., Amelinckx et al., 1985), it is possible to derive positional information about the minority atoms by reference to either bright- or dark-field images that highlight the

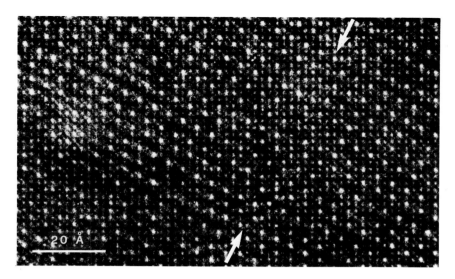

FIGURE 11.20 High-resolution image showing a region of the binary alloy $Au_{74}Mn_{21}$. Note the planar fault (arrow), which is best seen by looking at a small angle to the page. (From Terasaki et al., 1986.)

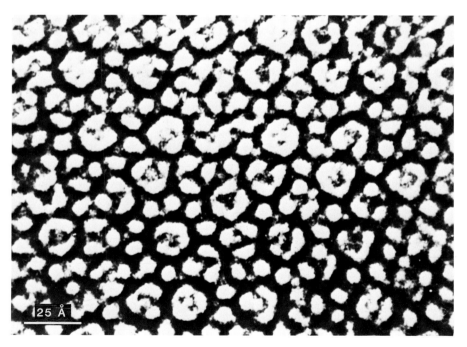

FIGURE 11.21 High-resolution image of Al_6Mn quasicrystal at 200 kV viewed along a fivefold symmetry axis. Note the absence of translational symmetry and the localized centers of pentagonal symmetry. (From Bursill and Peng Ju Lin, 1985.)

positions of these minority atomic columns under appropriate conditions (i.e., specimen thickness, defocus, etc.). Analysis has also shown (Terasaki et al., 1986) that different atomic columns can be differentiated, even in thin-specimen regions, once atomic resolution is achieved (see Figure 11.20).

Other intermetallic compounds develop complex crystal structures based on tetrahedral-close-packed phases, which are given designations such as μ-, σ-, Laves-, and M-phases (e.g., Stenberg and Andersson, 1979). Even at the level of 2.5-Å resolution, HRTEM is proving to be invaluable for characterizing these materials and elucidating the nature of commonly occurring structural defects (e.g., Ye et al., 1985).

Finally, fascinating fivefold symmetries have recently been observed (e.g., Bursill and Peng, 1985) in quenched alloys of the type Al–M (where M is a transition metal such as manganese or iron); an example is shown in Figure 11.21. The crystallography of these quasi crystals is currently intriguing scientists around the world because crystal structures based on five-fold symmetry do not completely fill space. Many groups are working feverishly to derive satisfactory structural models based on high-resolution images, since HRTEM is the only technique capable of providing localized information about the structure.

11.9 Radiation damage

A fundamental condition for high-resolution imaging is that high electron-current density at the sample (typically $\simeq 1$ to $10 \, \text{A}/\text{cm}^2$) is unavoidable if acceptable image statistics are to be achieved in reasonable times. Depending on the susceptibility of the sample to electron-induced radiation effects, such doses can have serious consequences for maintaining the structural integrity of the sample during observation. In this section, we review briefly the types of damage mechanisms and then consider the different ways in which they eventually become apparent in metals, semiconductors, and ceramics. Recognition of such damage by the microscopist is important so that erroneous image interpretation based on damage-induced image detail may be avoided.

11.9.1 Types of processes

The two distinct mechanisms for radiation damage are knock-on displacement, when the incident electron has sufficient energy to eject atoms from their normal lattice sites by momentum transfer, and radiolysis, when there are interactions between incident electrons and the atomic electrons. Knock-on displacement affects all materials above particular characteristic threshold energies, whereas radiolysis does not occur in metals because electronic excitations are effectively delocalized by the conduction band.

The displacement threshold for knock-on damage has been measured for

TABLE 11.1
Typical Threshold Energies for Knock-on Atomic Displacement

Material	Threshold (keV)	Material	Threshold (keV)
C (graphite)	130–140/220	GaAs–Ga	250
C (diamond)	330	–As	250
Al	140–160	CdS–Cd	290
Si	120–190	–S	120
Ge	370	Al_2O_3–Al	180
Fe	370	–O	370
Cu	400		
W	1300		

many materials (e.g., Urban, 1979; Hobbs, 1984), and some typical beam energies are listed in Table 11.1. The threshold increases with the atomic number of the material and depends on the specimen temperature and its orientation with respect to the incident-beam direction. For example, the threshold is 130 keV for graphite oriented with the beam parallel to the c-axis; it increases to about 220 keV for the beam perpendicular to c (Iwata and Nihara, 1966). In most compound semiconductors and oxides, a difference exists in the threshold energy for the two constituents—that is, selective displacement of one species can be anticipated when the incident-beam energy is somewhere between the two thresholds. Moreover, substantial beam-damage effects can be expected to occur for many compound materials during an extended observation period in any of the latest generation of medium-voltage (300 to 400 kV), high-resolution instruments. Finally, the rate of knock-on damage is greater at surfaces and defects, since the displacement energy is generally lower at these locations because of reduced interatomic bonding.

There are essentially four mechanisms for the radiolytic process (Hobbs, 1984):

1. Ionization of core electrons (displacement energy $T_D > 100$ eV).
2. Ionization of valence electrons ($T_D \simeq 10$ to 20 eV).
3. Promotion of valence electrons into bound electron-hole states ($T_D \simeq 5$ to 10 eV).
4. Plasmons ($T_D \simeq 25$ eV), although plasmons are delocalized and thus cannot cause atomic displacements.

The cross sections for these processes decrease with increasing beam energy. In materials where there are several different chemical species present, one element is often preferentially displaced from its normal lattice site by radiolysis. Recombination usually takes place since the ejected (interstitial) atoms are mobile within the lattice, but disproportionation of material at the surface commonly results from ejection (desorption) into the microscope vacuum. Some examples are described below.

11.9.2 Metals

Although metals are free of radiolytic processes, radiation damage due to knock-on displacement presents serious difficulties during high-resolution observations of low-atomic-number metals, since it is necessary to operate at 300 keV or more in order to attain atomic resolution. Once the threshold voltage for atomic displacement is exceeded, the likelihood of knock-on occurring is large, despite cross sections of $\simeq 20$ to 50 barns (b) ($1\,\text{b} \equiv 10^{-24}\,\text{cm}^2$), because of the high-primary-beam current (e.g., $10\,\text{A/cm}^2 \equiv 6 \times 10^{19}$ electrons/cm^2/s). Interstitials and vacancies are created in substantial numbers, and owing to their mobility at room temperature, aggregation at lattice defects or grain boundaries as well as separate precipitation as dislocation loops or stacking fault tetrahedra will occur (Kiritani, 1977). Whenever possible, the microscopist should endeavor to operate below the threshold voltage. Alternatively, every effort should be made to use the lowest possible beam current and electron optical magnification compatible with image viewing and recording with adequate signal/noise statistics.

11.9.3 Semiconductors

Semiconductors are covalent solids that are susceptible to radiolytic damage processes as well as to knock-on displacement. A breaking of bonds

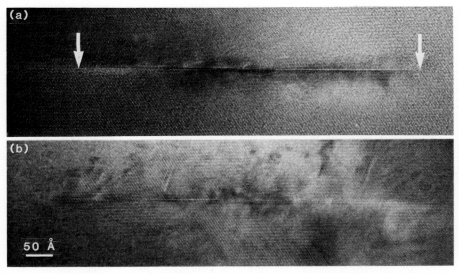

FIGURE 11.22 Radiation damage at a 60° dislocation in silicon. The dislocation is dissociated into partials (arrow) at nonequilibrium separation, with a stacking fault in between. (*a*) At the beginning of the observation. (*b*) After 5 min in the focused electron beam; electron dose $\simeq 5 \times 10^6$ electrons/Å^2.

between the atoms can lead to atomic rearrangements, resulting in a phase transition from crystalline to amorphous. For subthreshold imaging, the development of beam damage is not normally visible, presumably because of the mobility of the small defects and their recombination. However, damage occurs preferentially in the vicinity of lattice defects and at surfaces. Ehrlich and Smith (1986), for example, have proposed that the desorption of sulfur from the near-surface layers of CdS crystals has resulted in the observed accumulation of cadmium on the surface. The development of radiation damage in a thin silicon foil is more pronounced near the dissociated dislocation with nonequilibrium separation of the partials (Figure 11.22). In this case, the point defects had been produced by knock-on damage, with their subsequent diffusion to the dislocation being enhanced by radiolytic processes such as nonradiative electron-hole-pair recombination.

11.9.4 Ceramics

The types of damage that occur in ceramic materials depend on whether the ceramic has a covalent or ionic nature: If the ceramic is covalent, radiolytic processes are the dominant damage mechanism; if the ceramic is ionic, the

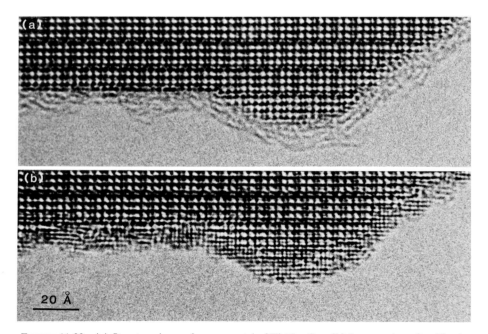

FIGURE 11.23 (a) Structure image from a crystal of $Ti_2Nb_{10}O_{29}$. (b) Same region after 30-min exposure to a 400-kV electron beam with current density of $\simeq 40$ A/cm^2. Note the development of metallization as indicated by the finely spaced fringes along the edge of the crystal (arrows), owing to electron-stimulated desorption of oxygen.

majority of the displaced atoms will eventually be relocated at their proper lattice sites owing to diffusion. In ionic materials, the rate of clustering—and therefore, the apparent extent of beam damage—can be reduced significantly by lowering the sample temperature, since this reduces defect diffusion and forestalls the precipitation of displaced atoms at dislocation loops and voids. In insulating solids, damage processes such as the transition from crystalline to amorphous have been observed to originate from a dislocation (Cherns et al., 1980). Close to the surface, preferential desorption can also be anticipated. Figure 11.23 shows 400-kV images of a $Ti_2Nb_{10}O_{29}$ crystal recorded before and after 30 min irradiation in the focused electron beam (current density $\simeq 40$ A/cm^2). Metallization of the near-surface layers due to the desorption of oxygen is apparent (Smith and Bursill, 1985).

11.9.5 Effects of ion implantation and annealing

Ion implantation is used nowadays in a variety of ways to modify a material or property in a particular manner—for example, to introduce a dopant into a semiconductor or to harden the surface of ceramics or metals. Implantation invariably results in damage to, or disruption of, the host material, and several groups have observed the characteristics of implantation and annealing by using HRTEM (e.g., Narayan and Holland, 1984; Carpenter et al., 1986). An important side issue concerning ion implantation is the recent study of ion-milling damage by Cullis et al. (1985), who showed that the well-known precipitation of dislocation loops in II–VI compound semiconductors caused by the usual milling process using Ar^+ could be overcome by milling with reactive I^+.

11.10 Complementary techniques

In several of the applications described earlier, we emphasized that conventional, large-aperture, bright-field structure imaging is not always the most effective method for obtaining useful structural information. In this section, we briefly outline two other imaging modes that are complementary to structure imaging, and we discuss some applications of these techniques. The size of the objective aperture is chosen in such a way that only the central beam and the elastic diffuse scattering around that beam are used to form the image, giving the well-known amplitude- or strain-contrast configuration that has been widely used for many years to characterize the structure of defects (e.g., Hirsch et al., 1977). Even where the strain contrast is effectively negligible—for example, with some end-on dislocations or gas bubbles—considerable image contrast can still be obtained.

11.10.1 Bright-field phase contrast

The phase-contrast technique is applicable to defects and precipitates that are not visible in an image at zero defocus but show good contrast in

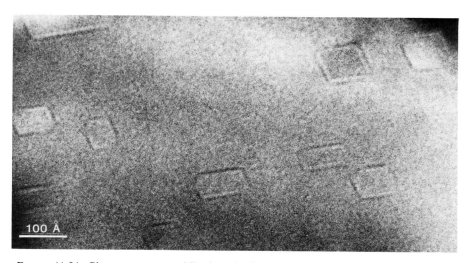

FIGURE 11.24   Phase contrast at voidite faces in diamond (400 kV). Note that the bright fringe is just inside the voidite face and the dark fringe is just outside, indicating a lower value for the mean inner potential for the voidite relative to that of diamond.

out-of-focus images. This contrast occurs because of the differences between the mean inner potential of the defect and that of the bulk crystal, as described by equation (2.37). An application of this effect is demonstrated in Figure 11.24, which shows an underfocused image of voidites in diamond. A detailed analysis (Barry, 1988b) established that the mean inner potential of these voidites is considerably less than that of diamond, which is consistent with recent results showing that these defects contain nitrogen gas under pressure (Hirsch et al., 1986). It is significant that a similar analysis applied to platelets in diamond led to incorrect estimates of the platelet width. It turns out that the height of the central fringe is far more sensitive to the defect structure than the central fringe width. Finally, note that, for quantitative analysis, it is necessary to compute the phase contrast directly by using multislice methods (Bursill et al., 1978).

11.10.2   Thickness contrast

Thickness contrast is applicable to defects and precipitates that are visible in zero-defocus images, and it occurs because the dynamical scattering from the defect differs from that of the host matrix. This contrast is optimized by using a small objective aperture around either the central (bright-field) beam or around a diffracted (dark-field) beam. The contrast is maximized, alternating either bright and dark, when the foil thickness minus one-half of the defect thickness is a fractional multiple of the extinction distance [i.e., $H - D_0/2 = (n + \frac{1}{2})\xi_g/2$] (Barry, 1988b). This expression can be used to

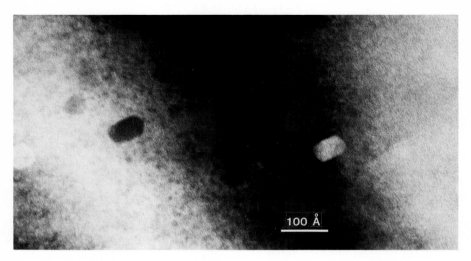

FIGURE 11.25 Thickness contrast from voidites at zero defocus (200 kV). The voidites are bright on the thin-crystal side of the dark thickness fringe and are dark on the thick-crystal side.

FIGURE 11.26 Weak-beam, dark-field image of voidites on a partially collapsed platelet in diamond. Here, $g = (1\bar{1}1)$ at 100 kV. Note that some of the larger voidites (arrow) have an annular appearance because the effective extinction distance is less than their diameter.

explain the variation in contrast of the voidites in diamond shown in Figure 11.25. In weak-beam, dark-field imaging, when the host crystal is tilted away from the exact Bragg condition, the effective extinction distance becomes small and the contrast for the smaller voidites is enhanced (Figure 11.26). This same method of emphasizing thickness contrast could be used to image very small ($\simeq 5$ Å) precipitates or even point defects, but surface roughness limits the detectability.

REFERENCES

Albu-Yaron, A., Barry, J. C., and Booker, G. R. (1984). A TEM study of the annealing of a polysilicon/silicon specimen with a thin oxide layer at the interface. In *Electron microscopy—1984*, eds. A. Csanady, P. Rohlich, and D. Szabo, 521.

Allpress, J. G., Hewat, E. A., Moodie, A. F., and Sanders, J. V. (1972). n-beam lattice images. I. Experimental and computed images from $W_4Nb_{26}O_{77}$. *Acta Crystallogr., Sect. A* **28**, 528.

Amelinckx, S., Van Tendeloo, G., and Van Landuyt, J. (1985). High-resolution studies of order and disorder in alloys. *Ultramicroscopy* **18**, 395.

Anstis, G. R., Hirsch, P. B., Humphreys, C. J., Hutchison, J. L., and Ourmazd, A. (1981). Lattice images of the cores of 30° partials in silicon. In *Microscopy of semiconducting materials—1981*, ed. A. G. Cullis, 15. The Institute of Physics, Bristol.

Barry, J. C. (1988a). Direct structure images of diamond platelets. *Philos. Mag.* (forthcoming).

——— (1988b). An analysis of phase and thickness contrast in high-resolution electron microscope images of platelets and voidites in diamond. *Philos. Mag.* (forthcoming).

———, Bursill, L. A., and Hutchison, J. L. (1983). Measurement of lattice displacement of {100} platelets in diamond. *Philos. Mag., Ser. A* **48**, 109.

———, Bursill, L. A., and Hutchison, J. L. (1985). On the structure of {100} platelet defects in type 1a diamond. *Philos. Mag., Ser. A* **51**, 15.

Bourret, A., and Desseaux, J. (1979). The low-angle [011] tilt boundary in germanium. I. High-resolution structure determination. *Philos. Mag., Ser. A* **39**, 405.

———, Desseaux, J., and D'Anterroches, C. (1981). Defect structure in CZ silicon and germanium studied by high resolution electron microscopy. In *Microscopy of semiconducting materials—1981*, ed. A. G. Cullis, 9. The Institute of Physics, Bristol.

———, Desseaux, J., and Renault, A. (1982). Core structure of the Lomer dislocation in germanium and silicon. *Philos. Mag., Ser. A* **45**, 1.

———, Thibault-Desseaux, J., D'Anterroches, C., Penisson, J. M., and DeCrecy, A. (1983). Are the core structures of dislocations and grain boundaries resolvable by HREM? *J. Microsc.* **129**, 337.

Bovin, J.-O., Wallenberg, L. R., and Smith, D. J. (1985). Structural rearrangements in small metal particles at atomic resolution. In *Electron microscopy and analysis—1985*, ed. G. J. Tatlock, 481. Adam Hilger, Bristol.

Bres, E. F., Barry, J. C., and Hutchison, J. L. (1984). A structural basis for the carious dissolution of apatite crystals of human tooth enamel. *Ultramicroscopy* **12**, 367.

Briscoe, N. A., and Hutchison, J. l. (1984). Surface morphology and voids in $ZnCrFeO_4$ spinel catalyst. In *Electron microscopy and analysis—1981*, ed. M. J. Goringe, 249. The Institute of Physics, Bristol.

Bursill, L. A., and Peng Ju Lin (1985). Penrose tiling observed in a quasi-crystal. *Nature* **316**, 50.

———, and Shen Guang Jun (1984). Visibility criteria for the imaging of small defects in crystals. *Optik* **66**, 251.

———, and Smith, D. J. (1984). Small and extended defect interactions in nonstoichiometric oxides. *Nature* **309**, 319.

———, and Wood, G. J. (1978). Electron-optical imaging of Ti_6O_{11} at 1.6-Å point-to-point resolution. *Philos. Mag. Ser. A* **38,** 673.

———, Barry, J. C., and Hudson, P. R. W. (1978). Fresnel diffraction at {100} platelets in diamond: An attempt at defect structure analysis by high-resolution (3-Å) phase-contrast microscopy. *Philos. Mag., Ser. A* **37,** 789.

———, Blanchin, M. G., and Smith, D. J. (1982). The nature and extent of disorder within rapidly-cooled $TiO_{1.9985}$. *Proc. R. Soc. London, Ser. A* **384,** 135.

———, Smith, D. J., and Peng Ju Lin (1985). Small-defect clusters in chromia-doped rutile. *J. Solid State Chem.* **56,** 203.

Carpenter, R. W., Vanderschaeve, G., Varker, C. J., and Wilson, S. R. (1986). Nucleation and growth of octahedral oxide precipitates in silicon: oxygen ion implantation. *Mater. Res. Soc. Symp. Proc.* **59,** 309.

Casanove-Lahana, M. J., Dorignac, D., and Jouffrey, B. (1982). Computed high resolution Guinier-Preston zone images in an Al–Cu alloy. In *Electron microscopy and analysis—1981,* ed. M. J. Goringe, 377. The Institute of Physics, Bristol.

Cherns, D. (1984). Finding the right problems: some HREM applications to interfaces. In *Proceedings of the 42nd annual meeting of the Electron Microscope Society of America,* ed. G. W. Bailey, 376. San Francisco Press, San Francisco.

———, Anstis, G. R., Hutchison, J. L., and Spence, J. C. H. (1982). Atomic structure of the $NiSi_2$/(111) Si interface. *Philos. Mag., Ser. A* **46,** 849.

———, Hutchison, J. L., Jenkins, M. L., and Hirsch, P. B. (1980). Electron irradiation induced vitrification at dislocations in quartz. *Nature* **287,** 314.

Clarke, D. R. (1979). On the detection of thin intergranular films by electron microscopy. *Ultramicroscopy* **4,** 33.

Cockayne, D. J. H., and Gronsky, R. (1981). Lattice fringe imaging of modulated structures. *Philos. Mag., Ser. A* **44,** 159.

———, Parsons, J. R., and Hoelke, C. W. (1971). A study of the relationship between lattice fringes and lattice planes in electron microscope images of crystals containing defects. *Philos. Mag.* **24,** 139.

Cook, J. M., O'Keefe, M. A., Smith, D. J., and Stobbs, W. M. (1983). The high-resolution electron microscopy of stacking defects in Cu–Zn–Al shape memory alloys. *J. Microsc.* **129,** 295.

Cowley, J. M., and Moodie, A. F. (1957). Fourier images. I. The point source. *Proc. Phys. Soc., London, Sect. B* **70,** 486.

Cullis, A. J., Chew, N. G., and Hutchison, J. L. (1985). Formation and elimination of surface ion milling defects in cadmium telluride, zinc sulphide and zinc selinide. *Ultramicroscopy* **17,** 203.

Dorignac, D., Casanove, M. J., Jagut, R., and Jouffrey, B. (1980). High-resolution Guinier-Preston zone images. In *Electron microscopy—1980,* eds. P. Brederoo and J. Van Landuyt, **4,** 174. Seventh European Congress on Electron Microscopy Foundation, Leiden.

Echigoya, J., Pirouz, P., and Edington, J. W. (1982). Preliminary studies of crystal defects in cadmium sulphide by high-resolution transmission electron microscopy. *Philos. Mag. Ser. A* **45,** 455.

Ehrlich, D. J., and Smith, D. J. (1986). Electron-beam-stimulated nonthermal crystallization of CdS surface layers: observations by real-time atomic-resolution electron microscopy. *Appl. Phys. Lett.* **48,** 1751.

Fields, P. M., and Cowley, J. M. (1978). Computed electron microscope images of atomic defects in f.c.c. metals. *Acta. Crystallogr. Sect. A* **34,** 103.

Gai, P. L., Goringe, M. J., and Barry, J. C. (1986). HREM image contrast from supported small metal particles. *J. Microsc.* **142,** 9.

Gibson, J. M. (1984). High resolution electron microscopy of interfaces between epitaxial thin films and semiconductors. *Ultramicroscopy* **14,** 1.

———, and Phillips, J. M. (1983). Analysis of epitaxial fluorite-semiconductor interfaces. *Appl. Phys. Lett.* **43,** 828.

———, McDonald, M. L., and Unterwald, F. C. (1985). Direct imaging of a novel silicon surface reconstruction. *Phys. Rev. Lett.* **55**, 1765.

Glaisher, R. W., and Spargo, A. E. C. (1984). Calculated diffuse scattering and imaging of a small defect in silicon. In *Electron microscopy and analysis—1983*, ed. P. Doig, 185. The Institute of Physics, Bristol.

Hashimoto, H. (1978–1979) Identification of atoms in crystals by the fine structure of their images. *Chem. Scripta* **14**, 23.

———, Takai, Y., Yokota, Y., Endoh, H., and Fukada, E. (1980). Direct observations of the arrangement of atoms around stacking faults and twins in gold crystals and the movement of atoms accompanying their formation and disappearance. *Jpn. J. Appl. Phys.* **19**, L1.

Hetherington, C. J. D., Barry, J. C., Bi, J. M., Humphreys, C. J., Grange, J., and Wood, C. (1985). High resolution electron microscopy of semiconductor quantum well structures. *Mater. Res. Soc. Symp. Proc.* **37**, 41.

Heuer, A., Kraus-Lanteri, S., Labun, P. A., Lanteri, V., and Mitchell, T. E. (1985). HREM studies of coherent and incoherent interfaces in ZrO_2-containing ceramics: a preliminary account. *Ultramicroscopy* **18**, 335.

Hirabayashi, M. (1980). Applications of high-voltage high-resolution electron microscopy: alloy structures. In *Electron microscopy—1980*, eds. P. Brederoo and J. Van Landuyt. **4**, 142. Seventh European Congress on Electron Microscopy Foundation, Leiden.

———, Hiraga, K., and Shindo, D. (1982). High-resolution imaging by 1-MV electron microscopy. *Ultramicroscopy* **9**, 197.

Hiraga, K. (1984). Application of high-resolution electron microscopy to the study of structure defects and grain boundaries in Si_3N_4 and SiC—a brief review. *Sci. Rep. Res. Inst. Tohoku Univ., Ser. A* **32**, 1.

———, Hirabayashi, M., Hayashi, S., and Hirai, T. (1983). High-resolution electron microscopy of chemically-vapour-deposited β-Si_3N_4-TiN composites. *J. Am. Ceram. Soc.* **66**, 539.

Hirsch, P. B., Pirouz, P., and Barry, J. C. (1986). Platelets, dislocation loops and voidites in diamond. *Proc. R. Soc. London*, Ser. A, **407**, 239.

———, Howie, A., Nicholson, R., Pashley, D. W., and Whelan, M. J. (1977). *Electron microscopy of thin crystals*. 2nd ed. Krieger, Huntington, N.Y.

Hobbs, L. W. (1984). Radiation effects in analysis by TEM. In *Quantitative electron microscopy*, ed. J. N. Chapman and A. J. Craven; 401. Scottish Universities Summer School in Physics, Edinburgh.

Hutchison, J. L. (1984). High resolution electron microscopy in the study of semiconducting materials. *Ultramicroscopy* **15**, 51.

———, and Briscoe, N. A. (1985). Surface profile imaging of spinel catalyst particles. *Ultramicroscopy* **18**, 435.

———, Booker, G. R., and Abrahams, M. S. (1981). Transmission high-resolution electron microscopy studies of silicon-sapphire epitaxial layer structures. In *Electron microscopy and analysis—1981*, ed. A. G. Cullis, 139. The Institute of Physics, Bristol.

———, Anstis, G. R., Humphreys, C. J., and Ourmazd, A. (1982). Atomic images of silicon and related materials: fact or artefact? In *Electron microscopy and analysis—1981*, ed. M. J. Goringe, 357. The Institute of Physics, Bristol.

Iijima, S., and Ichihashi, T. (1986). Structural instability of ultrafine particles of metals. *Phys. Rev. Lett.* **56**, 616.

———, and O'Keefe, M. A. (1979). Determination of defocus values using Fourier images for high-resolution electron microscopy. *J. Microsc.* **117**, 347.

Ishida, Y., Ichinose, H., Mori, M., and Hashimoto, M. (1983). Identification of grain boundary atomic structure in gold by matching lattice imaging micrographs with simulated images. *Trans. Jpn. Inst. Met.* **24**, 349.

Iwata, T., and Nihira, T. (1966). Atomic displacement in pyrolytic graphite by electron bombardment. *Phys. Lett.* **23**, 631.

Izui, K., Furuno, S., Nishida, T., Otsu, H., and Kuwabara, S. (1978). High-resolution electron

microscopy of images of atoms in silicon crystal oriented in (110). *J. Electron Microsc.* **27,** 171.
Kang, Z. L., Smith, D. J., and Eyring, L. (1986). Atomic imaging of oxide surfaces. I. General features. *Surf. Sci.* **175,** 684.
Kiritani, M. (1977). Electron radiation damage of metals and nature of point defects by high voltage electron microscopy. In *Point defects,* ed. M. Doyoma and S. Yoshida, 247. University of Tokyo Press, Tokyo.
Knowles, K. M. (1982a). A high-resolution electron microscope study of nickel-titanium martensite. *Philos. Mag. Ser. A* **45,** 357.
——— (1982b). A high-resolution electron microscope study of martensite and martensitic interfaces in titanium-manganese. *Proc. R. Soc. London, Ser. A* **380,** 187.
Komoda, T. (1964). On the observation of crossed lattice images by electron microscopy. *Optik* **21,** 93.
Krakow, K., Wetzel, J. T., Smith, D. A., and Trafas, G. (1985). Characterization of tilt boundaries by ultra high resolution electron microscopy. *Mater. Res. Soc. Symp. Proc.* **41,** 253.
Krivanek, O. L., Isoda, S., and Kobayashi, K. (1977). Lattice imaging of a grain boundary in crystalline germanium. *Philos. Mag.* **36,** 931.
———, Tsui, D. C., Sheng, T. T., and Kamgar, A. (1978). A high-resolution EM study of Si–SiO_2 interfaces. In *Proceedings of the international conference on the physics of SiO_2 and its interfaces,* ed. S. T. Pantelides, 356. Pergamon Press, New York.
Marks, L. D. (1984). Direct atomic imaging of solid surfaces. I. Image simulation and interpretation, *Surf. Sci.* **139,** 281.
———, and Smith, D. J. (1983). Surface structure imaging of small metal particles. *Nature* **303,** 316.
Merkle, K. L., Reddy, J. F., Wiley, C. L., Smith, D. J., and Wood, G. J. (1986). Atomic resolution studies of tilt grain boundaries in NiO. *Mater. Res. Soc. Symp. Proc.* **60,** 227.
Narayan, J., and Holland, O. W. (1984). Characteristics of ion-implanted damage and annealing phenomena in semiconductors. *J. Electrochem. Soc.* **131,** 2651.
Nelson, D. G. A., Wood, G. J., Barry, J. C., and Featherstone, J. D. B. (1986). The structure of (100) defects in carbonated apatite crystallites: a high resolution electron microscope study. *Ultramicroscopy* **19,** 253.
Olsen, A., and Spence, J. C. H. (1981). Distinguishing dissociated glide and shuffle set dislocations by high resolution electron microscopy. *Philos. Mag., Ser. A* **43,** 945.
Paus, K. C., Barry, J. C., Booker, G. R., Peters, T. B., and Pitt, M. G. (1985). Investigation of the early growth of epitaxial silicon-on-sapphire using high-resolution transmission electron microscopy. In *Microscopy of Semiconducting Materials—1985,* ed. A. G. Cullis, 35. The Institute of Physics, Bristol.
Penisson, J. M., and Bourret, A. (1979). High resolution study of [011] low-angle tilt boundaries in aluminium. *Philos. Mag. Ser. A* **40,** 811.
———, Gronsky, R., and Brosse, J. B. (1982). High resolution study of a $\Sigma = 41$ grain boundary in molybdenum. *Scripta Met.* **16,** 1239.
Salisbury, I. G., and Loretto, M. H. (1979). {113} loops in electron-irradiated silicon. *Philos. Mag., Ser. A* **39,** 317.
Saxton, W. O., and Smith, D. J. (1985). The determination of atomic positions in high resolution electron micrographs. *Ultramicroscopy* **18,** 39.
———, Smith, D. J., and Erasmus, S. J. (1983). Procedures for focussing, stigmating and alignment in high resolution electron microscopy. *J. Microsc.* **130,** 187.
Self, P. G., Bhadeshia, H. K. D. H., and Stobbs, W. M. (1981). Lattice spacings from lattice fringes. *Ultramicroscopy* **6,** 29.
Sinclair, R. (1977). The importance of electron diffraction to transmission electron imaging. *Trans. Am. Cryst. Assoc.* **13,** 101.
———, Ponce, F. A., Yamashita, T., and Smith, D. J. (1983). High resolution electron microscopy of II–VI compound semiconductors. In *Electron microscopy and analysis—*

1983, ed. A. G. Cullis, S. M. Davidson, and G. R. Booker, 103. The Institute of Physics, Bristol.
Smith, D. J. (1985). Atomic resolution surface dynamics by electron microscopy. *J. Vac. Sci. Technol. Sect. B* **3**, 1563.
―――― (1986). High-resolution electron microscopy in surface science. In *Chemistry and physics of solid surfaces*, Vol. 6, ed. R. Vanselow and R. Howe, Chap. 15. Springer-Verlag, Heidelberg.
――――, and Bursill, L. A. (1985). "Metallisation" of oxide surfaces observed by *in situ* high resolution electron microscopy. *Ultramicroscopy* **17**, 387.
――――, and Marks, L. D. (1985). Direct atomic imaging of solid surfaces. III. Small particles and extended Au surfaces. *Ultramicroscopy* **16**, 101.
――――, and O'Keefe, M. A. (1983). Conditions for direct structure imaging in silicon carbide polytypes. *Acta Crystallogr., Sect. A* **39**, 838.
――――, Bursill, L. A., and Blanchin, M. G. (1984). The structure of extended defect terminations in rutile. *Philos. Mag., Ser. A* **50**, 473.
――――, Bursill, L. A., and Jefferson, D. A. (1986). Atomic imaging of oxide surfaces. I. General features and surface rearrangements. *Surf. Sci.* **175**, 673.
――――, Bursill, L. A., and Wood, G. J. (1983). High resolution electron microscope study of tin dioxide crystals. *J. Solid State Chem.* **50**, 51.
――――, Bursill, L. A., and Wood, G. J. (1985). Non-anomalous high resolution imaging of crystalline materials. *Ultramicroscopy* **16**, 19.
――――, Camps, R. A., and Freeman, L. A. (1982). Atomic resolution in the high-voltage electron microscope. In *Electron microscopy and analysis—1981*, ed. M. J. Goringe, 381. The Institute of Physics, Bristol.
――――, Petford-Long, A. K., Wallenberg, L. R., and Bovin, J. O. (1986). Dynamic observations of atomic-level rearrangements in small gold particles. *Science* **233**, 872.
――――, Ponce, F. A., Yamashita, Y., and Sinclair, R. (1983). High-voltage high-resolution electron microscopy of compound semiconductors. In *Proceedings of the seventh international conference on high voltage electron microscopy,* eds. R. M. Fisher, R. Gronsky, and K. H. Westmacott, 31, Lawrence Berkeley Laboratory, Berkeley.
――――, Saxton, W. O., O'Keefe, M. A., Wood, G. J., and Stobbs, W. M. (1983). The importance of beam alignment and crystal tilt in high-resolution electron microscopy. *Ultramicroscopy* **11**, 263.
――――, Freeman, L. A., McMahon, R. A., Ahmed, H., Pitt, M. G., and Peters, T. B. (1984). The characterization of Si-implanted and electron-beam-annealed silicon-on-sapphire by high-resolution electron microscopy. *J. Appl. Phys.* **56**, 2207.
Stenberg, L., and Andersson, S. (1979). Electron microscope studies on a quenched Fe–Mo alloy. *J. Solid State Chem.* **28**, 269.
Stobbs, W. M., Wood, G. J., and Smith D. J. (1984). The measurement of boundary displacement in metals. *Ultramicroscopy* **14**, 145.
Suzuki, K., Takeuchi, S., Shino, M., Kanaya, K., and Iwanaga, H. (1983). Lattice image observations of defects in CdS and CdSe. *Trans. Jpn. Inst. Met.* **24**, 435.
Tan, T. Y., Foll, N., and Krakow, W. (1981). Intermediate defects in silicon and germanium. In *Microscopy of semiconducting materials—1981*, ed. A. G. Cullis, 1. The Institute of Physics, Bristol.
Tanaka, M., and Jouffrey, B. (1984). Dissociated dislocations in GaAs observed in high-resolution electron microscopy. *Philos. Mag. Ser. A* **50**, 733.
Terasaki, O., Smith, D. J., and Wood, G. J. (1986). The imaging of individual cation columns in f.c.c. mixed alloy systems. *Acta Crystallogr., Sect. B* **42**, 39.
Timsit, R. S., Waddington, W. G., Humphreys, C. J., and Hutchison, J. L. (1986). Examination of the Al/Al_2O_3 interface by high-resolution electron microscopy. *Ultramicroscopy* **18**, 387.
Tung, R. T., Gibson, J. M., and Poate, J. M. (1983). The growth of epitaxial $NiSi_2$ single crystals on silicon by the use of template layers. *Mater. Res. Soc. Symp. Proc.* **14**, 435.

Urban, K. (1979). Radiation-induced processes in experiments carried out *in-situ* in the high-voltage electron microscope. *Phys. Status Solidi a* **56,** 157.

Watanabe, D., and Terasaki, O. (1984). Long-period ordered structure of the Au-rich Au–Mn alloys. In *Phase transformations in solids,* ed. T. Tsakalakos, 231. North-Holland, New York.

Wood, G. J., Bursill, L. A., and Smith, D. J. (1983). An investigation of order/disorder in a chromia-doped rutile by high-resolution electron microscopy. *J. Microsc.* **129,** 263.

———, Stobbs, W. M., and Smith D. J. (1984). Methods for the measurement of rigid-body displacements at edge-on boundaries using high-resolution electron microscopy. *Philos. Mag., Ser. A* **50,** 375.

Ye, H. Q., Wang, D. N., and Kuo, K. H. (1985). Domain structures of tetrahedrally close-packed phases with juxtaposed pentagonal antiprisms. 2. Domain boundaries of the C-14 laves phase. *Philos. Mag., Ser. A* **51,** 839.

Yoshida, H., Hashimoto, H., Yokota, Y., and Takeda, M. (1984). High-resolution lattice images of G–P zones and solute clusters in Al–Cu and Cu–Be alloys. *Mater Res. Soc. Symp. Proc.* **21,** 131.

Zakharov, N. D., Pasemann, M., and Rozhanski, V. N. (1982). Observations of point defects in silicon by means of dark field lattice plane imaging. *Phys. Status Solidi a* **71,** 275.

Zemlin, F., Weiss, K., Schiske, P., Kunath, W., and Herrmann, K.-H. (1978). Coma-free alignment of high-resolution electron microscopes with the aid of optical diffractograms. *Ultramicroscopy* **3,** 49.

12

PRACTICAL HIGH-RESOLUTION ELECTRON MICROSCOPY

ONDREJ L. KRIVANEK

12.1 Introduction

This chapter deals with two subjects: how to attain the highest possible resolution with a given transmission electron microscope (TEM) and how to avoid experimental artifacts that may lead to misleading conclusions about the specimen structure or chemistry. It concentrates on the techniques of high-resolution imaging in the conventional transmission electron microscope (CTEM) and microanalysis with a small probe in the scanning transmission electron microscope (STEM). A broader survey of new structure-probing techniques is provided in Chapters 5 to 7 in this book, and a more general introduction to all aspects of experimental high-resolution electron microscopy can be found in recent books by Reimer (1984) and Spence (1981). The present chapter should be read in conjunction with Chapters 1 through 4, which introduce several key concepts of high-resolution electron microscopy and provide the necessary theoretical background.

The performance of high-resolution electron microscopes is now at the point where the lattice planes of practically any crystalline material can be resolved directly, and single, heavy atoms can be readily distinguished if placed on the featureless background of a low-Z-material. These capabilities are not new—Komoda first imaged the (200) and (220) planes in gold with spacings of 2.04 and 1.44 Å some two decades ago (Komoda, 1966), and Crewe and co-workers first imaged single atoms and identified clusters of a few thousand iron atoms by electron energy loss spectroscopy (EELS) over 15 years ago (Crewe, 1970). The past decade has witnessed more incremental advances. The number of high-resolution electron microscopes capable of reaching resolution around 2 Å on a routine basis has increased to several dozens, and the ease and convenience with which these instruments can resolve atomic-level detail has improved dramatically.

The high-resolution image is formed in two stages: First, the specimen scatters the incident-electron beam into the scattered beams, which become completely separated at the back focal plane of the objective lens, and then the lenses of the microscope bring the beams back into coincidence at the level of the image. The high-resolution image is therefore always simply an interference pattern, and hence, it is easily changed by minor adjustments of the many instrumental parameters that govern the way the scattered beams are made to coincide. Some of the interference patterns bear a close resemblance to the specimen structure and are called "structure images." Many more bear little resemblance to the specimen and can lead to erroneous conclusions about the specimen structure. Ironically, the increased resolution capabilities of HRTEM have made it much easier to produce the false images, and the experimenter needs to be on guard against the conditions that may produce them. Furthermore, even with the microscope correctly adjusted so that imaging artifacts are largely avoided, the required resolution is often close to the limits of the microscope being used, and it then becomes important to optimize the resolution by considering the precise effect of each experimental adjustment. This is equally true in analytical electron microscopy (AEM), where the signal is typically quite weak; and practical problems such as contamination and radiation damage often impose further limits.

To be able to adjust the microscope in the optimal way, it is necessary to understand the theory of image formation and how the theory relates to the practical aspects of obtaining high-resolution images. This knowledge is particularly important because many adjustments of the microscope produce effects that run contrary to simple intuition. As a practical example, it is always tempting to increase the brightness of a high-resolution image by increasing the beam current extracted from the filament, because this usually improves the visibility of the image on the viewing screen. However, increasing the beam current results in a larger energy spread of the beam and hence less coherent illumination; and the decreased coherence often causes a loss of resolution. Other adjustments run closer to intuition, but even with the simplest adjustments, a substantial error can make a whole experimental session completely worthless. Similarly, a malfunction in just one of the many crucial parts of a microscope may make it impossible to obtain good results until the problem is diagnosed and fixed. It is therefore useful to review the aspects that affect high-resolution imaging in practice one by one, and this approach is adopted in this chapter.

12.2 Instrumentation

The instrumental requirements for high-resolution imaging in the CTEM have been recently reviewed by Hermann (1983), and the instrumental requirements for an analytical STEM are covered in several review articles (e.g., Crewe, 1971), as well as in more recent conference proceedings (e.g., Colliex, 1985). We shall therefore concentrate on reviewing only those aspects of the instrumentation that pertain directly to the discussion of the optimum operation of the microscopes considered further on in this chapter.

Different high-resolution microscopes can be usefully compared by their theoretically attainable point-to-point resolution, which measures the microscope's ability to faithfully image the specimen structure. Another criterion of the microscope performance is the line resolution, which measures the microscope's ability to combine two beams so that they produce an interference pattern and principally tests the mechanical stability and coherence of illumination of the microscope. For bright-field imaging, the point-to-point resolution is usually taken as one over the extent of the main contrast-transfer interval at the optimum defocus for phase-contrast imaging. This is given by

$$r_{bf} = 0.67\lambda^{3/4} C_s^{1/4}, \qquad (12.1)$$

where λ is the electron wavelength and C_s is the coefficient of spherical aberration of the objective lens (Scherzer, 1949; Spence, 1981; see also Chapter 1). This well-known expression shows that the resolution performance of a microscope can be improved in two fundamental ways: by increasing the accelerating energy E_0 of the microscope to reduce λ, and by decreasing the coefficient of spherical aberration of the objective lens C_s. There are also several practical factors that contribute to the attainable resolution but are not covered by this simple expression. They are discussed later in this chapter.

Table 12.1 lists commercial TEMs that are capable of approaching or surpassing 2 Å point-to-point resolution in bright-field imaging. The table makes it clear that, although there are now several CTEMs capable of reaching this performance, the STEMs as yet do not have a single entry in the table. It is also worth noting that the small differences in resolution of

TABLE 12.1
Microscopes Approaching and Surpassing 2 Å point-to-point resolution in bright-field imaging (May 1986)

Microscope[a]	Maximum operating voltage (kV)	C_s (mm)	C_c (mm)	Point-to-point resolution (Å) Theoretical	Point-to-point resolution (Å) Demonstrated[b]	Line resolution demonstrated (Å)	Small-probe-forming capabilities ($d < 50$ Å)
Cambridge HREM	600	2.5	2.7	1.8	1.8	0.72	✓
Hitachi H 800	200	1.0	1.2	2.3	2.3	0.72	
Hitachi HU 1250	1200	2.5	$2.6(C_c^i) 3.5(C_c^v)$[c]	1.2	1.6	1.02	
Hitachi H 9000	300	0.9	1.5	1.9	1.9	1.02	
ISI 002A	120	0.3	0.6	2.2	2.2	1.02	✓
ISI 002B	200	0.4	0.8	1.9	2.0	1.02	✓
JEOL 2000 EX	200	0.9	1.2	2.3	2.3	1.02	
JEOL 4000 EX	400	1.0	1.7	1.7	1.7	1.02	
JEOL HAREM	500	1.0	1.4	1.6	1.8	1.02	
JEOL ARM	1000	2.3	3.4	1.3	1.8	1.02	
Philips 430ST	300	1.1	1.1	2.0	2.0	1.02	✓

[a] If various versions of a microscope are available, the one with the lowest objective-lens aberrations is cited here.
[b] Approximate figures only.
[c] C_c^i and C_c^v are the coefficients of chromatic aberration applicable to changes in the objective lens current and the high voltage, respectively.

the different microscopes listed can be significant, since even a 10 per cent improvement in the resolving power is likely to make new materials of increasingly smaller lattice plane spacings accessible to direct imaging. The table further shows that, whereas instruments designed for operation at primary voltages up to 400 kV are able to reach their theoretical resolution limit, higher-voltage instruments usually cannot reach this limit. Their resolution is, instead, constrained by electrical and mechanical instabilities. The 2 Å-resolution figure is somewhat arbitrary, although it is approximately the spacing of the close-packed planes in most metals and insulators. A table listing microscopes capable of achieving better than 3.5 Å point-to-point resolution, which is sufficient for high-resolution studies of most semiconductors and many ceramics and minerals, would include practically all the transmission electron microscopes currently in production.

Listing microscopes according to the point-to-point resolution as expressed by equation (12.1) neglects other important aspects of the instrumental performance. Chief among these considerations are the brightness and stability of the electron gun and the performance of the detectors. The gun design has now largely stabilized with LaB_6 guns, which give brightness about 10 times greater than conventional guns using heated tungsten filaments, dominant in high-resolution CTEM, and field-emission guns, whose brightness is another two orders of magnitude higher still, used in STEM. The CTEM detector of choice is still photographic film, although promising new approaches based on recording media in which the arrival of the high-energy electrons causes excitations that are subsequently read by a scanned laser beam are beginning to appear (Fujiyoshi, 1986). Furthermore, the resolution of fiber optically coupled TV cameras, which have been able to detect single electrons for some time (Hermann et al., 1971), has recently begun to approach the resolution of photographic film (Kraus et al., 1986). Such cameras make the task of adjusting the HRTEM far easier and permit the dynamic recording of changes in the specimen. They are therefore becoming increasingly common in all HRTEM installations. In analytical STEMs, the large variety of detectors used is still rapidly evolving, and the ultimate goal of detecting every possible analytical signal with 100 per cent efficiency is far from achieved.

The design of the electron optical column of the microscope is now fairly stabilized. Some of the more recent innovations are best embodied in one of the microscopes listed in Table 12.1, the ISI 002B (Figure 12.1). The objective lens of the 002B has a spherical-aberration coefficient of only 0.4 mm. This enables the 002B to achieve a theoretical point-to-point resolution of 1.9 Å at 200 kV—i.e., a performance similar to microscopes whose primary voltage is two times higher. The very low C_s is achieved by a pole-piece gap of 1.2 mm and a bore diameter of 0.8 mm. A 3-mm specimen can nevertheless be examined in the microscope at up to $\pm 10°$ tilt. This is made possible by the placement of the objective aperture, which enters through a transverse hole in the bottom pole piece (Figure 12.2), and by the

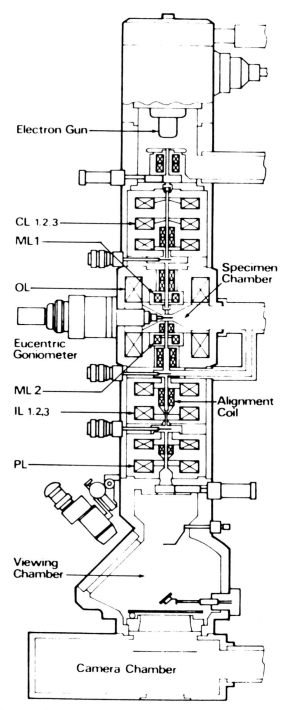

FIGURE 12.1 Cross section of the ISI 002B electron microscope. The symbols CL, ML, OL, IL, and PL stand for condenser lens, minilens, objective lens, intermediate lens, and projector lens.

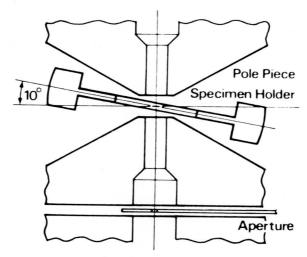

FIGURE 12.2 The placement of the specimen in the 1.2-mm-wide gap of the objective lens of the ISI 002B.

design of the specimen holder, which leaves the central part of the specimen disk as the only object that needs to be located in the intense-field region of the lens. The column of the microscope is almost symmetrical about the midplane of the objective lens, with minilenses in the upper and lower bores of the objective lens, three condenser lenses, three intermediate lenses, and a projector lens. The minilenses give complete control of the critical upper and lower parts of the objective-lens field, so that one can, for instance, form a 1-nm probe on the specimen and observe its sharp image at a magnification of 1.9×10^6 on the microscope-viewing screen and do this for a range of specimen heights. Another interesting point of the design is that the separation of the objective-lens pole pieces is adjustable from outside the vacuum (with the lens switched off), which permits increased maximum specimen tilt at the cost of decreased resolution.

A microscope incorporating the electron-optical-design innovations of the 002B, operating at a primary energy of 500 keV and possessing a bright, stable field-emission gun, would achieve about 1.4 Å point-to-point resolution in both conventional and scanning transmission electron microscopy. Such a microscope would be difficult to improve with established technologies. Since existing CTEMs already achieve 1.6 Å point-to-point resolution, the best prospect for a significant further improvement in the resolution lies with aberration correctors, particularly the new type of corrector utilizing a long sextupole (Crewe and Kopf, 1979). One can also expect that the current trend to increased computerization of the microscopes will approach the stage where the computer takes over all of the routine aspects of the instrumental operation. Although this will not increase the attainable performance as such, it will make microscopy more readily accessible to nonexperts; and it will also make the whole electron microscopy field increasingly quantitative by providing easy-to-use, on-line, data-processing capabilities.

12.3 Adjustment of the CTEM for high-resolution imaging

At the start of any high-resolution-imaging session, it is necessary to check the basic alignment of the electron microscope, such as the centering of the gun for optimum brightness and the alignment of the imaging lenses, which enables the image to remain centered on the viewing screen as the magnification is varied. The corresponding alignment procedures are usually different for each type of electron microscope and are typically described in detail in the microscope's operating manual provided by the manufacturer. A generally applicable outline of these basic alignments is given in an article by Self (1983). In the discussion that follows, we shall assume that the basic alignments have already been performed and concentrate, instead, on the final adjustments typically performed just before a high-resolution image is taken.

Although the theoretically attainable resolution of a CTEM as defined by expression 12.1 depends only on C_s and λ, in practice, there are many factors that influence the microscope resolution. The contrast-transfer-function description of the imaging process, introduced in Chapter 1, makes it possible to separate the effects of the various resolution-limiting factors, but it is only valid if the following two approximations can be made:

1. The imaging properties of the microscope must be the same for the whole image; i.e., the way an object is imaged by the microscope must be independent of the position of the object within the field of view. This is called the isoplanatic approximation (Lenz, 1971). In light optics, where the imaging characteristics typically vary markedly across the image, the isoplanatic approximation is usually inapplicable. In high-resolution electron microscopy, fortunately, the size of the examined-specimen region is so small that the imaging is nearly the same across the whole image field; i.e., the image of an atom will be the same no matter where the atom is located within the field of view. The image amplitude can then be represented by replacing each scattering center at the specimen by the point-spread function $t(\mathbf{r})$ at the level of the image, where the point-spread function describes the way the electrons originating from each scattering center are spread into a small patch by the microscope optics.

2. The image intensity must be linearly related to the amplitude of the electron wave function at the exit surface of the specimen. This essentially amounts to the weak-phase-object approximation (Chapter 1) and is valid when the main (unscattered) beam dominates the scattered beams. In this

case, only the pairwise interference between the main beam and each scattered beam needs to be considered; the interference among the scattered beams themselves can be neglected. The main beam then acts in the same way as the reference beam in holography, and the image intensity is linearly related to the amplitude of the electron wave function at the specimen. In practice, the weak-phase-object approximation fails badly except in regions right next to the edge of the specimen. Provided that one concentrates on the ultrathin regions, however, the approximation provides the theoretical basis for the unusually simple image interpretation whereby, at the optimum (Scherzer) defocus, atoms simply appear dark. The simple description of the imaging process made possible by the weak-phase-object approximation is also invaluable in providing guidelines for how to set up the microscope. We shall base most of this chapter on the approximation.

Combining these two approximations leads to a particularly simple representation of the imaging process as

$$I(\mathbf{r}) = t(\mathbf{r}) * \psi_s(\mathbf{r}), \tag{12.2}$$

where $\mathbf{r}$ is the image-plane coordinate $[\mathbf{r} = (x, y)]$ scaled such that the magnification of the image is one, $I(\mathbf{r})$ is the image intensity, $t(\mathbf{r})$ is the point-spread function, $*$ denotes a convolution product (Bracewell, 1978), and $\psi_s(\mathbf{r})$ is the amplitude of the scattered part of the electron wave function at the specimen-exit surface.

Expression (12.2) can be transformed into a form convenient for computation by using the fact that the Fourier transform of a convolution product is a multiplication (Bracewell, 1978), which leads to

$$I(\mathbf{r}) = FT^{-1}[T(\mathbf{U})\Psi_s(\mathbf{U})]. \tag{12.3}$$

Here, $\mathbf{U}$ is the reciprocal-spatial-frequency coordinate $[\mathbf{U} = (U, V)$, where U and V are the reciprocal-space counterparts of the image coordinates x and y], which can be taken to represent the coordinate system in the back focal plane of the objective lens where the diffraction pattern is located; $T(\mathbf{U})$ is the Fourier transform of the point-spread function; $\Psi_s(\mathbf{U})$ is the Fourier transform of $\psi_s(\mathbf{r})$; and FT^{-1} denotes the reverse Fourier transform. Equation (12.3) shows that a high-resolution electron microscope can be thought of as a device that transmits the spatial frequencies contained in the specimen to the image with various weights but otherwise unaltered. This is a consequence of the linear relationship between the image and the weak specimen potential. It means that the imaging process in the electron microscope can be described in terms of the frequency response in essentially the same way as the performance of an audio amplifier or of any linear transmission line. The function that describes the frequency response, $T(\mathbf{U})$, is known as the contrast transfer function.

Expressions (12.2) and (12.3) are applicable in all imaging configurations in which the unscattered main beam dominates the image, i.e., in all bright-field imaging of thin specimens. The bright-field images that show the closest resemblance to the specimen structure are, however, obtained when the electron beam enters the objective lens along the optic axis—i.e., in axial bright-field imaging. The phase-contrast transfer function $T(\mathbf{U})$ is then determined by the physical apertures that may be present in the microscope, attenuation effects due to incoherence of illumination and mechanical and other instabilities, and the phase distortion of the electron wave:

$$T(\mathbf{U}) = A(\mathbf{U})E(\mathbf{U}) 2 \sin \chi(\mathbf{U}). \tag{12.4}$$

Here, $A(\mathbf{U})$ is the aperture function, which is 1 inside and 0 outside the objective aperture; $E(\mathbf{U})$ is an envelope function describing the attenuation of contrast transfer due to a number of factors discussed below; and $\chi(\mathbf{U})$ is the aberration function, which describes the phase shift experienced by the beam due to the spatial frequency U—i.e., by the beam scattered by the specimen through an angle λU. Note that if there is no objective aperture, $A(\mathbf{U}) = 1$ everywhere. The aberration function is given by

$$\chi(U) = \pi \, \Delta f \lambda U^2 + \tfrac{1}{2}\pi C_s \lambda^3 U^4, \tag{12.5}$$

where Δf is the defocus of the objective lens and $U = |\mathbf{U}|$.

12.3.1 Illumination

Equations (12.4) and (12.5) show that if there were no objective aperture $[A(\mathbf{U}) = 1]$ and no contrast attenuation $[E(\mathbf{U}) = 1]$, the contrast transfer function would oscillate rapidly between $+2$ and -2 all the way to the highest spatial frequency $\mathbf{U}$ admitted through the bore of the objective-lens pole piece. In practice, various resolution-limiting factors attenuate the contrast transfer to zero after typically one to five oscillations. This imposes a practical limit on the resolution, which can be even more severe than the resolution limit owing to the objective lens described by the equation (12.1). For weakly scattering specimens, the attenuation of the contrast-transfer function is conveniently described by the envelope function $E(\mathbf{U})$.

The resolution-limiting factors include unavoidable effects, of which the principal one is the incoherence of illumination (energy and angular spread). Other contributing factors—such as instabilities of the high voltage and of the lens-current supplies, influence of stray magnetic fields, specimen drift and vibration, and finite resolution of the recording medium—can, in principle, be eliminated. Provided that the half angle of the angular spread of the illumination is less than about 5 mrad and the energy spread is less than a few electronvolts, the envelope function due to the incoherence of illumination can be approximated by a product of partial envelopes due to each of these contributions (Frank, 1973). The effect of instrumental stabilities can be similarly expressed by a partial envelope, yielding

$$E(U) = E_t(U)E_s(U)E_i(U), \tag{12.6}$$

where the subscripts t, s, and i refer to the temporal incoherence (defocus spread), spatial incoherence (angular spread), and the instrumental instabilities. For a Gaussian spread of the defocus with a full width at half maximum (FWHM) equal to δ, the temporal incoherence envelope is given by (Frank, 1973)

$$E_t(U) = \exp\left[\frac{-(\pi\delta\lambda U^2)^2}{16 \ln 2}\right]. \quad (12.7)$$

The defocus spread δ is due to the energy spread of the incident beam and the instabilities of the high voltage and the objective-lens power supplies. It is given by

$$\delta = C_c\left[\left(\frac{\Delta E}{E_0}\right)^2 + \left(2\frac{\Delta I}{I_0}\right)^2\right]^{1/2}. \quad (12.8)$$

Here, C_c is the coefficient of the chromatic aberration of the objective lens; ΔE is the FWHM of the total energy spread due to both the energy spread of the electron beam [which arises because of the spread of the velocities of the electrons emitted by the electron source and also the Boersch effect (Boersch, 1954; Loeffler, 1969)] and the high-voltage instabilities; E_0 is the average energy of the electron beam; ΔI is the FWHM of the objective-lens-current spread during the time of the image recording; and I_0 is the average current of the lens. The two terms inside the square brackets in (12.8) are added in quadrature because they represent independent, random fluctuations. In practice, the resolution-limiting effect of the fluctuations in the objective-lens current is usually negligible compared with the effect of the electron-energy spread.

For a Gaussian distribution of the illumination angles of FWHM α_0, the spatial-incoherence envelope is given by

$$E_s(U) = \exp\left\{\frac{-[\pi\alpha_0(\Delta fU + C_s\lambda^2 U^3)]^2}{4 \ln 2}\right\}. \quad (12.9)$$

The Gaussian distribution approximates the angular distribution when the illumination on the sample is defocused. When the illumination is fully focused—i.e., when the gun crossover is imaged onto the specimen—it is more appropriate to model the illumination by a sharply limited (top-hat) angular distribution of diameter α_0, giving

$$E_s(U) = \frac{2J_1[\pi\alpha_0(\Delta fU + C_s\lambda^2 U^3)]}{\pi\alpha_0(\Delta fU + C_s\lambda^2 U^3)}, \quad (12.10)$$

where J_1 is the Bessel function of the first kind.

There are three things to be noted about the expressions above:

1. The contrast-transfer attenuation due to energy spread is independent of defocus; the attenuation due to angular spread varies with defocus.

2. The spatial frequency U appears in (12.7) in the fourth power in the argument of the exponential function, which means that energy spread results in a fairly abrupt cut-off in contrast transfer at

$$U_{max} = 2 (\ln 2)^{1/4}(\pi\delta\lambda)^{-1/2}. \qquad (12.11)$$

3. The spatial-incoherence envelope reaches unity at U-values for which the term inside the parentheses in (12.9) and (12.10) becomes zero—i.e., at

$$U_1 = \left(\frac{-\Delta f}{C_s\lambda^2}\right)^{1/2}. \qquad (12.12)$$

This is also the spatial frequency at which the aberration function has a plateau (first derivative equal to zero) and the contrast transfer function usually has a broad maximum or minimum.

In other words, in axial-illumination imaging, the energy stability imposes an absolute limit on the resolution; the angular spread of illumination attenuates contrast transfer in a way that depends on the defocus and, for moderate spread magnitudes, has little effect on the broadest contrast-transfer interval. At spatial frequencies higher than U_1, the second term in expressions (12.9) and (12.10) dominates, and the falloff of the spatial-incoherence envelope is even more rapid than that of the temporal-incoherence envelope.

Term U_{max} in equation (12.11) is sometimes called the information-retrieval limit, and $1/U_{max}$ is the maximum retrievable resolution. Expressing δ in terms of the total energy spread (equation 12.8) and then rearranging (12.11), yields a condition for the maximum total energy spread if linear imaging up to a resolution of $1/U_{max}$ is to be achieved:

$$\Delta E < \frac{4 (\ln 2)^{1/2} E_0}{\pi C_c \lambda U_{max}^2}. \qquad (12.13)$$

For $C_c = 1$ mm and a resolution of 2 Å, this gives $\Delta E < 1.1$ eV at 100 kV primary energy and $\Delta E < 15$ eV at 500 kV. For $C_c = 1$ mm and a resolution of 1 Å, it gives $\Delta E < 0.28$ eV at 100 kV and $\Delta E < 3.7$ eV at 500 kV. This shows that the total-energy-spread requirements become more relaxed at higher operating voltages, especially when the aberration coefficients are similar to those of the lower-voltage microscopes. Nevertheless, practical experience shows that attaining the required stability of the accelerating voltage is rather difficult at voltages exceeding 500 kV, and that the present limit to the high-resolution performance of most 1 MeV microscopes comes precisely from high-voltage instabilities. At lower voltages, the high-voltage stability is usually adequate, but the energy spread can be a serious limitation. Its magnitude depends on the type of electron gun and the emission current. Typical values (determined by EELS) for an LaB_6 source are 1 eV energy spread at emission currents less than $2\,\mu A$, 2 to 3 eV at $10\,\mu A$, and about 5 eV at $30\,\mu A$. Attaining the best possible resolution therefore necessitates using emission currents of only 1 to 2 μA in a 100 kV microscope, 5 to 10 μA in a 200 kV microscope, and up to 30 μA in a 400 kV microscope. The small usable emission current at 100 kV means that it is particularly important to employ a bright source such as LaB_6 (or even a field-emission gun) to optimize the brightness. It is then also advisable to use a large illumination spot (weak first condensor lens), since a small illumination spot will lead to a further loss of usable brightness owing to the sizable aberrations of the illumination system.

The maximum tolerable spread of illumination angles is best judged by examining a selected-area electron diffraction (SAED) pattern obtained with a small SAED aperture and the illumination set up exactly the way it was for the imaging. First, however, one must make sure that the diffraction focus is correctly adjusted. This is done by fully focusing the illumination while in the image mode, switching over to diffraction, and adjusting the diffraction focus so that the diffraction pattern contains sharply focused disks. Switching back to imaging, defocusing the illumination to suit the imaging needs, inserting a SAED aperture, and switching once more into diffraction then reveals the actual value of the beam convergence. As a rough rule of thumb, the diameter of the central spot thus revealed should be smaller than about one-fifth the distance of the highest spatial frequency that one would like to contribute to the image from the central beam. In any case, the diffraction pattern should be recorded in this condition for each series of high-resolution micrographs so that the illumination-convergence value, which is necessary for an accurate simulation of computed images, can be determined.

The effects of the energy spread of illumination on high-resolution imaging are illustrated in Figure 12.3 for the particular case of the Philips 430ST electron microscope. The contrast transfer function that would be obtained at the optimum defocus (see next section) with perfectly coherent illumination is shown in Figure 12.3a, and the attenuating effect of the incoherence is shown by the envelope functions plotted for various values of energy and angular spread in Figures 12.3b and c, respectively. The attenuation results in a contrast transfer function whose main transfer interval is preserved but reduced in intensity and whose higher transfer intervals may be completely eliminated (Figure 12.3d).

This discussion is exactly applicable only to the case of axial bright-field imaging of weak phase objects and is further limited by the fact that Frank's envelope-function formulation is only valid for moderate values of incoherence (Frank, 1973). Nevertheless, it provides guidelines that are also roughly applicable to the way the illumination should be set up for bright-field imaging of thicker objects. It is also to some extent applicable to bright-field imaging in STEM, where the collector aperture plays the role of the TEM condenser aperture, but where the need to maximize the poor signal-to-noise ratio of the images usually means that rather large collection angles are used.

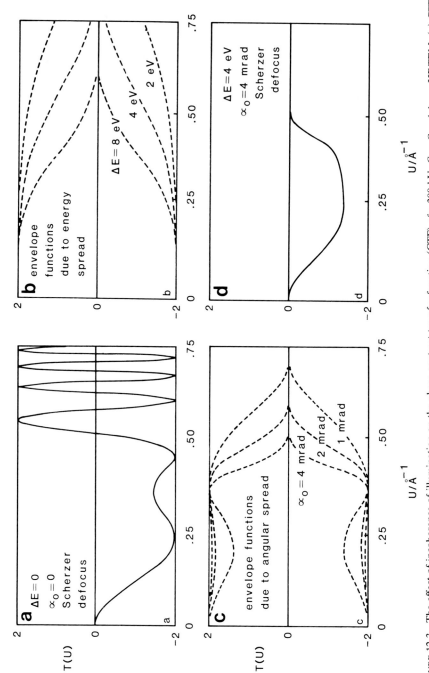

FIGURE 12.3 The effect of incoherence of illumination on the phase-contrast transfer function (CTF) of a 300-kV HRTEM. (a) CTF for Scherzer defocus and perfectly coherent illumination. (b) Envelope functions due to the energy spread of the illumination. (c) Envelope functions due to angular spread. (d) The resultant CTF for $\Delta E = 4$ eV and $\alpha_0 = 4$ mrad. All spread values are given as full widths at half maximum.

12.3.2 Adjustment of the objective lens

There are three important adjustments to be performed at the level of the objective lens: The lens must be optimally focused, its astigmatism must be compensated, and the illumination angle must be adjusted so that the electron beam impinges upon the specimen along a direction parallel to the optic axis of the lens. Whereas an error in the adjustment of the illumination is most likely to just lead to a slight decrease of the resolution, an error in the adjustment of the objective lens may lead to an image that actually contains sharper detail than the perfectly adjusted image, but that corresponds only poorly to the specimen structure and may even cause erroneous conclusions about the specimen. This makes the correct adjustment of the objective lens particularly crucial.

An essential prerequisite for correct alignment is that the magnetic field of the objective lens should not depart from cylindrical symmetry. This demands that the top and bottom pole pieces of the lens be parallel and centered on the same axis to within a few micrometers and that there be no magnetic inhomogeneities in the pole-piece material. If the objective lens has lost cylindrical symmetry, terms of odd power in U will appear in the aberration function (expression 12.5), and the cylindrical symmetry of the whole imaging process will also be lost. The usual way of diagnosing the misalignment is by reversing the polarity of the current in the objective lens and observing the resultant shift of the point about which the whole image rotates as the objective-lens current is varied by a small amount (the current center). With a well-aligned lens, the current center should shift typically less than $2\,\mu$m as the polarity is switched.

If the microscope provides no easy means for reversing the lens polarity, an indication of the presence of this type of misalignment can be obtained from the separation of the current center from the voltage center of the lens, where the voltage center is the point about which the image rotates when the high voltage of the microscope is varied by about 1 percent. The separation of the two centers should be less than about $1\,\mu$m. The separation is usually determined with the microscope in the imaging mode, at a magnification of about $30,000\times$, by first aligning the illumination direction so that one center falls on the center of the viewing screen and then determining the location of the other center. If the microscope is equipped with independent bright-field and dark-field channels, an alternative way to determine the separation of the current and voltage centers of a microscope is as follows: Select one of the two channels, produce an image at around $500,000\times$ magnification, and adjust the incident-beam direction until the current center is brought into the middle of the viewing screen. Next, switch to the other channel, and adjust the beam tilt until the voltage center is brought into the middle of the screen. Finally, switch over to the diffraction mode, and observe the angular separation of the voltage and current axes from the change in the position of the main beam that results when the microscope is switched between the bright-field and dark-field channels.

If the separation determined by the first method is more than about 1 μm, or if the separation determined by the second method is more than 1 mrad, the mechanical alignment of the objective-lens pole pieces or the homogeneity of the pole-piece material is not satisfactory. It then becomes advisable to either realign the pole pieces if they are alignable, or to replace them if they are prealigned. However, once the pole pieces are properly aligned, or once a good set of prealigned pole pieces is found, the problem is not likely to occur again.

12.3.2.1 Focusing. The intensity of the electron wave at the exit face of a thin phase object is uniform, since electrons are not absorbed in the specimen but only experience small angular deflections. If an electron microscope were focused so perfectly that all the electrons arrived at the image plane in places correspondingly exactly to where they exited from the thin specimen, the image would show no contrast whatever. In terms of the aberration function (equations 12.4 and 12.5), this would correspond to achieving $\chi(\mathbf{U}) = 0$ for all spatial frequencies $\mathbf{U}$, in which case $T(\mathbf{U}) = 0$ everywhere, and there is no phase contrast. In practice, some contrast is always preserved because of the spherical aberration of the objective lens, the removal of electrons from the beam by the microscope apertures, and the fact that no real specimen has the ideal attributes of the thin phase object. Nevertheless, there is a deep contrast minimum when the objective lens is focused a small distance (typically 200 to 300 Å) above the sample (in a conventional microscope in which the electrons travel downward and the coefficient of spherical aberration is positive). This is called the Gaussian defocus. It occurs $-0.3(C_s\lambda)^{1/2}$ above the exit surface of the specimen, and it corresponds to the defocus value that best cancels the contrast-producing effect of the spherical aberration. Changing the focus in either direction from the Gaussian defocus produces strong contrast maxima. Changing the defocus further still produces oscillatory contrast behavior with secondary contrast minima, which are much less pronounced than the main minimum, interspaced by secondary contrast maxima. The Gaussian defocus therefore constitutes a useful reference point to which all other defocus values can be easily related.

To obtain useful images of thin phase objects, one typically focuses the microscope onto a plane a few hundred angstroms higher than the Gaussian-defocus plane. The small defocus and the spherical aberration of the objective lens then cause the electrons to arrive outside the positions corresponding to where they exited from the specimen, and phase contrast is observed. The optimum defocus that produces strong contrast [large $T(\mathbf{U})$] over the broadest range of spatial frequencies is called the Scherzer defocus. It corresponds to the first contrast maximum, which is observed as the objective-lens current is weakened (underfocused) relative to the

TABLE 12.2
Defocus Values of the Objective Lens that Produce Broad Contrast-Transfer Intervals

Defocus value (Δf)	Center of main interval $\left[\left(\dfrac{-c_s \lambda^2}{\Delta f}\right)^{1/2}\right]$	Values for JEOL 4000EX at 400 kV		Transfer through main interval
		Defocus (Å)	Center of main interval (Å)	
$-(C_s\lambda)^{1/2}$	$C_s^{1/4}\lambda^{3/4}$	-404	2.57	Negative (black atoms)
$-(3C_s\lambda)^{1/2}$	$0.74 C_s^{1/4}\lambda^{3/4}$	-702	1.96	Positive (white atoms)
$-(5C_s\lambda)^{1/2}$	$0.65 C_s^{1/4}\lambda^{3/4}$	-906	1.72	Negative
$-(7C_s\lambda)^{1/2}$	$0.61 C_s^{1/4}\lambda^{3/4}$	-1071	1.58	Positive

Gaussian-defocus value. Further underfocusing leads to the secondary contrast maxima in which high spatial frequencies pass through broad contrast intervals and give strong image contrast, but the contrast due to small spatial frequencies (large specimen features) becomes difficult to interpret.

The underfocus values that produce broad contrast-transfer intervals and the locations of the centers of these intervals are listed in Table 12.2. In the so-called generalized contrast-transfer theory (Hawkes, 1980), where the units of defocus and the reciprocal-spatial frequency U are chosen so that the theory's results become independent of the microscope parameters, the quantity $-(C_s\lambda)^{1/2}$ is called defocus of 1 Scherzer, and the progression of optimum defocus values is simply $-1^{1/2}$, $-3^{1/2}$, $-5^{1/2}$, and so on.

The main contrast-transfer intervals can be broadened slightly by increasing the underfocus by a small amount, although this has the undesirable side effect that the transfer strength at the center of the main interval becomes weaker. A typical compromise involves changing the progression of optimum defocus values to $-1.5^{1/2}$, $-3.5^{1/2}$, $-5.5^{1/2}$, and so on, in which case the centers of the contrast-transfer intervals correspond to 90 per cent contrast transfer. Phase-contrast transfer functions calculated for the modified progression of optimum defocus values and typical values of energy spread and angular spread of the illumination are plotted in Figure 12.4.

As can be seen in Figure 12.4, the increase in the resolution owing to the shift of the main transfer interval to higher spatial frequencies has to be paid for by a more complicated contrast transfer that includes zero crossings and both positive and negative contrast transfer at lower spatial frequencies. These defects in the contrast transfer function can be partly corrected by *a posteriori* image processing (Maréchal and Croce, 1953) of a single micrograph. The simplest form of the processing works by changing the sign of every second contrast interval. This gives a contrast transfer function that contains no contrast reversals but still has contrast-transfer gaps. A more

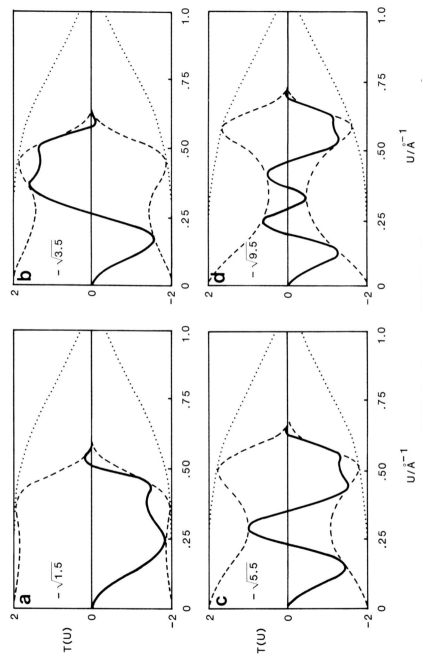

FIGURE 12.4 Contrast transfer functions for a 300-kV, $C_s = C_c = 1.1$ mm HRTEM at different values of defocus. (*a*) 570 Å (Scherzer defocus). (*b*) −870 Å ($-\sqrt{3.5}$ generalized defocus). (*c*) −1091 Å ($-\sqrt{5.5}$). (*d*) −1435 Å ($-\sqrt{9.5}$). The envelope functions shown are due to an energy spread of 2 eV (dotted line) and an angular spread of 2 mrad (dashed line).

complete correction can be obtained by combining a series of micrographs recorded at different defocus values, in which case the transfer gaps can be eliminated (Saxton, 1980). Nevertheless, the correction becomes complicated if the specimen is not very thin and nonlinear image terms are important, as is, unfortunately, the case in many practical specimens. For such specimens, it is advisable not to attempt to correct the contrast transfer function but simply work with the best contrast transfer function that the microscope can produce. The use of higher underfocus is therefore mostly unsuitable for imaging of large-unit-cell materials such as many complex oxides, where the relatively large spacings need to be transferred faithfully. In the imaging of small-unit-cell materials such as metals, there are usually no larger spacings to worry about. If the imaging is to be performed with a microscope whose basic point-to-point resolution is not quite sufficient to resolve the lattice planes of interest, the use of the higher underfocus is an attractive solution.

Provided that the specimen is stable in the electron beam, it is a usual practice to record a through-focus series of micrographs, in which the defocus is varied by a small amount about the best guess at the optimum value. The necessary sampling density can be estimated by requiring that there be n micrographs taken per each contrast-transfer reversal at the highest spatial frequency U_{max} to be recorded. A contrast reversal means a change in χ by one π, and equation (12.5) therefore gives

$$\Delta f = \frac{1}{n\lambda U_{max}^2} \quad (12.14a)$$

for the defocus step Δf, or in terms of the smallest spacing that we want to image, d_{min},

$$\Delta f = \frac{d_{min}^2}{n\lambda}. \quad (12.14b)$$

Taking $n = 3$ is usually sufficient. Indeed, one does not want to waste photographic material by selecting a defocus step smaller than the defocus spread δ, which is typically around 100 Å in a 100-kV microscope and 50 Å in 300 to 400-kV microscopes. Suitable defocus-step sizes are therefore around 100 to 200 Å in 100-kV microscopes, and 50 to 100 Å in 300 to 400-kV microscopes.

Estimating the optimum defocus setting prior to taking the photograph can only be learned while operating the microscope. The usual technique involves first setting the lens to the Gaussian defocus by finding the contrast minimum in a thin amorphous part of the specimen, such as the amorphous-carbon support film or the amorphous contamination at the edge of the specimen. The next step is to adjust the defocus by as many clicks of the objective-lens-current control as necessary to reach the estimated optimum value, where the calibration of the current control in terms of the defocus changes had been obtained previously.

When a crystalline specimen of known spacing is being examined, the defocus can be determined with improved precision from the contrast maxima and minima shown by the appropriate lattice fringes in the thinnest parts of the specimen (i.e., right at the specimen edge, where one has a good chance of finding something approaching a weak phase object). The maxima and minima occur when the aberration function $\chi(U)$ reaches values of $n\pi/2$, where n is either odd (maxima) or even (minima; see Chapter 1). The ambiguity in the value of n, which is a consequence of the recurring nature of lattice-fringe contrast (Fourier images), can usually be removed by observing the Fresnel fringe at the specimen edge, since this fringe disappears at a defocus close to zero. All of these steps are made much easier if an efficient TV is fitted to the microscope, which considerably improves the visibility of the high-resolution image above what is attainable by direct observation of the microscope-viewing screen through binoculars. It is even possible for a computer, which is interfaced to the TV camera, to determine the defocus value on-line with better precision than a human operator, as discussed later.

12.3.2.2 Astigmatism adjustment. The usual approach involves adjusting the astigmatism by making the contrast in an image of an amorphous specimen disappear equally in all directions as the Gaussian defocus is reached. Again, the adjustment can only be learned properly by operating the microscope, and a computer interfaced to a TV camera can perform the adjustment better than a human operator.

12.3.2.3 Alignment of the illumination direction. Provided that the mechanical alignment of the objective-lens pole pieces is satisfactory, the alignment of the illumination direction can, in principle be corrected simply by adjusting the incident-beam tilt so that the beam enters the objective lens along its axis. The problem, however, is that the optic axis of the lens is not easy to find. It is the center of symmetry of the aberration function $\chi(U)$, and it corresponds to neither the current center of the lens nor the voltage center. $\chi(U)$ cannot be measured directly, but it is revealed as the phase shift experienced by one beam traversing the objective lens relative to the phase shift of another beam. In order to determine the position of the optic axis accurately, one therefore has to analyze the detailed appearance of a high-resolution image, which is very sensitive to the relative phase shifts.

The most clearly visible effect of a small tilt of the incident beam on a high-resolution image looks the same as a change in astigmatism plus a change in defocus, and these "apparent" astigmatism and defocus changes can be compensated for by changing the real astigmatism and defocus in the opposite way. Another readily visible effect of the beam tilt is a shift of the image, which also depends on the defocus and C_s. A less visible but far more important effect of the beam tilt is the appearance of the so-called axial coma in the image (Krivanek, 1975, 1978; Zemlin et al., 1978; Smith et al., 1983), which distorts the image without affecting the strength of the contrast transfer of the different spatial-frequency components. These and other effects can be worked out by considering the phase difference between the main beam, which passes a distance $\mathbf{U}_0$ away from the optic axis instead of directly on the axis, and the scattered beam $\mathbf{U}$, which passes a distance $\mathbf{U} - \mathbf{U}_0$ away from the axis (McFarlane, 1975; Krivanek, 1975, 1978; Zemlin et al., 1978; Smith et al., 1983). The important point that emerges from the analysis is that, although small beam misalignment (less than about 1 mrad) cannot be detected from a single micrograph except under rather special conditions (Heinemann, 1971; Krivanek, 1976a), in large-unit-cell materials, the axial coma due to a beam misalignment of 1 mrad can produce significant differences between the image symmetry and the symmetry of the specimen.

The change in the image appearance due to the tilt of the incident beam is illustrated in Figure 12.5. The micrograph in Figure 12.5a was recorded after the microscope was aligned by using the method described below, and Figure 12.5b was recorded after a change in the incident-beam tilt by 1.2 mrad. No correction was made for the resultant changes in the apparent defocus and astigmatism, so the principal differences between the two images are actually the result of the change of apparent defocus. However, whereas the aligned micrograph shows nearly perfect glide symmetry, the misaligned micrograph shows, upon a careful examination, small departures from the glide symmetry. These are due to the axial coma and cannot be eliminated except by correcting the beam tilt.

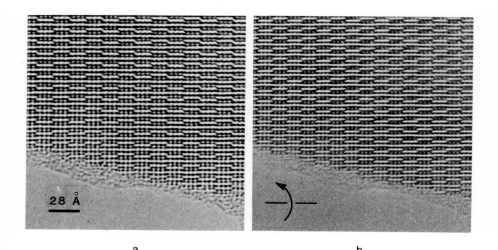

FIGURE 12.5 Micrographs of $Ti_2Nb_{10}O_{29}$ illustrating the effect of beam tilt. (*a*) Incident beam axially aligned. (*b*) Incident beam tilted by 1.2 mrad about the axis indicated. Micrograph taken on the Cambridge HRTEM at 500 kV. (Courtesy of D. J. Smith.)

The apparent astigmatism, defocus, and image shift owing to the beam tilt cannot be measured from a single micrograph alone, and one therefore needs to compare several images to determine the presence of the beam misalignment. Most strategies for aligning the incident-beam direction use the fact that the apparent astigmatism and defocus changes alter as U_0^2 (Krivanek, 1978), which means that adding an additional tilt α to a perfectly aligned main beam first in one direction and then in the opposite direction will produce two identical-looking images; whereas if there is a small amount of misalignment to start with, the images will be different. Hence, the adjustment requires a beam wobbler, which tilts the beam by plus or minus an adjustable amount about the normal angle of the main beam, at a frequency that allows the operator to assess each image appearance at a high magnification—i.e., about 1 s in each position—and provides an instantaneous switchover between the two. With the wobbler in operation, the microscopist adjusts the tilt of the main beam until the two images look exactly the same. The adjustment is performed for two perpendicular directions (x and y) of the beam wobble, and it is completed when one cannot tell, from the image appearance, that the beam direction has been switched by the wobbler (Zemlin, 1979; Smith et al., 1983). The optimum value of the additional beam tilt α is around $\frac{2}{3}\lambda/d$, where d is the point-to-point resolution of the microscope. Too small a value of α results in poor sensitivity of the adjustment; too large a value makes the adjustment difficult to perform. The adjustment works best with an amorphous film such as amorphous carbon, which must be thin enough so that its structure looks the same when projected along the two test directions (i.e., thickness $t \ll d/2\alpha$, where d is the image resolution and α is the magnitude of the additional tilt). The adjustment becomes easier if it is performed on a specimen area containing an easily recognizable feature such as the specimen edge or a small adsorbed particle, so that it is easy to keep track of the same image area. As one converges on two similar-looking images, it is also helpful to adjust the objective-lens defocus to minimize the image movement.

This procedure works well as long as there is no interaction between the beam-deflection coils and the final image other than through the effects of the beam tilt. With some microscopes, unfortunately, a small fraction of the field from the illumination-deflection coils leaks inside the objective lens below the sample, causing a spurious image deflection when the beam direction is switched. It then becomes impossible to get the two partial images to coincide exactly, unless one also makes the beam wobbler deflect the final image by a small amount.

12.3.2.4 Complete alignment of the objective lens. The adjustments of the objective lens defocus and astigmatism and of the beam tilt are not independent, so setting up the objective lens actually becomes a search for the optimum value in five dimensions (defocus, astigmatism x and y, beam tilt x and y), and it is important to consider the order in which the adjustments are performed. The apparent astigmatism depends sensitively on the beam tilt, so adjusting the astigmatism should always follow adjusting the tilt. Fortunately, the presence of a small amount of astigmatism has no effect on the beam-tilt adjustment other than a small image shift (Krivanek, 1978; Smith et al., 1983). Hence, one can usually roughly adjust the defocus to about the Scherzer value, adjust the beam tilt, readjust defocus to Gaussian (minimum contrast), and complete the adjustment by eliminating the astigmatism. If the lens was completely out of adjustment, several iterations may be necessary. Subsequent small changes in defocus can be made without upsetting the other parameters. Large changes in the objective-lens current necessitated by a change in the operating voltage of the microscope or by a change in the specimen height by several tens of micrometers require that the complete alignment be performed anew.

12.3.2.5 Specimen tilt. The final adjustment of the specimen tilt should be performed once the incident-beam direction has been adjusted. However, in the absence of a doubly eucentric, tilting specimen holder or at least a precise z-motion stage, a major change in the specimen orientation leads to a significant change in the specimen height and a major change in the objective-lens current; and this necessitates a realignment of the lens. Hence, it is best to first tilt the specimen into roughly the desired orientation, perform the objective-lens alignment, and then carry out the final specimen tilting. If the microscope can form a small probe so that higher-order Laue zone (HOLZ) lines (Steeds, 1979) become visible in the diffraction pattern, the final adjustment of the specimen tilt can, in principle, by performed with a precision of up to 0.1 mrad. However, small-probe capabilities are not implemented in most high-resolution microscopes; and in any case, the HOLZ lines are only clearly visible in specimen areas whose thickness is too large for high-resolution imaging. The specimen tilt is therefore normally adjusted by equalizing the strength of the plus and minus Bragg beams in the zone-axis diffraction pattern. The sensitivity of the adjustment is improved if one concentrates on higher-order Bragg beams, and adjustment to a precision of better than 2 mrad is not difficult. For specimen thicknesses less than 50 Å, this level of crystal mistilt produces virtually no observable effects in the image (Smith et al., 1983).

12.3.3 *Aligning the rest of the microscope column*

The first intermediate image formed by the objective lens has a typical magnification of 100 to 200×. The remaining lenses of the microscope column operate on this magnified image, so their aberrations, distortions, and lack of alignment are much less important. The most significant exception to this occurs in low-dose imaging of periodic biological objects, where it is important to record images extending over large specimen areas and image distortion caused by the intermediate lenses interferes with the image analysis. The convenience of having the image expand about the center of the screen as one steps through the magnification ranges justifies spending the time to align the microscope column well; but since the alignment is not critical to the high-resolution performance, it need not be done daily.

One important point is that a large change in the current of the first post-objective lens (called either the diffraction lens or the first intermediate lens) changes the height of the plane on which this lens is focused. Retaining a focused final image then requires that the height of the first intermediate image formed by the objective lens be adjusted by changing the objective-lens current, and this usually leads to significant changes in the astigmatism of the objective lens and, sometimes, the tilt of the lens axis. Since it is desirable to be able to change the magnification between about 200,000× and the top magnification without having to readjust the alignment of the objective lens, the current in the first intermediate lens should be kept constant over this range; and the magnification changes should be implemented by varying the current in the lenses further down the column. Several modern microscopes now avoid changes in the first-intermediate-lens current at the top magnifications. Modifying the other microscopes to keep the intermediate lens current constant is usually a simple matter of changing the program that governs the imaging lenses.

12.3.4 Recording the high-resolution image

A frequently expressed concern is that one focuses the image on a focusing screen that is at a different height inside the projector chamber than the photographic film, and that the defocus of the image on the film will therefore be different from the defocus of the image on the screen. This is certainly correct, but the defocus difference is only $\Delta f = \Delta z/M^2$, where Δz is the difference in the height of the focusing screen and of the photographic film and M is the magnification. At a magnification of 100,000× and $\Delta z = 10$ cm, the defocus difference is therefore just 0.1 Å. The difference is even less at higher magnifications, so correcting for this effect at any magnification used in high-resolution imaging is never required.

In recording the image, one must take into account the nature of the information desired from the micrograph and the resolution of the photographic medium. The usual resolution of a photographic emulsion is around 50 line pairs per millimeter, meaning that 2-Å detail can be captured at any magnification greater than 100,000×. The use of such low magnifications is appropriate for two applications where only the periodic information in the micrograph is of interest: low-dose imaging of periodic objects and the optical-diffractogram analysis of the properties of the objective lens described below. In both cases, the low magnification means that the illumination intensity can be reduced. For low-dose imaging, this is absolutely essential; for the optical-diffractogram application, it permits one to use weaker but more coherent illumination achieved by lowering the beam current and decreasing the condensor-aperture size. This leads to a greater extent of the envelope functions, giving diffractograms extending to higher spatial frequencies and containing many rings.

In most other high-resolution imaging applications, however, the primary interest lies in the nonperiodic component of the image such as the precise image detail at a lattice defect. Here, in order to see the nonperiodicity clearly, one must consider the image shot noise (i.e., the $\sqrt{N}$ statistical variation in the number of electrons arriving at each point in the image), which must be as small as possible. In practice, high-resolution images are recorded on photographic film with about 10^3 to 10^4 electrons per resolution element (pixel). These doses give a random variation in the intensity of each image pixel of 3 and 1 percent, respectively, provided that the recording medium records the incoming electrons with a detector quantum efficiency (DQE) close to 1, which is the case for most photographic emulsions. Variation of 3 per cent can be more than the contrast variation from pixel to pixel. To ensure that the observed contrast is actually due to the specimen and not the image shot noise (which can play unscientific tricks on the eyes of the unsuspecting observer), the number of electrons per pixel should therefore be made as high as possible. The best way to achieve this without saturating the photographic emulsion is to use rather high electron optical magnifications of the order of 1,000,000×, or to use a finer-grained (higher-resolution) but less sensitive emulsion at magnifications of the order of 500,000×.

The shot noise is even more serious when one is observing the image in real time through a television camera, since the integration in this case is typically reduced to the response time of the eye—i.e., about 0.1 s. This is 50 times less than the typical 5-s exposure on film, and the shot noise is therefore typically some 7 times worse than in the recorded images. The noise can give the impression that the object under observation is changing with time or, to put it more poetically, "is dancing under the electron beam." Indeed, a computer simulation of a sequence of high-resolution images of an unchanging specimen that explicitly includes the image shot noise (Wood, 1985) shows a striking resemblance to some of the recent TV pictures of atomic motion.

In summary, practical high-resolution imaging is a matter of achieving the optimum compromise between several conflicting requirements: The illumination should be made as coherent as possible by decreasing the emission current and the illumination aperture sizes; the exposure time should be minimized to prevent image smearing owing to specimen drift; and yet the signal-to-noise ratio in the image should be kept as high as possible by maximizing the number of electrons recorded per resolution element. Fortunately, the optimum is a fairly shallow one, so a small error in most of the adjustments is not too serious. Hence, anyone prepared to spend a little time analyzing the practical consequences of the adjustments can become an efficient HRTEM operator.

12.3.5 Diffractogram analysis

There are several methods available for determining each of the important imaging parameters from a recorded micrograph or a series of such micrographs. For instance, defocus and the coefficient of spherical aberration C_s of the objective lens can be determined by comparing the displacement of corresponding bright-field and dark-field images (Budinger and Glaeser, 1976); and the coefficient of chromatic aberration can be determined from the defocus change when the accelerating voltage is changed by a small amount. The most general method of image analysis, however, involves examining the power spectrum of a recorded micrograph of an amorphous object and deducing the important parameters from the contrast transfer function, which the power spectrum reveals. This is the optical-diffractogram analysis pioneered by Thon (1965, 1971) in which a micrograph of a thin amorphous specimen such as amorphous carbon is placed in an optical bench and its power spectrum is obtained by forming the micrograph's Fraunhofer diffraction pattern (or the micrograph is digitized and a power spectrum is obtained by a computer). The resultant diffractogram shows the strength of the contrast transfer at different spatial frequencies, which enables all the important imaging parameters to be determined.

The diffractogram intensity is given by

$$D(U) = c \, |I(U)|^2, \qquad (12.15)$$

where c is a constant dependent on the exposure of the diffractogram and other factors, $I(U)$ is the Fourier transform of the image intensity, and we have assumed that the diffractogram scale is calibrated in terms of the reciprocal spatial frequency U of the specimen. The calibration is most easily done if the specimen contains a strong periodicity of a known spacing that also shows up in the diffractogram, as, for instance, when an amorphous-carbon film is partially covered by small evaporated-gold crystallites. Assuming once more that the specimen is a thin phase object, and using expressions (12.4) and (12.5), we obtain

$$D(U) = c' \, |A(U)|^2 \, |E(U)|^2 \, |\sin \chi(U)|^2 \, |\Psi_s(U)|^2. \qquad (12.16)$$

The recorded diffractogram therefore contains information on the aperture function, the envelope function, and the $\sin^2$ of the aberration function. The diffractogram intensity is also weighed by the square of the electron-wave-function amplitude in reciprocal space—that is, by the intensity of the scattering distribution as revealed in the diffraction pattern. The ideal specimen for the diffractogram analysis would have a diffraction pattern showing the same intensity at all spatial frequencies U, which would be the

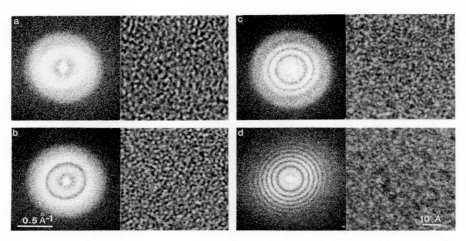

FIGURE 12.6 Micrographs and diffractograms from amorphous germanium recorded at different defocus values. (a) Scherzer defocus. (b) Generalized defocus $-\sqrt{3.5}$, (c) $-\sqrt{5.5}$. (d) $-\sqrt{15}$. The micrographs were recorded in a Philips 430ST at 300 kV. The contrast transfer extends out to 1.95 Å in (a), 1.8 Å in (b), 1.7 Å in (c), and 1.5 Å in (d).

case if the specimen consisted of a random assembly of point scatterers. On the other hand, for a crystalline specimen whose diffraction pattern consists of a series of delta-function spikes, the information on the imaging parameters obtainable by diffractogram analysis would be very incomplete. In practice, one uses amorphous specimens that contain all spatial frequencies but whose scattering intensity falls off at higher angles, as can be seen in the diffraction pattern. Amorphous carbon is the usual sample for diffractogram analysis, but it has very little diffracted intensity between $(1.6 \text{ Å})^{-1}$ and $(1.2 \text{ Å})^{-1}$ (Kakinoki et al., 1960). This may explain the paucity of convincing experimental diffractograms showing contrast transfer beyond $(1.6 \text{ Å})^{-1}$.

A diffractogram consists of a series of bright rings, ellipses, or hyperbolas, depending on the defocus, the astigmatism, and the incident-beam tilt, as shown in Figures 12.6, 12.7, and 12.8. It enables the microscopist to determine all of the important imaging parameters.

FIGURE 12.7 Micrographs and diffractograms of amorphous carbon illustrating various instrumental misadjustments. (a) Well-aligned, stigmated, with no specimen drift. (b) Moderate amount of astigmatism ($c_a = 140$ Å). (c) Large amount of astigmatism ($c_a = 800$ Å). (d) Stigmated, with about 3-Å specimen drift during exposure. (e) Stigmated, with 5-Å specimen drift. (f) Graphitized carbon used for calibration of diffractogram scale showing rings corresponding to the 3.44 and 2.03-Å spacings. Micrographs (a) and (f) were recorded at Scherzer defocus; micrographs (b), (d), and (e) were recorded at $-\sqrt{5}$ generalized defocus; and micrograph (c) was recorded at Gaussian defocus. All micrographs were taken with the Philips 430ST at 300 kV.

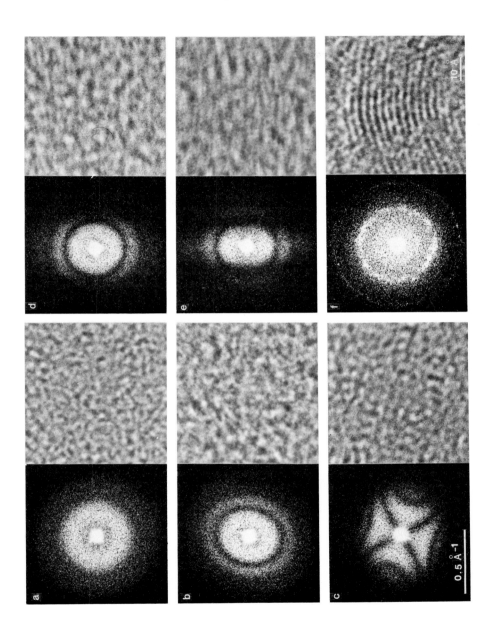

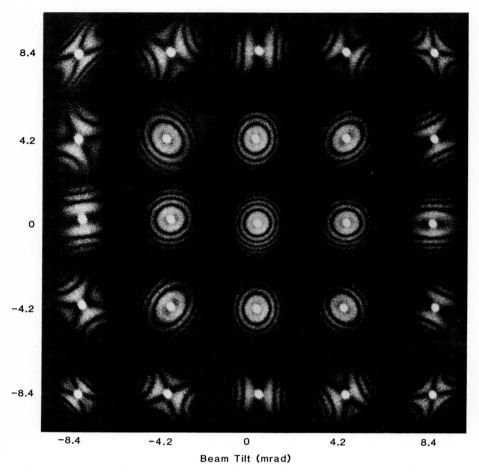

FIGURE 12.8 Diffractograms showing the effect of the tilt of the incident beam on the contrast transfer. Intentional beam tilt is proportional to the distance of the diffractogram from the center of the tableau. Micrographs were taken on the Siemens 102 at 125 kV.

12.3.5.1 Determination of the objective-lens defocus, astigmatism, and coefficient of spherical aberration. A series of experimental micrographs and diffractograms, taken at various defocus values of the objective lens, is shown in Figure 12.6. The diffractograms correspond closely to the computed contrast transfer functions shown in Figure 12.4. Note that a diffractogram corresponding to a micrograph taken near a generalized underfocus of $-\sqrt{(2n+1)}$ normally shows n bright rings, and that the main contrast-transfer interval (the broadest ring visible in each diffractogram) moves out to higher spatial frequencies as the underfocus increases. Also note that the image changes produced by the defocus variation are subtle and hard to interpret quantitatively, but the diffractograms change in a manner that is easy to understand and quantify.

The defocus value is obtained from the radii of the bright and dark diffractogram rings as follows: The intensity maxima of the bright rings arise when $\sin(\chi(U)) = 1$, and the intensity minima of the dark rings arise when $\sin \chi(U) = 0$. This means that the maxima occur when

$$\chi(U) = \frac{n\pi}{2} \quad n \text{ odd}, \quad (12.17\text{a})$$

and the minima occur when

$$\chi(U) = \frac{n\pi}{2} \quad n \text{ even}. \quad (12.17\text{b})$$

Hence, the location of the rings is fully determined by the value of the aberration function (equation 12.5), which depends only on the defocus Δf and the spherical aberration coefficient C_s. It is then conversely possible to determine Δf and C_s from the radius of the rings of a diffractogram, as long as the diffractogram contains at least two (and preferably many more) rings.

The most convenient way to determine Δf and C_s is to first rearrange equation (12.5) so that the experimental data—that is, the radii of the diffractogram rings—can be plotted in such a way that one obtains a straight line whose slope depends only on C_s and whose intercept with the ordinate depends only on Δf (Krivanek, 1976b). This is achieved by inserting the condition for the bright and dark rings (12.17) into the left side of (12.5), and rearranging the result slightly:

$$\frac{n}{U^2} = C_s \lambda^3 U^2 + 2 \Delta f \lambda. \quad (12.18)$$

This expression shows that plotting n/U^2 as a function of U^2 will result in a straight line whose slope is equal to $C_s \lambda^3$, and whose intercept with the ordinate is equal to $2 \Delta f \lambda$. The values of U^2 are determined from the diffractogram once it has been calibrated in reciprocal angstroms or nanometers by using a discrete reflection due to lattice planes of a known spacing also recorded in the micrograph.

The values of n are assigned starting at 1 for the intensity maximum of the bright ring closest to the origin, 2 for the first dark ring, 3 for the second bright ring, and so on, when the micrograph was taken at an overfocus, or assigning -1, -2, -3, and so on, when it was taken at moderately large underfocus. The assignment becomes slightly more complicated close to Scherzer defocus, where the first ring corresponds to $n = -1$, but subsequent rings (if visible) have $n = 0, 1, 2$, and so forth, and at other small underfocus values for which n decreases all the way to the broadest ring and then starts increasing. However, a mistake in the assignment is quickly discovered since the points will then not even approximate a straight line. It is usually also helpful to draw a reference line with a slope of $C_s\lambda^3$ by using the manufacturer's value for C_s of the microscope and making sure that the diffractogram lines are approximately parallel to it.

The precision of the analysis depends on the quality of the input data. Since the effect of spherical aberration becomes particularly strong at high spatial frequencies, the determination of C_s with good precision demands that the diffractogram rings extend to the highest values of U possible. In practice, this means that the best data are recorded at fairly large underfocus of the order of $-(9C_s\lambda)^{1/2}$, at which point it becomes possible to determine C_s to an accuracy of a few percent (Kuzua and Hibino, 1981).

The analysis can be simplified further by realizing that, in a plot of n/U^2 as a function of U^2, all the points have to lie on a set of hyperbolas in which there is one hyperbola for each value of n. Hence, it is possible to produce a reference graph containing the hyperbola set and to enter the diffractogram data by marking a point for each diffractogram ring at the determined U^2-value on the hyperbola corresponding to the appropriate n (Krivanek, 1976b). Another variant of the method is to enter the U^2 and n-values into a computer and use a least-squares fit to obtain the slope and the intercept of the straight line. A listing of a program that will do this is included in Appendix 1 of the book by Spence (1981). This makes the analysis convenient; but because the program generates no graphic output, it is not as easy to spot an incorrect assignment of the n-values as with the graphic method.

Micrographs and diffractograms revealing moderate and large astigmatism are shown in Figures 12.7b and c, respectively. Note that small to moderate astigmatism gives quasi-elliptical instead of round diffractogram rings. Large astigmatism may cause the defocus to be positive in one direction and negative in the perpendicular direction. In this case, the diffractogram shows a characteristic pattern of curves resembling hyperbolas (Figure 12.7c).

When the diffractogram shows that there was astigmatism, the analysis becomes slightly more complicated. First, it is necessary to find the two lines of mirror symmetry that are contained in all astigmatic diffractograms. In diffractograms showing the quasi ellipses, the mirror lines are simply the major and minor axes of the ellipses. In diffractograms showing the quasi hyperbolas, the mirror lines bisect the angles between the asymptotes of the quasi hyperbolas. In both cases, the mirror lines correspond to the directions of the largest and the smallest defocus. The defocus is then determined as before, but for the two directions separately. The coefficient of astigmatism C_a is equal to one-half the difference between the two extreme defocus values; the defocus is equal to their average.

12.3.5.2 Determination of the energy spread and angular spread of the illumination and of specimen drift or vibration. The energy spread and angular spread of the incoming electron beam, the specimen drift and vibration, and the other instabilities all attenuate the contrast transfer, but they do not change the radii of the bright and dark rings visible in the diffractograms. In other words, the illumination incoherence and instrumental instabilities primarily affect only the envelope function $E(U)$. Determining their magnitude from high-resolution micrographs therefore involves experimentally determining $E(U)$.

The energy spread and the angular spread of the illumination can be determined by analyzing the variation of the envelope function of a whole through-focus series of micrographs taken with axial illumination (Frank, 1978). However, both the energy spread and the angular spread can be determined more easily by other methods. The energy spread can be measured by attaching an electron-energy-loss spectrometer to the high-resolution electron microscope and determining the energy width of the zero-loss peak. The angular spread can be determined by analyzing a diffraction pattern recorded in the microscope with the illumination set up the way it was when the image was taken, as described in section 12.3.1. Diffractogram analysis is therefore used only rarely to determine these two parameters. If a spectrometer is not available and it is desirable to measure the energy spread (as, for instance, when there is reason to suspect that high-voltage instabilities are degrading the resolution), this can also be conveniently accomplished by recording micrographs with highly tilted illumination and analyzing the resultant diffractograms by using the methods of Parsons and Hoelke (1974).

The effect of specimen drift is discerned in diffractograms (see Figures 12.7d and e): Uniform drift by a distance **d** during the time of the exposure of the micrograph gives rise to an envelope function $E_i U)$ of the form

$$E_i(U) = \frac{\sin(\pi \mathbf{U} \cdot \mathbf{d})}{\pi d}, \qquad (12.19)$$

where **U** and **d** are both vectors. The right side of the equation shows that if there is specimen drift during the exposure, the diffractogram will become modulated by a series of bands whose profile resembles the diffraction pattern of a single slit, with a characteristic form of one main maximum and weaker side maxima of one-half the width of the main maximum. The main maximum forms a band of width $2/d$, and the drift direction is perpendicular to the bands. Large drift (greater than the resolution of the micrograph) produces diffractograms in which the main band and usually two secondary bands are clearly visible (Figure 12.7e). Smaller drift results in weakening of the outer diffractogram intensity in the direction of the drift and is not always readily distinguishable from other causes. Random vibrations of the specimen also weaken the outer diffractogram intensity and cannot always be distinguished from other effects. Real-time observation through a good TV system is therefore invaluable when mechanical instability of the specimen is suspected, but quantifying the effect is usually easiest from a diffractogram.

12.3.5.3 Determination of the incident-beam tilt with respect to the optic axis. As a rule, a single diffractogram does not reveal the angle between the main beam and the optic axis. One important exception does occur, however: When a micrograph of an amorphous object is taken at large underfocus [over $-(10C_s\lambda)^{1/2}$], with very small energy spread but sizable angular spread of the illumination, the maximum of the spatial-incoherence envelope $E_s(U)$ becomes directly visible in its diffractogram (Krivanek, 1976a). When the illumination is perfectly aligned, the maximum takes the form of a broad ring of the same radius as the main contrast-transfer interval [$U = (\Delta f/C_s\lambda^2)^{1/2}$], and is not readily distinguishable. When the illumination is misaligned, even by a small amount, the maximum splits into two complementary rings that are displaced from the center of the diffractogram by $\pm\frac{3}{2}U_0$, where U_0 is the tilt of the main beam. The ring radius stays the same, enabling C_s to be determined very accurately.

Under more general conditions, the tilt only becomes apparent in a series of diffractograms taken at different incident-beam tilts and put together in a diffractogram tableau, as shown in Figure 12.8 (Zemlin et al., 1978; Krivanek, 1978). In such a tableau, the diffractograms taken at plus and minus a given intentional beam tilt will only look the same if the beam tilt for the central diffractogram is zero. For instance, in the tableau of Figure 12.8, the diffractograms above and below the central diffractogram are nearly identical, showing that the incident-beam alignment in the y-direction was nearly perfect. The diffractograms to the left and to the right of the central one are slightly different (the rings have different radii), showing that the alignment in the x-direction was not perfect. Determining the apparent defocus for several diffractograms taken at different beam tilts and plotting the defocus as a function of the tilt will result in a parabola, whose center is the real axis of the objective lens and whose width is determined by C_s (Krivanek, 1978).

12.3.6 Automatic alignment

The alignment of the microscope can be handled on-line by a computer, and the results can surpass the accuracy attainable even by the most experienced human operators (Saxton et al., 1983). The necessary hardware consists of two links between the computer and the microscope (Figure 12.9): an image pickup system and computer interface enabling the computer to "see" the image produced by the microscope, and a microscope-control interface enabling the computer to set the important imaging parameters.

Before launching the computer alignment, the microscope operator needs to select the magnification and illumination conditions, find a suitable homogeneous area on an amorphous specimen, and perform a rough adjustment of the focus and astigmatism. The computer then takes over and finds the optimum defocus and adjusts the astigmatism and the beam tilt to zero. It proceeds by varying each parameter over a predetermined range

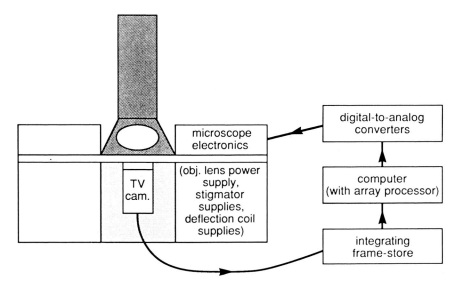

FIGURE 12.9 Block diagram of a computer control system for microscope autoalignment.

and analyzing the image appearance at each step. Although one might expect that the imaging parameters would be determined most accurately by obtaining and analyzing the image power spectrum essentially as described above, the computer alignment schemes implemented so far have simply analyzed the variation in the image contrast as each parameter is varied. The contrast is worked out either by cross-correlating pairs of images recorded at each microscope setting (Saxton et al., 1983) or by recording one image at each setting and determining the contrast after a low- and a high-spatial-frequency filter have been applied to the image (Krivanek and Wood, 1984). The first method takes advantage of the fact that the cross-correlation product gives a measure of the image contrast in which the effects of the image shot noise and irregularities of the photographic emulsion have been eliminated (Frank and Al-Ali, 1975). The second method eliminates most of the shot noise by filtering out spatial frequencies higher than the resolution limit and also removes the very low spatial frequencies that contain spurious intensities due to slow variation of image intensity across the whole image field.

Provided that the beam tilt and astigmatism are roughly adjusted, the image will show a well-defined contrast minimum at the Gaussian defocus and a maximum at the Scherzer defocus (Frank and Al-Ali, 1975). When the defocus is set to the minimum-contrast value and the astigmatism and beam tilt are varied, zero astigmatism and beam tilt correspond to a contrast minimum. If the defocus is set to the Scherzer value, zero astigmatism and beam tilt produce a contrast maximum. The alignment algorithm then consists of finding these minima or maxima by evaluating the contrast at each new value of the parameter being varied, and fitting the experimentally determined variation by a parabola, whose center marks the searched value. Complete alignment requires that all five parameters be adjusted one after the other; and since some 20 images are explored for each parameter, the total number of images that need to be acquired and processed is large. With efficient television systems and array processors, the whole alignment can nevertheless be accomplished in less than 1 min (Saxton et al., 1983).

The chief limitations of the autoalignment are that the algorithm will not converge if the microscope is initially badly misaligned and that a thin amorphous specimen is an absolute necessity. The first limitation could be overcome by incorporating different strategies for the initial search for the optimum parameters—for example, the computer could look initially for a contrast maximum and start the final optimization after the maximum is found. Extending the method to nonuniform or even crystalline samples will probably prove to be more difficult, although, in principle, an approach based on detecting the shift of the image as the beam tilt is varied should work on any sample.

12.3.7 Tilted-illumination imaging

Tilting the illumination by an angle such that the optic axis of the objective lens lies halfway between the scattered beam of interest and the main beam makes the image of the corresponding lattice planes completely insensitive to defocus variation and also decreases the loss of contrast due to the angular spread of illumination. The insensitivity to defocus variation is illustrated in Figure 1.10, which shows that two beams traveling at equal angles to the optic axis suffer an identical phase shift when the defocus changes. The phase difference between the beams and the position of the fringes that result from their interference are therefore not affected by defocus variation. Hence, the intensity of the fringes remains the same even as the electron energy is varied by several electron-volts.

In an optical diffractogram, the main beam and all the beams traveling at the same angle to the optic axis fall on the so-called achromatic circle (Hoppe, 1974; Parsons and Hoelke, 1974), which is centered on the optic axis and which can extend to spatial frequencies as high as $(1 \text{ Å})^{-1}$ even in a 3 Å point-to-point resolution microscope. It then becomes possible to transfer rather high spatial frequencies into the image, resulting in high lattice resolutions. This is the basis of the tilted-illumination-imaging technique introduced by Dowell (1963), which was at one time widely used to obtain information about specimen spacings smaller than the point-to-point resolution of the microscope. The main value of such images lies in revealing what spacings are present in a particular specimen region (Sinclair et al., 1976); but since the phases of the spatial frequencies transferred are invariably distorted, the images can never be taken as representative of the specimen structure. Computer simulation of the images could, in principle, resolve this problem; but the images are so sensitive to variations in the beam tilt, astigmatism, and specimen thickness and orientation (Cockayne and Gronsky, 1981) that these parameters are never known with enough precision to enable meaningful simulations to be carried out.

12.3.8 Nonlinear effects

Most of the preceding discussion applies only to very thin specimens. This is because the thin-phase-object approximation accounts adequately only for the contrast observed near the edge of a crystalline specimen, in a region extending from the very edge of the crystal to wherever the contrast begins to look different. The crystal thickness in this region is typically less than 20 to 100 Å, depending on the average density of the material and the primary voltage of the microscope. The weak-phase-object approximation should never be used to model the image contrast outside this region. It could, in fact, be completely misleading if applied to an electron micrograph in which the specimen edge cannot be seen.

For thicker crystals, one must consider explicitly the so-called nonlinear terms, which arise by pairwise interaction of scattered beams. In most crystals thicker than 100 Å, the nonlinear terms dominate the linear terms, which arise through the interference of the scattered beams with the unscattered (main) beam. The transfer of each nonlinear term is attenuated by a transmission cross-coefficient (O'Keefe, 1979; Ishizuka, 1980) that depends on the spatial frequency of each of the two beams (U_1 and U_2) as well as the energy and angular spread of illumination. The magnitude of the cross-coefficient is, in fact, identical to the magnitude of the envelope function for a spatial frequency $U_1 - U_2$ with the main beam tilted to U_2, and the equations determining it resemble equations (12.7), (12.9), and (12.10).

The most notable consequence of the nonlinear terms is that pairwise interference between two beams traveling at the same angle to the optic axis is not affected by the energy spread of the beam or by high-voltage or objective-lens-current instabilities. This is because the two beams then fall on the achromatic circle discussed in the previous section. Since the energy spread normally limits linear transfer to not much better than 2 Å in axial imaging, finer fringes in high-resolution images are usually of nonlinear origin. A supply of suitable pairs of beams is ensured by the fact that, in axial imaging, $+g$ and $-g$-beams automatically travel at the same angle to the axis. Interference with other beams traveling at similar angles can result in intricately detailed images whose resolution appears to surpass the resolution limit given by equation (12.1), but whose relation to the specimen structure is, at best, rather tenuous.

An interesting example of nonlinear terms dominating the image is shown in the image of a gold crystallite in Figure 12.10. Image detail inside the

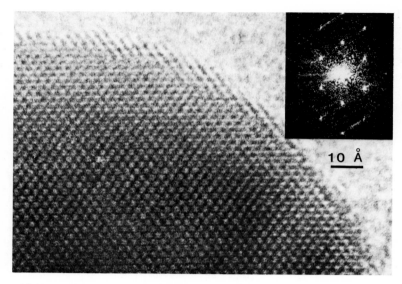

FIGURE 12.10 An image of a (011) projection of a gold particle recorded at a generalized defocus of $-\sqrt{3.5}$. The dumbells seen inside the particle are an artifact of nonlinear imaging—in this case, mostly the interference between the (200) and (−200) beams. The insert shows the optical diffractogram from the particle. The micrograph was taken on the Philips 430ST at 300 kV.

crystallite resembles the images of (011) silicon obtained by Izui and co-workers (1979) that were originally taken as evidence that the atomic structure of the silicon crystal could be resolved at a level surpassing the basic point-to-point resolution of their microscope by more than two times. In gold, such an image would be even more striking, but there is a problem: The projected structure of the gold crystal does not contain the double columns (dumbbells) seen in the micrograph. Instead, the dumbbells are a result of the interference of the (200)- and (111)-type beams from the [011] gold in exactly the same manner as discussed for silicon images by Rez and Krivanek (1980). The value of such images therefore lies mostly in demonstrating the mechanical stability of the microscope (freedom from specimen vibration and drift) and the quality of the shielding against stray magnetic fields.

12.4 Analytical electron microscopy

In contrast to conventional high-resolution electron microscopes, which utilize elastically scattered electrons and provide only images and diffraction patterns, analytical electron microscopes (AEMs) are capable of many different types of operation and of utilizing many different signals. Each signal can also be used in two different ways: spectroscopy when the probe is held stationary and the variation of the signal intensity with energy (and/or scattering angle) is explored, and imaging or mapping when the signal, or a combination of signals, is used to modulate the microscope cathode-ray tube (CRT) display as the electron probe is scanned across the sample. A listing of the analytical signals is provided in Chapter 7 of this book, and the whole analytical electron microscopy field has been reviewed in several recent books (Hren et al., 1983; Reimer, 1984; Williams, 1984). In the short space available here, it is not possible to cover the whole analytical microscopy field satisfactorily, and we concentrate instead on the factors that determine the spatial resolution attainable in analytical microscopy.

The spatial resolution at which an electron microscope operates is determined primarily by four factors:

1. Instrumental resolution (probe size in an AEM).
2. The spreading of the beam within the sample.
3. The size of the interaction region that gives rise to the detected signal.
4. The signal-to-noise ratio of the signal.

In conventional high-resolution imaging the first factor is nearly always dominant; but in analytical electron microscopy, the remaining three factors are often very important. We consider each one in turn.

12.4.1 Instrumental resolution

Although analytical electron microscopy can be performed in a fixed-beam microscope by using inelastically scattered electrons and a few other signals, most AEMs operate in the scanning mode where the microscope forms a small probe and scans it across the sample. The scanning instruments can be divided into the so-called dedicated scanning transmission electron microscopes and the hybrid TEM/STEMs. The dedicated STEMs follow the original STEM design of Crewe (1971) and are not capable of forming images with a fixed beam. The TEM/STEMs are basically conventional transmission microscopes with added scanning coils and are equally able to form images with a fixed beam or a scanning beam. The STEM performance of the two types of microscope should, in principle, be the same. In practice, the dedicated STEMs usually have a better vacuum and brighter and more stable field-emission guns, which gives them an edge in high-resolution analytical performance over the TEM/STEMs.

By consequence of the theory of reciprocity (see Chapter 1) the resolution for any STEM operating mode that has an equivalent conventional mode is the same as it would be in a CTEM with the equivalent electron optical configuration and parameters. However, the scanning image is recorded serially (pixel by pixel), while the conventional image is recorded in parallel (whole image at the same time). This means that if the number of electrons contributing to the two types of images is to be the same, the STEM gun brightness must be n times higher than the CTEM gun brightness, where n is the number of pixels (image elements) in the scanning image, or the scanning image acquisition time must be n times longer. The difference between the brightness of the LaB_6 guns that are now common in CTEMs and that of the field-emission guns used in high-resolution STEMs is about 10^3 ($10^6\,A/cm^2/sr^2$ versus $10^9\,A/cm^2/sr^2$ at $100\,keV$ primary energy). The typical number of pixels in a scanning image is 10^5 to 10^6. Hence, the dedicated-STEM-gun brightness is about 100 to 1000 times too low to produce images with the same number of electrons as the CTEM in the same exposure time. The lower number of contributing electrons means that the scanning images tend to be noisier than the conventional ones, and it is the main reason why the quality of bright-field images obtained in the CTEMs is usually superior to those obtained in the STEMs. In dark-field imaging, the flexible detector geometry and the possibity of operating with a large convergence of the incident beam in a STEM results in approximately equal performance of the two types of microscope.

For all AEM operating modes other than bright-field and dark-field imaging with elastically scattered electrons, the comparison with conventional imaging becomes less useful, and one needs to know, instead, the precise size and shape of the electron probe. This depends on several factors (Crewe et al., 1975; Mory et al., 1985):

1. Demagnification of the electron source.
2. Diffraction limit imposed by the STEM objective aperture.
3. Spherical aberration of the objective lens.
4. Instrumental instabilities.

Depending on which factor dominates, the probe can take on different shapes, as shown in Figure 12.11. In practice, the oscillating "tail" of the

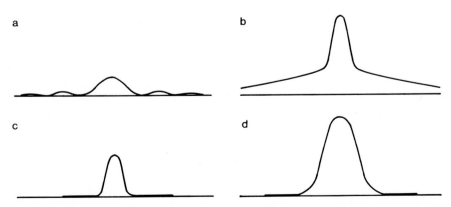

FIGURE 12.11 Different types of electron probes. (*a*) Diffraction-limited. (*b*) Limited by spherical aberration. (*c*) Optimum compromise between (*a*) and (*b*). (*d*) Limited by source size.

diffraction-limited probe (see Figure 12.11a) tends to be smeared out by the limited spatial coherence owing to the finite size of the demagnified image of the source. The intense tail of the spherical-aberration-limited probe means that there are many more electrons in the tail than in the peak, which remains sharp. This is particularly undesirable for microanalysis of inhomogeneous specimens where the signal from a small particle or a sharp interface produced by the peak can be completely masked by the signal from the matrix produced by the tail. The presence of the tail is not always obvious from an image, since it only produces a grey background, which can be eliminated by adjusting the electronic grey-level control. One way to detect the condition is to form a dark-field image or a spectroscopic line scan across an abrupt edge of the specimen and to see whether the image or the spectrum intensity goes to zero at a distance outside the specimen equal to one or two times the claimed resolution. In TEM/STEM instruments, it is usually possible to produce a highly magnified image of the probe on the final viewing screen, and this method is best for determining the probe shape.

The angular half-width of the objective aperture (condenser aperture in a TEM/STEM), which gives the optimum compromise between the diffraction-limited probe and the spherical-aberration-limited probe, is given by

$$\alpha_0 = 1.27 \lambda^{1/4} C_s^{-1/4}. \quad (12.20)$$

Using the optimum aperture and an optimum defocus of $\Delta f = -0.75 \lambda^{1/2} C_s^{1/2}$ results in a probe that contains 70 per cent of the electron intensity within a diameter of

$$d_{\text{inst}(70\%)} = 0.66 \lambda^{3/4} C_s^{1/4} \quad (12.21)$$

(Mory et al., 1985). Expression (12.21) resembles expression (12.1) for the CTEM point-to-point resolution. It does not account for the finite size of the electron source, which must be included by convoluting the theoretical probe shape with the demagnified electron source. Practically all STEMs enable the operator to choose the demagnification of the electron source, which is reduced to about 50 Å in diameter by the action of the accelerating field in the typical field-emission STEM, even before the electron beam enters the demagnifying lenses. A large demagnification means a better resolution but also a loss of the current in the probe in proportion to the demagnification squared. The loss of intensity can be tolerated better in bright-field or dark-field imaging with elastically scattered electrons, where the signal is strong, than in X-ray and electron-energy-loss spectroscopies, where the cross sections are smaller and the signal is weak. Hence, one tends to use large demagnifications, of the order of 50×, for bright-field, dark-field, and Z-contrast imaging (Crewe, 1971), and smaller demagnifications, of the order of 10 to 20×, for microanalysis (Mory et al, 1985).

The effect of chromatic aberration on the probe is usually negligible in the

STEM, mainly because a cold field-emission gun produces a very small energy spread (typically 0.25 eV). Mechanical instabilities deteriorate the resolution in exactly the same way as in the TEM, and stray AC magnetic fields, particularly in the sensitive region around the field-emission tip, have to be carefully shielded out.

12.4.2 Beam spreading

As it traverses the sample, the electron beam spreads owing to scattering and defocusing. In AEM, this causes an increase in the size of the analyzed volume, which must be explicitly considered. The spreading due to scattering, both inelastic and elastic, becomes particularly important in thicker samples of heavier materials. It is described by $d \sim t^{3/2} f(\rho)$, where d is the increase in the width of the beam at a specimen depth t and f is a function of the specimen density ρ. At 100 keV primary energy, the spreading reaches about 500 Å in a 1000-Å-thick foil of gold (Newbury, 1982). It is, however, much less at lower thicknesses and in lighter materials, reaching just 1.6 Å in a 100-Å-thick foil of carbon (Williams, 1984); and it is reduced even further at higher primary voltages. The deterioration of the attainable spatial resolution due to beam spreading is also reduced in EELS analysis, where the large-angle scattering events that deflect the electrons far from their original trajectories cause these electrons to be intercepted by the collection aperture in front of the spectrometer, thus excluding them from contributing to the collected spectrum. On the whole, provided that it is possible to prepare suitably thin foils from the sample of interest, beam spreading can usually be reduced to levels where it does not limit the attainable resolution.

12.4.3 Delocalization

Because of the wave nature of electrons, a scattering event that changes the momentum of a fast electron by Δp can take place with the electron passing any distance up to Δx away from the scattering center, where $\Delta x \, \Delta p \geq h$ and h is Planck's constant. If the change in the momentum is small, the electron can, for instance, pass several tens of angstroms from an atom and still ionize it. The scattering will then become delocalized, and the size of the interaction region may exceed the resolution of the microscope. In this case, the spatial resolution attainable will be determined by the scattering process and not by the microscope, and there will be no point in trying to optimize or improve the instrumental resolution. This result has not yet been reached for images formed with elastically scattered electrons as in conventional high-resolution imaging, since the characteristic scattering angles for elastic scattering and the corresponding momentum change are large, and the size of the interaction region is always less than 1 Å. Inelastic scattering, however, is characterized by much smaller scattering angles, particularly when the energy transfer is small. The possibility that the spatial resolution of AEM analysis will be determined by the scattering process itself therefore always needs to be carefully considered.

The width of the interaction region d_{del} for a scattering event in which a fast electron of velocity v loses energy ΔE is approximately given by (Howie, 1979)

$$d_{del} \simeq \frac{2hv}{\Delta E}. \qquad (12.22)$$

This shows that, at 100 keV primary energy ($v = 0.55c$, where c is the speed of light) and an energy loss of 100 eV, $d_{del} \simeq 40$ Å; and at $\Delta E = 20$ eV, $d_{del} \simeq 100$ Å.

The loss of resolution in inelastic images owing to delocalization has been observed experimentally in a high-resolution STEM (Isaacson et al., 1974; Colliex et al., 1981). An example of the effect is provided in Figure 12.12, which shows images and an energy-loss spectrum of uranium clusters recorded in a dedicated field-emission STEM at a primary energy of 100 keV. The image in Figure 12.12b was recorded with inelastically scattered electrons of energies just below the uranium $O_{4,5}$ edge and shows a resolution of about 40 Å. The image in Figure 12.12c was recorded with inelastically scattered electrons of energies just above the uranium $O_{4,5}$ edge and shows a resolution of about 30 Å. Both images were recorded at the same microscope operating condition as the dark-field elastic image in Figure 12.2a, which shows a resolution of about 6 Å, indicating that the diameter of the electron probe was about 6 Å. Neither image is statistically very noisy. The poor resolution of the images in Figures 12.12b and c is therefore not caused by statistical noise (see the next section) but is simply a result of the delocalized nature of the inelastic scattering.

The distance that an electron passes from the center of the interaction region is inversely related to the angular deflection that it experiences, that is, the electrons that pass the closest have the largest probability of being scattered by a large angle. In EELS, this can be used to decrease the loss of resolution resulting from delocalization by selectively examining only the scattering events that produced large angular deflections (Taftø and Krivanek, 1981). Excluding a large percentage of the electrons scattered by small angles will, however, decrease the signal strength to the point where the signal noise begins to limit the resolution, and concentrating on the electrons scattered by large angles will also increase the importance of beam spreading.

Since the emission of any analytical electron microscopy signal always involves an inelastic-scattering event in which the necessary energy is transferred to the sample, the delocalization of inelastic scattering imposes a fundamental resolution limit on all the analytical information that can be obtained in an AEM by any technique. It is therefore particularly unfortunate that the delocalization is especially serious for low-energy

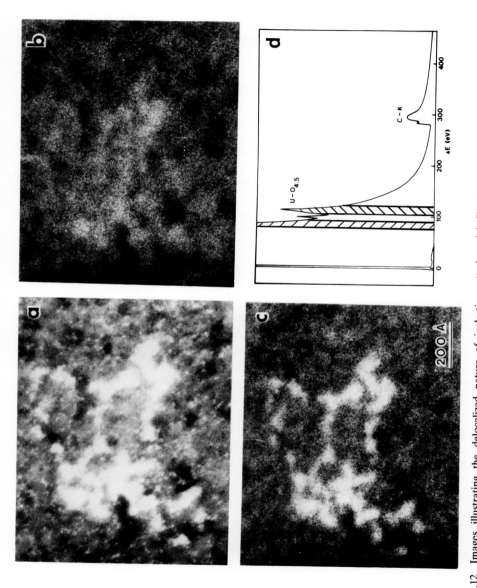

FIGURE 12.12 Images illustrating the delocalized nature of inelastic scattering. (a) Elastic dark-field image of uranium clusters on a thin amorphous-carbon film showing a resolution of about 6 Å. (b) Energy-filtered image ($\Delta E = 80$ to 90 eV). (c) Energy-filtered image ($\Delta E = 105$ to 115 eV). (d) Energy-loss spectrum from one of the uranium clusters showing the energy windows used for images (b) and (c).

transitions, for which the cross sections are much higher than for higher-energy transitions, and that the effect becomes worse at higher primary energies (see equation 12.22), which improve the instrumental resolution but increase the electron velocity. However, recent calculations by Kohl and Rose (1985) and experimental results by Scheinfein and Isaacson (1986) raise the possibility that the effect may not be as serious as first thought, and the number of practical situations for which delocalization has been shown to be the main resolution-limiting influence has so far been relatively small.

A much larger decrease in resolution arises when an intermediate particle carries the excitation-energy quantum from the original scattering center to a different region of the specimen where the de-excitation and emission of the detected analytical signal takes place. This effect is encountered in several spectroscopies and may limit the attainable resolution to several hundreds of angstroms even in very thin foils. Examples are cathodoluminescence and electron-beam-induced current (EBIC) spectroscopies of semiconductors, where the electrons or holes carrying the excitation energy typically diffuse over lengths comparable to sample thickness for thin samples, thereby eliminating any possibility that these spectroscopies could be performed with atomic-scale resolution. In X-ray and Auger spectroscopies in which the excitation and de-excitation usually take place within the same atom, the effect is not important.

12.4.4 Statistical noise

Detecting a particular feature in an image or a spectrum is subject to the condition that the signal S be recognizably stronger than the background noise B, usually expressed as

$$\frac{S}{B} \geq k, \qquad (12.23)$$

where k is chosen between 3 and 5 (Rose, 1948). The irreducible background noise is determined by the statistical variation (shot noise) in the number N of events detected in each resolution element—that is, $B = \sqrt{N}$. The signal S is given by $S = cN$, where c is the contrast. The minimum contrast that can be detected without a loss of resolution then becomes

$$c_{\min} = \frac{k}{\sqrt{N}} \qquad (12.24)$$

Features showing contrast $c < c_{\min}$ are only detectable in a two-dimensional image if their projected area is bigger than

$$d_n^2 = \frac{d_{\text{inst}}^2 k^2}{c^2 N}, \qquad (12.25)$$

where d_{inst} is the instrumental resolution. In bright-field high-resolution imaging, the usual number of electrons per pixel is 10^3 to 10^5. For $k = 4$, the minimum detectable contrast then becomes 1.3 per cent ($N = 10^5$) to 13 per cent ($N = 10^3$).

For other signals whose intensities are much less, such as X-ray emission, the resolution decreases markedly. Thus, a precipitate that has a concentration of one element 40 per cent higher than the surrounding matrix ($c = 0.4$) will only be detectable in an X-ray map acquired with one X-ray per pixel (on average) if it is larger than $10d_{inst}$. At the same time, the map will show many smaller sharp features due to the random variation in the number of the X-rays, and these features could be potentially mistaken for meaningful specimen features. Fortunately, the validity of all features at the threshold of visibility can be checked by acquiring a second map of the same area under identical conditions. If a particular feature fails to reappear, it was probably just noise.

The number of detected events is proportional to the gun brightness and to the detector efficiency. In situations where the resolution is limited by noise, an improvement by a factor of 4 in either the gun brightness or the detector efficiency will improve the attainable resolution by $2\times$, and they both therefore need to be carefully maximized. The resolution-decreasing effect of weak signals can be further mitigated by increasing the recording time until the instrumental instabilities such as specimen drift or contamination begin to limit the resolution (or the operator runs out of patience). With samples sensitive to radiation damage, this route is not available, and the statistical precision of the results that can be recorded with less than "lethal" dose nearly always limits the attainable resolution to values significantly worse than the instrumental resolution. When the specimen is periodic, the noise can be reduced—and the resolution improved—by averaging the signal over many unit cells (Unwin and Henderson, 1975). A simple variant of this idea can be used in the microanalysis of planar interfaces, where one can acquire a two-dimensional concentration map instead of a one-dimensional concentration profile and average the data in the direction of the interface.

12.4.5 Optimum adjustment of the AEM

The range of experimental situations encountered in analytical electron microscopy is so large that it is impossible to give a uniform prescription for optimizing the resolution, but a few guidelines can nevertheless be established:

1. For quantitative analysis, it is essential to do a one-time calibration of the probe-forming performance of the AEM used. Measure the total current in the electron probe of the AEM instrument over a range of source demagnifications, illumination-aperture sizes, total emission currents, and operating voltages, concentrating in particular on the range of operating

conditions normally used, and determine the probe diameter over the same range of instrumental conditions. In a TEM/STEM, the current can usually be determined by converting the exposure meter reading to electron density by using the manufacturer's calibration. In AEMs equipped with energy-loss spectrometers with single-electron detection capabilities, beam intensities up to about 0.1 pA can be measured in the electron-counting mode (provided that the slits are wide open, so that the whole beam enters the detection system) and up to about 1 nA if it is also possible to measure the anode current of the photomultiplier tube (PMT) and if the conversion factor between the electron-counting and the current-measuring modes (gain change) is known. For beam currents greater than 1 nA, the electrostatically isolated drift tube incorporated in most energy-loss spectrometers provides an efficient Faraday cup when the spectrometer is switched off. The probe diameter can be determined experimentally from dark-field images of the edge of an amorphous-carbon film or other suitable test object, or theoretically, with several experimental checks.

2. Estimate the attainable resolution as given by delocalization, the beam spreading due to the thickness of the specimen, the expected signal for a given incident-electron dose, and the expected contrast level. Choose a microscope condition with a probe about 2× smaller than the delocalization or the beam spreading, whichever is smaller, and work out the expected resolution limit due to the statistical noise d_n (equation 12.25). If d_n proves to be the main limitation, increase the beam current by increasing the probe size until the resolution limits due to the probe size and the statistical noise are about equal. If d_n is not a limitation, decrease the probe size, thereby improving the resolution in other signals that can be collected simultaneously; or leave the probe size as it is, thereby improving the detection limit for low-contrast features.

3. If the required instrumental resolution estimated in step 2 turns out to be at or below the limit imposed by the instrument, it may be necessary to optimize the instrumental resolution by performing the STEM equivalents of all the instrumental alignments as described in the CTEM part of this chapter. In most analytical situations, this is, however, not the case, and it is then more important to perform the rough estimates described above simply to make sure that the selected instrumental-operation regime is well matched to the problem at hand. Consistent use of this approach will also point out which resolution limit is most frequently encountered on your AEM, and this will show which aspects of the AEM's performance should be improved to provide better resolution most effectively.

12.5 Conclusion

The high-resolution electron microscope is not the only instrument capable of imaging atoms in real space or of analyzing the composition of matter on an atomic scale. The competing instruments are, in chronological order of their development, the field-ion microscope (FIM) (Müller and Tsong, 1969), the scanning tunneling microscope (STM) (Young et al., 1972; Binnig and Rohrer, 1985), and possibly also the recently proposed atomic-force microscope (AFM) (Binnig et al., 1986). The demonstrated spatial resolutions of FIM, HREM, and STM are all around 2 Å, with the exception of the STM resolution perpendicular to the studied surface, which is about 0.1 Å. Atom-probe FIMs reliably determine the chemical identity of single atoms, and the STMs can analyze the electronic states of a single surface atom. These capabilities are superior to present-day AEMs, which are only just beginning to approach single-atom sensitivity (Shuman and Somlyo, 1986).

The chief advantage of high-resolution electron microscopy is that it probes matter with high-energy electrons, which penetrate inside the specimen and reveal the atomic structure underneath the surface. By comparison, the FIM, STM, and AFM can only examine surfaces. In view of the large number of unresolved bulk structures and defects, which are not periodic and therefore cannot be solved by diffraction techniques, it is clear that high-resolution electron microscopy will continue to have an important role to play in the future.

Since high-resolution microscopes rely on electron optics to produce the high spatial resolution, they are necessarily more complicated than the other instruments, which have only one or two parameters that are critical. High-resolution electron microscopy is also unique in that a badly misadjusted electron microscope will produce images that bear practically no relationship to the specimen structure; the other techniques mostly produce either the correct results or no results at all. However, high-resolution electron microscopy can also be very simple: high-resolution images of thin specimens, taken at the optimum defocus in a correctly adjusted microscope, always show dark atoms on a light background. The fact that the imaging can be made so simple is precisely what makes the study of the correct adjustments extremely worthwhile. The reward for mastering the adjustments so that they become routine should be the excitement of discovery in the purest sense: Seeing a hitherto unknown structure clearly resolved for the first time.

REFERENCES

Binnig, G., and Rohrer, H. (1985). The scanning tunneling microscope. *Sci. Am.* **253** (2) (August 1985), 50.

———, Quate, C. F., and Gerber, Ch. (1986). Atomic force microscope. *Phys. Rev. Lett* **56**, 930.

Boersch, H. (1954). Experimentelle Bestimmung der Energieverteilung in thermisch ausgelösten Elektronenstrahlen. *Z. Phys.* **139**, 115.

Bracewell, R. N. (1978). *The Fourier transform and its applications.* McGraw-Hill, New York.

Budinger, T. F., and Glaeser, R. M. (1976). Measurement of focus and spherical aberration of an electron microscope objective lens. *Ultramicroscopy* **2**, 37.

Cockayne, D. J. H., and Gronsky, R. (1981). Lattice fringe imaging of modulated structures. *Philos Mag., Ser. A* **44,** 159.

Colliex, C., ed. (1985). Proceedings of the STEM special session. *J. Microsc. Spectrosc. Electron.* **10,** 313.

———, Krivanek, O. L., and Trebbia, P. (1981). Electron energy loss spectroscopy in the electron microscope: A review of recent progress. *Inst. Phys. Conf. Ser.* **61,** 183.

Crewe, A. V. (1970). Current state of high-resolution scanning electron microscopy. *Science* **168,** 1338.

——— (1971). High-intensity electron sources and scanning electron microscopy. In *Electron microscopy in materials science,* ed. U. Valdrè, 163. Academic Press, New York and London.

——— and Kopf, D. (1979). A sextupole system for the correction of spherical aberration. *Optik* **55,** 1.

———, Langmore, J. P., and Isaacson, M. S. (1975). Resolution and contrast in the scanning transmission electron microscope. In *Physical techniques of electron microscopy and microanalysis,* ed. B. Siegel and D. Bearman, 47. Wiley, New York.

Dowell, W. C. T. (1963). Das elektronenmikroscopie Bild von Netzebenscharen und sein Kontrast. *Optik* **20,** 535.

Frank, J. (1973). The envelope of electron microscopic transfer functions for partially coherent illumination. *Optik* **38,** 519.

——— (1976). Determination of the source size and energy spread from electron micrographs using the method of Young's fringes. *Optik* **44,** 379.

——— and Al-Ali, L. (1975). Signal-to-noise ratio of electron micrographs obtained by cross-correlation. *Nature* **256,** 376.

Fujiyoshi, Y. (1986). Private communication.

Hawkes, P. W. (1980). Units and conventions in electron microscopy, for use in ultramicroscopy. *Ultramicroscopy* **5,** 67.

Heinemann, K. (1971). In-situ measurement of the objective lens data of a high-resolution electron microscope. *Optik* **34,** 113.

Hermann, K.-H. (1983). Instrumental requirements for high-resolution imaging. *J. Microsc.* **131,** 67.

———, Krahl, D., Kübler, A., Müller, K.-H., and Rindfleisch, V. (1971). Image recording with semiconductor detectors and video amplification devices. In *Electron microscopy in materials science,* ed. U. Valdrè, 236. Academic Press, New York.

Hoppe, W. (1974). Towards three-dimensional electron microscopy at atomic resolution. *Naturwissenschaften* **61,** 239.

Howie, A. (1979). Image contrast and localized signal selection techniques. *J. Microsc.* **117,** 11.

Hren, J. J., Goldstein, J. I., and Joy, D. C., eds. (1983). *Introduction to analytical electron microscopy.* 2nd ed. Plenum, New York.

Isaacson, M., Langmore, J. P., and Rose, H. (1974). Determination of the nonlocalization of the inelastic scattering of electrons by electron microscopy. *Optik* **41,** 92.

Ishizuka, K. (1980). Contrast transfer of crystal images in the TEM. *Ultramicroscopy* **5,** 55.

Izui, K., Fununo, S., Nishida, T., and Otsu, H. (1979). High-resolution structure images and their application to defect studies. *Chem. Scripta* **14,** 99.

Kakinoki, J., Katada, K., Hanawa, T., and Ino, T. (1960). Electron diffraction study of evaporated carbon films. *Acta Crystallogr.* **13,** 171.

Kohl, H., and Rose, H. (1985). Theory of image formation by inelastically scattered electrons in the electron microscope. *Adv. Electron. Electron Phys.* **65,** 173.

Komoda, T. (1966). Fourier images of a gold crystal lattice. *Jpn. Appl. Phys.* **5,** 1120.

Kraus, B., Krivanek, O. L., Swann, N. T., Ahn, C. C., and Swann, P. R. (1986). The performance of Newvicon and CCD real-time EM observation systems. In *Proceedings of the 11th international electron microscopy congress* **1,** 455. Japanese Society of Electron Microscopy, Tokyo.

Krivanek, O. L. (1975). Ph.D. diss., University of Cambridge, England.

———— (1976a). Studies of the envelope of the EM contrast transfer function. In *Proceedings of the 6th European electron microscopy congress,* 263. TAL International Publishing Co., Jerusalem.

———— (1976b). Method for determining the coefficient of spherical aberration from a single electron micrograph. *Optik* **45,** 97.

———— (1978) EM contrast transfer functions for tilted illumination imaging. In *Proceedings of the 9th international electron microscopy congress,* **1,** 168. Microscopical Society of Canada, Toronto.

———— and Wood, G. J. (1984). Unpublished results.

Kuzua, M., and Hibino, M. (1981). The effect of defocus on the determination of the spherical aberration coefficient from a micrograph of an amorphous phase object. *J. Electron Microsc.* **30,** 114.

Lenz, F. A. (1971). Transfer of image information in the electron microscope. In *Electron microscopy in materials science,* ed. U. Valdrè, 540. Academic Press, New York.

Loeffler, K. H. (1969). Energy-spread considerations in electron-optical instruments. *Z. Angew. Phys.* **27,** 145.

McFarlane, S. C. (1975). Imaging of amorphous specimens in a tilted beam electron microscope. *J. Phys. C* **8,** 2819.

Maréchal, A., and Croce, P. (1953). Un filtre des fréquences spatiales pour l'amélioration du contrast des images optiques. *C.R. Acad. Sci.* **237,** 607.

Mory, C., Tence, M., and Colliex, C. (1985). Theoretical study of the characteristics of the probe for a STEM with a field emission gun. *J. Microsc. Spectrosc. Electron.* **10,** 381.

Müller, E. W., and Tsong, T. T. (1969). *Field ion microscopy, principles and applications.* Elsevier, New York.

Newbury, D. E. (1982). Beam broadening in the analytical electron microscope. In *Microbeam analysis 1982,* ed. K. J. F. Heinrich, 79. San Francisco Press, San Francisco.

O'Keefe, M. A. (1979). Resolution-damping functions in nonlinear images. In *Proceedings of the 37th annual EMSA meeting,* 556. Claitors, Baton Rouge.

Parsons, J. R., and Hoelke, C. W. (1974). Electron microscopy of plasmons. *Philos. Mag.* **30,** 135.

Reimer, L. (1984). *Transmission electron microscopy.* Springer-Verlag, Heidelberg.

Rez, P., and Krivanek, O. L. (1980). Imaging of atomic columns in [110] silicon. In *Proceedings of the 38th annual EMSA meeting,* 170. Claitor's, Baton Rouge.

Rose, A. (1948). Television pickup tubes and the problem of noise. *Adv. Electron.* **1,** 131.

Saxton, W. O. (1980). Computer techniques for image processing in electron microscopy. *Adv. Electron. Electron Phys. Suppl.* **10,** 289.

————, Smith, D. J., and Erasmus, S. J. (1983). Procedures for focussing, stigmating and alignment in high-resolution electron microscopy. *J. Micros.* **130,** 187.

Scheinfein, M., and Isaacson, M. (1986). Electronic and chemical analysis of fluoride interface structures at subnanometer spatial resolution. *J. Vac. Sci. Technol. Ser. B* **4,** 326.

Scherzer, O. (1949). The theoretical resolution limit of the electron microscope. *J. Appl. Phys.* **20,** 20.

Self, P. (1983). Outline of the alignment of a TEM. *EMSA Bull.* **13** (2), 79.

Shuman, H., and Somylo, A. P. (1987) Electron energy loss analysis of near-trace element concentrations of calcium. *Ultramicroscopy* **21,** 23.

Sinclair, R., Gronsky, R., and Thomas, G. (1976). Optical diffraction from lattice images of alloys. *Acta Metall.* **24,** 789.

Smith, D. J., Saxton, W. O., O'Keefe, M. A., Wood, G. J., and Stobbs, W. M. (1983). The importance of beam alignment and crystal tilt in high-resolution electron microscopy. *Ultramicroscopy* **11,** 263.

Spence, J. C. H. (1981). *Experimental high-resolution electron microscopy.* Clarendon Press, Oxford.

Steeds, J. W. (1979). Convergent beam electron diffraction. In *Introduction to analytical electron microscopy*, ed. J. J. Hren, J. I. Goldstein, and D. C. Joy, 387. Plenum, New York.

Taftø, J., and Krivanek, O. L. (1981). Characteristic energy losses from channelled 100-kV electrons. *J. Nucl. Instrum. Methods* **194,** 153.

Thon, F. (1965). Elektronenmikroskopie Untersuchungen an dünnen Kohlefolien. *Z. Naturforsch.* **20a,** 154.

────── (1971). Phase contrast electron microscopy. In *Electron microscopy in materials science*, ed. U. Valdrè, 570. Academic Press, New York and London.

Unwin, P. N. T., and Henderson, R. (1975). Molecular structure determination by electron microscopy of unstained crystalline specimens. *J. Mol. Biol.* **94,** 425.

Williams, D. B. (1984). *Practical analytical electron microscopy in materials science*. Philips Electronic Instruments, Mahwah, N.J.

Wood, G. J. (1985). Unpublished results.

Young, R., Ward, J., and Scire, F. (1972). The topografiner: An instrument for measuring surface microtopography. *Rev. Sci. Instrum.* **43,** 999.

Zemlin, F. (1979). A practical procedure for alignment of a high-resolution electron microscope. *Ultramicroscopy* **4,** 241.

────── , Weiss, K., Schiske, P., Kunath, W., and Hermann, K.-H. (1978). Coma-free alignment of high-resolution electron microscopes with the aid of optical diffractograms. *Ultramicroscopy* **3,** 49.

13
SURFACES
KATSUMICHI YAGI

13.1 Surface-sensitive methods

The meaning of *surface* depends on the individual scientist. Some are concerned with the surfaces met in practical problems such as tarnishing, blistering, scratching, coating, and fracture; and others may consider only the clean and well-defined surfaces that can be formed and maintained only in ultrahigh-vacuum (UHV) conditions. For imaging of surfaces in the former sense, scanning electron microscopy (SEM) and replica transmission electron microscopy (TEM) have been used (Laird, 1974). However, surfaces in the latter sense have become of great practical importance in semiconductor-device technology and catalysis.

The study of surfaces from this point of view is one of the most rapidly developing scientific fields in the past decade. In this chapter, we describe the imaging of surfaces at the level of single atomic layers under well-defined conditions by use of ultrahigh-vacuum conventional TEM (UHV–CTEM), which has been developed very recently. Several reviews on surface imaging have been given in the past (Yagi, 1980; Howie, 1981; Venables, 1981, 1982; Cowley, 1982a, 1982b; Yagi et al., 1982; Takayanagi and Yagi, 1983). Some of them (Venables, 1981; Cowley, 1982a, 1982b) mainly describe observations by scanning microscopy, which is not emphasized here.

Many different techniques for studying surfaces have been developed recently, such as low-energy electron diffraction (LEED), photoemission spectroscopy (UPS and XPS), ion channeling, neutral-atom scattering, and so on (Prutton, 1982). However, information obtained by these methods is averaged over the area covered by the probes. They cannot therefore directly characterize inhomogeneous surface processes, whose information is very important in surface science.

Various kinds of surface microscopy have been developed, as shown in Table 13.1. One type uses a magnification system with or without lenses; in another type, a microprobe is formed and a scanning system is used. In the latter case, there are various kinds of surface microscopy, depending on the species of the probe and the detected signals. The things we want to know and to do with surface microscopy are summarized in Table 13.2. The methods that can be used for each purpose are also indicated by number in Table 13.1.

The first step in surface microscopy is to observe surface topography on

TABLE 13.1
Imaging Methods

Magnification system	Microprobe plus scanning system
With lenses	(7) Scanning electron microscopy
(1) Optical microscopy	(low and high energy)
(2) CTEM	(8) Scanning Auger microscopy
(3) Photoemission electron microscopy	(9) STEM and SREM
(4) Low-energy, electron-reflection microscopy	(10) Scanning tunneling microscopy
Without lenses	
(5) Field-emission microscopy	
(6) Field-ion microscopy	

TABLE 13.2
Information Obtained by Surface Microscopy

Information	Method
Surface topography of atomic scale (monatomic steps)	(1) (2) (4) (6) (9) (10)
Texture of the structure (domains of surface structures)	(2) (3) (4) (6) (7) (8) (9) (10)
Characterization of the local structure	
Composition	(6) (8) ((2) (9))
Structure identification	(2) (4) (9) (10) ((7))
Electronic structure	(3) (4) (5) (9) (10) ((2) (8))
Structure analyses	(2) (6) (9) (10) ((4))
Surface dynamic processes (heterogeneous processes) (phase transition, adsorption processes, step growth)	(1) (2) (3) (4) (5) (6) (7) (8) (9) (10)

an atomic scale such as single atom-high steps, which are the main defect on surfaces and play an important role during surface dynamic processes. Also, one can see textures of surface structures, such as domain structures of reconstructed and absorbate systems. Structure analysis within the periodic unit cell is an important topic in surface science, and microscopic methods can be used for that also. One of the most important applications of surface microscopy is the observation of surface dynamic processes *in situ*, whereby heterogeneous surface reactions may be clearly seen. With the help of the previously mentioned capabilities, surface microscopy tells us what is happening on the surface and where and how it takes place.

Before going into the details of surface studies by UHV CTEM, we will summarize briefly the recent progress with several other surface-imaging methods.

13.1.1. Scanning electron microscopy (conventional SEM)

Generally, scanning electron microscopy does not reveal surface structures at a monolayer level. However, to increase surface sensitivity, progress has been made in two directions. One technique is to decrease the incident-electron energy to below 1000 eV. These slow electrons do not penetrate more than 50 Å below the surface. Using a field-emission gun, Ichinokawa et al. (1984) have obtained a resolution of 100 nm at an accelerating voltage of 200 V, and this resolution is expected to be improved on in the future. They succeeded in distinguishing between an area of clean Si (111) 7 × 7 surface and that covered by a monolayer of Au.

The other technique invented by the Venables' group in Sussex (Futamoto et al., 1985) is to bias the specimen. An example is shown in Figure 13.1, where areas of clean Si surface and those covered by 5 monolayer (ML) and 0.5 ML of Ag are clearly constrasted at $E_p = 30$ keV. Scanning Auger microscopy (Venables, 1981, 1982), whose resolution has been improved down to 30 nm, was also used to determine the concentration of Ag in the patched areas.

One shortcoming in SEM studies is the lack of surface-crystallographic

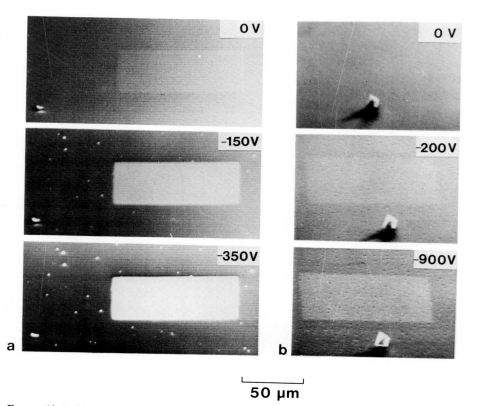

FIGURE 13.1 Biased SEM images of Si (111) surface. 5 ML and 0.5 ML of Ag were deposited in patches in (a) and (b), respectively. Note that the area covered with 0.5 ML of Ag shows contrast only in biased images. (Courtesy of Dr. J. A. Venables.)

information in the images because the secondary-electron yield is not strongly altered by diffraction effects. The Kikuchi patterns of back-scattered electrons are, however, useful in certain cases (Harland et al., 1981), but it is not certain that they reflect the crystallography of surface structures.

13.1.2 Photoemission electron microscopy (PEEM)

The intensity of photoemission electrons and their energy spectrum are determined by surface electronic structures (Prutton, 1982). Therefore, photoemission electron microscoy (PEEM) can reveal the spatial distribution of the work function and local, electronic surface states resulting from nonuniform absorption of materials. PEEM has a rather long history, but only recently were UHV–PEEM instruments constructed (Beamson et al., 1981; Bethge et al., 1982). Two types of PEEM images were conceived: electron-energy integrated images and electron-energy selected ones. An example of the former is shown in Figure 13.2, where differences of

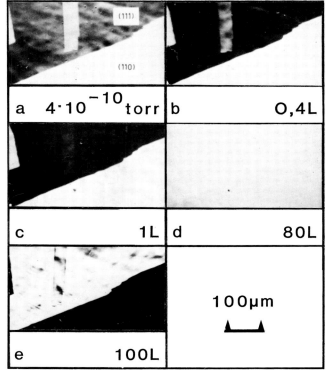

FIGURE 13.2 A series of PEEM images that recorded oxygen adsorption processes on clean Ni (110) and (111) surfaces. Oxygen exposure is shown in the Langmuir (1×10^{-6} torr · s) unit. (Courtesy of Prof. H. Bethge.)

572 HIGH-RESOLUTION TRANSMISSION ELECTRON MICROSCOPY

oxidation processes on various nickel surfaces are clearly seen (Bethge et al., 1982). The authors claimed that changes of photoemission intensity due to oxidation on different Miller planes are much stronger than those of Auger electrons. From consideration of the important roles of ultraviolet photoelectron spectroscopy (UPS) and soft-X-ray photoelectron spectroscopy (XPS) in surface studies, it is apparent that energy-selected images contain much additional information (Beamson et al., 1981).

The resolution of PEEM is limited to about 10 nm, which results from a spread of the photoelectron emission angle. Angle-resolved PS is a powerful tool for structure analysis (Fadley, 1985). However, it may be difficult to make angle-resolved and energy-selective PEEM measurements because of a lack of intensity. The availability of synchrotron-radiation light sources offers exciting possibilities in this regard. An advantage of PEEM is the small amount of damage to the surfaces caused by the incident irradiations; hence, it can safely be used for surfaces of organic crystals and biological materials.

13.1.3 Scanning tunneling microscopy (STM)

Scanning tunneling microscopy (STM), initiated by Young et al. (1972) as a topografiner, has recently been dramatically improved in its spatial resolution to the point where it can show surface unevenness on an atomic scale (Binnig and Rohrer, 1983). An atomic-scale-size tip is scanned over the

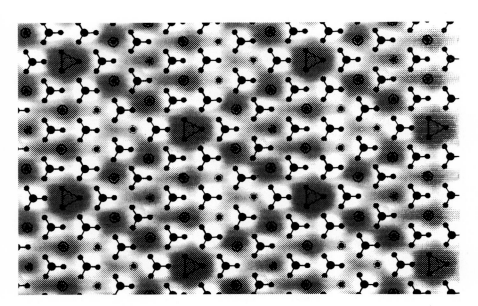

FIGURE 13.3 STM image of the Si(111) 7 × 7 structure (brightness represents height of the scanning tip.) The adatom-vacancy model is superposed. (Courtesy of Drs. Rohrer and Binnig.)

surface, and the tip height is controlled with an accuracy of better than 0.01 nm to get a constant tunneling current between the tip and the surface. Thus, a surface relief map on an atomic scale is obtained. The first such image was obtained for the Si (111) 7 × 7 structure and is shown in Figure 13.3 (Binnig et al., 1983a). Assuming a simple topographic interpretation, Binnig and Rohrer's group proposed an adatom-vacancy model, which is also shown in Figure 13.3. They have also produced images of the Au (110) 2 × 1 structure (Binnig et al., 1983b). These pioneering results have led many others to construct similar instruments and have stimulated much theoretical work on the detailed interpretation of STM images. There has been enormous progress in STM between the time when this manuscript was written (January 1985) and the time when it was finally ready for the printer. One of the most important developments is that the tunneling spectroscopy and microscopy were combined together to get images of local electronic states in the case of semiconductor surfaces. This was successfully done in the case of Si (111) 7 × 7, which proved that the DAS (dimer-adatom-stacking fault) model for the 7 × 7 structure proposed by Takayanagi et al. (1985) is correct (Hamers et al. 1986). Characterization of the surfaces and tips at the same time by other methods such as SEM, reflection electron microscopy (REM), FIM, and HRTEM is also in progress.

13.1.4 *Field-ion microscopy (FIM)*

Field-ion microscopy (FIM) has dominated surface imaging at atomic resolution. Although the range of materials to be studied is limited, even after the more recent improvements of the imaging techniques, it provides the ultimate in chemical analysis of surfaces, when combined with a mass analyzer to form the atom-probe FIM (Müller and Tsong, 1969).

13.1.5 *Low-energy, electron-reflection microscopy (LEERM)*

The idea of the low-energy, electron-reflection microscope was introduced about twenty years ago (Turner and Bauer, 1966). However, it is only recently that it has been used for clean-surface problems. A diffracted beam is used for imaging, so surface-crystallographic information is included in the image contrast. A resolution of 20 nm has been obtained with an incident-beam energy (at the specimen) of 20 eV. The apparatus was designed to allow various operating modes in addition to the LEERM mode (Telieps, 1984; Telieps and Bauer, 1985): LEED (spatial resolution of 1 to 2 μm with the use of a fine-focused electron beam from a field emitter), PEEM, mirror electron microscopy, and thermal emission.

13.1.6 *A short history of CTEM for surface studies*

13.1.6.1 Transmission mode.
An indirect approach—the decoration and carbon-extraction replica techniques—and a direct approach—viewing sur-

TABLE 13.3
TEM Observations of Surface Steps

Authors	Surface	Imaging mode
Cherns (1974)	Au (111)	DF[b]
Spence (1975)	Au (111)	DF
Kambe and Lehmpfuhl (1975)	MgO (001)	DF
Iijima (1977)	C_G (0001)	BF[b]
Lehmpfuhl and Uchida (1979)	MgO (111)	BF, DF
Iijima (1981)	Si (111)	BF, DF
Tanishiro et al. (1981a)	Ag (111)[a]	DF
Lehmpfuhl and Takayanagi (1981)	Ag (111)[a]	BF
Takayanagi (1982)	Ag (111)[a]	DF
Klaua and Bethge (1983)	Au (111)	BF, DF
Kodaira et al. (1983)	Ag (001)[a]	BF, DF

[a] Clean surfaces
[b] Bright-field (BF) and dark-field (DF) images

faces of thin film—were developed. In the former approach, the preferential nucleation of metals vacuum-deposited along surface atomic steps on substrate crystals is used (Bethge, 1962). This approach is applicable only to alkali halides, silver, and a few other substances that can be dissolved readily. Changes of step configurations of vacuum and air-cleaved surfaces during annealing have been observed and analyzed to get information on microscopic surface processes (Bethge, 1982). The technique is not general because it is indirect and applicable to only a limited range of materials.

The direct method started with Cherns's (1974) observation of surface atomic steps on Au (111) surfaces in the dark-field image using so-called forbidden reflections. The possibility was also shown in a calculation on MO_3 by Goodman and Moodie (1974). Dynamical effects on these images, convergent-beam patterns, and the effects of stacking faults were first discussed by Lynch (1971) and Spence (1975). Later reports on surface-step observation are summarized in Table 13.3. Proper imaging conditions were explored to increase contrast and to decrease exposure time for image recording not only in dark-field but also in bright-field conditions (Lehmpfuhl and Takayanagi, 1981; Takayanagi, 1983). Except for those observations marked by a footnote, ordinary vacuum microscopes were used for specimens prepared outside the microscopes and transferred through the air. Therefore, the conclusions could not be extended to clean-surface problems.

13.1.6.2 Reflection mode. Ruska (1933) proposed REM for the observation of surfaces of bulk specimens. Images were formed by use of electron beams diffusely scattered in the forward direction, so they were sometimes called low-loss images. Images in the early days showed only silhouettes of the surfaces (Menter, 1952–1953), so REM has been replaced by SEM and replica techniques.

Advances of REM were made by Halliday and Newman (1960), who proposed imaging with use of Bragg-reflected beams in the RHEED (reflection high energy electron diffraction) patterns from the surface because these beams contain surface-crystallographic information. Cowley's group in Arizona also noticed this and explored the technique of high-resolution REM using a conventional microscope (Nielsen and Cowley, 1976). They also made a UHV-scanning REM apparatus, where diffracted beams could be monitored during scanning (Cowley et al., 1975). However, owing to a lack of *in situ* specimen-treatment techniques, they could not apply their methods to clean-surface problems at that time. The first success of UHV–REM was reported for clean Si (111) surfaces in 1980 (Osakabe et al., 1980).

13.1.7 STEM and SREM

STEM (scanning TEM) and SREM (scanning REM) are equivalent to TEM and REM (Chapter 7), respectively. They will be discussed in section 13.5.

13.2 Experimental techniques for surface CTEM

13.2.1 Vacuum system

An improvement in the vacuum of electron-microscope columns results in improved high-voltage stability because of a reduction of microdischarges in the gun chamber, longer lifetime of bright-electron sources, and less specimen contamination. The last feature is important in HRTEM (high-resolution TEM), CBED (convergent beam electron diffraction), and analytical electron microscopy (AEM), and now, 10^{-8}-torr electron microscopes are commercially available.

However, such a vacuum level is still too poor to be used for surface studies. So that the 10^{-9} to 10^{-10}-torr level is reached, several modifications are necessary, including a reduction of the number of O-rings, introduction of liner tubes along electron paths, UHV pumps (ion pumps, sublimation pumps, turbos, and/or cryopumps), use of low-out-gas materials, introduction of a baking systems, and UHV-tight mechanical feedthroughs and goniometers. However, electron microscopes are very complicated vacuum systems, so a liquid He cryogenic shroud around the specimen is indispensable. A noteworthy feature is that all four groups, including ours (Takayanagi et al., 1978; Wilson and Petroff, 1983; Aseef et al., 1984; Gibson et al., 1985), where the Si (111) 7×7 structure was produced by *in situ* cleaning in CTEMs, have used a liquid He cryoshroud. Very recently, a commercial UHV electron microscope was developed with a vacuum level of 10^{-9} torr without using liquid nitrogen traps (Takayanagi et al., 1986).

576 HIGH-RESOLUTION TRANSMISSION ELECTRON MICROSCOPY

13.2.2 In situ *specimen treatment*

Two approaches for cleaning surfaces in the microscope are possible. One is to incorporate the specimen-treatment techniques into the UHV microscope so that all the specimen treatment is done *in situ*. The other is to make a UHV specimen-preparation chamber and a UHV specimen-transfer system (Ambrose, 1976). In either case, we need the standard techniques; sublimation or desorption by flash heating, vacuum deposition, ion sputtering and annealing, vacuum cleavage, and oxidation-reduction.

Figure 13.4 shows schematically the specimen chamber of our UHV electron microscope (Takayanagi et al., 1978). Notice that a specimen on either the heating TEM or the REM holder, which is introduced into the

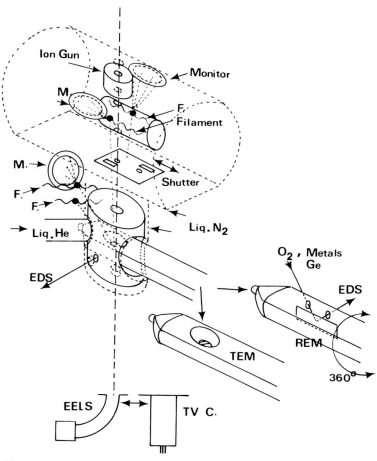

FIGURE 13.4 The arrangement of equipment for surface TEM–REM experiments. (For details, see the text.)

microscope through a double O-ring and differentially pumped goniometer, is surrounded by a liquid-helium cryogenic shroud cooled to 20 K. The shroud is also surrounded by a liquid-nitrogen-cooled shroud. An evaporator, which has two filaments, in the top-entry specimen chamber is surrounded by another liquid-nitrogen-cooled shroud. A pair of quartz oscillators is introduced to monitor the evaporation rate from each filament. An ion-sputter gun of discharge type is positioned just above the evaporator (Morita et al., 1980). At the side entry-holder level, two evaporation filaments are incorporated for *in situ* deposition during REM observations (Tanishiro et al., 1981b). Again, a quartz-thickness monitor is located behind the filaments. A gas-inlet system is also located at this level (Shimizu et al., 1985) to observe oxidation and reduction processes in the REM and TEM modes.

The specimen chamber is evacuated by a turbomolecular pump ($270 \, l \, s^{-1}$), an ion pump ($20 \, l \, s^{-1}$), and a titanium-sublimation pump; and the base pressure is 10^{-8} torr. The local pressure around the specimen may be at the 10^{-10}-torr level after operation of the liquid-nitrogen- and helium-cooled shrouds. A reduction of oxygen partial pressure by orders of magnitude after operation of the liquid-helium cryoshroud was estimated from the observation of the much slower oxidation rate for iron-evaporated films at 550°C.

13.2.3 Analytical techniques

Energy-dispersive, X-ray–spectroscopy (EDS) and electron-energy-loss-spectroscopy (EELS) techniques can be easily combined with the surface-observation techniques. However, their applications to practical problems are rather rare. This point will be discussed in sections 13.4.1 and 13.5.2. Auger-electron spectroscopy (AES), which is most widely used in surface studies, is generally hard to accommodate because of a strong magnetic field around the specimen. Only one paper (see Poppa, 1975) has reported AES combined with CTEM. Owing to a low collection efficiency, the signal-to-noise ratio was not as good as for ordinary Auger scans.

13.3 TEM studies of surfaces

13.3.1 Image-formation process

Figure 13.5 schematically illustrates the processes for the imaging of surfaces in the transmission mode. As shown in Figure 13.5a, a surface structure, either a reconstructed or adsorbate structure, is seen through a bulk-structure thin film (plan view mode). Here, a ($\sqrt{3} \times \sqrt{3}$) R30° surface structure on a trigonal-crystal surface is assumed for the drawing. In the transmission-electron-diffraction (TED) pattern in Figure 13.5b, diffraction spots due to the bulk (solid circles) and the surface (open circles) structures

578 HIGH-RESOLUTION TRANSMISSION ELECTRON MICROSCOPY

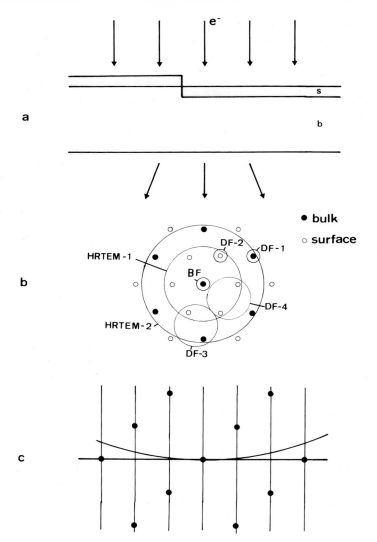

FIGURE 13.5 Surface-imaging process in the transmission mode. (*a*) Surface structure on the top of a thin film. (*b*) TED pattern and various kinds of aperture positions for surface-structure imaging. (*c*) The Ewald sphere and the reciprocal spots and rods. TED patterns can give intensities of (hk) rods with the third indices equal to zero.

are formed. Figure 13.5*c* shows a cross section of reciprocal-lattice points and rods of the thin film with the surface structure parallel to the electron beam and illustrates how the Ewald sphere cuts them.

Scattering of electrons in thin-surface layers is kinematic, but it is dynamic in the film. However, if the film is thin and/or is set so that no strong Bragg reflections are operating, then double diffraction of the surface and the bulk reflections is negligibly small. This has been shown experimentally in cases of the Si (111) 7×7 (Takayanagi et al., 1985) and Au (001) 28×5 structures. In the former case, intensity distributions of the 7×7 spots are not periodic with the projected bulk period, which would be true if dynamical effects were strong. In the latter case, the TED pattern shows rather strong diffraction spots from the top single layer and negligibly weak double diffraction spots (Takayanagi, 1984). This is quite different from the case of LEED patterns, where strong multiple diffraction does occur (Van Hove et al., 1981). A dynamical-diffraction calculation in the case of the Si (111) 7×7 structure by Spence (1983) showed that surface-reflection intensity shows only minor thickness oscillation for symmetric incidence, which clearly indicates that a kinematic-diffraction treatment for surface reflections is valid at least for certain specimen and Bragg conditions. Very recently, Tanishiro et al. (1986) calculated the TED intensities in the case of Si (111) 7×7 structure based on the DAS model (section 13.1.3) and showed that in certain conditions the dynamical effect can be avoided in the analysis.

Images of surfaces can be taken by using the imaging conditions shown in Figure 13.5b, namely, the bright-field image (BF), dark-field images of a bulk reflection (DF-1) (under strong- or weak-beam conditions) or of a surface reflection (DF-2), dark-field images of two or more of them (DF-3 and DF-4), HRTEM-1 due to surface reflections, and HRTEM-2 due to the surface and bulk reflections. The BF can show surface atomic steps under certain conditions, but generally image contrast of the surface structure is weak and it has not been used widely. The DF-1 can show surface steps very clearly. For practical purposes, such as to observe surface-dynamic processes *in situ,* very weak beam conditions (such as using forbidden reflections) are not useful except the cases where surface structures or adsorbate structures induce lattice distortions in the substrates (Yagi et al., 1979). Therefore, the proper Bragg conditions for imaging must be explored in each case. The DF-2 is useful to identify the surface regions that have the corresponding surface structure or one type of domain when equivalent orientational surface-structure domains exist. The DF-4 can also be used for the same purpose, but it also shows lattice fringes of the surface structure and their shift at out-of-phase boundaries. For DF-2 and DF-4, we need relatively long exposure times. In DF-3, moire fringes due to surface and bulk structures appear, which give us details of the local orientation relation and lattice parameters of the surface structure with respect to the bulk structure, as in the case of moire fringes between two overlapping crystals. Advantages of using DF-3 are the short exposure time for recording and the relatively strong contrast of surface structures, since the image is given by $\psi_b(\mathbf{r})^* \cdot \psi_s(\mathbf{r}) + \psi_b(\mathbf{r}) \cdot \psi_s(\mathbf{r})^*$, where $\psi_b(\mathbf{r})$ and $\psi_s(\mathbf{r})$ are scattered-electron-wave amplitudes from the bulk and surface structures, respectively. HRTEM-1 and HRTEM-2 will be discussed later.

From these considerations, the following possibilities of surface-structure analysis by the TEM–TED methods are concluded:

1. *Structure analysis from TED intensity measurement.* The fact that reflected intensities from the surface structure are kinematic means that they are proportional to squares of corresponding Fourier coefficients of the projected potential of surface structure $\phi_s(\mathbf{r})$ along the beam direction z:

$$I(hk) = \left| \int \int \left[\int \phi_s(\mathbf{r})\, dz \right] e^{2\pi i(hx+ky)}\, dx\, dy \right|^2. \quad (13.1)$$

This equation is for the case where incident electrons are perpendicular to the surface and $I(hk)$ is the intensity of an (hk)-rod with the third index $l = 0$. Notice that these parts of a reciprocal lattice cannot be observed by LEED or RHEED. From $I(hk)$, a surface structure projected along z is analyzed. Equation (13.1) can be extended to inclined-incidence cases, so that the (hk) rod with $l \neq 0$ can be observed. Thus, TED intensity can be used for three-dimensional structure analysis of surface structures, and the so-called DAS model was concluded for the Si (111) 7×7 structure (Takayanagi et al. 1985). The TED intensity can also be used to test previously proposed models or trial models (e.g., see McRae, 1983 Takayanagi et al., 1985). Note that the intensity given by equation (13.1) for normal incidence should be zero and that no surface reflections are observed if the surface reconstruction involves only atomic displacements normal to surface.

2. *Structure analysis by HRTEM of surfaces.* The real-space expression of the equation (13.1) shows that HRTEM-1 under optimum defocus conditions should tell us the arrangement of atoms in the surface structure. Thus, it can be used for structure analysis. HRTEM-2 may show the arrangement of the surface atoms with respect to the bulk lattice. Image simulations have shown such possibilities (Takayanagi, 1982; Spence, 1983). Observation of surfaces in profile by using HRTEM will be discussed in section 13.3.4.

3. *Structure analysis from DF-3.* As mentioned earlier, DF-3 gives details of the unit-cell size of the surface structure and its orientation on the bulk lattice. This is useful especially when the surface structure is incommensurate with the bulk lattice (see Figure 13.8 shown later in the chapter). In some cases, image-contrast analysis of DF-3 reveals strains produced by interfacial-misfit dislocations between the absorbed layers and bulk lattice, which could hardly be analyzed by LEED.

An important point in TEM–TED studies is that the dynamical-diffraction theory (Chapters 3 and 4) has been well developed. Thus, information obtained can be made quantitative by comparing the observed images with the calculated ones.

13.3.2 Surface-structure analysis by TEM–TED

Table 13.4 summarizes examples of surface-structure analysis done by the TEM–TED method. The recent analysis of the Si (111) 7 × 7 structure is a typical case where LEED analysis is too complex to be used because of a large unit cell, while TED analysis is rather simple and straightforward (Takayanagi et al., 1985).

As an example of TEM–TED analysis, the case of Au (111) reconstructed-surface structure is described (Tanishiro et al., 1981a). Figure 13.6 shows a dark-field image (DF-3) and a TED pattern of Au (111) platelets grown on MoS_2 *in situ* in the UHV microscope. In Figure 13.6a, three regions (A, B, C) with different directions of fringes (spaced about 6.3 nm) are domains of the surface reconstruction. This is concluded from the facts that directions of the fringes change during gold deposition and the fringes disappear by monolayer deposition of foreign metals (Figure 13.7). In the TED pattern, extra spots a, b, and c are seen in addition to spots from MoS_2 (M) and gold Au platelets (Au) of bulk structure. It was found that the fringe directions are perpendicular to the lines in reciprocal space from the $(10, 0)_{Au}$ spot to the extra spots (a, b, c), and the fringe spacings are inversely proportional to the distances between these spots (note the DF-3 condition). Thus, unidirectional contraction of a close-packed surface layer of Au by about 3.4 percent along one of the three $\langle 1\bar{1}0 \rangle$ directions on the (111) surface was concluded. The structure is denoted by 22 × 1. This was also concluded from an LEED study (Van Hove et al., 1981). An

TABLE 13.4
Surface-Structure Analyses by TEM–TED

Si (111) 7 × 7	TED[a]
Au (111) 22 × 1	TEM, TED[b]
Au (001) 28 × 5	TEM, TED[c]
Ag (111) twist–Pb	TEM, TED[d]
Au (111) twist–Pb	TEM, TED[e]
Au (110) 2 × 1	HRTEM*,[f,g]
Au (001) 28 × 5	HRTEM*,[g]
Si (111) 5 × 5–Ge	TED[h]
Si (001) 2 × 1	TED[i]

[a] Takayanagi et al. (1985).
[b] Tanishiro et al. (1981a).
[c] Takayanagi (1984).
[d] Takayanagi (1981).
[e] Takayanagi (1982).
[f] Marks and Smith (1983).
[g] Hasegawa et al. (1986).
[h] Kajiyama et al. (1986).
[i] Nakayama et al. (1986).
* Profile imaging (section 13.3.4).

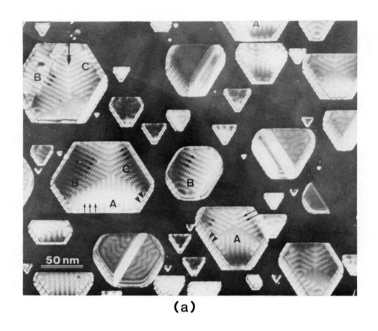

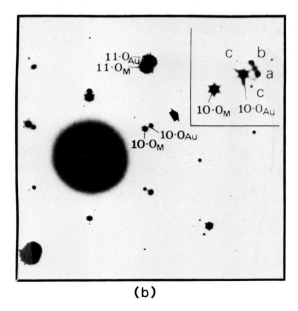

FIGURE 13.6 (a) Reconstructed-surface-structure image of Au (111) platelets grown on MoS$_2$. (b) A corresponding diffraction pattern, which shows surface-structure spots a, b, and c due to reconstructed-structure domains A, B, and C in (a).

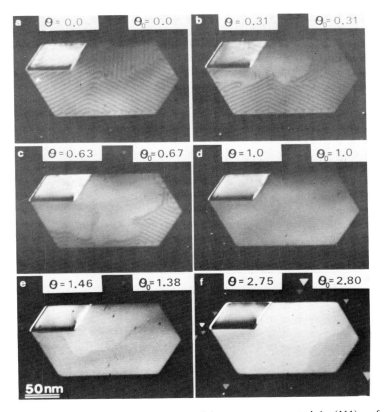

FIGURE 13.7 A layer-by-layer growth process of Ag on a reconstructed Au(111) surface. θ is the amount of deposit in ML units, which agrees with the fractional area θ_0 of fringe-free region from (a) to (d). In (e) and (f), second and third layers are growing, respectively. (Courtesy of *Surface Science*.)

important finding from the TEM–TED analysis is, however, that the contraction is not uniform but is similar to the contraction formed by an array of extended interfacial-misfit dislocations. Thus, the surface atoms are displaced from bulk positions not only along the contraction direction but also along the direction perpendicular to it. This was concluded not only from the contrast analysis of the image but also from the relative-intensity measurement of the extra spots in TED patterns and their analysis. This information is hard to obtain from LEED studies.

13.3.3 Observations of surface-dynamic processes

Table 13.5 summarizes possible observations of surface-dynamic processes, depending on the *in situ* techniques of specimen treatment. A few examples of observations are given below.

TABLE 13.5
Possible Observations of Surface-Dynamic Processes

Surface-structure-phase-transition processes
  Between two ordered phases
  Between ordered and disordered phases
    (including surface melting and roughening)
Adsorption processes
  Vapor deposition (metals and semiconductors)
  Gas adsorption and reaction (oxidation and reduction)
Step motions
  Due to sublimation
  Due to deposition (crystal growth)
  Due to deformation (slip trace)
Others
  Surface diffusion
  Bulk diffusion
  Ion sputtering (radiation damages)
  Catalytic reactions (small-particle surfaces)
  Surface-structure-domain growth

Three different kinds of observations of adsorption processes are shown in Figures 13.7, 13.8, and 13.9. In Figure 13.7, adsorption of silver on Au(111) 22×1 is shown. From Figures 13.7a to d, we see that 1 ML deposition of silver removed the surface reconstruction of gold (Tanishiro et al., 1981a). During this process, a fractional area θ_0 of fringe-free regions is equal to the amount of deposit θ in ML units. Also, later growth takes place by layerwise growth. Not the growing terrace edges (steps) of second and third layers of silver in Figures 13.7e and f, respectively.

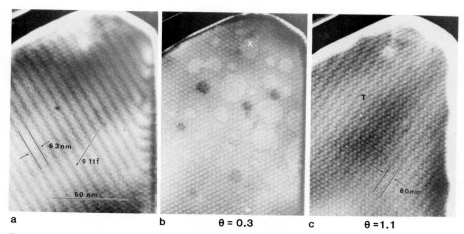

FIGURE 13.8 A sequence showing the growth process of Pb on a reconstructed Au(111) surface. (a) A reconstructed structure image, which disappears in (b) by 0.3 ML deposit of Pb. (c) Moire fringes between close-packed monolayers of Pb (twist structure) and the Au lattice. (courtesy of Dr. K. Takayanagi.)

For lead growth on Au (111) 22 × 1 in Figure 13.8, the reconstruction is removed only by 0.3 ML deposition (Figure 13.8b). At 0.8 ML, moire fringes due to the lead monolayer and gold lattice (DF-3 condition in Figure 13.5) are seen, and their spacing and orientation change. Figure 13.8c shows the moire fringe at 1.1 ML. The changes are explained by an increase in atom density of the lead monolayer and an increase of twist angle (misorientation angle) between lead and gold close-packed, two-dimensional lattices.

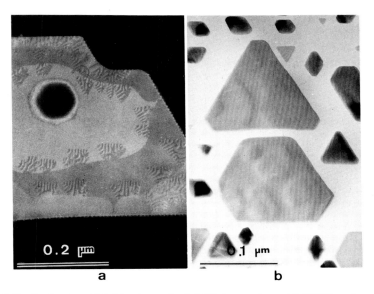

FIGURE 13.9 Single-atom high-island growth. (a) Of Ag on stepped Pd (111) surface. (b) Of Pd on Ag. Misfit dislocations are formed between single-atom high Ag islands and Pd substrate in (a) but not in (b).

Figure 13.9 shows cases of single-atom, high-nucleus formation. In Figure 13.9a, showing silver adsorption on palladium, single-atom high nuclei with misfit dislocations formed along the lower sides of the steps (Yagi et al., 1985). Note the similarity of the image intensity in the silver nuclei (newly formed on palladium terraces) and preexisting palladium terraces of the same height (silver is next to palladium in the periodic table). In Figure 13.9b, showing the reversed case (palladium growth on silver), single-atom high nuclei have no misfit dislocations (Yagi et al., 1979). The nuclei have strong contrast even in the bright-field image, resulting from the strain of the silver lattice produced by the palladium nuclei at the peripheries of the nuclei, which are expanded by 4.8 percent to be coincident with the silver lattice.

13.3.4 HRTEM of surfaces

Two modes of HRTEM of surfaces have been discussed, as shown in Figure 13.10: observations of the surface (A) through the film (plan view mode) and edge-on observations of the surface (B) (profile imaging mode). Image simulations showed that, with proper specimen and imaging conditions, both modes are useful. UHV–HRTEM—with a point-to-point resolution of around 0.17 nm at 1000 kV (Yagi et al., 1983), provision for a heating holder, a liquid-helium cryoshroud, and an *in situ* evaporator—has already been developed (Kodaira et al., 1983). Using these techniques in the transmission mode, atomic arrangements of clean Ag (111) and (001) surfaces have been imaged by Takayanagi et al. (1983) (HRTEM-1 in Figure 13.5). HRTEM-2 of Au (110) surfaces of small particles was done in Cambridge in the profile imaging mode (Marks, 1983; Marks and Smith,

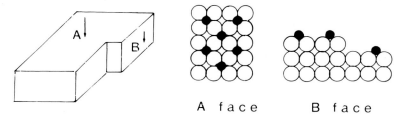

FIGURE 13.10 Two modes of an HRTEM study of surfaces.

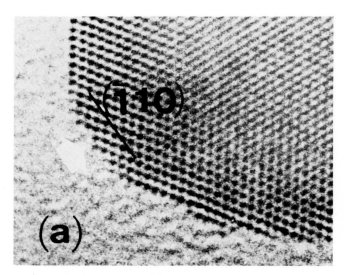

FIGURE 13.11 An HRTEM image of an Au particle epitaxially grown on a (110) cleavage surface of NaCl taken by 1-MV HRTEM. The {111} faceting of the (110) surface of Au is shown.

SURFACES

1983), although, in this case, the specimens were transferred to the microscope through the air and not observed in UHV conditions. They concluded the missing-row model from an image similar to that in Figure 13.11, for the 2×1 reconstruction of an Au(110) surface under the assumption that reconstruction formed in the vacuum chamber remains through their specimen treatment in air and observations. The profile imaging in clean conditions has been applied recently to Si(111) and (311) surfaces (Gibson et al., 1985) and Au(001) and (110) surfaces (Hasegawa et al., 1986a, 1986b).

13.4 REM studies of surfaces

13.4.1 Image-formation processes

Imaging geometry in REM–RHEED is schematically shown in Figure 13.12. An electron beam, which hits a bulk-crystal surface with a small

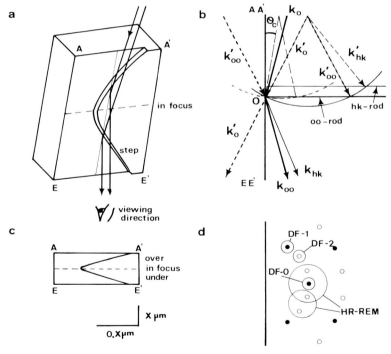

FIGURE 13.12 A schematic illustration of imaging conditions in REM. (a) The electron beam direction and the viewing direction (c) the obtained image and scaling marker for a REM micrograph (a foreshortened image of a surface). (b) Refraction and diffraction geometry in the reflection mode. A prime means wave vectors in the crystal; θ_c is a critical angle for the total reflection in the crystal. (d) A RHEED pattern (solid and open circles are diffraction spots from the bulk and surface structures, respectively) and various aperture positions for surface imaging.

glancing angle (Figure 13.12a) forms a RHEED pattern (Figure 13.12d) on a back focal plane. Owing to the small glancing angle, most of electrons are reflected at surface regions of the specimen (penetration depth is less than several nanometers); so the dynamical-scattering effect is strong, and bulk reflections (solid circles) as well as surface reflections (open circles) are surface-structure sensitive. Note that, owing to a refraction effect, (hk)-rods with l smaller than a critical value $(1 \sim 2)$, which is determined by a value $2k\theta_c$ as shown in Figure 13.12b, do not appear in the RHEED pattern. Here, θ_c is a critical angle for the total reflection, and k is the electron wave number.

Since the imaging beam makes a small glancing angle at the surface, REM images are foreshortened, as shown in Figure 13.12c, and are in focus only along a line perpendicular to the beam direction (note also scale marks). Images may be taken with the objective-lens aperture at the positions of the specular spot (DF-0), one of the bulk reflections (DF-1), surface reflections (DF-2), and two or more of them (HR-REM). Since RHEED is surface-structure sensitive, each type of image is also. The DF-1 and DF-2 are similar to DF-1 and DF-2 in Figure 13.5, respectively. HR-REM is like DF-3 and DF-4 in Figure 13.5 and can give lattice fringes of the surface structure.

Two kinds of image contrast in REM, Bragg contrast and Fresnel contrast, have been discussed (Osakabe et al., 1981a). The former is seen at in-focus regions, where

$$I(\mathbf{r}) = I(\mathbf{r}_s) = \left| \sum A_h(\mathbf{r}_s) * \frac{J_1(\pi u_0 r_s)}{\pi u_0 r_s} * \mathscr{F}(e^{i\chi(h)}) \right|^2 \qquad (13.2)$$

is recorded on a photographic plate (see Figure 13.13, position P). Here, $A_h(\mathbf{r}_s)$ is the complex scattering amplitude of **h**-reflection at the surface $\mathbf{r}_s$, and $J_1(\pi u_0 r_s)$ and $\chi(\mathbf{h})$ are given by equations (1.7) and (1.9), respectively; but for discussions of low-resolution images, they can be ignored, and images are mainly determined by $|A_h(\mathbf{r}_s)|^2$, which we call Bragg contrast.

The latter contrast appears for out-of-focus regions (position Q in Figure 13.13). Ignoring the aperture and aberration effects, image contrast is given in

$$I(\mathbf{r}) = \left| \sum \int A_h(\mathbf{r}_s) \cdot P(\mathbf{r}, \mathbf{r}_s) \, d\mathbf{r}_s \right|^2. \qquad (13.3)$$

Here, $P(\mathbf{r}, \mathbf{r}_s)$ is a propagation function. Images are determined not only by $|A_h(\mathbf{r}_s)|$ but also by the phase $\alpha_h(\mathbf{r}_s)$ of the scattering amplitude through the propagation from $\mathbf{r}_s$ to $\mathbf{r}$. Thus, we call an additional contrast produced by the propagation the Fresnel contrast.

Clearly to know $A_h(\mathbf{r}_s)$ is the first step for the image analysis. For perfect- and flat-crystal surfaces, several approaches have been proposed (Collela, 1972; Collela and Menadue, 1972; Moon, 1972; Masud and Pendry, 1976; Maksym and Beeby, 1981, 1984; Ichimiya, 1983). However, there has been

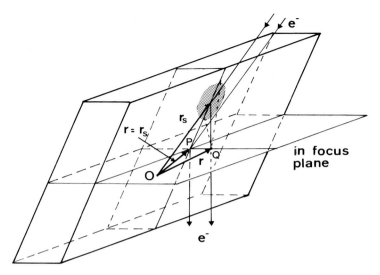

FIGURE 13.13 Fresnel-diffraction-contrast formation process for an out-of-focus position Q. The shaded area corresponds to the first Fresnel zone on the surface from Q.

no general RHEED theory for imperfect crystals—hence, no general theory for REM-image contrast of imperfect surfaces. Shuman (1977) first attempted calculations of the Bragg contrast of dislocations and stacking faults emergent at the surface with a column approximation, which assumes that the reflected intensity from a part of distorted areas around the dislocation is the same as that of the distorted area. Following him, Osakabe et al. (1981a) calculated a rocking curve for the specular reflection from the Si (111) surface, as shown in Figure 13.14a. From the lattice misorientation around a dislocation and the curve, the image profile of the dislocation was calculated as shown in Figure 13.14b and c, which are image profiles for crystal settings such that the perfect-crystal regions are at the exact Bragg setting (A in Figure 13.14a) and off-Bragg setting (B in Figure 13.14a), respectively.

Note that, because of the narrow Bragg width in the rocking curve in the reflection case, image widths of dislocations are much larger ($\simeq 500$ nm) than those in TEM ($\simeq 20$ nm), and REM is sensitive to lattice distortion, a distortion of 10^{-4} being detectable. Figure 13.15 shows screw-dislocation images under the two conditions (A and B in Figure 13.14a), which agree well with Figures 13.14b and c, respectively.

Step-image contrast was also analyzed in terms of a lattice distortion along the step and compared with the observations. Recently, Kawamura and Maksym (1985) applied the RHEED theory to stepped surfaces and

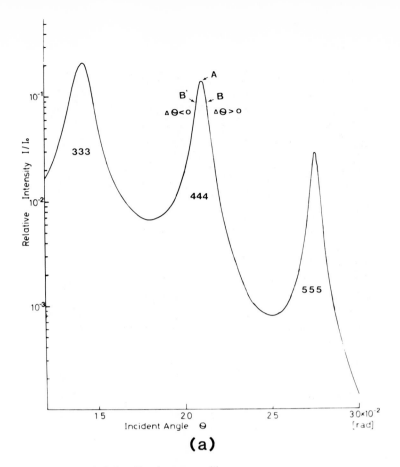
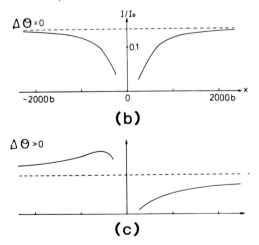

FIGURE 13.14 (a) Intensity variation of the specular spot as a function of incident angle θ for the Si(111) surface for 100-keV electrons. (b) and (c) Screw-dislocation-image profiles on Si(111) at different Bragg conditions, A and B in (a), respectively. The horizontal axis x is expressed in units of the magnitude of the Burgers vector (0.31 nm).

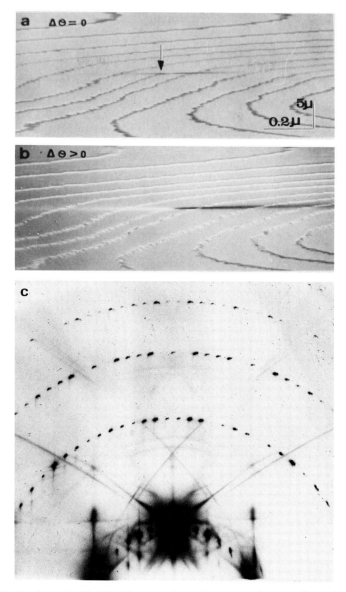

FIGURE 13.15 (*a*) and (*b*) REM images of atomic steps and screw dislocation on a clean Si(111) 7 × 7 surface at different Bragg conditions, which correspond to Figures 13.14*b* and *c*, respectively. (*c*) A corresponding 7 × 7 RHEED pattern.

interpreted the intensity-oscillation behavior* of RHEED spots recently observed during growth of III–V semiconductors by molecular-beam epitaxy. They assumed a periodic array of steps perpendicular or parallel to the beam direction. The calculation showed that the intensity behavior is different for two cases, which are clearly related to the REM-image-constant behavior in the two cases. In their calculation, a large unit cell was assumed, similar to the method of periodic continuation (see Chapter 8).

The clear Fresnel contrast was first pointed out experimentally by Osakabe et al. (1981a) and later discussed in detail by Cowley and Peng (1985). Apparently, when $|A_h(\mathbf{r}_s)|$ and/or $\alpha_h(\mathbf{r}_s)$ changes appreciably in the area shaded in Figure 13.13 (the area corresponds to the first Fresnel zone from the point Q). In this case, we have to know $A_h(\mathbf{r}_s)$. In addition to this, a geometrical phase factor produced by surface unevenness also causes the Fresnel contrast. If we have a surface partially shifted by $\Delta \mathbf{r}_s$ ($\mathbf{r}_s$) due to slight lattice distortion, which does not cause appreciable change in $|A_h(\mathbf{r}_s)|$ and $\alpha_h(\mathbf{r}_s)$, it causes phase shift $\delta(\mathbf{r}_s) = 2\pi \mathbf{h} \, \Delta \mathbf{r}_s(\mathbf{r}_s)$ of the **h**-reflected wave.

In the normal-imaging condition, $h = (3 \sim 4)/d$ (d is a lattice spacing of low-index lattice plane), and the value of δ is about $\frac{1}{3} \sim \frac{1}{2}$ rad, even for $|\Delta \mathbf{r}_s|$ of $d/50$, which is large enough to produce Fresnel contrast. Steps and dislocation images for out-of-focus conditions were qualitatively explained by the effect of δ.

For REM-image-contrast analyses, a few remarks should be made. The first is the fact that reflected intensity changes sensitively with incident polar and azimuthal angles. This means that we need a parallel electron beam to get fine details of surface images. However, because of the forefield of the objective lens (it is very strong in the case of the so-called C–O lens), the incident angle of electrons changes appreciably from place to place in the illuminated area. Thus, even when a large area of a perfect crystal is illuminated, the REM image is bright only in a restricted surface area whose shape is similar to that of the corresponding diffraction spots in RHEED. Care is needed on this point for the image-contrast analyses.

The second remark relates to an effect of inelastic-scattering processes. Energy analysis of reflected electrons from flat and clean surfaces showed pronounced surface-plasmon-excitation losses for small-glancing-angle incidence (Figure 13.16) (Yagi, 1982; Krivanek et al., 1983); so in REM, one of the most important resolution-limiting factors is chromatic aberration. It was concluded that the loss process takes place when electrons travel in the vacuum close to the surface. The excitation probability agrees well with a theoretical prediction (Howie, 1983). A change of energy does not affect Bragg conditions so much; but a slight change of momentum of incident electrons owing to the surface-plasmon excitation in the loss-diffraction process results in a change of incident angles, so images due to these

---

* The oscillation is now used to monitor a completion of one-layer growth and is used to control the thickness of overgrowth precisely. This is the first case where electron-diffraction equipment is used for a manufacturing-process control.

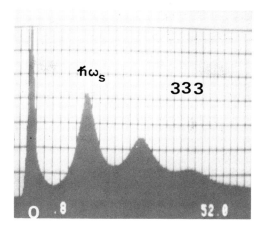

FIGURE 13.16 An energy-loss spectrum of the (333) specular spot from a clean and flat part of Si (111) surface ($E_p = 80$ kV). A strong surface-plasmon-loss peak and its multiples are shown.

electrons are expected to be different from those of no-loss electrons. However, an energy loss after diffraction (diffraction loss) does not affect the contrast. Changes of the image contrast of dislocations and steps by a slight change of objective-aperture positions to the higher- or lower angle side of the specular-reflection positions (Osakabe et al., 1981a) is attributed to the loss-diffraction process. Thus, an image with a lower- (higher-) angle-aperture position should be similar to an image taken at a larger (smaller) incident angle from the Bragg conditions, as actually is observed (Figure 7 of Osakabe et al., 1981a; Figure 10 of Osakabe et al., 1980).

The third remark concerns HR-REM (see Figure 13.12). No experimental and theoretical studies of lattice fringes in REM have been reported until recently. Imaging of lattice fringes is accompanied by many complex problems, such as foreshortening (which depends on exit angles for each reflection), out-of-focus conditions, inelastic-scattering contributions, and geometrical complexity (Tanishiro et al., 1986).

The fourth remark concerns surface undulation and its effect on REM and RHEED. If we have large undulations so that transmission of electrons through protruding parts takes place, REM shows TEM-like images similar to large-angle, wedge-shaped-crystal images (see Figure 2 of Russel and Woods, 1976), and RHEED shows spotty patterns similar to TED (Figure 3 of the same reference). But if we have almost flat, perfect surfaces, an image should have no contrast (see the large flat areas of Figure 13.15a and b); and a sharp spotty pattern is seen in RHEED (Figure 13.15c). Owing to the perfection of the surface, the reciprocal-lattice rods should be thin; and because of the refraction effect, the Ewald sphere should cut the rods in the zeroth Laue zone with finite angles (see Figure 13.12b). These two facts are

TABLE 13.6
REM–RHEED Studies of Surfaces

Surface topography (morphology)
  Steps
  Dislocations emergent at the surface
  Surface-structure domains [orientational and
    translational (out-of-phase domains)]

Surface-dynamic processes
  Phase transition
    Si (111) $7 \times 7 \leftrightarrow 1 \times 1$
    Si (111) $5 \times 1 \leftrightarrow 1 \times 1$ Au
  Adsorptions
    Au, Ag, Sn, Ge, Sn, Cu, Al on Si (111) 7×7
    Au on Si (001) 2×1, Au on Pt (111)
  Oxidation: O_2 on Si (111) 7×7

[Applications of REM to other field of science.]

the reasons for a spotty pattern. This is also apparent from the fact that diffraction-spot width has a reciprocal relation to the surface image (no contrast). In the intermediate case of surface undulation, a local misorientation from a flat face is less than a Bragg angle, bright and dark modulations of reflected intensity are seen in the image, and RHEED shows a streak pattern (see Figure 5 of Yagi et al., 1980; Figure 7 of Nielsen and Cowley, 1976). The streak has a reciprocal relationship with the intensity modulation in REM. The origins of the modulation in this case were discussed by Nielsen and Cowley (1976): a geometrical factor and a dynamical-diffraction effect caused by refraction. Another reason for the streak in RHEED is the small sizes of the surface-structure domains even on a flat surface. A typical case is Si (111) $\sqrt{3}_d$–Au and $\sqrt{3}_s$–Au (see section 13.4.2.2). In the former structure, the domain size is very small in REM, and diffuse streaks are seen in RHEED; in the latter structure, large domains and sharp spots are observed. Very recently, Peng and Cowley (1986) developed a multislice formulation of the many beam dynamical theory to be applied to the REM images of a surface with a surface step.

Table 13.6 summarizes REM studies of surfaces. One group of studies involves characterization of surface morphology such as the observation of steps, dislocations emergent on the surface, and surface-structure domains. Two types of domain have been observed, as in the case of three-dimensional crystals: orientational and translational (out-of-phase) domains.

13.4.2 Observations of surface-dynamic processes

13.4.2.1 Phase transitions. A typical and well-known example of surface-phase-transition processes is the reversible transition between the Si (111) 7×7 and 1×1 structures at about 830°C studied by LEED (Bennett and

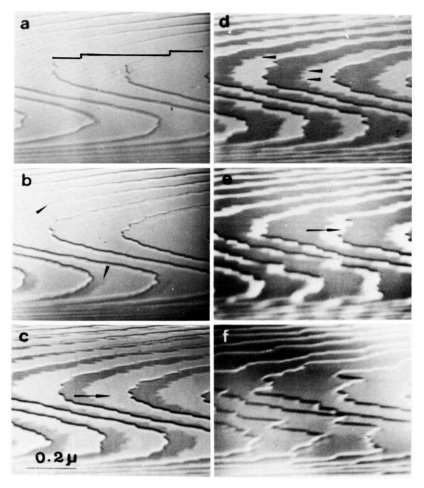

FIGURE 13.17 A sequence of REM images, which shows a phase-transition process from the Si (111) 1 × 1 structure to the 7 × 7 structure (darker regions on the terraces) on cooling. (For details, see the text.) (Courtesy of *Surface Science*.)

Webb, 1981). Osakabe et al. (1981b) studied it by REM. Figure 13.17 shows a sequence of REM images (DF-0 in Figure 13.12) obtained on cooling from 830°C. In Figure 13.17a, the surface is in the 1 × 1 high-temperature phase. Terraces are higher on the right-hand side of the steps. On cooling, dark regions (arrowheads) appear along the outer edge of the steps (Figure 13.17b) and expand toward the other side of steps on the terraces (see arrows in Figures 13.17c and e). The DF-3 images by the 7 × 7 spots showed that the dark regions are of the 7 × 7 structure. The phase transition is completed when the phase boundaries reach the inner edges of the steps. The reverse process was observed on heating. The arrowheads in Figure 13.17d show thin, bright lines that are out-of-phase boundaries in the 7 × 7 structure. The details of the physics of the phase transition were discussed by Osakabe et al. (1981b) and Tanishiro et al. (1983). Here, the following should be noted:

1. Because of a strong dynamical effect in REM, strong contrast between the different surface-structure regions and contrast of the out-of-phase boundaries are seen in the DF-0 image of the bulk reflection as well as in DF-3 of the surface reflections.
2. Heterogeneous-surface processes are clearly seen.

13.4.2.2 Adsorption processes. Apparent from REM studies was that many adsorption processes are heterogeneous, being affected by the surfaces steps. For example, Figure 13.18 shows the initial-growth stage of

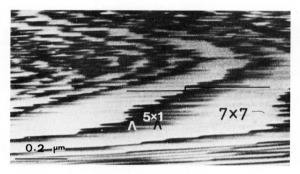

FIGURE 13.18 A REM image of the initial stage of Au adsorption (5 × 1 structure) on an Si (111) 7 × 7 surface at 650°C. Preferential formation of the adsorbate structure at the outer terrace edges along the steps is shown.

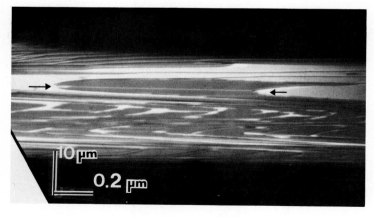

FIGURE 13.19 A domain structure of the Si (111) 5 × 1 Au surface. Arrows indicate an out-of-phase boundary.

the Si(111) 5×1 Au structure at 650°C (Tanishiro et al., 1981b). The absorbate structure is preferentially formed along the upper side of terrace edges. In the later stage, it expands over the terraces and covers the terraces entirely. An interesting finding, which is particular to surface microscopy, is that, during the adsorption, the steps move in a direction so as to expand the upper terraces. The reason is not clear yet since no conclusive structure model of the 5×1 structure is available.

Figure 13.19 shows domains of the 5×1 structure, which can take three equivalent orientations on the Si(111) surface. They are seen with bright and dark and intermediate contrast (in the micrograph, domains with dark contrast are narrow). Most of the domain boundaries are steps. In one type of domain, an out-of-phase boundary indicated by arrows is also seen (Yagi et al., 1980). In the later state of Au adsorption, the 5×1 structure transforms gradually to the $(\sqrt{3} \times \sqrt{3})_d$ structure (small domain size and diffuse streaks owing to thick rods are observed in REM and RHEED), and it transforms suddenly to the $(\sqrt{3} \times \sqrt{3})_s$ structure (sharp spots).

13.4.2.3 Oxidations processes. Figure 13.20 shows the initial stage of oxidation of the Si(111) 7×7 surfaces at about 670°C. Hollows with the depth of one atomic layer nucleate and grow in the central part of the terraces. The fact that the formation of hollows is depressed along the steps suggests the following process: Vacancies are formed all over the terraces owing to mono-oxide formation and sublimation, and they meet each other to nucleate into the hollows. The vacancies formed at places close to the preexisting steps annihiliate there, and the hollows are not formed there. At

FIGURE 13.20 Hollow formation (0.31 Å deep) at central regions on the terrace at the initial stage of oxidation of the Si(111) 7×7 surface at 670°C.

598 HIGH-RESOLUTION TRANSMISSION ELECTRON MICROSCOPY

this stage, the 7 × 7 structure spots remain in RHEED. In the later stage, the surface transforms into the 1 × 1 structure, and oxide covers the surface. Details of the growing process of the hollows and the temperature dependence of the oxidation process are given by Shimizu et al. (1985).

This example again shows that surface electron microscopy is a powerful technique for revealing the details of surface-dynamic processes.

13.4.3 Applications of REM to other fields of sciences

REM's capability of surface characterization can be used in various fields of science. This is because we can see step structures of surfaces even after clean surfaces are covered by structureless adsorbates, as in the cases of air-exposed Si (111) surfaces and the gold-deposited Si (111) 7 × 7 surface at room temperature. One application is to use REM for characterization of surface morphology of as-grown crystals made by various techniques. These surfaces are considered to be clean and flat when they are formed. This was attempted by Hsu (1983) and Hsu and Cowley (1983) on surfaces of single crystal spheres of platinum and gold formed by heating the metal wires in

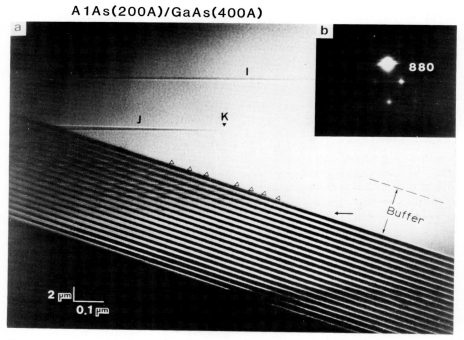

FIGURE 13.21 A REM image of a (110) cleavage plane of an AlAs (200 Å)/GaAs (400 Å) superlattice grown on GaAs with GaAs buffer layers. (Courtesy of Dr. N. Yamamoto and *Jpn. J. Appl. Phys.*)

air. Step configurations of (111) surfaces, slip traces, and vicinal surfaces were characterized.

Another example is to characterize bulk-crystal structures indirectly through an observation of surfaces of cross sections of the crystals. The first attempt was made for superlattices formed by alternative depositions of $Ga_{1-x}Al_xAs$ and GaAs (Yamamoto and Muto, 1984); see Figure 13.21. This is an extension of early trials of REM observations of cleavage surfaces of III to V semiconductors (Yamamoto and Spence, 1983; Hsu et al., 1984). The superlattice is cleaved together with the base crystal; and a periodic arrangement of two kinds of layers, which forms a quantum-well structure, is clearly contrasted. The origins of the contrast are structure-factor difference and lattice strains at the interface owing to small-lattice mismatch ($10^{-3} \sim 10^{-4}$).

13.5 STEM, SREM, and microanalysis of the surface

13.5.1 SREM and STEM

As mentioned in Chapter 7, STEM and SREM of surfaces are equivalent to TEM and REM of surfaces by reciprocity. Three types of experimental approach are possible. One is to use CTEM in the STEM mode. The second is to use a dedicated, high-resolution STEM instrument commercially available with probe diameter of 0.5 nm or less at 100 kV. The third is to construct a special UHV apparatus for SREM (or STEM) with a probe diameter of $10 \sim 20$ nm with the use of field-emission guns.

The CTEM approach to STEM has not been done. HR–STEM approaches to SREM or to surface studies have been reported by Cowley (1982a, 1982b), Cowley and Kang (1983), and Cowley and Neumann (1984); but owing to lack of an *in situ* specimen-preparation technique, it has not been applied to clean-surface problems. However, it showed the importance of nanometer probes for the characterization of local structures and showed details of interactions of electrons with surfaces involving diffraction, refraction, energy losses, and surface channeling. For details, refer to papers by Cowley (1982a, 1982b) and Cowley and Kang (1983).

The UHV–SREM approach sacrifices spatial resolution in order to get space around the specimen for *in situ* specimen-treatment devices and analyzers. As pointed out by Yagi (1980), to get images similar to REM by CTEM, the acceptance angle of the detector should be made small (10^{-4} rad), since REM images by CTEM are sensitive to incident-beam directions. A recent success of SREM by Ichikawa et al. (1984) is based on their effort to reduce their effective detecting angle. A beam diameter of 20 nm with a beam current of 10 nA at 20 kV is one of their operating conditions. They succeeded in observing steps, dislocations, and absorption processes as in REM. The Arizona group has also improved the detecting system and specimen-treatment devices for their SREM apparatus, developed ten years ago (Cowley et al., 1975), and has succeeded in observing domains of adsorbate structure formed by gold and silver deposition on clean Si (111) 7 × 7 surfaces (Elibol et al., 1985). The beam diameter is 7 nm at 20 kV, and the lowest operating voltage is 0.5 kV. By replacing a specimen holder, STEM-mode observations are possible. They introduced a vacuum-cleavage device, and Bennett et al. (1985) succeeded in observing domains of the Si (111) 2 × 1 structure, which is characteristic of the cleaved surface.

An advantage common to all of the scanning approaches is easy image processing, such as correction of the foreshortening and dynamic focusing (reduction of the out-of-focus effect).

13.5.2 Surface analysis

Analytical techniques are important for characterization of surfaces. The importance of combining them with surface-imaging methods lies in the fact that only in this case can we know where we are analyzing and what kind of place it is. Possible methods are Auger-electron spectroscopy, electron-energy-loss spectroscopy, energy-dispersive, X-ray analysis, microdiffraction, and cathodoluminescence. In general, UHV–SREM instruments are easily combined with these techniques because of the large space around the specimen.

EELS in the REM mode has already been mentioned. Microdiffraction at the surface of small crystals has been discussed by Cowley and Spence (1981). EDS in the TEM mode can detect monolayer adsorbates; but at present, it is difficult to get quantitative data, as required, for example, for surface ALCHEMI (atom location by channeling electron microanalysis) (see Chapter 7). EDS in the reflection mode was studied by Ino et al. (1980)., who observed peaks quantitatively from a monolayer of adsorbate, which shows a possibility of ALCHEMI in the reflection mode similar to standing-wave methods in X-ray experiments. Lately, Hasegawa et al. (1985) developed a technique to enhance the signal from adsorbate atoms on the surface.

As mentioned in section 13.2.3, combining Auger electron spectroscopy with CTEM is difficult. However, it is easily introduced in UHV–SREM instruments (Ichikawa et al., 1984). Horio and Ichimiya (1985) used RHEED–AES to analyze the Si (111) ($\sqrt{3} \times \sqrt{3}$)–Ag structure (a kind of ALCHEMI). Surface cathodoluminescence in the TEM mode may be difficult, but it may be applied in the REM mode.

13.6 Concluding remarks

In this chapter, the importance and usefulness of surface electron microscopy for obtaining monolayer-level structure information with high spatial resolution are described by showing recent progress, with emphasis on the CTEM approach.

One of the future challenges in surface CTEM is high-resolution, surface-structure imaging, and we are now exploring this field. Recent progress in this direction has been very rapid (see Bovin et al. 1985; Iijima and Ichihashi, 1985; Gibson et al., 1985; Hasegawa et al., 1986a; 1986b; Smith, 1986). Gibson et al. and Hasegawa et al. used UHV microscopes for their observations.

Further development of specimen-preparation techniques to meet various requirements for materials, specimen-preparation chambers and transfer methods, and analytical techniques are also important. Also, the image theory for the REM mode should be developed.

Surface observations in the scanning mode and by other types of surface-imaging methods are now developing very quickly. As in ordinary surface studies, it is important to combine the various analytical techniques with the imaging techniques.

REFERENCES

Ambrose, B. K. (1976). A device for transferring specimens from an ultrahigh vacuum chamber into the tilting goniometer of JEOL 100B electron microscope. *J. Phys. E* **9**, 382.

Aseef, A. L., Latyshev, A. V., and Stenin, S. I. (1984). Investigation of structure transformation on clean surface of silicon and germanium by means of ultrahigh vacuum reflection electron microscopy. *Proc. European Reg. Cong. Electron Microscopy (Budapest)*, 1215.

Beamson, G., Porter, H. Q., and Turner, D. W. (1981). Photoelectron spectromicroscopy. *Nature* **290**, 556.

Bennett, P., and Webb, M. B. (1981). The Si (111) 7×7 to 1×1 transition. *Surf. Sci.* **104**, 74.

———, Ou, H., Elibol, C., and Cowley, J. M. (1985). Domain structure of the Si (111) 2×1 surface studied by reflection electron microscopy. *J. Vac. Sci. Technol., Sect. A* **3**, 1634.

Bethge, H. (1962). Oberflachenstructrum and Kristallbaufehler im electron mikroskopischen Bild, untersucht am NaCl. II. *Phys. Status Solidi* **2**, 775.

——— (1982). Studies of surface morphology on an atomic scale. In *Interfacial aspects of phase transformation,* ed. B. Mutaftschiev, 669. Reidel Publishing Company, Boston.

———, Gerth, D., and Matern, D. (1982). Photoemission microscopy and surface analysis. *Proc. 10th Int. Cong. Electron Microscopy (Hamburg)*, 69.

Binnig, G., and Rohrer, H. (1983). Scanning tunneling microscopy. *Surf. Sci.* **126**, 236.

———, Rohrer, H., Gerber, Ch., and Weibel, E. (1983a). 7×7 reconstruction on Si (111) resolved in real space. *Phys. Rev. Lett.* **50**, 120.

———, Rohrer, H., Gerber, Ch., and Weibel, E. (1983b). (111) faces as the origin of reconstructed Au (110) surfaces. *Surf. Sci.* **131**, L379.

Bovin, J. O., Wallenburg, R., and Smith, D. J. (1985). Image of atomic clouds outside the surface of gold crystal by electron microscopy. *Nature* **317**, 47.

Cherns, D. (1974). Direct resolution of surface atomic steps by transmission electron microscopy. *Philos. Mag.* **30**, 549.

Collela, R. (1972). n-Beam dynamical diffraction of high-energy electrons at glancing incidence. General theory and computational methods. *Acta Crystallogr., Sect. A* **28**, 11.

———, and Menadue, J. F. (1972). Comparison of experimental and n-beam calculated intensities for glancing incidence high-energy electron diffraction. *Acta Crystallogr., Sect. A* **28**, 16.

Cowley, J. M. (1982a). Microdiffraction, STEM imaging and ELS at crystal surfaces. *Ultramicroscopy* **9**, 231.

——— (1982b). Surface energies and surface structure of small crystals studied by use of a STEM instrument. *Surf. Sci.* **114,** 587.

———, and Kang, Z.-C. (1983). STEM imaging and analysis of surfaces. *Ultramicroscopy* **11,** 131.

———, and Neumann, K. D. (1984). The alignment of gold particles on MgO crystal faces. *Surf. Sci.* **145,** 301.

———, and Peng, L. M. (1985). The image contrast of surface steps in reflection electron microscopy. *Ultramicroscopy* **16,** 59.

———, and Spence, J. C. H. (1981). Convergent beam electron microdiffraction from small crystals. *Ultramicroscopy* **6,** 359.

———, Albain, J. L., Hembree, G. G., Nielsen, P. E. H., Koch, F. A., Landry, J. D., and Shuman, H. (1975). System for reflection electron microscopy and electron diffraction at intermediate energies. *Rev. Sci. Instrum.* **46,** 826.

Elibol, C., Ou, H., Hembree, G. G., and Cowley, J. M. (1985). Improved instrument for medium-energy electron diffraction and microscopy. *Rev. Sci. Instrum.* **56,** 1215.

Fadley, C. S. (1984). Angle-resolved X-ray photoelectron spectroscopy. In *Progress in surface science,* vol. 16, ed. S. G. Davison, p. 275. Pergamon Press, Elmsford, N.Y.

Futamoto, M., Hanbuchen, M., Harland, C. J., Jones, G. W., and Venables, J. A. (1985). Visualization of submonolayers and surface topography by biased secondary electron imaging. *Surf. Sci.* **150,** 4306.

Gibson, J. M., McDonald, M. L., And Unterwald, F. C. (1985). Direct imaging of a novel silicon surface reconstruction. *Phys. Rev. Letts.* **55,** 1765.

Goodman, P., and Moodie, A. F. (1974). Numerical evaluation of N-beam wave functions in electron scattering by the multislice method. *Aca Crystallogr., Sect. A* **30,** 280.

Halliday, J. S., and Newman, R. C. (1960). Reflection electron microscopy using diffracted electrons. *Br. J. Appl. Phys.* **11,** 158.

Hamers, R. J., Tromp, R. M., and Demuth, J. E. (1986). Surface electronic structure of Si (111)–7 × 7 resolved in real space. *Phys. Rev. Letts.* **56,** 1972.

Harland, C., Akhter, P., and Venables, J. A. (1981). Accurate microcrystallography at high spatial resolution using electron back scattering patterns in a FEG–SEM. *J. Phys. E* **14,** 175.

Hasegaswa, S., Ino, S., Yamamoto, Y., and Daimon, H. (1985). Chemical analysis of surfaces of total-reflection-angle X-ray spectroscopy. In RHEED experiments (RHEED–TRAXS). *Jpn. J. Appl. Phys.* **24,** L387.

Hasegawa, T., Kobayashi, K., Igarashi, N., Takayanagi, K., and Yagi, K. (1986a). Atomic resolution TEM images of Au (001) reconstructed surface. *Jpn. J. Appl. Phys.* **25,** L366.

———, Ikarashi, N., Kobayashi, K., Takayanagi, K., and Yagi, K. (1986b). High resolution profile images of the reconstructed structures of Au (001), (110), and (111) surfaces. *Proc. 11th Int. Cong. on Electron Microscopy (Kyoto),* 1345.

Horio, Y., and Ichimiya, A. (1985). Surface study of Si (111) $\sqrt{3} \times \sqrt{3}$–Ag by incident beam rocking AES method. *Surf. Sci.* **164,** 589.

Howie, A. (1981). Electron microscopy of surface structure and reactions. In *Electron microscopy and analysis—1981,* ed. J. M. Goringe. Chap. 9, 419. Institute of Physics, Bristol.

———, (1983). Surface reactions and excitations. *Ultramicroscopy* **11,** 141.

Hsu, T. (1983). Reflection electron microscopy of vicinal surfaces of fcc metals. *Ultramicroscopy* **11,** 167.

———, and Cowley, J. M. (1983). Reflection electron microscopy of fcc metals. *Ultramicroscopy* **11,** 239.

———, Iijima, S., and Cowley, J. M. (1984). Atomic and other structures of cleaved GaAs (110) surfaces. *Surf. Sci.* **137,** 551.

Ichikawa, M., Doi, T., Ichihashi, M., and Hayakawa, K. (1984). Observation of surface microstructures by microprobe reflection high-energy electron diffraction. *Jpn. J. Appl. Phys.* **23,** 913.

Ichimiya, A. (1983). Many-beam calculation of reflection high-energy diffraction (RHEED). Intensities by the multislice method. *Jpn. J. Appl. Phys.* **22,** 176.

Ichinokawa, T., Kinoshita, S., Kemmochi, M., Ikeda, N., and Ishikawa, Y. (1984). Surface analysis by low-energy scanning electron microscope in ultrahigh vacuum. *Proc. European Reg. Cong. Electron Microscopy (Budapest),* 571.

Iijima, S. (1977). High-resolution electron microscopy of phase object: Observation of small holes and steps on graphite crystals. *Optik* **47,** 437.

────── (1981). Observation of atomic steps of (111) surface of a silicon crystal using bright-field electron microscopy. *Ultramicroscopy* **6,** 41.

──────, and Ichihashi, T. (1985). Motion of surface atoms on small gold particles revealed by HREM with real-time VTR system. *Jpn. J. Appl. Phys.* **24,** L125.

Ino, S., Ichikawa, T., and Okada, S. (1980). Chemical analysis of surface by fluorescent X-ray spectroscopy using RHEED–SSD method. *Jpn. J. Appl. Phys.* **19,** 1451.

Kajiyama, K., Takayanagi, K., Tamishiro, Y., and Yagi, K. (1986). Transmission electron diffraction and microscopy of the 5×5 and 7×7 reconstructed structures of Ge deposited on Si (111). *Proc. 11th Int. Cong. on Electron Microscopy (Kyoto),* 1341.

Kambe, K., and Lehmpfuhl, G. (1975). Weak-beam technique for electron microscopic observation of atomic steps on thin single-crystal surface. *Optik* **42,** 187.

Kawamura, T., and Maksym, P. A. (1985). RHEED from stepped surfaces and its relation to RHEED intensity oscillations observed during MBE. *Surf. Sci.* **161,** 12.

Klaua, M., and Bethge, H. (1983). Images of atomic steps on ultrathin Au films by TEM. *Ultramicroscopy* **11,** 125.

Kodaira, Y., Takayanagi, K., Kobayashi, K., and Yagi, K. (1983). An evaporator for *in situ* studies of surfaces and thin-film growth in an UHV 1-MV high-resolution electron microscope. *Proc. 7th Int. Cong. High Voltage Electron Microscopy (Berkeley)* 103.

Krivanek, O. L., Tanishiro, Y., Takayanagi, K., and Yagi, K. (1983). Electron energy loss spectroscopy in glancing reflection from bulk crystals. *Ultramicroscopy* **11,** 215.

Laird, C. (1974). Electron microscopy. In *Characterization of solid surfaces,* ed. P. F. Cane and G. B. Larrabee, 75. Plenum, New York.

Lehmpfuhl, G., and Uchida, Y. (1979). Dark-field and bright-field techniques for electron microscopic observation of atomic steps on MgO single-crystal surfaces. *Ultramicroscopy* **4,** 275.

──────, and Takayanagi, K. (1981). Electron microscopic contrast of atomic steps on fcc metal crystal surfaces. *Ultramicroscopy* **6,** 195.

Lynch, D. (1971). Out-of-zone effects on Au (111) convergent beam patterns. *Acta Crystallogr. Sect. A* **27,** 399.

McRae, E. G. (1983). Surface stacking sequence and (7×7) reconstruction at Si (111) surface. *Phys. Rev. B.* **28,** 2305.

Maksym, P. A., and Beeby, J. L. (1981). A theory of RHEED. *Surf. Sci.* **110,** 423.

──────, and Beeby, J. L. (1984). Calculation of MEED intensities in the 5–10-keV electron energy range. *Surf. Sci.* **140,** 77.

Marks, L. D. (1983). Direct imaging of carbon-covered and clean gold (110) surface. *Phys. Rev. Lett.* **51,** 1000.

──────, and Smith, D. J. (1983). Direct surface imaging in small metal particles. *Nature* **303,** 316.

Masud, N., and Pendry, J. B. (1976). Theory of RHEED. *J. Phys. C* **9,** 1833.

Menter, J. W. (1952–1953). Direct examination of solid surfaces using a commercial electron microscope in reflection. *J. Inst. Met.* **81,** 163.

Moon, A. R. (1972). Calculation of reflected intensities for medium- and high-energy electron diffraction. *Z. Naturforsch.* **27,** 390.

Morita, E., Takayanagi, K., Kobayashi, K., Yagi, K., and Honjo, G. (1980). An ion-sputtering gun to clean crystal surfaces in an ultrahigh vacuum electron microscope. *Jpn. J. Appl. Phys.* **19,** 1981.

Muller, E. W., and Tsong, T. T. (1969). *Field ion microscopy, principles and applications.* Elsevier, New York.
Nakayama, T., Takayanagi, K., Tanishiro, Y., and Yagi, K. (1986). Transmission electron diffraction analysis of the Si (001) 2×1 reconstructed surface. *Proc. 11th Int. Cong. on Electron Microscopy (Kyoto),* 1343.
Nielsen, P. E. H., and Cowley, J. M. (1976). Surface imaging using diffracted electrons. *Surf. Sci.* **54,** 340.
Osakabe, H., Tanishiro, Y., Yagi, K., and Honjo, G. (1980). Reflection electron microscopy of clean and gold-deposited (111) Si surfaces. *Surf. Sci.* **97,** 393.
―――, Tanishiro, Y., Yagi, K., and Honjo, G. (1981a). Image contrast of dislocations and atomic steps on (111) silicon surface in reflection electron microscopy. *Surf. Sci.* **102,** 424.
―――, Tanishiro, Y., Yagi, K., and Honjo, G. (1981b). Direct observation of the phase transition between the 7×7 and 1×1 structures of clean (111) silicon surfaces. *Surf. Sci.* **109,** 353.
Peng, L. M., and Cowley, J. M. (1986). Dynamical diffraction calculation for RHEED and REM. *Acta. Crystallogr. Sect. A* **42,** 545.
Poppa, H. (1975). Studies of thin-film nucleation and growth by transmission electron microscopy. In *Epitaxial growth A,* ed. J. W. Matthews, 215. Academic Press., New York.
Prutton, M. (1982). *Surface physics.* 2nd ed. Clarendon Press, Oxford.
Ruska, E. (1933). Die electronenmikroskopische abbildung elektronenbestrahlter Oberflachen. *Z. Phys.* **83,** 492.
Russel, G. J., and Woods, J. (1976). A specimen holder for reflection electron microscopy using a JEM 120 or 7A electron microscope. *J. Phys. E* **9,** 98.
Shimizu, N., Tanishiro, Y., Kobayashi, K., Takayanagi, K., and Yagi, K. (1985). Reflection electron microscope study of initial stage of oxidation of Si (111) 7×7 surfaces. *Ultramicroscopy* **18,** 1981.
Shuman, H. (1977). Bragg diffraction imaging of defects at crystal surfaces. *Ultramicroscopy* **2,** 361.
Smith, D. J. (1986). High-resolution electron microscopy in surface science. In *Chemistry and physics of solid surfaces,* Vol. VI, ed. R. Vansen and R. Howe, 413. Springer-Verlag, Berlin.
Spence, J. C. H. (1975). Single-atom contrast In *Developments in electron microscopy and analysis,* ed. J. A. Venables, 257. Academic Press, London.
――― (1983). High-energy transmission electron diffraction and imaging studies of the Si (111) 7×7 surface structure. *Ultramicroscopy* **11,** 117.
Takayanagi, K. (1981). Ultrahigh vacuum transmission electron microscopy on the monolayer condensation process of lead on silver (111) surface. *Surf. Sci.* **104,** 527.
――― (1982). High-resolution surface study by *in situ* UHV transmission electron microscopy. *Ultramicroscopy* **8,** 145.
――― (1984). Surface structure imaging by electron microscopy. *J. Microsc.* **136,** 287.
――― (1983). Transmission electron microscopy of monatomic surface layers. *Jpn. J. Appl. Phys.* **22,** L4.
―――, and Yagi, K. (1983). Monatom high-level electron microscopy of metal surfaces. *Trans. Jpn. Inst. Met.* **24,** 337.
―――, Tanishiro, Y., Takahashi, M., and Takahashi, S. (1985). Structure analysis of Si (111) 7×7 by UHV transmission electron diffraction and microscopy. *J. Vac. Sci. Technol.,* **A3,** 1502.
―――, Yagi, K., Kobayashi, K., and Honjo, G. (1978). Techniques for routine UHV *in situ* electron microscopy of growth processes of epitaxial thin films. *J. Phys. E* **11,** 441.
―――, Kobayashi, K., Kodaira, Y., Yokayama, T., and Yagi, K. (1983). Surface studies by high-resolution imaging. *Proc. 7th Int. Cong. High Voltage Electron Microscopy (Berkeley),* 47.
―――, Tanishiro, Y., Takahashi, M., Motoyoshi, H., and Yagi, K. (1982). UHV-transmission

electron diffraction study on (111) Si 7 × 7 surface. *Proc. 10th Int. Cong. Electron Microscopy (Hamburg)*, II, 285.

———, Tanishiro, Y., Kobayashi, K., Yamamoto, N., Yagi, K., Ohi, K., Kondo, Y., Hirano, H., Ishibashi, Y., Kobayashi, H., and Harada, Y. (1986). A new ultra-high vacuum and high resolution electron microscope for in-situ surface study. *Proc. 11th Int. Cong. on Electron Microscopy (Kyoto)*, 1337.

Tanishiro, Y., Takayanagi, K., and Yagi, K. (1983). On the phase transition between the 7 × 7 and 1 × 1 structures of silicon (111) surface studied by reflection electron microscopy. *Ultramicroscopy* **11**, 95.

———, Takayanagi, K., and Yagi, K. (1986). Observations of lattice fringes of the Si (111) 7 × 7 structure by reflection electron microscopy. *J. Microsc.* **142**, 211.

———, Kanamori, H., Takayanagi, K., Yagi, K., and Honjo, G. (1981a). UHV transmission electron microscopy in the reconstructed surface of (111) gold. *Surf. Sci.* **111**, 395.

———, Takayanagi, K., Kobayashi, K., and Yagi, K. (1981b). *In situ* reflection electron microscope study of metal deposition on clean Si (111) surface. *Acta Crystallogr., Sect. A* **37**, C-300.

———, Takayanagi, K., Takahashi, S., Takahashi, M., and Yagi, K. (1986). Validity of kinematical approximation in transmission electron diffraction for structure analysis of Si (111)–7 × 7 reconstructed surface, *Proc. 11th Int. Cong. on Electron Microscopy* **2**, 1339.

Telieps, W. (1984). An analytical reflection and emission UHV surface electron microscope. *Proc. European Reg. Cong. Electron Microscopy (Budapest)*, 593.

———, and Bauer, E. (1985). An analytical reflection and emission UHV surface electron microscope. *Ultramicroscopy* **17**, 57.

Turner, G., and Bauer, E. (1966). An ultrahigh vacuum electron microscope and its application to work function studies. *Proc. 6th Int. Cong. on Electron Microscopy (Kyoto)*, ed. R. Uyeda, 163. Maruzen, Tokyo.

Van Hove, M. A., Kostner, R. J., Stair, P. C., Biberian, J. P., Kesmodel, L. L., Bartos, I., and Somorjai, G. A. (1981). The surface reconstruction of the (100) faces of irridium, platinum and gold. I. Experimental observations and possible structure model. *Surf. Sci.* **103**, 189.

Venables, J. A. (1981). Electron microscopy of surfaces. *Ultramicroscopy* **17**, 81.

——— (1982). Analytical electron microscopy in surface science. In *Chemistry and physics of solid surfaces*, Vol. 4, ed. R. Vanselow and R. Howe, 123. Springer-Verlag, Berlin.

Wilson, R. J., and Petroff, P. M. (1983). A cryopump and sample holder for clean reconstructed surface observations in a transmission electron microscope. *Rev. Sci. Instrum.* **54**, 1534.

Yagi, K. (1980). UHV transmission and reflection microscopy for surface studies. *38th Ann. Proc. Electron Microscopy Soc. Amer*, ed. G. W. Bailey, 290. Claitors', Baton Rouge.

——— (1982). Surface studies by ultrahigh vacuum transmission and reflection electron microscopy. In *Scanning electron microscopy*, Vol. 4, ed. O. Johari, 1421. SEM. Inc., Chicago.

———, Takayanagi, K., and Honjo, G. (1982). *In situ* UHV electron microscopy of surfaces. In *Crystals, growth, properties and applications*, Vol. 7, ed. H. C. Freyhardt, 47. Springer-Verlag, Berlin.

———, Kobayashi, K., Tanishiro, Y., and Takayanagi, K. (1985). *In situ* electron microscopy of initial stage of metal growth on metals. *Thin Solid Films* **126**, 95.

———, Osakabe, N., Tanishiro, Y., and Honjo, G. (1980). Reflection electron microscope study of Au and Ag deposited (111) and (001) Si surfaces. *Proc. 4th Int. Conf. Solid Surfaces (Cannes)*, I, 1007.

———, Takayanagi, K., Kobayashi, K., and Nagakura, S. (1983). Structure images taken by a high-resolution 1-MV electron microscope. *Proc. 7th Int. Cong. High Voltage Electron Microscopy (Berkeley)*, 11.

———, Takayanagi, K., Kobayashi, K., Osakabe, N., Tanishiro, Y., and Honjo, G. (1979). Surface study by an UHV electron microscope. *Surf. Sci.* **86,** 174.

Yamamoto, N., and Muto, S. (1984). Direct observation of $Al_xGa_{1-x}As/GaAs$ superlattices by REM. *Jpn. J. Appl. Phys.* **23,** L804.

———, and Spence, J. C. H. (1983). Surface imaging of III–V semiconductors by reflection electron microscopy and internal potential measurement. *Thin Solid Films* **104,** 43.

Young, R., Ward, J., and Scire, F. (1972). The topografiner: An instrument for measuring surface microtopography. *Rev. Sci. Instrum.* **43,** 999.

14

HIGHLY DISORDERED MATERIALS

A. HOWIE

14.1 Significance and outline nature of the problem

The structure and scattering properties of highly disordered materials is a subject of at least transitory interest to any electron microscopist who employs them either as high-resolution test objects or as specimen supports. Amorphous thin films frequently do act as the kind of broad band-scattering objects required to produce good diffractograms for correction of astigmatism as well as determination of defocus, spherical aberration, or other transfer properties within the weak-phase-object approximation as described in Chapter 2. As specimen supports, however, they generally fail to provide, at high resolution, the homogeneous, featureless, or low-contrast background that is required and might naively be expected. Bright-field images are typically characterized by a very irregular "fringey" appearance and dark-field images by a speckle pattern of bright spots on a low-intensity background (e.g., see Figures 14.5 and 14.12).

In the phase-object approximation, which is valid for small scattering angles (Chapter 2), these noisy-image effects can evidently be ascribed to varying degrees of alignment, possibly purely by chance, of groups of atoms along or close to the beam direction. Depending on the support film thickness, it may be difficult or impossible to image satisfactorily or even to detect the small particles or other supported objects that may be of primary interest. An acute example of these difficulties occurs in the case of the typical heterogeneous catalyst containing small clusters of heavy atoms embedded in a light-atom support, which, being frequently thicker with a much less random structure than the ideal amorphous support film, thus generates a very irregular background intensity.

Catalyst supports, of course, fall into a category where features of the disordered structure, such as the atomic-scale porosity of the support, are themselves of consuming scientific interest. In this respect, catalysts therefore belong to a rapidly increasing list, including, for example, both oxide and metallic glasses, amorphous semiconductors, a great variety of disordered carbons, amorphous intergranular phases in ceramics, oxide coatings, and overgrowths on semiconductors. In all these cases amorphous or highly disordered structures play a vital role in the technical performance of materials and devices, where the maximum possible information is required about local structure, composition, and other properties. Elliott

(1984) has recently published a useful introduction to the structure and physical properties of amorphous materials.

As a source of high-resolution information about such highly disordered structures, there is as yet no serious rival to the classical diffraction methods and their more recent refinements such as extended X-ray absorption fine structure (EXAFS). These lead, as indicated in section 14.2, to a convenient statistical description of the structure in terms of a radial-distribution function, which is often adequate to determine first-neighbor numbers and distances. Probably the most significant achievement of electron microscopy so far (see section 14.3) has been to supplement this high-resolution statistical data with information about the large, real-space structural fluctuations that are frequently visible on the medium-range (>30 Å) scale in micrographs of disordered materials.

Even at such modest resolution levels, two peculiar difficulties in the study of highly disordered specimens become apparent. First, the observed structure is often a most sensitive function of specimen preparation and history. Comparisons between materials produced in different laboratories, and perhaps investigated in different ways, can be very difficult, and it becomes essential for adequate characterization to measure as wide a variety of properties as possible on the same set of specimens.

The second difficulty concerns the need for a detailed model of the proposed structure for which image predictions can be made and tested in detail against observations and that can also account for other properties of interest. Even when the model does not extend to the atomic level, it can be a formidable test to ascertain whether the proposed three-dimensional structure is fully consistent with the two-dimensional projected images observed.

As just indicated and as discussed in more detail in sections 14.4 and 14.5, the problem of extracting reliable structural information from high-resolution electron micrographs of disordered specimens is greatly complicated by noise originating from purely random atomic alignments as well as by instrumental artifacts. Some progress has been made in the special case of substitutional disorder in alloys; but in the case of amorphous materials, the task of deducing and adequately describing three-dimensional structures on the basis of projected-image information becomes truly formidable. To overcome this, new imaging techniques and theory will most certainly be required. The study of highly disordered materials is thus currently one of the most challenging and open-ended applications of high-resolution electron microscopy.

14.2 Classical diffraction and structural description

14.2.1 Radial-distribution function

As indicated in Chapter 3, the kinematical, scattered-intensity data $I(\mathbf{q})$ from a highly disordered specimen consists of diffuse halos superimposed on falling background. This can be processed to yield statistical information about interatomic distances via a pair-correlation function or radial-distribution function (RDF). If we assume that the structure is statistically homogeneous, we can write, for a sample of N identical atoms at positions r_j,

$$I(\mathbf{q}) = |f(q)|^2 \sum_{i,j} \exp(i\mathbf{q} \cdot (\mathbf{r}_i - \mathbf{r}_j))$$

$$= N |f(q)|^2 \left(1 + \frac{F(\mathbf{q})}{q}\right). \tag{14.1}$$

The interference function $F(\mathbf{q})$ defined in this way is constructed to emphasize the structure-sensitive features of the data. If we further assume that the structure is on average isotropic, we obtain

$$F(q) = q \sum_{i \neq j} \exp[i\mathbf{q} \cdot (\mathbf{r}_i - \mathbf{r}_j)] \tag{14.2}$$

$$= q \int \rho(r) \exp(i\mathbf{q} \cdot \mathbf{r}) \, d\tau$$

$$= 4\pi \int_0^\infty \rho(r) \sin(qr) \, r \, dr, \tag{14.3}$$

where $\rho(r)$ is the RDF and gives the probability, per unit small element of volume, that an atom will be found at distance r from another atom taken to be at the origin. Clearly, only variations in $\rho(r)$, not its average value ρ_0, are significant in equation (14.3). We can thus invert the equation to read

$$\rho(r) - \rho_0 = \frac{1}{r} \int_0^\infty F(q) \sin qr \, dq. \tag{14.4}$$

Suitable data (ideally extending out to at least $q = 15 \, \text{Å}^{-1}$) can be obtained from X-ray or neutron scattering as well as by electron scattering. For light atoms such as carbon and silicon, Stenhouse and Grout (1978) pointed out the importance of correcting the X-ray and electron-scattering amplitudes to allow for bonding effects. Of course, the electron-scattering data can, in principle, be collected in the electron microscope; but the specimens must be very thin to avoid multiple-scattering effects, and energy filtering should be employed to remove inelastic scattering.

It is usually advisable in any given case to examine both $F(q)$ and $\rho(r)$, comparing them with predictions from any model structure, since each is sensitive to different features. The turbostratic structure of vitreous carbon, for instance, with layers stacked locally parallel at a separation of $d \simeq 3.54 \, \text{Å}$ gives rise (Wignall and Pings, 1974) to a prominent peak at $q = 2\pi d^{-1}$ in the $F(q)$-function. The peaks found in the $\rho(r)$-function correspond, however, only to interatomic distances within a single layer, since the ordering between different layers is confined to the layer separation d and does not impose any lateral registry between adjacent layers. The bonding in this form of carbon thus seems to be predominantly trigonal.

The classical RDF analysis can be extended to deal with cases where two or more different atomic species are present. Partial-distribution functions for the various pairs of atoms present can be derived if scattering data for different relative values of the atomic-scattering amplitudes can be obtained. This goal can be achieved by utilizing, with synchrotron radiation, the anomalous X-ray–scattering effect, by comparing X-ray with neutron scattering, or by carrying out neutron scattering on samples composed of different isotopes. Extended X-ray–absorption fine structure (EXAFS), which exhibits weak oscillations in absorption above the characteristic edge (see Chapter 7), can also be analysed to give useful information about distances to various shells of neighboring atoms. Although the EXAFS data obtained generally omits some of the significant low-q end of the range available from scattering data, the environment of each species present can be conveniently probed by tuning to its characteristic edge.

14.2.2 Interpretation of RDF data and structure modeling

Gaskell (1979, 1983) and Elliott (1984) have reviewed the scattering data and its interpretation for a wide range of amorphous solids. Analysis of peak positions and areas in the $\rho(r)$ (RDF) data gives valuable information about averaged, near-neighbor distances and numbers, particularly when only one atomic species is present. Supplementary information about the symmetry of the first-neighbor arrangement may be available from nuclear magnetic resonance or Mössbauer measurements. An important fact to emerge in many cases is that the first nearest-neighbor distance and numbers are often very similar to the crystalline phase; the distribution of second and further neighbors is, however, broadened and altered. Amorphous carbon, which exists in a variety of forms (Robertson, 1986), is a remarkable exception in having an average first-neighbor distance intermediate between that of graphite (1.41 Å) and tetrahedrally bonded diamond (1.55 Å). This has led to various proposals for the structure as a mixture of three-fold and four-fold coordinated regions. In most other cases, however, the main uncertainties lie in the arrangements of second and further neighbors. The second neighbor in the tetrahedrally bonded, amorphous semiconductors can, for instance, be analyzed in terms of the variations in the bond angle β, but peaks in $\rho(r)$ at higher distances are controlled in a more complex way by factors such as the distribution of the dihedral angle ψ shown in Figure 14.1.

In an attempt to resolve some of these problems, a variety of models have been considered for different amorphous materials. At one extreme lie polycrystalline or microcrystalline models, and at the other lie models such as the random-network structures for covalently bonded, amorphous semiconductors or the dense, random-packed, hard-sphere (Bernal, 1960)

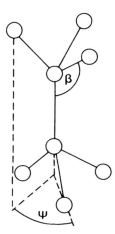

FIGURE 14.1 Bond angle β and dihedral angle ψ in a tetrahedrally bonded network.

models for metallic glasses. An intermediate position is occupied by models based on structural units, such as SiO_2 tetrahedra or trigonal prisms, with irregular packing constrained by corner or edge sharing between neighboring units. The units are often derived from crystalline structures of similar composition.

For detailed calculations, any of these models should ideally consist of several hundred or even several thousand atoms arranged with specified coordinates in a quasi-spherical cluster. They can be created mechanically by ball-and-spoke construction but can also be generated in the computer, starting from some small noncrystalline seed. A given structure can then be modified in the computer by allowing the atoms to relax to minimum-energy positions, using some assumed form for the bonding potential. This also allows recrystallization energies to be estimated. Although models have been built for individual grain boundaries (e.g., Krivanek et al., 1977), few if any models have actually been built with a microcrystallite structure.

Apart from allowing comparisons to be made between computed and observed scattering data, structural models provide additional information about other parameters such as the dihedral-angle distribution mentioned previously and ring statistics—that is, relative numbers of five-membered, six-membered, seven-membered, and so on, rings in covalently bonded networks. It seems to be impossible, for instance, to fit the observed scattering data for a–Ge and a–Si with models based only on the six-membered rings found in the diamond or wurtzite crystal structures; considerable numbers of five- and seven-membered rings must be used. In amorphous III–V semiconductors such as Ga–As, however, the occurrence of odd-membered rings implies Ga–Ga or As–As bonds. Photoemission experiments showing that the atomic-core, energy-level positions are very close to those in the crystal suggest that the number of such wrong bonds in the amorphous structures must be small. EXAFS data, however, does indicate that the environment of arsenic in amorphous Ga–As is different from the environment of gallium, which is the same as in the crystal. The data are consistent with about 20 per cent of As–As bonds.

14.3 Medium-scale structure

Conventional transmission electron micrographs reveal that many amorphous specimens do not possess the homogeneous structure assumed in the simple RDF analysis but can contain noncrystalline domains, troughs, voids, networks of pores, or variations in composition on the scale of 30 Å or more. An example is shown in Figure 14.2. In some cases, the effects of these local fluctuations can be observed in the small-angle region of the scattering data $I(q)$, and these can be analyzed to yield statistical information about the dimensions of voids, pores, or precipitates that may be present. In other cases, the structure can be markedly anisotropic over local areas, which can easily be resolved in the electron microscope. In the absence, however, of any related variation in density or composition, there

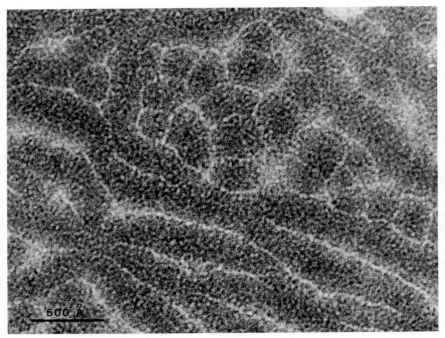

FIGURE 14.2 Bright-field image of the pore structure in silicon deposited on rock salt by the silane process. (Courtesy of Mistry, 1979.)

might well be no small-angle-scattering effect, and diffraction data would fail to reveal anything unless it could be obtained from selected areas sufficiently small to yield some anisotropic texture or other azimuthal structure in the diffuse rings. In these situations, which include cases of amorphous films containing a very small fraction of crystalline materials, microscopic images can be much more informative than the conventional, wide-beam diffraction patterns. An extreme form of anisotropy is the columnar structure that frequently develops parallel to the direction of deposition in evaporated amorphous films. Since this structures extends over the whole sample, it can be detected as an anisotropy in diffraction patterns when highly oblique deposition is used. The structure can also be resolved in micrographs, particularly near the edges of holes in the specimen where overlap effects are reduced. Recent examples and discussions of the significance of these various medium-scale structures in amorphous materials, together with further references, are given in Chen et al. (1981) and Elliott (1984).

Partially graphitized carbon affords a particularly rich variety of striking structures on the medium scale (Robertson, 1986). A typical example of a partially graphitized catalyst support is shown in Figure 14.3. Commercial carbon fibres for strong-materials applications can have graphite planes roughly parallel to the outer surfaces. However, the carbon fibres produced as unwanted deposits in disproportionation reactions in chemical reactors appear to have a hollow core around which the graphite planes run at a more or less constant angle in a spiral structure (Audier et al., 1981). The stresses so generated can cause the whole fibre to twist into a second spiral on a still larger scale. Further examples of carbon structures include the interleaved ribbon structure of polymeric carbons and bird's-nest structure of carbon blacks. Detailed information about the relation between such medium-scale structure and structure at the near-atomic level can often be obtained by dark-field microscopy, whereby local "crystal" orientations can be determined. High-resolution, bright-field microscopy can also be a powerful method but is restricted to cases where little overlap occurs between regions of different orientation. In all cases, it is important to ensure that the structure proposed to account for the diffraction and imaging data is fully consistent in three dimensions. It has been pointed out, for instance (Howie, 1983), that high-resolution, bright-field images of the polymeric-carbon-ribbon structure should show many more regions with terminating lattice fringes than are observed in images of the bird's-nest structure. Some examples of terminating fringes can, in fact, be seen in Figure 14.3 (also see Figure 9.8c,d).

In some cases, the complicated image detail of high-resolution micrographs can be irrelevant or even a positive nuisance in the visualization of medium-scale structure. It may then be preferable to operate the microscope in a lower-resolution mode, particularly since much lower and therefore less damaging flux densities can be used. An alternative that can

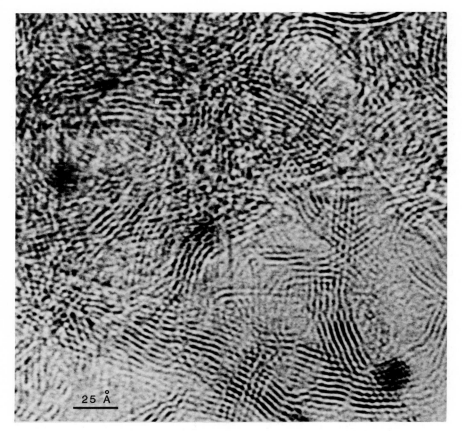

FIGURE 14.3 Platinum-catalyst particles on a partially graphitized support. (Courtesy of D. J. Smith.)

be useful when medium-scale and high-resolution features are to be correlated is to take a through-focal series. Medium-scale structure can then be observed (Mistry, 1979) at high defocus values (positive or negative) by cutting out the high-resolution detail by optical-bench methods or by computer-image processing. Gibson and Dong (1980) were able to observe systematic differences between "dry" and "wet" SiO_2 films by concentrating on low-resolution-image features not affected by the spherical-aberration term in the phase-shift expression and therefore displaying a simple contrast reversal between overfocus and underfocus imaging.

Several of the techniques of scanning transmission electron microscopy (STEM), such as diffraction and small-angle scattering from selected areas combined with localized X-ray emission or electron-energy-loss spectroscopy (Patterson et al., 1983), directly address problems of medium-scale

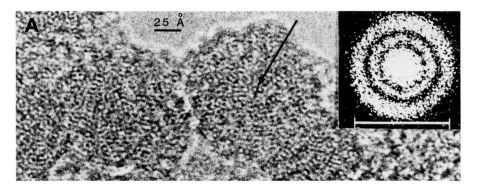

FIGURE 14.4 Bright-field image of solution-precipitated (Ludox) silica. (Courtesy of Gaskell and Mistry, 1979.)

variations in structure or composition. Shifts or shape changes in core-energy losses relative to crystal values can throw light on bonding in the disordered state (Taftø and Kampas, 1985). It may also be possible to image pores and voids by using the valence-loss electron signal. Surface troughs could, in principle, be distinguished from internal pores by using the secondary electron signal. Several of these STEM techniques also have relevance for high-resolution imaging (see section 14.6).

It is important to revise or refine the interpretation of the classical scattering data in the light of a sufficiently precise model of the medium-scale structure. In the case of graphitized carbons, for instance, the data can be reinterpreted in terms of ribbon lengths and widths as well as in terms of layer d-spacing and its relation to the degree of graphitization. Reliable procedures of this kind are particularly useful because they yield additional statistical data of a more quantitative nature than the micrograph can easily provide. They are likely to be even more valuable in very highly disordered specimens like silica gels (see Figure 14.4), where Gaskell and Mistry (1979) have demonstrated the intriguing but still qualitative information available at various levels of resolution. For such cases, the construction of atomic models on a scale large enough to incorporate the inhomogeneities observed would be a daunting task but probably a highly rewarding one.

14.4 High-resolution, bright-field imaging

14.4.1 Image theory and properties

The simple kinematical theory of image formation (see Chapter 3) provides an instructive and useful basis for the discussion and interpretation of the images of amorphous materials (for a review, see Howie, 1978). We consider a thin specimen consisting of atoms j at positions $\mathbf{r}_j$ relative to an origin in the midplane of the specimen. The bright-field image intensity at a point $\mathbf{r} = (x, y, \Delta f)$ in some defocus plane at distance Δf from the same midplane is

$$I_B(\mathbf{r}) = 1 + \left(\frac{\lambda A_k}{2\pi^2}\right) \sum_j \sum_{k_i} f_j(q) \cos\left[\mathbf{q} \cdot (\mathbf{r} - \mathbf{r}_j)\right] T(u), \tag{14.5}$$

where $\mathbf{q} = \mathbf{k}_i - \mathbf{k}_0$, the summation includes points $\mathbf{k}_i$ on a two-dimensional grid accepted by the objective aperture, and A_k is the unit-cell area of this grid. Function $T(u)$ is the transfer function including the usual $\sin\chi(u)$ factor as well as an envelop function that allows for the energy spread (temporal incoherence) and angular spread (spatial incoherence) of the illumination. Frequently, the minimum change in defocus Δf resulting in any detectable change in the image exceeds the specimen thickness, so we can make the weak-phase-object approximation and ignore the different heights z_j of the atoms in the specimen.

Equation (14.5) allows us to compute the image expected for any prescribed structure under specific imaging conditions. The computations required can be rather laborious for atomic clusters greater than about 30 Å in diameter, so strict comparison with results obtained for typical thin-film specimens of thickness $t \simeq 100$ Å is difficult. Nevertheless, the availability of such results can play a valuable role in identifying imaging artifacts or errors of interpretation. The computed images also demonstrate directly the importance of chance alignments of atoms and verify the predictions of simple theories of the image-intensity statistics (for references, see Howie, 1978). Of course, the kinematical theory is a poor approximation in all but the thinnest samples. More accurate images of amorphous specimens can, in principle, be computed on the basis of the slice method (Chapters 3 and 4) applied to a periodic structure whose unit cell can be an amorphous cluster of several hundred atoms. The extra computational labor involved is very considerable, and present results (Bursill et al., 1982) indicate that the errors in the kinematical theory in the case of light-atom amorphous structures are not serious at thicknesses up to perhaps 30 Å.

The presence of the $T(u)$-factor in equation (14.5) shows that the image is a sensitive function of defocus and other instrumental conditions. Before indulging in any image interpretation, therefore, one should be sure that the image obtained is as reliable and free from instrumental artifacts as possible. This can be checked by examining the two-dimensional Fourier transform of I_B, whose intensity can be obtained on an optical diffractogram or by computer processing of the image. If we rewrite the image intensity in the form

$$I_B(\mathbf{r}) = 1 + \sum_{k_i} C(\mathbf{q}) \cos(\mathbf{q} \cdot \mathbf{r} + \phi_\mathbf{q}), \tag{14.6}$$

where $\phi_\mathbf{q} = -\phi_{-\mathbf{q}}$, and if we compare equations (14.1), (14.5), and (14.6), we see that the diffractogram intensity or image power spectrum is

$$|C(\mathbf{q})|^2 = \left(\frac{\lambda A_k}{2\pi^2}\right)^2 I(q) |T(u)|^2. \tag{14.7}$$

Ideally, the diffractogram should be axially symmetric with dark rings at radii corresponding to the zeros of $|T(u)|$ where no information is transferred to the image. The missing information in these regions can be obtained from a series of images taken with different defocus values. Figure 14.5 shows a through-focal series in amorphous arsenic taken by Stobbs and Smith (1980). In a number of cases (for references, see Howie, 1983), the various members of such through-focal series have been digitized and processed by computer to yield a final image free of the zeros and sign reversal of the $T(u)$-function, thus reflecting in a more faithful way the structure of the specimen.

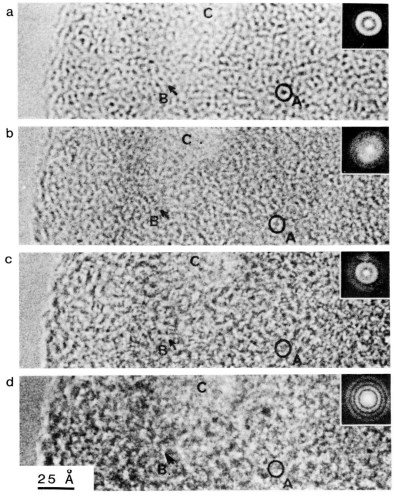

FIGURE 14.5 Bright-field, through-focal series of amorphous arsenic. (Courtesy of Stobbs and Smith, 1980.)

14.4.2 Image testing and assessment

Comparing equations (14.6) and (14.7), we see that the essential new information that the high-resolution image conveys about the specimen is the phases ϕ_q of the different Fourier components. If these phases were completely random, it would be impossible to extract more information than is already available from classical scattering data, which gives $I(q)$. In such cases, high-resolution imaging is a waste of time.

In practice, the assessment of bright-field images for nonrandom features tends to be based on a rather subjective judgement about the degree of "structure" present as measured, for example, by the typical area of the patches of coherent fringes. Clearly, however, this is a function of the image power spectrum, in that images with a narrow power spectrum will tend to look more structured than those with a broad power spectrum. The test for nonrandomness can therefore be made a little less subjective (Krivanek et al., 1976) by comparing the observed image with an image that has the same power spectrum but completely random phases $\phi'_q = -\phi'_{-q}$. Such a random image can be simulated in a computer or can be generated from the original image by phase-randomization procedures (either in the computer or in the optical bench). To preserve the power spectrum as required, the new phases ϕ'_q, which replace ϕ_q in equation (14.6), must still satisfy the relation $\phi'_q = -\phi'_{-q}$, and this is most easily achieved in optical processing by employing an optical bench that transmits only one-half of the diffraction pattern. In Figure 14.6, (from Smith et al., 1981a), the image from an amorphous-carbon film (Figure 14.6a) has been phase-randomized (Figure 14.6b) to yield a result that is evidently less structured, having fewer regions with long, unbroken fringes. It should, however, be possible to devise more quantitative tests for nonrandomness in the phase ϕ_q and even to analyze

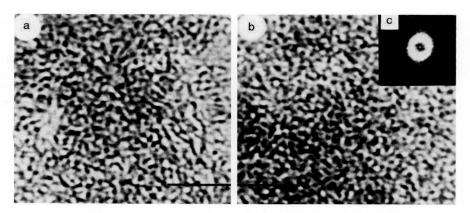

FIGURE 14.6 Images of an amorphous-carbon film. (a) Original. (b) Phase-randomized. More structure is visible in (a). The power spectrum is shown in (c). (Courtesy of Smith et al., 1981a).

the nature of nonrandomness in terms of phase correlations by more detailed analysis of digitized images. Recently, Fan and Cowley (1985) have obtained interesting indications of local order by superimposing the Patterson functions constructed from different small parts of the image after relative rotation of these to maximize angular cross-correlations. Such procedures may be particularly useful now that high-voltage, high-resolution electron microscopes are available with transfer functions extending out to or beyond the first peak in the $F(q)$-function of typical amorphous materials.

14.4.3 Projection effects and object reconstruction

Equation (14.5) shows that the intensity in a bright field image takes the form of an amplitude superposition of projected atomic images. This is because the scattered wave points $\mathbf{k}_i$ all lie more or less on a plane in reciprocal space normal to the optic axis z—that is, the phase-object approximation is valid. In most cases, the atomic correlations in the specimen we wish to analyze, which are responsible for the nonrandom phases $\phi_\mathbf{q}$, extend only for atomic separations up to <10 Å. For specimen thicknesses much greater than this value, therefore, the images will tend to be dominated by the purely random overlap effect from atoms separated by large distances in the beam direction. This is in line with the observation that the most structured regions of an image from an amorphous material can often be found in the thinnest areas near the edges of the specimen.

The measurement of specimen thickness in local regions is clearly very important if images of amorphous specimens are to be reliably assessed. In the conventional microscope, one can monitor the bright-field image obtained when a small objective aperture is employed, since its intensity decreases with increasing thickness because of increased scattering outside the aperture, and the relevant cross sections can be estimated with reasonable accuracy (Mistry, 1979). In theory (Saxton et al., 1977), the mean-square variation in image intensity is proportional in thin specimens to the thickness, but this seems to be less reliable as a way of measuring local thickness. Further methods for thickness determination are, of course, available in STEM, such as inelastic-scattering or backscattering intensity measurements, but these have not as yet been systematically applied to amorphous materials.

In principle, the projection problem could be overcome by taking a number of stereo images of the same area, but this is, unfortunately, rather difficult when the image is such a highly sensitive function of defocus, making it far from easy to recognize an area of the image with adequate precision. So far, the main approach (Smith et al., 1981a) has been to concentrate on specimen areas that are as thin as possible to minimize statistical-overlap effects. The presence of significant structure can be checked by phase-randomization tests. Computations based on equation

(14.5) have indicated, for instance, that even a region as highly structured as a 15-Å crystal may not be visible when immersed in amorphous surroundings in a film more than about 60 Å thick. Unfortunately, there is evidence (Smith et al., 1981a) that some additional complication (possible anomalous specimen vibration or thermal diffuse scattering) can destroy the high-resolution, image-contrast detail in extremely thin ($t \simeq 10$ Å), unsupported films.

14.4.4 Results of high-resolution, bright-field imaging

In view of these difficulties, it is, perhaps, not surprising that very few definitive results have as yet emerged from high-resolution, bright-field imaging of amorphous specimens. Evaporated-carbon films seem to be systematically more structured than evaporated or sputtered a–Si or a–Ge films, an observation probably related to the presence of a large fraction of trigonal sp^2-bonds in the carbon case. So far, however, no consistent difference has been found between evaporated-carbon films and the interesting, so-called hard-carbon films, which are prepared by glow-discharge methods and which have strikingly different optical, electrical, and mechanical properties. Important differences in optical and electrical properties are also found in silicon films produced by glow-discharge methods when compared with films produced by sputtering or evaporation. These are attributed to the presence of large amounts of hydrogen. A systematic comparison of these specimens by high-resolution microscopy would undoubtedly be an interesting and challenging project.

The advent of high-voltage, high-resolution electron microscopy has enabled the techniques of bright-field imaging developed at 100 keV to be extended to the resolution region below 3 Å, thus making accessible the prominent interatomic distances in many more amorphous materials. Interesting fringe structures have, for instance, been resolved near the edges of small islands of metallic glass (Gaskell et al., 1979) and in amorphous silicon at considerable distances from the clean silicon crystal substrate used (Philips et al., 1987). These advances in resolution have not, however, so far overcome the basic projection problem and difficulties of interpretation described earlier. More positive results have been obtained on several heterogeneous-catalyst specimens (Smith et al., 1981b; White et al., 1983), where the ability to resolve the crystal lattice on very small particles carried on the typical catalyst support can be very effective in identifying their nature and morphology. An example of platinum particles on a highly structured carbon support was shown above in Figure 14.3. Similarly, Figure 14.7 shows an amorphous GeO_2 specimen in which small, partially ordered regions can be identified as tiny Ge crystals (Levi, Phillips, and Smith, private communication).

Before concluding this section, we should take some note, however brief and inadequate, of the great explosion of interest in quasi crystals of

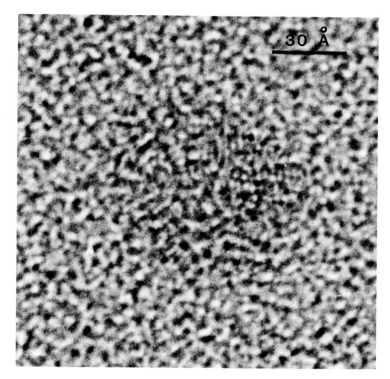

FIGURE 14.7 Bright-field image of an amorphous GeO_2 film showing anomalous region, probably a small Ge crystal. (Courtesy of Levy, Phillips, and Smith, private communication.)

icosahedral symmetry that has developed recently. Considerable excitement was caused when Schechtman et al. (1984) reported the observation of a metallic Al–Mn phase with long-range orientational order but with icosahedral symmetry, which is inconsistent with lattice periodicity. Although the diffraction patterns cannot be indexed to any Bravais lattice, the spots are as sharp as normal crystalline spots. Subsequent investigations (e.g., Levine and Steinhardt, 1985) have shown that such diffraction patterns can result from quasi-periodic structures composed of two different rhombohedra (Mackay, 1982) that fill space in a Penrose-tiling pattern and that have some remarkable scaling, self-similarity properties. Whether or not these structures can form a satisfactory basis to describe the actual atomic arrangement is not clear, since the positions of the atoms within the rhombohedra have yet to be determined. High-resolution, bright-field images may play an important role here and have been obtained in several laboratories. Figure 14.8, taken at 500 keV by Knowles et al. (1985) from a fivefold symmetry axis in $Al_{86}Mn_{14}$, illustrates in detail several of the theoretically predicted features of the quasi-crystal structure. Quantitative interpretation of such

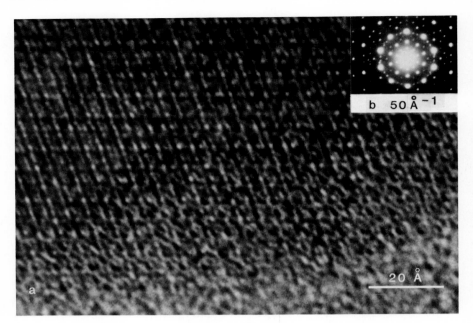

FIGURE 14.8 High-resolution, bright-field image from a fivefold axis of $Al_{86}Mn_{14}$. Inset is the electron diffraction pattern. (Courtesy of Knowles et al., 1985.)

images depends, however, even more strongly than usual on accurate knowledge of the specimen thickness, since dynamical-diffraction effects can generate extra spots, leading to a new diffraction pattern of reduced scale but similar symmetry and therefore not automatically distinguishable from the kinematical one. Although the practical significance of these fascinating structures is at present dubious, the intellectual challenge they pose is undeniable.

14.5 High-resolution, dark-field imaging

14.5.1 Dark-field-image theory and properties

The kinematical theory is a useful basis for the discussion of dark-field as well as of bright-field images from amorphous materials. Assuming tilted illumination with a single plane wave $\mathbf{k}_0$, the coherent dark-field intensity at a point $\mathbf{r} = (\mathbf{\rho}, \Delta f)$, where $\mathbf{\rho} = (x, y)$, is given by the expression

$$I_D(\mathbf{r}) = \left| \sum_j a_j(\mathbf{\rho} - \mathbf{\rho}_j, \Delta f, \mathbf{k}_0) \exp(-i\mathbf{Q} \cdot \mathbf{r}_j) \right|^2, \qquad (14.8)$$

where the various scattering vectors are shown in Figure 14.9. The

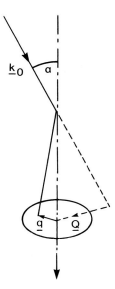

FIGURE 14.9 Schematic diagram for conventional, tilted, dark-field imaging. The scattering vector $\mathbf{k}_i - \mathbf{k}_0 = \mathbf{Q} + \mathbf{q}$ has a mean value $\mathbf{Q}$.

atomic-amplitude functions that superimpose are defined by

$$a_j(\mathbf{\rho} - \mathbf{\rho}_j, \Delta f, \mathbf{k}_0) = \left(\frac{i\lambda A_k}{4\pi^2}\right) \sum_{\mathbf{q}} f_j(Q) \exp\left[i\mathbf{q} \cdot (\mathbf{\rho} - \mathbf{\rho}_j) - i\chi(\mathbf{u})\right], \quad (14.9)$$

where the summation includes all waves $\mathbf{k}_i - \mathbf{k}_0 - \mathbf{Q} + \mathbf{q}$ collected by the objective aperture.

Coherent, dark-field images of amorphous materials can, like bright-field images, be dominated by statistical effects, since the atomic amplitudes that superimpose in equation (14.8) can be seen to arise in equation (14.9) from a sum of plane waves with random or nearly random phases. These random influences produce a speckle effect, with many small, bright spots against a very low-intensity background. In the fully random situation, the size of the spots in the coherent, dark-field image is independent of focus and cannot be taken as indicative of local, ordered regions such as diffracting domains or microcrystals. The intensity at any point has an exponential probability distribution whose most probable value is zero, corresponding to zero image amplitude. Because of the nonlinear nature of the dark-field imaging, positive or negative fluctuations in the amplitude given in equation (14.8) will both give rise to bright spots. In practice, because of the finite range of illumination angles, most dark-field images are not fully coherent, and this modifies the intensity distribution as well as the focal dependence of the bright spots (Gibson and Howie, 1979).

The dependence of the dark-field intensity on the square of the atomic-amplitude contributions means that the simple linear properties of bright-field imaging (in the kinematical approximation) are lost, together with the well-understood transfer theory so useful for detecting and correcting imaging artifacts. These disadvantages can be far outweighed, however, by the ability to employ, in dark-field imaging, selected portions of reciprocal space where the scattering is sensitive to particular features of the specimen structure. Despite its low intensity, the scattering required can thus contribute to the image with good contrast. This principle is quite familiar in the field of crystal-defect imaging and is employed, for example, in the weak-beam method. Applications to disordered alloys are discussed in section 14.5.2.

A further significant advantage of dark-field imaging, which we explore in more detail in section 14.5.3, is the ability to use much larger scattering angles than in bright-field. Recalling that the phase-object (projection) approximation (see Chapter 2) depends for its validity on small scattering angles, it can be appreciated that high-angle, dark-field methods may offer hope for overcoming the projection difficulties that plague the bright-field imaging of amorphous materials. Note that, although the atomic amplitude a_j given by equation (14.9) can be taken to be independent of the depth z_j of each atom in the sample, a phase term involving the z_j-coordinates does occur in equation (14.8) and will be significant if the illumination angle α in Figure 14.9 is large, giving **Q** an appreciable component along the optic axis.

14.5.2 *Dark-field imaging of disordered alloys*

The investigation of short-range order in alloys has a number of features in common with the study of the structure of amorphous materials, although the problem in some respects is a simpler one because the disordered atoms still lie on a lattice. Fourier techniques can be applied to the diffuse intensity in the diffraction pattern, which, in simple cases, has a peak near the position of superlattice spots in the ordered structure. This leads, by arguments similar to those given in section 14.2, to values of short-range-order parameters (Cowley, 1981a) that describe the statistical order around any lattice site. Some success (e.g., see Moss and Walker, 1974) has been obtained in relating these parameters via a simplified thermodynamic argument to an interatomic, pairwise ordering potential.

In the attempt to set up a real-space picture of short-range order, controversy has arisen from time to time between descriptions based on purely statistical ideas or perhaps Fourier concepts like concentration waves and real-space concepts like partially ordered clusters or domains (for discussion, see Stobbs and Chevalier, 1978). Any attack on these problems by electron microscopy has to confront the various difficulties of high-resolution and projection effects already described. Although bright-field,

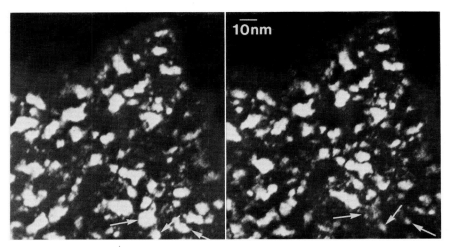

FIGURE 14.10 Stereo images of ordered microdomains in Cu–Pt. Arrows indicate regions where contrast changes occur. (Courtesy of Chevalier and Stobbs, 1979.)

structure-imaging methods can be applied and are particularly successful when there is a considerable degree of order, difficulties arise when the disorder is high or when the atoms involved differ only slightly in atomic-scattering factor. High-resolution, tilted, dark-field methods can then be employed to advantage by using the diffuse-scattered intensity to form the image. In some cases, such as CuPt, it has been possible (Chevalier and Stobbs, 1979) to demonstrate impressively and unequivocably the presence of microdomains by dark-field stereo imaging. An example is shown in Figure 14.10. In other cases, such as Ni_4Mo, however (Stobbs and Chevalier, 1978), the evidence points against microdomains.

14.5.3 High-angle, incoherent, dark-field imaging

The usefulness of relatively high-angle electron scattering in high-resolution imaging was dramatically demonstrated by the work of Crewe and coworkers (1975) when they employed the so-called Z-contrast method to produce images of single, heavy atoms supported on thin-carbon films. This STEM technique (see Chapter 2) depends mainly on the signal collected by an annular detector working typically in the 20 to 100 mrad range of scattering angles. Unwanted contrast effects due to local variations in the thickness of the support film can be greatly reduced by displaying the ratio of the annular-detector signal to the valence-loss ($\simeq 20$ eV) signal, which can be collected simultaneously in a spectrometer working close to axis in the instrument. The Z-contrast method has also proved to be quite useful in imaging small, heavy-atom catalyst particles on charcoal or alumina supports (Treacy et al., 1978; Howie et al., 1982). Figure 14.11 shows an

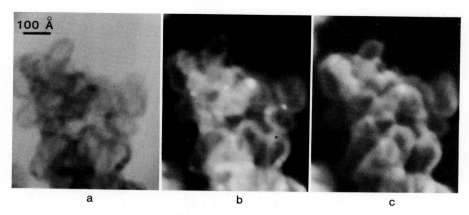

FIGURE 14.11 STEM images of platinum-catalyst particles in a charcoal support. (*a*) Bright-field image. (*b*) Annular-detector, dark-field image. (*c*) Secondary electron image. (Courtesy Bosch and Imeson, private communication.)

annular-detector image of platinum particles on a carbon support compared with bright-field and secondary-electron STEM images of the same area.

The bright images of heavy atoms in the annular-detector image are explained by the single-atom-scattering theory since the Rutherford-scattering, differential cross section, at any scattering angle, is proportional to Z^2. At smaller scattering angles, this effect, however, can be swamped by Bragg reflection or, in highly disordered media, chance alignments of the atoms in the support. The suppression of the coherent effects in the annular-detector signal can be seen to arise because of two factors. First, at the scattering angles used, the Debye-Waller factor greatly reduces the Bragg intensity, and the signal is mainly the diffuse intensity typical of independently scattering atoms (e.g., see Howie et al., 1982). Second, the annular detector collects scattered electrons over a range of directions lying between two cones, an imaging mode that can be seen, thanks to the reciprocity principle (see Chapter 2), to be equivalent to dark-field imaging in the conventional microscope using hollow-cone illumination. Of course, the annular-detector semiangle in the STEM is typically a good deal larger that the illumination semiangle α in Figure 14.9 that could be achieved in practical hollow-cone illumination.

The effects on the conventional, dark-field image of using a range of incident-wave directions $\mathbf{k}_0$ in Figure 14.9 can be computed for any specific structure by integrating the intensity $I_D(\mathbf{r}, \mathbf{k}_0)$ in equation (14.8) over $\mathbf{k}_0$ (Gibson and Howie, 1979). This assumes that the different, incident, plane-wave components are mutually incoherent (a good approximation, in most cases). Theory and experiment both indicate (Gibson and Howie, 1979) that, with increasing angular spread of the illumination, the dark-field-imaging mode becomes more and more incoherent. The speckle size

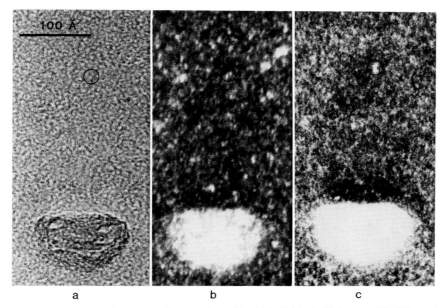

FIGURE 14.12 Image of an amorphous carbon film. (a) Bright-field image. (b) Coherent, dark-field image. (c) Incoherent, hollow-cone, dark-field image. (Courtesy of Gibson and Howie, 1979.)

depends on focus (as is observed in practical, dark-field images), and the intensity probability distribution changes to a form where the most probable intensity level is no longer zero. More significantly, the interference-overlap effects between the amplitude contributions a_i, a_j from different atoms are reduced. In the case of hollow-cone illumination, it appears that interatomic-interference effects are confined to pairs of atoms where $\rho_i \simeq \rho_j$ and $|z_i - z_j| < \lambda/8 \sin^2 \alpha/2$. The purely statistical speckle due to widely separated pairs of atoms that complicates coherent, dark-field images is highly reduced, therefore, in large-angle, hollow-cone or STEM annular-detector images. Residual intensity peaks in these incoherent images are probably a direct consequence of structural correlations between closely separated pairs of atoms. Although some qualitative evidence is available (see Figure 14.12) to support these ideas, further experiments have yet to be done to assess them more critically.

14.6 Other imaging modes

A number of other imaging modes can be conveniently applied in STEM and offer promise for structural studies of highly disordered solids. Figure 14.11c shows an example of a secondary-electron image obtained from a typical catalyst sample. The secondary electrons have an escape depth of perhaps 20 Å and thus show surface topography, troughs, and so on, but do not show pores or catalyst particles in the interior of the sample. This mode gives a better picture of the medium-scale morphology of the support than either the bright-field or annular-detector image do.

Microdiffraction patterns can be obtained from $\simeq 10$-Å-diameter areas in STEM. From our knowledge of the speckle structure in coherent, dark-field images of amorphous solids, we can deduce, by an application of the reciprocity theorem, that, as the probe in STEM is scanned across the specimen, the intensity at each point in the diffraction pattern will fluctuate rapidly and irregularly in a way not simply related in general to the intensity fluctuations at other points in the pattern. Microdiffraction patterns thus have a spotty appearance instead of the diffuse rings characteristic of the diffraction patterns from large volumes. In principle, therefore (Cowley, 1981b), they contain a great deal of information about the local structure. Much of the pattern comes, moreover, from relatively high scattering angles and might therefore be useful in overcoming the structural ambiguities arising from the projection problems that affect conventional, bright-field images. Unfortunately, however, the patterns will still be complicated by statistical noise similar to that found in high-resolution images. It will probably be necessary to study an enormous number of microdiffraction patterns and apply pattern-recognition techniques or other processing methods on an automatic basis to make full use of all this information.

As an initial step in automatic, on-line processing of STEM microdiffraction data, the simplest procedure would be to measure the correlation in the signals received in two detectors defining the different scattering vectors $\mathbf{q}_1$ and $\mathbf{q}_2$, indicated in Figure 14.13. As the focused electron probe is scanned across the specimen, a detector placed at the point $\mathbf{q}$ in the microdiffraction pattern will collect a time-varying signal $J(\mathbf{q}, t)$. For two such detectors, we could then study the correlation function

$$C(\mathbf{q}_1, \mathbf{q}_2) = \frac{\langle J(\mathbf{q}_1, t) J(\mathbf{q}_2, t) \rangle}{\langle J(\mathbf{q}_1, t) \rangle \langle J(\mathbf{q}_2, t) \rangle}, \qquad (14.10)$$

where $\langle \ \rangle$ denotes a time average during continuous scanning. Defined in this way, C would have a value of 1 for completely uncorrelated signals and a value respectively greater or less than 1 for correlated or anticorrelated signals. An obvious procedure would be to take equal, mean scattering angles—that is, $q_1 = q_2 = q$—corresponding, perhaps, to a maximum in the interference function $F(q)$ of equation (14.1). One could then investigate C as a function of ψ, the azimuthal angle between the two detectors (Figure 14.13). Then, $C(\psi)$ would be a maximum at $\psi = 0$ (corresponding to complete correlation), falling off rapidly with increasing ψ, at a rate determined by the semiangle β of the probe-forming illumination, to a much lower background value close to unity. In a sample with completely random structure, the individual intensity levels collected by point detectors would

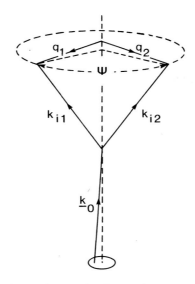

FIGURE 14.13 Schematic diagram for correlated-scattering measurements in STEM.

obey the exponential-intensity statistics mentioned in section 14.5.1 and characteristic of coherent, dark-field imaging. This would give the results $C(0) = 2$, $C(\psi) = 1$ for $\psi > \beta$. The presence of nonrandom features—in particular, any structural unit—would produce oscillations in C as a function of ψ.

A pilot experiment of this kind was carried out by Clark et al. (1983), using laser-beam illumination focused and scanned across a thin, liquid-crystal film that contained small domains with a hexagonal structure. The $C(\psi)$-function showed pronounced structure with peaks at $\psi = 60°$ and 120°. For a typical amorphous thin film, the correlation effects would probably be much smaller. Preliminary computations have been made by Howie et al. (1986), using amorphous clusters of several hundred atoms and parallel illumination. The intensities received by two detectors, separated by a fixed, azimuthal angular difference of ψ, were then computed (by Fourier transforms) and the function $C(\psi)$ obtained by averaging the intensity product over all possible orientations of the cluster. As a function of ψ, $C(\psi)$ was found to have variations of a few percent. The correlations from such an isolated cluster may be only a rough guide to the results that might be obtainable from amorphous, thin films; however, they can be related to a theory developed by Kam (1977) to describe the scattering of radiation from identical, randomly oriented particles. Kam's theory depends on expansion of the scattered intensity from a single cluster in terms of spherical harmonics. A simple expression is then obtained for the correlation function $C(\psi)$ in terms of $P_l(\cos \psi)$ Legendre functions.

Although these ideas have yet to be put to detailed practical test in the high-resolution imaging of amorphous or other disordered materials, they do have the merit of offering some hope of surmounting the two most serious difficulties suffered by existing work. In the first place, since high-scattering angles are readily available in the diffraction mode, the projection problem can be overcome. In the second place, they aim to produce a function that is a convenient statistical measure of angular correlations in the structure. In terms of conveying relatively simple quantitative information to supplement RDF data, this would be vastly better than the kind of high-resolution electron micrograph, which, looking more or less like random noise to the nonexpert eye, can so far only be interpreted qualitatively and with great difficulty.

14.7 Conclusions

Electron microscopy has made significant contributions to our knowledge of the medium-range structure of highly disordered materials. In providing high-resolution, atomic-scale information, particularly about highly amorphous materials, existing techniques encounter severe difficulties, and although progress continues to be made, they do not as yet rival classical diffraction methods. Some of the new imaging methods available in STEM, particularly those involving high-angle scattering and statistical processing of the data, may change this situation, however.

Finally, the difficulty of obtaining completely reproducible samples of such highly disordered material in different laboratories poses a challenge to electron microscopists and to their colleagues in other disciplines. In such a situation, it is essential that electron-microscopy techniques be integrated more effectively into other laboratory procedures so that the practice of studying, in the microscope, samples of the same material whose other properties have been measured as thoroughly as possible is a matter of routine custom and not the very rare phenomenon that it is today.

REFERENCES

Audier, M., Oberlin, A., Oberlin, M., Coulon, M., and Bounetain, L. (1981). Morphology and crystalline order in catalytic carbons. *Carbon* **19**, 217.

Bernal, J. D. (1960). Geometry of the structure of monatomic liquids. *Nature* **185**, 68.

Bursill, L. A., Mallinson, L. G., Elliott, S. R., and Thomas, J. M. (1982). Computer simulation and interpretation of E. M. images of amorphous materials. *J. Phys. Chem.* **85**, 3004.

Chen, C. H., Phillips, J. C., Tai, K. L., and Bridenbaugh, P. M. (1981). Domain microscopy in chalcogenide thin films. *Solid State Commun.* **38**, 657.

Chevalier, J.-P. A. A., and Stobbs, W. M. (1979). The state of local order in quenched CuPt. *Acta Metall.* **27**, 285.

Clark, N. A., Ackerson, B. J., and Hurd, A. J. (1983). Multidetector scattering as a probe of local structure in disordered phases. *Phys. Rev. Lett.* **50**, 1459.

Cowley, J. M. (1981a). *Diffraction physics.* 2nd ed. North-Holland, Amsterdam.

——— (1981b). Electron microdiffraction and microscopy of amorphous solids. In *Diffraction studies of non-crystalline substances,* ed. I. Hargittai and W. J. Orville Thomas, 849. Akademici Kiado, Budapest.

Crewe, A. V., Langmore, J. P., and Issacson, J. S. (1975). Resolution and contrast in the scanning transmission electron microscope. In *Physical aspects of electron microscopy and microbeam analysis,* eds. B. M. Siegel and D. R. Beaman, 47. Wiley, New York.

Elliott, S. R. (1984). *Physics of amorphous materials.* Longman, London.

Fan, G. W., and Cowley, J. M. (1985). Autocorrelation analysis of high-resolution electron micrographs of near-amorphous thin films. In *43rd EMSA Proceedings,* ed. G. W. Bailey, 60. San Francisco Press, San Francisco.

Gaskell, P. H. (1979). On the structure of simple inorganic amorphous solids. *J. Phys. C* **12,** 4337.

——— (1983). Models for the structure of amorphous metals. In *Glassy Metals. II.* ed. H. Beak and H. J. Guntherodt, 5. Springer-Verlag, New York.

———, and Mistry, A. B. (1979). High-resolution transmission electron microscopy of small amorphous silica particles. *Philos. Mag. A* **39,** 245.

———, Smith, D. J., Catto, C. J. D., and Cleaver, J. R. A. (1979). Direct observation of the structure of a metallic glass. *Nature* **281,** 465.

Gibson, J. M., and Dong, D. W. (1980). Direct evidence for pores in "dry" SiO_2 by HREM. *J. Electrochem. Soc.* **127,** 2722.

———, and Howie, A. (1979). Investigation of local structure and composition in amorphous solids by high-resolution electron microscopy. *Chem. Scripta* **14,** 109.

Howie, A. (1978). High-resolution electron microscopy of amorphous thin films. *J. Non-Cryst. Solids* **31,** 41.

——— (1983). Problems of interpretation in high-resolution electron microscopy *J. Microsc.* **129,** 239.

———, McGill, C. A., and Rodenburg, J. M. (1986). Intensity correlations in microdiffraction from amorphous materials. *J. Phys. (Paris) Colloque C* **9,** 59.

———, Marks, L. D., and Pennycook, S. J. (1982). New imaging methods for catalyst particles. *Ultramicroscopy* **8,** 163.

Kam, Z. (1977). Determination of macromolecular structure in solution by spatial correlation of scattering fluctuations. *Macromolecules* **10,** 927.

Knowles, K. M., Greer, A. L., Saxton, W. O., and Stobbs, W. M. (1985). High-resolution electron microscopy of an Al–Mn alloy exhibiting icosahedral symmetry. *Philos. Mag., Ser. B* **52,** L31.

Krivanek, O. L., Gaskell, P. H., and Howie, A. (1976). Seeing order in amorphous materials. *Nature* **262,** 454.

———, Isoda, S., and Kobayashi, K. (1977). Lattice imaging of a grain boundary in crystalline Ge. *Philos. Mag.* **36,** 931.

Levine, D., and Steinhardt, P. J. (1985). Quasicrystals. *J. Non-Cryst. Solids* **75,** 85.

Mackay, A. (1982). De nive quinquangula (On the pentagonal snowflake). *Sov. Phys. Crystallogr.* **26,** 517.

Mistry, A. B. (1979). High-resolution electron microscopy of amorphous thin films. Ph.D. diss., University of Cambridge, England.

Moss, S. C. and Walker, R. H. (1974). Screening singularities and Fermi surface effects in the diffuse scattering from alloys. *J. Appl. Cryst.* **8,** 96.

Patterson, A. M., Long, A. R., Craven, A. J., and Chapman, J. N. (1983). Structural studies of sputtered aSi:H films in a high-resolution analytical STEM. *J. Non-Cryst. solids* **59,** 225.

Phillips, J. C., Bean, J. C., Wilson, B. A. and Ourmazd, A. (1987). Bragg diffraction from amorphous silicon. *Nature* **325,** 121.

Robertson, J. (1986). Amorphous carbon. *Adv. in Phys.* **35,** 317.

Saxton, W. O., Howie, A., Mistry, A. B., and Pitt, A. (1977). Fact and artefact in HREM. In *Developmens in electron microscopy and analysis—1977,* ed. D. L. Misell, 119. Institute of Physics, Bristol.

Shechtman, D., Blech, I., Gratias, D., and Cahn, J. W. (1984). Metallic phase with long-range orientational order and no translation symmetry. *Phys. Rev. Lett.* **53,** 1951.

Smith, D. J., Stobbs, W. M., and Saxton, W. O. (1981a). Ultra-high-resolution electron microscopy of amorphous materials. *Philos. Mag.* B **43,** 907.

———, Fisher, R. M., and Freeman, L. A. (1981b). Resolution electron microscopy of metal-intercalated graphite. *J. Catal.* **72,** 51.

Stenhouse, B. J., and Grout, P. J. (1978). Diffraction intensities and the structure of amorphous carbon. *J. Non-cryst. Solids* **27,** 247.

Stobbs, W. M., and Chevalier, J.-P. A. A. (1978). The classification of short-range order by electron microscopy. *Acta. Metall.* **26,** 233.

———, and Smith, D. J. (1980). Observations of the structure of amorphous arsenic by high-resolution electron microscopy. *J. Microsc.* **119,** 29.

Taftø, J., and Kampas, F. J. (1985). Evidence of chemical ordering in amorphous hydrogenated silicon carbide. *Appl. Phys. Lett.* **46,** 949.

Treacy, M. M. J., Howie, A., and Wilson, C. J. (1978). Z-contrast of platinum and palladium catalysts. *Philos. Mag., Ser.* A **38,** 569.

White, D., Baird, T., Fryer, J. R., Freeman, L. A., Smith, D. J., and Day, M. (1983). Electron microscope studies of platinum/alumina reforming catalysts. *J. Catal.* **81,** 119.

Wagnall, G. D., and Pings, C. J. (1974). The structure of vitreous carbon from wide-angle and small-angle X-ray diffraction. *Carbon* **12,** 51.

INDEX

aberration-free focus (AFF), 485, 496
aberration function, 527, 543. *See also* phase-contrast transfer function
aberrations, 13, 45. *See also* astigmatism
  astigmatism, 13, 45
  chromatic, 13, 24, 97
  spherical, 13, 14, 45, fig. 1.6, 543, 547
absorption. *See also* inelastic scattering
  absorption effects, 173, 177
  absorption function, 9, 86, 118
  phenomenological, 251
achromatic circle, 552
Al-Si ordering, 320, 346, 347
albite, 346
ALCHEMI (atom location by channeling enhanced microanalysis), 219, 365
  axial channeling, 224, 367
  interstitial sites, 224
  localization of X-ray generation, 224, 367
  planar electron channeling, 367
  surface, 600
  temperature reduction, 224
aliasing, 255. *See also* multislice method
alignment, 302. *See also* beam tilt
  alignment techniques, 483
  by computer, 550
  crystal, 482
alkali feldspars, 346
alkaline-earth fluorides, 490
alloys, 479. *See also* metals
  disordered, 624
  spinodally decomposed, 479
aluminum, 493
amorphous materials, 269. *See also* highly disordered materials
  calculation of diffraction amplitudes, 269
  films, 480
  semiconductors, 610
amphiboles, 311, 312, 316, 317, figs. 9.1–9.5
  anthophyllite, 358
amplitude contrast, 16, 510

analytical electron microscopy (AEM), 56, 383, 391, 555, chap 7
  microanalysis, 132, 138, 146, 469
angle-resolved photoelectron spectroscopy (ARPES), 190
anorthite, 346
anthophyllite, 358
antigorite, 335, 342, 348, 358, figs. 9.16, 9.17
apatite (biological), 491
aperture function, 12, 277, 283, 527, 543
appearance-potential spectroscopy (APS), 190
apuanite, 338
artifacts
  heating by the electron beam, 364
  imaging, 367, 483
  radiation damage, 363
astigmatism, 277, 302, fig. 12.7
  adjustment of, 537
  apparent, 537
  correction of, 302
  determination of, 548
  large, 548
atom-pair images (dumbell images), 485
atomic scattering factor, 61
"atomic clouds", 238
atomic disorder, 103
  medium-range, 104
  short-range, 104
Auger-electron spectroscopy (AES), 190, 577
automated focusing, 237. *See also* alignment; computer control of electron microscopes
axial channeling, 367. *See also* ALCHEMI

bastnaesite-synchysite, 337
beam alignment, 483, 484, fig. 11.6. *See also* alignment

For acronyms see list on pages xviii and xix. Spelled-out versions for most acronyms are listed in the index.

beam divergence (= beam convergence), 27, 278, fig. 8.8
beam interaction. *See also* radiation damage
  knock-on, 388
  radiolysis, 388
beam tilt, 253, 302, fig. 8.13. *See also* alignment
beryl, 320, fig. 9.9
Bethe theory, 154
binding energy, 194
biopyribole, 311, 312, 314–16, 354, 360, figs. 9.1–9.6
biopyribole reactions, 315
biotite, 322
Bloch waves, 208. *See also* dispersion surface
  Bloch-wave picture, 177
  formulation, 245
block structures, 408, 427
  disorder in, 429
  mechanisms of reduction, 410, 429, 431, fig. 10.15
  nonstoichiometry in, 429
  oxidation of, 432
  review of, 427
  structural characteristics, 427
  types of Nb_2O_5, 429
blocking, 155, 164. *See also* channeling
Boersch effect, 528
bonding
  in amorphous carbon films, 620
  in disordered materials, 610–12, 615, fig. 14.1
  scattering amplitude corrections, 609
Born series, 40
bornite, 348, 364
Borrmann effect, 86, 220
boundary conditions, 116
Bragg's law, 62, 113
bright field. *See also* imaging; dark field
  images, 579
  phase contrast, 510, 511
broadband-imaging conditions, 285
bronzoids, ITB type, 416
brucite units. *See also* chlorite; humite
Burgers circuits, 494. *See also* dislocations

cadmium telluride, 482
calaverite, 348
calcite, 347
carbon
  amorphous films, 620
  bird's nest structure, 613, fig. 14.3

electron energy-loss near-edge structure, 206, 216
  partially graphitized, 318, fig. 9.8, 506, 613, 615
  polymeric structure, 613
carbonates
  bastnaesite-synchysite, 337
  calcite, 347
  dolomite, 347
carlosturanite, 354
catalysts, 433, 607, 620, figs. 10.32,10.33
  Ge particles, 620
  GeO_2 particles, 620
  Pt particles, 620
cathodoluminescence, 190, 224, fig. 7.13
  diffusion length, 226
  dislocations in silicon, 228
  from individual dislocations, 227
  in diamond, 227
  infrared, 228
  liquid-helium temperatures, 228
  spatial resolution, 226
cation ordering, 220
CdS, 509
centrifugal-barrier effect, 199
ceramics, 490–92, 501, 509
cerium oxides, 440–44, fig. 10.35
chain silicates, 325
  amphiboles, 311, 312, 316, 317, figs. 9.1–9.5
  anthophyllite, 358
  carlosturanite, 354
  chesterite, 311, 312
  double chains, 311, 316
  enstatite, 330
  jimthompsonite, 311, 312, 358
  pyribole, 312, 316
chain-periodicity faults, 336
chalcogenides, formation of, 451–53, fig. 10.42
channeling, 164, 212, 214, 221. *See also* ALCHEMI
chemical analysis, HRTEM techniques, 391. *See also* analytical electron microscopy
chemical change, electron-beam-induced, 364, 387–90. *See also* radiation damage
chemical disorder, 365, 398, figs. 10.8,10.10
chemical equilibrium, deviation from, 378
chemical mapping, 167
chemical reactions
  Au-In, Au-Sn, 455–68, figs. 10.45–10.50,10.52,10.53
  chalcogenide, 451–53, fig. 10.42
  $Cr-Cr_2O_3$, 447, 448, fig. 10.40

INDEX 635

crystallization or precipitation, 453–55, figs. 10.43,10.44
Pb-Bi-S, 449, 450
chemical shifts, 200, 202
chemical twinning, 332, 340. *See also* intergrowth: structures; polysomatism
chemical-vapor deposition, 492
chesterite, 312
chlorite, 322, 327, 328, 368, figs. 9.10,9.11, 9.13
  disorder in chlorites, 328
chlorite-like layers, 357
chromatic aberration, 278. *See also* aberrations
chromite spinel, 221
chrysotile, 335, 348, 358
clay minerals, 334
  kaolinite, 334
  mixed-layer clay, 334
  rectorite, 335
  smectite, 334
coherence, 13
  finite sources, 26
  imaging of amorphous materials, 616, 623
  partial, 24
column approximation, 100, 121, fig. 3.24, 268, 589
coma, axial, 537, 538
computer control of electron microscopes, 238. *See also* alignment
computer interfacing, 238
contrast, minimum detectable, 562
contrast transfer function (CTF), 20, 284, 526, 528, 534. *See also* phase-contrast transfer function
  envelope of, 530, 543, 549, fig. 12.3
  maximum of, 529, 530, 533, 551
  voltage center of, 532
convergent beam electron diffraction (CBED), 67, 88, 92, 228, 269, figs. 3.18,3.20
  coherent interference effects, 69, 70, fig. 3.8
  pattern, fig. 3.6
  space group, 229
  symmetry, fig. 3.19
  Tanaka method, 229
convolution integral, 42
copper, twin boundary, 496
cordierite, 320, fig. 9.9
core electron, 202, 507
core exciton, 200
core hole, 194
  lifetime, 203

correlation function, in microdiffraction, 628–30
Cr_2O_3, 447
crystal alignment, 482. *See also* alignment
crystallographic shear, 312, 394, 401, 420–33, fig. 10.5
  biopyriboles, 312
  definition of, 394
  two-dimensional, 427
  $\{102\}$ CS phase in WO_{3-x}, 394
  $\{103\}$ CS phase in WO_{3-x}, 396

dark field, 344, fig. 9.18. *See also* bright field; images
  beam stop, fig. 1.13
  central stop, 31, 53
  displaced aperture, fig. 1.13
  high resolution, 31, 53, fig. 1.13, 362
  image, 30, 52, 579
  of amorphous materials, 623, fig. 14.12
  pyrrhotite, 362
  STEM, 34
Debye–Waller factor, 102, 145, 174, 176, 250
  and Z-contrast, 626
decomposition reactions, 411, 434, 444–46, 449, figs. 10.15,10.39,10.41
  ternary sulfides, 449
deconvolution, 179. *See also* electron energy-loss spectroscopy
  logarithmic, 140
defects (see chaps. 9–11)
  computation of diffraction amplitudes from, 267
  dislocations, 498
  extended, 105
  formation, 386
  in $WO_{3-\delta}$, 398, 391–400
  planar, 314, 478, 494–97
  returning, 267. *See also* periodic continuation
  types of, 386
defocus
  apparent, 537
  determination of, 543, 547, fig. 12.6
  Gaussian, 533, 536, 551
  optimum, 536
  Scherzer, 526, 533, 551
delocalization, 367, 558, 559, 563. *See also* electron energy-loss spectroscopy
detector, efficiency of, 522, 562
diamond, 485
  platelets, 482, 494, 496, 511, fig. 11.3
  platelets, nitrogen-fretwork model, 496

diamond (*contd.*)
  platelets, zigzag model, 496
  voidites, 511, fig. 11.24
dielectric constant, 193
dielectric theory of energy loss, 173
diffraction contrast, 163, 164, 510. *See also* bright field
diffraction methods, high-resolution, 447. *See also* microdiffraction
diffractogram
  analysis of, 541, 543, 544
  amorphous materials, 616, 617
  tableau, 550, fig. 12.8
diffuse scattering, 99, 104, fig. 3.26, 146
diffusion length, 226
digenite, 364
digital acquisition of images, 238. *See also* computer interfacing
dipole approximation, 197, 207
dipole-selection rule, 194
dislocations, 498. *See also* cathodoluminescence; weak-beam images
disorder. *See also* chaps. 9–11, 14
  block structures, 427–33, figs. 10.28,10.29,10.31
  chlorites, 328
  dynamical diffraction, 104
  intergrowth disorder, 332, 333
  random intergrowth, 358
  rare earth oxides, 444–46, fig. 10.39
  runs-probability test, 360
  site-occupancy disorder, 365
  stacking disorder, 325
  structural disorder, 316
  $TiO_{2-\delta}$, 420–27, figs. 10.23,10.25,10.26
  V_nO_{2n-1}, 427, fig. 10.27
  WO_{3-x}, 391–400, figs. 10.10,10.11
  zeolites, 433–438, figs. 10.32,10.34
disordered alloys, 624
dispersion equations, 116. *See also* Bloch waves
dispersion relation, 148
dispersion surface, 80, 133
displacement threshold, 183. *See also* radiation damage
displacive planar faults, 316
dolomite, 347
domains, in disordered alloys, 624, 625, fig. 14.10
double chains, 311, 316
dry-cell batteries, 350
dynamical diffraction, 109–111, fig. 3.1
  Bethe theory, 78
  Bloch-wave formulation, 78, 79–86, 109, 116

  Born series, 110
  difference equations, 101, 120
  disorder, 104
  extended defects, 105
  in quasi crystals, 621, 622
  matrix formulation, 78
  multiple scattering, 254
  multislice formulation, 78, 86, 121, fig. 3.17
  scattering matrix formulation, 79, 120
  semi-reciprocal formulation, 79, 120
  symmetry, 88
  two-beam approximation, 118, figs. 3.14,3.15,4.1
dynamical scattering, 58–60, 109
dynamically forbidden reflections, 293

e-plagioclase, 346
edge-on boundaries, 486
effective source size, 26
elastic scattering. *See also* dynamical diffraction; dynamical scattering
electron
  interaction constant, 18, fig. 3.10
  refractive index, 38
  relativistic mass, 109
  relativistic wavelength, 109
  scattering amplitude, 7, 61, 111
  transmission through matter, 43
  wave equation, 38
  wavelength, 6, fig. 3.10
  waves, 38
electron beam
  chemical effects, 387
  spreading of, 558, 563
electron channeling, 367. *See also* ALCHEMI; channeling
electron gun
  brightness of, 522, 556, 562
  field emission, 556
  lanthanum hexaboride, 529, 556
electron optics, 3
  lenses, 6
electron probe, shape of, 556, 557, 563
electron energy-loss filters, 167
electron energy-loss spectroscopy (EELS), 135, 190
  in ALCHEMI, 213, fig. 7.7
  in reflection electron microscopy (REM), 592, 600
  orientation effects in, 211, 213
electron energy-loss near-edge structure (ELNES), 132, 198
  boron nitride, 206

INDEX

carbon, 210
chemical shifts, 206
density of states, 209
energy width, 203
excited-state lifetime, 202
graphite, 206
isolated atom, 200
local atomic coordination, 199
momentum-conservation requirement, 208
"muffin tin" potential, 209
nickel oxide, 206
screening, 205
site symmetry, 206
titanium dioxide, 206
transition dioxide, 206
transition metals, 204
"white lines", 205
electronic excitations, 506
emission current, 529
energy-dispersive X-ray spectroscopy (EDS), 132, 190, 600. *See also* analytical electron microscopy
in reflection electron microscopy (REM), 600
in reflection high-energy electron diffraction (RHEED), 600
energy-filtered diffraction patterns, 163
energy-filtered microdiffraction, 170
enstatite, 330, 334, 358
envelope function, 26, 27, 36, 49, 125. *See also* transfer function
for beam divergence, 284
for chromatic aberration, 284
limitations of, 292
evaporation, *in situ*, 503
Ewald sphere, 64, 66, 133, 135, figs. 3.4,5.2
excitation error, 251, fig. 8.2
excitons, 202
extended crystallographic-shear defects, 394, 398, 421, 497, figs 10.8,10.10,10.23, 10.25–10.27,10.29,11.15. *See also* crystallographic shear
extended defects, 105
extended electron-loss fine structure (EXELFS), 132, 193
channeling effects, 198
logarithm deconvolution, 197
parallel-recording detectors, 197
extended X-ray absorption fine-structure (EXAFS), 132, 190, 610
amorphous materials, 608, 610
of amorphous Ga-As, 612
extinction distance, 147. *See also* dynamical diffraction

feldspar, 221, 334, 346
Al-Si ordering, 320, 346, 347
albite, 346
alkali feldspars, 346
anorthite, 346
e-plagioclase, 346
microcline, 347
orthoclase, 347
sanidine, 347
fibrous habit, 354
anthophyllite, 358
carlosturanite, 354
chrysotile, 335, 348, 358
field-emission gun, 26, 33, 570
field-ion microscopy, 573, 579
first Born approximation, 40, 110
focal-series restoration, 486
forbidden reflection, 574
Fourier images, 30, 51, fig. 2.3, 480–82, 537
self-image, 482
Fourier transform, 43, 71, 111, 254, 256, 257, 275
convolution theorem, 42
fast, 303
multiplication theorem, 43
Franck-Hertz experiment, 191
Fraunhofer diffraction, 42, fig. 1.4
pattern, 11
Fresnel diffraction, 42, 122, fig. 2.2
Fresnel fringe, 537
Friedel's law, 88

GaAs, 499
garnet, 367
Gaussian function, 145
germanium, 487, 498, 499
dislocation, 30° partial, 498
dislocations, 479
grain boundary, high-angle, 487
grain boundary, low-angle, 487
glass. *See* highly disordered materials
goethite, 334
gold, 485, 486, 496, 502, fig. 11.14
grain boundaries, 486
aluminum, 493
germanium, 488, fig. 11.5
molybdenum, 494, fig. 11.12
nickel oxide, 492, fig. 11.11
graphite, 318, 507, fig. 9.8. *See also* carbon
grey scale, 295. *See also* image simulations
Guinier-Preston zones, 496

heating by the electron beam, 364

INDEX

high-resolution transmission electron microscopy (HRTEM)
　chemical information from, 379
　combination of X-ray analysis with, 391
　in preparative chemistry, 431
　in situ reactions, 464
　video recording, 464
higher-order Laue zones, 259
highly disordered materials, 607–32
　amorphous semiconductor bonding, 611, 612
　columnar structure, 613
　medium-scale structure, 612–15
　metallic glasses, 620
historical account, applications to chemistry, 380–82
hollandite-romanechite, 350
homologous series, W_nO_{3n-1}, 394
Howie–Whelan difference equations, 101
humite, 337
hydroxyapatite (HAP), 491

I-beam diagrams, figs. 9.2, 9.7
illumination
　incoherence of, 527–30, 549
　tilt of, 537, 539, 550
image
　false, 520, 554
　nonperiodic, 542
　processing of, 534
　recording of, 541
　shot noise of, 542, 561
image calculation, 367, chap. 8
image formation
　Abbe theory, 10, 28, 44, fig. 1.4
　crystal defects, 99
　crystals, 93, 124
　dark-field images, 30, 52
　incoherent, 46
　lattice fringes, 29, 36
　partial coherence, 48
　periodic objects, 28, 50
　two-beam approximation, 93
　weak phase objects, 46
image intensifier, 238
image processing, 299, 363, 390
image simulations, 105, chap. 8
　quantitative comparisons, 440
imaging
　bright field, 527, 529, 530, 557. *See* bright field
　dark field, 344, 557. *See* dark field
　dark-field method, 362

HRTEM, 275
　in real time, 542
　lattice-fringe images, 309
　linear, 282
　low dose, 540, 541
　nonlinear, 286, 553
　tilted-illumination, 552
　Z-contrast, 557
imaging artifacts, 367, 483
impact parameter, 160
in situ
　Au_3In formation, 466
　deposition, 577
　reaction mechanism, 364, 466
　technique, 464
　treatment, 364, 576, 583
incoherent-imaging factors, 125
incommensurate modulated structures, 342, 344, 346
independent Bloch-wave approximation, 155, 217
indium film, oxidation of, 460
inelastic scattering, 9, 56, 129, fig. 5.1
　angular distribution of, 138
　Bethe ridge, 138
　Bethe theory, 137
　Bloch waves, 133, 149
　Born approximation, 137
　conservation of energy, 132
　conservation of momentum, 132
　cross section, 138
　crystal momentum, 133
　density of states, 137
　dielectric formulation, 138
　diffraction contrast, 149
　dispersion relation, 133, 135
　dynamical, 147, 154
　ionization, 138
　lattice images, 148
　localization, 160
　matrix element, 138
　mean free path, 147
　momentum transfer, 133
　multiple scattering, 146
　multislice algorithm, 155
　nuclear recoil, 150
　optical properties, 138
　oscillator strength, 137, 138
　reflection mode, 138
　relativistic effects, 146, 147
　removal of multiple inelastic scattering, 140
　resonance error, 150
information retrieval limit, 259

inner-shell excitations, 135, fig. 5.3. *See also* inelastic scattering
InP, 490
instabilities, instrumental, 527, 528, 529, 549, 558
interaction constant, 77, fig. 3.10
 relativistic, 248
interband transitions, 146. *See also* inelastic scattering
interface, 486–94
 aluminum/aluminum oxide, 493
 martensite, 493
 metals, 493, 494
 metal silicide/silicon, 489
 Si (100)/SiO$_x$, 488, fig. 11.6
 silicon on sapphire, 488
interference lattice images, 485
intergrowth
 disorder, 332, 333
 rare earth oxides, 445
 structures, 333, 357
 tungsten bronze (ITB), 417, 418
intraband transitions, 146. *See also* inelastic scattering
inverse photoelectron spectroscopy (IPES), 190
ion milling, 362
 damage, 510
ionization, 132
isoplanatic approximation, 525

jimthompsonite, 311, 312, 358, fig. 9.1

kaolinite, 334
Kikuchi lines, 102, 135, 147, 154, 155, 163, 170
kinematic approximation, 19
kinematical diffraction, 111–15
 amorphous materials, 72
 intensities, 70
 limitations, 74
kinematical scattering, 60–74, fig. 3.2
 crystal defects, 115
 geometry, 62
 intensities, 113
kinematical theory, in amorphous structure imaging, 615
knock-on, 388. *See also* radiation damage
Koopman's theorem, 202
Kramers–Kronig relations, 138, 140
krennerite, 349
kulkeite, 357

lattice fringes, 94, 126
 defects, 94
 images, 309
 three-beam case, 96
 two-dimensional, 97
lattice-fringe imaging
 applications, 478
 limitations, 478
lattice images in STEM, 169, fig. 6.3
Laue zones
 higher order (HOLZ), 66
 zero order (ZOLZ), 66
layer silicates. *See* micas; sheet silicates
lens transfer function, 24. *See also* transfer function
leucophoenicite, 337, 338
lifetime, 200
lifetime broadening, 203
light optics, 12, 17, 20, 35
 lenses, 3, 6
 microscopy, 3
line defects, 478, 498, 499
lizardite, 335, 358
localization, 132, 154, 160, 163, 164
 of X-ray generation, 367
 scattering angle dependence, 214, fig. 7.7
low-energy electron diffraction (LEED), 195
low-energy electron-reflection microscopy (LEERM), 573
ludwigite, 340, fig. 9.15

magnetic-sector analyzer, 167
manganese oxides, 350, 358, figs. 9.20–9.23
 dry-cell batteries, 350
 hollandite-romanechite, 350
 nsutite, 350
 todorokite, 353, 358
mechanical instabilities, 277. *See also* vibration, effects on HRTEM images
metal alloys, 504–6
 quasi crystals, 506, figs. 11.21,14.8
metallic glass. *See* highly disordered materials
metals, 493, 494, 501, 504
metastable remnants, 310
meteorites, 335, 348
micas, 311, 327, 367, 368, figs. 9.10–9.12
 biotite, 322
microanalysis. *See* analytical electron microscopy
microcline, 347

microdiffraction, 169, 228, fig. 7.15
  coherent, 232, fig. 7.16
  combined with HRTEM, 230, fig. 7.14
  correlation effects in, 628
  focus needed to form most compact probe, 231
  incoherent, 232
  from highly disordered materials, 628
  lattice imaging in STEM, 169, 232
  optical system, 235
  Ronchigram, 233
  small particles, 235
  stacking faults, 235
  to determine fault vectors, 234, fig. 7.17
  twin boundaries, 235
microscope
  atomic-force, 564
  commercial, 521, 522
  field-ion, 564
  scanning transmission, 555
  scanning tunneling, 564
microscope column,
  alignment of, 525, 540
  construction of, 522, fig. 12.1
microsyntaxy, 337
millerite, 364
mineral definition, 356
mineral nomenclature, 356
mineralogy, 308
minilens, 524
mixed layer clay, 334
mixed-layer compounds, 337
modulated structures, 340, 346
moiré fringes, 585
molybdenum, 494
  grain boundary, 494, fig. 11.12
Mott formula, 111
"movies", 238
multilayers, $GaAs/Ga_{1-x}Al_xAs$, 489, 490, fig. 11.8
multiple energy-loss effects, removal, 177. *See also* deconvolution
multiple energy-loss processes, 194
multiple scattering. *See* dynamical diffraction
multiple-quantum-well devices, 489
multislice method, 245–59
  basis of, 246
  consistency tests, 257
  fast-Fourier transform, 254
  iteration, 247, 253
  propagation function, 247, 251
  transmission function, 247, 248

near-edge X-ray absorption fine structure (NEXAFS), 132
neutron diffraction, 141
nickel oxide, bicrystals, 492, fig. 11.11
noise amplification, 180
nonintegral diffraction patterns, 344
nonstoichiometry, 332, 340, 384–87, 429
norbergite, 337. *See also* humite
nsutite, 350, fig. 9.21
nuclear recoil, 152

objective lens
  alignment, 537, 538, 539
  current center of, 532, 537
  focusing of, 533
  optic axis of, 537, 550
  polepieces, alignment of, 532, 533
  voltage center of, 532, 537
octacalcium phosphate, 491, fig. 11.9
olivine, 221, 337
optical absorption coefficient, 138
optical diffractogram, 483. *See also* diffractogram
optical microscopes, 3
optical potential, 174
ordered domains, CuPt, 625, fig. 14.10
ordered intergrowth, 359, 445, fig. 10.39
ordered iron vacancies, 344, fig. 9.18
ordered structures, 359
orientation effect, in energy-loss spectra, 217. *See also* ALCHEMI; electron energy-loss spectroscopy
orthoclase, 347
overlapping orders, 170
oxidation state, 202
oxysulfides, 338
  apuanite, 338
  schafarzikite, 338
  versiliaite, 338

parallel detection of electron diffraction patterns, 238
Parseval's theorem (in electron energy-loss spectroscopy), 55
partial coherence, 24
Patterson function, 114
  for amorphous images, 619
pendulum solution, 84
pentagonal-tunnel structures, 396
  block structure, 408
  in the W-Nb-O system, 407
  reactions, 410

periodic continuation, 265
phase contrast, 15, 16
  Schlieren technique, 17
  transfer function, 20, 36
  Zernike method, 17
phase definition, minimum size, 360
phase grating, 248, 251
phase problem, 69
phase randomization, 618
phase-contrast transfer function (PCTF), 47. *See also* transfer function
phase-object approximation (POA), 9, 43, 47, 87, fig. 1.3. *See also* weak-phase-object approximation
  in amorphous samples, 619, 624
  limitation, 77
phonons, 142, 143
  dispersion relation, 145
  Einstein model, 144
  emission, 152. *See also* Umklapp scattering, 132
photoelectric effect, 161, 193
photoelectron, 194
photoelectron wave vector, 194
photoemission electron microscopy (PEEM), 571
pinakiolite, 340, fig. 9.15
planar defects, 314, 478, 494–97
plasmons, 130, 142, 507
  Bragg scattering, 143
  cross section, 142
  dispersion, 143
  double-plasmon process, 143
  microanalytical technique, 142
  profile imaging, 142
  small particles, 142
  thickness dependence, 143
platinum, 503
point defects, 500–2
  ceramics and metals, 501
  semiconductors, 500
point-spread function, 526. *See also* transfer function
Poisson statistics, 178
polycrystal, 337
polysomatic series, 312
  biopyribole, 311, 312, 314–16, 354, 360, figs. 9.1–9.6
  humite, 337
polysomatism, 312, 333
polytypism, 325, 331
  chlorites, 328
  ITB bronzoids, 418, fig. 10.20

micas, 327
power spectrum. *See* diffractogram
praseodymium oxide, 440–46, figs. 10.37,10.39
preparative techniques for HRTEM of minerals, 362
  ion milling, 362
  ultramicrotomy, 362
projected potential, 249, 259, 300
projected-charge-density approximation (PCDA), 16
projection effects in amorphous specimens, 619, 624
propagation function, 122, 123, 251. *See also* multislice method
pumpellyite, 355, fig. 9.25
pyribole, 312, 316
pyrophyllite, 334
pyrosmalite, 331
pyroxene, 311, 312, 316, 317, 325, 330, 334, 358, figs. 9.1–9.3
pyroxenoids, 335, fig. 9.14
  pyroxmangite, 336
  rhodonite, 336
pyroxmangite, 336
pyrrhotite, 342, 362, fig. 9.18

quasi crystals, 506, 620, figs. 11.21,14.8. *See also* amorphous materials

radial-distribution function, 74, 608, 609
  interpretation, 610, 611
  partial distribution function, 610–12
radiation damage, 182, 238, 387–90, 506–10, 562, fig. 6.8
  alloys, 184
  ceramic, 509
  displacement energies, 184
  electron channeling, 183
  interstitials, 183
  ion implantation, 510
  ions, 184
  knock-on displacement, 183, 388, 506
  metals, 508
  on surfaces, 186
  quartz, 185
  radiolysis, 183, 388, 389, 506, 507
  semiconductors, 508
  silicates, 363

radiation damage (*contd.*)
  vacancies, 183
  videorecording, 186
radiolysis. *See also* radiation damage
random intergrowth, 358
rare-earth oxides, 436–46
  bastnaesite-synchysite, 337
  beam-induced reactions in, 444
  defects in, 439
  fluorite-related, 436
  homologous series, 438
  oxidation in the electron beam, 445
  structural models, 440
reaction mechanisms, 310, 314, 455–68, figs. 9.6,10.52
reaction types,
  Al-Si ordering, 320, 346, 347
  replacement reactions, 315
  spinodal, 322
real-time recording, 237. *See also* videorecording in HRTEM
reciprocal lattice, 63, 112
reciprocity principle, 33, 55, 91, 169
  and annular dark-field imaging, 626
  in elastic scattering, 215
  in microdiffraction, 628
reconstructed surface, 2x1 {100} Au, 485, 486
rectorite, 335
reflection electron microscopy (REM), 574
  Fresnel contrast, 592
  image of screw dislocation, 589, fig. 13.15
  image step, 594, figs. 13.15,13.17
  lattice fringe, 588, 593
  scanning, 575, 599
  step image, 589
  study of surface, 587–99
  superlattice, 599
  vicinal surface, 599
reflection high-energy electron diffraction (RHEED), 575, 587
  theory, 589
refraction effect, 588
relaxation energy, 200, 202
replacement reactions, 315
replica technique, 573
resolution, 34
  information limit, 36
  line, 520, 552
  loss of, by delocalization, 559
  optical diffraction measurement, 35
  optimization of, 520, 562
  performance, 521, 524
  point-to-point, 35, 520–22

Rayleigh criterion, 12, 46
resolution increase, 180
resolvable distance, 47
scanning transmission microscopy, 555
Scherzer, 21, 35, 47
reversibility in chemical reactions, lack of, 387
rhodium, 503
rhodonite, 336
rigid-body displacements, 486
rodlike defects, 499
runs-probability test, 360
Rutherford scattering, 132
  and Z-contrast, 626
rutile, 497, 501, fig. 11.15
  chromia-doped, 497
  CS planes, 421, 497
  interstitial cation, 499
  platelets in, 423, fig. 10.26

sanidine, 347
scanning electron microscopy (SEM), 568, 569, 575, 599
scanning transmission electron microscopy (STEM), 5, 33, 54, 269, 555
  annular detector, fig. 2.4, 55
  dark field, 55
  detector configurations, 56
  finite detector, 55
  instrument, fig. 2.4, 555
  lattice imaging, 169
  resolution of, 555
  systems, fig. 1.2
scattering factors, 250
scattering matrix, 44
scattering power in reciprocal space, 63
schafarzikite, 338
Scherzer limit, 484–86
Scherzer optimum defocus, 22, 35
Schottky barrier heights, 489
secondary electron, 194, fig. 14.11
  catalyst imaging, 627
  escape depth, 627
selenides, formation of, 451–53, fig. 10.42
sellaite, 337. *See also* humite
semiconductors, 486–90, 500, 508
serpentine, 363, figs. 9.10,9.26
  antigorite, 335, 342, 348, 358, figs. 9.16,9.17
  chrysotile, 335, 348, 358
  lizardite, 335, 358
  sheets in serpentine, 354
shape transform, 113

sheet silicates, 333
    chlorite, 322, 327, 328, 333, 368
    chlorite-like layers, 357
    chrysotile, 335, 348, 358
    clay minerals, 334
    kaolinite, 334
    kulkeite, 333
    micas, 311, 327, 333, 367, 368
    pyrophyllite, 334
    pyrosmalite, 331. *See also* clay minerals;
        serpentine
    sheets in serpentine, 354
    talc, 358
    talc-like layers, 357
short-range order, 320, 624
    Ni_4Mo, 625
Si-Al-O-N, 477
SiC. *See* silicon carbide
sign conventions, 39
silica, wet and dry films, 614
silicate chains, 314
silicon, 485, 488, 490, 499, 500, fig. 11.22
    on insulator, 488
    on sapphire, 488
    reconstructed surface, 504
    split $\langle 100 \rangle$ self-interstitial, 500
silicon carbide (SiC), 492, fig. 11.10
    3C, 492
    α-polytypes, 492
silicon nitride, 491, 492
silver, 502
simulated images, comparison with observations, 391. *See also* image simulations
single chains, 311
single-electron excitations, 142, 146
single-loss spectrum, 178
site-occupancy disorder, 365
slice method, amorphous specimens,
    application to, 616
small particles, 502–4, fig. 11.19
    rearrangements, 503
smectite, 334
solid state chemistry
    future HRTEM techniques, 468
    general review, 380
    scope of, 378
solid-state chemical synthesis, role of
    HRTEM in 283, 431
solid-state reactions, 310
    in WO_{3-x}, 400
    mechanisms of, 400
solid-state synthesis, 382, 431
spatial frequency, 526
specimen

drift of, 549
tilt of, 540
specimen preparation, 235, 362. *See also*
    preparative techniques for HRTEM of
    minerals
    ultramicrotomy, 362
speckle, 607, 626, 627
    in dark-field images, 623
specular reflection, 593
spinel catalyst, 502
spinels, 366
spinodal mechanism, 322
    bornite, 348, 364
spread function, 11, 15, 19, 45, fig. 1.8. *See
    also* transfer function
stacking disorder, 325
stacking fault tetrahedra, 508
stacking sequences, 328
standing wave, 220
strain-contrast, 510. *See also* diffraction
    contrast
structural disorder, 316
structure amplitude, 71
structure analysis, electron diffraction, 72
structure factors, electron, 249
structure image, 480–82
sudden approximation, 161
sulfides
    bornite, 348, 364
    digenite, 364
    millerite, 364
    pyrrhotite, 342, 362
    ternary, 449
supercell, 265. *See also* periodic
    continuation
superlattice, 342
    spots, and disorder scattering, 624
surface
    adsorption process, 596
    Au (001) 28×5, 579, 587
    Au (110) 2×1, 573, 587
    Au (111) 22×1, 581
    Au $\sqrt{3} \times \sqrt{3}$, 597
    Au 5×1, 597
    dynamic process, 569, 583, 594–97
    HRTEM, 580
    microscopy, 568–73
    out-of-phase boundary, 596, fig. 13.19
    phase transition, 594, fig. 13.17
    plan view, 577, 586, fig. 13.5
    profile imaging, 469, 502–4, 586
    Pt (111), 598
    Si (111) 2×1, 600
    Si (111) 5×1, 597

surface (*contd.*)
　Si (111) 7 × 7, 570, 573, 579, 587, 595, 597
　Si (311), 587
　steps, 568, 574, 584
　structure analysis, 581–83
　structure phase transition, 584, tab. 13.5
　ultra-high-vacuum transmission electron microscopy, 468
　well-defined, 568
surface plasmons, 142
surface structure, 263
sursassite, 355, fig. 9.25
sylvanite, 349
symmetry
　determination, 90
　type I, 90
　type II, 91

Takagi triangle, 164
takeuchiite, 355
talc, 358
talc-like layers, 357. *See also* chlorite; kulkeite
tellurides
　calaverite, 348
　formation of, 452
　krennerite, 349
　sylvanite, 349
temperature factor. *See* Debye–Waller factor
terbium oxide, 444, 446, 469, 502
thermal diffuse scattering, 102
thermal vibrations, 102
thickness contrast, 511
thickness measurement, in amorphous films, 619
thin films
　Au-In reactions, 455, 458
　Au-Sn reactions, 455
　chalcogenide formation in, 451
　chemical reactions in, 446, 450
　contamination layers, 456
　Cu-Se reaction, 452
　diffusion couples, 450, 455
　HRTEM studies of, 446
　in situ reactions in, 464, 446
　low-temperature preparations, 452
　micro-techniques, 447
　new structure in, 452
　precipitation in, 455
　reaction analysis, 458
　reaction specimens for, 451
　surface reactions, 447

through-focal series, 486, 536
tight-binding approximation, 151, 155
tilted illumination, 478
tilted-beam, 362
tin dioxide, 480, 483, 497, figs. 11.2,11.6
　(011)-type twin, 497
$TiO_{2-\delta}$ phases, 420
　crystallographic shear in, 420
　homologous series, 420
　mechanism of oxygen loss, 425
　paired CSP in, 421
　platelet defects in, 423
　point defects in, 420
　slightly reduced, 420
　small defects in, 424
　vanadium doped, 422
titanium niobate, 483, figs. 11.4,11.23
todorokite, 353, 358, figs. 9.22,9.23
tooth enamel, 491
transfer function, 12, 19, 47, figs. 1.8,1.9.
　See also spread function
　imaging of amorphous materials, 616
transistors, bipolar, 488
transition radiation, 143
transmission cross-coefficient, 553
transmission electron diffraction (TED), 577
　from surface structure, 580
transmission function, 8, 248. *See also* multislice method
triple chains, 311, 316
tungsten bronzes, 412–20
　intergrowth, 414
　occurrences, 414
tungsten trioxide, 499
TV camera, 522
twin boundaries, 496, fig. 11.14

ultra-high vacuum, 504
　electron microscope, 575
　pump, 575
ultramicrotomy, 362. *See also* specimen preparation
ultraviolet photoelectron spectroscopy (UPS), 190
Umklapp, 133, 152
uncertainty principle, 4, 162
unitary test, 257. *See also* multislice method: consistency tests
uranium oxide, surface-profile image, 502

valence electrons, 507
vanadium fluoride bronzes, 413

INDEX

versiliaite, 338
vibration, effects on HRTEM images, 278
videorecording in HRTEM, 237
   for demonstration purposes, 238
virtual inelastic scattering, 175
V_nO_{2n-1}, homologous series, 427

Wadsley defect, 381
   an element of structure, 394
   definition of, 394
weak-beam images, 227
weak-beam-imaging technique, 99
weak-beam-object approximation (WPOA), 18, 46, 78, 525, 526, 553. *See also* phase-object approximation
   imaging, 284, 485
   imaging theory, 19
   transfer function, figs. 1.10, 1.11, 8.9
   validity, 18, 47
weathering products, 334
WO_3, doped, 402–12
   comparison of Ta-W-O and Nb-W-O systems, 403
   crystal habits, 402
   CS found, 404
   homologous series, 394
   (Nb, Ta, W)$O_{3-\delta}$ systems, 404
   preparation, 402, 404
   reduction of, 393
   reviews of, 392
   shear type, 402
   structure, 393, fig. 10.4
   the {102} CS phase, 394
   the {103} CS phase, 396
   the $W_{18}O_{49}$ phase, 397
   the $W_{24}O_{68}$ phase, 396
   (W, Nb)$O_{3-\delta}$, 402
   (W, Ta)$O_{3-\delta}$, 402

X-ray absorption, near-edge structure (XANES), 132, 190
X-ray photoelectron spectroscopy (XPS), 190

yttrium-aluminum garnet detector, 238

Z-contrast, of catalysts, 625
zeolites, 363
   beam sensitivity of, 433
   characterization of, 433
   dealumination of, 434
   defects in, 434
   diffraction from, 434
   intergrowth in, 434
zirconia, tetragonal to monoclinic transformation, 491
ZrO_2 (see zirconia)